ENCYCLOPEDIA

of

Nuclear Magnetic Resonance

ENCYCLOPEDIA

of

Nuclear Magnetic Resonance

Editors-in-Chief

David M. Grant
University of Utah, USA

Robin K. Harris
University of Durham, UK

Volume 8

Tis – Z
Indexes

JOHN WILEY & SONS
Chichester • New York • Brisbane • Toronto • Singapore

Baffins Lane, Chichester
West Sussex PO19 1UD, England

Telephone National (01243) 779777
International (+44) 1243 779777

Other Wiley Editorial Offices

John Wiley & Sons, Inc., 605 Third Avenue,
New York, NY 10158-0012, USA

Jacaranda Wiley Ltd, 33 Park Road, Milton,
Queensland 4064, Australia

John Wiley & Sons (Canada) Ltd, 22 Worcester Road,
Rexdale, Ontario M9W 1L1, Canada

John Wiley & Sons (SEA) Pte Ltd, 37 Jalan Pemimpin #05-04,
Block B, Union Industrial Building, Singapore 2057

Library of Congress Cataloging-in-Publication Data

Encyclopedia of nuclear magnetic resonance / editors-in-chief, David
M. Grant, Robin K. Harris.
p. cm.
Includes bibliographical references and index.
ISBN 0-471-93871-8 (alk. paper)
1. Nuclear magnetic resonance—Encyclopedias. I. Grant, David M.
II. Harris, Robin Kingsley.
QC762.E53 1996
538′.362—dc20 95-23825
CIP

British Library Cataloguing in Publication Data

A catalogue record for this book is available from the British Library

ISBN 0 471 93871 8 (set)

Chemical structures produced in ChemDraw by Synopsys, Leeds
Data Management and Typesetting by BPC Techset, Exeter
Printed and bound in Great Britain by BPC Wheatons, Exeter

This book is printed on acid-free paper responsibly manufactured from sustainable forestation,
for which at least two trees are planted for each one used for paper production.

Preface

Nuclear magnetic resonance is now one of the most versatile tools available to the scientific community. Based solidly in the laws of physics, this area has contributed not only to investigations of molecular-level structures and dynamics, important in modern chemistry and biology, but also to the study of magnetic resonance images arising from macroscopic distribution of materials such as are found in the human body. The power of NMR resides in the wide variety of experiments made possible by an almost unlimited choice of radiofrequency pulse sequences that may be used to perturb and then to observe nuclear spins. NMR now applies to all states of matter and to virtually any material. The method is isotope-specific, but nearly every element in the Periodic Table provides a suitable magnetic nucleus. Multinuclear operations combine to provide complementary information on microscopic nuclear environments.

The current situation traces to two groups of physicists who in 1945 made simultaneous discoveries of the phenomenon, though, as Volume 1 of this Encyclopedia describes, those crucial experiments by the Purcell and Bloch groups had a number of important antecedents. Since then, there have been many *revolutions* in technique; seldom has a scientific field manifested the capacity for such outstanding new discoveries and novel applications as experienced by nuclear magnetic resonance across the whole gamut of scientific effort. One may point particularly to the use of very high magnetic fields, to the application of Fourier transform principles, to the development of imaging possibilities, to the establishment of multinuclear and multi-dimensional methods, and to the accommodation of a variety of high-resolution spectral techniques that make the study of solids feasible.

The first two decades of work firmly established the underlying principles of NMR and foreshadowed many of the subsequent developments, even though the full ramifications and ultimate complexities of the subject are still unfolding in this fifth decade since the discovery of NMR. Much of the significant early work is summarized admirably in the seminal books by Andrew (1955), by Pople, Schneider, and Bernstein (1959), by Abragam (1961), and by Slichter (1963), which approached NMR from a variety of viewpoints. Since then, the subject has expanded so much, and the applications have become so numerous and wide-ranging in character, that no single volume can hope to give a comprehensive treatment of NMR. The output of research papers, reviews, and books has been enormous and testifies to the vitality of the discipline.

The 50th anniversary of the discovery of NMR appeared to us and to the publishers of this work to be an appropriate time to attempt an overview, and an Encyclopedia with a historical component seemed to be an appropriate format to commemorate the event. The undertaking has been enormous and fraught with obvious challenges in choice of both topics and contributors. As extensive as the work has become, we recognized from its inception that only representative contributions would fit within the page limitations of the project. However, we have been very encouraged by the enthusiasm and dedicated hard work of a great number of people, in particular the six editors (Ted Becker, Sunney Chan, Jim Emsley, Bernie Gerstein, Tom Farrar, and Ian Young), who have accepted the responsibilities for major components of this work. They have recruited a veritable army of expert contributors, and without their serious commitment to this project its completion would not have been possible. To a significant extent we have relied upon the judgment of the contributors to select appropriate material for inclusion in the Encyclopedia. Doubtless, some topics remain that are only represented in a cursory manner, but we believe the seven technical volumes present both a sensible and a reasonably comprehensive selection of topics from physics, chemistry, materials science, geology, biology, and medicine (not to mention some relevant aspects of agricultural, clinical, and industrial technologies).

Volume 1, under the direction of Ted Becker, is unique. It seemed appropriate at the 50-year commemoration of NMR to celebrate also the contribution of the human spirit and intellect by producing a history of the subject and its workers. The intent has been to preserve some of the anecdotal material that is often left unreported in the technical literature but is so important to a new discovery and its ultimate exploitation. In this the personal recollections of about 200 prominent practitioners of the subject are indispensable. We are especially delighted that many of the pioneers from the early days of NMR as well as representative innovators of recent significant advances have responded to the invitation to participate in this unusual volume. We know that they would want their reflections to represent the thousands of dedicated investigators who have contributed to this monumental half-century of scientific endeavor.

We hope this Encyclopedia commends itself to the widespread NMR community and that readers everywhere find that the details of the various techniques and applications, all in one set of volumes, provide a valuable reference collection. Articles span the full range from introductory to advanced levels, and we trust that all, from novice to expert, will find their needs addressed in this work.

As well as being extremely indebted to the Section Editors and to the contributors, who have set the tone for this enterprise, we are grateful to our distinguished International Advisory Board for their counsel and help, to Dr Colin Drayton who conceived and commissioned the project, and to Dr Ernest Kirkwood and the publishing team at Wiley. The contributions by our secretaries, Louise Trapier and Karen

Widdrington, assisted greatly in the project's accomplishments and relieved many of the burdens associated with this effort. Finally, we express sincere appreciation to our respective wives, Reva and Maureen, who have had evenings and weekends interrupted by meetings, transatlantic telephone calls, and husbands sometimes distracted and perplexed by the various crises that were common during the time that this publication was in preparation.

David M. Grant
University of Utah
Robin K. Harris
University of Durham

Foreword

In order to cover properly the range of areas mentioned in the Preface, five Section Editors with specialized knowledge of different aspects of NMR were chosen to oversee the technical articles. These articles appear in Volumes 2–7 in alphabetical order of their first word. Careful choice of that first word was essential so that it is one a reader might reasonably be expected to seek in order to provide the desired information. Trivial first words were therefore avoided and, in many cases, groups of articles with a common theme will appear together in the Encyclopedia. Moreover, although the articles are self-contained, cross-referencing to other articles is frequently provided, both by the use of ***Bold Italics*** in the text and especially in the form of a short list of related topics at the end of articles.

Brief biographical details of authors are also given at the end of each article.

A particular topic can be found either directly, by turning to the appropriate place in the Encyclopedia, or from the Subject Index in Volume 8.

A strong attempt has been made to obtain consistent usage of terminology and symbols throughout the Encyclopedia. However, variations have been accepted for the historical volume and also for medical imaging and related articles (where different usages have grown up in some instances). For the convenience of readers, a list of relevant symbols is given after this Foreword. A list of terms and acronyms for imaging and medical articles is given in a Glossary immediately before the Subject Index in Volume 8.

In general, all equations (except in Volume 1) are written in SI form. The conventions favored by international bodies such as IUPAC and IUPAP are followed for quantities and units.

An attempt has been made to keep the use of abbreviations and acronyms within bounds. Those commonly used are defined on the front endpaper. Additional symbols and acronyms are defined in the relevant articles. Further discussion regarding conventions and nomenclature, together with a list of nuclear spin parameters, is to be found in the technical article ***Nuclear Spin Properties & Notation***.

The following systematic outline groups the articles under the five technical areas used in commissioning the encyclopedia. A full alphabetical listing of the articles appears after the symbols.

1 INORGANIC APPLICATIONS; POLYMER AND LIQUID CRYSTALLINE SOLUTIONS; QUADRUPOLAR NUCLEI; ONE- AND MULTI-DIMENSIONAL SPECTROSCOPY OF SOLUTIONS (J. W. Emsley)

1.1 Inorganic Applications

1.2 Polymers in the Liquid State

1.3 Liquid Crystalline Samples

Liquid Crystalline Samples: Deuterium NMR
Liquid Crystalline Samples: Diffusion
Liquid Crystalline Samples: Relaxation Mechanisms
Liquid Crystalline Samples: Spectral Analysis
Liquid Crystalline Samples: Structure of Nonrigid Molecules
Liquid Crystals: General Considerations
Liquid Crystals: Mixed Magnetic Susceptibility Solvents
Lyotropic Liquid Crystalline Samples
Multiple Quantum Spectroscopy in Liquid Crystalline Solvents
Spinning Liquid Crystalline Samples
Structure of Rigid Molecules Dissolved in Liquid Crystalline Solvents
Two-Dimensional NMR of Molecules Oriented in Liquid Crystalline Phases

1.4 Methods for 1D Spectra of Liquid Samples

Recording One-Dimensional High Resolution Spectra
Vicinal ^{1}H,^{1}H Coupling Constants in Cyclic π-Systems

1.5 Methods for 2D Spectra of Liquid Samples

COSY Spectra: Quantitative Analysis
COSY Two-Dimensional Experiments
Coupling Constants Determined by ECOSY
Heteronuclear Shift Correlation Spectroscopy
Multidimensional Spectroscopy: Concepts
NOESY
Relayed Coherence Transfer Experiments
ROESY
Selective Hartmann–Hahn Transfer in Liquids
Selective NOESY
Two-Dimensional *J*-Resolved Spectroscopy
Two-Dimensional Methods of Monitoring Exchange

1.6 Methods for 3D and 4D Spectra of Liquid Samples

Homonuclear Three-Dimensional NMR of Biomolecules
Three-Dimensional HMQC–NOESY, NOESY–HMQC, & NOESY–HSQC
Three- & Four-Dimensional Heteronuclear Magnetic Resonance
TOCSY in ROESY & ROESY in TOCSY

1.7 General Methods for Liquid Samples

Analysis of High Resolution Solution State Spectra
Analysis of Spectra: Automatic Methods
Chemical Exchange Effects on Spectra
Chemically Induced Dynamic Nuclear Polarization
Composite Pulses
Concentrated Solution Effects
Decoupling Methods
Double Resonance
Field Cycling Experiments
Flow NMR
Heteronuclear Assignment Techniques
INADEQUATE Experiment
INEPT
Magnetic Equivalence
Magnetic Field Induced Alignment of Molecules
Multiple Quantum Spectroscopy of Liquid Samples
Phase Cycling
Polarization Transfer Experiments via Scalar Coupling in Liquids
Quadrupolar Interactions
Radiofrequency Pulses: Response of Nuclear Spins
Relaxation of Quadrupolar Nuclei Measured via Multiple Quantum Filtration
Selective Pulses
Shaped Pulses
Spin Echo Spectroscopy of Liquid Samples
Steady-State Techniques for Low Sensitivity & Slowly Relaxing Nuclei

1.8 Diffusion and Flow

Diffusion & Flow in Fluids
Electrophoretic NMR

1.9 Applications to Unusual Samples

Colloidal Systems
Electrolytes
Geological Applications
Molten Salts
Terrestrial Magnetic Field NMR

2 PHYSICS APPLICATIONS; SOLID METHODS; SOLID-STATE APPLICATIONS (B. C. Gerstein)

2.1 Electron–Nuclear Interactions

Dynamic Nuclear Polarization & High-Resolution NMR of Solids
Electron–Nuclear Hyperfine Interactions
Electron–Nuclear Multiple Resonance Spectroscopy
Knight Shift
Laser Devices in NMR: Routes to Chaos
Optically Enhanced Magnetic Resonance
Polarization of Noble Gas Nuclei with Optically Pumped Alkali Metal Vapors
Ruby NMR Laser
SQUIDs

2.2 Heterogeneous Catalysis; Zeolites; Silicates; Supported Metals

Adsorbed Species: Spectroscopy & Dynamics
Brønsted Acidity of Solids
Chemical Exchange on Solid Metal Surfaces
Diffusion in Porous Media
Hydrogen Molecules
Microporous Materials & Xenon-129 NMR
Molecular Sieves: Crystalline Systems
Reactions in Zeolites
Silica Surfaces: Characterization
Silicon-29 NMR of Solid Silicates
Supported Metal Catalysts
Vanadium Catalysts: Solid State NMR

2.3 Polymers and Biopolymers

Polymer Blends: Miscibility Studies

Polymer Dynamics & Order from Multidimensional Solid State NMR
Polymer Physics
Polysaccharide Solid State NMR
Solid Biopolymers
Solid Polymers: Shielding & Electronic States

2.4 Materials Science; Coals; Cokes; Ceramics and Glasses

Amorphous Materials
Amorphous Silicon Alloys
Ceramics
Ceramics Imaging
Coal Structure from Solid State NMR
Cokes
Diamond Thin Films
Fast Ion Conductors
Ferroelectrics & Proton Glasses
Fossil Fuels
Intercalation Compounds
Wood & Wood Chars

2.5 Quadrupolar Nuclei; Nutation; Double Rotation; Dynamic Angle Spinning; Shielding, Dipolar, and Electric Field Gradient Tensors

Chemical Shift Tensor Measurement in Solids
Chemical Shift Tensors in Single Crystals
Deuterium NMR in Solids
Dipolar & Indirect Coupling Tensors in Solids
Dipolar Spectroscopy: Transient Nutations & Other Techniques
Double Rotation
Dynamic Angle Spinning
High Speed MAS of Half-Integer Quadrupolar Nuclei in Solids
Inorganic Solids
Line Narrowing Methods in Solids
Magic Angle Turning & Hopping
Nutation Spectroscopy of Quadrupolar Nuclei
Overtone Spectroscopy of Quadrupolar Nuclei
Quadrupolar Nuclei in Glasses
Quadrupolar Nuclei in Solids
Two-Dimensional Powder Correlation Methods

2.6 High-Resolution Solid-State NMR; Magic Angle Spinning; CRAMPS; REDOR; TEDOR; Line Narrowing

Double Rotation
Dynamic Angle Spinning
Echoes in Solids
HETCOR in Organic Solids
Homonuclear Recoupling Schemes in MAS NMR
Line Narrowing Methods in Solids
Magic Angle Spinning
Magic Angle Spinning Carbon-13 Lineshapes: Effect of Nitrogen-14
Magic Angle Spinning: Effects of Quadrupolar Nuclei on Spin-1/2 Spectra
Magic Angle Turning & Hopping
Magnetic Susceptibility & High Resolution NMR of Liquids & Solids
Nutation Spectroscopy of Quadrupolar Nuclei
REDOR & TEDOR
Rotating Solids
Sideband Analysis in Magic Angle Spinning NMR of Solids
Spectral Editing Techniques: Hydrocarbon Solids
Two-Dimensional Powder Correlation Methods
Variable Angle Sample Spinning

2.7 Solid-State Physics; Superconductors; Metals; Fast Ionic Conductors

Amorphous Materials
Amorphous Silicon Alloys
Diffusion in Rare Gas Solids
Diffusion in Solids
Dynamic Spin Ordering of Matrix Isolated Methyl Rotors
Fast Ion Conductors
Ferroelectrics & Proton Glasses
High Temperature Superconductors
Hydrogen–Metal Systems
Hydrogen Molecules
Incommensurate Systems
Knight Shift
Laser Devices in NMR: Routes to Chaos
Low Spin Temperature NMR
Metallic Superconductors
Metals: Pure & Alloyed
Nuclear Ferromagnetism & Antiferromagnetism
Optically Enhanced Magnetic Resonance
Ortho–Para Hydrogen at Low Temperature
Phase Transitions & Critical Phenomena in Solids
Polarization of Noble Gas Nuclei with Optically Pumped Alkali Metal Vapors
Polymer Physics
Polymorphism & Related Phenomena
Quantum Exchange
Quantum Optics: Concepts of NMR
Reorientation in Crystalline Solids: Propeller-Like R_3M Species
Ruby NMR Laser
Spin Diffusion in Solids
Thermodynamics of Nuclear Magnetic Ordering
Ultraslow Motions in Solids
Ultrasonic Irradiation & NMR
Zero Field NMR

2.8 Specialized Techniques; High-Pressure NMR; Squids

Force Detection & Imaging in Magnetic Resonance
Gas Phase Studies of Chemical Exchange Processes
Gases at High Pressure
Geometric Phases
High-Pressure NMR
Instrumentation for the Home Builder
Kinetics at High Pressure
Knight Shift
Thermometry
Well Logging
Wide Lines for Nonquadrupolar Nuclei
Zero Field NMR

2.9 Theory; Rudimentary NMR; Average Hamiltonian; Multiple Quantum NMR

Average Hamiltonian Theory
Chemical Exchange on Solid Metal Surfaces
CRAMPS
Cross Polarization in Rotating Solids: Spin-1/2 Nuclei
Cross Polarization in Solids
Double Quantum Coherence
Floquet Theory
Internal Spin Interactions & Rotations in Solids
Multiple Quantum Coherence in Spin-1/2 Dipolar Coupled Solids
Multiple Quantum Coherences in Extended Dipolar Coupled Spin Networks
Multiple Quantum NMR in Solids
Nuclear Spin Properties & Notation
Rudimentary NMR: The Classical Picture

3 BIOLOGICAL APPLICATIONS (S. I. Chan)

3.1 Biological Macromolecules

Biological Macromolecules
Biological Macromolecules: NMR Parameters
Chemical Shifts in Biochemical Systems
Vicinal Coupling Constants & Conformation of Biomolecules

3.2 Applications of Isotopes

Biosynthesis & Metabolic Pathways: Carbon-13 and Nitrogen-15 NMR
Biosynthesis & Metabolic Pathways: Deuterium NMR
Enzymatic Transformations: Isotope Probes
Oxygen-17 NMR: Applications in Biochemistry
Oxygen-18 in Biological NMR
Tritium NMR in Biology

3.3 Biological Structures in Solution

Biological Macromolecules: Structure Determination in Solution
Distance Geometry
Nucleic Acid Structures in Solution: Sequence Dependence
Protein Structures: Relaxation Matrix Refinement
Relaxation Matrix Refinement of Nucleic Acids
Structures of Larger Proteins, Protein–Ligand, & Protein–DNA Complexes by Multi-Dimensional Heteronuclear NMR

3.4 Biological Structures in the Solid State

Membrane Proteins
Membranes: Deuterium NMR
Polysaccharide Solid State NMR
Rotational Resonance in Biology
Time-Resolved Solid-State NMR of Enzyme–Substrate Interactions

3.5 Biological Dynamics

Dynamics of Water in Biological Systems: Inferences from Relaxometry
Enzyme-Catalyzed Exchange: Magnetization Transfer Measurements
Hydrogen Exchange & Macromolecular Dynamics
Molecular Motions: T_1 Frequency Dispersion in Biological Systems
Motional Effects on Protein Structure: Acyl Carriers
Nucleic Acid Flexibility & Dynamics: Deuterium NMR
Protein Dynamics from NMR Relaxation
Protein Dynamics from Solid State NMR
Selective Relaxation Techniques in Biological NMR
Slow & Ultraslow Motions in Biology

3.6 Peptides and Proteins

Amino Acids, Peptides & Proteins: Chemical Shifts
Bacterial Chemotaxis Proteins: Fluorine-19 NMR
Calcium-Binding Proteins
Calmodulin
Carbonic Anhydrase
Cobalt(II)- & Nickel(II)-Substituted Proteins
Copper Proteins
Enzymes Utilizing ATP: Kinases, ATPases and Polymerases
Haptens & Antibodies
Heme Peroxidases
Hemoglobin
Iron–Sulfur Proteins
Metallothioneins
Mitochondrial Cytochrome *c*
Muscle Proteins
Myoglobin
Nonheme Iron Proteins
Pancreatic Trypsin Inhibitor
Peptide Hormones
Peptide & Protein Secondary Structural Elements
Peptides & Polypeptides
Phage Lysozyme: Dynamics & Folding Pathway
Proteases
Protein Hydration
Proteins & Protein Fragments: Folding
SH2 Domain Structure
Synthetic Peptides
Thioredoxin & Glutaredoxin
Transferrins
Zinc Fingers

3.7 Nucleic Acids

DNA: A-, B-, & Z-
DNA–Cation Interactions: Quadrupolar Studies
DNA Triplexes, Quadruplexes, & Aptamers
Drug–Nucleic Acid Interactions
Nucleic Acids: Base Stacking & Base Pairing Interactions
Nucleic Acids: Chemical Shifts
Nucleic Acids: Phosphorus-31 NMR
Nucleic Acids: Spectra, Structures, & Dynamics
Ribosomal RNA
RNA Structure & Function: Modified Nucleosides

3.8 Membranes

Bacteriorhodopsin & Rhodopsin
Bilayer Membranes: Deuterium & Carbon-13 NMR
Bilayer Membranes: Proton & Fluorine-19 NMR

Filamentous Bacteriophage Coat Protein
Gramicidin Channels: Orientational Constraints for Defining High-Resolution Structures
D-Lactate Dehydrogenase
Lipid Polymorphism
Lipoproteins
Membrane Lipids of *Acholeplasma laidlawii*
Membranes: Carbon-13 NMR
Membranes: Phosphorus-31 NMR
Phospholipid–Cholesterol Bilayers
Sonicated Membrane Vesicles

3.9 Carbohydrates

Carbohydrates & Glycoconjugates
Glycolipids
Polysaccharides & Complex Oligosaccharides

3.10 Other Systems and Methods

Agriculture & Soils
Biological Systems: Spin-3/2 Nuclei
Bioreactors and Perfusion
Cell Suspensions
Half-Filter Experiments: Proton–Carbon-13
Plant Physiology
Water Signal Suppression in NMR of Biomolecules

4 INSTRUMENTATION; ORGANIC APPLICATIONS; RELAXATION TOPICS; THEORY (T. C. Farrar)

4.1 Relaxation Times

4.1.1 Experimental

Carbon-13 Relaxation Measurements: Organic Chemistry Applications
Coupled Spin Relaxation in Polymers
Gas Phase Studies of Intermolecular Interactions & Relaxation
Nuclear Overhauser Effect
Paramagnetic Relaxation in Solution
Relaxation: An Introduction

4.1.2 Theory

Brownian Motion & Correlation Times
Liouville Equation of Motion
Quantum Tunneling Spectroscopy
Relaxation of Coupled Spins from Rotational Diffusion
Relaxation Effects of Chemical Exchange
Relaxation Mechanisms: Magnetization Modes
Relaxation Processes in Coupled-Spin Systems
Relaxation Processes: Cross Correlation & Interference Terms
Relaxation Theory: Density Matrix Formulation
Relaxation Theory for Quadrupolar Nuclei
Relaxation of Transverse Magnetization for Coupled Spins
Rotating Frame Spin–Lattice Relaxation Off-Resonance
Spin–Rotation Relaxation Theory

4.2 Instrumentation

Probe Design & Construction
Probes for High Resolution
Probes for Special Purposes
Rapid Scan Correlation Spectroscopy
Shimming of Superconducting Magnets
Solid State Probe Design
Spectrometers: A General Overview
Stochastic Excitation

4.3 Chemical Shifts

4.3.1 Experimental

Carbon & Nitrogen Chemical Shifts of Solid State Enzymes
Chemical Shift Scales on an Absolute Basis
Isotope Effects on Chemical Shifts & Coupling Constants
Proton Chemical Shift Measurements in Biological Solids

4.3.2 Theory

Chemical Shift Tensors
Dynamic Frequency Shift
Electric Field Effects on Shielding Constants
Semiempirical Chemical Shift Calculations
Shielding Calculations: IGLO Method
Shielding Calculations: LORG & SOLO Approaches
Shielding Calculations: Perturbation Methods
Shielding: Overview of Theoretical Methods
Shielding in Small Molecules
Shielding Tensor Calculations
Shielding Theory: GIAO Method

4.4 Organic Applications

Coupling Through Space in Organic Chemistry
Heterocycles
Isotope Effects in Carbocation Chemistry
Micellar Solutions & Microemulsions
Natural Products
Organic Chemistry Applications
Quantitative Measurements
Vitamin B_{12}

4.5 Data Processing

Computer Assisted Structure Elucidation
Data Processing
Fourier Transform & Linear Prediction Methods
Fourier Transform Spectroscopy
Maximum Entropy Reconstruction

4.6 Spin Coupling

Indirect Coupling: Semiempirical Calculations
Indirect Coupling: Theory & Applications in Organic Chemistry
Stereochemistry & Long Range Coupling Constants

4.7 Solvent Studies

Hydrogen Bonding
Indirect Coupling: Intermolecular & Solvent Effects

4.8 Carbon-13 Studies

Carbon-13 Relaxation Measurements: Organic Chemistry Applications
Carbon-13 Spectral Simulation
Carbon-13 Studies of Deuterium Labeled Compounds
Two-Dimensional Carbon–Heteroelement Correlation

4.9 Field Gradient Experiments

Diffusion Measurements by Magnetic Field Gradient Methods
Field Gradients & Their Application
Radiofrequency Gradient Pulses

5 BIOMEDICAL APPLICATIONS; IMAGING PRINCIPLES AND APPLICATIONS (I. R. Young)

5.1 General Overview

EPR and In Vivo EPR: Roles for Experimental & Clinical NMR Studies
High Field Whole Body Systems
Imaging: A Historical Overview
Imaging using the Electronic Overhauser Effect
In Vivo ESR Imaging of Animals
Low-Field Whole Body Systems
Midfield Magnetic Resonance Systems

5.2 Image Formation Methods and Sensitivity

Image Formation Methods
Image Quality & Perception
Projection–Reconstruction in MRI
Selective Excitation in MRI
Sensitivity of the NMR Experiment
Sensitivity of Whole Body MRI Experiments
Spin Warp Data Acquisition
Spiral Scanning Imaging Techniques
Wavelet Encoding of MRI Images
Whole Body MRI: Strategies Designed to Improve Patient Throughput

5.3 Techniques of Imaging/Spectroscopy In Vivo

Brain: Sensory Activation Monitored by Induced Hemodynamic Changes with Echo Planar MRI
Chemical Shift Imaging
Complex Radiofrequency Pulses
Echo-Planar Imaging
Functional MRI: Theory & Practice
Inversion Recovery Pulse Sequence in MRI
Localization by Rotating Frame Techniques
Lung and Mediastinum: A Discussion of the Relevant NMR Physics
Multi-Echo Acquisition Techniques Using Inverting Radiofrequency Pulses in MRI
Multiple Quantum Coherence Imaging
Partial Fourier Acquisition in MRI
Spatial Localization Techniques for Human MRS
Tissue Water & Lipids: Chemical Shift Imaging & Other Methods
Trabecular Bone Imaging
Water Suppression in Proton MRS of Humans & Animals
Whole Body Magnetic Resonance: Fast Low-Angle Acquisition Methods

5.4 Machine and Equipment-Related Topics

Birdcage & Other High Homogeneity Radiofrequency Coils for Whole Body Magnetic Resonance
Cardiac Gating Practice
Coils for Insertion into the Human Body
Cryogenic Magnets for Whole Body Magnetic Resonance Systems
Gradient Coil Systems
Health & Safety Aspects of Human Magnetic Resonance Studies
Image Segmentation, Texture Analysis, Data Extraction, & Measurement
Multifrequency Coils for Whole Body Studies
Patient Life Support & Monitoring Facilities for Whole Body MRI
Radiofrequency Systems & Coils for MRI & MRS
Refrigerated & Superconducting Receiver Coils for Whole Body Magnetic Resonance
Resistive & Permanent Magnets for Whole Body MRI
Surface & Other Local Coils for In Vivo Studies
Whole Body Machines: NMR Phased Array Coil Systems
Whole Body Magnetic Resonance Spectrometers: All-Digital Transmit/Receive Systems
Whole Body MRI: Local & Inserted Gradient Coils

5.5 Artifacts

Eddy Currents & Their Control
Functional Neuroimaging Artifacts
Pulsatility Artifacts due to Blood Flow & Tissue Motion & Their Control
Respiratory Artifacts: Mechanism & Control
Selective Excitation Methods: Artifacts
Spin Warp Method: Artifacts
Susceptibility Effects in Whole Body Experiments
Whole Body Magnetic Resonance Artifacts

5.6 Measurement Techniques and Problems

pH Measurement In Vivo in Whole Body Systems
Quantitation in Whole Body MRS
Relaxation Measurements in Imaging Studies
Relaxation Measurements in Whole Body MRI: Clinical Utility
Relaxometry of Tissue
Surface Coil NMR: Quantification with Inhomogeneous Radiofrequency Field Antennas
Temperature Measurement for In Vivo MRI
Whole Body Machines: Quality Control

5.7 Radiofrequency Intensive Techniques and Aspects

Bioeffects & Safety of Radiofrequency Electromagnetic Fields
Linewidth Manipulation in MRI
Magnetization Transfer Contrast: Clinical Applications
Magnetization Transfer & Cross Relaxation in Tissue
Magnetization Transfer between Water & Macromolecules in Proton MRI

5.8 Contrast Agents

5.9 Flow and Flow-Related Phenomena

5.10 Clinical Studies

5.10.1 General

5.10.2 Head, Neck, and Brain

5.11 Spectroscopy

5.11.1 Human Studies

5.11.2 Nonhuman Studies

5.12 Microscopy and Solid-State Imaging

Symbols for NMR and Related Quantities

(i) Roman Alphabet

Symbol	Meaning
a or A	Hyperfine (electron–nucleus) coupling constant
$A_q^{(l,m)}$	The mth component of an irreducible tensor representing the nuclear spin operator for an interaction of type q
$\boldsymbol{B}$	Magnetic induction field (magnetic flux density)
$\boldsymbol{B}_0$	Static magnetic field of an NMR spectrometer
$\boldsymbol{B}_1$, $\boldsymbol{B}_2$	Radiofrequency magnetic fields associated with ν_1, ν_2
$\boldsymbol{B}_\mathrm{L}$	Local magnetic field (components $B_{x\mathrm{L}}$, $B_{y\mathrm{L}}$, $B_{z\mathrm{L}}$) of random field or dipolar origin
c	Coefficient in linear expansion of wavefunctions
$\mathbf{C}$	Spin–rotation interaction tensor
C_X	(i) Natural abundance of nuclide X, expressed as %; (ii) spin–rotation coupling constant of nuclide X
$\mathbf{D}$	Dipolar interaction tensor
D^C	Nuclear receptivity relative to that of carbon-13
D^P	Nuclear receptivity relative to that of the proton
$\boldsymbol{E}$	Electric field
$\hat{\boldsymbol{F}}_\mathrm{G}$	Nuclear spin operator for a group, G, of nuclei (conponents $\hat{F}_{\mathrm{G}x}$, $\hat{F}_{\mathrm{G}y}$, $\hat{F}_{\mathrm{G}z}$, $\hat{F}_{\mathrm{G}+}$, $\hat{F}_{\mathrm{G}-}$)
F_G	Magnetic quantum number associated with $\hat{\boldsymbol{F}}_\mathrm{G}$
g	Nuclear or electronic g factor (Landé splitting factor)
G	Magnetic field gradient amplitude
H_{ij}	Element of matrix representation of $\hat{\mathcal{H}}$
$H_{a,b,\ldots}^{(k)}$	The kth term in the Magnus expansion associated with interactions $a, b, \ldots$
$\hat{\mathcal{H}}$	Hamiltonian operator (in energy units); subscripts indicate the nature of the operator
i	$\sqrt{-1}$
$\hat{\boldsymbol{I}}_j$	Nuclear spin operator for nucleus j (components $\hat{I}_{jx}$, $\hat{I}_{jy}$, $\hat{I}_{jz}$)
$\hat{I}_{j+}$, $\hat{I}_{j-}$	'Raising' and 'lowering' spin operators for nucleus j
I_j	Magnetic quantum number associated with $\hat{\boldsymbol{I}}_j$
I	Moment of inertia
$\mathbf{J}$	Indirect coupling tensor
^nJ	Nuclear spin–spin coupling constant through n bonds (in Hz). Further information may be given by subscripts or in parentheses. Normally subscripts are only used for algebraic symbols for nuclei in spectral analysis cases, e.g. J_AX. Parentheses are used for indicating the species of nuclei coupled, e.g. $J(^{13}\mathrm{C}, {}^1\mathrm{H})$ or, additionally, the coupling path, e.g. J(POCF). The nucleus of higher mass should be given first
$J(\omega)$	Spectral density at angular frequency ω. Subscripts to J may be used to indicate the transition quantum change (0, 1 or 2)
^nK	Reduced nuclear spin–spin coupling constant (see the notes concerning ^nJ)
m_e	Mass of the electron
m_j	Eigenvalue of $\hat{I}_{jz}$ (magnetic component quantum number)
m_T	Total magnetic component quantum number for a spin system (eigenvalue of $\Sigma_j \hat{I}_{jz}$)
m_T(X)	Total magnetic component quantum number for X-type nuclei
$\boldsymbol{M}_0$	Equilibrium macroscopic magnetization of a spin system in the presence of $\boldsymbol{B}_0$
M_x, M_y, M_z	Components of a macroscopic magnetization
M_n	Moment of a spectrum (M_2 = second moment, etc.)
n_α, n_β	Populations of the α and β spin states
N	Total number of nuclei of a given type in the sample
$\boldsymbol{P}$	Angular momentum
P	Transition probability
$\mathbf{q}$	Electric field gradient tensor (principal components q_{xx}, q_{yy}, q_{zz}) (see also V)
Q	Nuclear quadrupole moment
r	(i) General symbol for distance; (ii) general symbol for spin state (as $\langle r|$ or $|r\rangle$)
R	Dipolar coupling constant, $(\mu_0/4\pi)\gamma_1\gamma_2(\hbar/2\pi)r^{-3}$
R_1^X	Spin–lattice (longitudinal) relaxation rate for nucleus X
R_2^X	Spin–spin (transverse) relaxation rate for nucleus X
$R_{1\rho}^\mathrm{X}$	Spin–lattice relaxation rate in the rotating frame for nucleus X
$R_{j\phi}$	Rotation by angle ϕ about axis j
s	General symbol for spin state (as $\langle s|$ or $|s\rangle$)
S	(i) Signal height; (ii) electron (or, occasionally, nuclear) spin; cf. I

Symbol	Meaning
T	Temperature
T_c	Coalescence temperature for an NMR spectrum
T_1^X	Spin–lattice (longitudinal) relaxation time of the X nucleus (further subscripts refer to the relaxation mechanism)
T_2^X	Spin–spin (transverse) relaxation time of the X nucleus (further subscripts refer to the relaxation mechanism)
T'_2	Inhomogeneity contribution to the dephasing time for M_x or M_y
T_2^*	Total dephasing time for M_x or M_y; $(T_2^*)^{-1} = T_2^{-1} + (T'_2)^{-1}$
$T_{1\rho}^X$	Spin–lattice relaxation time of the X nucleus in the frame of reference rotating with $\boldsymbol{B}_1$
T_d	Pulse delay time (in FT NMR)
T_{ac}	Acquisition time (in FT NMR)
T_p	Period for repetitive pulses (= interpulse time = $T_{ac} + T_d$ if τ_p is negligible)
$T_q^{(l,m)}$	The mth component of an irreducible tensor representing the strength of an interaction of type $\boldsymbol{q}$
u	In-phase (dispersion mode) signal
U_{1Q}	Single-quantum propagator
U_{2Q}	Double-quantum propagator
$U_\lambda(t,0)$	Propagator from time 0 to time t associated with the interaction λ
$U_\lambda^{-1}(t,0)$	Inverse of $U_\lambda(t,0)$
v	Out-of-phase (absorption mode) signal
$V_{\alpha\beta}$	Elements of cartesian field gradient tensor (see also $\boldsymbol{q}$)
W_0, W_1, W_2	Relaxation rates between energy levels differing by 0, 1, and 2 (respectively) in m_T; especially, but not uniquely, for systems of two spin-$\frac{1}{2}$ nuclei

(ii) Greek Alphabet

Symbol	Meaning
α	Nuclear spin wavefunction (eigenfunction of $\hat{I}_z$) for a spin-$\frac{1}{2}$ nucleus
α_A^2	The s-character of hybrid orbital at atom A
α_E	The Ernst angle (for optimum sensitivity)
β	Nuclear spin wavefunction (eigenfunction of $\hat{I}_z$) for a spin-$\frac{1}{2}$ nucleus
γ_X	Magnetogyric ratio of nucleus X
δ_k	The kth time interval in a pulse sequence
δ_X	Chemical shift (for the resonance) of nucleus of element X (positive when the sample resonates to high frequency of the reference). Usually in ppm. Further information regarding solvent, references, or nucleus of interest may be given by superscripts or subscripts or in parentheses
Δn	Population difference between nuclear states (Δn_0 at Boltzmann equilibrium)
$\Delta\delta$	Change or difference in δ
$\Delta\nu_{1/2}$	Full width (in Hz) of a resonance line at half-height
$\Delta\boldsymbol{\sigma}$	Anisotropy in $\boldsymbol{\sigma}$ ($\Delta\boldsymbol{\sigma} = \sigma_\parallel - \sigma_\perp$)
$\Delta\chi$	(i) Susceptibility anisotropy ($\Delta\chi = \chi_\parallel - \chi_\perp$); (ii) difference in electronegativities
ϵ_0	Permittivity of a vacuum
ζ	Anisotropy in shielding, expressed as $\sigma_{zz} - \sigma_{iso}$
η	(i) Nuclear Overhauser enhancement; (ii) tensor asymmetry factor (e.g. in e^2qQ/h); (iii) viscosity
θ	Angle, especially for that between a given vector and $\boldsymbol{B}_0$
κ	The skew of a tensor
$\boldsymbol{\mu}$	(i) Magnetic dipole moment (component μ_z along $\boldsymbol{B}_0$); (ii) electric dipole moment
μ_0	Permeability of a vacuum
μ_B	Bohr magneton
μ_N	Nuclear magneton
ν_c	Carrier frequency of the radiation
ν_j	Larmor precession frequency of nucleus j (in Hz)
ν_0	(i) Spectrometer operating frequency; (ii) Larmor precession frequency (general, or of bare nucleus)
ν_1	Frequency of 'observing' rf magnetic field
ν_2	Frequency of 'irradiating' rf magnetic field
Ξ_X	Resonance frequency for the nucleus of element X in a magnetic field such that the protons in TMS resonate at exactly 100 MHz
$\rho(x,y,z)$	Spin density at point x, y, z
$\boldsymbol{\rho}$	Density matrix
$\hat{\rho}$	Density operator
ρ_{ij}	Element of matrix representation of $\hat{\rho}$
$\boldsymbol{\sigma}$	Shielding tensor
σ_j	(Isotropic) shielding constant of nucleus j. Usually in ppm. Subscripts may alternatively indicate contributions to σ
$\sigma_\parallel$, $\sigma_\perp$	Components of σ parallel and perpendicular to a symmetry axis
$\hat{\sigma}$	Reduced density operator
τ	Time between rf pulses (general symbol)

τ_A	Pre-exchange lifetime of molecular species A
τ_c	Correlation time, especially for molecular tumbling
τ_d	Dwell time
τ_e	Electronic correlation time
τ_{null}	Recovery time sufficing to give zero signal after a 180° pulse
τ_p	Pulse duration
τ_{sc}	Correlation time for relaxation by the scalar mechanism
τ_{sr}	Correlation time for spin–rotation relaxation
$\tau_{\parallel}$, $\tau_{\perp}$	Correlation times for molecular tumbling parallel and perpendicular to a symmetry axis
χ	(i) Magnetic susceptibility; (ii) electronegativity; (iii) nuclear quadrupole constant ($=e^2qQ/h$)
ω_j, ω_0, ω_1, ω_2, ω_c	As for ν_j, ν_0, ν_1, ν_2, ν_c but in rad s^{-1}
Ω	The span of a tensor
$\mathbf{\Omega}_1$, $\mathbf{\Omega}_2$	The strength of a rf magnetic field, expressed in angular frequency units (as of precession) for a nucleus of magnetogyric ratio γ ($\mathbf{\Omega}_1 = -\gamma \boldsymbol{B}_1$, $\mathbf{\Omega}_2 = -\gamma \boldsymbol{B}_2$)

Contents

VOLUME 4

VOLUME 6

VOLUME 7

VOLUME 8

Tis

Tissue Behavior Measurements Using Phosphorus-31 NMR

Simon D. Taylor-Robinson
Hammersmith Hospital, London, UK

&

Claude D. Marcus
Hôpital Robert Debré, Reims, France

1 INTRODUCTION

Phosphorus-31 MRS provides a noninvasive method of assessment of mobile phosphorus-containing compounds. A typical in vivo ^{31}P MR spectrum contains seven resonances (Figure 1). Phospholipid cell membrane precursors, adenosine monophosphate (AMP) and glycolytic intermediates (sugar phosphates) contribute to the phosphomonoester (PME) peak. Phospholipid cell membrane degradation products and endoplasmic reticulum contribute to the phosphodiester (PDE) peak. Information on tissue bioenergetics can be obtained from inorganic phosphate (Pi), phosphocreatine (PCr) and the three nucleoside triphosphate resonances. A measurement of intracellular pH (pHi) can be calculated from the chemical shift of the Pi peak.

Pathological processes involving hypoxia and ischemia are particularly amenable to ^{31}P MRS assessment using absolute or relative quantitation of the PCr and Pi peaks. The PCr/Pi ratio has been the most commonly used marker of tissue energy reserve under these conditions. Phosphorus-31 MRS may also

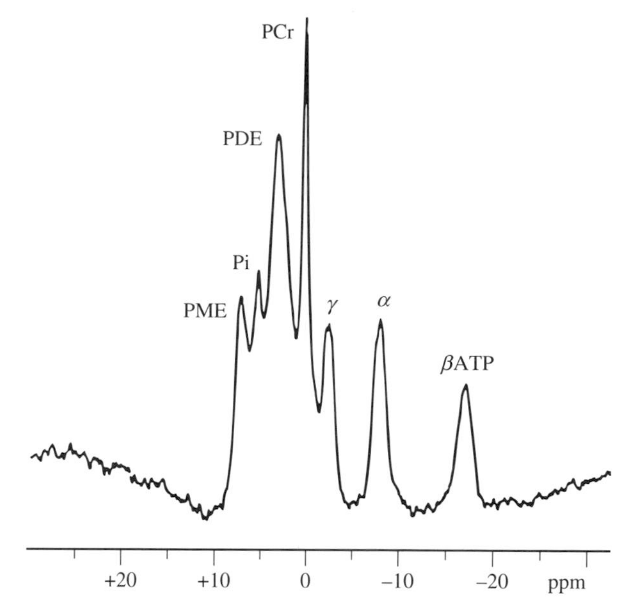

Figure 1 An unlocalized ^{31}P MR spectrum from the head of a healthy adult volunteer. There are seven resonances. PME = phosphomonoester, Pi = inorganic phosphate, PDE = phosphodiester, PCr = phosphocreatine, ATP = adenosine triphosphate

provide an indication of the viability of isolated donor organs prior to transplantation.

In this article, we discuss the role of in vivo ^{31}P MRS in the examination of tissue bioenergetics in human studies.

2 ^{31}P MRS AND CELLULAR ENERGY STATUS

The energy metabolism of each cell is dependent on the synthesis and utilization of compounds which contain high-energy phosphate bonds such as ATP and PCr.[1] ATP is present in all cells, but PCr is limited to those tissues containing creatine and the enzyme creatine kinase (CK), such as skeletal muscle and brain. ATP has a pivotal role in cellular bioenergetics. The demand for ATP is usually reasonably constant under normal resting conditions. The hydrolysis of ATP to ADP (adenosine 5′-diphosphate) and Pi releases potential energy from high-energy phosphate bonds for all activities involved in maintaining intracellular homeostasis and the specialized functions which may be unique to each cell type. Phosphocreatine acts as an energy reservoir in tissues such as muscle and brain. The enzyme CK splits PCr to provide an energy source for ATP resynthesis (Figure 2). Oxidative phosphorylation, which involves an electron transport chain in the mitochondrial membrane, is the process which provides most of the ATP for each cell under conditions of adequate oxygen supply. In a range of situations the requirements for ATP cannot be met by oxidative phosphorylation in the mitochondria. For example, oxidative phosphorylation is impaired in hypoxia and may be inadequate in normal exercising muscle. The shortfall in ATP production is met by glycolysis in the cytoplasm. Lactic acid may accumulate as a consequence of this process. Phosphocreatine is utilized under these conditions and as PCr falls, Pi increases, but any reduction in ATP is minimized because of the buffering effect of CK. Only when the PCr pool

For References see p. 4770

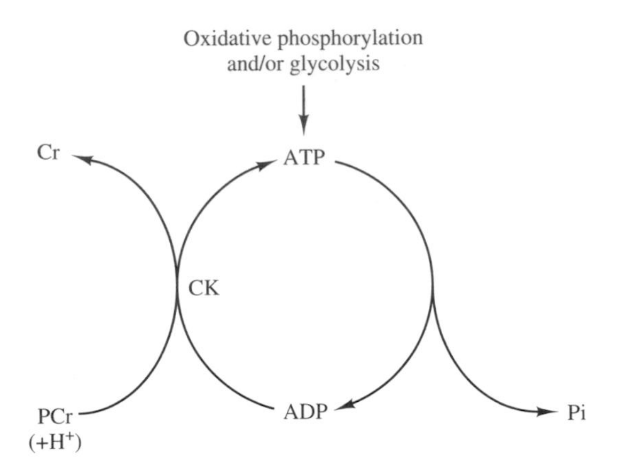

Figure 2 The interrelationship between ATP and PCr. Cr = creatine, CK = creatine kinase, ADP = adenosine diphosphate

is completely consumed do the tissue ATP levels fall appreciably, leading to a rise in both ADP and Pi.

Mitochondrial oxidative metabolism may also be affected by changes in pHi,[2] but the rate of ATP biosynthesis remains constant under normal conditions. The concentration of ADP is a major factor in the rate of ATP production.[2,3] The majority of ADP is not directly detectable using NMR methods because of tissue binding. The ability of a particular tissue to synthesize ATP may be calculated from the phosphorylation potential, given by equation (1)

$$[\mathrm{ATP}]/[\mathrm{ADP}][\mathrm{Pi}] \tag{1}$$

A further indication of cellular energy status is given by equation (2):

$$[\mathrm{PCr}]/[\mathrm{Cr}][\mathrm{Pi}] \tag{2}$$

Absolute or relative concentrations of tissue PCr, Pi, and ATP can be obtained using ^{31}P MRS. The rate of ATP biosynthesis, the phosphorylation potential, ADP concentrations, and the kinetics of CK may be calculated from ^{31}P MRS measurements[2,3] (see also ***Enzymes Utilizing ATP: Kinases, ATPases and Polymerases*** and ***Enzyme-Catalyzed Exchange: Magnetization Transfer Measurements***). The PCr/Pi and PCr/ATP ratios are commonly used MR indices of cellular energy status or 'bioenergetic reserve', as they reflect equations (1) and (2).

The oxidative phosphorylation pathway can be assessed using the PCr/Pi ratio. Glycolytic intermediates contribute to the PME peak and therefore an indirect assessment of glycolysis may be obtained from quantitation of this resonance. This is of particular importance in assessment of glycolytic disorders in muscle and in dynamic studies of liver metabolism.

3 ^{31}P MRS AND INTRACELLULAR pH

Phosphorus-31 MRS can be used to measure intracellular pH from the chemical shift of the Pi peak with reference to the PCr resonance in tissues such as brain and muscle where PCr is present. In tissues such as liver where PCr is absent the reference used is αATP. The MR signal from Pi is thought not to represent the total intracellular levels of Pi. It is unclear why the remainder is not detected, but it may be bound in the mitochondria. It is not known whether there are pH differences between the cytoplasm and the mitochondria. Split Pi resonances representing intracellular compartmentation have not been observed in human ^{31}P MRS in vivo.

Different body tissues are more susceptible to ischemia and hypoxia than others. In normal exercising muscle, anaerobic glycolysis may take place with the accumulation of lactic acid as a normal sequence of events. The large PCr reservoir in muscle ensures an energy source for these anaerobic reactions. The accumulation of lactate in the brain is more likely to be of pathological consequence because cerebral function is particularly sensitive to hypoxic insults. The measurement of intracellular acidosis may be used as a marker of hypoxia–ischemia and cellular dysfunction in conditions such as stroke or birth asphyxia. Under these circumstances, the PCr/Pi ratio is reduced. Lactate accumulation leads to a reduction in pHi and a change in the chemical shift of Pi. Intracellular pH may change with time: for example, an intracellular alkalosis can develop in ischemic brain tissue of stroke patients over an extended time period.[4] The underlying mechanisms behind such pH changes are not known. However, the MR measurement of pHi may be used to discriminate between diseased and healthy tissue in combination with indices of bioenergetic reserve such as PCr/Pi.

4 ^{31}P MRS AND PHOSPHOLIPID METABOLISM

The PME and PDE peaks are multicomponent. Phosphoethanolamine (PE) and phosphocholine are cell membrane precursors and contribute to the PME peak with signal from glycolytic intermediates and AMP. Glycerophosphorylethanolamine (GPE) and glycerophosphorylcholine (GPC), which are cell membrane degradation products, contribute to the PDE resonance. Phosphoenolpyruvate and endoplasmic reticulum form other contributory factors. The relative contributions of these compounds to the PME and PDE peaks may change with disease. In situations where there is rapid cell turnover, the PME resonance may be elevated due to an increase in PE and phosphocholine. Similarly, under conditions of rapid cell death, after tumor embolization or chemotherapy, for example, an increased contribution of GPE and GPC to the PDE peak may be expected. Phosphorus–proton decoupling can be used to resolve further these resonances in vivo.

5 ^{31}P MRS AND TRANSPLANT ORGAN VIABILITY

Organ transplantation is a steadily expanding surgical field. Increased patient survival rates have been achieved because of improved operative techniques, anti-rejection chemotherapy, and methods for harvesting and preserving donor organs.

The success of any transplant procedure is dependent on the quality of the donor graft and this is a reflection of organ storage methods. The use of more physiological preservation fluids has led to increased storage times, allowing donor organs to be transported between transplant centers. Despite these advances, tissue damage, caused by cold preservation, is an important factor in patient morbidity and mortality. Phosphorus MRS provides a noninvasive assessment of the viability of the isolated donor organ prior to transplantion. The standard indices of bioenergy reserve such as PCr/ATP and PCr/Pi ratios may be appropriate markers in heart transplantation, but the ^{31}P MR spectra from healthy liver and kidney contain no appreciable PCr. Therefore, other indices such as the PME/Pi ratio have to be employed. Adenosine 5′-triphosphate begins to degenerate to ADP and Pi immediately each organ has been harvested. With time, ADP further degenerates to AMP which contributes to the PME peak. The PME/Pi ratio may reflect the ability of the isolated organ to rephosphorylate AMP to ATP on transplantation. Specific transplant studies are considered later in this chapter.

6 CLINICAL APPLICATIONS

The development of whole body magnets has allowed clinical ^{31}P MRS studies to be undertaken in patients with a variety of pathological processes, facilitating comparisons with healthy volunteers and offering the possibility of disease monitoring in response to treatment. MR measurements of bioenergetic reserve such as the PCr/Pi ratio have been proposed as predictors of outcome in hypoxic and/or ischemic conditions such as birth asphyxia. A review of the role of ^{31}P MRS in some of the major disease processes follows.

6.1 ^{31}P MRS and the Adult Brain (See Also *Brain MRS of Human Subjects*)

6.1.1 Stroke

Stroke is the most common adult neurological condition and is a prominent cause of mortality in developed countries. Early diagnosis of ischemia facilitates more appropriate treatment, and delay may result in an irreversible loss of neuronal function. Ischemia and the consequent tissue hypoxia result in a depletion in PCr, ATP levels being maintained initially. The acute stages of stroke are characterized by a decreased PCr/Pi ratio in the ^{31}P MR spectrum and an intracellular acidosis.[5] These changes may be detectable before MRI changes become evident.[5] A combination of ^{31}P MRS and MRI may aid early diagnosis, and help to monitor the brain's response to treatment.

Persistent cerebral ischemia results in irreversible cell damage and neuronal death. A reduction in phosphorus signal has been seen in patients with chronic stroke, consistent with a reduction in viable cells in the area of infarction.[4] The pHi has been noted to change with time, resulting in a rebound intracellular alkalosis which may persist.[4] The reasons underlying this change remain unclear.

6.1.2 Transient Ischemia

A transient ischemic attack is defined as a reversible neurological deficit that lasts for 24 h or less. The diagnosis is therefore retrospective and treatment is aimed at preventing recurrence. One study of two patients suggested that the total phosphorus signal was reduced in the absence of MRI changes.[4] This may be due to ischemia, but remains difficult to explain.

6.1.3 Epilepsy

Temporal lobe epilepsy may be unresponsive to standard antiepileptic therapy and a small number of patients require surgical resection of the epileptogenic focus. The definition of the pathological area needs careful preoperative planning. MRI studies have been used to obtain hippocampal volumes[6] but the role of ^{31}P MRS has been limited. A reduced PME, probably as a result of underlying hippocampal sclerosis, has been noted in some studies. An increased Pi and an unexplained intracellular alkalosis have been noted in most studies,[4,7] all of which have involved relatively small numbers of patients.

6.1.4 Alzheimer's Disease

The results of ^{31}P MRS studies have been disappointing. Bottomley and colleagues[8] found no changes in either metabolite ratios or absolute concentrations of metabolites.

6.1.5 Multiple Sclerosis

This condition is common in temperate climates and is characterized by episodes of focal neurological deficit which relapse and remit over a period of many years. The classic histological lesions are plaques of demyelination. MRI has revolutionized the diagnosis of this condition, but the results of ^{31}P MRS studies are less clear-cut. In one study[9] there was a decrease in the PCr/ATP ratio in some patients with active disease, whereas in another study[10] there was an increase in the PCr/ATP ratio with disease activity.

6.1.6 Other Cerebral Conditions

Decreased PCr/Pi has been observed in the cerebral ^{31}P MR spectra from patients with migraine and mitochondrial cytopathies. This suggests an alteration in bioenergetic reserve.

Chronic hepatic encephalopathy is defined as the neuropsychiatric impairment observed in patients with cirrhosis of the liver. The results of some ^{31}P MRS studies have suggested altered brain energy metabolism in these patients (see also ***Hepatic & Other Systemically Induced Encephalopathies: Applications of MRS***).

6.2 ^{31}P MRS and Pediatric Brain Studies

The normal ^{31}P MR spectrum from a healthy neonatal brain is significantly different from adult spectra. The PME signal is much larger in the neonate and this varies with gestational age. Azzopardi and colleagues[11] found that the PME is smaller and the PDE larger in the healthy full term infant than the healthy preterm infant, related most probably to the changes in membrane lipids with myelin formation. The PCr/Pi ratio increases with gestational age, indicating an increased phosphorylation potential with brain development. The measured intracellular pH appears not to vary with the gestational age of healthy newborn infants.

6.2.1 Hypoxic–Ischemic Encephalopathy

Birth asphyxia of the newborn infant has been extensively investigated using ^{31}P MRS. This condition is almost unique because there are well-defined MR indices of prognosis. Reduced PCr/Pi ratio has been correlated with outcome.[12]

For References see p. 4770

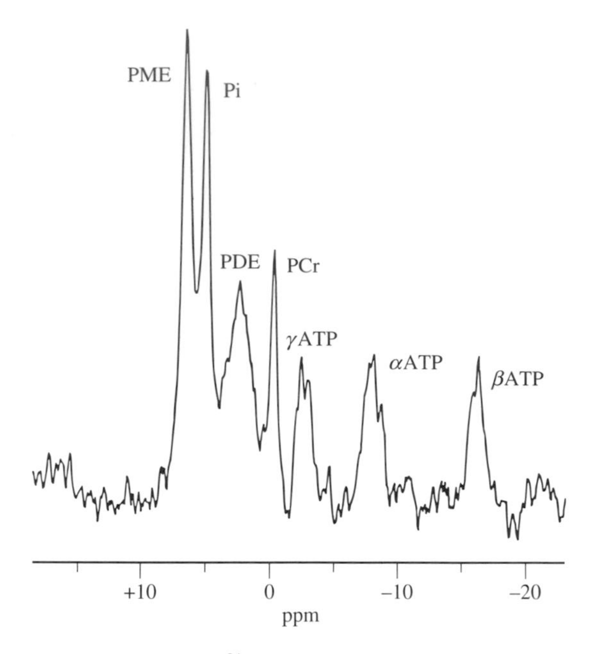

Figure 3 An unlocalized ^{31}P MR spectrum from the head of a birth asphyxiated neonate. The PCr peak is reduced and the Pi peak is elevated

Spectra obtained in the initial 24 h after birth may be normal, but a fall in PCr and a rise in Pi (reduced PCr/Pi ratio) may develop over the ensuing hours and days (Figure 3). This reflects defective oxidative phosphorylation as a result of birth trauma. Intracellular pH tends to rise in a delayed response to the hypoxic–ischemic insult and there may also be a reduction in ATP levels. The metabolite ratios tend to return to normal within 2 weeks in neonates who recover. The reduction in PCr/Pi ratio is proportional to the degree of subsequent neurodevelopmental impairment and to reduced cranial growth in the first year of life.[12] In the severest cases, where the neonates subsequently die, the PCr and ATP may be almost undetectable. The Pi often rises out of proportion to the reduction in PCr. This delayed or secondary response to ischemic injury is poorly understood, but may be partly due to the toxic effects of neurotransmitters such as glutamate, which may induce mitochondrial membrane disruption through the generation of free radicals.[13] Phosphorus-31 MR spectroscopy may be utilized to monitor the effectiveness of treatment designed to prevent this secondary energy failure. The development of suitable therapeutic regimens is an area of current and future research. However, the PCr/Pi ratio is already being used as a predictor of patient outcome and may give the pediatrician insight into planning future management decisions.

6.3 Myocardial Metabolism

Heart muscle is well supplied with oxygen and mitochondria. It relies on glycolysis to a much smaller extent than skeletal muscle, which has a higher concentration of glycolytic enzyme and fewer mitochondria. The regulation of cardiac metabolism and its relation to mechanical function has been widely studied in animals and humans[14,15] (see also ***Cardiovascular NMR to Study Function*** and ***Heart Studies Using MRS***).

6.3.1 Measurements in Normal Myocardium

Some of the major problems encountered in cardiac studies are related to signal contamination of the myocardium from the chest wall and from blood circulating through the cardiac chambers. The chest wall contains about four times more PCr than the normal myocardium. Signal from blood usually includes appreciable amounts of ATP and PDE, and an intense 2,3-diphosphoglycerate (2,3-DPG) resonance, which may obscure the Pi and PME resonances. Under such circumstances the pHi and the PCr/Pi ratio cannot be measured.[16]

Inorganic phosphate and pHi measurements still remain difficult in humans. Localization techniques may be used to minimize signal contamination from chest wall or blood. Correction for residual blood contamination and saturation effects may also be made.

The PCr/ATP ratio is the most frequently used index of bioenergy reserve in cardiac studies. Results from human and animal studies are comparable.[17] This ratio remains relatively constant during the cardiac cycle and in exercise. It is also highly reproducible. The PDE/ATP ratio is difficult to measure accurately because of interference from the 2,3-DPG signal. The available data show considerable biological variation in human studies. Different examination techniques may complicate this matter. The pHi of normal human myocardium is pH 7.15 ± 0.03.[18]

6.3.2 Measurements in Myocardial Diseases

Spectral abnormalities in human myocardium mainly reflect ischemic conditions.[19] The PCr/ATP ratio decreases significantly during hand grip exercise[20] in patients with a high degree of coronary artery stenosis. This ratio is also reduced in advanced stages of heart failure, in left ventricular hypertrophy, or dilated cardiomyopathy.[21] An increase in normal PCr/ATP ratio has been shown during treatment for heart failure.[17] Apart from blood contamination, an increased PDE/ATP ratio might reflect an accumulation of cell membrane degradation products in patients with a decreased left ventricular ejection fraction.[22]

6.3.3 Cardiac Transplantation

Advances in operative technique and in antirejection therapy have led to cardiac transplantation being used as a viable treatment for patients with end-stage heart disease. Endomyocardial biopsies with histological grading are the gold standard for detection of rejection. No reliable noninvasive alternative is available at present. Phosphorus-31 MRS has been studied in this context.

Animal models of cardiac transplantation have demonstrated an early decrease in PCr/ATP and PCr/Pi ratios.[23] A parallel rise in PDE/ATP was found in two studies which preceded the onset of histologically detectable rejection by 1 or 2 days.[23,24] The increase in the PDE peak measurements may be due to immunological or reperfusion injury. The abnormalities of high-energy phosphate during acute, severe rejection are reversible with antirejection therapy.[25]

In human studies no correlation was found between the decrease in the PCr/ATP ratio measurements and the biopsy grading. Severe rejection may therefore not be distinguishable from mild or moderate rejection by ^{31}P MRS alone.[26] The discrepancy between animal and human studies may be explained by the different time courses for MRS examinations and the rather more heterogeneous human population and study conditions.

6.4 ^{31}P MRS and the Kidney

Spectral localization is required because there are considerable regional differences in renal function and metabolism. The cortex is dependent on oxidative phosphorylation and the inner medulla is more reliant on glycolysis for energy requirements. Localization has proved difficult owing to the anatomical position and the displacement on respiratory excursion. No large-scale studies have been undertaken in the renal failure patients, but transplanted kidneys are positioned in a relatively superficial, static position in the anterior pelvis and are therefore much more amenable to MRS examination.

The normal ^{31}P MR spectrum does not contain a large PCr signal arising from the kidney. There is a relatively large PME peak compared with liver and muscle. Urinary Pi may contribute to the PDE resonance, but this only becomes significant when the collecting ducts are distended through obstructive disease.[27]

6.4.1 *The Isolated Donor Kidney*

Phosphorus-31 MRS can be used to assess the viability of isolated donor organs, kept in physiological preservation fluid, prior to transplantation. An alternative index of tissue energy reserve has to be used because the renal PCr pool is small. Adenosine 5′-triphosphate degrades to ADP and Pi fairly rapidly, but provided there is sufficient cellular AMP, rephosphorylation to ATP should be possible. Adenosine monophosphate resonates in the PME region of the spectrum and therefore the PME/Pi ratio has been used as an indicator of viability. Human studies have positively correlated the PME/Pi ratio with subsequent postoperative renal function.[28] The best renal function was observed in kidneys where ATP was seen in the donor organ spectra.

6.4.2 *The Transplanted Kidney*

Phosphorus-31 MRS may be used to investigate renal failure, graft rejection, or organ viability postoperatively. The PME/ATP ratio may be slightly higher in the transplanted kidney than in healthy volunteers. Animal studies have shown an elevated Pi/ATP ratio with renal failure or ischemia due to poor graft viability. Organ rejection or renal dysfunction due to cyclosporin toxicity may produce a similar picture.[27] Such changes are nonspecific, but perhaps may be used to monitor the effectiveness of antirejection chemotherapy.

6.5 ^{31}P MRS and Liver

The position of the liver renders it much more amenable to MRS investigation. The MR spectrum contains no PCr signal arising from the liver itself. Most studies have concentrated on changes in PME/ATP and PDE/ATP ratios with disease, reflecting changes in hepatic phospholipid and carbohydrate metabolism. The Pi/ATP ratio remains relatively constant. Clinical interest has focused on spectral changes under conditions such as alcoholic liver disease, cirrhosis, primary and secondary liver tumors, and the dynamic changes in metabolites following infusions of fructose, alanine, and alcohol. This subject is discussed in detail in another article (see ***Liver: in vivo MRS of Humans***). Tissue behavior measurements have been used to measure the effectiveness of chemoembolization therapy for hepatic cancers. The resulting ischemia can be assessed using PME/Pi or PME/ATP ratios, both of which fall after successful treatment.[29]

The PME/Pi ratio has been used to assess the viability of isolated donor organs prior to transplantation in a similar fashion to the renal studies already mentioned.[30]

6.6 ^{31}P MRS in Tumors (See Also *Animal Tumor Models*; *Liver: in vivo MRS of Humans*; and *Brain Neoplasms in Humans Studied by Phosphorus-31 MRS*)

The difficulty with reporting metabolic changes in tumors, as reflected by ^{31}P MRS, is that the results are variable, depending on the size, location, and precise histology. In human tumors, the typical metabolite characteristics include lower PCr and higher PME and PDE levels.[31] Variations in the PME and PDE levels may reflect different rates of membrane synthesis, catabolism, or metabolic turnover.

The PCr/Pi ratio is reduced in most high-grade tumors of the brain, reflecting an increased demand for ATP in rapid growth. The measured pHi is often found to be alkaline under these circumstances.[32] The reasons for this intracellular alkalinization remain unclear.

Treatment such as radiotherapy or chemotherapy may induce hypoxia, ischemia, or necrosis in tumors. These changes may lead to an increase in Pi[33] and acidosis.[34] The early decrease in PME seems to be a sensitive indicator of changes in the phospholipid metabolism of the cell membrane.[31] However, because of the wide interindividual variability, changes should be related to initial measurements performed in patients before the start of therapy.

6.7 ^{31}P MRS and Muscle

Phosphorus-31 MRS allows investigation of muscle bioenergetics at rest, during exercise, and during the recovery period. Normal muscle has a low metabolic rate and a high energy capacity at rest, reflected in a high PCr/Pi ratio. During exercise the ATP levels are maintained at the expense of the PCr pool. Anaerobic glycolysis results in lactate accumulation and a reduced pHi. In the recovery phase PCr regenerates and the PCr/Pi ratio rises to preexercise levels. The observed pHi also returns to normal, because oxidative phosphorylation then provides the bulk of the energy requirements. Phosphorus-31 MRS may be utilized as a screening test in patients with exercise intolerance. The various muscle pathologies are discussed in a separate article (see ***Peripheral Muscle Metabolism Studied by MRS***). We briefly consider some of the major conditions where tissue behavior measurements are important.

6.7.1 *Mitochondrial Myopathies*

Mitochondrial myopathies, where there is defective oxidative phosphorylation, are characterized by a reduced resting energy state and a decreased PCr/Pi ratio, which falls rapidly to very low levels on exercise.[35] The capacity for exercise is

For References see p. 4770

reduced and the resynthesis of PCr after exercise is impaired, because this is dependent on mitochondrial function. Therefore there is a prolonged recovery phase before the PCr/Pi ratio reaches resting levels. Intracellular acidosis is often observed, which may become marked on exercise.

6.7.2 *Glycolytic Disorders*

Specific enzyme deficiencies in the glycolytic pathway block the production of lactic acid and therefore the pHi does not fall in anaerobic exercise as it does in normal muscle. This failure of intracellular acidification is characteristic.[35] During exercise an accumulation of glycolytic intermediates can be measured indirectly from the ^{31}P MR spectrum, because sugar phosphates contribute to the PME resonance.

7 CONCLUSIONS

In vivo ^{31}P MRS provides a noninvasive assessment of tissue bioenergetics and phospholipid metabolism. Comparisons may be made between healthy and diseased tissue. Measurements of the PCr/Pi ratio and pHi may provide insights into pathogenic mechanisms and in the future, the efficacy of therapeutic intervention in cardiac and cerebral hypoxic–ischemic states. Energy reserves at rest and during exercise may be monitored in muscle disease, and the PME/Pi ratio is a noninvasive indicator of viability in isolated donor organs prior to transplantation. In vivo ^{31}P MRS remains predominantly a research tool, but it has proved to be clinically useful in birth-asphyxiated babies and in studies of muscle disease.

8 RELATED ARTICLES

Animal Tumor Models; Brain MRS of Human Subjects; Brain Neoplasms in Humans Studied by Phosphorus-31 MRS; Cardiovascular NMR to Study Function; Enzyme-Catalyzed Exchange: Magnetization Transfer Measurements; Enzymes Utilizing ATP: Kinases, ATPases and Polymerases; Focal Brain Lesions in Human Subjects Investigated Using MRS; Heart Studies Using MRS; Hepatic & Other Systemically Induced Encephalopathies: Applications of MRS; Liver: in vivo MRS of Humans; Peripheral Muscle Metabolism Studied by MRS; Quantitation in Whole Body MRS; Whole Body Studies: Impact of MRS.

9 REFERENCES

1. E. E. Conn, P. K. Stumpf, G. Bruening, and R. H. Doi, 'Outlines of Biochemistry', 5th edn., Wiley, New York, 1987.
2. S. Nioka, B. Chance, M. Hilberman, H. V. Subramanian, J. S. Leigh, Jr., R. L. Veech, and R. E. Forster, *J. Appl. Physiol.*, 1987, **62**, 2094.
3. B. Chance, J. S. Leigh, Jr., J. Kent, K. McCully, S. Nioka, B. J. Clark, J. M. Maris, and T. Graham, *Proc. Natl. Acad. Sci., U.S.A.*, 1986, **83**, 9458.
4. J. W. Hugg, G. B. Matson, D. B. Twieg, A. A. Maudsley, D. Sappey-Marinier, and M. W. Weiner, *Magn. Reson. Imaging*, 1992, **10**, 227.
5. J. Kucharczyk, M. Moseley, J. Kurhanewicz, and D. Norman. *Invest. Radiol.*, 1989, **24**, 951.
6. C. R. Jack, F. W. Sharbrough, C. K. Twomey, G. D. Cascino, K. A. Hirschorn, W. R. Marsh, A. R. Zinsmeister, and B. Scheithauer. *Radiology*, 1990, **175**, 423.
7. J. W. Hugg, K. D. Laxxer, G. B. Matson, A. A. Maudsley, C. A. Husted, and M. W. Weiner. *Neurology*, 1992, **42**, 2011.
8. D. G. M. Murphy, P. A. Bottomley, J. A. Salerno, C. DeCarli, M. J. Mentis, C. L. Grady, D. Teichberg, K. R. Giacometti, J. M. Rosenberg, C. J. Hardy, M. B. Schapiro, S. I. Rapoport, J. R. Alger, and B. Horwitz. *Arch. Gen. Psychiatry*, 1993, **50**, 341.
9. T. A. D. Cadoux-Hudson, A. Kermode, B. Rajagopalan, D. Taylor, A. J. Thompson, I. E. C. Ormerod, W. I. McDonald, and G. K. Radda. *J. Neurol. Neurosurg. Psychiatry*, 1991, **54**, 1004.
10. J. M. Minderhoud, E. L. Mooyaart, R. L. Kamman, A. W. Teelken, M. C. Hoogstraten, L. M. Vencken, E. J's. Gravenmade, and W. van den Burg. *Arch. Neurol.*, 1992, **49**, 161.
11. D. Azzopardi, J. S. Wyatt, P. A. Hamilton, E. B. Cady, D. T. Delpy, P. L. Hope, and E. O. R. Reynolds. *Pediatr. Res.*, 1989, **25**, 440.
12. D. Azzopardi, J. S. Wyatt, E. B. Cady, D. T. Delpy, J. Baudin, A. L. Stewart, P. L. Hope, P. A. Hamilton, and E. O. R. Reynolds. *Pediatr. Res.*, 1989, **25**, 445.
13. J. S. Wyatt, A. D. Edwards, D. Azzopardi, and E. O. R. Reynolds. *Arch. Dis. Child.*, 1989, **64**, 953.
14. C. B. Higgins, M. Saeed, M. Wendland, and W. M. Chew. *Invest. Radiol.*, 1989, **24**, 962.
15. S. Schaefer. *Am. J. Cardiol.*, 1990, **66**, 45F.
16. A. de Roos, J. Doornbos, S. Rebergen, P. van Rugge, P. Pattynama, and E. E. van der Wall. *J. Radiol.*, 1992, **14**, 97.
17. S. Neubauer, T. Krahe, R. Schindler, M. Horn, H. Hillenbrand, C. Entzeroth, H. Mader, E. P. Kromer, G. A. J. Riegger, K. Lackner, and G. Ertl. *Circulation*, 1992, **86**, 1810.
18. A. de Roos, J. Doornbos, P. R. Luyten, L. J. M. P. Oosterwaal, E. E. van der Wall and J. A. den Hollander. *J. Magn. Reson. Imaging*, 1992, **2**, 711.
19. S. Schaefer, G. G. Schwartz, J. R. Gober, B. Massie, and M. W. Weiner. *Invest. Radiol.*, 1989, **24**, 969.
20. R. G. Weiss, P. A. Bottomley, C. J. Hardy, and G. Gerstenblith. *N. Engl. J. Med.*, 1990, **323**, 1593.
21. H. Sakuma, K. Takeda, T. Tagami, T. Nakagawa, S. Okamoto, T. Konishi, and T. Nakano. *Am. Heart J.*, 1993, **125**, 1323.
22. W. Auffermann, W. M. Chew, C. L. Wolfe, N. J. Tavares, W. W. Parmley, R. C. Semelka, T. Donnelly, K. Chatterjee, and C. B. Higgins. *Radiology*, 1991, **179**, 253.
23. R. C. Canby, W. T. Evanochko, L. V. Barrett, J. K. Kirklin, D. C. McGiffin, T. T. Sakai, M. E. Brown, R. E. Foster, R. C. Reeves, and G. M. Pohost. *J. Am. Coll. Cardiol.*, 1987, **9**, 1067.
24. C. D. Fraser, Jr., V. P. Chacko, W. E. Jacobus, R. L. Soulen, G. M. Hutchins, B. A. Reitz, and W. A. Baumgartner. *Transplantation*, 1988, **46**, 346.
25. P. McNally, N. Mistry, J. Idle, J. Walls, and J. Freehally. *Transplantation*, 1989, **48**, 1068.
26. P. A. Bottomley, R. G. Weiss, C. J. Hardy, and W. A. Baumgartner. *Radiology*, 1991, **181**, 67.
27. M. D. Boska, D. J. Meyerhoff, D. B. Twieg, G. S. Karczmar, G. B. Matson, and M. W. Weiner. *Kidney Int.*, 1990, **38**, 294.
28. P. N. Bretan, Jr., N. Baldwin, A. C. Novick, A. Majors, K. Easley, T. Ng, N. Stowe, P. Rehm, S. B. Streem, and D. R. Steinmuller. *Transplantation*, 1989, **48**, 48.
29. D. J. Meyerhoff, G. S. Karczmar, F. Valone, A. Venook, G. B. Matson and M. W. Weiner. *Invest. Radiol.*, 1992, **27**, 456.
30. R. F. E. Wolf, R. L. Kamman, E. L. Mooyaart, E. B. Haagsma, R. P. Bleichrodt, and M. J. H. Slooff. *Transplantation*, 1993, **55**, 949.
31. W. Negendank. *NMR Biomed.*, 1992, **5**, 303.
32. W.-D. Weiss, W. Heindel, K. Herholz, J. Rudolf, J. Burke, J. Jeske, and G. Friedman. *J. Nucl. Med.*, 1990, **31**, 302.

33. A. Schilling, B. Gewiese, G. Berger, J. Boese-Landgraf, F. Fobbe, D. Stiller, V. Gallkowski, and K. J. Wolf. *Radiology*, 1992, **182**, 887.
34. M. W. Dewhirst, H. D. Sostman, K. A. Leopold, H. C. Charles, D. Moore, R. A. Burn, J. A. Tucker, J. M. Harrelson, and J. R. Oleson. *Radiology*, 1990, **174**, 847.
35. Z. Argov and W. J. Bank. *Ann. Neurol.*, 1991, **30**, 90.

Biographical Sketches

Claude D. Marcus. *b* 1957. M.D., 1986, Reims, France. Chef de clinique assistant des hopitaux, 1986–1990; praticien hospitalier temps plein centre hospitalier universitaire de Reims, France, 1990–present. Introduced to NMR by Professor D. Doyon (CHU Kremlin-Bicêtre, France). Titulaire du Diplome d'Université d'imagerie par resonance magnétique, 1988. Research interests include clinical applications of NMR to brain and breast diseases.

Simon. D. Taylor-Robinson. *b* 1960. M.B., B.S., 1984, London; M.R.C.P., 1989, (UK). Introduced to NMR by G. M. Bydder. Honorary lecturer, medicine, Royal Free Hospital and School of Medicine, London, 1992–94. Research fellow and Honorary Senior Registrar at NMR Unit, Royal Postgraduate Medical School, Hammersmith Hospital, London, 1992–94. Senior Registrar, Division of Gastroenterology, Department of Medicine, Royal Postgraduate Medical School, Hammersmith Hospital, London, 1994–present. Approx. 14 publications. Research interests include clinical application of NMR to liver disease and transplant organ viability.

Tissue & Cell Extracts MRS

Patrick J. Cozzone, Sylviane Confort-Gouny, & Jean Vion-Dury

Centre de Résonance Magnétique Biologique et Médicale, Faculté de Médecine, Marseille, France

1 INTRODUCTION

Biochemistry has been based for a long time on the identification of molecules involved in the complex architecture and organization of living systems. Prerequisites to the identification of these molecules are their extraction from cells, their chemical and structural analysis, and their quantitation. High-resolution magnetic resonance spectroscopy (MRS) of tissue and cell extracts has been playing a growing role over the past 15 years in the identification of key molecules participating in biochemical processes (metabolism). In the 1990s, studies by MRS of extracts have generated more than 100 publications per year and are often considered as a required complement to in vivo and ex vivo experiments.

MRS of extracts is gradually developing into a major analytical technique. It combines the most advanced magnetic resonance technology (spectrometers) and methods (pulse sequences) with the methods of biochemical extraction and separation based on chemical and physicochemical properties of molecules. In most cases, the MR spectrometer analyses complex mixtures of compounds in a way which is similar to the detector of a gas chromatograph or the chamber of a mass spectrometer. In some cases, chromatographic separation and subsequent analysis of compounds by MRS have been combined online in the same apparatus as in the coupling of a MR spectrometer with a high-performance liquid chromatograph.

There are many compelling reasons for an MR spectroscopist to analyze tissue and cell extracts. The first pertains to the assignments of resonances in spectra recorded ex vivo and in vivo on organs, animals, and humans, accounting for about 20% of publications on the subject. The second reason of interest is the description and monitoring of normal and pathological metabolism in cells and organs (respectively 34%

For References see p. 4775

and 21% of publications), essentially using isotope labeling. Attempts at diagnosing pathologies based on the analysis of human tissue extracts are being made (6% of publications). Finally, the field of pharmacology (bioconversion of drugs, pharmacokinetics, effect of drugs on metabolism) benefits significantly from the MRS analysis of extracts (19% of publications).

In this chapter, we are not aiming to cover exhaustively the field of MRS of extracts. Instead of presenting a comprehensive review, we will try to illustrate what is unique to MRS of tissue and cell extracts in the general context of analytical techniques and modern studies of metabolism.

2 TECHNICAL ASPECTS

The extraction of compounds to be analyzed by MRS has to meet two requirements. It must be *selective* in order to separate molecules of interest from unwanted compounds or cellular substructures (cell debris, membranes etc.). Also, the extraction procedure has *to avoid/limit the degradation and chemical transformation* of molecules. Treatment by perchloric acid is a widely used general method of extraction of water-soluble metabolites.[1] This method requires neutralization and allows removal of salts and proteins by centrifugation (pellet). Perchloric extraction usually follows freeze-clamping of tissues to stop metabolism. It is routinely conducted at a temperature below 5 °C in order to avoid/limit enzymatic or chemical degradation of extracted metabolites. In some cases, inhibitors of enzymatic activity can be added in the supernatant. Other methods are more specific of particular metabolites. For example, extraction by NaOH is used to obtain glycogen and extraction of lipids requires an organic phase such as a 2:1 mixture of chloroform and methanol.[2]

Extracts have to be stored at very low temperature (deep freezer at −80 °C or liquid nitrogen) to limit degradation. Progressive thawing is also critical since chemical modifications of metabolites may occur during this process.

It should be kept in mind that extraction procedures may give a distorted image of the biological reality. Aggressive procedures such as extraction by strong acids or organic solvents

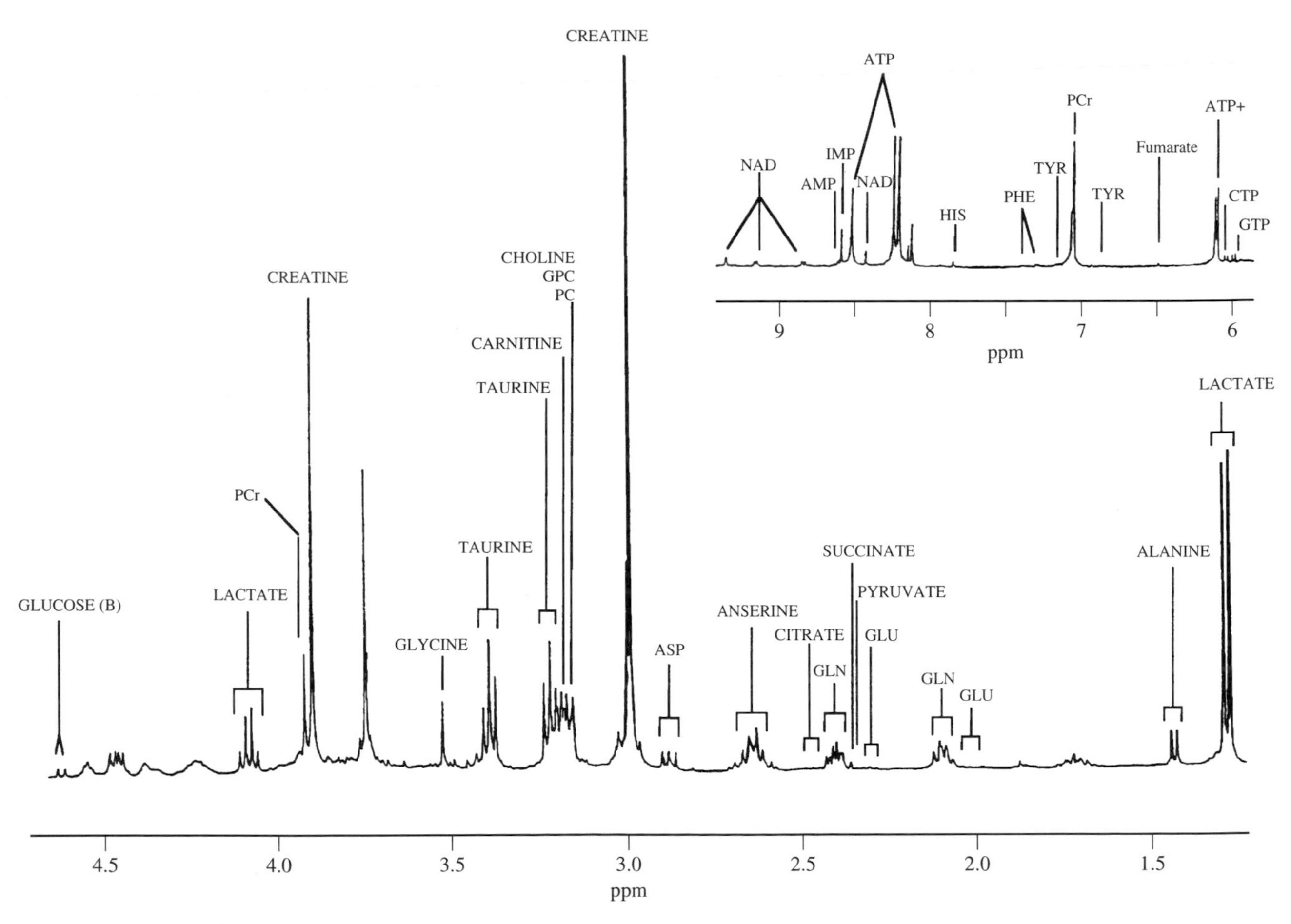

Figure 1 Proton MR spectrum (400 MHz) of a perchloric extract of rat gastrocnemius muscle obtained after freeze-clamping. The supernanant is lyophilized in order to eliminate water. Then the metabolites are dissolved in D_2O. Main Figure: aliphatic region of the spectrum (0–5 ppm). Inset: aromatic region of the spectrum (5–12 ppm). Abbreviations: PC, phosphorylcholine; GPC, glycerophosphorylcholine; PCr, phosphocreatine; ASP, aspartate, GLN, glutamine; GLU, glutamate; NAD, nicotinamide-adenine dinucleotide; AMP, adenosine monophosphate; IMP, inosine monophosphate; HIS, histidine; PHE, phenylalanine; TYR, tyrosine; CTP, cytidine 5′-triphosphate; GTP, guanosine 5′-triphosphate; ATP, adenosine triphosphate (Reproduced by permission of Elsevier from C. E. Mountford, W. B. Mackinnon, M. Bloom, E. E. Burnell, and I. C. P. Smith, *J. Biochem. Biophys. Methods*, 1984, **9**, 323)

modify very significantly the physicochemical environment of metabolites, as compared with the one they experience in the integer cell. As an example, the ionic environment is modified at very low pH (perchloric acid) or when protons and water molecules are lost in organic phases.

Finally, all metabolites are not equal with respect to extraction procedures. They display wide variations in chemical stability, binding properties, conformational rearrangement etc. Hence, one should exert caution when comparing metabolic profiles obtained by in vivo MRS with results obtained on extracts by in vitro MRS. The comparison can be at best misleading if appropriate control experiments are not run since the information offered by extracts is, in essence, different from the information obtained by localized in vivo MR spectroscopy.

MRS of extracts does not require customized MR spectrometers. Standard high-field spectrometers (4.7–18 T) can be used with either narrow-bore or wide-bore magnets. No specific developments in the hardware or software are necessary and MRS of extracts can be performed at any MR laboratory with a basic knowledge of NMR techniques.

3 TISSUE EXTRACTS FOR RESONANCE ASSIGNMENTS

Specific assignment of the resonances recorded in vivo on an MR spectrum remains a critical issue, and a growing one, in consideration of the multiplicity of new applications of MRS in metabolic, pharmacological, and medical studies. Assignment is undoubtedly facilitated by the analysis of extracts using data banks (chemical shifts, coupling patterns, coupling constants), pH titration, and selective addition of authentic compounds to extracts.[3] Also, one can use the whole arsenal of pulse sequences and multinuclear approaches that the chemists have developed in high-resolution MRS of organic compounds.[4–6] With the limitation already mentioned, in vitro MRS of extracts provides the best basis for the assignments of in vivo and ex vivo spectra. (See Figure 1.)

Extracts reflect a ‘frozen situation’ in cell metabolism which is often an advantage. Spectral resolution is always superior in vitro vs. in vivo and ex vivo by several orders of magnitude and facilitates the interpretation of spectra recorded on living cells, tissues, and organs. Interestingly, assignments of resonances in extracts have revealed the presence of unexpected compounds in plants, yeasts, bacterias, and human tissues,[7,8] the identification of which has required chromatographic separation and chemical structure identification by standard analytical techniques, including high-resolution MRS. In these instances, MRS of extracts has directly contributed to a better understanding of metabolic events occurring in the cell.

4 TISSUE EXTRACTS IN METABOLIC STUDIES

MRS of tissue extracts has played a major role in the study of metabolism of cells and organs. Two strategies have been used. The first strategy is ‘passive’ and consists of the detection and dosage of the metabolites present in the tissue extract reflecting, as a snapshot, the metabolic status of the living tissue at the time of extraction. Quantitation can be achieved with respect to wet weight of tissue, protein content, or an internal reference (e.g., an endogenous compound of known concentration). Proton and ^{13}C MR spectra bear information on the intermediary metabolism (amino acids, lipids, Krebs cycle intermediates, etc.). Phosphorus-31 MR spectra carry different information depending on the type of extraction. Aqueous extracts contain phosphorylated molecules of energy metab-

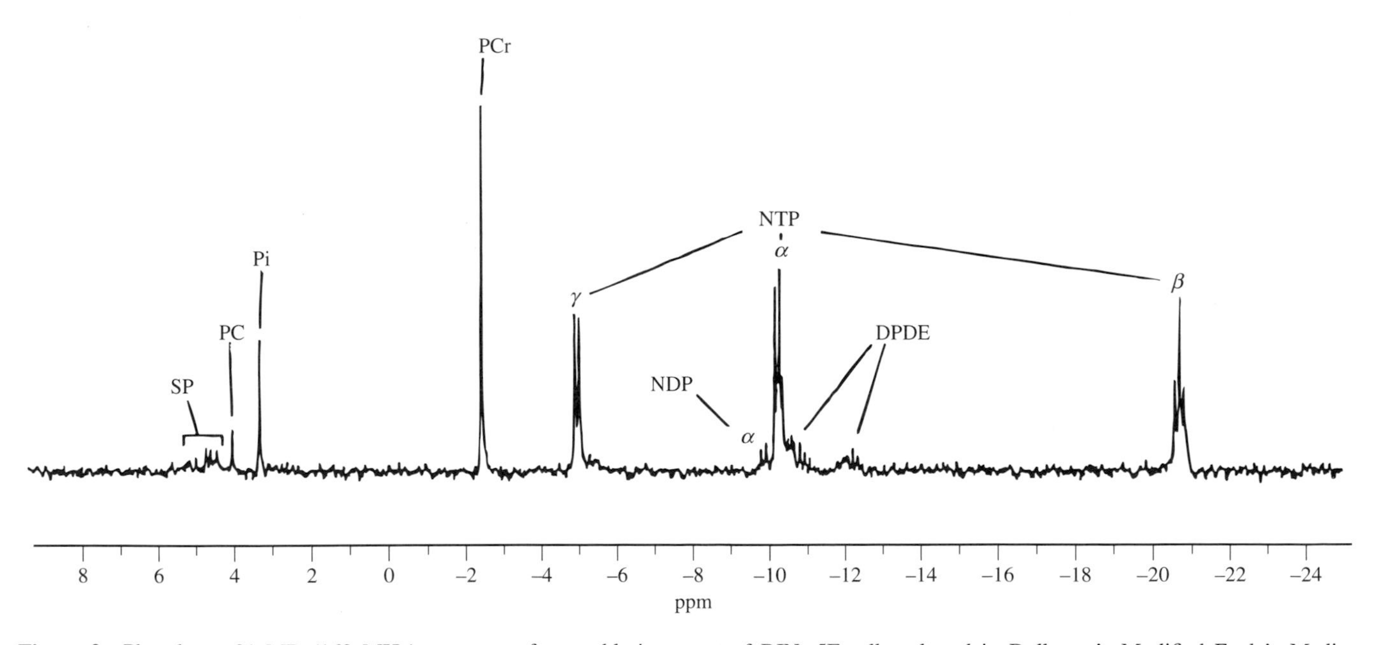

Figure 2 Phosphorus-31 MR (162 MHz) spectrum of a perchloric extract of RINm5F cells cultured in Dulbecco's Modified Eagle's Medium containing glucose. This spectrum (sum of 4800 scans, with complete proton decoupling) corresponds to the perchloric extract of 6×10^8 RINm5F cells cultured in flasks. Abbreviations: PC, phosphorylcholine; SP, sugar phosphate; PCr, phosphocreatine; DPDE, diphosphodiesters; NTP, nucleoside triphosphates; NDP, nucleoside diphosphates (Reproduced by permission of the Radiological Society of North America from E. J. Dellkatny, P. Russell, J. C. Hunter, R. Hancock, K. H. Atkinson, C. Van-Haaften-Day, and C. E. Mountford, *Radiology*, 1993, **188**, 791)

For References see p. 4775

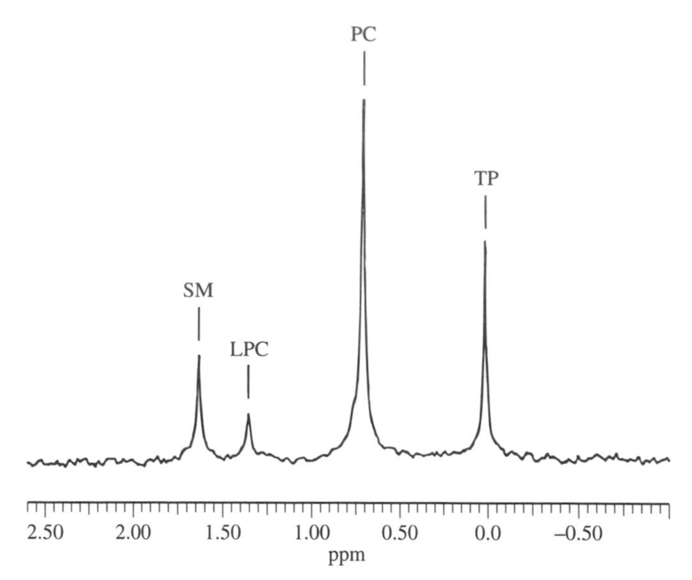

Figure 3 Phosphorus-31 MR (162 MHz) spectrum of a control plasma lipid extract. A line broadening of 4 Hz was applied. Abbreviations: SM, sphingomyelin; PC, phosphatidylcholine; LPC, lysophosphatidylcholine; TP, triethyl phosphate (Reproduced by permission of Academic Press from B. Kunnecke, E. J. Delikatny, P. Russell, J. C. Hunter, and C. E. Mountford, *J. Magn. Reson. B*, 1994, **104**, 135)

olism (ATP, creatine phosphate, phosphomonoesters and phosphodiesters, inorganic phosphate, etc.). (See Figure 2.) The polar head groups of many phospholipids can be assayed on lipidic extracts. In fact, ^{31}P MRS provides a viable alternative to the quantitative assay of phospholipids. (See Figure 3.)

The second strategy is 'active' and involves the perfusion of a selectively ^{13}C-enriched substrate prior to the freeze of metabolic activity and subsequent extraction. By choosing appropriately both the enriched molecule and the chemical site of the enrichment on the molecule, specific metabolic pathways can be singled out and monitored and metabolic fluxes can be quantitated under a variety of conditions, in relation to pathological state, cell differentiation,[9] cell growth, aging etc. This approach is widely used on perfused organs and cultured cells. In these experiments, extracts are prepared and analyzed at

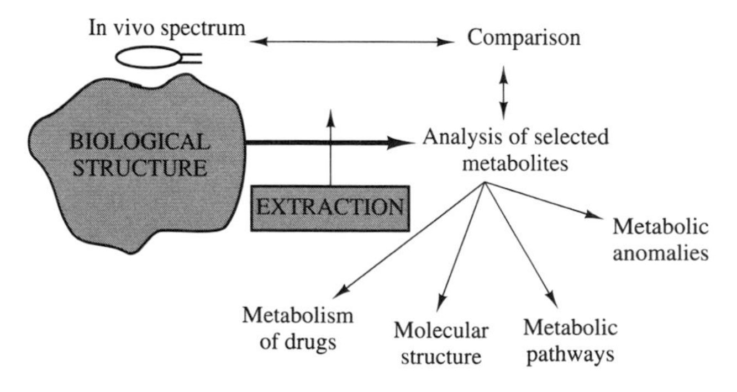

Figure 4 Synthetic scheme summarizing the position of extracts in analysis of biological activity (Reproduced by permission of Williams & Wilkins from C. L. Lean, R. C. Newland, D. A. Ende, E. L. Bokey, I. C. P. Smith, and C. E. Mountford, *Magn. Reson. Med.*, 1993, **30**, 525)

specific times to follow up the conversion and biodistribution of the ^{13}C marker. Basic metabolic events have been documented such as the phosphorylation status of liver under normal conditions[10] or in the presence of ethanol, the metabolism of acetate,[11] or the redox state of the brain,[12,13] etc.

Cellular compartmentation can also be studied from a metabolic standpoint and particularly interesting information has been obtained in brain tissue to shed additional light on the respective role of glial and neuronal cells.[14]

Particular attention has been devoted to the study of metabolic disorders in relation to cancer. The reader is referred to the abundant literature on this subject[15,16] which illustrates the large extent to which MRS of extracts has become an indispensable analytical tool to study normal and pathological metabolism in cells and tissues.

5 APPLICATIONS TO PHARMACOLOGY

Besides the study of cell metabolism, and naturally occurring metabolites, MRS of extracts can document the metabolic fate and the metabolic impact of xenobiotics. Bioconversion of drugs can be readily studied with unsurpassed accuracy. The same MRS analysis can detect, identify (structure), and quantitate the metabolites to which a drug is converted and the kinetics of their appearance.[17]

Prototype studies have been conducted on antimitotic, antifungal, and antiviral drugs.[18,19] The analysis of drug metabolites benefits from the use of selective isotope labeling and multinuclear (^{31}P, ^{19}F, ^{13}C, ^{2}H) and multidimensional MRS experiments.

A second pharmacological application of MRS of extracts is the evaluation of the metabolic changes induced by the administration of a drug. As an example, the impact of the different interleukins on cell metabolism is not well known. However, obtaining additional information is critical to develop, adapt, and optimize anticancer treatment using this new family of molecules. MRS plays a key role in evaluating the effect of these peptides on the general metabolism of the patient toward eliciting the adequate immune response to the presence of a cancer.[20] Drug resistance, another new problem of public health, is currently tackled by MRS of extracts in bacterial and eukaryotic cells, animal models, and even patients in an attempt to establish some metabolic basis to the alarming decrease in therapeutic efficacy of a number of antibiotic and antimitotic drugs.[21,22]

6 NEW DIRECTIONS: TOWARD MEDICAL DIAGNOSIS BY MRS OF TISSUE EXTRACTS

Tissue characterization in medicine is essentially based on the methods developed in pathology and cytology with a large role devoted to optical and electronic microscopy, and the use of selective dyes and immunological labeling. Biochemical studies of excised tissues is not common at the hospital, with the exception of blood and some body fluids. The analysis of tissue biopsies is a growing area of application for the MRS of extracts. Muscle and brain biopsies have been studied but most of the work done to date has focused on human tissues biopsies: breast cancer,[23] uterus,[24] colon.[25] As a matter of fact, the knowledge of the metabolic features of a cancer tumor is of the utmost interest in diagnosis, prognosis and therapeutic decisions.[26] Specific metabolic profiles are observed from extracts of malignant tumors with deviations in the metabolism of amino acids and lipids/phospholipids.[24,27] (See Figure 3.)

The wealth of information that is available on the spectra (usually proton spectra) of biopsy extracts has prompted new strategies for data processing and analysis, based on pattern recognition, neural networks, and chemometrics.[28,29] This integrated and automated approach to spectral processing should facilitate very significantly the acceptance of MRS of biopsy extracts in routine medical diagnosis.

7 CONCLUSIONS

Obviously, MRS of extracts is not 'the MRS of the poor' that scientists and clinicians use when they do not have access to sophisticated methods of in vivo spectroscopy. It is making a definite contribution to the understanding of normal and pathological cell metabolism when it is submitted to a variety of perturbations. MRS of extracts is a cost-effective, highly informational, easy-to-use analytical procedure which is now entering the field of clinical biology by providing unique information that the medical profession can readily use for diagnosis, treatment, and prognosis. (See Figure 4.)

8 RELATED ARTICLES

Body Fluids; Cells and Cell Systems MRS; Whole Body Studies: Impact of MRS.

9 REFERENCES

1. M. Leach, L. Le Moyec, and F. Podo, in 'Magnetic Resonance Spectroscopy in Biology and Medicine', eds. J. De Certaines, W. M. M. J. Bovée, and F. Podo, Pergamon Press, Oxford, 1992, Chap. 18.
2. J. Folch, M. Lees, and G. H. S. Stanley, *J. Biol. Chem.*, 1957, **226**, 497.
3. O. Kaplan, Van Zijl, and J. S. Cohen, in 'NMR Basic Principles and Progress', eds. P. Diehl, E. Fluck, H. Günther, R. Kosfeld, and J. Seelig, Springer, Berlin, 1992, Vol. 28, Chap. 1.
4. G. Navon, T. Kushnir, N. Askenasy, and O. Kaplan. in 'NMR Basic Principles and Progress', eds. P. Diehl, E. Fluck, H. Günther, R. Kosfeld, and J. Seelig, Springer, Berlin, 1992, Vol. 27, Chap. 10.
5. C. Arus, Y. C. Chang, and M. Barany, *Physiol. Chem. Phys. Med. NMR*, 1985, **17**, 23.
6. M. Barany, C. Arus, and Y. C. Chang, *Magn. Reson. Med.*, 1985, **2**, 289.
7. N. De Tomasi, S. Piacente, F. De Simone, C. Pizza, and Z. L. Zhou, *J. Nat. Prod.*, 1993, **56**(10), 1669.
8. F. H. Kormelink, R. A. Hoffmann, H. Gruppen., A. G. Voragen, J. P. Kamerling, and J. F. Vliegentart. *Carbohydr. Res.*, 1993, **249**(2), 369.
9. J. P. Galons, J. Fantini, J. Vion-Dury, P. J. Cozzone, and P. Canioni. *Biochimie*, 1989, **71**, 949.
10. R. A. Iles, A. N. Stevens, J. R. Griffiths, and P. G. Morris. *Biochem J.*, 1985, **229**, 141.
11. S. R. Williams, E. Proctor, K. Allen, D. G. Gadian, and H. A. Crockard. *Magn. Reson. Med.*, 1988, **7**, 425.
12. O. Ben-Yosef, S. Badar-Goffer, P. G. Morris, and H. S. Bachelard. *Biochem. J.*, 1993, **291**, 915.
13. C. Remy, C. Arus, A. Ziegler, E. Sam Lai, A. Moreno, Y. Le Fur and M. Décorps. *J. Neurochem.*, 1994, **62**(1), 166.
14. J. Urenkak, S. R. Williams, D. G. Gadian, and M. Noble. *J. Neurosci.*, 1993, **13**(3), 981.
15. C. E. Mountford, C. L. Lean, and W. B. Mackinnon. *Annu. Rep. NMR Spectrosc.*, 1993, **27**, 174.
16. M. Czuba and I. C. P. Smith. *Pharmacol. Ther.*, 1991, **50**, 147.
17. J. Vion-Dury, S. Confort-Gouny, and P. J. Cozzone, in 'Human Pharmacology', eds. H. Kuemmerle, T. Shibuya, and J. P. Tillement, Elsevier, Amsterdam, 1991, Chap. 13.
18. M. O. F. Fasoli, D. Kerridge, P. G. Morris, and A. Torosantucci. *Antimicrob. Agents Chemother.*, 1990, **34**(10), 1996.
19. R. A. Vere-Hodge, S. J. Darlinson, and S. A. Readshaw. *Chirality*, 1993, **5**(8), 577.
20. E. Prioretti., F. Belardelli., G. Carpinelli, M. Di Vito, D. Woodrow, J. Moss, P. Sestelli, W. Fiers, I. Gresser and F. Podo. *Int. J. Cancer*, 1988, **42**, 582.
21. A. Ferretti, L. L. Chen, M. Di Vito, S. Barca, M. Tombesi, M. Cianfriglia, A. Bozzi, R. Srom, and F. Podo. *Anticancer Res.*, 1993, **13**, 867.
22. G. L. May, L. C. Wright, M. Dyne, W. B. Mackinnon, R. M. Fox, and C. E. Mountford. *Int. J. Cancer*, 1988, **42**, 728.
23. T. A. D. Smith, C. Bush, C. Jameson, J. C. Titley, M. O. Leach, D. E. V. Wilman, and V. R. McCready. *NMR Biomed.*, 1993, **6**, 318.
24. C. E. Mountford, E. J. Delikatny, M. Dyne, K. T. Holmes, W. B. Mackinnon, R. Ford, J. C. Hunter, I. D. Truskett, and P. Russell. *Magn. Reson. Med.*, 1990, **13**, 324.
25. A. Moreno, M. Rey, J. M. Montane, J. Alonso, and C. Arus. *NMR Biomed.*, 1993, **6**, 111.
26. D. G. Gadian, T. E. Bates, S. R. William, J. D. Bell, S. J. Austin, and A. Connelly. *NMR Biomed.*, 1991, **4**, 85.
27. M. Kriat, J. Vion-Dury, S. Confort-Gouny, R. Favre, P. Viout, M. Sciaky, H. Sari, and P. J. Cozzone. *J. Lipid Res.*, 1993, **34**, 1009.
28. S. L. Howells, R. J. Maxwell, and J. R. Griffiths. *NMR Biomed.*, 1992, **5**, 59.
29. S. L. Howells, R. J. Maxwell, A. C. Peet, and J. R. Griffiths. *Magn. Reson. Med.*, 1992, **28**, 214.

Biographical Sketches

Patrick J. Cozzone. *b* 1945. Ph.D., 1971, University of Marseille, MBA, University of Chicago. Professor of Biochemistry, University of Marseille, 1975–90; Professor of Biophysics, Faculty of Medicine, Marseille, 1990–present; Director, Centre de Résonance Magnétique Biologique et Médicale (CRMBM), 1986–present.

Sylviane Confort-Gouny. *b* 1957. Ph.D., 1984, University of Grenoble, MRS instrumentation. Research engineer, Centre National de la Recherche Scientifique; in charge at Centre de Résonance Magnétique Biologique et Médicale, Marseille, of methodological developments and clinical transfers of MRS preclinical protocols, 1986–present.

Jean Vion-Dury. *b* 1956. M.D. 1983, School of Medicine of Marseille, Ph.D., 1988, University of Marseille. Assistant Professor of Biophysics, Faculty of Medicine, Marseille, 1994–present. NMR research at the Centre de Résonance Magnétique Biologique et Médicale, 1986–present.

Tissue NMR Ex Vivo

Ian C. P. Smith
National Research Council of Canada, Winnipeg, Canada

&

Carolyn E. Mountford
University of Sydney, NSW, Australia

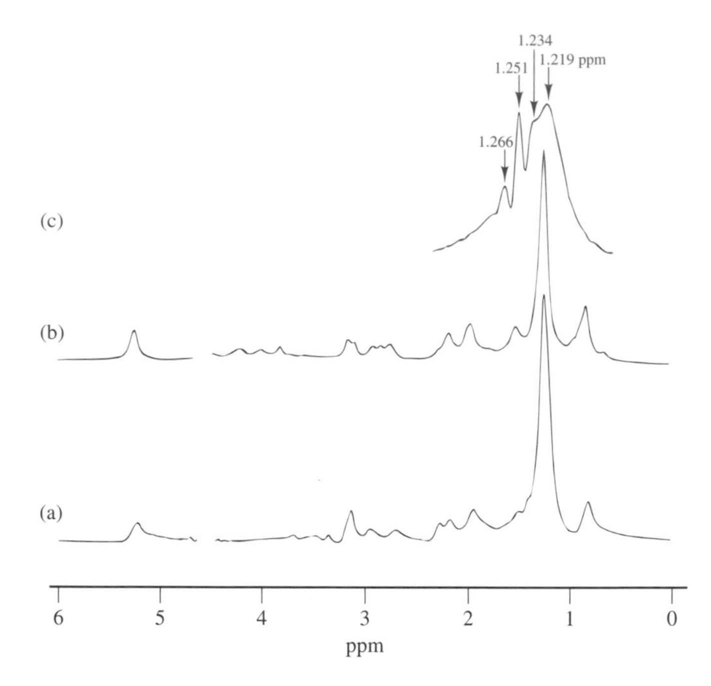

Figure 1 Proton 400 MHz NMR spectra of the mammary adenocarcinoma cell line 'J clone'.[3] (a) A solid tumor excised from a Fisher rat. (b) 'J clone' cells (1×10^8). (c) A Lorentzian–Gaussian resolution enhancement applied to (b), expanded scale

1 INTRODUCTION

NMR of tissue specimens is now an established, powerful adjunct to histopathology for classification of human cancers. The initial study that provided insight into its potential for the early detection and diagnosis of cancer was reported in 1980 for mouse thymus.[1] Spectral changes preceded the cellular changes observed by histopathology, suggesting that the tissue NMR method could detect preinvasive changes prior to any morphological manifestation. The concept that information on the biological outcome of a tumor was also present in the NMR spectrum came from a study of a rat mammary model for adenocarcinoma metastasis.[2] Spectra from the excised tumor closely resembled those from suspensions of the same cells grown in vitro[3] (Figure 1). The T_2 relaxation time for the strong peak at 1.3 ppm correlated well with the metastatic potential of this tumor in the rat. It was this same model, over a decade later, that showed the acute sensitivity of the NMR method for detecting microfoci of malignant cells in lymph nodes.[4]

This early success with a rat model led us to attempt similar studies with human tissues. Human colorectal tumors yielded spectra with some parameters similar to those obtained from rat mammary adenocarcinoma. Over 90% of the human colorectal tumors gave long T_2 relaxation values for the resonance at 1.3 ppm, implying a high metastatic potential.[5] Patient follow-up showed no false positives; that is, no patients whose tumor generated only a short T_2 relaxation value for the resonance at 1.3 ppm developed secondary tumors.[6] Many of those with long T_2 relaxation values did develop secondary tumors.

From these studies it was apparent that the strong signals from fat were superimposed upon those of potentially diagnostic species present in the spectra from colorectal biopsies, a problem now known to exist also for lymph node and breast tissues. The implementation of two-dimensional (2D) NMR spectroscopy was crucial to determine the pathology of human tumors.[7,8] The individual chemical species useful for the various pathological diagnoses were assigned, and some of the biological events they were reporting were identified.[9]

2 METHODS

All successful studies of viable tissues have relied heavily on careful sample handling prior to ^{1}H MRS experiments.[9,10] Tissues are placed into sterile tubes immediately after excision, immersed in liquid nitrogen, and stored at −70 °C. Either a glass wool platform supports the sample surrounded by buffer, or the specimen is inserted into a capillary containing buffer, which in turn is inserted into an NMR tube containing the same buffer and a reference. This latter method of sample positioning has several advantages including measurable volume for quantification and easier homogeneity adjustment.[10]

Time constraints imposed by the limited viability of cell and tissue specimens dictate careful selection of acquisition parameters for 2D NMR spectroscopy. In order to obtain a pathological assessment of the sample used for NMR, time in the instrument must be minimized. Thus, the trade-off between signal strength, spectral resolution, and sample viability must be addressed for each tissue type. Cells and tissues contain a variety of biochemical species at a range of concentrations, with T_2 relaxation times ranging from <100 ms to >1 s. The choice of window functions during processing has a dramatic effect on the resolution, sensitivity, and appearance of the 2D spectra.[11] To obtain all available information from one data set on a tissue specimen, the 2D data file should be processed using a number of different window functions.[11] It is imperative that histological examination be undertaken on the NMR specimen itself, rather than relying upon the routine hospital report.

We describe below studies on tissue from three organs, each representing a solution to different technical/clinical problems: (1) the distinction between preinvasive and invasive cancers; (2) the presence of significant levels of fat, obscuring signals of interest, and (3) the detection of cellular abnormalities which are not morphologically manifest.

3 CERVIX

Cervical epithelium has preinvasive states well documented by histopathology. Independent clinical studies in each of our laboratories have shown that ^1H NMR can distinguish between preinvasive lesions, including carcinoma in situ (CIS), and frankly invasive cervical cancer with $p < 0.0001$ (Student *t*-test).[8,10,12] The ^1H NMR acquisitions are completed in 15 min, and the tissue is not subject to the vagaries of sampling since the entire punch biopsy specimen is examined. Two different approaches were used by our respective laboratories to extract diagnostic information from the ^1H NMR data. In one, manual measurements of peak heights were performed. Typical spectra of preinvasive and invasive cervical epithelium are characterized by intense lipid resonances at 0.9, 1.3, and 5.2 ppm. In contrast, spectra of preinvasive lesions are free of signals from high-resolution lipid.[12] A second spectral region which provides diagnostic information lies between 3.4 and 4.2 ppm. These resonances

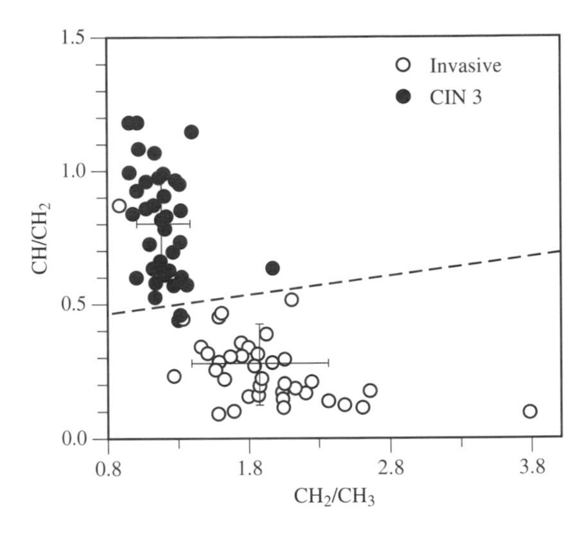

Figure 2 Peak height ratios from one-dimensional ^1H NMR spectra.[12] Severely dysplastic epithelium (CIN3) is compared with invasive carcinoma. The cross bars show the mean ± standard deviation

are more prevalent in spectra from preinvasive than invasive specimens. The resonance intensities, a ratio of the strongest resonance between 3.4 and 4.2 ppm and the methylene resonance at 1.3 ppm, is plotted against the ratio of the methylene to methyl resonances at 1.3 and 0.9 ppm, respectively (Figure 2). The separation between the mean ratio values for CIS and invasive cancer is significant. The *p* values obtained when comparing invasive cancer to the low,

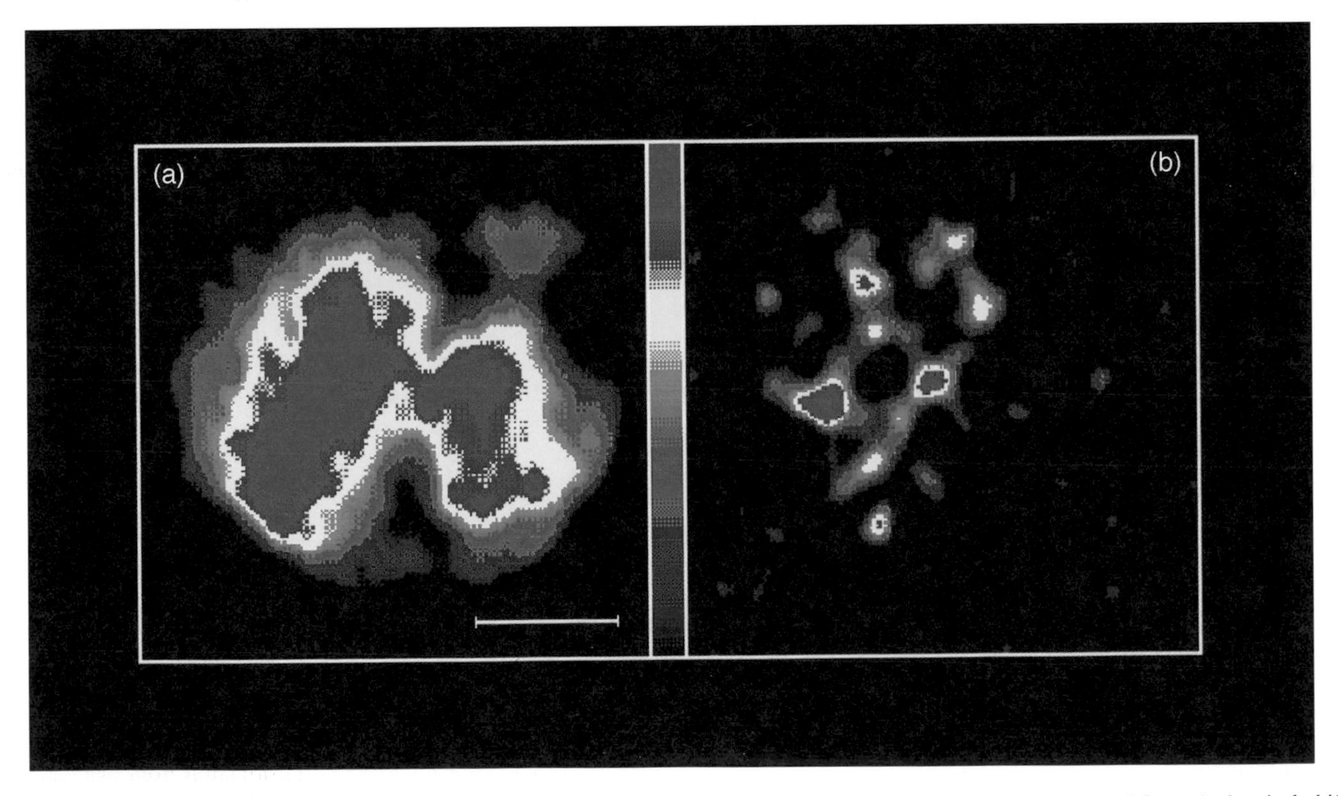

Figure 3 Microimaging of cervical biopsies. Water image (a) (excitation at 4.8 ppm) and lipid-based (b) (excitation at 1.3 ppm) chemical shift images of two human cervical punch biopsies. Right of each image: low-grade intraepithelial neoplasia (CIN1). Left of each image: invasive adenocarcinoma.[16] The bar shown in (a) represents 1 mm

For References see p. 4779

moderate, or severe levels of dysplasia are all <0.0001, with a specificity of 0.94 and sensitivity of 0.98.[12]

Alternatively, a mathematical treatment of the data sets may be used. A variety of multivariate analyses, such as linear discriminant analysis, neural nets, and genetic programming were applied to these data. Using a consensus approach for a series of 150 spectra of cervical tissue, spectra were allocated to characteristic groups. Groupings were then compared with the data from histopathology,[13] and yielded sensitivities and specificities greater than 0.99 for the various levels of dysplasia.

Chemical shift imaging (CSI) has the potential to revolutionize the detection of cancers since it can provide information on the biochemical composition and distribution, the biological potential, and the spatial location of neoplasms. We have used CSI as described by Dixon[14] with the pulse sequence of Hall et al.,[15] where the image is constituted from a single specific frequency. Having established a correlation with histopathology that the presence of a lipid spectrum signifies the presence of malignant cells, we considered that a chemical shift image based on the intense methylene signal at 1.3 ppm should distinguish adenocarcinoma from both normal tissue and preinvasive lesions.

The water-based and lipid-based images of two human cervical punch biopsies are shown in Figure 3.[16] The water-based image [(a), excitation at 4.8 ppm] and lipid-based image [(b), lipid excitation at 1.3 ppm] were collected simultaneously over 2 h at 8 T. The left biopsy sample is from an invasive adenocarcinoma, whilst the right-hand one is from a low-grade squamous dysplasia. In the water-based image (a) both biopsies are observed as distinct components with equal intensities. In contrast, in the lipid-based images, only the left (malignant) biopsy shows substantial accumulation of MR-visible lipids. Histological assessment revealed that the lipid-rich areas detected by CSI corresponded to foci of malignant cells whilst the lipid-deficient areas corresponded to normal cervical stroma. Thus, we have shown that with prior knowledge of the chemical changes associated with specific pathologies of the human uterine cervix, proton CSI can provide a spatial map of the diseased areas.[16]

4 COLON

The colon presents the technical difficulty of naturally high levels of fat,[6] necessitating either 2D NMR spectroscopy,[6] or T_2 relaxation filters[17] to examine the diagnostic resonances present in the pathologically abnormal specimens. The 2D ^{1}H–^{1}H COSY spectra of normal colorectal mucosa and an invasive adenocarcinoma are compared in Figure 4. The diagnostic parameters include choline, choline-based metabolites, and other metabolites, and altered lipid profiles. Tissues histopathologically classified as normal, while remaining distinct from the malignant spectral profile, were found to fit into two categories, one of which had some of the spectral characteristics of malignancy. These results indicate that ^{1}H MRS identifies colorectal mucosa whose abnormality is not morphologically manifest.[17]

The colorectal cancer biopsy program has relied heavily on the correlation of biological and genetic information with the NMR data from cultured cell lines.[18] It was shown that 2D NMR data for cell surface fucosyl residues unambiguously identify the various preinvasive states as well as a feature unique to the carcinoma in situ.[19] These pathological states, identifiable by NMR, include the distinction between malignant tumors and normal tissue; carcinoma in situ and frankly invasive adenocarcinoma; the extents of dysplasia in preinvasive polyps.

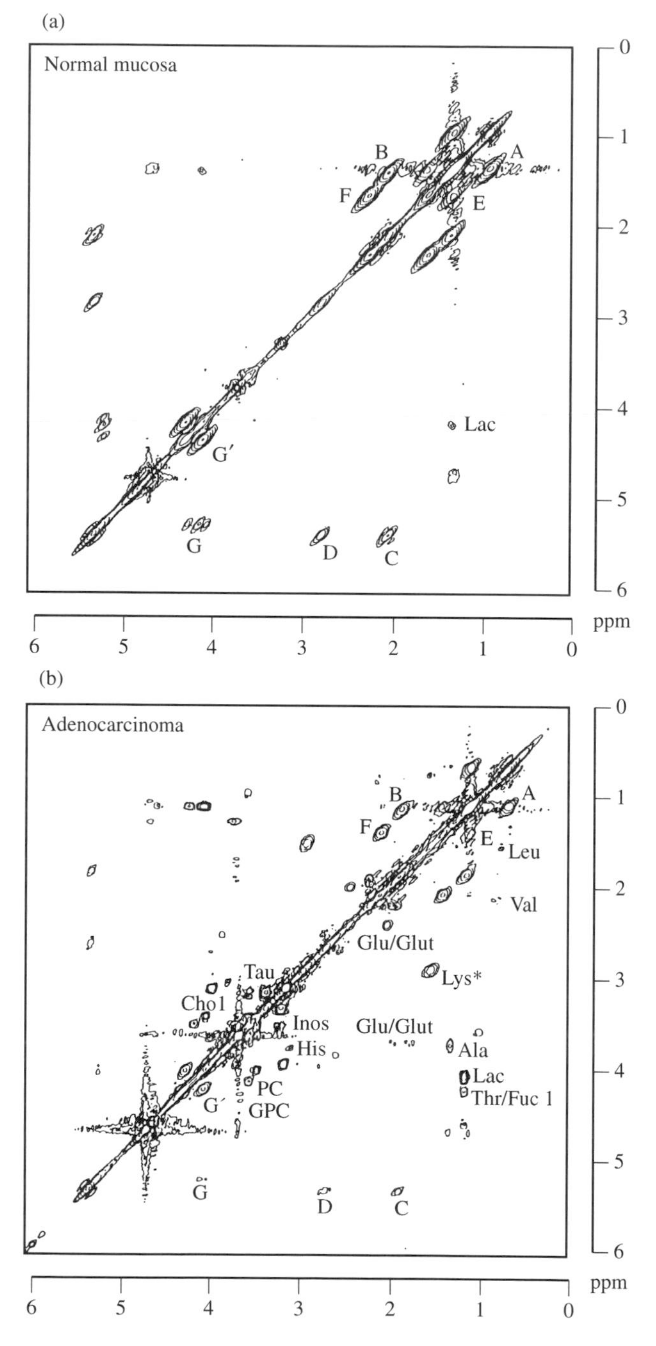

Figure 4 Proton 360 MHz NMR symmetrized COSY spectra of (a) normal colonic mucosa, (b) carcinoma, in phosphate-buffered saline/ D_2O, 37 °C.[17] Lac = lactate, Leu = leucine, Vol = valine, Glu/Glut = glutamic acid/glutamate, Lys* = lysine, Tau = taurine, Cho 1 = choline, Inos = inositol, His = histidine, Ala = alanine, Thr = threonine, Fuc I = fucose, PC = phosphatidyl choline, GPC = glycerophosphorylcholine

5 THYROID

We include data for the thyroid because it is the first organ where NMR has distinguished between benign adenoma and carcinoma which are morphologically identical under the light microscope.[20,22] Histological diagnosis of follicular thyroid cancer relies upon visualization of capsular invasion or the patients having demonstrated secondary tumors. The morphological similarity of malignant and benign follicular cells leads to surgery solely for diagnostic purposes. One-dimensional NMR spectroscopy clearly distinguishes normal thyroid tissue from morphologically distinct invasive papillary carcinoma, based on differences in the relative intensities of resonances at 1.7 and 0.9 ppm,[20] demonstrating the potential to distinguish genuinely benign follicular adenomas from follicular carcinomas. The MR parameters separated follicular neoplasms into two categories, each of which was directly comparable with either normal thyroid or papillary carcinoma. The patients with clinically proven or histologically invasive follicular carcinoma were categorized by ^{1}H MR together with papillary cancers, as were a number of patients with clinically or morphological atypical neoplasms. It was subsequently established that the same diagnostic information could be obtained from fine needle aspiration biopsies, without recourse to surgery.[21] Multivariate analysis undertaken in a blind study delineated 5 out of 6 patients with follicular carcinoma (identified clinically by the presence of secondary tumors).[22] A clinical trial has commenced in Australia where patients with normal NMR profiles from fine needle aspiration biopsies are given the option not to undergo surgery for diagnostic purposes.[23]

6 CONCLUSIONS

It has now been established that the tissue NMR parameters allow pathological and in some cases clinical classification of human cancers. More importantly, we have demonstrated that ^{1}H NMR can discern changes to the cellular chemistry in human tissues prior to their histological manifestation. This has significant implications for identifying the extent of cellular abnormality in the preinvasive states, subjects with a predisposition to cancer, and pathologies that have to date eluded the light microscope. The natural extension to CSI of excised biopsies, now completed for cervix and thyroid, indicates a very promising future for the in vivo diagnosis of human cancer by ^{1}H NMR. Clearly those organs that require 2D spectroscopy to identify the diagnostic parameters are in abeyance until 2D NMR in vivo becomes practical.

7 RELATED ARTICLES

. Body Fluids; Cells and Cell Systems MRS; Kidney, Prostate, Testicle, & Uterus of Subjects Studied by MRS; Tissue & Cell Extracts MRS.

8 REFERENCES

1. C. E. Mountford, G. Grossman, P. A. Gatenby, and R. M. Fox, *Br. J. Cancer*, 1980, **41**, 1000.
2. C. E. Mountford, L. C. Wright, K. T. Holmes, W. B. Mackinnon, P. Gregory, and R. M. Fox, *Science*, 1984, **226**, 1415.
3. C. E. Mountford, W. B. Mackinnon, M. Bloom, E. E. Burnell, and I. C. P. Smith, *J. Biochem. Biophys. Methods*, 1984, **9**, 323.
4. C. E. Mountford, C. L. Lean, R. Hancock, S. Dowd, W. B. Mackinnon, M. H. N. Tattersall, and P. Russell, *Invasion Metastasis*, 1993, **13**, 57.
5. C. E. Mountford, G. L. May, P. G. Williams, M. H. N. Tattersall, P. Russell, J. K. Saunders, K. T. Holmes, R. M. Fox, I. R. Barr, and I. C. P. Smith, *Lancet*, 1986, **i**, 651.
6. I. C. P. Smith, E. J. Princz, and J. K. Saunders, *J. Can. Assoc. Radiol.*, 1990, **41**, 32.
7. K. J. Cross, K. T. Holmes, C. E. Mountford, and P. E. Wright, *Biochemistry*, 1984, **23**, 5895.
8. C. E. Mountford, E. J. Delikatny, M. Dyne, K. T. Holmes, W. B. Mackinnon, R. Ford, J. C. Hunter, I. D. Truskett, and P. Russell, *Magn. Reson. Med.*, 1990, **13**, 324.
9. C. E. Mountford, C. L. Lean, W. B. Mackinnon, and P. Russell, *Annu. Rep. NMR Spectrosc.*, 1993, **27**, 173.
10. A. C. Kuesel, T. Kroft, J. K. Saunders, M. Préfontaine, N. Mikhael, and I. C. P. Smith, *Magn. Reson. Med.*, 1992, **27**, 349.
11. E. J. Delikatny, W. E. Hull, and C. E. Mountford, *J. Magn. Reson.*, 1991, **94**, 563.
12. E. J. Delikatny, P. Russell, J. C. Hunter, R. Hancock, K. H. Atkinson, C. van Haaften-Day, and C. E. Mountford, *Radiology*, 1993, **188**, 791.
13. R. L. Somorjai, A. E. Nikulin, A. C. Kuesel, M. Préfontaine, N. Mikhael, and I. C. P. Smith, *Proc. Soc. Magn. Reson. Med.*, 1992, **1**, 56.
14. W. T. Dixon, *Radiology*, 1984, **153**, 189.
15. L. D. Hall, S. Sukumar, and S. L. Talagala, *J. Magn. Reson.*, 1984, **56**, 275.
16. B. Kunnecke, E. J. Delikatny, P. Russell, J. C. Hunter, and C. E. Mountford, *J. Magn. Reson.*, 1994, **10**, 135.
17. C. L. Lean, R. C. Newland, D. A. Ende, E. L. Bokey, I. C. P. Smith, and C. E. Mountford, *Magn. Reson. Med.*, 1993, **30**, 525.
18. C. L. Lean, W. B. Mackinnon, E. J. Delikatny, R. H. Whitehead, and C. E. Mountford, *Biochemistry*, 1992, **31**, 11 095.
19. D. A. Ende, C. L. Lean, W. B. Mackinnon, P. Chapuis, R. Newland, P. Russell, E. L. Bokey, and C. E. Mountford, *Proc. Soc. Magn. Reson. Med.*, 1993, **2**, 1033.
20. P. Russell, C. L. Lean, L. Delbridge, G. L. May, S. Dowd, and C. E. Mountford, *Am. J. Med.*, 1994, **96**, 383.
21. C. L. Lean, P. Russell, L. Delbridge, G. L. May, S. Dowd, and C. E. Mountford, *Proc. Soc. Magn. Reson. Med.*, 1993, **1**, 71.
22. R. L. Somorjai, N. Pizzi, A. Nikulin, R. Jackson, C. E. Mountford, P. Russell, C. L. Lean, L. Delbridge, and I. C. P. Smith, *Proc. Soc. Magn. Reson. Med.*, 1995, **33**, 257.
23. L. Delbridge, C. L. Lean, P. Russell, G. L. May, S. Roman, S. Dowd, T. S. Reeve, and C. E. Mountford, *World J. Surg.*, 1994, **18**, 512.

Biographical Sketches

Carolyn E. Mountford. *b* 1950. NZCS, 1971, Auckland, M.Sc., 1976, crystallography, Oxford; D.Phil., 1978, Oxford. Introduced to NMR by R. J. P. Williams. Postdoctoral work with P. Coleman, R. M. Fox, and M. H. N. Tattersall, University of Sydney 1978–1980. Mentor in application of NMR to biological membranes, 1980–present, I. C. P. Smith. Director, Membrane NMR Unit, Ludwig Institute for Cancer Research (Sydney Branch) 1982–86. Director, Membrane MR Unit, Department of Cancer Medicine, University of Sydney, 1987–present. Approx. 100 publications. Research specialty: applications of proton NMR for detection and diagnosis of cancer and the immune response.

Ian C. P. Smith. *b* 1939. B.Sc. 1961, M.Sc. 1962, Manitoba; Ph.D. Cambridge, 1965 (Research Director, Alan Carrington); Fil. Dr. (H.C.)

Stockholm, 1986; D.Sc. (H.C.) Winnipeg, 1990. Introduced to NMR by W. G. Schneider, NRC, Ottawa, 1960. Postdoctoral work with H. M. McConnell, Stanford, 1965–66 and R. G. Shulman, Bell Labs, 1966–67. Research Officer, Institute for Biological Sciences, NRC, 1967–87; Director-General, 1987–91; Director-General, Institute for Biodiagnostics, NRC, 1992–present. Approx. 380 publications. Research specialty: applications of modern spectroscopic and computational methods in medicine, with emphasis on magnetic resonance spectroscopy.

Tissue Perfusion in MRI by Contrast Bolus Injection

Johannes C. Böck & Roland Felix
Freie Universität Berlin, Germany

1 INTRODUCTION

The success of clinical magnetic resonance imaging (MRI) is based on its superb anatomic and contrast resolution, in particular in the central nervous system. However, MRI was initially unable to contribute to the field of perfusion imaging because of two factors: the lack of MRI contrast media and insufficient temporal resolution of the pulse sequences.

Contrast media are now available. The first agent in clinical use was gadolinium–DTPA, a well-tolerated ionic paramagnetic contrast agent developed by Schering AG, Berlin and first used in humans at our Institution in 1983.[1] After intravenous injection, gadolinium–DTPA is distributed in the extracellular space with the notable exception of the central nervous system, where the extravasation of gadolinium–DTPA into the interstitial space is effectively prevented by the intact blood/brain barrier. No such barrier prevents leakage of gadolinium–DTPA in the normal myocardium.

The limitation of temporal resolution has been overcome by the development of pulse sequences such as gradient echo and echo planar imaging. These pulse sequences allow for acquisition times ranging from 0.5–20 images per second, sufficient to follow the first passage of a contrast bolus through an organ.

In 1986, Haase et al.[2] published a report which pioneered contrast enhanced perfusion MRI. In an experimental model, they combined the intravenous bolus injection of the first available paramagnetic contrast agent, gadolinium–DTPA, and a new rapid image acquisition strategy 'fast low angle shot' (FLASH). When a series of images was acquired in the same slice position after contrast bolus injection ('FLASH MR movie') a reversible signal intensity decrease was observed in the brain which apparently coincided with the first pass of the contrast medium through the brain.

Unfortunately, the interpretation of the contrast kinetic is limited because of the physical properties of currently available contrast agents, which distribute either in the intravascular space (in the intact brain) or, in addition, in the interstitial space (in the myocardium but also in cerebral pathologies with disruption of the blood/brain barrier). From a rigorous mathematical standpoint only relative cerebral blood *volume* can be calculated on the basis of concentration–time curves measured in the brain[3]—cerebral blood *flow* can only be estimated.[3,4] Since gadolinium–DTPA is not a strictly intravascular contrast medium in other organs than the brain, both blood flow and volume cannot be derived from gadolinium–DTPA kinetics measured in the myocardium and the kidneys.

By contrast, organ blood flow can be quantitatively assessed using diffusible contrast media such as deuterium oxide[5] or $H_2{}^{17}O$.[6] However, these techniques will not be addressed in depth in this article since clinical experience is still lacking, probably because the methodology requires an injection either directly into the tissue or into the feeding artery. A less invasive approach, intravenous injection, would require the measurement of the concentration–time curve in the feeding artery and deconvolution of the arterial and tissue curves.[7] Such techniques offer the same advantages as positron emission tomography in that blood flow can be measured in absolute terms (mL min^{-1} per 100 g of tissue).

2 PRINCIPLES

Conventional contrast enhanced MRI using paramagnetic agents such as gadolinium–DTPA relies on dipole–dipole interactions between the nuclear spins of local protons and unpaired electrons of the contrast molecules. These interactions shorten the T_1 relaxation times and thereby increase signal intensity in the interstitial space *minutes* (the time required for contrast extravasation) after injection. If, however, the contrast is rapidly injected as a bolus, and if an appropriate (susceptibility-sensitive) pulse sequence is used, then a signal intensity *decrease* is observed within *seconds* after injection. The underlying mechanism is the compartmentalization of the contrast material in the vasculature (at least in the brain) with consecutively high contrast concentration gradients between the intravascular and interstitial compartments. (Positive) susceptibility is the property of paramagnetic substances to produce an intrinsic magnetic field when exposed to an extrinsic magnetic field. Because the magnetic susceptibility of gadolinium–DTPA is different to that of biological tissues, the contrast concentration gradients translate into regional (intravoxel) magnetic field gradients.

Intravoxel magnetic field inhomogeneity broadens the resonance frequency spectrum with subsequent dephasing and signal loss. Since gradient echo pulse sequences do not refocus static magnetic field inhomogeneities, signal intensity is directly affected by this process, in particular when the echo time is long (on the order of 20 ms). Spin echo pulse sequences do refocus *static* inhomogeneities and are therefore only sensitive to dephasing of mobile spins (diffusion of water molecules); again, the sensitivity increases with the echo time.

The signal intensity decrease is reversed as the contrast bolus passes the brain (first pass); however, a second (third, etc.) passage of the contrast (systemic contrast recirculation) is usually observed *before* the baseline signal intensity has been reached.

For further computations (e.g. blood volume), indicator dilution theory requires a concentration–time curve which can be computed for every voxel on the basis of the measured signal intensities by logarithmic transformation:[8]

$$c(t) = -\ln[S(t)/S(t_0)]/(k_2 TE) \quad (1)$$

(c = concentration, S = signal intensity, t = time, t_0 = time of injection, k_2 = constant depending on system and tissue properties, TE = echo time).

The postprocessing algorithm is based on the hypothesis that the area under the concentration–time curve recorded in every voxel reflects regional cerebral blood volume.[9] This assumption is correct for the first pass of the contrast bolus and requires numerical elimination of recirculating contrast. This is achieved with sufficient accuracy by a nonlinear least-squares fit procedure using a γ-variate or some other algebraic function as a model for the cerebral contrast transit in the absence of systemic contrast recirculation. Finally, a parameter image is generated in which the brightness of each voxel reflects the area of the concentration–time curve and thus regional cerebral blood volume.

In the myocardium, the contrast extravasates into the interstitial space. This already occurs to a substantial degree during the first pass;[10] the area under the concentration–time curve therefore does not correspond to a well-defined biologic compartment, but reflects the intravascular and a variable portion of the interstitial compartment. Myocardial blood volume can therefore not be derived from the concentration–time curve of small molecules[11] like gadolinium–DTPA. The use of a macromolecular agent with a molecular weight above 20 000 Da such as gadolinium–DTPA–polylysine[12] would be required for this purpose, but macromolecular agents are not yet available for clinical use.

As already mentioned, there is no rigorous theoretical concept for the construction of parameter images from concentration–time curves which reflect regional cerebral or myocardial blood flow. Some approaches are based on mean transit time concepts although the correct regional mean transit times cannot be determined using intravascular contrast agents.[3,4] In fact, the mean transit time calculated from the tissue concentration–time curve alone (without consideration of the arterial input function) overestimates the true organ mean transit time (which is on the order of only a few seconds in the brain) by more than 100%. Correction for the dispersion of the contrast bolus outside of the brain (finite injection duration, passage from the peripheral vein to the brain) would require

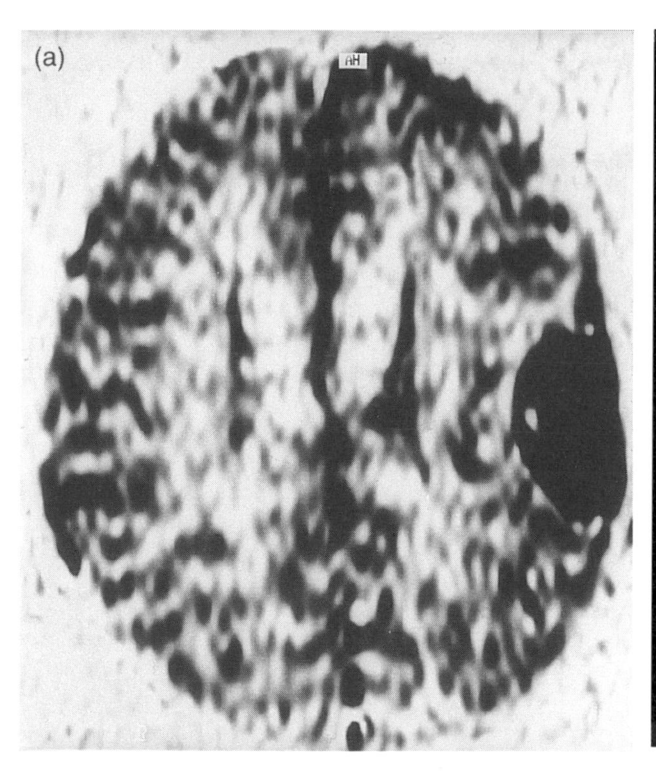

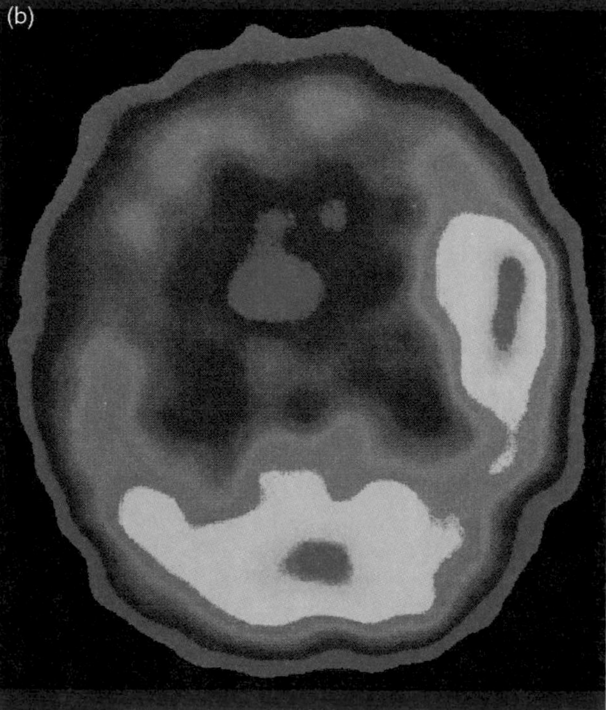

Figure 1 Sixty-one year old man with focal seizures of the right arm and speech arrest. Left parietal mass with homogeneous hyperperfusion pattern documented by MRI and confirmed by the radionuclide reference technique. Histology: meningiomatous and angiomatous meningioma. (a) T_2*-weighted first-pass study (subtraction image); (b) radionuclide study (HMPAO–SPECT)

For References see p. 4783

numerical deconvolution of the tissue curves with the arterial input function. Although the arterial input function can be measured[13] it is not the correct input function for every voxel. Finally, since the volume of distribution of gadolinium–DTPA-type contrast material is restricted, the inverse of the (correct) mean transit time would not represent blood flow in terms of [mL min^{-1} per 100 g tissue] but blood flow per unit blood volume [mL min^{-1} per ml blood]. Despite these limitations, some investigators have used transit time approaches with some success.[14, 15]

Another approach is to calculate a subtraction image where the image acquired at the time of maximum cerebral contrast concentration is subtracted from an image acquired before any contrast has reached the brain on a voxel-by-voxel basis [Figure 1(a)].[16] The underlying theoretical concept is valid under certain experimental conditions[17] and requires the consideration of the arterial input function. Fortunately, the arterial input function can be neglected as long as the image contains its own reference (e.g. the normal contralateral hemisphere), since the arterial input function is identical for all voxels under considerations in this concept. Subtraction images appear to reflect regional cerebral blood flow as assessed by an established radionuclide reference technique (HMPAO–SPECT) in the vast majority of patients with cerebral infarcts[16] and in all patients with homogeneous intracranial tumors [Figures 1(a) and (b)].[18] Discrepant results between the radionuclide method and first-pass subtraction images were found in inhomogeneous tumors (vital tumor/necrosis/edema). However, the discrepancy did not appear to result from the inability of MRI to depict adequately regional cerebral blood flow in these tumors but from the lower spatial resolution of SPECT (voxel volume = 1500 mm^3) versus MRI (voxel volume = 50 mm^3).

3 CEREBRAL PERFUSION IMAGING

Clinical experience with contrast enhanced MR perfusion imaging is steadily increasing. A number of recent reports focus on cerebral ischemia, intracranial tumors, and arteriovenous malformations.

In cerebral infarcts, regional cerebral blood flow is usually decreased but a hyperperfusion pattern is identified in some cases.[16, 19] While the prognostic significance of the hyperperfusion pattern in subacute infarcts has yet to be established, one might speculate that evidence of reperfusion indicates a better clinical outcome.

The grade of malignancy of intraaxial tumors appears to be correlated with the perfusion pattern. Low-grade astrocytomas are characterized by decreased blood volume when compared with contralateral gray matter. High-grade astrocytomas show either a homogeneous increase of blood volume or signs of marked inhomogeneity. These findings parallel—to a certain degree—the extent of the blood/brain barrier disruption as demonstrated by contrast enhancement on T_1-weighted (or Cranial Computerized Tomography—CCT) images. However, it is important to remember that the blood/brain barrier and regional cerebral perfusion are two distinct features which obey different underlying regulating and disturbing mechanisms.[20]

Multiple biopsies of intraaxial tumors have shown that high-grade tumors are composed of tissues with different histological grading. It is obvious that biopsies should be obtained from the most malignant part of the tumor. These areas are often characterized by increased metabolism, increased tumor blood flow, and volume. Perfusion imaging should therefore be considered for optimal biopsy guiding.[21] The differentiation of radiation necrosis after therapy of high-grade intraaxial tumors from recurrent tumor is cumbersome because both present with mass effect, necrosis and blood/brain barrier disruption. However, initial clinical results indicate that blood volume is homogeneously decreased in radiation necrosis, while an inhomogeneous pattern with areas of high regional blood volume can be demonstrated in recurrences.[21]

Arteriovenous malformations are quite easily demonstrated by MRI, including magnetic resonance angiography. However, due to inherent limitations of magnetic resonance angiography, the success of interventional therapy by embolization tends to be overestimated. In a recent study[22] we showed that contrast enhanced perfusion imaging is by far the most sensitive MRI technique to demonstrate residual pathologic vasculature after embolization. Perfusion imaging is also highly useful to demonstrate reopening of previously occluded pathways and is therefore part of our standard protocol in the evaluation and follow-up of arteriovenous malformations.

4 MYOCARDIAL PERFUSION IMAGING

Initial experience with gadolinium–DTPA at our Institution demonstrated increased enhancement in acute infarcts but no significant enhancement in subacute and chronic infarcts.[23] Since spin echo pulse sequences with an acquisition time of more than 5 min were used in these studies, the enhancement did not reflect blood flow but contrast extravasation into the interstitial space which is mainly a function of microvascular permeability. Other MR contrast media such as dysprosium, manganese, macromolecular gadolinium compounds, and superparamagnetic agents have been successfully studied.[24,25]

Atkinson and co-workers performed bolus injection experiments using a T_1-weighted pulse sequence;[26] they described a transient signal intensity increase during the first pass of the contrast material in normal myocardium of experimental animals and volunteers, but no signal intensity change after experimental occlusion of the left coronary artery. In a subsequent study, similar effects were seen in patients with coronary artery disease.[27] No attempt was made in these studies to quantify the effects in terms of myocardial blood flow.

Most recently, Wilke et al.[15] have used a kinetic model involving the mean transit time concept to obtain blood flow estimates in an experimental model. Blood flow was measured by a radiomicrosphere reference technique. A correlation coefficient of $r = 0.89$ was found between the MRI estimate of flow (the inverse of mean transit time) and the reference technique. It must be noted, however, that changes in myocardial blood volume are not accounted for in that model. Since myocardial blood volume is highly variable, myocardial blood flow estimates based on mean transit time alone are likely to be much less reliable in a clinical setting.

5 PERFUSION IMAGING OF OTHER ORGANS

Next to the brain and heart, perfusion imaging of the kidneys might be of clinical interest. Imaging of renal blood flow is quite complex because flow and function (=excretion) are tightly related. Gadolinium–DTPA is freely filtrated in the glomerulus and neither secreted nor reabsorbed in the tubular system.[28] Due to water reabsorption in the tubules, the concentration of gadolinium–DTPA increases by up to a factor of 100 from the initial plasma concentration. This wide range of concentrations is unique in the kidneys and has important effects on the observed signal intensity: at lower concentrations, signal intensity increases due to T_1 shortening; at higher concentrations, signal intensity decreases due to T_2 shortening. Obviously, this nonlinearity complicates the interpretation of signal intensity versus time curves measured in the kidneys. Nevertheless, three phases can be distinguished after contrast bolus injection: a vascular, a tubular, and a ductal phase.[29] Due to the complexity of both renal physiology and the contrast mechanisms, assessment of renal blood flow will remain one of the most challenging tasks in MR imaging of organ blood flow.

The methodology discussed in this article also applies to perfusion imaging of other tissues. It must be noted, however, that the quality of the primary data (the concentration versus time curve) depends on the contrast concentration obtained during the first pass after bolus injection; tissues with low regional blood flow and volume are therefore unlikely to be candidates for the assessment of perfusion by contrast bolus experiments.

6 FUNCTIONAL IMAGING OF THE HUMAN CORTEX

Functional imaging of the brain refers to the detection of stimulated cortical activity. The stimulus may be an optical, motor, or sensory paradigm; even ideation of motion or a visual pattern can act as a stimulus. Cortical stimulation entails alterations of regional cerebral blood flow, blood volume, oxygenation, and metabolism which can be detected by MRI. Classically a domain of nuclear medicine, in particular of positron emission tomography, MRI has been very successful in detecting evidence of cortical activation.[30] The methodology relies upon the above-described contrast bolus experiment which is performed under resting conditions and during stimulation. The computed blood volume images are subtracted from one another; the subtraction image represents areas of locally altered perfusion due to cortical stimulation.

7 RELATED ARTICLES

Blood Flow: Quantitative Measurement by MRI; Brain: Sensory Activation Monitored by Induced Hemodynamic Changes with Echo Planar MRI; Functional Neuroimaging Artifacts; Gadolinium Chelate Contrast Agents in MRI: Clinical Applications; Gadolinium Chelates: Chemistry, Safety, & Behavior; Susceptibility Effects in Whole Body Experiments.

8 REFERENCES

1. W. Schörner, R. Felix, M. Laniado, L. Lange, H. J. Weinmann, C. Claussen, W. Fiegler, U. Speck, and E. Katzner, *Fortschr. Röntgenstr.*, 1984, **140**, 493.
2. A. Haase, D. Matthaei, W. Hänicke, and J. Frahm, *Radiology*, 1986, **160**, 537.
3. N. A. Lassen, *J. Cerebr. Blood Flow Metab.*, 1984, **4**, 633.
4. R. M. Weisskoff, D. Chesler, J. L. Boxerman, and B. R. Rosen, *Magn. Reson. Med.*, 1993, **29**, 553.
5. J. J. H. Ackerman, C. S. Ewy, S. G. Kim, and R. A. Shalwitz, *Ann. N. Y. Acad. Sci.*, 1987, **508**, 89.
6. K. K. Kwong, A. L. Hopkins, J. W. Belliveau, D. A. Chesler, L. M. Porkka, R. C. McKinstry, D. A. Finelli, G. J. Hunter, J. B. Moore, B. R. Rosen, and R. G. Barr, *Magn. Reson. Med.*, 1991, **22**, 154.
7. N. A. Lassen and W. Perl, 'Tracer Kinetic Methods in Medical Physiology', Raven Press, New York, 1979.
8. B. R. Rosen, J. W. Belliveau, J. R. Vevea, and T. J. Brady, *Magn. Reson. Med.*, 1990, **14**, 249.
9. L. Axel, *Radiology*, 1990, **177**, 679.
10. J. A. Rumberger and M. R. Bell, *Invest. Radiol.*, 1992, **27**, S40.
11. J. M. Canty, R. M. Judd, A. S. Brody, and F. J. Klocke, *Circulation*, 1991, **84**, 2071.
12. M. Saeed, M. F. Wendland, T. Masui, A. J. Connolly, N. Derugin, R. C. Brasch, and C. B. Higgins, *Radiology*, 1991, **180**, 153.
13. W. H. Perman, M. H. Gado, K. B. Larson, and J. S. Perlmutter, *Magn. Reson. Med.*, 1992, **28**, 74.
14. J. W. Belliveau, B. R. Rosen, G. J. Hunter, E. Tasdemiroglu, R. MacFarlane, P. Boccalini, L. M. Hamberg, M. A. Moskowitz, and S. H. Simon, *Stroke*, 1993, **24**, 444.
15. N. Wilke, C. Simm, J. Zhang, J. Ellermann, X. Ya, H. Merkle, G. Path, H. Lüdemann, R. J. Bache, and K. Ugurbil, *Magn. Reson. Med.*, 1993, **29**, 485.
16. J. C. Böck, B. Sander, J. Hierholzer, M. Cordes, J. Haustein, W. Schörner, and R. Felix, *Fortschr. Röntgenstr.*, 1992, **156**, 382.
17. N. A. Mullani and K. L. Gould, *J. Nucl. Med.*, 1983, **24**, 577.
18. J. C. Böck, B. Sander, J. Hierholzer, J. Haustein, M. Scholz, K. H. Radke, W. Schörner, W. Lanksch, and R. Felix, *Fortschr. Röntgenstr.*, 1992, **157**, 378.
19. R. R. Edelman, H. P. Mattle, D. J. Atkinson, T. Hill, J. P. Finn, C. Mayman, M. Ronthal, H. M. Hoogewoud, and J. Kleefield, *Radiology*, 1990, **176**, 211.
20. J. C. Böck, B. Sander, K. H. Radke, J. Haustein, C. Zwicker, and R. Felix, *Adv. MRI Contrast*, 1993, **1**, 76.
21. B. R. Rosen, J. W. Belliveau, A. J. Aronen, D. Kennedy, B. R. Buchbinder, A. Fischman, M. Gruber, J. Glas, R. M. Weisskoff, M. S. Cohen, F. H. Hochberg, and T. J. Brady, *Magn. Reson. Med.*, 1991, **22**, 293.
22. J. C. Böck, H. P. Molsen, B. Sander, and R. Felix, *Fortschr. Röntgenstr.*, 1992, **157**, 471.
23. H. W. Eichstaedt, R. Felix, F. C. Dougherty, M. Langer, W. Rutsch, and H. Schmutzler, *Clin. Cardiol.*, 1986, **9**, 527.
24. A. D. Watson, S. M. Rocklage, and M. J. Carvlin, in 'Magnetic Resonance Imaging', 2nd edn. ed. D. D. Stark and W. G. Bradley, Mosby Year Book, St. Louis, MO, 1992, Vol. 1, Chap. 14.
25. C. B. Higgins, G. Caputo, M. F. Wendland, and M. Saeed, *Invest. Radiol.*, 1992, **27**, S66.
26. D. J. Atkinson, D. Burstein, and R. Edelman, *Radiology*, 1990, **174**, 757.
27. W. J. Manning, D. J. Atkinson, W. Grossman, S. Paulin, and R. R. Edelman, *J. Am. Coll. Cardiol.*, 1991, **18**, 959.
28. C. H. Lorenz, T. A. Powers, and C. L. Partain, *Invest. Radiol.*, 1992, **27**, S109.
29. P. L. Choyke, J. A. Frank, M. E. Girton, S. W. Inscoe, M. J. Carvlin, J. L. Black, H. A. Austin, and A. J. Dwyer, *Radiology*, 1989, **170**, 713.

30. J. W. Belliveau, D. N. Kennedy, R. C. McKinstry, B. R. Buchbinder, R. M. Weisskoff, M. S. Cohen, J. M. Vevea, T. J. Brady, and B. R. Rosen, *Science*, 1991, **254**, 716.

Biographical Sketches

Roland Felix. *b* 1938. M.D., 1962, Munich. 1962–78; Departments of Surgery, Freiburg; Gynecology, Hamburg; Medicine, Mainz; Physiology, Bonn; Medicine, Kiel; Radiology, Bonn. Professor of Clinical Radiology and Chairman, Strahlenklinik und Poliklinik (Department of Diagnostic Radiology, Nuclear Medicine, and Radiooncology), Virchow-Klinikum, Medizinische Fakultät der Humboldt-Universität zu Berlin, 1978–present. Approximately 900 publications. Research interests cover contrast enhanced CT and MRI, telecommunication, and radiooncology including therapeutic hyperthermia.

Johannes C. Böck. *b* 1957. M.D., 1987, Göttingen. Introduced to MRI by R. Felix in 1989. Research fellow, Departments of Surgery, Radiology and Nuclear Medicine, University of California, San Francisco, 1987–89; Strahlenklinik und Poliklinik, Virchow-Klinikum, Medizinische Fakultät der Humboldt-Universität zu Berlin, 1989–present. Approximately 60 publications. Research interests include cerebral perfusion imaging by NMR, MR contrast media, MR angiography, and pulmonary imaging by NMR.

Tissue Water & Lipids: Chemical Shift Imaging & Other Methods

W. Thomas Dixon

Emory University, Atlanta, GA, USA

1 INTRODUCTION

Most signal contributing to images in vivo comes from water and from lipid methylene protons which resonate at a frequency about 3.4 ppm lower. Other spins are so much rarer they are not seen without special effort. Spectroscopic imaging sensitive to these additional, minor lines as well is discussed in ***Chemical Shift Imaging***. Because more measurements are required to determine more unknowns, imaging methods that can determine pixel intensities at the chemical shifts of many rarer resonances are slower or provide coarser spatial resolution than those dealing with only a two-line spectrum. Methods described in this article typically produce images with the same spatial resolution as standard, chemical shift blind imaging methods. Because images are made from echoes, only water and lipid lines with reasonable T_2 values contribute. Lipid droplets are readily seen but membrane lipids are not. The contribution of intracellular water to most images has been questioned.[1] This article explains why and how this basic, two-line spectrum is manipulated in imaging. The subject has been reviewed more thoroughly before.[2,3]

2 MANIPULATION TYPES AND USES

What water and lipid manipulations are desired and why? When lipids are a nuisance in images, fat suppression is needed; when lipid is to be studied, water suppression may be useful; when fat/water interfaces, displacement of water by fat or of fat by water are to be detected, an image of water signal minus lipid signal is very sensitive.

2.1 Lipid Suppressed Images

Lipids signals are often a nuisance. Because lipid is abundant, has high proton density, and repolarizes before the water in most tissues, it is the dominant source of signal in many images. A small fraction of the lipid signal in the wrong place often degrades images of less intense but medically important, watery regions. Optic nerves are surrounded by fat and provide an example of the benefit of suppressing fat. Signal increases in the spinal chord, brain, optic nerves, and retinas following injection of Gd–DTPA are a sign of pathology. Ordinarily, a barrier around blood vessels in these organs prevents Gd–DTPA and many other substances from escaping into the surrounding tissue. However, when this barrier is damaged, Gd in the tissue enhances longitudinal relaxation and therefore increases signal intensity. The patient shown in Figure 1 has autoimmune optic neuritis affecting the right side with an apparently normal left side. The increased intensity of the right optic nerve, relative to the left side or to the brain, is easily seen when fat is suppressed, but not when the nerve is surrounded by high intensity fat.

Fat signal gets out of control in several ways. The most important are probably spreading of the lipid signal in the phase-encoding direction when fat moves (heart beat, breathing, wiggling), Gibbs's ringing, and the 'chemical shift artifact'. This artifact is an offset in apparent position of lipid in the slice selection and read gradient directions caused by the water–lipid chemical shift difference. Gibbs's ringing and the chemical shift artifact obscure small structures near a large amount of fat. Cartilage in joints and the optic nerve are examples. Moving fat produces long-range phase-encoding artifacts which are a serious problem in the abdomen and thorax. These can, for example, interfere with the detection of low contrast masses in the liver. Side lobes or other slice selection problems, quadrature errors, lack of dynamic range in signal detection,

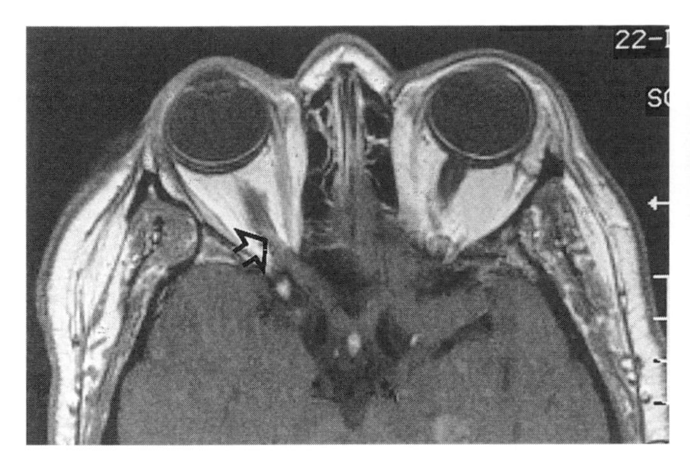

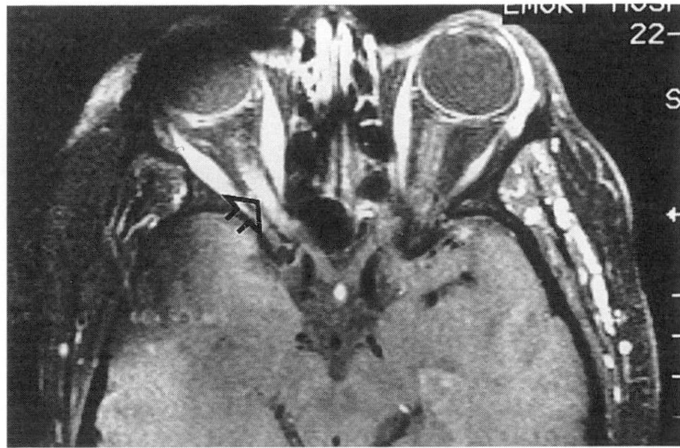

Figure 1 T_1-Weighted images of the optic nerves. The right optic nerve (arrow, left side of image), but not the left optic nerve, is enhanced in intensity by Gd–DTPA. Saturation of the surrounding fat by a selective pulse made this easier to detect in the image at right. (Courtesy of Dr. Susan B. Peterman)

instrument instability, and coil tuning variations caused by breathing can also spread lipid signal to lipid-free regions of an image.

Fat suppression methods are not perfect. Generally, partial water suppression is a more serious image flaw than is incomplete or nonuniform fat suppression; however, both flaws may be serious in quantitative imaging. Examples of quantitative studies in organs which accumulate lipid are measuring the fat content of the heart[4] or the water relaxation rates of the liver.[5] Water relaxation measurements may be distorted by partial water suppression and will be by any lipid signal, whether it originates within the region of study or elsewhere.

2.2 Water Suppressed Images

Most methods suitable for fat suppression can suppress water instead, and some easily produce an image pair, one of fat, the other of water. Replacement of cancer by fat is a favorable prognostic indicator. Lipid signal from heart muscle may indicate myocardial ischemia or infarction.[4] For some nutritional studies, quantitation of adipose tissue is needed. Water suppression is useful here.

2.3 Opposed Images

Lipid images by themselves can be hard to interpret because interesting organs and landmarks for finding them are lipid-free and invisible. However, the presence of fat can be a useful source of contrast or mark the distinction between muscles or other organs. Images in which the phase of the fat signal is opposite to that of the water, 'opposed images', have the chemical shift information of a lipid image but show all anatomic structures. These images can be made on purpose, and in gradient echo imaging occur incidentally when certain echo times are chosen. Usually images are presented as absolute values. At fat/water interfaces in absolute value opposed images, a line of dark pixels occurs where the water and fat signals within the pixels cancel. A dark outline can make it easy to see a lymph node or an adrenal gland surrounded by fat of similar intensity. Of course, if the absolute value is not taken, or rather the effort is made to preserve the sign, all organs are easily distinguished from fat.

Liver can be diffusely infiltrated by fat. Opposed images are more sensitive to infiltration than water-suppressed or fat-suppressed images because they show both the increase in fat signal and decrease in water signal caused by displacement. Decreasing liver intensity this way can aid detection of metastases in liver[6] because the intensity of the metastases is not affected.

Some applications simply require any manipulation that identifies lipids. For example, in leukemia, fatty bone marrow in the spine is replaced by water but reappears after successful treatment.[7]

3 METHODS

The following outline orders some of the many published fat–water manipulation methods by principle of operation. Many pulse sequences combining different methods of operation have been demonstrated also.

- Longitudinal relaxation (Section 3.1):
  - Single inversion (STIR)
  - Double inversion
- J Coupling (Section 3.2):
  - Classical multiple quantum sequences
  - Multiple echo fast imaging sequences
- Chemical shift (Section 3.3):
  - Selective rf pulses
    - rf pulses without gradients (selective excitation; selective suppression
    - rf pulses with gradients
  - Phase methods
    - Hahn and gradient–echoes
    - gradient–echoes only

3.1 T_1 Based Fat Suppression

The T_1-based, short inversion time inversion–recovery, STIR,[8] is a very simple fat suppression method. An inversion–recovery image is made with TI chosen to null signal from

For References see p. 4788

undesired fat. With a long *TR*, this is (ln 2)T_1. With proper interleaving of pulses, multislice IR imaging can be quite time efficient.

Because this method is not sensitive to shimming and is hardly sensitive to flip angle variations, it can give uniform suppression over a large region. Adipose tissue contains some water and varies regionally in composition. Temperature and relaxation rates also vary, especially in the skin; therefore, suppression is not completely uniform. Either the lipid signal or the combined lipid and water signal from adipose tissue can be nulled with the proper *TI*; the difference is slight.

Objections to STIR for fat suppression are that it partially suppresses water, by a different amount in different tissues, and it fixes *TI* so that T_1-weighting cannot be effectively adjusted. Figure 2 illustrates this for a typical STIR sequence. Assuming relaxation rates for fat and most other tissues are around 4 s^{-1} and 1 s^{-1}, respectively, this sequence gives essentially no T_1-weighting for most tissues and most tissue magnetizations are just over one-half their equilibrium values.

Following a second inversion pulse, tissues with two different T_1 values can recover to zero magnetization simultaneously. Two inversion times must be adjusted to align the two zero crossings. In angiograms, pictures of flowing blood, this has been done[9] to eliminate both fat and one other tissue (muscle or brain) from images.

3.2 *J* Coupling

Lipids have *J*-coupled spins while water does not. Classic multiple quantum pulse sequences have been combined with imaging to produce lipid images.[10] More recently, *J* coupling has been proposed as an important reason that fat is more intense in certain fast images than in conventional images having the same nominal *TE*. These fast images (RARE) collect several lines of *k*-space in closely spaced Hahn echoes following each excitation [***Multi-Echo Acquisition Techniques Using Inverting Radiofrequency Pulses in MRI***]. *J* coupling reduces the intensity of coupled spins less when more 180° pulses are used to produce an echo at a given time. This has led to the production of lipid images by subtracting fast images made with the same echo time but different numbers of refocusing pulses.[11] Water intensity is much less dependent on the number of refocusing pulses used. A threshold can be set to identify voxels containing fat, and postprocessing then used to produce either water or lipid images.[11]

J coupling methods are immune to shimming problems and are temperature-independent. They require that two images be made to obtain any chemical shift information. This takes additional time and increases sensitivity to motion, even if acquisitions of the two images are interleaved. Water images will necessarily be T_2-weighted and lipid images will have an intensity which may depend in a complicated way on lipid chemistry.

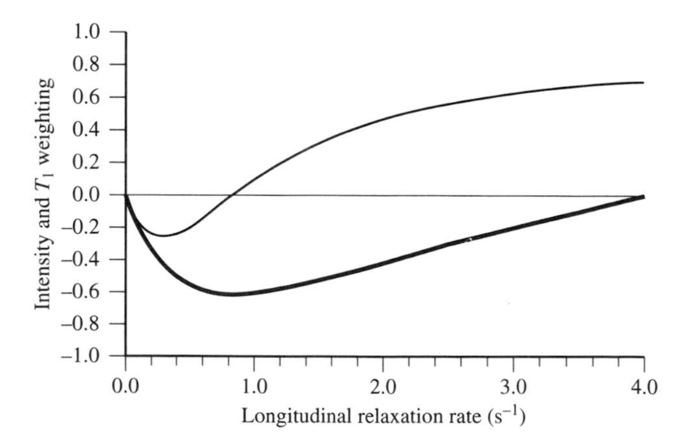

Figure 2 Intensity and T_1-weighting in STIR images. Signal intensity (bold line) and T_1-weighting (fine line) versus longitudinal relaxation rate. Weighting is defined here as 100-times the intensity change caused by a 1% rate change. Weighting and intensity are measured in units of equilibrium magnetization. (Pulse sequence *TR*/*TI*/*TE* = 2500/175/negligible ms)

3.3 Chemical Shift Based Methods

For most of our concerns, the spectrum of tissues can be approximated as two δ functions 3.4 ppm apart. An approximation that cannot be made is that the water Larmor frequency is everywhere the same. Shimming is by far the most serious problem with chemical shift based lipid manipulation schemes. Even if the magnet is perfectly shimmed when empty, the susceptibility of the subject distorts the field by an amount comparable to the 3.4 ppm separation between fat and water resonances. To aid shimming, the subject can be surrounded by a material with a susceptibility near that of water, but near cavities such as sinuses, airways, lungs, or bowel gas, this does not help.

3.3.1 *Chemically Selective rf Pulses without Gradients*

A simple approach to chemical shift imaging is to saturate either the water or the fat magnetization before a standard imaging sequence.[12] This can be done with a soft, frequency selective pulse of about 90° in the absence of a gradient, followed by a spoiling gradient, then the rest of the imaging sequence. As an alternative to the extra chemical shift selective pulse, one of the pulses in a multipulse imaging sequence can be made chemically rather than spatially selective. A chemical shift selective saturation pulse can be applied during an inversion–recovery sequence at the time of the zero crossing of fat. This combined method has been compared with some other chemically selective pulse methods.[13] Of these shift selective pulse methods, those with pulses selective for the desired signal are harder to incorporate in multislice methods than those selectively exciting the unwanted species.

In the presence of shimming errors, an rf pulse intended to suppress fat may suppress water instead in regions with a lower field and suppress nothing at all in higher field regions. Such pulses can give good results in relatively small regions that are not near odd-shaped air spaces and which can be placed at the magnet isocenter, for example the knee. A suppression pulse combined with surface coil reception is useful in examining the spine. The surface coil's lack of sensitivity to moving fat in the front of the body prevents phase-encoding artifacts, so the suppression pulse need only be effective near the spine. Magnet shimming is in general much better now than when selective pulse methods were put forward. Originally only of academic interest, selective pulses are now included in commercial, clinical software for fat saturation.

3.3.2 *rf Pulses with Gradients*

Angiograms, or images of flowing blood, can be made from a stack of thin, highly T_1-weighted images. Most tissues are dark because there is little time for recovery between acquisitions. Blood flows into the slice fully polarized if it moves more than a slice thickness in *TR*. Fat limits the smallest vessels that can be seen because it recovers magnetization rapidly and has an intensity comparable to the blood in small vessels. Veins present another problem in images of arteries. Veins usually run along arteries but are less often of interest. Both problems can be solved simultaneously by saturating a thick slab on one side of the slice being imaged.[14] This is done using a soft pulse with a weak gradient. The slab is positioned to saturate venous blood upstream of the slice but arterial blood downstream, thus removing veins but not arteries from the angiogram. The region of fat saturated by the extra pulse is not exactly the same as the region of water saturated because of the Larmor frequency difference. The gradient direction and magnitude for saturating the slab are chosen so that fat in the region to be imaged is saturated, but upstream blood is not. The shimming effect on this fat suppression method depends on the gradient strength. Excited slices are not perfectly flat but are warped like potato chips. Consecutive slices will nest perfectly if made with the same gradient and the same rf pulse (except for a frequency shift) but otherwise will overlap or leave gaps where none are intended.

3.3.3 *Phase Based Methods*

These methods are based on the difference between a gradient echo and Hahn's original spin echo, better called a Hahn, rf recalled, or rf echo to avoid confusion (Figure 3). Pixel phase in the image is the spin phase at the time of the gradient echo. At a Hahn echo, spin phase is independent of chemical shift and shimming. Therefore, when these echoes coincide, water and fat have the same phase in the reconstructed image. When they do not, phase accumulates at the Larmor frequency from the Hahn echo until the gradient echo. (For simplicity, the model uses a frame rotating at the instrument frequency.) Because the water–fat Larmor frequency difference is known, images with any desired phase difference between lipid and water can be produced by choosing the two echo times carefully.

Availability of starting images with known water–lipid phase shifts creates an urge to Fourier transform them for presentation. Transforms with one,[15] two,[16,17] three,[18] four,[19] and more[20] starting images have been used. Using more starting images allows solving more equations (by FT) for more unknowns. Formally, the output of a one-point FT is the same as its input. This is appropriate for opposed images as discussed before. The output of a two-point FT is the sum and the difference of the input points. Given images with water and lipid in phase and 180° opposed, a two-point FT produces separate, quantifiable images of lipid and of water. Additional unknowns of interest that can be imaged by phase-based methods include regional B_0 variation and the lipid linewidth.

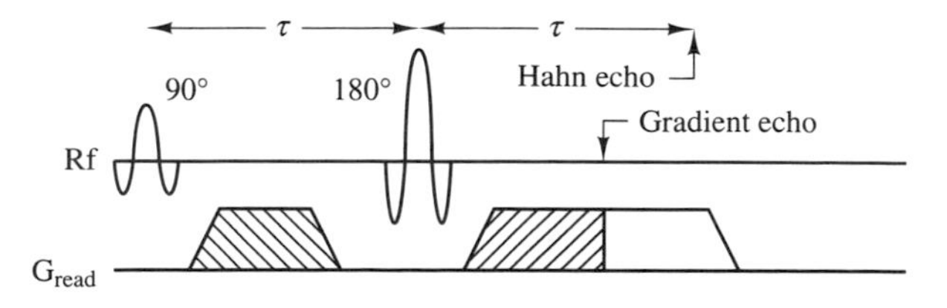

Figure 3 Hahn and gradient echoes in a spin echo imaging sequence

Phase-based manipulations are most straightforward when refocusing pulses produce Hahn echoes. This allows images with different phase relations to be made with the same gradient echo time. In these, regions differ because of chemical shift differences but not because of T_2 differences. Nevertheless, phase-based fat suppression has been done with gradient echoes alone in the absence of refocusing pulses.[21]

The sign of a signal is often a question in IR imaging, especially when contrast agents are employed. In images produced by phase-based fat–water manipulations, the sign is always a question and occasionally a problem. If the phases and the signs of signals were ideal, separation of fat (F) from water (W) images given in phase (Si) and opposed (So) images would be unambiguous.

$$\begin{array}{lcl} \mathrm{Si} = \mathrm{W} + \mathrm{F} & \rightarrow & \left.\begin{array}{l} 2\mathrm{W} = \mathrm{Si} + \mathrm{So} \\ 2\mathrm{F} = \mathrm{Si} - \mathrm{So} \end{array}\right\} \\ \mathrm{So} = \mathrm{W} - \mathrm{F} & & \end{array} \quad (1)$$

However, if only absolute value images are available,

$$\begin{array}{lcl} 2\mathrm{W} = |\mathrm{Si}| + |\mathrm{So}| & \text{or} & \left.\begin{array}{l} 2\mathrm{F} = |\mathrm{Si}| + |\mathrm{So}| \\ 2\mathrm{W} = |\mathrm{Si}| - |\mathrm{So}| \end{array}\right\} \\ 2\mathrm{F} = |\mathrm{Si}| - |\mathrm{So}| & & \end{array} \quad (2)$$

The second solution applies where F > W. Simple-minded addition of images to produce a water image and subtraction to produce lipid images incorrectly reports a 60:40 mixture of lipid and water signal as 40:60 and so forth. The impact of this twofold ambiguity varies. Adipose tissue appears in the water instead of in the lipid image. Quantitative studies of fatty livers are affected while searches for liver metastases are not. Wherever the absolute value reverses the sign of the opposed image, misclassification occurs.

In principle, use of the original, complex images rather than absolute value images prevents the twofold ambiguity in fat signal fractions. In practice, images made with gradient and Hahn echoes which do not coincide have complicated phases depending on things other than chemical shift, shimming being the most important. The sign problem has been solved by noting that chemical shift often changes suddenly from one voxel to the next but $\boldsymbol{B}_0$ always changes gradually. In a phase map, fat/water boundaries can produce buttes or mesas superimposed on slowly rolling hills produced by shimming or other instrumental errors. A global view allows identification of region boundaries and therefore correction of signs in absolute value and absolute value-derived images. This phase unwrapping shows that one region is primarily lipid and another primarily water, but another method, usually inspection, is needed to show which is which.

When the phase of a particular pixel corresponds to a frequency 3.4 ppm off reference, that could indicate its signal comes from lipid rather than water. Alternatively B_0 could be 3.4 ppm low at that location. No local method using information from that voxel alone can distinguish chemical shift differences from B_0 differences, regardless of what pulse sequence is used or how many basis images are available. Even though the fat signal fraction ambiguity can only be

For References see p. 4788

solved by improved reconstruction methods, not by improved pulse sequences, desire to solve this ambiguity has led to new pulse sequences. Pulse sequence changes can make phase unwrapping easier. For example, increasing the phase difference between water and lipid from 180° to 360° removes the chemical shift induced cliffs from the opposed image phase map while doubling the height of the pattern of rolling hills caused by shimming and other variables.[19,22] The resulting $\boldsymbol{B}_0$ map helps unwrap the opposed image and can be used to adjust the instrument. Using comparison images of a phantom with uniform chemical shift also helps solve the sign problem.[23]

A less elaborate solution has been used to determine fat fractions in large regions of interest in vertebral bone marrow of leukemia patients.[7] Examination of the real opposed image determines whether signals from the marrow and from the intervertebral disks have the same or opposite signs. The disks contain no fat and have a positive signal. Knowing the sign of the opposed image signal removes the twofold ambiguity in fat signal fraction.

4 OTHER USES

Manipulations described here can be used for substances other than fat. T_1-Based fat suppression has been combined with frequency selective water suppression pulses[24,25] to image silicone prostheses. Silicone images have also been produced using opposed images in which delays were chosen to produce water/lipid, methylene/silicone, methyl proton phases of 0°, 0°, 180°.[26] Fluorinated blood substitutes are difficult to image because multiple resonances with large shift differences frustrate position determination in both slice selection and read directions. Pulse sequence[27] and reconstruction changes[28,29] can help. Imaging the sugar distribution in grapes[30–32] is a challenging application because glucose and fructose spectral lines lie closer to water than do lipid lines.

5 RELATED ARTICLES

Breast MRI; Chemical Shift Imaging; Inversion Recovery Pulse Sequence in MRI; Multi-Echo Acquisition Techniques Using Inverting Radiofrequency Pulses in MRI; Respiratory Artifacts: Mechanism & Control; Selective Excitation Methods: Artifacts; Water Suppression in Proton MRS of Humans & Animals; Whole Body Magnetic Resonance Artifacts.

6 REFERENCES

1. M. Neeman, K. A. Jarrett, L. O. Sillerud, and J. P. Freyer, *Cancer Res.*, 1991, **51**, 4072.
2. J. H. Simon and J. Szumowski, *Invest. Radiol.*, 1992, **27**, 865.
3. L. Brateman, *Am. J. Roentgenol.*, 1986, **146**, 971.
4. A. Bouchard, M. Doyle, P. E. Wolkowicz, R. Wilson, W. T. Evanochko, and G. M. Pohost, *Magn. Reson. Med.*, 1991, **17**, 379.
5. C. S. Poon, J. Szumowski, D. B. Plewes, P. Ashby, and R. M. Henkelman, *Magn. Reson. Imaging*, 1989, **7**, 369.
6. D. D. Stark, J. Wittenberg, M. S. Middleton, and J. T. Ferrucci, *Radiology*, 1986, **158**, 327.
7. E. L. Gerard, J. A. Ferry, P. C. Amrein, D. C. Harmon, R. C. McKinstry, B. E. Hoppel, and B. R. Rosen, *Radiology*, 1992, **183**, 39.
8. G. M. Bydder and I. R. Young, *J. Comput. Assist. Tomogr.*, 1985, **9**, 659.
9. W. T. Dixon, M. Sardashti, M. Castillo, and G. P. Stomp, *Magn. Reson. Med.*, 1991, **18**, 257.
10. C. L. Dumoulin and D. Vatis, *Magn. Reson. Med.*, 1986, **3**, 282.
11. R. T. Constable, R. C. Smith, and J. C. Gore, *J. Magn. Reson. Imaging*, 1993, **3**, 547.
12. P. A. Bottomley, T. H. Foster, and W. M. Leue, *Proc. Natl. Acad. Sci. USA*, 1984, **81**, 6856.
13. E. Kaldoudi, S. C. R. Williams, G. J. Barker, and P. S. Tofts, *Magn. Reson. Imaging*, 1993, **11**, 341.
14. M. Doyle, T. Matsuda, and G. M. Pohost, *Magn. Reson. Med.*, 1991, **21**, 71.
15. J. K. T. Lee, J. P. Heiken, and W. T. Dixon, *Radiology*, 1985, **156**, 429.
16. W. T. Dixon, *Radiology*, 1984, **153**, 189.
17. D. D. Blatter, A. H. Morris, D. C. Ailion, A. G. Cutillo, and T. A. Case, *Invest. Radiol.*, 1985, **20**, 845.
18. G. H. Glover and E. Schneider, *Magn. Reson. Med.*, 1991, **18**, 371.
19. G. H. Glover, *J. Magn. Reson. Imaging*, 1991, **1**, 521.
20. R.E. Sepponen, J. T. Sipponen, and J. L. Tanttu, *J. Comput. Assist. Tomogr.*, 1984, **8**, 585.
21. J. Szumowski and D. B. Plewes, *Magn. Reson. Med.*, 1988, **8**, 345.
22. H. N. Yeung and D. W. Kormos, *Radiology*, 1986, **159**, 783.
23. J. A. Borrello, T. L. Chenevert, C. R. Meyer, A. M. Aisen, and G. M. Glazer, *Radiology*, 1987, **164**, 531.
24. B. Pfleiderer, J. L. Ackerman, and L. Garrido, *Magn. Reson. Med.*, 1993, **29**, 656.
25. S. Mukundan, Jr., W. T. Dixon, B. D. Kruse, D. L. Monticciolo, and R. C. Nelson, *J. Magn. Reson. Imaging*, 1993, **3**, 713.
26. E. Schneider and T. S. Chan, *Radiology*, 1993, **187**, 89.
27. H. K. Lee, O. Nalcioglu, and R. B. Buxton, *Magn. Reson. Med.*, 1991, **21**, 21.
28. L. J. Busse, R. G. Pratt, and S. R. Thomas, *J. Comput. Assist. Tomogr.*, 1988, **12**, 824.
29. H. K. Lee and O. Nalcioglu, *J. Magn. Reson. Imaging*, 1992, **2**, 53.
30. J. M. Pope, H. Rumpel, W. Kuhn, R. Walker, D. Leach, and V. Sarafis, *Magn. Reson. Imaging*, 1991, **9**, 357.
31. J. M. Pope, R. R. Walker, and T. Kron, *Magn. Reson. Imaging*, 1992, **10**, 695.
32. B. A. Goodman, B. Williamson, and J. A. Chudek, *Magn. Reson. Imaging*, 1993, **11**, 1039.

Biographical Sketch

W. Thomas Dixon. *b* 1945. B. Mech. Engr., 1967 Ph.D., 1980, physics, (with O. Lumpkin), University of California, San Diego. Postdoctoral work at Washington University/Monsanto with J. Schaefer. Approx. 35 publications. Research areas: cross polarization, magic angle spinning, proton imaging.

TOCSY in ROESY & ROESY in TOCSY

J. Schleucher, J. Quant, S. J. Glaser, & C. Griesinger
Universität Frankfurt, Frankfurt, Germany

1 INTRODUCTION

Homonuclear TOCSY[1] and ROESY[2] experiments are essential tools for the identification of spin systems and for the measurement of internuclear distances, respectively. (ROESY has been reviewed by Bull;[3] see also ***ROESY*** in this Encyclopedia.) TOCSY experiments rely on the transfer of polarization via J coupling, ROESY experiments rely on dipolar relaxation. Cross peaks due to slow chemical exchange are present in both experiments. The success of TOCSY and ROESY depends critically on the performance of the mixing sequence used to accomplish polarization transfer via the desired interaction. It will be shown that because of the complicated mixing sequences used, cross relaxation and J transfer cannot easily be separated. In fact, longitudinal and transverse cross relaxation and J transfer are intimately connected, and it is the aim of this article to describe handy tools to estimate the contribution of either of these interactions for a given pulse sequence. Unwanted interactions must be suppressed, since they lead to artifacts that can hamper the interpretation of the spectra. For example, transfer via J couplings alters the intensity of cross peaks in ROESY, and cross relaxation interferes with weak cross peaks in TOCSY. The offset dependence of TOCSY transfer and the offset dependence of the weight of transverse and longitudinal cross-relaxation rates in CW ROESY is illustrated in Figure 1. Hartmann–Hahn transfer along the diagonal cannot be avoided in any ROESY experiment. Hartmann–Hahn transfer along the antidiagonal is a more serious drawback, which can be overcome in more sophisticated ROESY experiments. The offset dependence of the net cross-relaxation rate

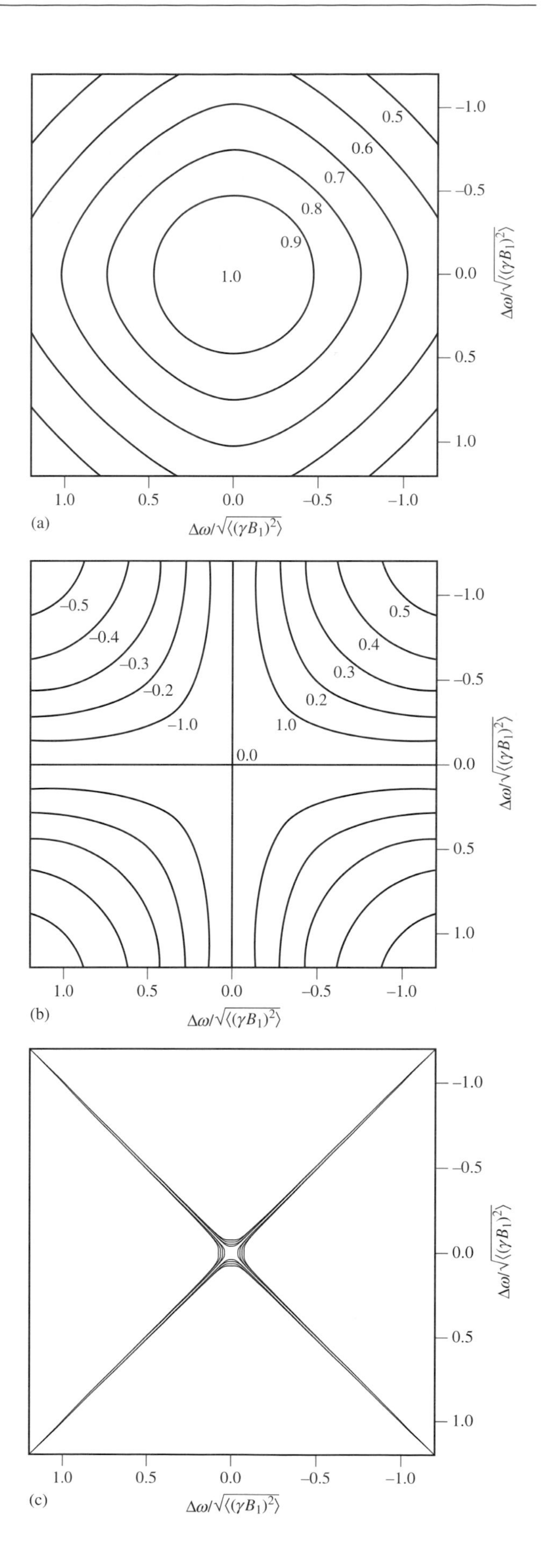

Figure 1 Simulation of the offset dependences of the weights of transverse (a) and longitudinal (b) cross relaxation and Hartmann–Hahn transfer (c) in CW ROESY.[11] In (c) the largest possible Hartmann–Hahn transfer is shown as a contour plot

For References see p. 4803

Table 1 Relations between J Transfer, Cross Relaxation, and Chemical Exchange

	TOCSY	NOESY	ROESY	EXSY
(a) Large Molecules ($\omega\tau_c >> 1$)				
J transfer		Zero for weakly coupled spins; optimally suppressed for strongly coupled spins	Suppression of J transfer and of longitudinal cross relaxation are conflicting goals	Good suppression of J transfer and both types of cross relaxation necessary and possible
Cross relaxation	Zero (clean TOCSY)	σ_{NOE}		
Exchange	Fully operative	Fully operative	Fully operative	
(b) Medium Sized Molecules ($\omega\tau_c \approx 1$)				
J transfer		Zero for weakly coupled spins; optimally suppressed for strongly coupled spins	Suppression of J transfer and maximization of transverse cross relaxation are conflicting goals	Application of NOESY yields complete suppression of cross relaxation and J transfer
Cross relaxation	Reduced transverse cross relaxation cannot be suppressed	Not applicable, since $\sigma_{NOE} = 0$		
Exchange	Fully operative	Fully operative	Fully operative	
(c) Small Molecules ($\omega\tau_c << 1$)				
J transfer		Zero for weakly coupled spins; optimally suppressed for strongly coupled spins	Suppression of J transfer has priority since longitudinal and transverse cross relaxation have the same sign	Application of NOESY suppresses J transfer in weakly coupled spin systems; cross relaxation cannot be suppressed
Cross relaxation	Small cross-relaxation rates do not compete efficiently	σ_{NOE}		
Exchange	Fully operative	Fully operative	Fully operative	

σ_{net} requires offset corrections to be applied if quantitative information is to be extracted from ROESY spectra.

Therefore, separation of coherent J transfer (TOCSY) and transfer via cross relaxation is an important goal in the development of multiple pulse sequences used in these experiments. In this article, we give a summary of theoretical tools used to analyze the performance of multiple pulse sequences in this respect. We discuss approaches to obtain cross-relaxation-free TOCSY spectra and suppression of coherent transfer and longitudinal cross relaxation in ROESY spectra. Relations between J transfer, cross relaxation, and chemical exchange for molecules of different sizes are collected in Table 1 (since cross-relaxation rates are strongly dependent on molecular weight, the possibility of separating cross relaxation from the other interactions is considered separately for small, medium sized, and large molecules). It is obvious from this table that for molecules of any size, longitudinal cross relaxation (NOE) can be observed without contributions from ROE or J transfer. Suppression of cross relaxation in TOCSY spectra is a problem for large molecules only. Suppression of TOCSY in ROESY spectra is necessary for molecules of any size. Chemical exchange cannot be suppressed in any experiment, but can be identified by its sign in ROESY spectra.

2 THEORY

2.1 Average Hamiltonian Theory

An arbitrary multiple pulse sequence can be represented formally by a time-dependent Hamiltonian $\hat{\mathcal{H}}(t)$. $\hat{\mathcal{H}}(t)$ can generally be represented by a series of constant Hamiltonians that act for finite intervals of time. Each Hamiltonian introduces a unitary transformation of the density matrix that is given by its propagator, and the overall effect of the multiple pulse sequence is given by the product of these propagators. Since the product of unitary transformations is again a unitary tranformation, the product of propagators can always be written as a single propagator. The Hamiltonian corresponding to this propagator is called the effective Hamiltonian $\mathcal{H}_{eff}$ of the multiple pulse sequence. Average Hamiltonian theory[4–6] yields criteria indicating whether the effective Hamiltonian can be expressed as an 'average Hamiltonian' and whether the effective and the average Hamiltonian are the same. It also shows how the average Hamiltonian for a given $\hat{\mathcal{H}}(t)$ can be obtained. Average Hamiltonian theory is most useful if $\hat{\mathcal{H}}(t)$ fulfills two conditions:

1. $\hat{\mathcal{H}}(t)$ is periodic with a cycle time τ_{cyc}, i.e., $\hat{\mathcal{H}}(t + n\tau_{cyc}) = \hat{\mathcal{H}}(t)$;
2. the state of the spin system is only sampled at times $n\tau_{cyc}$.

These conditions are generally fulfilled for multiple pulse sequences used for TOCSY and ROESY. $\hat{\mathcal{H}}(t)$ is time-dependent during each cycle of the multiple pulse sequence, and the state of the spin system is interrogated only at multiples of the cycle times, i.e., at the end of the mixing time.

To apply average Hamiltonian theory to an arbitrary multiple pulse sequence, we first note that $\hat{\mathcal{H}}(t)$ can be decomposed into a constant part $\hat{\mathcal{H}}^0$, the Hamiltonian of the spin system in the absence of any rf irradiation, and a time-dependent part $\hat{\mathcal{H}}^1(t)$, which describes the action of the multiple pulse sequence:

$$\hat{\mathcal{H}}(t) = \hat{\mathcal{H}}^0 + \hat{\mathcal{H}}^1(t), \quad \hat{U}(t) = T\exp\left\{ -\mathrm{i}\int_0^t [\hat{\mathcal{H}}^0 + \hat{\mathcal{H}}^1(t')]\,\mathrm{d}t' \right\} \quad (1)$$

The propagator $\hat{U}(t)$ has been expressed as a time integral, as a generalization of the product of two propagators. The Dyson time-ordering operator[7,8] T formally expresses the fact that in the operator product, factors with different time arguments must be arranged in order of decreasing time from left to right. The propagator $\hat{U}^1(t)$ of $\hat{\mathcal{H}}^1(t)$ takes the form

$$\hat{U}^1(t) = T\exp\left[-\mathrm{i}\int_0^t \hat{\mathcal{H}}^1(t')\,\mathrm{d}t' \right] \quad (2)$$

Multiple pulse sequences with the property $\hat{U}^1(t) = 1$ for multiples of the cycle time ($t = n\tau_{cyc}$) are called cyclic sequences. By definition, such sequences leave the state of the spin system unchanged after completion of a cycle. For such sequences and times $t = n\tau_{cyc}$, it is a good approach to try to split $\hat{U}(t)$ into a product of two terms:

$$\hat{U}(\tau_{cyc}) = \hat{U}^1(\tau_{cyc})\hat{\tilde{U}}(\tau_{cyc}) \quad (3)$$

By definition of cyclic sequences ($\hat{U}^1(\tau_{cyc}) = 1$), (τ_{cyc}) alone describes the time evolution of the spin system for times $t = n\tau_{cyc}$. (τ_{cyc}) is the propagator of the toggling-frame Hamiltonian $\hat{\tilde{\mathcal{H}}}^0(t)$, which is defined by transforming $\hat{\mathcal{H}}^0$ into the time-dependent coordinate system defined by the action of $\hat{\mathcal{H}}^1(t)$:

$$\hat{\tilde{\mathcal{H}}}^0(t) = [\hat{U}^1(t)^{-1}]\hat{\mathcal{H}}^0\hat{U}^1(t) \quad (4)$$

Looking at the time evolution of a density matrix $\hat{\sigma}(t)$, we see that this definition of $\hat{\tilde{\mathcal{H}}}^0(t)$ indeed yields a simplified description.[5] $\hat{\sigma}(t)$ can be expressed as

$$\hat{\sigma}(t) = \hat{U}^1(t)\hat{\tilde{\sigma}}(t)[\hat{U}^1(t)]^{-1} \quad (5)$$

Differentiation yields [note equation (2) and $\dot{\hat{U}}^1(t) = -\mathrm{i}\hat{\mathcal{H}}^1(t)\hat{U}^1(t)$]:

$$\begin{aligned}\dot{\hat{\sigma}}(t) &= \dot{\hat{U}}^1(t)\hat{\tilde{\sigma}}(t)[\hat{U}^1(t)]^{-1} + \hat{U}^1(t)\dot{\hat{\tilde{\sigma}}}(t)[\hat{U}^1(t)]^{-1} \\ &\quad + \hat{U}^1(t)\hat{\tilde{\sigma}}(t)\{[\hat{U}^1(t)]^{-1}\}^{\cdot} \\ &= -\mathrm{i}\hat{\mathcal{H}}^1(t)\hat{U}^1(t)\hat{\tilde{\sigma}}(t)[\hat{U}^1(t)]^{-1} + \hat{U}^1(t)\dot{\hat{\tilde{\sigma}}}(t)[\hat{U}^1(t)]^{-1} \\ &\quad + \mathrm{i}\hat{U}^1(t)\hat{\tilde{\sigma}}(t)[\hat{U}^1(t)]^{-1}\hat{\mathcal{H}}^1(t) \\ &= -\mathrm{i}[\hat{\mathcal{H}}^1(t), \hat{U}^1(t)\hat{\tilde{\sigma}}(t)[\hat{U}(t)]^{-1}] + \hat{U}^1(t)\dot{\hat{\tilde{\sigma}}}(t)[\hat{U}^1(t)]^{-1} \end{aligned} \quad (6)$$

Inserting the definition of $\hat{\mathcal{H}}(t)$ [equation (1)] and equation (5) into the Liouville–von Neumann equation, we obtain for $\dot{\hat{\sigma}}(t)$

$$\dot{\hat{\sigma}} = -\mathrm{i}[\hat{\mathcal{H}}^0 + \hat{\mathcal{H}}^1(t), \hat{U}^1(t)\hat{\tilde{\sigma}}(t)[\hat{U}^1(t)]^{-1}] \quad (7)$$

Equating equation (6) and equation (7), the commutators involving $\hat{\mathcal{H}}^1(t)$ cancel, and we obtain,

$$\hat{U}^1(t)\dot{\hat{\tilde{\sigma}}}(t)[\hat{U}^1(t)]^{-1} = -\mathrm{i}[\hat{\mathcal{H}}^0, \hat{U}^1(t)\hat{\tilde{\sigma}}(t)[\hat{U}^1(t)]^{-1}] \quad (8)$$

Multiplying this equation by $[(\hat{U}^1(t)]^{-1}$ from the left and by $\hat{U}^1(t)$ from the right yields

$$\begin{aligned}\dot{\hat{\tilde{\sigma}}}(t) &= -\mathrm{i}\{[\hat{U}^1(t)]^{-1}\hat{\mathcal{H}}^0\hat{U}^1(t)\hat{\tilde{\sigma}}(t) - \hat{\tilde{\sigma}}(t)[\hat{U}^1(t)]^{-1}\hat{\mathcal{H}}^0\hat{U}^1(t)\} \\ &= -\mathrm{i}[[\hat{U}^1(t)]^{-1}\hat{\mathcal{H}}^0\hat{U}^1(t), \hat{\tilde{\sigma}}(t)] = -\mathrm{i}[\hat{\tilde{\mathcal{H}}}^0(t), \hat{\tilde{\sigma}}(t)]\end{aligned} \quad (9)$$

The formal solution of this equation is [note that $(0) = \hat{\tilde{\sigma}}(0)$]

$$\hat{\tilde{\sigma}}(t) = \hat{U}(t)\hat{\tilde{\sigma}}(0)[\hat{\tilde{U}}(t)]^{-1} = \hat{\tilde{U}}(t)\hat{\sigma}(0)[\hat{\tilde{U}}(t)]^{-1} \quad (10)$$

where

$$\hat{\tilde{U}}(t) = T\exp\left[-\mathrm{i}\int_0^t \hat{\tilde{\mathcal{H}}}^0(t')\,\mathrm{d}t' \right] = \exp(-\mathrm{i}\hat{\mathcal{H}}_{eff}t) \quad (11)$$

Inserting equation (10) into equation (5), we see that the propagator given in equation (3) is a valid decomposition of $\hat{U}(t)$.

As has been mentioned, (τ_{cyc}) yields a complete description of the time evolution of the spin system under cyclic sequences. $\hat{\mathcal{H}}_{eff}$ is the effective Hamiltonian. In general, the effective Hamiltonian has the form

$$\hat{\mathcal{H}}_{eff} = \hat{\mathcal{H}}_{eff}^{lin} + \hat{\mathcal{H}}_{eff}^{bil} + \hat{O}(\geq 3) \quad (12)$$

The terms that are linear in spin operators are collected in $\hat{\mathcal{H}}_{eff}^{lin}$:

$$\hat{\mathcal{H}}_{eff}^{lin} = \sum_k \sum_\mu c_{k\mu}\hat{I}_{k\mu} \quad (13)$$

with $\mu = x$, y, or z. With the help of the coefficients $c_{k\mu}$, the effective offset of spin k may be defined as

$$\Omega_k^{eff} = \sqrt{\sum_\mu |c_{k\mu}|^2} \quad (14)$$

Terms that are bilinear in spin operators are summarized in $\hat{\mathcal{H}}_{eff}^{bil}$, which may be used to define effective coupling constants J_{kl}^{eff}. All terms that contain products of three or more spin operators are represented in equation (12) by $\hat{O}(\geqslant 3)$.

$\hat{\mathcal{H}}_{eff}$ can be expanded with the help of the Magnus expansion.[5,9] The zeroth-order term in the Magnus expansion is the average Hamiltonian $\bar{\mathcal{H}}$, which is given by

For References see p. 4803

$$\hat{\bar{H}} = \frac{1}{\tau_{\mathrm{cyc}}} \int_0^{\tau_{\mathrm{cyc}}} \hat{\tilde{\mathcal{H}}}^0(t')\,\mathrm{d}t' \quad (15)$$

2.2 Hartmann–Hahn Transfer

Let us consider a system of two coupled spins $\frac{1}{2}$. The Hamiltonian for this system can be written as a sum of the Zeeman part $\hat{\mathcal{H}}_Z$ and a coupling part $\hat{\mathcal{H}}_J$:

$$\begin{aligned}\hat{\mathcal{H}}_0 &= \hat{\mathcal{H}}_Z + \hat{\mathcal{H}}_J \\ &= \Omega_1\hat{I}_{1z} + \Omega_2\hat{I}_{2z} + 2\pi J(\hat{I}_{1x}\hat{I}_{2x} + \hat{I}_{1y}\hat{I}_{2y} + \hat{I}_{1z}\hat{I}_{2z})\end{aligned} \quad (16)$$

Assuming the spin system to be prepared at $t = 0$ with spin 1 polarized along z and spin 2 saturated, the initial density operator can be represented as $\hat{\rho}(0) = \hat{I}_{1z}$. For the analysis of Hartmann–Hahn transfer, it is more convenient to decompose $\hat{\mathcal{H}}_0$ into the terms $\hat{\mathcal{H}}_0'$ and $\hat{\mathcal{H}}_0''$:

$$\hat{\mathcal{H}}_0' = (\Omega_1 - \Omega_2)\frac{\hat{I}_{1z} - \hat{I}_{2z}}{2} + 2\pi J(\hat{I}_{1x}\hat{I}_{2x} + \hat{I}_{1y}\hat{I}_{2y}) \quad (17a)$$

$$\hat{\mathcal{H}}_0'' = (\Omega_1 + \Omega_2)\frac{\hat{I}_{1z} + \hat{I}_{2z}}{2} + 2\pi J\hat{I}_{1z}\hat{I}_{2z} \quad (17b)$$

$\hat{\mathcal{H}}_0''$ does not influence the time evolution of the density operator, and can be ignored, since it commutes with both $\hat{\mathcal{H}}_0'$ and $\hat{\rho}(0) = \hat{I}_{1z}$.

The evolution of $\hat{\rho}(t)$ due to the remaining operator $\hat{\mathcal{H}}_0'$ is strongly dependent on the ratio of the coupling constant J with the difference $\Delta\Omega = (\Omega_1 - \Omega_2)$. The evolution of $\hat{\rho}(t)$ is presented in Figure 2 with numerical simulations for the cases $|\Delta\Omega| >> |2\pi J_{12}|$ [(a): weak coupling limit], $|\Delta\Omega| \approx |2\pi J_{12}|$ [(b): strong coupling], and $|\Delta\Omega| << |2\pi J_{12}|$ [(c): Hartmann–Hahn limit].

In the weak coupling limit (Figure 2a), the density operator of the spin system does not evolve, so the polarization of spin 1 is invariant.

In the case of strong coupling (Figure 2b), the density operator evolves in an oscillatory manner. Starting from $\hat{\rho}(0) = \hat{I}_{1z}$, terms $\hat{I}_{2z}$ as well as $\hat{I}_{1y}\hat{I}_{2x} - \hat{I}_{1x}\hat{I}_{2y}$ and $\hat{I}_{1x}\hat{I}_{2x} + \hat{I}_{1y}\hat{I}_{2y}$ are created periodically. Thus, a part of the initial polarization of spin 1 ($\hat{I}_{1z}$) is periodically transferred to spin 2 ($\hat{I}_{2z}$).

In the Hartmann–Hahn limit (Figure 2c), only the terms $\hat{I}_{2z}$ and $\hat{I}_{1y}\hat{I}_{2x} - \hat{I}_{1x}\hat{I}_{2y}$ are formed, starting from $\hat{\rho}(0) = \hat{I}_{1z}$. The characteristic property of this limiting case is the complete transfer of polarization between the coupled spins after a time $1/2J$.

In the general case, the rate and maximum efficiency of transfer via scalar coupling in multiple pulse sequences is a complicated function of the exact sequence.[10,11] Such cases can be described using effective offsets $\Omega_{\mathrm{eff},1}$ and $\Omega_{\mathrm{eff},2}$ as well as an effective coupling constant J_{12}^{eff} (cf. equations (12)–(14)). Fulfillment of the Hartmann–Hahn condition $|\Delta\Omega| << |2\pi J_{12}|$ is a necessary condition for the occurrence of Hartmann–Hahn transfer between two spins. Under the influence of a multiple pulse sequence, the Hartmann–Hahn condition takes the form $|\Delta\Omega_{\mathrm{eff}}| << |2\pi J_{12}^{\mathrm{eff}}|$. For spins that have similar chemical shift, corresponding to cross peaks along the diagonal of a 2D spectrum, $\Delta\Omega_{\mathrm{eff}}$ can be expressed as $\Delta\Omega_{\mathrm{eff}} = (\Delta\Omega_{\mathrm{eff}}/\Delta\Omega)\,\Delta\Omega \approx (\partial\Omega_{\mathrm{eff}}/\partial\Omega)\,\Delta\Omega = \lambda\,\Delta\Omega$. Thus, the slope λ of Ω_{eff} as a function of Ω gives a measure for the occurrence of Hartmann–Hahn transfer near the diagonal. λ is identical to the

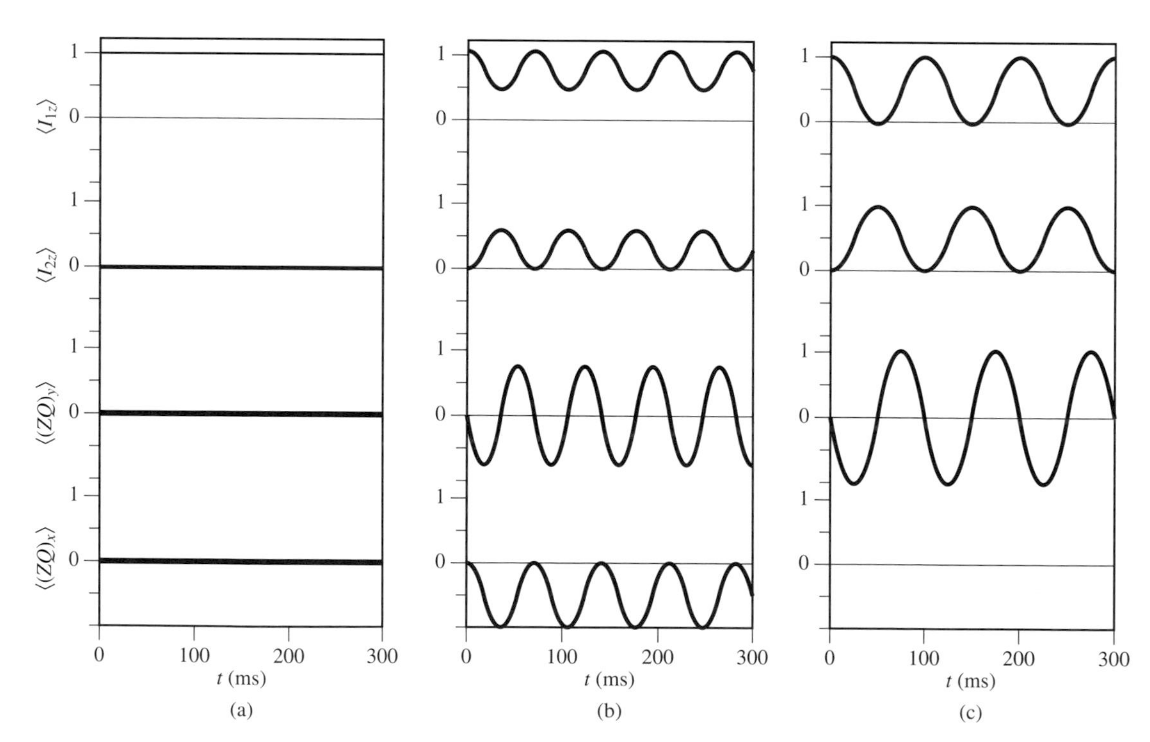

Figure 2 Time evolution of the initial density operator $\hat{\rho}(0) = \hat{I}_{1z}$ in a two-spin system with a coupling of 10 Hz under the influence of the Hamiltonian $\hat{\mathcal{H}}_0$. Curves from top to bottom show the expectation values of $\hat{I}_{1z}$, $\hat{I}_{2z}$, $\hat{I}_{1y}\hat{I}_{2x} - \hat{I}_{1x}\hat{I}_{2y} = (\mathrm{ZQ})_y$ and $\hat{I}_{1x}\hat{I}_{2x} + \hat{I}_{1y}\hat{I}_{2y} = (\mathrm{ZQ})_x$. (a) In the weak coupling limit ($|\Delta\nu| = 1000$ Hz $>> |J_{12}|$), the polarization of $\hat{I}_{1z}$ is invariant. (b) In the case of strong coupling ($|\Delta\nu| = 10$ Hz $= |J_{12}|$), the polarization $\hat{I}_{1z}$ is partly transferred to $\hat{I}_{2z}$. (c) In the Hartmann–Hahn limit ($|\Delta\nu| = 0$ Hz $<< |J_{12}|$), transfer of $\hat{I}_{1z}$ to $\hat{I}_{2z}$ is complete for odd multiples of $1/2J$

scaling factor introduced by Waugh[12] in the context of heteronuclear decoupling.

2.3 Invariant Trajectory Approach

Consider an uncoupled spin k subject to the action of an arbitrary multiple pulse sequence consisting of a basic sequence that is repeated during the mixing time of an experiment. The trajectory followed by a magnetization vector during the basic sequence is called invariant[13] if the orientations of the vector at the beginning and end of the basic sequence are the same. By definition, magnetization that moves on this trajectory does not experience a net rotation during a complete basic sequence. All other magnetization components experience a net rotation after each basic sequence, and are dephased by rf inhomogeneity during the repetitions of the basic sequence. The extent of dephasing depends on the effective field produced by the multiple pulse sequence, the mixing time, and the rf inhomogeneity of the probe. In this case, only the behavior of magnetization moving on the invariant trajectory need be considered in order to understand the properties of the multiple pulse sequence. The method of invariant trajectories allows an easy calculation of the effective cross relaxation rate σ_{eff} that is active during a multiple pulse sequence. For a pair of spins with identical offsets, σ_{eff} is given by

$$\sigma_{\mathrm{eff}} = w_{\mathrm{l}}\sigma_{\mathrm{NOE}} + w_{\mathrm{t}}\sigma_{\mathrm{ROE}} \tag{18}$$

where the weights w_{l} and w_{t} of longitudinal and transverse cross relaxation are given by the average squared longitudinal and transverse components of the invariant trajectory:[13]

$$\begin{aligned} w_{\mathrm{l}} &= \frac{1}{\tau_{\mathrm{cyc}}}\int_0^{\tau_{\mathrm{cyc}}} n_{0,z}^2(t)\,\mathrm{d}t = \langle n_{0,z}^2\rangle \\ w_{\mathrm{t}} &= \frac{1}{\tau_{\mathrm{cyc}}}\int_0^{\tau_{\mathrm{cyc}}} (n_{0,x} + n_{0,y})^2(t)\,\mathrm{d}t = \langle n_{0,(x,y)}^2\rangle \end{aligned} \tag{19}$$

Thus, the effective cross relaxation during a multiple pulse sequence is a linear combination of the longitudinal and transverse cross relaxation rates σ_{NOE} and σ_{ROE}, which are properties of the sample, with the weights $w_{\mathrm{l}}(\Omega)$ and $w_{\mathrm{t}}(\Omega)$, which are properties of the sequence. As an example, invariant trajectories of the phase-alternating ROESY sequence $(\pi_x\pi_{-x})_n$ for two offsets are shown in Figure 3. From the x, y, and z components of the invariant trajectories of this sequence (top of Figure 3), the weights of longitudinal and transverse cross relaxation can be obtained (bottom).

2.4 Connection between the Invariant Trajectory Approach and AHT

For basic sequences that create a nonzero effective field $B_{\mathrm{eff},k}$ (spin lock sequences), the orientation of the invariant trajectory is given by the orientation of $B_{\mathrm{eff},k}$.

However, there is also a more fundamental connection between invariant trajectories and effective fields. In the following discussion, we shall show that for any multiple pulse sequence, there is a relationship between the time-averaged z

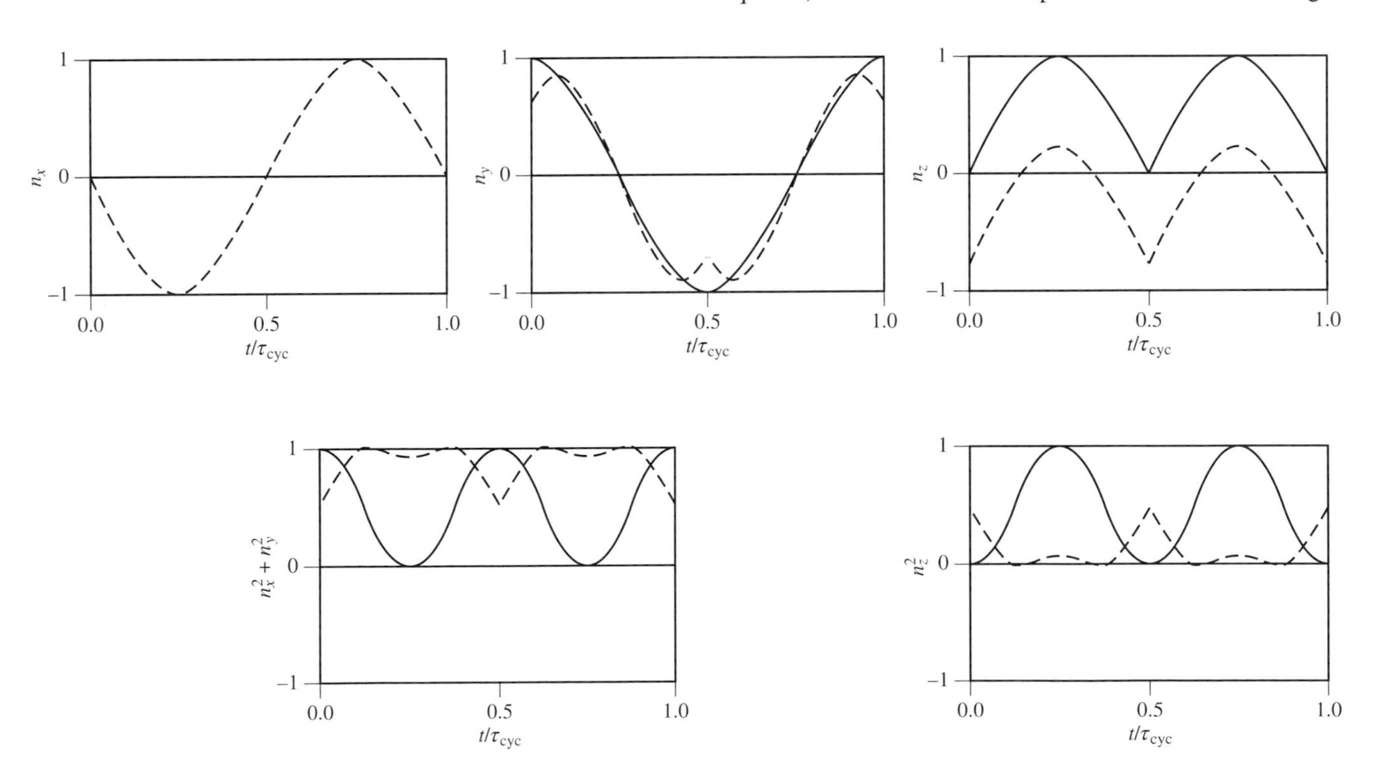

Figure 3 Invariant trajectories for the sequence $(\pi_x\pi_{-x})_n$ for an on-resonant spin (solid lines) and for an off-resonant spin with offset $\Omega = \gamma B_1$ (dashed lines). The x, y, and z components, together with the squares of the longitudinal and transverse components, are shown. For off-resonant spins, the sequence locks in the (y,z) plane. The weights w_{t} and w_{l} of transverse and longitudinal cross relaxation are calculated as the average square transverse and average square longitudinal components of the invariant trajectories. For the on-resonant spin, $\langle n_x^2 + n_y^2\rangle = \langle n_z^2\rangle = 0.5$, and for the off-resonant spin, $w_{\mathrm{t}} = \langle n_x^2 + n_y^2\rangle = 0.88$ and $w_{\mathrm{l}} = \langle n_z^2\rangle = 0.12$

For References see p. 4803

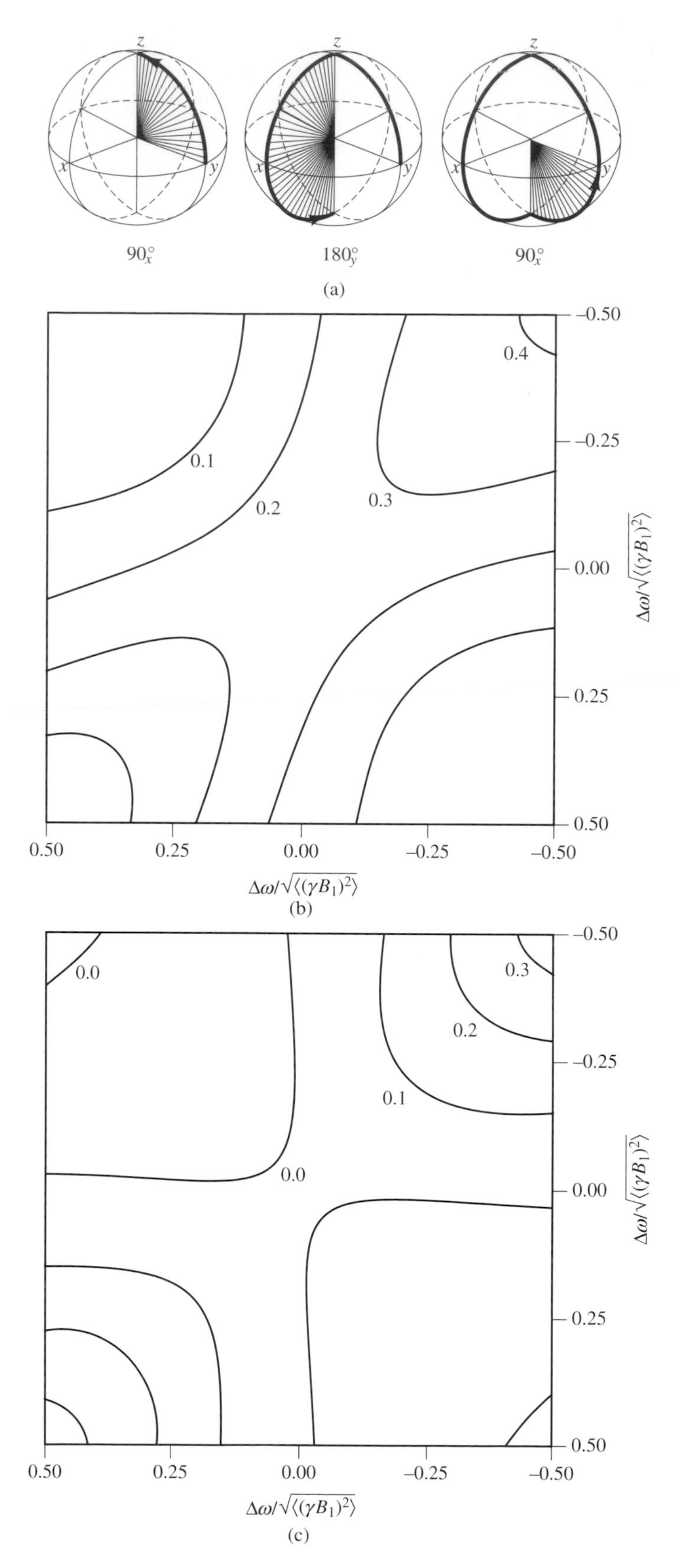

Figure 4 (a) Invariant trajectory starting from y magnetization for the $90^\circ_x 180^\circ_y 90^\circ_x$ composite pulse of the MLEV sequence for on-resonant irradiation. In (b) and (c), the offset dependence of the effective cross-relaxation rate σ_{eff} is shown in units of σ_{eff} in the spin diffusion limit. On-resonance, longitudinal and transverse cross relaxation have equal weights ($w_l = w_t = \frac{1}{2}$). Therefore, a net cross-relaxation rate of $\frac{1}{4}\sigma_{\mathrm{ROE}}$ is observed for two on-resonant spins during this sequence for a molecule in the spin diffusion limit (b). Introduction of delays of length $\Delta = \tau_{90}$ in CLEAN MLEV,[19] $90^\circ_x \Delta 180^\circ_y \Delta 90^\circ_x$, results in suppression of cross relaxation on-resonance (c)

component of the invariant trajectory $\langle n_{0,z}\rangle$ and the slope λ of the effective field as a function of the offset.[14] This result is central to the development of ROESY sequences, since the weight of longitudinal cross relaxation is proportional to $\langle n_{0,z}^2\rangle$, while λ is a measure for the occurrence of Hartmann–Hahn transfer along the diagonal of the 2D spectrum.

In order to calculate $\lambda = \partial\Omega_{\rm eff}/\partial\Omega$ as the limit $\lim_{\Delta\Omega\to 0}\Delta\Omega_{\rm eff}/\Delta\Omega = \lim_{\Delta\Omega\to 0}(\Omega_{\rm eff} - \Omega'_{\rm eff})/\Delta\Omega$, the effective propagator $\hat{U}(\tau_{\rm cyc})$ after one cycle of duration $\tau_{\rm cyc}$ of a repetitive multiple pulse sequence is calculated:

$$\hat{U}(\tau_{\rm cyc}) = \prod_j \hat{U}_j = \prod_j \exp(-i\hat{\mathcal{H}}_j\tau_j) = \exp(-i\hat{\mathcal{H}}_{\rm eff}\tau_{\rm cyc}) = \exp[-i\Omega_{\rm eff}\tau_{\rm cyc}\hat{I}_0(0)] \tag{20}$$

$\hat{U}(\tau_{\rm cyc})$ is given by the time-ordered product of individual pulse propagators $\hat{U}_j$, which represent the propagators for individual pulses of duration τ_j and Hamiltonian $\hat{\mathcal{H}}_j$. $\hat{\mathcal{H}}_{\rm eff}$ is the effective Hamiltonian for one cycle. For a system consisting of a single spin-$\frac{1}{2}$, the effective Hamiltonian $\hat{\mathcal{H}}_{\rm eff}$ may be written as $\Omega_{\rm eff}\hat{I}_0(0)$, where $\Omega_{\rm eff}$ is the effective chemical shift and $\hat{I}_0(0)$ defines the invariant trajectory[13] at time t = 0:

$$\hat{\mathcal{H}}_{\rm eff} = \Omega_{\rm eff}\hat{I}_0(0) \tag{21}$$

To calculate the effective chemical shift $\Omega'_{\rm eff}$ at the offset $\Omega' = \Omega + \Delta\Omega$, the $\hat{\mathcal{H}}_j$ are modified by $\hat{\mathcal{H}}_1 = \Delta\Omega\,\hat{I}_z$: $\hat{\mathcal{H}}'_j = \hat{\mathcal{H}}_j + \hat{\mathcal{H}}_1 = \hat{\mathcal{H}}_j + \Delta\Omega\,\hat{I}_z$. The resulting propagator,

$$\hat{U}'(\tau_{\rm cyc}) = \exp(-i\hat{\mathcal{H}}'_{\rm eff}\tau_{\rm cyc}) = \exp[-i\Omega'_{\rm eff}\tau_{\rm cyc}\hat{I}'_0(0)] \tag{22}$$

may be divided into two factors:[4,5]

$$\hat{U}'(\tau_{\rm cyc}) = \hat{U}(\tau_{\rm cyc})\hat{U}_1(\tau_{\rm cyc}) \tag{23}$$

with

$$\hat{U}_1(\tau_{\rm cyc}) = \exp(-i\hat{\mathcal{H}}_{1,\rm eff}\tau_{\rm cyc}) = T\exp\left\{-i\left[\int_0^{\tau_{\rm cyc}}\hat{\tilde{\mathcal{H}}}_1(t)\,dt\right]\right\} \tag{24}$$

where the toggling frame Hamiltonian is given by $\hat{\tilde{\mathcal{H}}}_1(t) = \hat{U}^{-1}(t)\,\Delta\Omega\,\hat{I}_z\hat{U}(t)$. In zeroth-order average Hamiltonian theory, equation (24) simplifies to

$$\hat{U}_1(\tau_{\rm cyc}) = \exp(-i\hat{\mathcal{H}}_{\rm eff,1}\tau_{\rm cyc}) = \exp\left\{-i\left[\int_0^{\tau_{\rm cyc}}\hat{\tilde{\mathcal{H}}}_1(t)\,dt\right]\right\} \tag{25}$$

This simplification is justified in the limit $\Delta\Omega \to 0$. Expansion of $\hat{U}'(\tau_{\rm cyc})$, equation (23), according to the Baker–Campbell–Haussdorff formula to first order yields

$$\hat{U}'(\tau_{\rm cyc}) = \exp\{-i(\hat{\mathcal{H}}_{\rm eff}\tau_{\rm cyc} + \hat{\mathcal{H}}_{\rm eff,1}\tau_{\rm cyc}) - \tfrac{1}{2}[\hat{\mathcal{H}}_{\rm eff}, \hat{\mathcal{H}}_{\rm eff,1}]\tau_{\rm cyc}^2\} \tag{26}$$

Comparing equations (22) and (26), we find

$$\hat{\mathcal{H}}'_{\rm eff} = \Omega'_{\rm eff}\hat{I}'_0(0) = (\hat{\mathcal{H}}_{\rm eff} + \hat{\mathcal{H}}_{\rm eff,1}) - \tfrac{1}{2}i[\hat{\mathcal{H}}_{\rm eff}, \hat{\mathcal{H}}_{\rm eff,1}]\tau_{\rm cyc} \tag{27a}$$

$$= \Omega_{\rm eff}\hat{I}_0(0) + \Omega_{\rm eff,1}\hat{I}_1(0) - \tfrac{1}{2}i\Omega_{\rm eff}\Omega_{\rm eff,1}[\hat{I}_0(0), \hat{I}_1(0)]\tau_{\rm cyc} \tag{27b}$$

where $\hat{I}_1(0)$ is defined as $\hat{\mathcal{H}}_{\rm eff,1}/\Omega_{\rm eff,1}$

For $\Delta\Omega << \Omega_{\rm eff}$ and $\Delta\Omega\,\tau_{\rm cyc} << 1$, only terms parallel to $\hat{I}_0(0)$ contribute to the magnitude of the effective field $\Omega'_{\rm eff}$ at offset Ω'. Since the commutator $[\hat{I}_0(0),\hat{I}_1(0)]$ is perpendicular to $\hat{I}_0(0)$, the last term in equation (27b) may be neglected. To find the contribution of the second term of equation (27b), we calculate the projection of $\Omega_{\rm eff,1}\hat{I}_1(0)$ onto the direction of $\hat{I}_0(0)$. This is given by $\Omega_{\rm eff,1}\,{\rm Tr}\,\{\hat{I}_0(0)\hat{I}_1(0)\}$. Thus, the magnitude of the effective field $\Omega'_{\rm eff}$ at offset Ω' is given by

$$\Omega'_{\rm eff} = \Omega_{\rm eff} + \Omega_{\rm eff,1}{\rm Tr}\{\hat{I}_0(0)\hat{I}_1(0)\} \tag{28}$$

For $\Delta\Omega \to 0$, the slope λ can be expressed as

$$\lambda = \frac{\Omega'_{\rm eff} - \Omega_{\rm eff}}{\Delta\Omega} = \frac{\Omega_{\rm eff,1}}{\Delta\Omega}{\rm Tr}\{\hat{I}_0(0)\hat{I}_1(0)\} \tag{29}$$

According to equation (25), $\Omega_{\rm eff,1}\hat{I}_1(0) = \hat{\mathcal{H}}_{\rm eff,1}$ is given by

$$\Omega_{\rm eff,1}\hat{I}_1(0) = \frac{1}{\tau_{\rm cyc}}\int_0^{\tau_{\rm cyc}}\hat{\tilde{\mathcal{H}}}_1(t)\,dt = \frac{1}{\tau_{\rm cyc}}\int_0^{\tau_{\rm cyc}}\hat{U}^{-1}(t)\Delta\Omega\hat{I}_z\hat{U}(t)\,dt \tag{30}$$

Inserting equation (30) into equation (29), we find

$$\begin{aligned}\lambda &= \frac{1}{\tau_{\rm cyc}}{\rm Tr}\left\{\hat{I}_0(0)\int_0^{\tau_{\rm cyc}}\hat{U}^{-1}(t)\hat{I}_z\hat{U}(t)\,dt\right\}\\ &= \frac{1}{\tau_{\rm cyc}}\int_0^{\tau_{\rm cyc}}{\rm Tr}\{\hat{I}_0(0)\hat{U}^{-1}(t)\hat{I}_z\hat{U}(t)\}\,dt\\ &= \frac{1}{\tau_{\rm cyc}}\int_0^{\tau_{\rm cyc}}{\rm Tr}\{\hat{U}(t)\hat{I}_0(0)\hat{U}^{-1}(t)\hat{I}_z\}\,dt\\ &= \frac{1}{\tau_{\rm cyc}}\int_0^{\tau_{\rm cyc}}{\rm Tr}\{\hat{I}_0(t)\hat{I}_z\}\,dt\\ &= \frac{1}{\tau_{\rm cyc}}\int_0^{\tau_{\rm cyc}}n_{0,z}(t)\,dt = \langle n_{0,z}\rangle\end{aligned} \tag{31}$$

where $n_{0,z}(t)$ is the z component of the invariant trajectory $\hat{I}_0(t)$. The general result of equation (31) can be stated as follows: the slope of $\Omega_{\rm eff}$ with respect to Ω is identical to the average z component of the invariant trajectory. Equation (28) relates the parameter λ central to the theory of spin decoupling to the invariant trajectory approach. We shall return to this result in Section 3.2.

3 SEPARATION OF HARTMANN–HAHN TRANSFER AND CROSS RELAXATION

3.1 Cross-Relaxation Compensated TOCSY Experiments

The TOCSY experiment is used to achieve transfer among spins in scalar-coupled spin systems. To achieve efficient TOCSY transfer, the effective fields experienced by the spins must be matched over the offset range of interest ($\lambda << 1$). Several multiple pulse sequences have been proposed to achieve this goal.[1,11,15–18] During these multiple pulse sequences, longitudinal and transverse cross-relaxation rates $\sigma_{\rm NOE}$ and $\sigma_{\rm ROE}$ are active and contribute to the effective cross relaxation rate $\sigma_{\rm eff}$ with the weights $w_{\rm l}$ and $w_{\rm t}$ [equation (18)]. Since $\sigma_{\rm NOE}$ and $\sigma_{\rm ROE}$ have the same sign for small molecules, it follows that cross-relaxation contributions cannot be sup-

For References see p. 4803

pressed near the diagonal of TOCSY spectra of small molecules, since, whatever w_t and w_l might be, the terms in equation (18) will always add to yield a nonvanishing σ_{eff}. Fortunately, cross-relaxation rates of such molecules are small, so that artifacts due to cross relaxation are rarely observed in the relatively short mixing times used.

As shown in Figure 4(a), y magnetization of on-resonant spins entering the $90^\circ_x180^\circ_y90^\circ_x$ composite pulse of a MLEV sequence spends equal amounts of time along longitudinal and transverse directions; therefore, $w_t = w_l = \frac{1}{2}$. Since $\sigma_{NOE} = 0$ for medium sized molecules, a cross relaxation contribution $\frac{1}{2}\sigma_{ROE}$ results for TOCSY spectra obtained with this mixing sequence. This contribution cannot be suppressed, since this would require $w_t = 0$, which cannot be realized for any multiple pulse sequence. In the case of large molecules ($\sigma_{ROE}/\sigma_{NOE} = -2$), cross-relaxation rates are large, and cross relaxation can there-

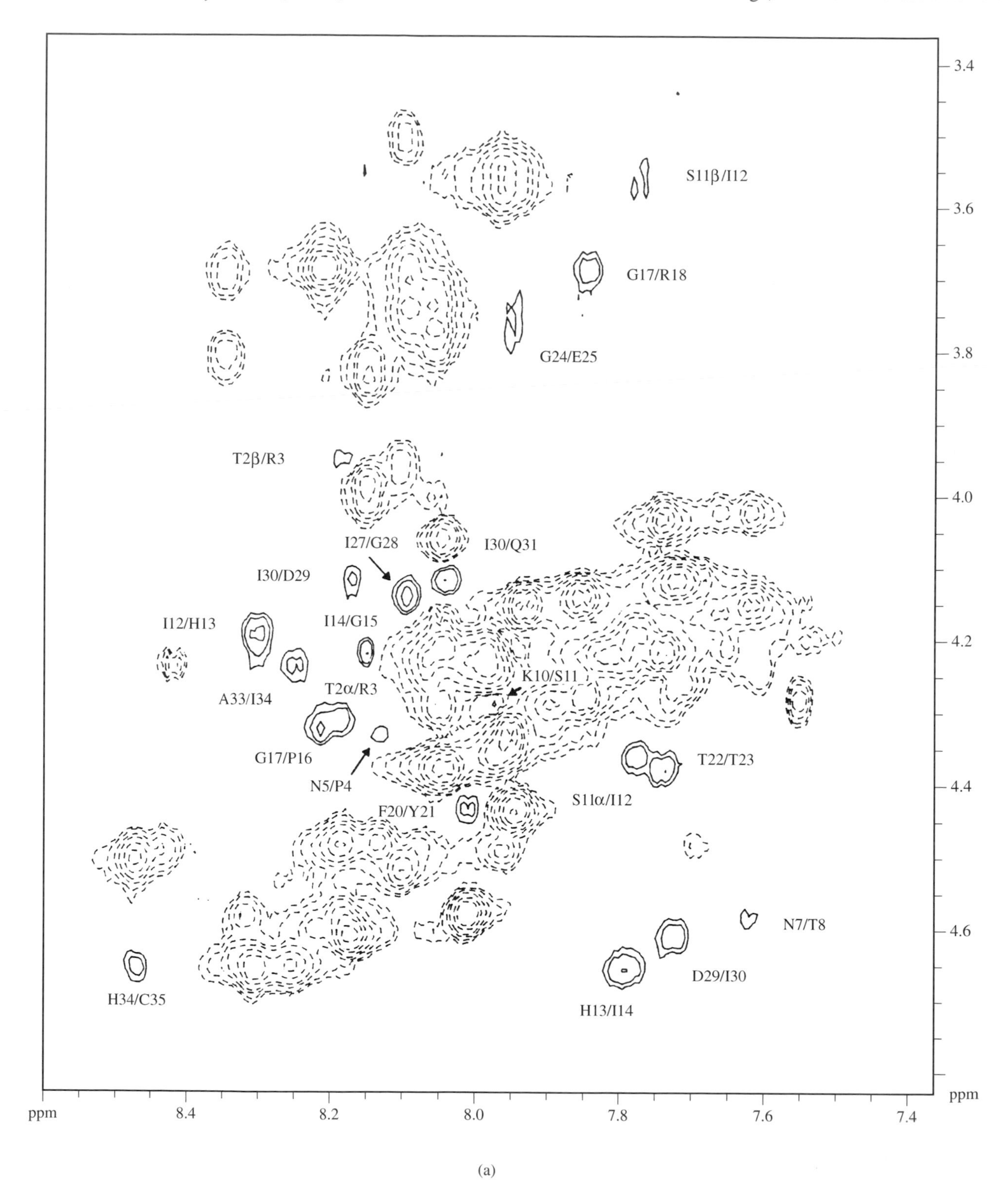

(a)

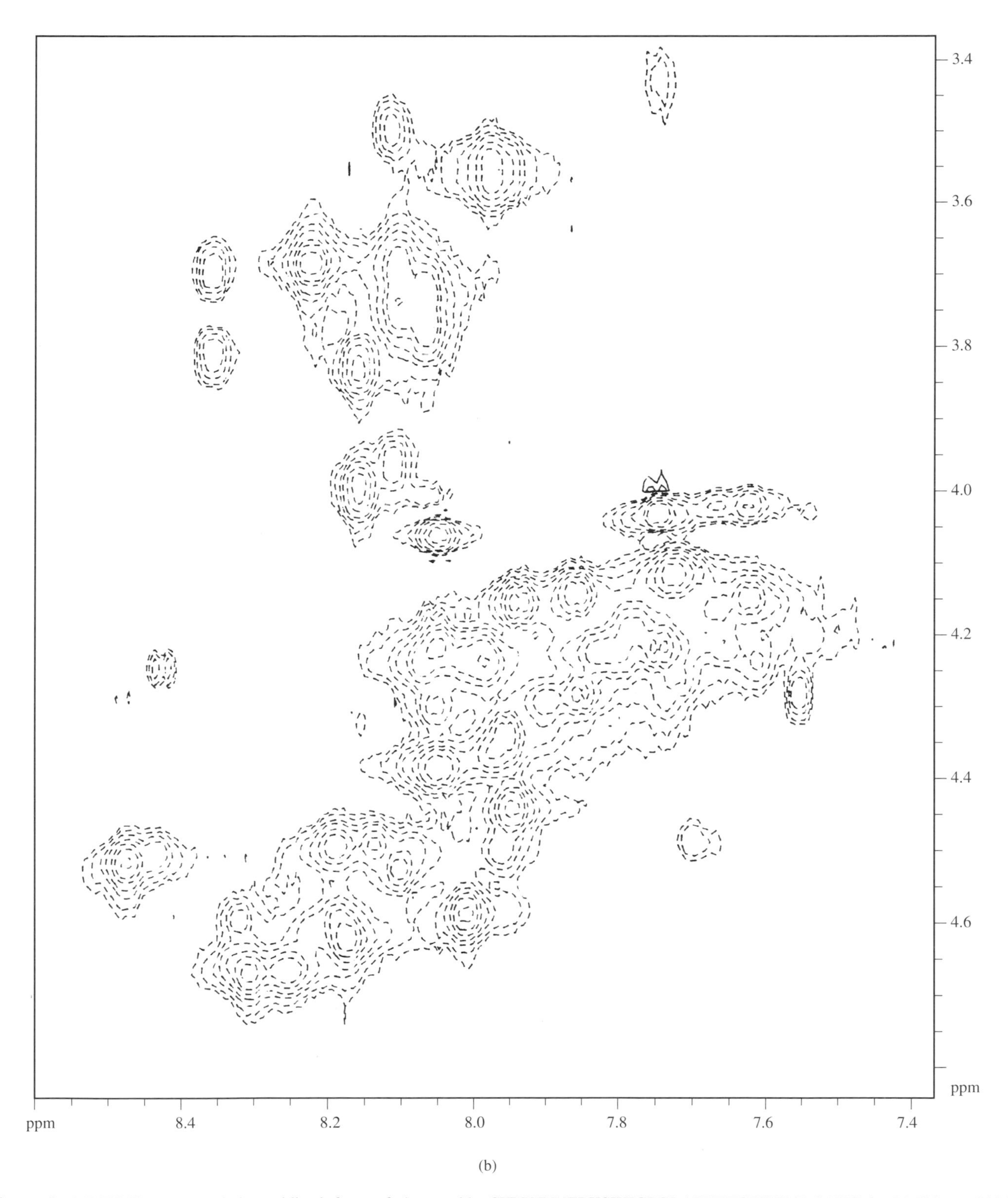

(b)

Figure 5 MLEV-17 spectra of the oxidized form of the peptide CTRPNNNTRKSIHIGPGRAFYTTGEIIGDIRQAHC in DMSO at 24 °C, recorded using a mean rf field strength $\gamma B_1/2\pi$ = 6.79 kHz at 600 MHz. Cross peaks with the same sign as the diagonal peaks (coherent transfer) are plotted as dashed lines. Cross peaks with opposite sign (incoherent transfer) are plotted as solid lines. Only the cross-relaxation peaks are assigned. (a) In the uncompensated TOCSY experiment, strong cross peaks occur owing to cross relaxation. (b) In the TOCSY spectrum recorded using an MLEV-17 expansion of an optimized shaped 90° pulse,[21] the cross-relaxation peaks have disappeared

For References see p. 4803

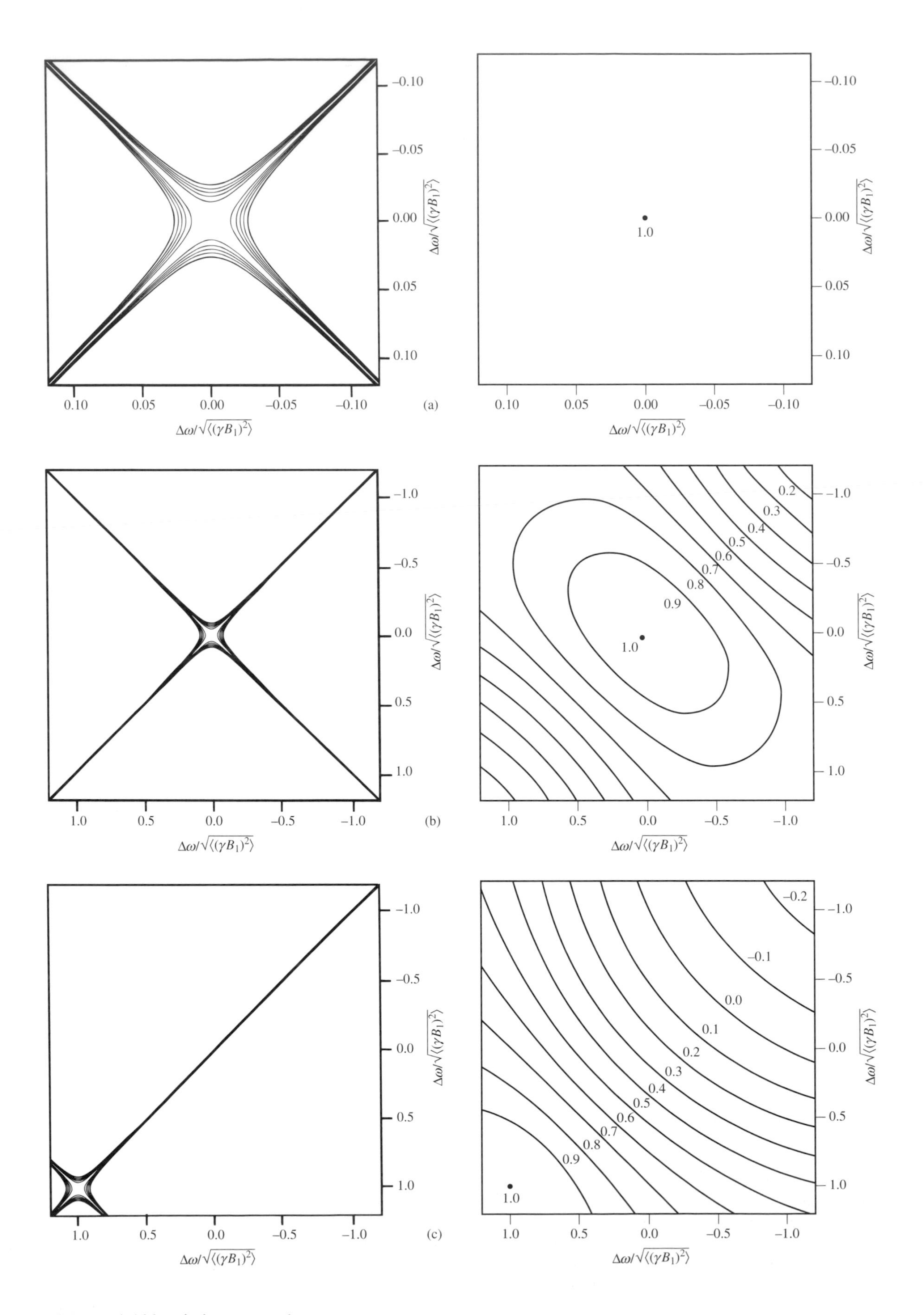

(a)
(b)
(c)
0.10
0.05
0.00
−0.05
−0.10
1.0
0.5
0.0
−0.5
−1.0
0.9
0.8
0.7
0.6
0.5
0.4
0.3
0.2
0.1
−0.1
−0.2
$\Delta\omega/\sqrt{\langle(\gamma B_1)^2\rangle}$

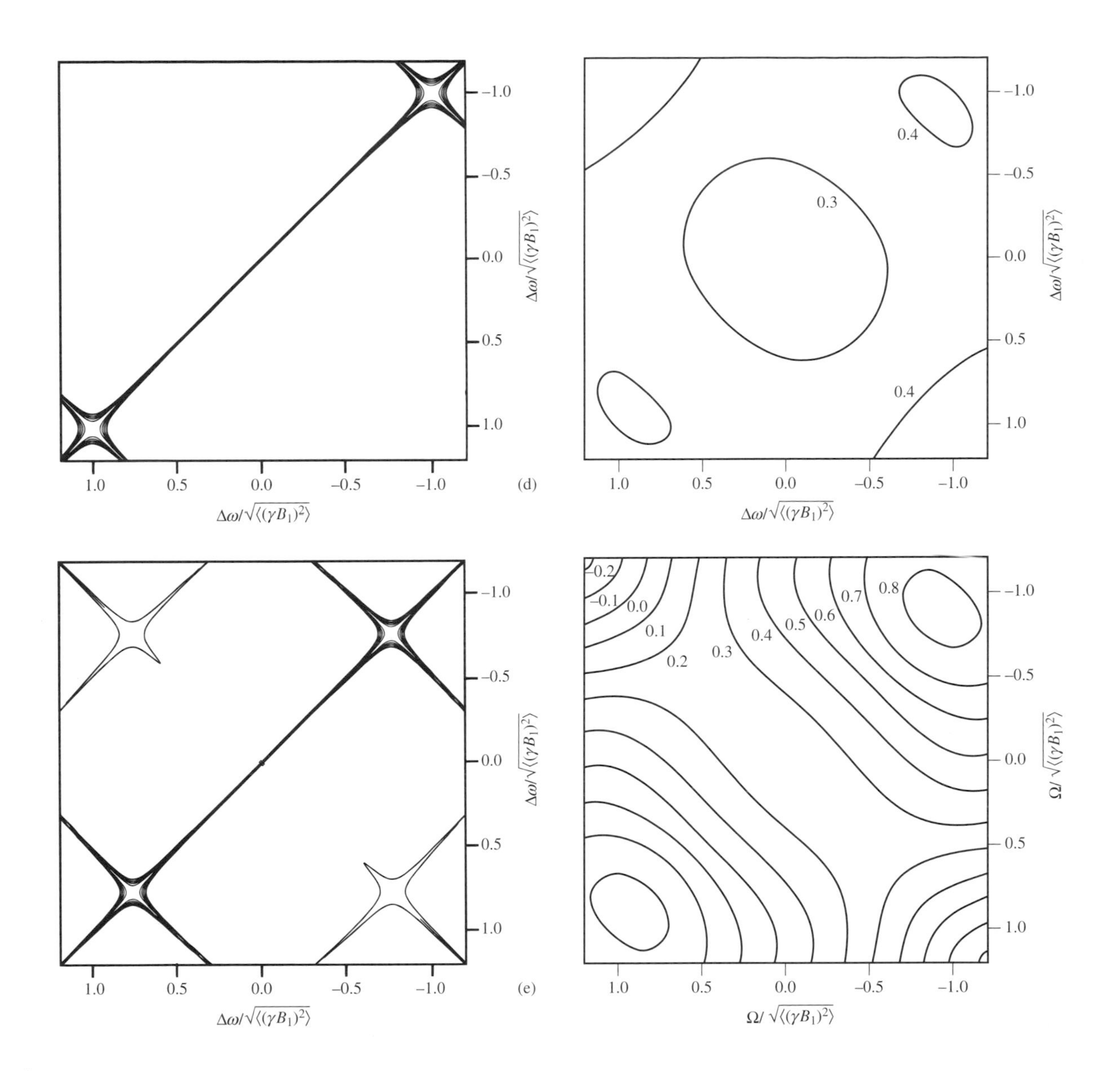

Figure 6 Simulations of Hartmann–Hahn transfer (left) and net cross-relaxation rate σ_{net} (right) for a molecule in the spin diffusion limit in ROESY sequences. Application of a very strong spin lock field (a) would result in a maximal σ_{net}, but complete Hartmann–Hahn matching would make such spectra useless. Reducing the spin lock field strength (b) strongly attenuates TOCSY at the expense of a relatively moderate offset dependence of σ_{net}. TOCSY transfer along the antidiagonal, as observed in (b), is suppressed by off-resonant irradiation of the spin lock field (c). However, the offset dependence of σ_{net} is much more pronounced in this case. Alternating application of two spin lock fields at opposite ends of the spectrum in JS ROESY (b) results in very smooth offset behavior, with TOCSY still being suppressed (d). Phase-alternating ROESY sequences (Figure 7a) show similar offset behavior for cross relaxation and coherent transfer (e). It is obvious that maximal transverse cross relaxation is achieved on-resonance at the cost of complete Hartmann–Hahn matching. Simulations of TOCSY show contour plots of the largest possible Hartmann–Hahn transfer amplitude in a two-spin system. The lowest level corresponds to a transfer amplitude of 0.1; the level spacing is 0.2

fore compete with TOCSY transfer, leading to artifacts. It follows from equation (19) that for $w_t \approx w_l$, the contribution $w_t\sigma_{ROE}$ of transverse cross relaxation outweighs $w_l\sigma_{NOE}$, and a residual ROE contribution is therefore observed in TOCSY spectra of large molecules. The net cross-relaxation rate of this sequence for molecules in the spin diffusion limit is displayed in Figure 4(b) in units of σ_{ROE}. On resonance, ROESY cross peaks of one-quarter maximum intensity are expected to appear. From the invariant trajectory of the $90^\circ_x 180^\circ_y 90^\circ_x$ composite pulse (Figure 4a), there follows a straightforward way to eliminate this contribution for large molecules. If delays of appropriate duration ($\Delta = \tau_{90}$ for $\sigma_{ROE}/\sigma_{NOE} = -2$) are introduced into the MLEV scheme at positions where the magnetization is longitudinal, w_t and w_l can be modified such that ROE and NOE contributions cancel each other. Figure 4(c) shows the offset dependence of the cross-relaxation rate in the so-called CLEAN TOCSY experiment[19] obtained in this way. Cross relaxation is completely suppressed on-resonance in this experiment, while residual ROE contributions remain along the diagonal of the 2D spectrum.

For References see p. 4803

While the introduction of delays is the most straightforward approach to suppress cross relaxation, and has been used frequently,[19,20] it suffers from an increase in the rf power dissipated in the sample (at constant average field strength γB_1), since the introduction of delays into any multiple pulse sequence lowers the duty ratio. This undesired effect is alleviated in more modern approaches to obtain cross relaxation compensated TOCSY mixing sequences. These approaches make use of crafted pulses to achieve a proper weighting of NOE and ROE during the multiple pulse sequence. Either the individual pulses of an existing TOCSY sequence are replaced by shaped pulses,[21] or TOCSY mixing sequences are optimized from scratch with the required weighting of NOE and ROE as a constraint,[22–24] resulting in sequences such as CITY[22] and TOWNY.[23] An example of a cross-relaxation compensated TOCSY spectrum, obtained using a MLEV mixing sequence consisting of shaped 90° pulses, is displayed in Figure 5.

An alternative approach to separate Hartmann–Hahn transfer and ROE was suggested by Ravikumar and Bothner-By.[25] It relies on the characteristic time evolution of coherent and incoherent magnetization transfer. However, the separation remains incomplete, since cross relaxation and coherent transfer both contain zero-frequency contributions.

3.2 Suppression of Hartmann–Hahn Transfer in ROESY Experiments

Using equation (31), which relates λ to $\langle n_{0,z}\rangle$, a relation between λ and the weight of longitudinal cross relaxation $w_l = \langle n_{0,z}^2\rangle$ can be derived for cross peaks near the diagonal. As the inequality $\langle n_{0,z}^2\rangle \geqslant \langle n_{0,z}\rangle^2$ always holds, it follows immediately that

$$w_l \geq \lambda^2 \tag{32}$$

This inequality shows that the suppression of longitudinal cross relaxation (w_l = min.!) and of TOCSY transfer (λ = max.!) are conflicting goals. Suppression of longitudinal cross relaxation can be achieved only if TOCSY transfer is allowed, and vice versa—or, there is no ROESY without NOESY and without TOCSY. The best a sequence can do is

$$w_l = \lambda^2 \tag{33}$$

This is only possible if $n_{0,z}(t)$ is constant during τ_{cyc}, as is the case in CW ROESY.[2,26,27] The latter can be described analytically in the following way. The invariant trajectory is static and has the orientation

$$\boldsymbol{n}_0 = \cos\theta \boldsymbol{e}_z + \sin\theta \boldsymbol{e}_x, \quad \text{with } \tan\theta = \frac{\gamma B_1}{\Omega - \Omega_{rf}} \tag{34}$$

where γB_1 is the spin lock field strength, θ is the lock angle as measured off the z axis, and Ω_{rf} is an offset frequency of the CW irradiation. The weights for longitudinal (w_l) and transverse (w_t) cross relaxation along the diagonal are

$$w_l = \cos^2\theta, \quad w_t = \sin^2\theta \tag{35}$$

The effective chemical shift Ω_{eff} is given by

$$\Omega_{eff} = \sqrt{(\gamma B_1)^2 + (\Omega - \Omega_{rf})^2} \tag{36}$$

and the scaling $\lambda_{CW} = \partial\Omega_{eff}/\partial\Omega$ of the chemical shift of this sequence is

$$\lambda_{CW} = \frac{\partial\Omega_{eff}}{\partial\Omega} = \frac{\Omega - \Omega_{rf}}{\sqrt{(\gamma B_1)^2 + (\Omega - \Omega_{rf})^2}} = \cos\theta = \sqrt{w_l} \tag{37}$$

Thus, CW ROESY has the best properties with respect to the simultaneous suppression of NOE and TOCSY that are possible, since equation (33) is fulfilled. For any w_t, there is no better ROESY sequence with respect to the goal of suppressing longitudinal cross relaxation and TOCSY simultaneously. In practice, a compromise must be made between the efficiency of ROESY transfer (minimization of w_l) and the degree of suppression of coherent transfer (maximization of λ). The large variation in w_l and w_t over the spectral range in CW ROESY is a strong drawback of this sequence. From an experimental point of view, an additional requirement is that w_t be uniform over the spectral width and λ not be scaled beyond a certain degree over the spectral width. A number of approaches have been proposed to address this goal.

3.2.1 Sequences Composed of CW Irradiations

The application of a very strong spin lock field in a ROESY experiment yields a very smooth offset dependence in CW ROESY, as displayed in Figure 6(a) (left). However, the application of such strong fields is prohibitive because of serious Hartmann–Hahn transfer (Figure 6(a), right). Therefore, judicious choice of the spin lock field strength is important, as

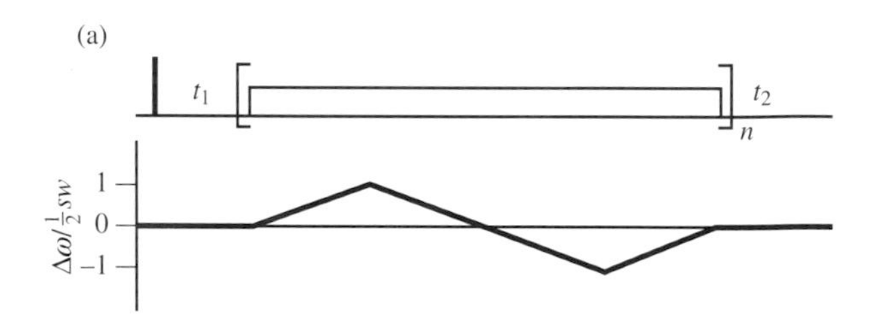

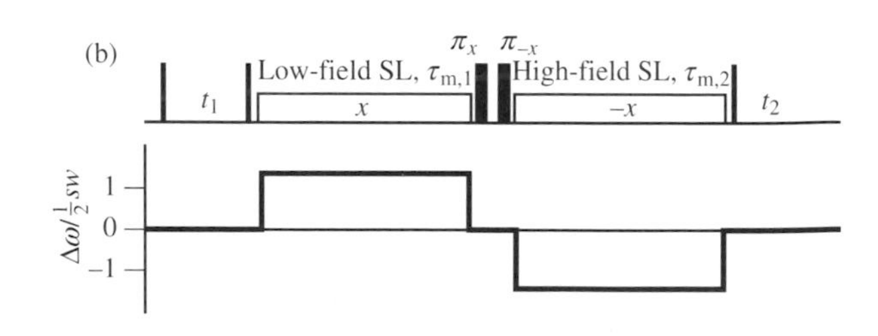

Figure 7 CW ROESY sequences with off-resonant irradiation spin locks. For each experiment, the pulse sequence (upper trace) and the position of the spin lock irradiation (lower trace) are given. (a) In the frequency swept variant,[29] the irradiation frequency is moved from the middle of the spectral region to the low-field edge, to the high-field edge, and back to the middle. (b) In the JS ROESY experiment, the mixing time consists of periods of low-field and high-field irradiation with a pair of 180° pulses mediating between the spin locks. The offset of the spin locks is generated by superimposing a phase gradient over a CW spin lock

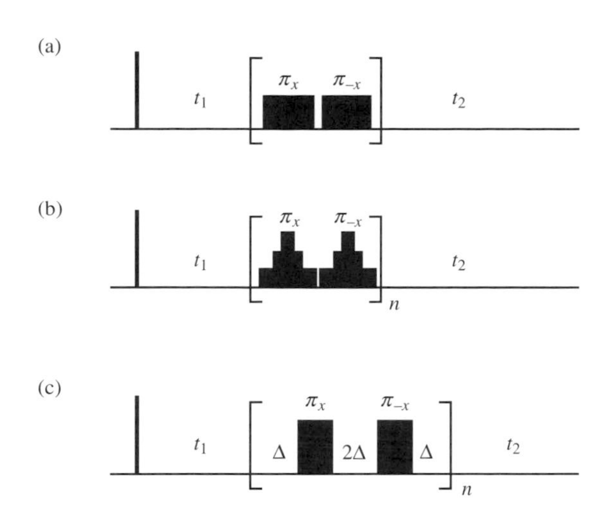

Figure 8 ROESY sequences employing spin locks consisting of phase-alternating pulses.[30,31]. The original sequence (a) can be modified by the use of shaped pulses (b) or by the introduction of delays (c) to increase the weight of transverse cross relaxation

shown in Figure 6(b).[26] Furthermore, coherent transfer is much more dependent on the position of the rf irradiation frequency than cross relaxation. Therefore, coherent transfer can be reduced by judicious placement of the irradiation position (Figure 6c).[26,28] However, optimal placement of the transmitter offset requires prior knowledge of the coupling networks that can lead to TOCSY transfer. Without such prior knowledge, the transmitter offset has to be placed outside the spectral region according to equation (37) in order to avoid Hartmann–Hahn matching between any two spins. However, this results in a strong dependence of the net cross-relaxation rate on Ω, which renders the interpretation of the spectra difficult. Still, the optimal behavior of CW irradiation with respect to TOCSY suppression renders sequences employing stretches of CW irradiation attractive.

Two related approaches have been developed to deal with the stated problem. In the first, proposed by Cavanagh and Keeler,[29] the transmitter offset is swept through the offset range of the spectrum as depicted in Figure 7(a). This method avoids continuous irradiation at any one frequency, and therefore avoids extended periods of Hartmann–Hahn matching. However, the offset dependence of the cross-relaxation rate of this sequence has not been published, and only qualitative in-

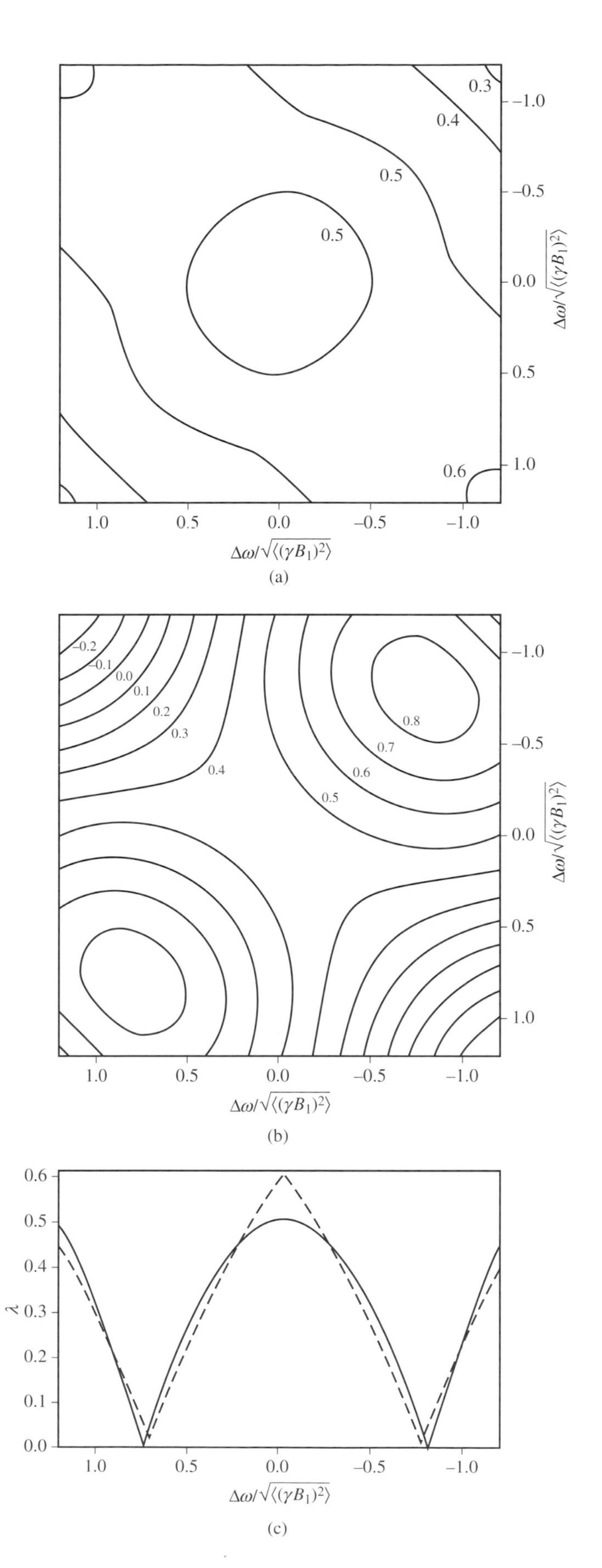

Figure 9 Simulations of the offset behavior of cross relaxation of JS ROESY and a ROESY employing a phase-alternating spin lock $(\pi_x^{3L}\pi_{-x}^{3L})_n$,[31] with parameters chosen to yield a net cross-relaxation rate of $0.476\sigma_{\mathrm{ROE}}$ on-resonance. For JS ROESY (a), an on-resonance lock angle $\theta_0 = 53.77°$ was used. The phase-modulating sequence (b) yields the same net cross-relaxation rate using the shaped pulse described by Hwang et al.[31] The offset dependence of cross relaxation is less pronounced in JS ROESY, making direct use of intensities feasible for distance calculations. (c) Simulations of λ as a function of Ω in the offset range $\pm 1.2\sqrt{\langle(\gamma B_1)^2\rangle}$. Near on-resonance, JS ROESY suppresses TOCSY better, while the phase-alternating sequence performs better towards the edges of the spectrum

For References see p. 4803

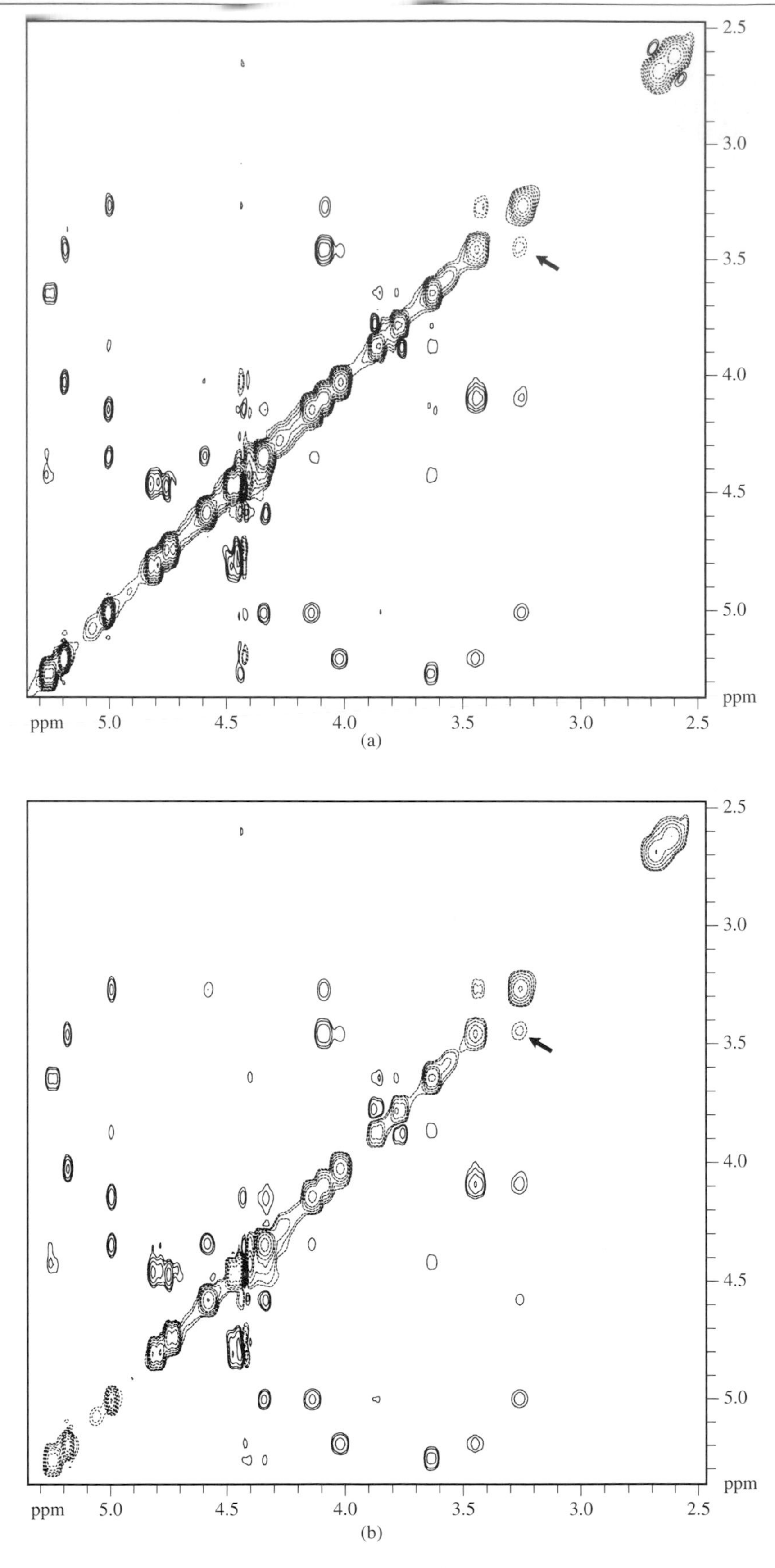

Figure 10 ROESY spectra of a disaccharide obtained with the ROESY sequences of Figures 7(b) (a) and Figure 8(b) (b). Positive (negative) levels are plotted as dashed (solid) lines. Both pulse sequences show similar ROE cross-peak intensities (solid contour lines). An artifact due to Hartmann–Hahn transfer between geminal protons is visible in both spectra (marked with an arrow). In contrast, no TOCSY transfer is observed between the geminal protons at 2.6 and 2.7 ppm, although they are even more strongly coupled than the protons mentioned above. This is probably due to the oscillatory behavior of TOCSY transfer, which vanishes for the chosen mixing time for the pair of geminal protons at 2.6 and 2.7 ppm. The example shows that a careful analysis of ROESY spectra is mandatory for signals close to the diagonal

terpretations of such spectra appear feasible at this time. Furthermore, the programming of the frequency sweep is rather demanding from an experimental point of view.

The second approach to modify CW ROESY to obtain ROESY spectra with suppression of TOCSY transfer—the use of JS ROESY ('jump symmetrized') sequences—makes use of only two irradiation positions placed outside the low- and high-field edges of the spectrum [Figure 7(b)]. Since the directions of the spin lock axes differ for the high-field and the low-field spin lock for all offsets (except for $\Omega = 0$), a compensating pulse has to be applied that rotates the magnetization into the new lock direction for every offset. In the pulse sequence of JS ROESY shown in Figure 7(b), this is achieved by the pair of 180° pulses between the two off-resonant spin locks. The field strength of these pulses is related to the spin lock field strength and to the on-resonance lock angle θ.[14] The averaging of the offset dependences of the two stretches of CW irradiation results in a very smooth offset behavior and good suppression of Hartmann–Hahn transfer, as shown in Figure 6(d). Close to on-resonance, this sequence achieves optimal suppression of TOCSY, since λ does not differ from the optimum value [given by equation (37)] at any time. Further off-resonance, the performance of the sequence is reduced, since the λ value for a given offset may differ grossly for high-field and low-field CW irradiation, and must not be averaged between the extended stretches of CW irradiation. The smaller of the two λ values (of the low- and high-field irradiations) gives a measure of the occurrence of TOCSY.

3.2.2 $\pi_x\pi_{-x}$ *ROESY*

A second approach to obtain ROESY spectra with suppression of TOCSY transfer is the use of multiple pulse sequences optimized for this goal. Only one group of sequences has so far been proposed.[30,31] These consist of a series of phase-alternating 180° pulses $(\pi_x\pi_{-x})_n$ (Figure 8), where each individual 180° pulse can be rectangular (which might include delays) or shaped. Sequences using pulses with flip angles other than 180° can also be used, and have been exploited to suppress cross relaxation in exchange spectra of macromolecules.[32] For large ($\omega\tau_c >> 1$) or intermediate sited ($\omega\tau_c \approx 1$) molecules, the unmodified mixing scheme $(\pi_x\pi_{-x})_n$ yields a net ROE rate of only $\frac{1}{4}\sigma_{ROE}$ or $\frac{1}{2}\sigma_{ROE}$ on resonance,[30] since $w_l = w_t = \frac{1}{2}$. It is therefore of interest to increase the weight of transverse cross relaxation. This can be achieved using shaped pulses (Figure 8b), or by the introduction of delays into the sequence (Figure 8c).

In the following, ROESY sequences will be compared that yield equal net cross-relaxation rates on-resonance and that dissipate the same power in a sample. Figure 9 shows the offset behavior of cross relaxation of JS ROESY [(a): $\theta = 53.77°$] and of a phase-alternating ROESY employing a shaped 180° pulse to increase w_t. (b):[31]

The offset dependence of the cross relaxation of the JS ROESY sequence is somewhat less pronounced than that of the phase-alternating sequence.

In Figure 9(c), the scaling λ is plotted as a function of the offset for both sequences. The minimal λ of the two spin locks is plotted for JS ROESY, since λ may not be averaged in the case of extended stretches of CW irradiation. On-resonance, λ is larger for JS ROESY than for the phase-alternating sequence. This result is a direct consequence of the fact that the scaling of λ in CW sequences is minimal at given w_t. For larger offsets, the phase-alternating sequence yields a somewhat better suppression of TOCSY (larger λ). Although λ can serve as a measure of TOCSY transfer along the diagonal only, this reasoning shows that it appears to be worthwhile to search for optimized ROESY sequences that minimize the loss in λ on-resonance while achieving maximum λ off-resonance.

Figure 10 shows a comparison of JS ROESY and the phase-alternating ROESY sequence. The spectra differ only marginally. Coherent transfer is well suppressed, except for the two strongly coupled geminal protons H-6′, H-6″. For more details, see the figure legend.

4 REFERENCES

1. L. Braunschweiler and R. R. Ernst, *J. Magn. Reson.*, 1983, **53**, 521.
2. A. A. Bothner-By, R. L. Stephens, J. M. Lee, C. D. Warren, and R. W. Jeanloz, *J. Am. Chem. Soc.*, 1984, **106**, 811.
3. T. E. Bull, *J. Magn. Reson.*, 1988, **80**, 470; T. E. Bull, in 'Progress in Nuclear Magnetic Resonance Spectroscopy', eds. J. W. Emsley, J. Feeney, and L. H. Sutcliffe, Pergamon Press, Oxford, 1992, Vol. 24, p. 377.
4. U. Haeberlen and J. S. Waugh, *Phys. Rev.*, 1968, **175**, 453.
5. Haeberlen, U. 'High Resolution NMR in Solids', Academic Press, New York, 1976, pp. 47–69.
6. R. R. Ernst, G. Bodenhausen, and A. Wokaun, 'Principles of Nuclear Magnetic Resonance in One and Two Dimensions', Oxford University Press, Oxford, 1987.
7. F. J. Dyson, *Phys. Rev.*, 1949, **75**, 486, 1736.
8. R. P. Feynman, *Phys. Rev.*, 1951, **84**, 108.
9. W. Magnus, *Commun. Pure Appl. Math.*, 1954, **7**, 649.
10. J. S. Waugh *J. Magn. Reson.*, 1986, **68**, 189.
11. S. J. Glaser and G. P. Drobny, in 'Advances in Magnetic Resonance', ed. W. S. Warren, Academic Press, New York, 1990, Vol. 14, p. 35.
12. J. S. Waugh, *J. Magn. Reson.*, 1982, **50**, 30.
13. C. Griesinger and R. R. Ernst, *Chem. Phys. Lett.*, 1988, **152**, 239.
14. J. Schleucher, J. Quant, S. J. Glaser, and C. Griesinger, submitted.
15. A. Bax and D. G. Davis, *J. Magn. Reson.*, 1985, **65**, 355.
16. A. J. Shaka, J. Keeler, and R. Freeman, *J. Magn. Reson.*, 1983, **53**, 313.
17. S. P. Rucker and A. J. Shaka, *Mol. Phys.*, 1989, **68**, 509.
18. M. Kadkhodaie, O. Rivas, M. Tan, A. Mohebbi, and A. J. Shaka *J. Magn. Reson.*, 1991, **91**, 437.
19. C. Griesinger, G. Otting, K. Wüthrich, and R. R. Ernst, *J. Am. Chem. Soc.*, 1988, **110**, 7870.
20. J. Cavanagh and M. Rance, *J. Magn. Reson.*, 1992, **96**, 670.
21. U. Kerssebaum, R. Markert, J. Quant, W. Bermel, S. J. Glaser, and C. Griesinger, *J. Magn. Reson.*, 1992, **99**, 184.
22. J. Briand and R. R. Ernst, *Chem. Phys. Lett.*, 1991. **185**, 276.
23. M. Kadkodaei, T.-L. Hwang, J. Tang, and A. J. Shaka, *J. Magn. Reson., Ser. A*, 1993, **105**, 104.
24. J. Quant, Diploma Thesis, University of Frankfurt, 1992.
25. M. Ravikumar and A. A. Bothner-By, *J. Am. Chem. Soc.*, 1993, **115**, 7537.
26. A. Bax and D. G. Davis, *J. Magn. Reson.*, 1985, **63**, 207.
27. C. Griesinger and R. R. Ernst, *J. Magn. Reson.*, 1987, **75**, 261.
28. D. Neuhaus and J. Keeler, *J. Magn. Reson.*, 1986, **68**, 568.
29. J. Cavanagh and J. Keeler, *J. Magn. Reson.*, 1988, **80**, 186.
30. T.-L. Hwang and A. J. Shaka, *J. Am. Chem. Soc.*, 1992, **114**, 3157.
31. T.-L. Hwang, M. Kadkhodaei, A. Mohebbi, and A. J. Shaka, *Magn. Reson. Chem.*, 1992, **30**, 24.

32. J. Fejzo, W. M. Westler, S. Macura, and J. L. Markley, *J. Magn. Reson.*, 1991, **92**, 20.

Acknowledgements

This work was supported by the Fonds der Chemischen Industrie. J.S. and J.Q. thank the Graduiertenkolleg Chemische und Biologische Synthese von Wirkstoffen at the Institute for Organic Chemistry (Gk Eg 53/3-3) for support. J.Q. gratefully acknowledges a scholarship of the Fonds der Chemischen Industries. S.G. thanks the DFG (Gl 203/1-1, Gl 203/1-2) and the Dr. Otto Röhm Stiftung for support. The sample of TFPI-2 was a gift from Novo Nordisk A/S. We thank Dr T. Keller, Dr W. Bermel, and Dr R. Kerssebaum, BRUKER Analytische Meßtechnik, for continuous support.

Trabecular Bone Imaging

Felix W. Wehrli

University of Pennsylvania Medical School, Philadelphia, PA, USA

1 INTRODUCTION

Bone is a composite material consisting of an inorganic phase—calcium apatite, $Ca_{10}(PO_4)_6(OH)_2$, corresponding to about 65% of total volume—and an organic phase—essentially collagen—accounting for most of the remaining 35%. From an architectural point of view, bone can be subdivided into cortical and trabecular, the latter providing most of the strength of the axial skeleton (e.g., the vertebral column) and the portions of the appendicular skeleton near the joints. Trabecular bone is made up of a three-dimensional network of struts and plates, the trabeculae, which are on the order of 100–150 μm in width and spaced 300–1000 μm apart.

Like engineering materials, trabecular bone derives its mechanical strength from its inherent elastic properties, its volume density, and its structural arrangement. Bone is constantly renewed through a process called 'bone remodeling', a term referring to a dynamic equilibrium between bone formation and bone resorption, controlled by two essential types of cells: the osteoblasts—bone forming cells—and the osteoclasts—bone resorbing cells. During bone formation, osteoblasts eventually become imbedded in bone, turning into osteocytes, which presumably act as piezoelectric sensors transmitting signals to the osteoblasts to induce bone formation. Since the seminal work of Wolff,[1] it has been known that bone grows in response to the forces to which it is subjected. Therefore, weightlessness and physical inactivity are well-known factors inducing bone loss.

The most common pathologic process leading to bone loss is osteoporosis.[2] Among the various etiologies, postmenopausal osteoporosis, which results from increased osteoclast activity, is the most frequent form of the disease, afflicting a substantial fraction of the elderly female population. The most common clinical manifestations are fractures of the hip and vertebrae. If detected early, calcium supplements and estrogen replacement are effective forms of therapeutic intervention. Further, the development of drugs inhibiting osteoclastic activity is in progress.

Bone mineral density is the most widely invoked criterion for fracture risk assessment, typically measured by dual X-ray absorptiometry (DEXA), which is based on the measurement of the attenuation coefficient in a quantitative radiographic procedure, or by quantitative computerized tomography (QCT). Whereas both methods measure bone mineral density with sufficient precision, neither provides information on the properties or structural arrangement of the bone. NMR, on the other hand, has the potential to probe structure as well as chemical composition of bone, both relevant to biomechanical competence.

2 DIRECT DETECTION OF BONE MINERAL BY ^{31}P NMR

The difficulties of detecting phosphorus in the solid state in vivo are considerable, but do not seem insurmountable. The problems are symptomatic of high-resolution NMR in the solid state in general: a combination of long T_1 (on the order of minutes) and short T_2 (on the order of 100 μs), as well as additional line broadening by anisotropic chemical shift and dipolar coupling, all of which are addressed in detail elsewhere in this Encyclopedia.

Brown et al.[3] first demonstrated the feasibility of quantitative analysis by solid state ^{31}P spectroscopy in human limbs as a means of measuring bone mineral density noninvasively.

Imaging adds an additional level of complexity, since the short lifetime of the signal demands short gradient duration, a requirement that can only be reconciled with gradients of high amplitude and large slew rates, which are both difficult to achieve in large sample volumes. Ackerman et al.[4] produced one-dimensional spin echo images at 7.4 T with echo times on the order of 1 ms and flip-back 180° rf pulses as a means to restore the longitudinal magnetization, inverted by the phase reversal pulse. More recently, the same group of workers reported two-dimensional images of chicken bone at 6 T by means of a combination of back-projection and ^1H–^{31}P cross polarization for sensitivity enhancement, with echo times as short as 200 μs.[5] In the cross-polarization technique, the ^{31}P magnetization is derived from that of dipolar-coupled protons, which have shorter T_1 and thus permit shorter pulse sequence recycling times. Making use of the dependence of the cross-polarization rate on the proton–phosphorus dipolar coupling, the same workers showed that different phosphate species can be distinguished, demonstrating the presence of a minor HPO_4^{2-} species in immature chicken bone.[6] While this work is at an early stage, it clearly has unique potential for nondestructive assaying of the chemical composition and its age-related changes of bone parameters that might in part explain the increased fragility of bone in older individuals.

3 IMPLICATIONS OF BONE DIAMAGNETISM ON NMR LINE BROADENING

Another approach toward assessing the properties of trabecular bone—specifically, as its architectural arrangement is concerned—exploits the diamagnetic properties of bone mineral. By virtue of the higher atomic number of its elemental composition (i.e., calcium and phosphorus), mineralized bone is more diamagnetic than marrow constituents in the trabecular marrow cavities, which consist mainly of water and lipids (i.e., oxygen, carbon, and hydrogen).

It is well known that near the interface of two materials of different magnetic susceptibility, and depending on the geometry of the interface, the magnetic field is inhomogeneous. Among the first to investigate these effects systematically were Glasel and Lee,[7] who studied deuteron relaxation of beads of different size and susceptibility, suspended in ^2H_2O. Specifically, they showed that the linewidth $1/\pi T_2^*$ scaled with $\Delta\chi$, the difference in volume susceptibility between the beads and deuterium oxide. The transverse relaxation rate $1/T_2$ was found to increase linearly with reciprocal bead size, an observation that could be reconciled with diffusion in the induced magnetic field gradients. Similar phenomena were reported by Davis et al.[8] who measured proton NMR linewidths at 5.9 T in powdered bone suspended in various solvents and found T_2^* to decrease with decreasing grain size and thus increased surface-to-volume ratio.

One of the earliest studied magnetically heterogeneous systems in biology is lung tissue, where the local magnetic field distortions are caused by air in the alveoli, and some of the concepts described here have parallels in imaging pulmonary parenchyma[9] (see also ***Lung and Mediastinum: A Discussion of the Relevant NMR Physics***). Transverse relaxation enhancement from diffusion in intrinsic microscopic gradients has received increased attention in conjunction with the BOLD contrast phenomenon, resulting from physiologic variations of deoxyhemoglobin in capillary vessels during functional activation.[10,11] However, since trabeculae are considerably larger than the venules of the capillary bed (100–200 μm versus 10–20 μm), and diffusion of the protons in the marrow spaces is small (on the order of $10^{-5}\,cm^2\,s^{-1}$), diffusion-induced shortening of T_2 is expected to be negligible. Consequently, the effect of the susceptibility-induced inhomogeneous field is essentially line broadening.

Suppose there is a distribution $\Delta B(x,y,z)$ across the sample volume, such as an imaging voxel of dimension $\Delta x\,\Delta y\,\Delta z$; then the transverse magnetization $M_{xy}(t)$ can be written as

$$M_{x,y}(t) = \frac{M_0}{\Delta x\,\Delta y\,\Delta z}\mathrm{e}^{-t/T_2} \times \int_{\Delta x}\int_{\Delta y}\int_{\Delta z} \mathrm{e}^{\mathrm{i}\gamma\Delta B(x,y,z)t}\,\mathrm{d}x\,\mathrm{d}y\,\mathrm{d}z \qquad (1)$$

For a Lorentzian field distribution, the integral in equation (1) can be described as an additional damping term, yielding

$$M_{x,y}(t) = M_0\mathrm{e}^{-t/T_2}\mathrm{e}^{-t/T_2'} \qquad (2)$$

T_2' is thus the time constant for inhomogeneity-induced spin dephasing. While the line broadening, in general, of course, is not Lorentzian, it will be seen subsequently that the assumption of a single exponential time constant is often a valid approximation.

Note that in this article, the following definitions and notation will be used for the contribution from magnetic field inhomogeneity to the total dephasing rate $1/T_2^*$, irrespective of the notation in the original literature: $1/T_2' \equiv R_2' \approx \frac{1}{2}\gamma\Delta B_z = 1/T_2^* - 1/T_2$, with ΔB_z representing the full width at half maximum of the magnetic field histogram in the sampling volume such as the imaging voxel.

3.1 Susceptibility of Bone

The difference in volume susceptibility between calcium apatite and water is about -3.8×10^{-6} (in rationalized SI units of the volume susceptibility), measured by means of a vibrating sample magnetometer (J. C. Ford and F. W. Wehrli, unpublished work). Recently, the author determined the susceptibility of bone using a susceptibility matching technique[12] (F. W. Wehrli and S. L. Wehrli, unpublished work). For this purpose, powdered bone from bovine femoral head was suspended in a cylindrical sample tube with a coaxial capillary containing water and serving as a reference, both aligned along the axis of a superconducting magnet. Potassium hexacyanoferrate(II), $K_4Fe(CN)_6$, which is highly diamagnetic, was then added incrementally to the suspension. This operation resulted in a decrease in linewidth and an increase in the bulk magnetic susceptibility (BMS) shift. For concentric cylinders, the BMS shift is given as[13]

$$\Delta\nu = \frac{\gamma B_0}{2\pi}\frac{|\chi_\mathrm{d} - \chi_\mathrm{w}|}{3} \qquad (3)$$

with χ_w and χ_d being the volume susceptibilities of water and hexacyanoferrate(II) solution, respectively [Figure 1a]. Whereas the critical concentration at which the solution matched the susceptibility of bone could not be attained owing to limited

For References see p. 4810

solubility of $K_4Fe(CN)_6$, the matching concentration was determined by extrapolation of the line broadening–concentration curve, from which a BMS shift of -0.95 ± 0.13 ppm was obtained (corresponding to $\chi_d - \chi_w = -2.85$ ppm) [Figure 1b]. Hence, bone is less diamagnetic than calcium hydroxyapatite by about 25%, which agrees well with the composition of bone, which has a 30% organic (i.e., nonmineral) constituent.

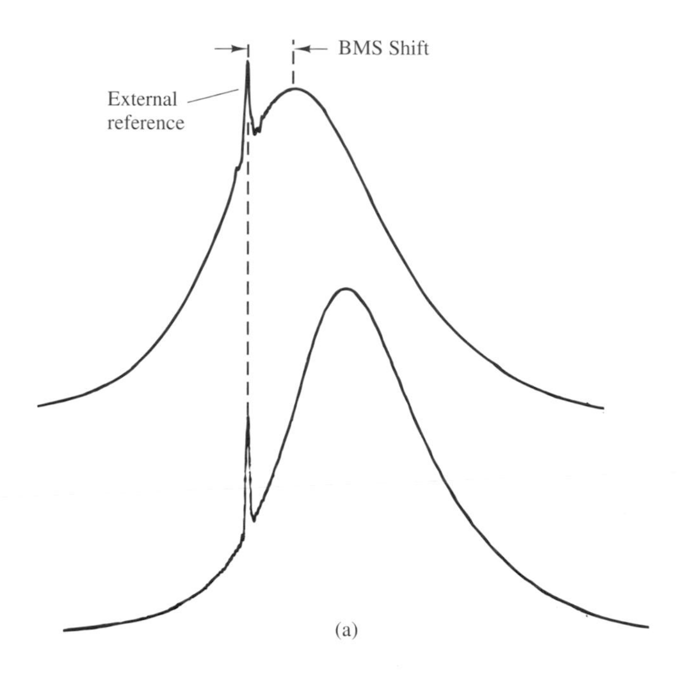

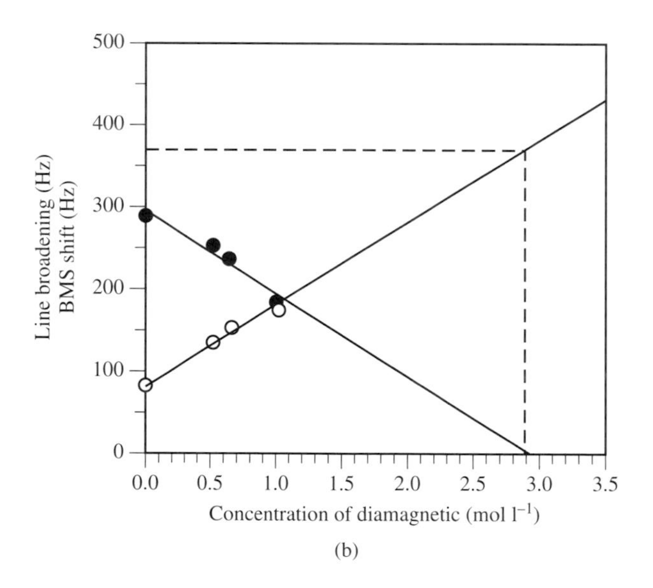

Figure 1 (a) 400 MHz proton spectrum from an aqueous suspension of ground bone from bovine tibia, showing susceptibility-broadened line ($\Delta\nu_{\frac{1}{2}} \approx 300$ Hz) and BMS shift relative to reference (top). The linewidth is reduced and the shift increased upon addition of 0.67 molar $K_4Fe(CN)_6$ owing to increased diamagnetism of the solution (bottom). (b) Line broadening (filled symbols) and diamagnetic shift (open symbols) versus concentration of $K_4Fe(CN)_6$, from experiments like those ones of Figure 1(a), affording differential susceptibility (see text for details)

3.2 Theoretical Considerations and Computer Modeling

Consider two adjoining materials of susceptibilities χ_m and χ_b. The induced magnetic surface charge density σ at some location on the interface between the two materials is then given as

$$\sigma = \Delta\chi \boldsymbol{B}_0 \cdot \boldsymbol{n} \tag{4}$$

where $\boldsymbol{n}$ is the unit vector normal to the interface and $\Delta\chi = \chi_m - \chi_b$. The additional field $\boldsymbol{B}_i(\boldsymbol{r})$ resulting from the magnetic charges at the phase boundary can be estimated from the Coulomb integral[14]

$$\boldsymbol{B}_i(\boldsymbol{r}) = \int \sigma(\boldsymbol{r}') \frac{\boldsymbol{r} - \boldsymbol{r}'}{|\boldsymbol{r} - \boldsymbol{r}'|^3} \, \mathrm{d}\boldsymbol{S}' \tag{5}$$

where the integration is over the surface S of the interface, with $\boldsymbol{r}'$ and $\boldsymbol{r}$ representing the locations of the source and field points, respectively.

The induced field is thus proportional to the difference in susceptibility of the two adjoining materials, the strength of the applied field, and the inverse square of the distance between source and field location. For an array of trabeculae, the field should be highly inhomogeneous, and a relationship is expected to exist between the magnetic field distribution within the volume of interest and the number density, thickness, and orientation of trabeculae.

Majumdar[15] evaluated the line broadening on the basis of a two-dimensional model of circular inclusions of radius r_s and susceptibility χ_2, immersed in a medium of susceptibility χ_1, by computing the field variation $\Delta B = B_z - B_0$ as a sum of contributions from N point discontinuities:

$$\Delta B = \sum_{i=1}^{N} \frac{4\pi(\chi_1 - \chi_2) r_s^3 (1 + 3\cos^2\theta_i)}{[3 + 4\pi(\chi_1 + 2\chi_2)] r_i^3} \tag{6}$$

where r_i and θ_i are the length and orientation of the position vector of the ith discontinuity from the location for which the induced field is calculated and r_s is the diameter of the perturber. The decay of the transverse magnetization was then computed for an assembly of random spins placed within the imaging pixel by evaluating the phase at each time step for a varying number density of inclusions. While not characteristic of the trabecular network, the model provides some insight in that it predicts that, for a susceptibility difference corresponding to calcium carbonate and water, the decay should be biexponential, approaching monoexponential behavior for echo times greater than 10 ms. The model also predicts R_2', the contribution to the total dephasing rate $1/T_2^*$, to increase linearly with particle density, assuming the particles to be uniformly distributed.

A more realistic three-dimensional model that resembles strut-like trabecular bone as found in the vertebrae[16] has been invoked to predict the line broadening behavior of protons in the marrow spaces, as a means to investigate the structural dependence of R_2'.[17] The model consists of a tetragonal lattice of interconnected parallelepipeds ('struts') of square cross-section, differing in susceptibility from the medium by $\Delta\chi$. When oriented so that the two parallel faces of each transverse strut are normal to the direction of the applied field $\boldsymbol{B}_0$, according to equation (4), these two faces will have uniform charge density $\sigma = \pm\Delta\chi B_0$. The other faces of the transverse (and longitudinal) struts will have $\sigma = 0$, and so do not contribute to $\boldsymbol{B}_i(\boldsymbol{r})$. If the field is oriented arbitrarily relative to the lattice, all faces will be uniformly polarized, with a charge density $\sigma = \Delta\chi B_0 \cos\alpha$ where α is the angle between $\boldsymbol{B}_0$ and the unit vector normal to any given polarized surface. An analytical expression exists for the induced field given by the integral of equation (5) for a rectangular lamina, and thus the total field at any one point in space can be calculated as a sum of contributions from all charge-bearing faces. In this manner, a histogram of the field for the unit cell of this lattice was obtained by randomly placing field points within the unit cell, and R_2' was calculated by fitting the Fourier transform of the field histogram to a decaying exponential. The model predicts nearly exponential decay within the experimentally practical range of echo times (about 10–50 ms). Further, R_2' is predicted to increase with both the number density of transverse struts and their thickness. Since the latter two quantities scale with material density, this finding appears unremarkable, since it would imply that R_2' merely measures bone mineral density. However, if both strut thickness and number density are varied in an opposite manner (so as to keep the material density constant), the model indicates that R_2' will increase as strut thickness decreases and number density increases. These predictions, which have been confirmed in analogous physical models,[18] underscore the importance of the distribution of the material, and suggest that different etiologies of bone loss (e.g., trabecular thinning as opposed to loss of trabecular elements) might be distinguishable.

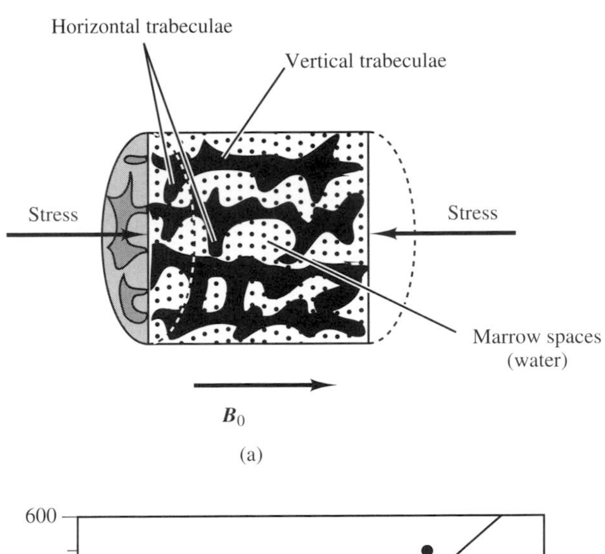

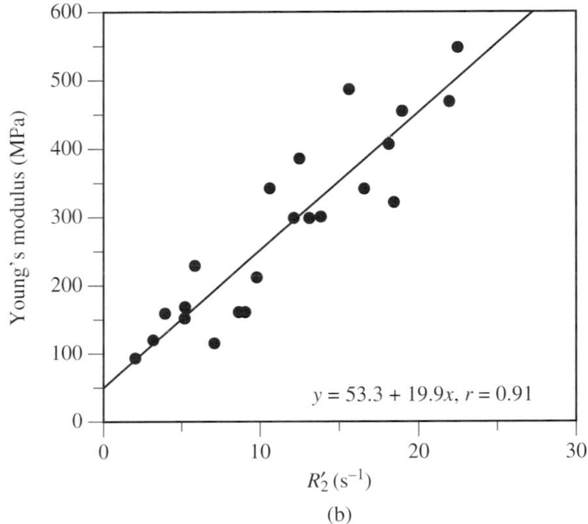

Figure 2 (a) Cross-section through a cylindrical trabecular bone specimen (schematic) used for R_2', structural, and stress analysis. The cylinder axis is parallel to the anatomic inferior–superior axis, aligned with the external field polarizing predominantly horizontal trabeculae, which cause line broadening of the proton resonances in the marrow spaces. Compressive loading is applied along the cylinder axis. (b) Young's modulus of elasticity obtained from compression tests in 22 cylindrical specimens from the lumbar vertebral bodies of 16 human subjects aged 24–86 years, plotted as a function of R_2' for the water protons in the intertrabecular spaces ($r = 0.91$, $p < 0.0001$). (Modified from Chung et al.[20])

4 RELATIONSHIP BETWEEN TRABECULAR ARCHITECTURE, LINE BROADENING, AND MECHANICAL COMPETENCE

Trabecular bone is well known to be anisotropic, with the orientation of the trabeculae following the major stress lines. In the vertebrae, for example, the preferred orientation of the trabeculae is along the body axis, in response to the compressive forces acting in this direction. The role of the horizontal trabeculae is to act as cross ties preventing failure by buckling. It has been shown that, during aging, horizontal trabeculae are lost preferentially.[19]

If the static magnetic field is applied parallel to the inferior–superior axis of the vertebrae, the horizontal trabeculae (i.e., those orthogonal to the field) are polarized, and thus are expected to be the principal cause of the susceptibility-induced line broadening. Chung et al.[20] measured the mean spacing of horizontal trabeculae (i.e., the reciprocal of horizontal trabecular number density) in cadaver specimens of human lumbar trabecular bone after bone marrow removal and suspension of the bone in water, using NMR microscopy and digital image processing. They found a positive correlation between water proton R_2', measured at 1.5 T, and mean number density of the horizontal trabeculae ($r = 0.74$, $p < 0.0001$).

The critical role of the horizontal trabeculae in the vertebrae in conferring compressive strength is illustrated with the correlation between R_2' measured with the polarizing field parallel to the body axis and Young's modulus of elasticity (stiffness) for compressive loading. Figure 2(a) shows the relationship between anatomic axis, trabecular orientation, the orientation of the applied magnetic field, and the direction of compressive loading. A strong association between stiffness and R_2' exists over a wide range of values, corresponding to trabecular bone of very different morphologic composition [Figure 2(b)]. From these data, it is inferred that a global measurement of R_2' in trabecular bone seems able to predict the compressive strength of this highly complicated structure.

For References see p. 4810

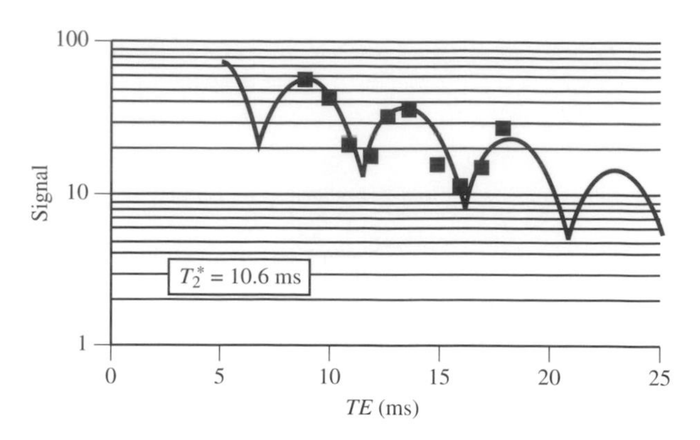

Figure 3 Region-of-interest signal from a 1 cm^2 region of lumbar vertebra 5, plotted versus gradient echo time in a typical subject. The solid line is a best fit to a two-component interferogram resulting from amplitude modulation between CH_2 lipid resonance and water in the bone marrow. (From Wehrli et al.[22])

5 IN VIVO QUANTITATIVE NMR FOR THE ASSESSMENT OF BONE DISEASES

5.1 Measurement and Data Analysis

Bone marrow has cellular (hematopoietic) and fatty components, with the relative fractions varying widely, depending on anatomic site and age. The major chemical constituents of the two types of marrow are water and fatty acid triglycerides. This chemical heterogeneity of bone marrow complicates in vivo measurement of T_2^*. A linewidth measurement by means of image-guided localized spectroscopy has the advantage of providing T_2^* for each spectral component.[21] However, image-based (non-spectrally resolved) techniques are generally preferred,[21–24] by means of either gradient echo or asymmetric spin echo techniques.[25] By collecting an array of images with incrementally stepped time for inhomogeneity dephasing (gradient echo delay or echo offset), the pixel amplitudes can be fitted to some model for signal decay. These methods are less sensitive to magnetic field inhomogeneity arising from effects unrelated to susceptibility-induced gradients, since the field across an imaging voxel of a few cubic millimeters is, in general, quite homogeneous.

The presence of multiple chemically shifted constituents causes an amplitude modulation that has the characteristics of an interferogram. The latter can be expressed as the modulus of the vector sum of the individual phase-modulated spectral components:[22,26]

$$I(t) = \left|\sum_{i=1}^{n} \boldsymbol{I}_i\right| = \left[\sum_{i=1}^{n}\sum_{j=1}^{n} I_{0i}\mathrm{e}^{-t/T_{2i}^*} I_{0j}\mathrm{e}^{-t/T_{2j}^*}\cos(\Delta\omega_{ij}t)\right]^{1/2} \quad (7)$$

where I_{0i} is the initial amplitude of the ith chemically shifted constituent, $\Delta\omega_{ij}$ is the chemical shift difference in rad s^{-1} between nuclei i and j, t is the dephasing time (e.g., the echo time TE in a gradient echo), and the summation is over all spectral components n. Typically, the most abundant spectral components are those of the CH_2 protons of fatty acids and of water, which are separated by δ = 3.35 ppm. It has been shown that $I(t)$ can be fitted to a two-component interferogram with T_2^* and fat and water signal amplitudes as adjustable parameters and assuming $T_{2,\mathrm{fat}}^* \approx T_{2,\mathrm{water}}^*$,[22] an example of which is illustrated in Figure 3.

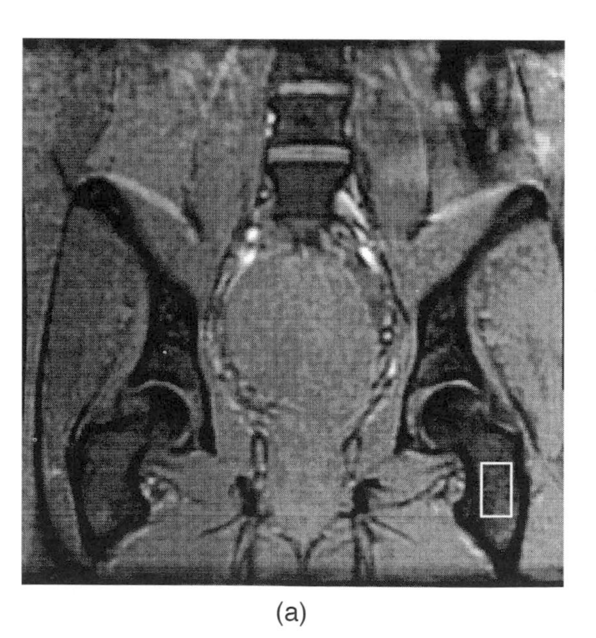

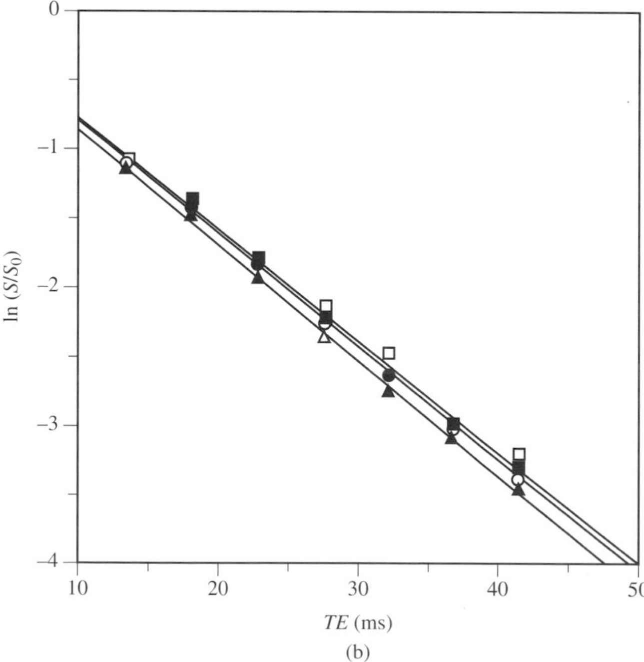

Figure 4 (a) First of a series of eight coronal gradient echo images for simultaneous measurement of T_2^* in the hip and lumbar spine, obtained by collecting 128 data samples every 4.65 ms, from a single gradient echo train so as effectively to demodulate the signal (see text for details). (b) Plot of signal measured in the trochanter [see region indicated in (a)] versus echo time, obtained from five successive scans. Solid lines are linear least-square fits, affording T_2^* = 12.55 ± 0.38 ms

Multiparameter curve fits are hampered by the difficulty of locating the global minimum, and require a relatively large number of images. One alternative is to suppress one spectral component,[24] which sacrifices some S/N and is relatively sensitive to global magnetic field inhomogeneity, demanding that the field across the sample volume vary less than the chemical shift. Another approach to suppress the modulation is to sample the interferogram at the modulation frequency, ideally in such a manner that the two components are in phase with one another.[27] This condition is satisfied for sampling at multiples of the modulation period $T = 2\pi/\gamma\delta B_0$, which is 4.65 ms at 1.5 T field strength.

Rather than computing T_2^* by fitting the mean signal from a two-dimensional array of pixels (often called the 'region of interest'), it is desirable to perform this calculation pixel by pixel for the generation of T_2^* maps. This, however, requires that misregistration be minimal between acquisitions, which may, for example, be achieved with a multiple echo pulse sequence that collects gradient echoes of the same polarity at multiples of the chemical shift modulation period. The precision achievable in vivo with this technique is illustrated with the data obtained from five separate scans in the same test subject in Figures 4(a) and (b). The data also show that chemical shift modulation is completely suppressed and that the decay is exponential to a high degree (Figure 4b).

The susceptibility-induced line broadening depends on the dimensions of the volume element across which the inhomogeneity is measured. In an image-based measurement, the smallest signal-producing volume element is the imaging voxel. If the voxel size decreases below the order of magnitude of the intertrabecular space then the likelihood of this voxel falling in a region sufficiently removed from the field gradients induced by trabeculae increases. As a consequence, the mean T_2' is expected to increase with decreasing pixel size. Majumdar and Genant[24] studied this effect in trabecular bone of various densities. They found that the T_2' histogram becomes wider and more asymmetric with decreasing pixel size. On the other hand, if the voxel size is large relative to the range of the gradients, T_2' becomes independent of voxel size, which is the case for pixels on the order of 1.5–2 mm. The distribution of T_2' as a function of image resolution is shown in Figure 5.

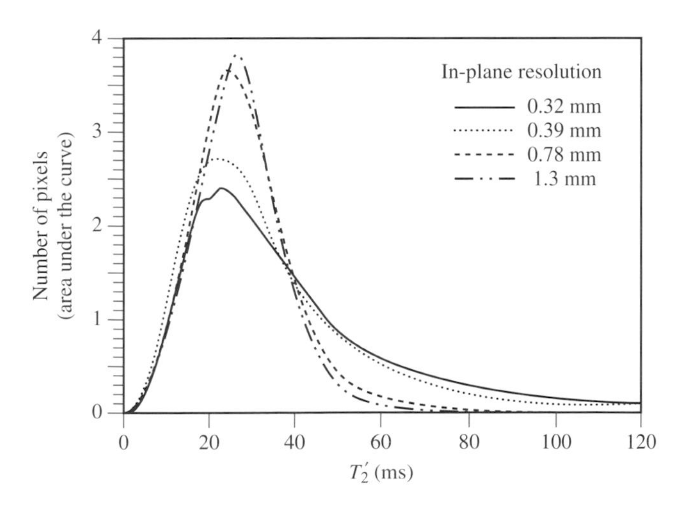

Figure 5 Smoothed T_2' histograms obtained from T_2' maps computed from axial images in the epiphysis of the distal femur, a site of dense trabeculation. As image resolution increases (smaller pixel size), the histogram broadens and becomes skewed. Means are found to vary between 41.7 ms at 0.32 mm pixel size and 25.7 ms at 1.25 mm pixel size. (Modified, with permission, from Majumdar and Genant[24])

5.2 Empirical Relationship with Bone Mineral Density

A possible relationship between bone mineral density (BMD) and the extent of inhomogeneity broadening measured by T_2^* was suggested by the work of Ford and Wehrli,[21] who found T_2^* in the distal femur to shorten with increasing density of trabeculation from the diaphysis (lowest trabeculation) toward the metaphysis and epiphysis of the bone (highest trabeculation). Majumdar et al.[23] determined R_2' in intact specimens of human vertebral trabecular bone after bone marrow removal and suspending the bone in saline. The measurements afforded a linear correlation between R_2' and BMD (in mg cm^{-3}), obtained from QCT, with a slope of 0.20 $\pm$ 0.02 s^{-1} mg^{-1} cm^3 ($r = 0.92$). More recently, the same group extended these studies in vivo in normal volunteers to the distal radius and proximal tibia.[24] Confirming earlier work, they found both BMD and R_2' to depend on the anatomic site of measurement. In excellent agreement with in vitro data, they measured 0.20 $\pm$ 0.01 s^{-1} mg^{-1} cm^3 ($r = 0.88$) for the combined data from both anatomic sites.

5.3 Clinical Studies in Patients with Osteoporosis

An early pilot study on a small group of patients with clinically established osteoporosis of the spine ($n = 12$) and an equal number of control subjects showed the former to have significantly reduced bone marrow $R_2^*(\equiv 1/T_2^*)$ in lumbar vertebra L5.[22] A follow-up study carried out in the author's laboratory, involving a larger cohort of patients and using more advanced methodology confirmed this observation.[28] This study was designed to explore whether image-based measurements of R_2^* may provide an index of the integrity of trabecular bone as a possible criterion for predicting fracture risk. Measurements included R_2^*, without correction for R_2, and DEXA BMD in the lumbar vertebral bodies. Patients were characterized as women with abnormally low bone mass (<90 mg cm^{-3} on QCT or <0.9 g cm^{-2} on DEXA), and or having roentgenographically proven thoracic vertebral fractures. Means and standard deviations of R_2^* and the result of the t test are listed in Table 1. Results indicated that in lumbar vertebrae L3, L4, and L5, osteoporotics had significantly lower R_2^* in conformance with the hypothesis that decreased trabecular plate density lessens the extent of susceptibility-induced line broadening. In addition, R_2^* was shown to correlate with DEXA BMD ($R = 0.53$, $p < 0.0005$), a finding that is in qualitative agreement with QCT BMD and R_2^* in the wrist and tibia.[24]

Table 1 R_2^* in the Lumbar Vertebrae of Patients ($n = 52$, Mean Age 59.1 ± 12.1 years) and Their Controls ($n = 51$, Mean Age 44 ± 12.7 years)

	R_2^* (s^{-1})		
	L3	L4	L5
Controls	67.2 ± 10.2	66.3 ± 9.93	66.7 ± 11.9
Osteoporotics	55.9 ± 10.5	56.2 ± 11.1	55.8 ± 10.2
p	<0.0001	<0.0001	<0.0001

For References see p. 4810

While it is too early to predict the effectiveness of NMR line broadening as a prognostic criterion, the method has potential for screening of the population at risk—mainly women before and after menopause, with hereditary or lifestyle-related risk factors. This would represent a new application of MRI, differing fundamentally from its current widespread diagnostic use. Whether such a scenario will materialize depends, besides effectiveness relative to established procedures, on procedure and equipment cost.

6 HIGH-RESOLUTION IMAGING OF TRABECULAR STRUCTURE

Whereas measurement of the induced magnetic field inhomogeneity provides structural information indirectly, high-resolution MRI at microscopic dimensions has the potential for nondestructive mapping of three-dimensional trabecular morphology, as an alternative to conventional microscopy from sections. In fact, bone is ideal in that it has nearly background intensity, which facilitates image segmentation into bone and marrow. In order to measure the histomorphometric parameters such as trabecular volume fraction, mean thickness, and intertrabecular spacing, a resolution on the order of 25–50 μm is required, which can easily be achieved in vitro from a 1 cm^3 specimen[20] with relatively standard microimaging hardware.

The background gradients from the susceptibility-induced fields can cause artifacts in the form of signal loss from intravoxel phase dispersion [cf. equation (1)]. Since these phase losses are recoverable—assuming diffusion to be negligible—spin echo detection is advantageous. Alternatively, the dephasing time should be held minimal, which can, for example, be achieved with projection–reconstruction techniques, as they were applied in microimaging of lung parenchyma.[29]

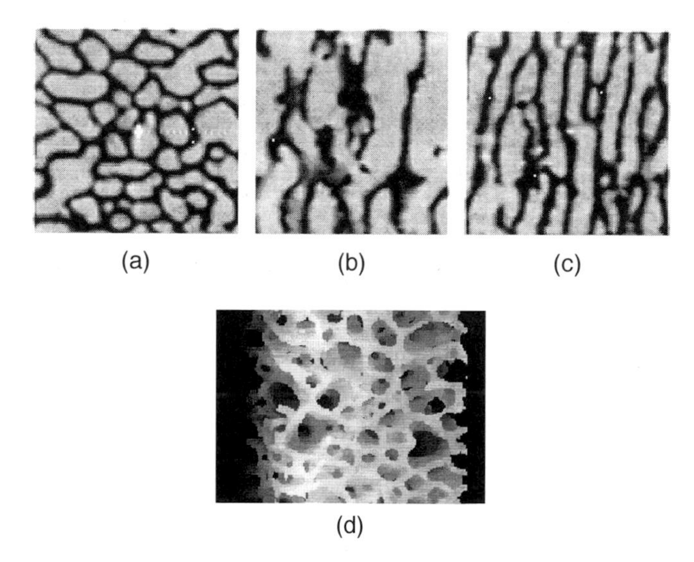

Figure 6 Images at microscopic dimension showing trabecular structure: (a)–(c) orthogonal views obtained from a 64^3 array of 3D spin echo array of images acquired at 9.4 T on a specimen of trabecular bone from bovine tibia (114 μm × 114 μm × 139 μm voxel size); (a) transverse and (b), (c) longitudinal sections. Note the preferential orientation of the trabeculae along the inferior–superior direction. (d) Shaded surface display of the same array of data, resolution-enhanced by means of a subvoxel tissue classification technique using Bayesian segmentation.[32]

The potential use of NMR microscopy as a histologic tool for assessing age-related changes in trabecular morphology and their agreement with histology is implied by recent work in animal models.[30] Since trabecular structure is anisotropic, three-dimensional (3D) imaging is advantageous, since it permits display and thus analysis in all three orthogonal planes. An example of such an application illustrating highly anisotropic trabecular bone is displayed in Figure 6, along with a 3D shaded surface display (H. Chung and F. W. Wehrli, unpublished). From data like those shown, the fabric tensor can be determined by measuring the mean intercept lengths between trabeculae in all spatial directions. Images at a resolution comparable to those shown can be achieved in vivo in circumstances permitting small closely coupled receive coils, as in the finger, by means of partial flip angle spin echo techniques.[31]

7 RELATED ARTICLES

Lung and Mediastinum: A Discussion of the Relevant NMR Physics; Oil Reservoir Rocks Examined by MRI; Susceptibility & Diffusion Effects in NMR Microscopy; Susceptibility Effects in Whole Body Experiments.

8 REFERENCES

1. J. Wolff, 'Das Gesetz der Transformation der Knochen', A. Hirschwald, Berlin, 1892.
2. B. Riggs and L. Melton, 'Osteoporosis', Raven Press, New York, 1988.
3. C. E. Brown, J. R. Allaway, K. L. Brown, and H. Battocletti, *Clin. Chem.*, 1987, **33**, 227.
4. J. L. Ackerman, D. P. Raleigh, and M. J. Glimcher, *Magn. Reson. Med.*, 1992, **25**, 1.
5. J. Moore, L. Garrido, and J. Ackerman, *Magn. Reson. Med.*, 1995, **33**, 293.
6. Y. Wu, M. Glimcher, C. Rey, and J. Ackerman, *J. Mol. Biol.*, 1994, **244**, 423.
7. J. Glasel and K. Lee, *J. Am. Chem. Soc.*, 1974, **96**, 970.
8. C. A. Davis, H. K. Genant, and J. S. Dunham, *Invest. Radiol.*, 1986, **21**, 472.
9. D. C. Ailion, T. A. Case, D. Blatter, A. H. Morris, A. Cutillo, C. H. Durney, and S. A. Johnson, *Bull. Magn. Reson.*, 1984, **6**, 130.
10. K. K. Kwong, J. W. Belliveau, D. A. Chesler, I. Goldberg, R. Weisskoff, B. Poncelet, D. N. Kennedy, B. Hoppel, M. Cohen, R. Turner, H. Cheng, T. Brady, and B. Rosen, *Proc. Natl. Acad. Sci. U.S.A.*, 1992, **89**, 5675.
11. S. Ogawa, D. W. Tank, R. Menon, J. M. Ellerman, S. G. Kim, H. Merkle, and K. Ugurbil, *Proc. Natl. Acad. Sci. U.S.A.*, 1992, **89**, 5951.
12. P. C. Lauterbur, B. V. Kaufman, and M. K. Crawford, in 'Biomolecular Structure and Function', ed. P. F. Agris. Academic Press, New York, 1978, p. 343.
13. S. Chu, Y. Xu, J. Balschi, and C. Springer, *Magn. Reson. Med.*, 1990, **13**, 239.
14. A. Morrish, 'The Physical Principles of Magnetism', Robert E. Krieger, Malabar, FL, 1983.
15. S. Majumdar, *Magn. Reson. Med.*, 1991, **22**, 101.
16. L. J. Gibson, *J. Biomech.*, 1985, **18**, 317.

17. J. C. Ford, F. W. Wehrli, and H. W. Chung, *Magn. Reson. Med.*, 1993, **30**, 373.
18. K. Engelke, S. Majumdar, and H. K. Genant, *Magn. Reson. Med.*, 1994, **31**, 380.
19. L. Mosekilde, *Bone Miner.*, 1990, **10**, 13.
20. H. Chung, F. W. Wehrli, J. L. Williams, and S. D. Kugelmass, *Proc. Natl. Acad. Sci. U.S.A.*, 1993, **90**, 10250.
21. J. C. Ford and F. W. Wehrli, *Magn. Reson. Med.*, 1991, **17**, 543.
22. F. W. Wehrli, J. C. Ford, M. Attie, H. Y. Kressel, and F. S. Kaplan, *Radiology*, 1991, **179**, 615.
23. S. Majumdar, D. Thomasson, A. Shimakawa, and H. K. Genant, *Magn. Reson. Med.*, 1991, **22**, 111.
24. S. Majumdar and H. K. Genant, *J. Magn. Reson. Imaging*, 1992, **2**, 209.
25. G. L. Wismer, R. B. Buxton, B. R. Rosen, C. Fisel, R. Oot, T. Brady, and K. Davis, *J. Comput. Assist. Tomogr.*, 1988, **12**, 259.
26. F. W. Wehrli, T. G. Perkins, A. Shimakawa, and F. Roberts, *Magn. Reson. Imaging*, 1987, **5**, 157.
27. J. C. Ford and F. W. Wehrli, *J. Magn. Reson. Imaging*, 1992, **2**(P), 103.
28. F. W. Wehrli, J. C. Ford, and J. G. Haddad, *Radiology*, 1995, **196**, 631.
29. S. L. Gewalt, G. H. Glover, L. W. Hedlund, G. P. Cofer, J. R. MacFall, and G. A. Johnson, *Magn. Reson. Med.*, 1993, **29**, 99.
30. R. D. Kapadia, W. B. High, H. A. Soulleveld, D. Bertolini, and S. K. Sarkar, *Magn. Reson. Med.*, 1993, **30**, 247.
31. H. Jara, F. W. Wehrli, H. Chung, and J. C. Ford, *Magn. Reson. Med.*, 1993, **29**, 528.
32. Z. Wu, H. Chung, and F. W. Wehrli, *Magn. Reson. Med.*, 1994, **31**, 302.

Biographical Sketch

Felix W. Wehrli. *b* 1941. M.S., 1967, Ph.D., 1970, chemistry, Swiss Federal Institute of Technology, Switzerland. NMR application scientist, Varian AG, 1970–79; Executive Vice President, Bruker Instruments, Billerica, 1979–82; NMR Application Manager, General Electric Medical Systems, Milwaukee, 1982–88. Currently Professor of Radiologic Science and Biophysics, University of Pennsylvania Medical School. Approximately 95 publications. Current research specialty: NMR imaging of biomaterials, specifically trabecular bone.

Transferrins

Claudio Luchinat & Marco Sola
University of Bologna, Bologna, Italy

1 INTRODUCTION

Transferrins are a family of metal binding glycoproteins which play a fundamental role in iron metabolism in vertebrates and invertebrates.[1] Transferrin from serum (Tf hereafter) acts as the principal iron carrier in plasma, binding the Fe^{3+} ion strongly and reversibly and shuttling it from the sites of metal uptake to those of iron–proteins biosynthesis and iron storage. Ovotransferrin from avian egg white (Otf), and lactoferrin (Ltf) found in many secretory fluids and in leukocytes, are mainly iron-sequestering proteins which exert a bacteriostatic and detoxifying function. Transferrins are monomeric proteins of about 700 residues with molecular weight close to 80 kDa. They are related by a sequence homology of about 50%. The X-ray structural information[2,3] indicates that the polypeptide chain is organized into two nearly equal-sized and about 40% homologous globular lobes connected by a short peptide segment, each containing one chain terminus (hence termed N-lobe and C-lobe), and one metal site (Figure 1). Each lobe is in turn structured into two domains. The metal lies at the bottom of the cleft between the two domains, about 10 Å from the protein surface. Both metals are bound by two tyrosines, one histidine, and one aspartate in a pseudooctahedral coordination which is completed by a bidentate carbonate ion hydrogen bonded to some protein groups (Figure 1).[2,3] X-ray data also indicate the absence of metal-coordinated water in the crystal.

The requirement of a concomitant (synergistic) binding of an anion (besides carbonate, carboxylate anions with a second donor group are also active) for specifically loading the metal in the two sites is a unique feature of the biochemistry of transferrins. In recent years X-ray crystallography and NMR have allowed the elucidation of several molecular details of the metal sites. It is apparent that the formation of the ternary complex at both sites proceeds through initial anion binding to the protein: the anion is most probably required to reduce the net positive charge in the binding cavity, due to the presence of a number of cationic residues, and for increasing the overall stability of the adduct by further anchoring the metal to the protein. Overall, it functions as a bridge between the metal and the protein. The synergistic anion appears to play a central role in the reactivity of the metal centers, acting as the key for the

For References see p. 4818

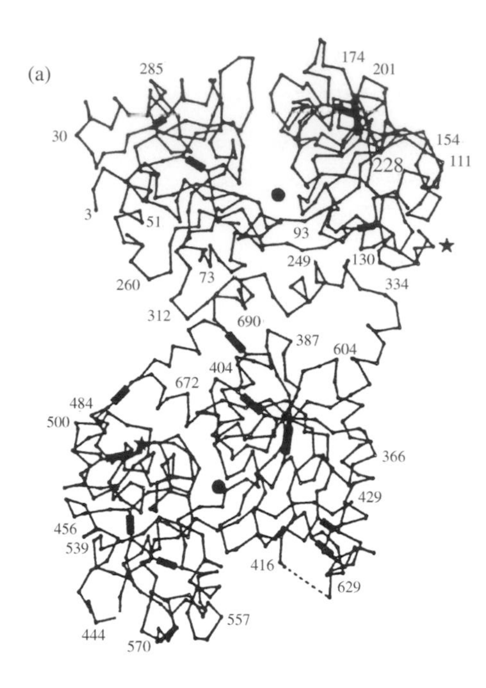

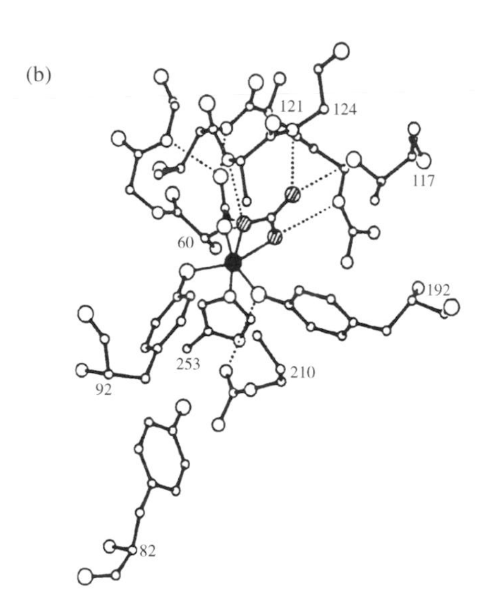

Figure 1 (a) C^{α} plot of the lactoferrin molecule showing its subdivision into two lobes, the N-lobe (top) and C-lobe (bottom). The iron atoms are shown as filled circles bound in the cleft between the two domains in each lobe. (b) The metal- and anion-binding site in lactoferrin (N-lobe). The carbonate anion is shown by hatched circles. Hydrogen bonds are indicated by dotted lines. (Reproduced by permission of Academic Press from B. F. Anderson, H. M. Baker, G. E. Norris, D. W. Rice, and E. N. Baker, *J. Mol. Biol.*, 1989, **209**, 711)

processes of metal uptake and release. The associated protein conformational changes are also being extensively studied, in particular the open–close equilibrium of the interdomain cleft at each lobe, with metal binding and removal occurring only in the open form.

The two metal sites of transferrins are, in general, spectroscopically slightly inequivalent and differ, although most often not dramatically, in metal binding affinity. The relative binding strength of the two sites is sensitive to pH, synergistic anion, and ionic composition of the medium, and may vary toward different metals. However, the C-site almost invariably shows a greater metal affinity than the N-site at acidic pH. The structural bases of these differences are still not unequivocally established (although the disulfide bridge content of the two lobes appears to be relevant in this respect). Their physiological relevance is also under investigation. The biological requirement for the structural 'doubleness' of transferrins is nowadays a central issue.

Transferrins can firmly bind a large number of transition, nontransition, and lanthanide ions. Several metals have served as spectroscopic probes for investigating the structural features of the metal sites, but some of the adducts also have a biological relevance since serum transferrin, being only about 30% saturated with iron in vivo, may function as an efficient carrier for metals other than iron.

2 THE ROLE OF NMR IN THE CHEMISTRY OF TRANSFERRINS: ACHIEVEMENTS

It should be recalled that for a protein of the size of transferrin full structural studies in solution (see ***Biological Macromolecules: Structure Determination in Solution***; ***Biological Macromolecules***; ***Biological Macromolecules: NMR Parameters***; ***Multidimensional Spectroscopy: Concepts***; ***Structures of Larger Proteins, Protein–Ligand, & Protein–DNA Complexes by Multi-Dimensional Heteronuclear NMR***) are not yet feasible. The NMR of transferrins was until recently mainly devoted to the elucidation of the molecular details of the ternary adducts at the metal sites with the following methodological approaches: (a) ^{13}C NMR of protein derivatives of diamagnetic metals with ^{13}C-enriched anions; (b) observation of protein-bound NMR active metal nuclei; (c) observation of ^{1}H NMR hyperfine-shifted resonances arising from the protons surrounding a bound paramagnetic metal (see ***Electron–Nuclear Hyperfine Interactions***; ***Cobalt(II)- & Nickel(II)-Substituted Proteins***; ***Electron–Nuclear Multiple Resonance Spectroscopy***); (d) NMRD profiles (see). Only in recent years, has one- and two-dimensional ^{1}H NMR at high field (see ***Multidimensional Spectroscopy: Concepts***) succeeded in handling, at least partially, the largely unresolved proton spectrum and proved to be of use in monitoring metal binding to the protein and the related conformational changes. In what follows we have outlined the contribution of NMR to the understanding of the main issues of the metallobiochemistry of transferrins.

2.1 Identification of the Metal-Binding Residues

The involvement of tyrosine residues in metal binding at both sites of transferrins, indicated by a number of spectroscopic and chemical modification studies, has been conclusively supported by ^{1}H NMR since the early 1970s from the shift induced in the corresponding resonances by Ga^{3+} binding, and their paramagnetic broadening upon Fe^{3+} binding (see ***Paramagnetic Relaxation in Solution***).[4] In addition, two histidines were also proposed as metal ligands from the detection of the C(2)H resonances which did not titrate with pH in the Ga^{3+} ternary complexes[5] of either intact or half molecules of Otf and human Tf[6,7] (Ga^{3+} has proved to be a good diamagnetic substitute for the Fe^{3+} ion, owing to the similarity of the ionic radius and to a similar binding affinity for transferrins). Metal binding of two histidines and two tyrosines was also consistent with the hyperfine shifted ^{1}H resonances in the mono- and bis-Co^{2+} derivatives of Otf.[8,9] The spread of the hyperfine shifted signals and their relatively long relaxation times[10] (see ***Paramagnetic Relaxation in Solution*** and ***Cobalt(II)- & Nickel(II)-Substituted Proteins***) suggested octahedral coordination around the metal ions.[8,9]

The X-ray structure later confirmed octahedral coordination and the presence of two tyrosine ligands, but showed that only one of histidine is involved in the metal binding site.[2,3] The identity of the additional nontitrating histidine in the Ga^{3+} derivative remains unknown, while the additional hyperfine shifted exchangeable proton in the spectrum of the Co^{2+} derivative could well arise, in retrospect, from one of the NH protons H bonded to the metal-coordinated carbonate ion.

Cadmium-113 NMR[11] (see ***Cadmium-113 NMR: A Surrogate Probe for Zinc & Calcium in Proteins***) and ^{205}Tl NMR data[12] (see ***Thallium NMR***) are in full agreement with the presence of only one histidine in the binding set. Aspartate coordination invariably went undetected, most probably due to the rather elusive ^{1}H NMR spectral changes in the crowded aliphatic region upon metal binding and the impossibility of metal NMR to discriminate it from other potential oxygen donors.

2.2 Exchangeable Protons in the Metal Sites

The issue of possible water coordination was tackled through water proton relaxation measurements. Early studies of water T_1 measurements in solutions of the Fe^{3+} adducts of Htf and Otf simply showed the accessibility of both sites to water molecules.[13,14] This was substantiated later by measurements of the field dependence of the longitudinal nuclear magnetic relaxation rates of water protons (NMRD profiles) on the iron adduct of Htf, also carried out for comparative purposes on the apoprotein and the diamagnetic analog containing Co^{3+}, which were interpreted as indicative of a water molecule only loosely bound to the metal in each site.[15,16] Analogously, one water molecule axially bound to the metal was also proposed for the Cu^{2+} derivative of Htf.[17] It was later shown that for slowly rotating protein complexes the anisotropic structure of the ligand fields of the complexed metals is not averaged out as it is for fast rotating aqua ions or small complexes.[18,19] The NMRD profiles for Cu^{2+} and VO^{2+} complexes of Htf were then evaluated with computations that included the anisotropic hyperfine interaction of metal ions with their own nuclei (see ***Electron–Nuclear Hyperfine Interactions***).[20] It was found that relaxation occurs through exchange of bulk water with a second coordination sphere water molecule. This was also confirmed for the Gd^{3+}–Htf complex[21] (see ***Gadolinium Chelates: Chemistry, Safety, & Behavior***) and demonstrated later by the X-ray structure of human Ltf.[3] Recent[22] more quantitative interpretation of the NMRD data for Fe^{3+}–Htf,[23] taking into account zero field splitting of the iron ion,[18,19,24] suggests the presence of slowly exchanging protons (possibly from the protein) and of fast exchanging water at about 4–5 Å from the metal ion. NMRD data on Cu^{2+}–transferrin were also instrumental in demonstrating rotation-independent electronic relaxation mechanisms for Cu^{2+} in solution.[25]

2.3 Synergistic Anions in the Metal Sites

Elucidation of the anion binding arrangement in the metal sites was obtained from a number of ^{13}C NMR data using ^{13}C-enriched anions (mainly carbonate and oxalate) and/or NMR-active nuclei loaded in the protein sites. The spectral parameters and resonance assignments to individual sites, where available, have been collected in Tables 1 and 2. The broadening beyond detection of the ^{13}C NMR resonance of isotopically-enriched bicarbonate bound to Fe^{3+}–Tf[26] allowed an estimate to be made of an upper limit of 9 Å for the Fe–^{13}C distance, which was later reduced to 4.5 Å[27] thanks to instrumental developments, consistent with the presence of the anion in the first coordination sphere of the metal. The latter study also indicated that the anion, most probably carbonate rather than bicarbonate as judged by the ^{13}C chemical shift value, was specifically bound in both metal binding sites of the diamagnetic Co^{3+} derivative and of the apoenzyme, giving rise in both cases to resonances in slow exchange with that of free bicarbonate (see ***Chemical Exchange Effects on Spectra***).[27] Specific anion binding to the apoprotein was taken as indicative that anion binding to the sites may precede metal binding,

Table 1 ^{13}C NMR Chemical Shifts for ^{13}C-enriched Synergistic Anions in Various Metal–Transferrin Derivatives[a]

Derivative	Chemical shift[b]	
Carbonate derivatives		
Co^{3+}–Htf[27]	169.5	169.3
Al^{3+}–Otf[30]	165.7 (N)	165.5(C)
Al^{3+}–Htf[29]	165.4	
Ga^{3+}–Otf[29]	166.3	
Ga^{3+}–Htf[29]	166.5	
Tl^{3+}–Htf[12]	165.98 (N)	166.22 (C)
Tl^{3+}–Otf[12]	165.92	
Sc^{3+}–Otf[38]	166.74 (N)	166.61 (C)
Sc^{3+}–Htf[38]	167.19	166.76
Zn^{2+}–Otf[29]	168.2	
Zn^{2+}–Htf[29]	168.5	
Cd^{2+}–Otf[11]	168.2	
Cd^{2+}–Htf[11]	168.2	
Oxalate derivatives		
	C-1	C-2
Al^{3+}–Otf[30]	168.42 (C)	165.33 (C)
	168.47 (N)	165.89 (N)
Al^{3+}–Htf[36]	168.58	166.07
	168.40	165.33
Al^{3+}–Ltf[36]	168.55	166.0
	168.24	166.0
Ga^{3+}–Otf[29]	168.4 (C)	165.7 (C)
	168.4 (N)	166.2 (N)
Tl^{3+}–Otf[12]	167.80 (C)	165.35 (C)
	167.91 (N)	165.99 (N)
Tl^{3+}–Htf[12]	167.88 (N)	165.62 (N)
Sc^{3+}–Otf[38]	170.52 (C)	167.00 (C)
	170.60 (N)	167.68 (N)
Zn^{2+}–Otf[29]	169.6	
Cd^{2+}–Htf[32]	168.5 (C)	169.9 (N)

[a]Htf: human serum transferrin; Otf: ovotransferrin; Ltf: human lactoferrin. Assignments to the N- and C-site are reported in parentheses, when available.

[b]In ppm from Me_4Si.

For References see p. 4818

Table 2 Spectral Parameters for Metal NMR Resonances and M–^{13}C Coupling Constants for Various Metal–Transferrin Derivatives[a]

Derivative	Chemical shift (ppm)[b]	$\Delta\nu_{\frac{1}{2}}$(Hz)	^{2}J(M,^{13}C) (Hz)
$^{27}Al^{3+}$–Otf–CO_3^{2-} [c] [30,36,37]	−2.3 (N)	140	
	−3.8 (C)	170	
$^{27}Al^{3+}$–Tf–CO_3^{2-} [c] [36]	−2.8	180	
$^{27}Al^{3+}$–Ltf–CO_3^{2-} [c] [36]	−4.3	200	
$^{27}Al^{3+}$–Otf–$C_2O_4^{2-}$ [c] [30]	−4.4 (N)	170	
	−5.7 (C)	200	
$^{205}Tl^{3+}$–Htf–CO_3^{2-} [d] [34,43]	2055 (C)	100	265
	2075 (N)	100	290
$^{205}Tl^{3+}$–Htf–CO_3^{2-} [e] [12]	2055 (C)		285
	2072 (N)		278
$^{205}Tl^{3+}$–Otf–CO_3^{2-} [e] [12]	2054 (C)		281
	2075 (N)		281
$^{205}Tl^{3+}$–Otf–$C_2O_4^{2-}$ [e] [12]	2100 (C)		32, 21[f]
	2103 (N)		28, 14[f]
$^{205}Tl^{3+}$–Htf–$C_2O_4^{2-}$ [e] [12]	–		27, 15[f,g]
$^{51}V^{5+}$–Htf–CO_3^{2-} [c] [39,40]	−529.5 (C)	200	
	−531.5 (N)	310	
$^{45}Sc^{3+}$–Otf–CO_3^{2-} [c] [38]	77 (C)	940	
	85 (N)	740	
$^{45}Sc^{3+}$–Htf–CO_3^{2-} [c] [38]	76	1020	
$^{45}Sc^{3+}$–Otf–$C_2O_4^{2-}$ [c] [38]	65 (C)	1030	
	70 (N)	1160	
$^{113}Cd^{2+}$–Otf–CO_3^{2-} [e] [11]	21.6	100	22
$^{113}Cd^{2+}$–Htf–CO_3^{2-} [e] [11]	11.7	100	20
$^{113}Cd^{2+}$–Htf–CO_3^{2-} [c] [54]	38.1 (N)	313	
	43.6 (C)	377	
$^{113}Cd^{2+}$–Otf–$C_2O_4^{2-}$ [e] [32]	54 (C)	150	
	– (N)[h]		
$^{113}Cd^{2+}$–Otf–Gly[e] [33]	139	320	

[a]Htf: human serum transferrin; Otf: ovotransferrin; Ltf: human lactoferrin. Assignments to the N- and C-site are reported in parentheses.
[b]The following references were used: ^{27}Al: 1 M $Al(NO_3)_3$ in D_2O; ^{205}Tl: Tl^{+} ion at infinite dilution; ^{51}V: $VOCl_3$; ^{45}Sc: 0.1 M $Sc(H_2O)_6^{3+}$; ^{113}Cd: 0.1 M $Cd(ClO_4)_2$ in D_2O. Assignment to the N- and C-site are reported in parentheses, when available.
[c]11.7 T.
[d]1.4 T.
[e]4.7 T.
[f]Inequivalent carbons.
[g]N-terminal half-molecule.
[h]Broadened beyond detection.

which is now indeed a widely accepted step in the mechanism of metal uptake of transferrins. A positively charged arginine residue was suggested as one of the anion binding groups for carbonate,[27] as later confirmed by X-ray data. The pH dependence of the ^{1}H NMR resonances of the histidine C(2)H protons in diamagnetic Ga(III) derivatives of Otf, Htf, and their half-molecules in the presence of dicarboxylic acids as synergistic anions indicated a histidine as the likely anion binding group in each site.[5–7,28] Interestingly, the X-ray structure of human Ltf allows for binding of anions larger than carbonate not to the arginine involved in the carbonate derivative, but to polar residues of a cavity beyond such a residue, which actually also includes a histidine residue.[3]

Further ^{13}C NMR studies of diamagnetic Al^{3+}–, Ga^{3+}– and Zn^{2+}–transferrin derivatives with carbonate and oxalate as the synergistic anion[29,30] were interpreted in terms of the anion bridging the metal and a positively charged group of the protein, in line with the 'interlocking-sites' model (Figure 2).[31] Oxalate bound in the Ga^{3+}–and Al^{3+}–Otf derivatives gave rise to two partially overlapped AB signal patterns.[29] The inequivalence of the two carbon atoms was assumed to arise from oxalate binding to the metal via only one carboxylate group, the other interacting with a positive residue of the site. Reference to ^{13}C chemical shift data of simple inorganic oxalate complexes allowed the assignment of the resonances of the protein-bound and metal-bound carbons of each oxalate molecule.[29,30] In the case of Zn^{2+}–[29] and Cd^{2+}–Otf–oxalate derivatives,[32] the effect of the metal on the shielding of the carbon atom is more similar to that of the protein residue, leading to the equivalence of the two carbon atoms (see ***Magnetic Equivalence***), hence to the observation of a single signal for each protein-bound anion.

Overall, although subject to a successive reevaluation (see below), these ^{13}C NMR data strongly supported the bridging arrangement of the synergistic anion prior to publication of the X-ray data.[2,3] Moreover, more recent ^{113}Cd NMR spectra of the ^{113}Cd–Otf derivative with glycine as the synergistic anion unequivocally indicated the binding of the amino group of the amino acid to the metal, fully in line with the interlocking sites model.[33]

Observation of the magnetic coupling between an $I = \frac{1}{2}$ metal and the carbon atom of the anion in both metal sites

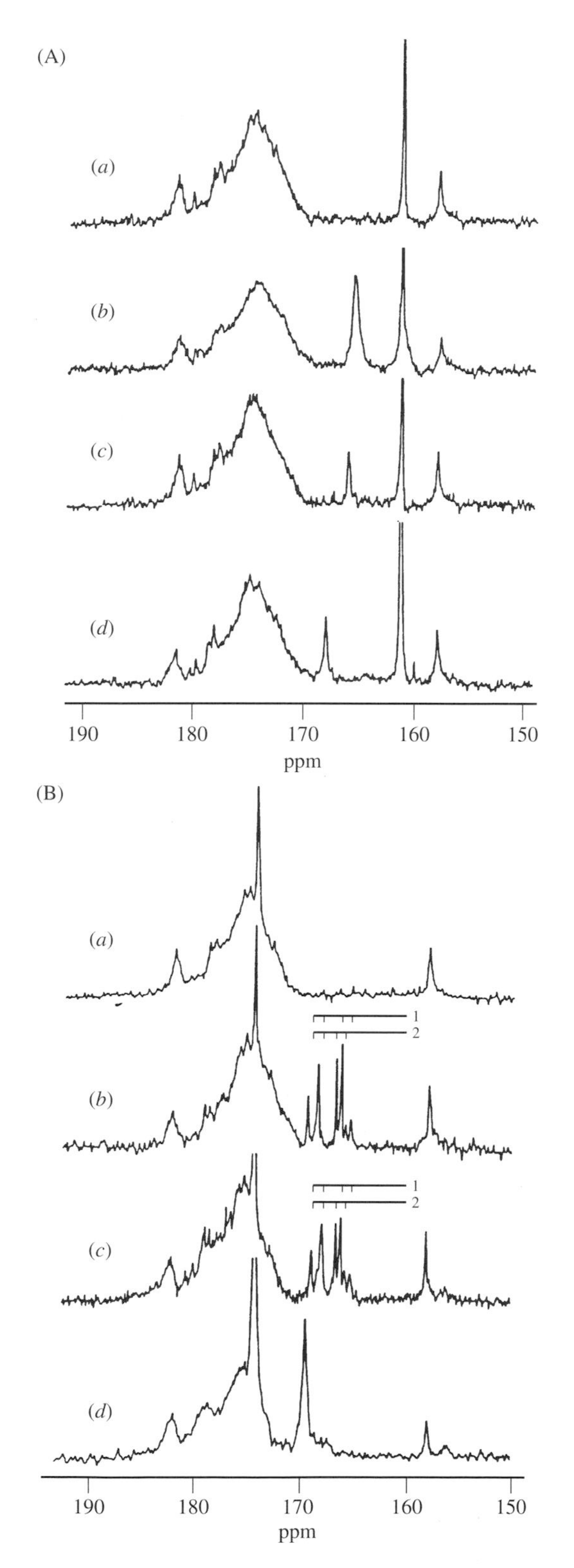

Figure 2 75.4 MHz ^{13}C NMR spectra of the carbonyl region of (A) Otf–carbonate–metal complexes and (B) Otf–oxalate–metal complexes in the presence of excess anion, pH 8 (protein concentration 1.5 mM): (*a*) apoprotein; (*b*) Al^{3+} derivatives; (*c*) Ga^{3+} derivatives; and (*d*) Zn^{2+} derivatives. (Reproduced by permission of the American Chemical Society from I. Bertini, C. Luchinat, L. Messori, A. Scozzafava, G. C. Pellacani, and M. Sola, *Inorg. Chem.*, 1986, **25**, 1782)

yielded conclusive spectroscopic evidence of a direct metal–ion linkage. In particular, this was achieved by using ^{205}Tl and ^{113}Cd ions loaded in one or both metal sites (Figure 3).[11,12,34] The ^{2}J coupling constants are collected in Table 2.

Thallium–transferrin derivatives allowed a reevaluation of the binding mode of oxalate, and led to further insight into the chemistry of the ternary adduct in the protein sites. The ^{13}C NMR signals of bound oxalate in each site of the $^{205}Tl^{3+}$ derivatives of human Tf and Otf are split into a doublet of doublets due to carbon–carbon and thallium–carbon coupling.[12] The comparable values of the ^{13}C–metal coupling constants for the two oxalate carbons (Table 2) indicate that the anion binds to the metal in a 1,2-bidentate mode at both sites. This is likely to be the case for all the oxalate adducts of transferrins with tripositive metals, as recently suggested from the similarity of the ^{13}C NMR spectral patterns for these adducts.[12] This view receives conclusive support from the X-ray structure of the Cu^{2+}–oxalate ternary adduct with the C-terminal half-molecule of Ltf,[35] which confirms the 1,2-bidentate bridging arrangement of the oxalate anion and indicates that the spectral inequivalence of the two carbon atoms in the ternary adducts with tripositive metals arises from differences in the protein residues H bonded to the two carboxylate groups. In the light of these results, the ^{13}C spectra of the $^{113}Cd^{2+}$–oxalate adduct of Otf, for which, unlike the carbonate derivative,[11] no ^{113}Cd–^{13}C coupling could be detected,[32] consistent with a longer metal–carbon distance, can also be interpreted in terms of the above anion arrangement.

Recent work on NMR of quadrupolar metal nuclei (^{51}V, ^{27}Al, and ^{45}Sc) (see ***Aluminum-27 NMR of Solutions***; ***Quadrupolar Transition Metal & Lanthanide Nuclei***) loaded in the sites of transferrins deserve a special mention.[12,30,36–40] For an $I = \frac{n}{2}$ nucleus (n = 3, 5, 7) firmly bound to a slowly rotating macromolecule, far from the extreme narrowing limit, it is shown that only the central $\frac{1}{2} \rightarrow -\frac{1}{2}$ nuclear spin transition can be observed (see ***Quadrupolar Interactions***) (Figure 4). Under these conditions the field dependence of the linewidth and chemical shift (see ***Dynamic Frequency Shift***) provides the values of the quadrupole coupling constant, χ, and of the rotational correlation time, τ_c, for the protein-bound metals, and hence yields information on the symmetry of the metal coordination environment. For example, independent of the synergistic anion, the Al^{3+} ion appears to be octahedrally coordinated in both sites, with a higher degree of symmetry in the N-site as compared to the C-site.[30] Moreover, the metal sites of the Sc^{3+}–Otf derivative show a higher symmetry than those in the homologous Al^{3+} derivative.[38]

Since detection of the resonances of these quadrupolar nuclei is facilitated by increasing the magnetic field and the molecular size of the protein, the technique is of general interest for probing the properties of the metal binding sites of large metalloproteins.

2.4 The (In)Equivalence of the Two Binding Sites

Early water proton relaxation measurements of Fe^{3+} binding to Otf, coupled with electrophoretic profiles, failed to detect the difference in iron binding ability of the two sites,[41] indicating a random distribution of iron in the two lobes.[14] Inequivalences can be detected with the aid of paramagnetic metals like Co^{2+} [8,9] and lanthanides,[42] taking advantage of the

For References see p. 4818

sensitivity of hyperfine shifts to even slight changes in the coordination sphere.[10] The advent of high-field Fourier transform spectrometers allowed the safe determination of site (in)equivalence simply from the presence of either one or two NMR signals (or signal patterns) due to the ^{13}C-labeled synergistic anion and/or NMR-active diamagnetic metals bound to the protein. A metal titration then generally indicated if sequential metal binding had occurred, and in some instances assignment of the NMR resonances to the individual sites allowed the site with higher metal affinity to be determined. This approach has been followed for the derivatives of Otf and Htf with Co^{3+}, Ga^{3+}, Al^{3+}, Tl^{3+}, Sc^{3+}, V^{5+}, Cd^{2+}, Zn^{2+} in combination with different synergistic anions, most often carbonate and oxalate. The NMR-active metals and the ^{13}C-labeled synergistic anion loaded in the two sites led to either overlapping peaks or distinct, but invariably closely spaced, resonances indicative of equivalent (within the resolution available) or slightly different chemical environments, respectively (Tables 1 and 2). The higher metal affinity of the C-site over the N-site at acidic pH was, in general, exploited to obtain resonance assignment to individual sites, although recently the use of proteolytically-obtained half-molecules has proved to be a safer means for such an assignment. This was evidenced for the derivatives of human Tf and Otf with Al^{3+}, Tl^{3+}, and Sc^{3+}, for which the metal resonances and the ^{13}C signals of the synergistic anion in the bimetallic derivative were invariably found to exactly superimpose the signals of the two one-sited half-molecules.[12,30,36–38]

These extensive NMR data concurred to indicate that the relative properties of the two sites for a given transferrin vary with the nature of the metal, the synergistic anion, pH, and medium composition, and may differ from those of other transferrins. The following selected examples are worth mentioning. Aluminum–27 and ^{13}C NMR showed that in the presence of carbonate the two aluminum metal sites of Otf are slightly inequivalent and that Al^{3+} binds preferentially to the N-site at pH 8, while the site preference is reversed with oxalate.[30] In contrast, spectroscopically equivalent sites are detected in the homologous derivatives of human Tf and Ltf.[36] At pH values slightly above neutrality in the presence of carbonate, the Tl^{3+} ion was shown at an early stage,[43] unlike Al^{3+}, to bind more strongly to the C-site of human Tf, and more recently to bind with the same affinity to both sites of Otf.[12] Moreover, Sc^{3+} showed no site preference toward Otf with either carbonate or oxalate as synergistic anions,[38] and site inequivalence in the Cd^{2+} derivative of Tf with oxalate appeared more pronounced than in the homologous Zn^{2+} derivative.[11,32]

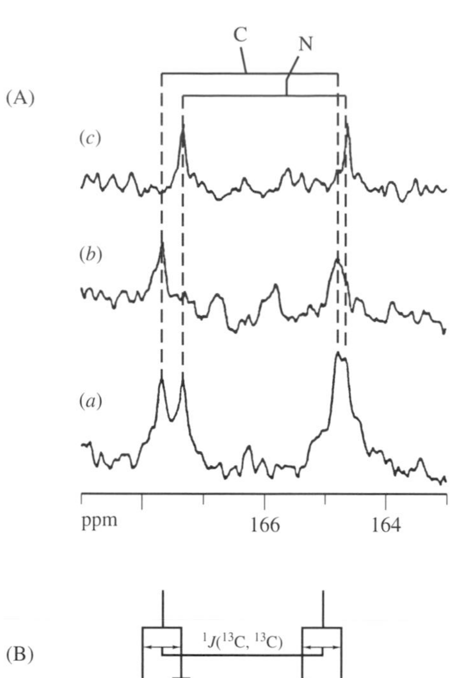

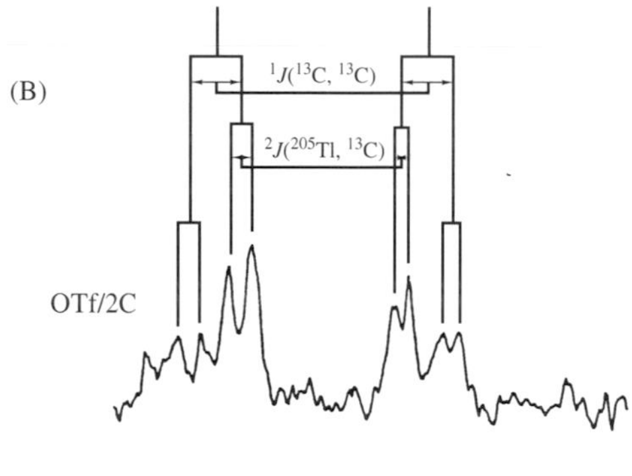

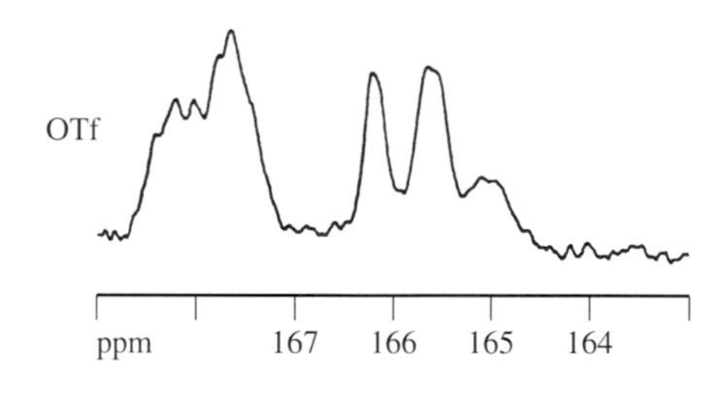

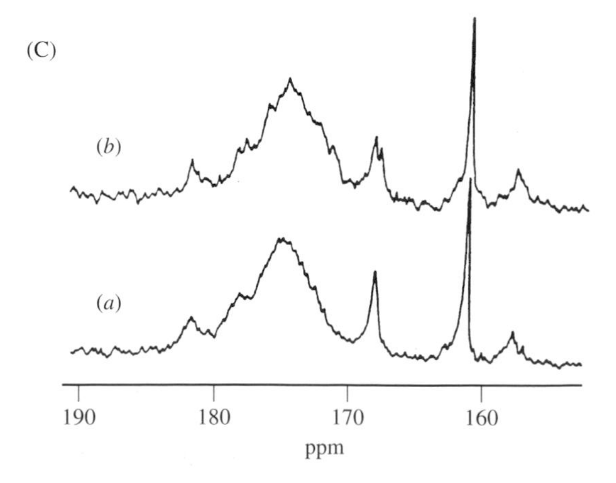

Figure 3 (A) Expanded carbonyl region of the 100.6 MHz ^{13}C NMR spectra of (*a*) Tl_2–Tf–CO_3^{2-} derivative; (*b*) Tl–Tf–carbonate derivative with Tl^{3+} loaded in the C-site; (*c*) Tl–Tf/2N–carbonate derivative. (Reproduced by permission of the American Chemical Society from J. M. Aramini, P. H. Krygsman, and H. J. Vogel, *Biochemistry*, 1994, **33**, 3304). (B) Expanded carbonyl regions of the 125.7 MHz ^{13}C NMR spectra of the Tl_2–Otf–oxalate derivative and of the Tl–Otf/2C–oxalate derivative. (Reproduced by permission of the American Chemical Society from J. M. Aramini, P. H. Krygsman, and H. J. Vogel, *Biochemistry*, 1994, **33**, 3304). (C) Carbonyl regions of the 50.3 MHz ^{13}C NMR spectrum of the Cd_2–Otf–carbonate derivatives containing (*a*) natural-abundance Cd^{2+} ion and (*b*) isotopically-enriched $^{113}Cd^{2+}$ ion. (Reproduced by permission of the American Chemical Society from M. Sola, *Inorg. Chem.*, 1990, **29**, 1113)

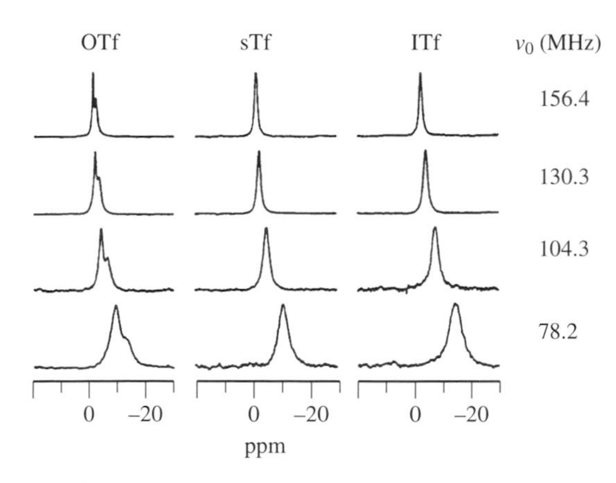

Figure 4 ^{27}Al NMR spectra of the Al_2 derivatives of Otf, human Tf, and human Ltf at different magnetic fields. (Reproduced by permission of the American Chemical Society from J. M. Aramini, M. W. Germann, and H. J. Vogel, *J. Am. Chem. Soc.*, 1993, **115**, 9750)

Although this overview demonstrates NMR to be a powerful tool for probing the differences in structural properties and binding affinity of the two metal sites, at present, as noted elsewhere,[44] it does not address unequivocally the question of the presence of a functional association between the lobes. The above-mentioned near-identity of the spectral properties of a given site in the half and intact transferrin molecules does not necessarily imply that the metal affinity of the sites remains unchanged upon proteolytic cleavage.

The application of resolution enhancement techniques, although allowing resolution of only a small fraction of the total proton resonances, particularly those belonging to relatively mobile portions of the protein, has recently enabled one- and two-dimensional ^1H NMR (see ***COSY Two-Dimensional Experiments***; ***Multidimensional Spectroscopy: Concepts***; ***NOESY***) to monitor the binding of diamagnetic metals to whole and half-transferrin molecules.[45–47] Low-frequency methyl peaks probably experiencing aromatic ring current shifts, glycan *N*-acetyl and sugar ring protons, Lys/Arg ϵ/δ-CH_2 protons, and His C-2,H resonances could be resolved in the apo form of human Tf.[47] The disappearance of some of these peaks and the rise of new resonances in a slow exchange regime upon metal titration of the apo form of whole human Tf and its N-lobe (HTf/2N hereafter) allowed the detection of the sequential binding of Al^{3+} and Ga^{3+} to Tf, with preferential binding of the former metal to the N-site (in the presence of carbonate, pH 8.8),[46] and of the latter to the C-site (with oxalate as a synergistic anion, pH 7.2).[45] Examples of one- and two-dimensional proton spectra are illustrated in Figure 5.[45]

Since X-ray data indicate that the structural features of the N-lobe are likely to be closely similar in the intact protein and in the isolated lobe,[2] the lack of a perfect match between the spectral features of the apo form and the Al^{3+} adducts of Tf and Tf/2N was interpreted as indicative of an interaction between the lobes in the intact protein.[46] This approach also allowed a determination of the apparent binding constant of oxalate to apo-HTf/2N ($\log K = 4.04 \pm 0.09$) from the shift of some resonances induced by the weak binding of the anion, as a consequence of the fast exchange between the free and anion bound species on the NMR timescale.[45] Moreover, the time course of the paramagnetic broadening of the proton resonances in the Ga^{3+}–oxalate–HTf/2N adduct upon Fe^{3+} addition allowed the kinetics of Ga^{3+} displacement ($k_{obs} = 0.162\ h^{-1}$ at 310 K) to be determined.[45]

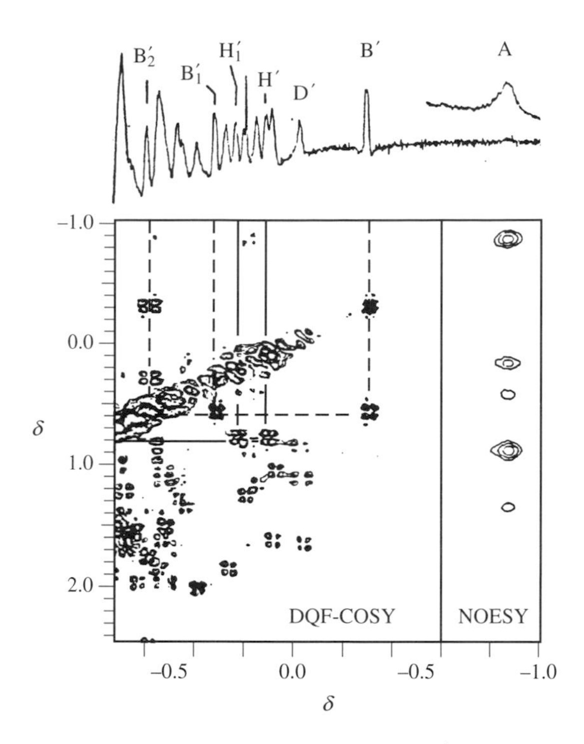

Figure 5 600-MHz phase-sensitive DQF-COSY ^1H NMR spectrum (subject to resolution enhancement) of the Ga^{3+}–Tf–oxalate adduct, together with a NOESY spectrum for the lowest frequency peak. (Reproduced by permission of the American Chemical Society from G. Kubal, A. B. Mason, S. U. Patel, P. J. Sadler, and R. C. Woodworth, *Biochemistry*, 1993, **32**, 3387)

2.5 Other Structural Features in Solution

Tentative assignments of the low-frequency methyl proton resonances to individual Leu, Val, and Ile residues were proposed for human Tf and its N-lobe through 2D COSY techniques (see ***COSY Two-Dimensional Experiments***) and calculation of the shifts induced by the ring currents of nearby aromatic groups (see ***Shielding: Overview of Theoretical Methods***).[45,46] The shifts of these resonances upon metal binding appear to be related to changes in their orientation toward the aromatic groups which reside in relatively mobile, yet structured hydrophobic domains possibly involved in conformational changes related to the closure of the interdomain cleft.[45] Unambiguous signal assignments and determinations of the precise movements of single residues are still unavailable, but much effort is being devoted toward this end.

Nonsynergistic anions like chloride, perchlorate, thiocyanate, and pyrophosphate are known to decrease the stability of various metal–transferrin derivatives, and, the chloride ion in particular, to affect the relative metal binding affinity of the sites.[48] They most probably bind to positively charged residues in the proximity of the metal site, altering the network of salt bridges and hydrogen bonds that stabilize the complex. The effect of perchlorate in decreasing the stability of the metal complexes was clearly monitored by the decrease of the ^{13}C

For References see p. 4818

and ^{113}Cd NMR resonances of the $^{113}Cd_2$–Otf–CO_3^{2-} adduct upon anion addition, due to dissociation of the complex.[11] Chlorine-35 and ^{14}N NMR linewidth measurements of chloride, perchlorate, and nitrate anion binding to human Tf in a fast exchange regime allowed the determination of a 2:1 stoichiometry for V^{5+} binding to the protein, since the formation of the ternary metal–carbonate complex at the sites determines the specific anion binding to the protein and hence leads to a decrease in quadrupolar relaxation (see ***Relaxation Theory for Quadrupolar Nuclei***).[49]

Proton NMR has also played an important role in the determination of the primary structure on the glycan moieties of a variety of transferrins and in the determination of solution conformation through 2D techniques.[50–53] Studies were carried out on glycopeptides obtained by proteolytic digestion or on the intact oligosaccharide chain detached from the protein by hydrazinolysis. Readers are referred to the original literature as far as the assignment procedures are concerned. However, it is worth mentioning that recent ^1H NMR studies on Ga^{3+} binding to human Tf indicate that metal binding alters the shift of the resonances of the *N*-acetylated sugar residues of both glycan chains, suggesting that the transmission of the perturbation to the carbohydrate chains may have significance for cell receptor recognition of the metal-loaded protein.[45]

3 THE ROLE OF NMR IN THE CHEMISTRY OF TRANSFERRINS: FUTURE PROSPECTS

The applicability of NMR techniques becomes less and less trivial with increase in molecular size. From this point of view, molecules of the size of transferrins represent a challenge. Multidimensional NMR, coupled with resolution enhancement techniques, of either intact and half-transferrin molecules, appear as a promising approach for detecting structural changes in the individual lobes upon metal uptake or release and ionization equilibria, and should be of use for monitoring the protein behavior toward the binding of a variety of metals and synergistic anions, under different conditions of pH and medium composition. The bottom line is to understand whether the two lobes undertake different roles in metal uptake and release processes, and whether they are functionally associated, and hence to address one of the most intriguing points of the biochemistry of transferrins related to their structural 'doubleness'. The advent of very high field spectrometers (750 MHz or higher) would greatly enhance the diagnostic power of this approach, together with the feasibility of NMR of quadrupolar metal nuclei which has proved highly sensitive to the structural features of the metal sites. The use of genetically-engineered proteins containing either amino acid substitutions, deletions, or labeled residues could also add essential information. Nitrogen-15 and/or ^{13}C selective and nonselective labeling is indeed providing a breakthrough to structural studies on medium sized proteins (see ***Heteronuclear Assignment Techniques***; ***Heteronuclear Shift Correlation Spectroscopy***).

Besides metal binding, NMR should also be a valuable means of monitoring the potential ability of transferrins to accommodate small molecules in the binding cleft.[3] Overall, the application of ^1H NMR to determine protein structure in solution, coupled with molecular dynamics calculations (although the size of transferrin would set this as a long-term project) appears to be the main research direction toward the elucidation of the structure–function relationships in this class of proteins.

4 RELATED ARTICLES

Calcium-Binding Proteins; Carbonic Anhydrase; Cobalt(II)- & Nickel(II)-Substituted Proteins; Copper Proteins; Field Cycling Experiments; Heme Peroxidases; Hemoglobin; Iron–Sulfur Proteins; Mitochondrial Cytochrome *c*; Myoglobin; Nonheme Iron Proteins; Proteases.

5 REFERENCES

1. D. C. Harris and P. Aisen, in 'Physical Bioinorganic Chemistry', ed. T. M. Loehr, VCH Publishers, New York, 1989, Vol. 2, p. 239.
2. R. Sarra, R. Garratt, B. Gorinski, H. Jhoti, and P. Lindley, *Acta Crystallogr.*, 1990, **B46**, 763.
3. B. F. Anderson, H. M. Baker, G. E. Norris, D. W. Rice, and E. N. Baker, *J. Mol. Biol.*, 1989, **209**, 711.
4. R. C. Woodworth, K. G. Morallee, and R. J. P. Williams, *Biochemistry*, 1970, **9**, 839.
5. B. M. Alsaadi, R. J. P. Williams, and R. C. Woodworth, *J. Inorg. Biochem.*, 1981, **15**, 1.
6. A. A. Valcour and R. C. Woodworth, *Biochemistry*, 1987, **26**, 3120.
7. R. C. Woodworth, N. D. Butcher, S. A. Brown, and A. Brown Mason, *Biochemistry*, 1987, **26**, 3115.
8. I. Bertini, C. Luchinat, L. Messori, R. Monnanni, and A. Scozzafava, *J. Biol. Chem.*, 1986, **261**, 1139.
9. L. Messori and M. Piccioli, *Inorg. Chim. Acta*, 1990, **174**, 137.
10. I. Bertini and C. Luchinat, 'NMR of Paramagnetic Molecules in Biological Systems', Benjamin/Cummings, Menlo Park, CA, 1986.
11. M. Sola, *Inorg. Chem.*, 1990, **29**, 1113.
12. J. M. Aramini, P. H. Krygsman, and H. J. Vogel, *Biochemistry*, 1994, **33**, 3304.
13. A. Wishnia, *J. Chem. Phys.*, 1960, **32**, 871.
14. P. Aisen, A. Leibman, and H. A. Reich, *J. Biol. Chem.*, 1966, **241**, 1666.
15. S. H. Koenig and W. E. Schillinger, *J. Biol. Chem.*, 1969, **244**, 3283.
16. S. H. Koenig and W. E. Schillinger, *J. Biol. Chem.*, 1969, **244**, 6520.
17. B. P. Gaber, W. E. Schillinger, S. H. Koenig, and P. Aisen, *J. Biol. Chem.*, 1970, **245**, 4251.
18. I. Bertini, C. Luchinat, M. Mancini, and G. Spina, in 'Magneto-Structural Correlations in Exchange-Coupled Systems', ed. D. Gatteschi, O. Kahn, and R. D. Willet, Reidel, Dordrecht, 1985, p. 421.
19. L. Banci, I. Bertini, and C. Luchinat, 'Nuclear and Electron Relaxation. The Magnetic Nucleus–Unpaired Electron Coupling in Solution', VCH, Weinheim, 1991.
20. I. Bertini, F. Briganti, S. H. Koenig, and C. Luchinat, *Biochemistry*, 1985, **24**, 6287.
21. P. B. O'Hara and S. H. Koenig, *Biochemistry*, 1986, **25**, 1445.
22. I. Bertini, O. Galas, C. Luchinat, L. Messori, and G. Parigi, *J. Phys. Chem.*, in press.
23. S. H. Koenig and R. D. Brown, III, in 'The Coordination Chemistry of Metalloenzymes', ed. I. Bertini, R. S. Drago and C. Luchinat, Reidel, Dordrecht, 1983, p. 19.
24. I. Bertini, C. Luchinat, M. Mancini, and G. Spina, *J. Magn. Reson.*, 1984, **59**, 213.

25. I. Bertini, C. Luchinat, R. D. Brown, III, and S. H. Koenig, *J. Am. Chem. Soc.*, 1989, **111**, 3532.
26. D. C. Harris, G. A. Gray, and P. Aisen, *J. Biol. Chem.*, 1974, **249**, 5261.
27. J. A. Zweier, J. B. Wooten, and J. S. Cohen, *Biochemistry*, 1981, **20**, 3505.
28. R. C. Woodworth, *J. Inorg. Biochem.*, 1986, **28**, 245.
29. I. Bertini, C. Luchinat, L. Messori, A. Scozzafava, G. C. Pellacani, and M. Sola, *Inorg. Chem.*, 1986, **25**, 1782.
30. J. M. Aramini and H. J. Vogel, *J. Am. Chem. Soc.*, 1993, **115**, 245.
31. G. W. Bates and M. R. Schlabach, *J. Biol. Chem.*, 1975, **250**, 2177.
32. M. Sola, *Eur. J. Biochem.*, 1990, **194**, 349.
33. G. Battistuzzi and M. Sola, *Biochim. Biophys. Acta*, 1992, **1118**, 313.
34. I. Bertini, L. Messori, G. C. Pellacani, and M. Sola, *Inorg. Chem.*, 1988, **27**, 761.
35. M. S. Shongwe, C. A. Smith, E. W. Ainscough, H. M. Baker, A. M. Brodie, and E. N. Baker, *Biochemistry*, 1992, **31**, 4451.
36. J. M. Aramini, M. W. Germann, and H. J. Vogel, *J. Am. Chem. Soc.*, 1993, **115**, 9750.
37. J. M. Aramini and H. J. Vogel, *Bull. Magn. Reson.*, 1993, **15**, 84.
38. J. M. Aramini and H. J. Vogel, *J. Am. Chem. Soc.*, 1994, **116**, 1988.
39. A. Butler and H. Eckert, *J. Am. Chem. Soc.*, 1989, **111**, 2802.
40. A. Butler and M. J. Danzitz, *J. Am. Chem. Soc.*, 1987, **109**, 1864.
41. P. Aisen, A. Leibman, and J. Zweier, *J. Biol. Chem.*, 1978, **253**, 1930.
42. L. Messori and M. Piccioli, *J. Inorg. Biochem.*, 1991, **42**, 185
43. I. Bertini, C. Luchinat, and L. Messori, *J. Am. Chem. Soc.*, 1983, **105**, 1347.
44. M. Sola, *Chemtracts—Inorg. Chem.*, 1993, **5**, 201.
45. G. Kubal, A. B. Mason, S. U. Patel, P. J. Sadler, and R. C. Woodworth, *Biochemistry*, 1993, **32**, 3387.
46. G. Kubal, A. B. Mason, P. J. Sadler, A. Tucker, and R. C. Woodworth, *Biochem. J.*, 1992, **285**, 711.
47. G. Kubal and P. J. Sadler, *J. Am. Chem. Soc.*, 1992, **114**, 1117.
48. N. D. Chasteen, *Adv. Inorg. Biochem.*, 1983, **5**, 201.
49. N. D. Chasteen, J. K. Grady, and C. E. Holloway, *Inorg. Chem.*, 1986, **25**, 2754.
50. S. W. Homans, R. A. Dwek, D. L. Fernandes, and T. W. Rademacher, *Biochim. Biophys. Acta*, 1983, **760**, 256.
51. H. Van Halbeek, L. Dorland, J. F. G. Vliegenthart, G. Spik, A. Cheron, and J. Montreuil, *Biochim. Biophys. Acta*, 1981, **675**, 293.
52. L. Dorland, J. Haverkamp, J. F. G. Vliegenthart, G. Spik, B. Fournet, and J. Montreuil, *Eur. J. Biochem.*, 1979, **100**, 569.
53. L. Dorland, J. Haverkamp, B. L. Schut, J. F. G. Vliegenthart, G. Spik, G. Strecker, B. Fournet, and J. Montreuil, *FEBS Lett.*, 1977, **77**, 15.
54. W. Kiang, P. J. Sadler, and D. G. Reid, *Magn. Reson. Chem.*, 1993, **31**, 5110.

Biographical Sketches

Claudio Luchinat. *b* 1952. Doctor in Chemistry, 1976, University of Florence. Introduced to NMR by I. Bertini. Faculty positions at University of Florence, 1981–86. Professor of Chemistry, University of Bologna, 1986–present. More than 200 publications and two books. Research interests: NMR of paramagnetic species, theory of electron and nuclear relaxation, biological applications of NMR, bioinorganic, biophysical and environmental chemistry.

Marco Sola. *b* 1957. B.S., 1984, University of Modena, Ph.D., 1987, University of Parma. Introduced to NMR by I. Bertini and C. Luchinat. Faculty in Chemistry, University of Modena, 1990–92. Associate Professor of Chemistry, University of Basilicata, 1992–1994. University of Bologna, 1994–present. More than 60 publications. Research interests: NMR of hydrolytic enzymes and metalloproteins involved in metal transport and electron transport, and NMR of model complexes.

Tritium NMR

Mark G. Kubinec & Philip G. Williams

Lawrence Berkeley National Laboratory, Berkeley, CA, USA

1 BACKGROUND

The fundamental magnetic properties of tritium (^3H) were first reported in 1947[1–3] in a series of NMR measurements which demonstrated the triton to be a spin-$\frac{1}{2}$ nucleus, with a positive magnetic moment about 6.66% larger than that of the proton. Despite the fact that these measurements established tritium as the most sensitive NMR-active nucleus, very little research was conducted over the next two decades. In short, we have found only 24 reports of tritium spectra in the first 30 years, and 14 of those were published in the period 1974–76. The lack of activity in the early years of tritium NMR development is readily rationalized—tritium is a radioactive nucleus and, with the limited sensitivity of early spectrometers, the quantities required to make the 1947 measurements were enormous. [No specific details of sample sizes were given, but 50 μL of HTO with 50% abundance of ^3H contains ca. 3000 GBq of radioactivity, and we estimate that the samples must have been of this order, i.e. 1850–18 500 GBq (50–500 Ci).] Hence, applications were limited to US National Laboratories, where facilities existed for obtaining and handling large quantities of tritium.

The first liquids ^3H NMR spectrum other than HTO was described in 1964,[4] and set several important precedents for the newly developing field: (i) samples required large amounts of tritium (ca. 370 GBq in this instance); (ii) instrumental modification for ^3H detection was simple; (iii) proton–tritium

For References see p. 4830

couplings were obvious and readily measured; and (iv) the labeling pattern (or percentage of ^{3}H in various positions) was accessible from the spectrum. The sample used in this experiment was ethylbenzene prepared by metal-catalyzed reduction of phenylacetylene with tritium gas, and interestingly, the integrated ratio of methylene to methyl tritium was 1:1.5, not 1:1 as predicted. This result established high-resolution ^{3}H NMR as the tool of choice for analysis of tritiated products. The 'nontextbook' labeling pattern would have been difficult to detect and almost certainly overlooked using the conventional methods of the time, and incorrect assumptions about the labeling pattern would have affected downstream uses of the product. Indeed, biochemical applications of tritium have been slow to recover from the poorly characterized materials used in the 1950s and 1960s, when ^{3}H NMR was not available for product analysis.

Such demonstrations of the power of ^{3}H NMR spectroscopy led to a renewed interest in technical development in the late 1960s and early 1970s. Fortunately, the resurgence coincided with advances in instrumentation that allowed many other low abundance and/or low sensitivity nuclei to be routinely observed. This revolution in NMR spectrometer sensitivity was crucial. Since tritium sensitivity is >75-times that of ^{13}C, the ability to collect ^{13}C spectra at natural abundance clearly implies that tritium spectra could be obtained at the very low ^{3}H concentrations necessary to ensure the technique is safe and accessible to laboratories around the world. During this period, the sample requirement of 370–3700 GBq was drastically reduced to the range of 3.7–3700 MBq, depending on the complexity of the signals and the S/N level required. Of course, tritium detection by NMR still suffers from the inherent insensitivity of the technique, and some perspective can be gained from the fact that liquid scintillation counting (LSC) techniques can detect tritium at a level 10^{-7} lower than the highest field spectrometers of today. Yet, the detailed information which NMR yields about the chemical and magnetic environment of the nuclei makes this loss of sensitivity easy to justify.

The most influential force in the promotion of ^{3}H NMR spectroscopy as a routine research tool came from a fruitful collaboration between the University of Surrey and Amersham International. The early published work (1971–76) concentrated on methodology and proof of principle, and in the late 1970s a host of applications were reported. A number of other groups adopted the ^{3}H NMR approach, and by 1985 the Surrey group was able to publish an excellent compilation of techniques and results, describing a mature field.[5] In this article, we will discuss the major features of ^{3}H NMR spectroscopy, and address the more recent, popular, and significant chemical applications.

2 SAFETY

Tritium is the very rare, but naturally occurring, radioactive isotope of hydrogen. It is a pure β-emitter and its 12.4 year halflife is convenient for most chemical or biological experiments. In Table 1 we list a variety of chemical and physical properties of tritium and elementary tritiated compounds.

As mentioned above, the large amounts of tritium necessary for early experiments made containment of tritiated NMR samples a point of major concern, and a number of convenient containment systems have evolved.[5] Since tritium is a 'soft' β emitter, i.e. the energy of the β particle associated with its radioactive decay has very little penetrating power (Table 1, E_{max} = 18.6 keV, 6 μm range in water), it is readily contained by a thin surface of plastic, Teflon, or glass. The major hazard is from ingestion, so care should be taken to keep samples sealed and clean on the exterior. We favor a double encapsulation approach using either a 3-mm glass tube inside a regular NMR tube, or a capped Teflon liner inside a sealed glass tube. These components are all commercially available.

Since the number of tritium nuclei needed for NMR detection exceeds that required for liquid scintillation counting by a factor of ca. 10^{7}, the monitoring of equipment or personnel for contamination is readily accomplished by standard radiochemical techniques. At the minimum level readily detectable by liquid scintillation counting (LSC) (37 Bq L^{-1}) in a conventional urine analysis, the radiation exposure for an individual is less than 0.001 mSv y^{-1}. The sensitivity of LSC thus provides

Table 1 General Physical and Chemical Data for Tritium, and Common Radioactivity Units[a]

Production	^{6}Li(n, α)^{3}H
Radiation	β(100%); range = 4.5–6 mm in air, 6 μm in H_2O
β Energy	E_{max} = 18.6 keV, E_{mean} = 5.7 keV
Radioactive halflife	12.43 y (4540 d)
Biological halflife	ca. 10 d
Decay product	^{3}He
Maximum specific activity	1064 GBq milliatom^{-1} (28.76 Ci milliatom^{-1})
Detection limit (in 24 h): LSC	0.037 Bq = 2.09 × 10^{7} atoms (ca. 0.1 fmol)
NMR	1064 kBq = 6.02 × 10^{14} atoms (ca. 1 nmol)
Diameter of tritium atom (approx.)	1.1 Å
Volume of 37 GBq (1 Ci) of tritium gas at standard temperature and pressure	0.385 mL
Radiation produced by 37 MBq (1 mCi) in a man (70 kg = 40 L, i.e. 25 μCi L^{-1})	0.044 mSv d^{-1} (= 4.4 mrem d^{-1}, or 1.6 rem y^{-1})
Density, triple point, and boiling point of isotopic water molecules	H_2O: 1.000 g mL^{-1}; 0.01 °C; 100.00 °C
	D_2O: 1.106 g mL^{-1}; 3.82 °C; 101.42 °C
	T_2O: 1.214 g mL^{-1}; 4.49 °C; 101.51 °C
Tritium content of pure T_2O at STP	117.4 TBq mL^{-1} (3173 Ci mL^{-1})
	96.7 TBq g^{-1} (2614 Ci g^{-1})

[a]37 gigabecquerels (GBq) = 1 curie (Ci) = 3.7 × 10^{10} disintegrations per second; 1 gray (Gy) = 10 000 erg g^{-1} = 100 rad (units of absorbed dose); 1 sievert (Sv) = 100 rem (units of radiation exposure).

a level of detection well below the average exposure from background, medical, and terrestrial sources (3 mSv y^{-1}) or the current recommendation for a US Department of Energy radiation worker (<20 mSv y^{-1}). In addition, tritium uptake in the body is cleared as tritiated water (HTO) with a 10 day biological halflife, and this period can be drastically reduced by increasing fluid intake or by diuretics (e.g. water, beer, or coffee).

The hazards associated with analyzing ^{3}H samples are real, but have been grossly overemphasized in the past. A rational assessment of the risks involved is in order. Tritium is readily contained and is physiologically benign at the levels used in most high-resolution NMR experiments. When making a benefit/hazard judgement of conducting research on a 2220 MBq (60 mCi) sample contained as described above, it is reasonable to remember that safety lights in aircraft contain up to 370 GBq (10 Ci) of tritium, a marine compass may have as much as 28 GBq (0.75 Ci), and commonly available night sights for rifles contain 2 GBq (50 mCi) of tritium. The spectroscopist may be at greater risk from the lock solvent or other chemical entity in the sample than from the radioactivity, e.g. in a 0.25 mL sample containing 2220 MBq of a tritiated compound in C_6D_6, there are 2 μmol of ^{3}H but 2.75 mmol of benzene. It is clearly less hazardous to handle a well-contained tritium sample than milligram quantities of neurotoxins and other biologically active compounds routinely analyzed in normal glass NMR tubes in many laboratories.

In summary, the handling and health physics of analyzing tritiated samples is an important but trivially solved consideration, allowing ^{3}H NMR spectroscopy to be readily executed in most laboratories having a pulsed Fourier transform NMR spectrometer.

3 FEATURES AND PARAMETERS

The characteristics of ^{3}H are compared with other common NMR active nuclei in Table 2. Tritium enjoys all the spectral advantages normally associated with spin-$\frac{1}{2}$ nuclei of high γ: narrow lines, high sensitivity, and good dispersion. A brief perusal of Table 2 indicates the strong similarity between triton and proton magnetic properties, with tritium having slightly higher sensitivity and dispersion, and similar relaxation characteristics. The similarity between nuclei extends to chemical shifts, allowing one to use the extensive library of proton chemical shifts when making tritium resonance assignments. Since tritium also has extremely low natural abundance, spectra are free of background signals and this, combined with the a priori knowledge of chemical shifts from analogous protons, makes assignment of tritium spectra particularly straightforward.

Many of the favorable properties of tritium are illustrated in the spectra in Figure 1, where a comparison of proton, deuterium, and tritium NMR spectra of labeled glucose is shown. The superior resolution and dispersion of tritium over deuterium, a spin-1 nucleus, is obvious. Note that this labeled molecule is low molecular weight, and the line broadening of deuterium signals for larger molecules is even more striking, hence eliminating deuterium NMR as a useful technique for the structural analysis of large molecules in solution. In addition, since the dominant relaxation mechanism for deuterium is quadrupolar, NOE determinations are impossible, and sensitivity in saturation transfer is reduced due to rapid spin–lattice relaxation. In contrast, tritium analysis greatly simplifies the proton spectrum in Figure 1(a), particularly around the solvent where a less fortuitous proton chemical shift (or sample temperature) would have completely obscured one of the glucose

Table 2 NMR Properties of ^{3}H and Common NMR Nuclei[a]

Nucleus	Natural abundance	Spin	γ	Resonant frequency	Relative sensitivity[b]
^{1}H	99.984	$\frac{1}{2}$	26.7510	300.13	1.00
^{2}H	0.0156	1	4.1064	46.07	9.65×10^{-3}
^{3}H	**$<10^{-16}$**	**$\frac{1}{2}$**	**28.5335**	**320.13**	**1.21**
^{13}C	1.10	$\frac{1}{2}$	6.7263	75.46	1.59×10^{-2}
^{15}N	0.37	$\frac{1}{2}$	−2.7116	30.40	1.04×10^{-3}
^{19}F	100	$\frac{1}{2}$	25.1665	282.23	0.83
^{31}P	100	$\frac{1}{2}$	10.8289	121.43	6.63×10^{-2}

Chemical shift (δ) range:	20 ppm
Isotope effects:	^{3}H primary, 1° $\approx$ 0.028 ppm; 9.0 Hz at 7.1 T
	^{3}H secondary, 2° $\approx$ 0.013 ppm; 4.2 Hz at 7.1 T
	^{2}H primary, 1° $\approx$ 0.020 ppm; 6.7 Hz at 7.1 T
Coupling constant (J) range:	0–20 Hz
	$J(\mathrm{T,H}) = J(\mathrm{H,H}) \times \gamma_T/\gamma_H$
	$J(\mathrm{T,T}) = J(\mathrm{H,H}) \times (\gamma_T)^2 \times (1/\gamma_H)^2$
T_1	0–10 s
T_2	0–10 s

[a]7.1 T field.
[b]Sensitivity given for equal numbers of nuclei in the same field.

For References see p. 4830

signals under the HDO peak. Even in H_2O, tritium spectra require no solvent suppression and are much more easily assigned than the generally more crowded proton analog.

To understand the quantities involved in a tritium NMR experiment, one must be aware of the common units and conversion factors which are listed in Table 1. One of the most important benchmarks is that complete replacement of hydrogen by tritium in one position of a molecule gives a specific activity (*SA*) of 1064 GBq $mmol^{-1}$ (28.76 Ci $mmol^{-1}$). Obviously, for lower percentages of tritium replacement the *SA* value is proportionately reduced and this relationship, combined with a knowledge of the total number of millimoles of labeled substrate, readily yields the amount of radioactivity involved in the experiment.

3.1 Chemical Shifts and Referencing

Early measurements established that the ratios of nuclear screenings among the hydrogen isotopes was very close to unity. This implies that the chemical shift for a triton is essentially identical to that of a proton or deuteron in the analogous molecular environment. Thus tritium spectra can be referenced to, say, monotritiated TMS and chemical shifts will be nearly identical to proton chemical shifts. Since the tritium/proton Larmor frequency ratio is very well determined for TMS[6] ($1.066\,639\,739 \pm 2 \times 10^{-9}$), in practice it is not necessary to add a tritiated reference material to a sample to achieve accurate referencing. Instead, a simple measurement of the *proton* frequency of a reference material and multiplication by the figure above gives the tritium resonance frequency for the reference. This procedure is known as ghost referencing. Some care must be taken in very precise work when using ghost referencing because more careful studies of the Larmor frequency ratio for a variety of tritium-labeled materials[6,7] have shown that nuclear screenings are not identical, but vary slightly with carbon–hydrogen (tritium) bond hybridization. Thus, when referencing tritiated methyl groups a TMS ghost may be appropriate, but a tritiated phenyl position is more accurately referenced by a phenyl proton and a slightly different Larmor frequency ratio. These differences are usually so small that they do not interfere with routine analysis, and are not discerned in a wide display (0–6 ppm), such as that in Figure 1.

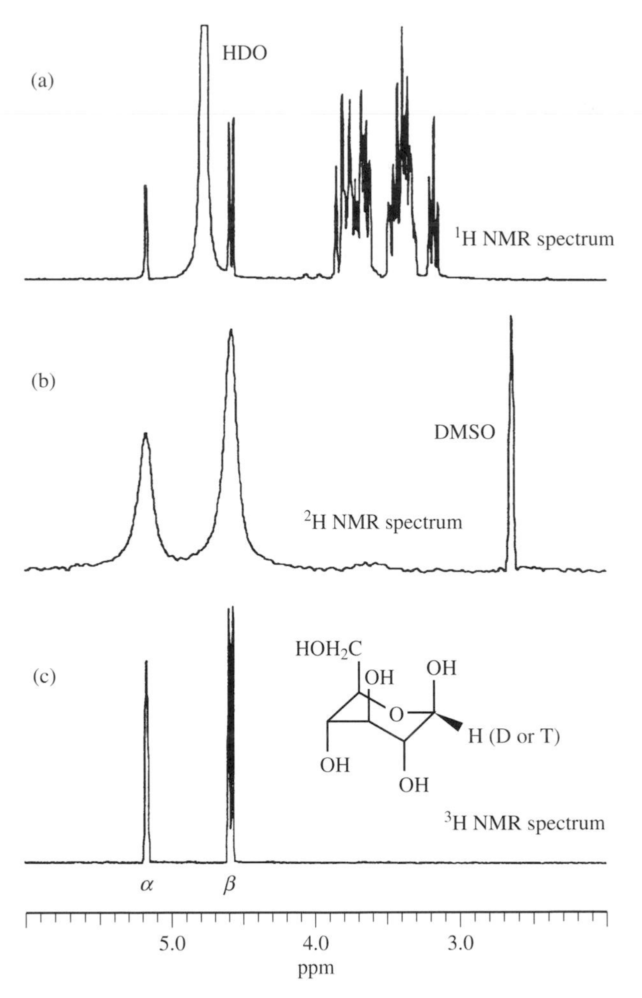

Figure 1 Hydrogen NMR spectra of D-glucose in water. (a) Proton spectrum of C-1 tritiated D-glucose; the large peak is due to HDO. (b) Proton decoupled deuterium NMR spectrum of C-1 deuterated glucose; DMSO is an integration marker. (c) Proton-coupled tritium spectrum of C-1 tritiated glucose, labeled by the same catalytic technique as in (b)

3.2 Coupling Constants and Isotope Effects

Coupling constants between tritium and other nuclei may be calculated from the analogous proton coupling constants. In particular, *J*(T,T) and *J*(T,H) are given by the relationships in Table 2, i.e. tritium–proton spin–spin coupling constants may be calculated as ca. 6.664% larger than the analogous proton constant. There is a small isotope effect on the coupling constant,[8] and measurements have been reported for tritium coupled to a number of other nuclei besides protons including carbon, tin, silicon, and boron.

Although there are isotope effects, measurement of a *J*(T,H) value is still an accurate way of predicting a proton–proton coupling constant, since the calculation involves division by 1.06664 rather than multiplication by 6.5144, as required by prediction from *J*(D,H). For isochronous protons (e.g. methyl protons having degenerate chemical shifts), measurement of the *J*(T,H) value readily yields a usually 'invisible' coupling constant (see Figure 2, below).

Besides the small isotope effects on nuclear screening and on coupling constants discussed above, there are also more commonly observed isotope effects on tritium chemical shifts caused by substitution of a heavy isotope in place of a proton. These are largest and therefore most easily observed when a geminal proton is replaced by a triton. This effect was first reported in analyses of highly tritiated methyl groups,[9] where the R–CT_3 singlet was resolved from the R–CT_2H doublet in the proton coupled tritium spectrum. The proton-decoupled spectrum was used to show that this primary isotope effect is ca. 0.025 ppm (i.e. 2.4 Hz at 96 MHz).

Figure 2 illustrates tritium isotope effects on chemical shifts and demonstrates how the methyl *J*(H,H) may be calculated from measured *J*(T,H) values, as discussed above. A sample of *m*-xylene was tritium-labeled by Raney nickel catalyzed exchange with T_2, which is known to give almost exclusive methyl tritiation. All three tritio-methyl species are produced, as shown in the proton-decoupled tritium spectrum in Figure 2(a), where these species are separated by the primary isotope

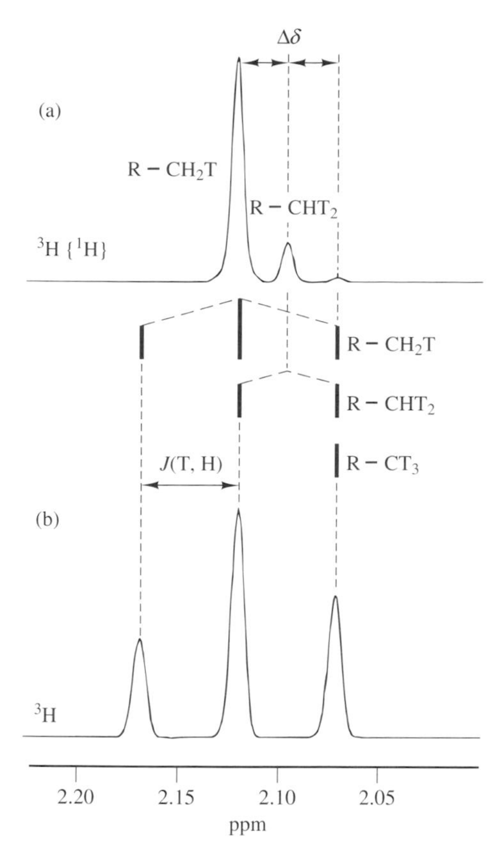

Figure 2 Deceptively simple tritium NMR spectra in C_6D_6 of *m*-xylene labeled by T_2 gas exchange over Raney nickel catalyst. (a) Proton-decoupled tritium spectrum showing peaks for R–CH_2T, R–CHT_2, and R–CT_3 species separated by $\Delta\delta = 0.025 \pm 0.001$ ppm (7.89 ± 0.21 Hz at 7.1 T). (b) Proton-coupled spectrum showing overlapped tritium multiplets separated by J(H,T) = 15.58 ± 0.21 Hz

effect ($\Delta\delta$) on the tritium chemical shift. At 320 MHz, tritium signals from methyl groups have a proton–tritium coupling constant which is double $\Delta\delta$, so that the proton-coupled tritium spectrum in Figure 2(b) has a deceptively simple appearance. The splitting patterns are shown on the figure, and our assignments are readily verified by integration of the signals in each of the spectra. The methyl J(H,H) for *m*-xylene methyl groups may be calculated from the observed J(T,H) value.

Secondary tritium isotope effects on chemical shifts have also been well quantified,[10] and the primary deuterium isotope effect on tritium chemical shifts is also easily measured (Table 2).

3.3 Tritium NMR Intensity and the NOE

One of the major advantages of tritium NMR spectroscopy is that the labeling pattern in a sample may be determined without the time-consuming and laborious degradative determinations previously required.[11] Indeed, a tritium spectrum gives nondestructive, quantitative information on relative tritium abundance at different sites in a molecule. Of course, care must be exercised in any measurement of NMR peak intensities so that accumulation of the data does not *differentially* affect the various peaks. When proton decoupling is used, for example, the tritium intensities will be affected by the Nuclear Overhauser Effect (NOE). Because of the frequent use of proton decoupling in tritium experiments, the importance of this effect has been the subject of several studies and discussions.[12] While the maximum theoretical NOE enhancement to tritons from proton irradiation is 47%, the observed enhancements vary from 10–35%. More importantly, there are situations where differential NOEs may prejudice tritium spectral intensities, such as a labeled molecule with a high abundance of tritium in a position without neighboring protons and a second tritiated position which has many relaxation partners, and integrals should only be compared when these issues have been considered.

3.4 Relaxation

The similarity of tritium and proton nuclei means that their pathways for relaxation are nearly identical. A feature of tritium NMR is that the simple spectra allow straightforward relaxation measurements. The analogous measurements in proton spectra can be difficult due to spectral crowding, solvent suppression, or the complexity of the physical system. Hence tritium relaxation can be the more useful probe of molecular interactions. A classic example is the use of proton and tritium relaxation measurements combined with homo- and heteronuclear saturation transfer experiments to study the relaxation of water molecules in the presence of a macromolecular suspension.[13] A rather simple suspension was created by the formation of egg phosphatidylcholine/cholesterol lipid bilayers in solution. In a lipid suspension the water can exist in two distinct environments, a 'free' state exhibiting bulk water properties, and a motionally hindered or 'bound' state that is associated with the lipid in a hydration shell. Nuclei in these populations relax differently. The free water has the familiar characteristics of small rapidly tumbling molecules in solution, while the bound water has an additional slow component of tumbling due to its association with the lipid. This lipid-induced hydrodynamic effect causes a more rapid relaxation in the bound state and the presence of two water populations with different relaxation rates makes the overall relaxation behavior of the water signal complex. In general, it will depend on a combination of processes including the hydrodynamic effect, chemical exchange between free and bound environments, and the dipolar interaction (cross relaxation).

In the investigation of these processes, tritium doped water was used to separate the exchange contribution from other relaxation processes in the heteronuclear tritium–proton experiments. The measurements made on this rather complicated system were thus simplified to an analysis of several single-line spectra using inversion–recovery and saturation transfer experiments. The results indicated that the dipolar interaction (in the spin diffusion limit) dominates the relaxation of the water, and two models of this limit were proposed which accurately predicted the results, including a rigid and dynamic description of the lattice in the lipid solution. A complete model of the relaxation behavior of water is very desirable since the difference in relaxation properties of free and bound water populations is the basis for contrast and tissue characterization techniques in magnetic resonance imaging.

For References see p. 4830

4 APPLICATIONS

We will limit the discussion here to chemical applications: (i) analysis of labeled materials for chemical and radiochemical purity; (ii) reaction mechanisms and tritium NMR techniques; (iii) specific activity measurements; (iv) solution conformation, stereochemistry, and optical purity; (v) proton exchange studies; and (vi) miscellaneous tritium NMR studies. Biological applications of ^{3}H NMR, e.g. ligand binding, metabolism studies, etc., are discussed elsewhere (see ***Tritium NMR in Biology***).

4.1 Analysis of Labeled Materials

The vast majority of reports involving ^{3}H NMR spectroscopy describe the products of tritium-labeling reactions, and in this application the technique is invaluable. Tritium NMR has been used extensively for the investigation of many tritium exchange techniques, including radiation-induced and acid-, base-, metal-, or zeolite-catalyzed reactions.[14] In general, these techniques rely on the replacement of a hydrogen atom by a tritium (i.e. exchange) in the native structure, and give low specific activity products which may contain tritium in one of several distinct sites.

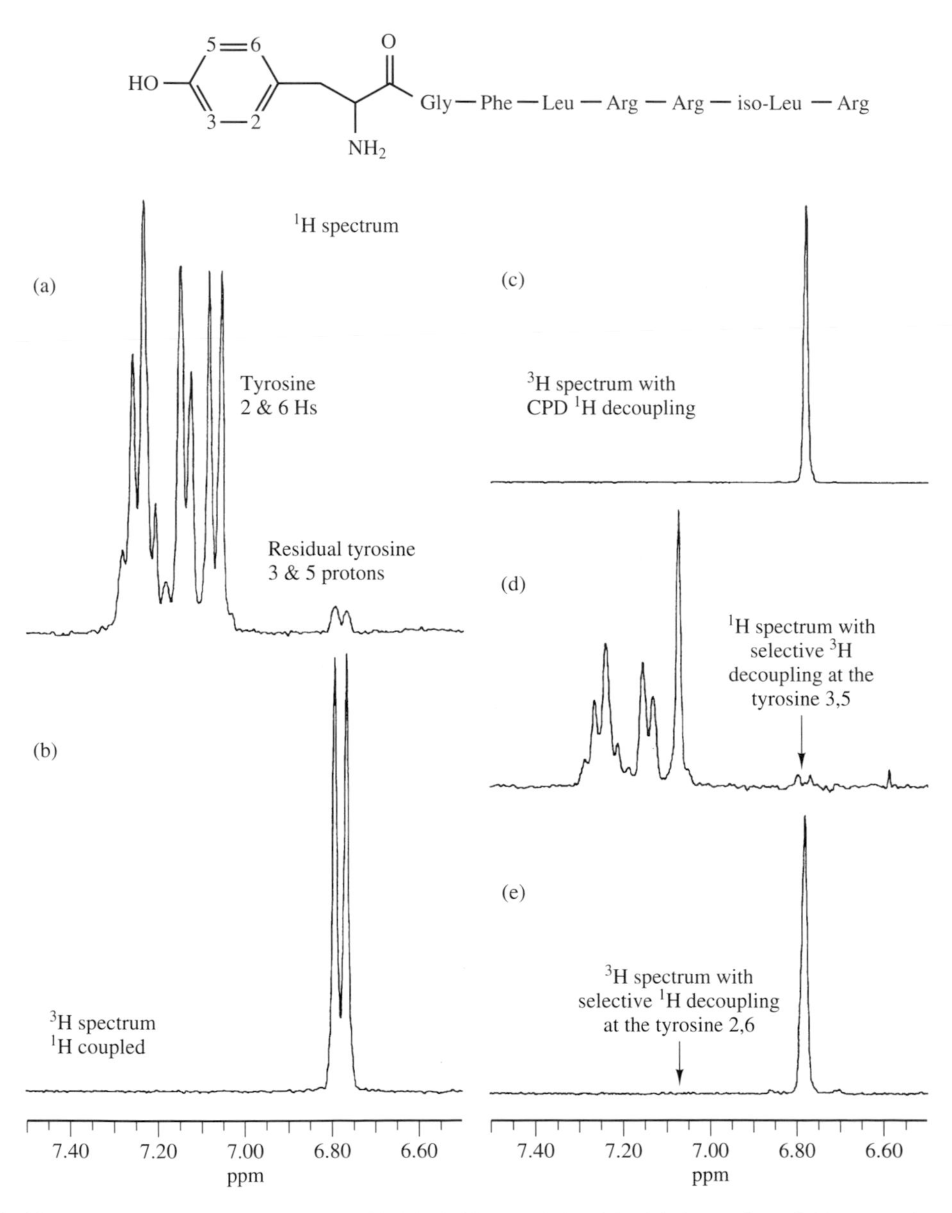

Figure 3 Proton and tritium NMR spectra in D_2O of a peptide labeled by catalytic tritio-dehalogenation of the appropriate diiodotyrosyl peptide. (a) Aromatic region of the proton spectrum showing signals from the phenylalanine residue and a clean doublet from the tyrosine 2 and 6 protons. Residual proton signals are obvious from the tyrosine 3 and 5 position. (b) Tritium spectrum showing the doublet of the tyrosine 3 and 5 tritons coupled to the 2/6 protons. (c) Proton-decoupled tritium spectrum showing the tritium singlet from the 3 & 5 tritium atoms. (d) Selectively tritium-decoupled proton spectrum: comparison with the spectrum in (a) shows the collapse of the tyrosine 2 & 6 doublet. (e) Selectively proton decoupled tritium spectrum, with irradiation only at the tyrosine 2 & 6 doublet

In contrast, high level tritium labeling usually involves a synthetic transformation such as dehalogenation, hydrogenation, or hydride reduction of a suitable precursor. Dehalogenation reactions are an attractive method for introducing tritium into a compound, but surprisingly these reactions rarely give quantitative replacement of halogen by tritium, even when 100% T_2 gas is used. Our experience is that 10–30% of product molecules will incorporate adventitious hydrogen (^1H) rather than ^3H. In general, the shorter the reaction time, the less hydrogen incorporated (i.e. the higher the *SA* value obtained), and it is advisable to use an iodo derivative as the substrate since the facility of halogen removal is I > Br > Cl > F. Also, 1 equiv. of organic base (e.g. triethylamine) enhances the reaction rate and neutralizes the acid formed.

The power of tritium NMR in the analysis of a tritiodehalogenation reaction product is best illustrated by an example. Figures 3(a)–(e) show a full proton and ^3H NMR analysis of a peptide labeled by catalytic dehalogenation of the analogous 3,5-diiodotyrosyl peptide. The aromatic portion of the proton spectrum of the tritiated material [Figure 3(a)] shows ca. 15% proton intensity at the tyrosine 3 & 5 chemical shift, with the usual tyrosine 2 & 6 and Phe aromatic signals. Note that the splitting of the tyrosine 2 & 6 doublet reflects *J*(T,H), *not* coupling to the 3 & 5 protons. The proton-coupled tritium spectrum in Figure 3(b) shows a clean doublet at the tyrosine 3 & 5 chemical shift for the incorporated tritium. Broadband decoupling of the protons yields the singlet tritium signal in Figure 3(c), as expected. Selective irradiation at the tritium tyrosine 3 & 5 frequency causes the tyrosine 2 & 6 doublet to collapse, as shown in the proton spectrum in Figure 3(d) [cf. Figure 3(a)]. Similarly, selective ^1H decoupling at the tyrosine 2 & 6 frequency yields a singlet tritium signal at the tyrosine 3 & 5 chemical shift [Figure 3(e)]. These experiments are readily performed and characterize the product beyond any doubt.

The other most common method for introducing high levels of tritium is catalytic hydrogenation of an unsaturated site on a suitable precursor. Interestingly, heterogeneous metal-catalyzed hydrogenation reactions often do not yield 'textbook' results, and it is remarkable that so few researchers are aware of these details, despite their repeated demonstration.[5] Two effects contribute to the complicated mixture of isotopomers obtained from these reactions:[10] (i) there is an inventory of hydrogen ($^1H_1$) in the catalyst, solvent, substrate, and reaction vessel which serves to **decrease** the maximum level of tritium which may be incorporated into the labeled material; and (ii) vinylic and allylic protons may undergo metal-catalyzed exchange prior to saturation of the multiple bond, thus leading to **increased** specific activity.[15] As a result, the products of a hydrogenation reaction often have (a) lower than theoretical incorporation of tritium, (b) tritium in positions remote from the original site of unsaturation, and (c) unequal amounts of tritium on each side of the original multiple bond.

When ^3H NMR is used to analyze the products of both heterogeneous and homogeneous catalysis of an identical reduction, the complexities of heterogeneous catalysis are clear. Such a comparison of tritiation techniques for the labeling of an unsaturated phospholipid molecule is shown in Figure 4. All the problems described above for heterogeneous catalysis complicate the spectrum shown in Figure 4(a), including small amounts of tritium incorporated at the positions adjacent to the

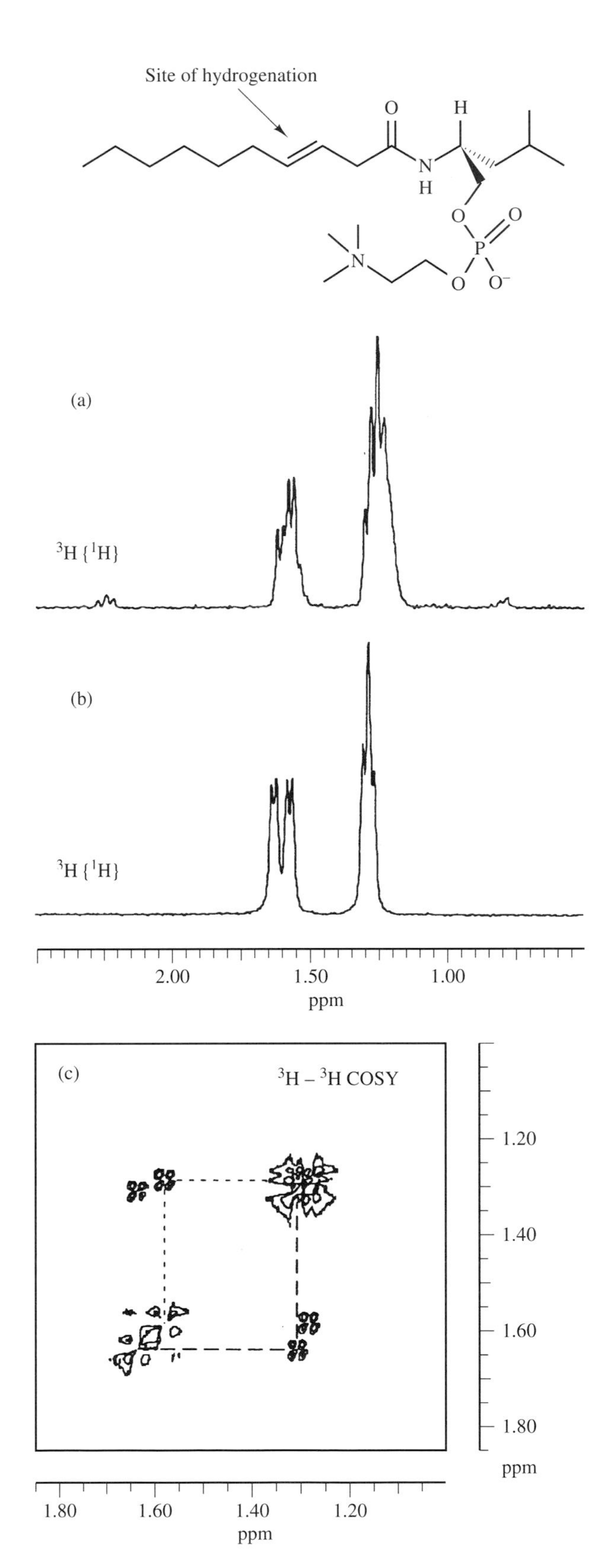

Figure 4 Proton-decoupled tritium NMR spectra of a labeled phospholipid in D_2O. (a) Product of a heterogeneous catalytic hydrogenation reaction. (b) Product from catalytic hydrogenation in the presence of Wilkinson's catalyst. (c) {^1H} ^3H–^3H COSY of the product in (b)

For References see p. 4830

double bond. In contrast, the product of the homogeneous tritiation using Wilkinson's catalyst [Figure 4(b)] shows clean addition of tritium, with a 1:1 ratio at the C-7 and C-8 positions. A ^{3}H–^{3}H COSY analysis of this product [Figure 4(c)] showed that the two expected diastereotopic products were obtained, in a 1:1 ratio.[16] Although this tritium analysis shows that Wilkinson's catalyst gives a very clean product, it is not applicable in all circumstances. Such mundane issues as sample solubility and recovery of the labeled material from the reaction mixture often affect the decision of which catalyst is used. Heterogeneous catalysts have the advantages that they are tolerant of a variety of solvent systems and are easily filtered away after the reaction. Wilkinson's catalyst is usually employed as a benzene solution and may not perform well in a binary mixture if the substrate is insoluble in benzene. In addition, Wilkinson's catalyst is designed to form an intimate mixture with the substrate, and postreaction separation steps need to be optimized for each new substrate.

4.2 Reaction Mechanisms and Tritium NMR Techniques

Tritium spectra are almost always simple, hence there has been no strong drive to apply more sophisticated NMR techniques to tritium analyses. However, the subtleties of exchange effects in heterogeneous catalysis and the complex products from enzymatic reactions often require careful interpretation, and such studies may be greatly enhanced by the use of modern multipulse approaches.

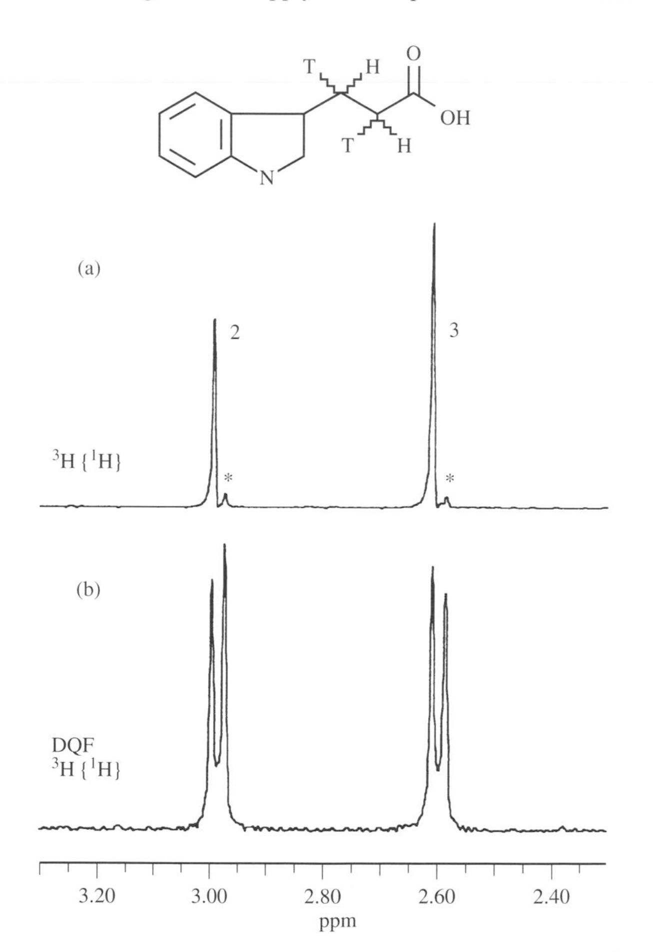

Figure 5 Tritium NMR spectra of tritiated 3-indolepropionic acid in CD_3OD. (a) Proton-decoupled tritium spectrum of the product of a heterogeneous catalytic hydrogenation reaction using 15% T/H. (b) Double quantum filtered proton-decoupled tritium spectrum of the same product as shown in (a). The observed $J(T,T) = 7.32 \pm 0.12$ Hz, suggesting a $J(H,H)$ value of ca. 6.4 Hz

The following techniques have been applied to tritium analyses over the past decade: double quantum filtering, DEPT, 2D homonuclear COSY, 2D tritium/proton correlation, 2D J-resolved, and 2D NOESY and EXSY experiments. The NOESY and EXSY approaches have been used for macromolecular structure studies and are discussed elsewhere (see ***Tritium NMR in Biology***). The pulse sequences we regard as most useful for analytical applications are the DQF and 2D J-resolved experiments. These have been used for detailed investigation of the 'nontextbook' results typically encountered in tritiation reactions, which are usually invisible in the corresponding reaction with $^{1}H_2$. Hence tritiation and tritium NMR spectroscopy provide an excellent approach for investigation of these imperfectly understood reaction mechanisms, as discussed below.

The research staff at Berkeley routinely use a mixture of 5–15% T_2 in H_2 for exploratory labeling reactions and in a training program. In the hydrogenation of 3-indoleacrylic acid using a 10% Pd/C catalyst and 15% T_2 in H_2 gas, simple statistics predict that 2.25% of labeled 3-indolepropionic acid product molecules will contain two ^{3}H atoms, 25.5% will have a tritium at one carbon or the other, and 72.25% will contain no tritium. The ^{1}H-decoupled tritium NMR spectrum in Figure 5(a) shows several features: (i) the large singlet signals arise from molecules containing one tritium atom, and it is obvious that unequal amounts of ^{3}H are incorporated across the double bond, and (ii) one signal from a small doublet can be discerned at the base of each singlet (asterisked), presumably arising from the doubly tritiated species. The unequal substitution pattern in the singly tritiated species is thought to arise from exchange at those carbons prior to saturation of the double bond, and tritium NMR represents the best method for the further study of the mechanism of such exchange.[15] The doubly tritiated species can be observed simply and directly by the application of a familiar multiple pulse technique, a double quantum filtered (DQF) experiment. The results of the DQF experiment on this molecule are remarkably clean, as shown in Figure 5(b). The spectrum is simple because the doubly tritiated species is the only one present with tritium homonuclear double quantum transitions, and is further simplified because all couplings to protons are heteronuclear and have been removed by broadband decoupling. This result has tremendous potential for the study of the mechanism of this type of addition and would not have been possible without the use of tritium NMR. Clearly this unique ability to separate labeled species and perform careful analysis of complex mixtures of isotopomers resulting from a theoretically simple tritiation reaction is essential for understanding the mechanism of these reactions.

Two-dimensional J-resolved ^{3}H NMR[10] can be an even more powerful tool for the analysis of complex isotopic mixtures. β-Methylstyrene was tritiated by catalytic reduction with 100% T_2 over 5% Pd/C catalyst, to give highly tritiated n-propylbenzene with ca. 40% of the tritium on the α-methylene carbon, 50% on the β-methylene, and 10% in the methyl group. In addition, all of the multiplets for these tritiated pos-

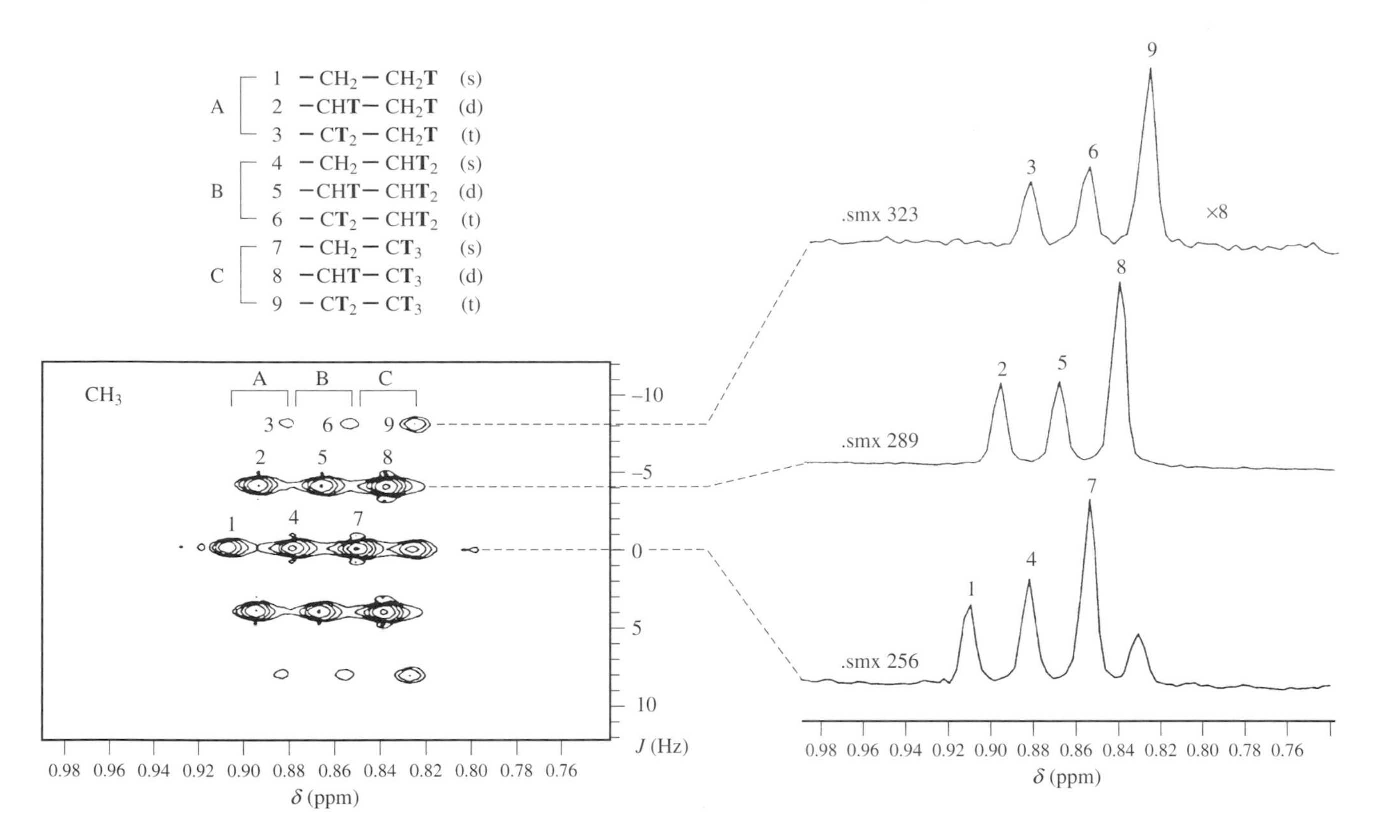

Figure 6 The methyl section (0.68–0.93 ppm) of the contour plot of the proton-decoupled tritium J-resolved spectrum of n-propylbenzene, labeled by catalytic tritiation of β-methylstyrene. Sine-bell window functions were applied in both dimensions, 2k × 512W transform, with magnitude calculation of peak intensities. The data were 'tilted' and 'symmetrized'

itions showed a complex structure, not at all like the simple doublet of doublets which might have been proposed assuming clean addition of a tritium atom to each end of the double bond.

The methyl region of the proton-decoupled tritium J-resolved analysis of this sample (Figure 6) provided most insight into the species giving rise to the various signals. There is a repeating and slightly overlapped singlet–doublet–triplet pattern, with each repetition moved successively to lower frequency by ^3H isotope effects. These first of these patterns (group A, Figure 6) is due to the species X–CH_2–CH_2T, singlet; X–CHT–CH_2T, doublet; and X–CT_2–CH_2T, triplet (species 1, 2, and 3), and the small change in chemical shift between these signals is the secondary tritium isotope effect induced by addition of a T at the β-CH_2. There are similar patterns for the –CHT_2 (B) and –CT_3 (C) methyl species, successively moved to lower frequency by the primary tritium isotope effect on the methyl chemical shift. Inspection of several slices along the ω_2 dimension (submatrix, or .smx plots shown in Figure 6) show the relative abundance of each species in this mixture, and analysis of ω_1 slices allows us to extract all the ^3H–^3H coupling constants. Similar detailed analysis of the α-CH_2 and β-CH_2 regions of the J-resolved spectrum is also possible, and evidence was found for almost all of the 35 statistically possible, tritium-containing isotopomers. While the theoretical product, Ph–CHT–CHT–CH_3, was the most abundant species, the presence of another 34 tritiated isotopomers makes it clear that the mechanism of heterogeneous metal-catalyzed hydrogenation reactions is not as simple as commonly thought.

4.3 Specific Activity Measurements

The molar specific activity (i.e. radioactivity per unit weight) of tritium-labeled molecules is an important property, and tritium NMR may be used to determine this quantity.[5] This approach is essential when the labeled material has no UV spectrum, or the high specific activity precludes weighing to determine sample concentration. An example of the latter was the determination of the *SA* value for a series of $NaBT_4$ preparations.[17] Boron has two NMR active isotopes (^{10}B, 20%, $I = 3$ and ^{11}B, 80%, $I = \frac{3}{2}$), so the proton spectrum of sodium borohydride in basic CD_3OD shows a large quartet ($Na^{11}BH_4$, $\delta = -0.168$ ppm, $J = 80.6$ Hz) and a small septet ($Na^{10}BH_4$, $\delta = -0.166$ ppm, $J = 26.98$ Hz). The proton-decoupled tritium spectrum of exchange labeled $NaBH(T)_4$ shows a similar spectrum for each isotopomer, i.e. $NaBH_3T$, $NaBH_2T_2$, $NaBHT_3$, $NaBT_4$, shifted to lower frequency for each additional T. The multiplicity (redundancy) of the peaks is useful since peak overlap does not occur for all the species, and the isotopomer ratios may be compared between large quartet signals and the smaller septets. The amount of unlabeled ($NaBH_4$) material present was estimated by comparison with $NaBH_3T$ signals in the proton spectrum. Knowing the mole ratio of all the isotopomers yields the *SA* value, and this property was compared for several production 'lots' of tritiated borohydride.

Even when determination of *SA* by UV spectroscopy and LSC counting, or by mass spectrometry, is possible, the NMR technique provides a useful cross check. In comparison to the sodium borohydride case above, a very simple example of

For References see p. 4830

using ^3H and ^1H NMR to determine specific activity is given in Figure 7. Pinoline (6-methoxy-1,2,3,4-tetrahydro-9*H*-pyrido[3,4-*b*]indole) was labeled by catalytic hydrogenation of the unsaturated 3,4-dihydro analog. The product consisted of a mixture of three isotopomers containing zero, one, or two tritium atoms per molecule. Obviously, at the chemical shift of the C-1-methylene, one of these species has only a proton spectrum, one has a tritium spectrum, and one [R–CHT, δ = 4.39 ppm, Figures 7(a) and (b)] has both a tritium and a proton NMR spectrum. In this case, the use of selective tritium [Figure 7(a)] or proton [Figure 7(b)] decoupling allows the observation of single lines for each of the isotopomers in the proton or tritium spectrum. Since the R–CHT signal occurs in both spectra, the molar ratio of the three species may be extracted from these spectra and the specific activity calculated.

4.4 Solution Conformation, Stereochemistry, and Optical Purity

Tritium NMR has been used to obtain the axial–equatorial conformational preference of tritium in [^{3}H]cyclohexane.[18] Tritium has the advantage of giving sharp and well-separated peaks that are well suited for high-accuracy integration at low temperatures, i.e. the peak separation in these experiments was ca. 150 Hz at 7.1 T, whereas the analogous deuterium experiment at 11.7 T gave a separation of only 36 Hz.

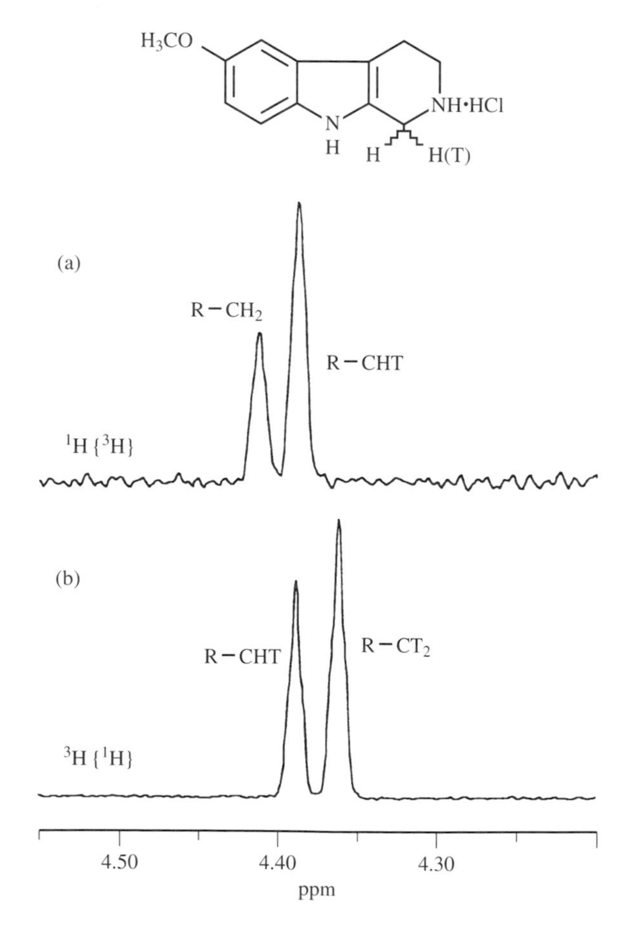

Figure 7 Proton and tritium NMR spectra of tritiated pinoline in CD_3OD. (a) Selectively tritium-decoupled proton spectrum. (b) Selectively proton-decoupled tritium spectrum

[^{3}H]Cyclohexane was made from cyclohexylmagnesium chloride in diethyl ether by the addition of HTO (5% tritium) followed by distillation of the tritiated product. A sample (ca. 0.9 GBq, 33 mCi) in CS_2 containing some CD_2Cl_2 for lock purposes was transferred into a standard thin-wall glass 5-mm NMR tube and ^3H {^1H} NMR spectra were obtained at 320 MHz (300-MHz ^1H frequency). Narrow lines (ca. 2 Hz at half-height) with good lineshapes were obtained at −88 °C by using the tritium FID as well as the ^2H lock signal for shimming. For accurate integration it is essential that the base of the peaks be narrow and symmetrical, and that the baseline is flat. A number of data sets, each consisting of the sum of 512 FIDs were acquired, several with the carrier at a higher frequency than the sample signals, some with the carrier at a lower frequency, and others with the carrier centered between the signals. The ^{13}C satellites of both the axial and equatorial ^3H signals ($\delta\nu$ = 0.471 ppm) were visible in each set of spectra and gave $J(^{13}C,^3H_{ax})$ = 131.1 Hz and $J(^{13}C,^3H_{eq})$ = 135.2 Hz. Integration of the spectra showed that tritium prefers the equatorial over the axial site by about 46.9 J mol^{-1} and a ^3H EXSY experiment showed that the axial and equatorial tritium atoms are undergoing (slow) exchange at −88 °C. Conformational analysis by ^3H NMR relaxation has also been used in enzyme ligand studies (see ***Tritium NMR in Biology***).

The absolute stereochemistry of a tritium atom on a carbon may be determined by tritium NMR if there is another chiral center in the molecule. If this is not the case on the parent molecule, it can often be created by esterification or some other derivatization. This feature was exploited in analysis of tritiated ethanol produced by enzymatic oxidation of 'chiral ethane', i.e. CH_3–CHDT, as part of a study of the mechanism of methane monooxygenase.[19] Tritium NMR analysis of the methylene region of the ethanol product showed signals from CH_3–CHT–OH and CH_3–CDT–OH, separated by the ^2H isotope effect on the tritium chemical shift [Figure 8(a)]. Whereas this spectrum gives information on the relative rate of removal of an H or D atom in the oxidation process, it gives no stereochemical data. A simple derivatization with (2*R*)-2-acetoxy-2-phenylethanoic (mandelic) acid yields ethyl mandelate, and the four possible species can be readily separated and quantitated by ^3H {^1H} NMR, as shown in Figure 8(b). The *absolute* stereochemistry was previously established by a known synthesis of the deuterium-substituted molecules and proton NMR analysis.

Tritium NMR has been used to determine the optical purity of labeled materials. The use of a lanthanide shift reagent[20] or Pirkle's alcohol[21] is a more general technique than the careful stereochemical determinations discussed above, and does not require the presence of an alcohol or amine functionality for derivatization to the appropriate ester or amide. This general approach is especially important for materials labeled at low tritium abundance because the optical properties of the bulk material may not be the same as the small subpopulation of labeled molecules.

4.5 Proton Exchange Studies

The NMR properties of tritium make it particularly well suited for in situ measurements of acid- and base-catalyzed proton exchange kinetics using tritium NMR spectroscopy.[22] Despite the great sensitivity and simplicity of this approach,[5] previous tritium NMR mechanistic studies of proton exchange

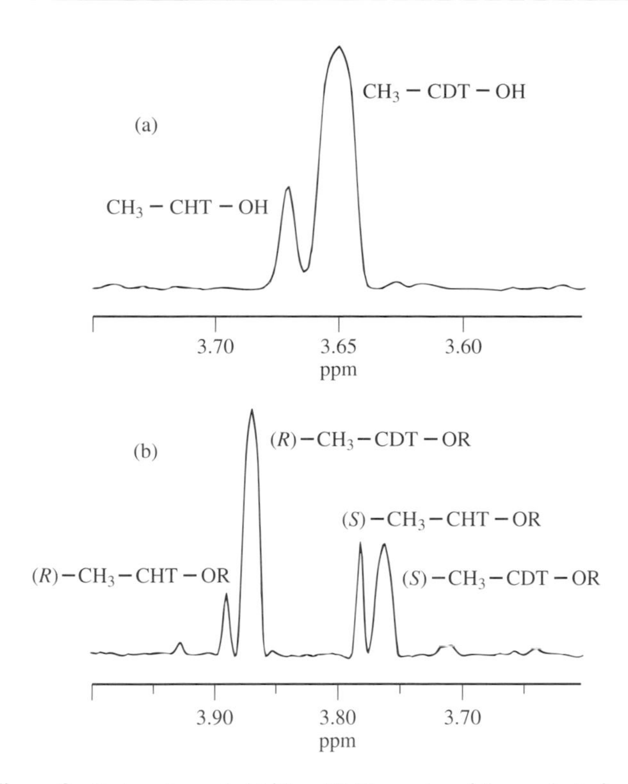

Figure 8 Proton-decoupled tritium NMR spectra of the products from enzymatic oxidation of chiral ethane. (a) Ethanol in H_2O. (b) Derivatized ethanol in C_6D_6, where R = (2*R*)-2-acetoxy 2-phenylethanoic acid

have been confined to a static approach, i.e. the analysis of aliquots of a kinetic experiment.[23] In most circumstances, it is desirable to perform one-tube, accurate, kinetic experiments, especially for compounds which are only available in small amounts or have several exchanging positions with similar rates. Use of deuterium NMR spectroscopy, hampered by its low sensitivity and resolution, is inherently limited to initial rates,[24] and if several positions are measured simultaneously, relative rates for the positions must lie within a small range. In contrast, tritium NMR spectroscopy allows one to measure the proton exchange kinetics of several compounds which may have similar rates or vary by a factor of 1000. Since tritium is a highly sensitive nucleus, the sample need not contain high levels of tritium for detection (especially compared with deuterium) and tritium therefore functions as a true tracer in the reaction kinetics. Only 20–180 MBq (0.5–5 mCi) per position (1.5×10^{-5}% incorporation) is necessary for an adequate S/N ratio. This low tritium concentration also ensures the absence of tritium–tritium coupling, and the high γ and excellent relaxation characteristics of the nucleus lead to sharp resonance lines. The integrity of the reactant mixture may be monitored by proton NMR spectroscopy at any stage of the experiment. An inverse-gated proton-decoupling scheme may be used to ensure that single peaks are observed for each magnetic environment and that the relative peak integrals are not affected by NOE build-up. The quality of the kinetic data obtained by this technique suggests that it may be readily applied to other chemical proton-transfer systems.

As an example, this technique has been applied to the measurement of detritiation rates for the characterization of isotope effects. The Swain–Schaad relationship predicts that the deuterium and tritium isotope effects on the rate of a reaction are related as follows: $(k_H/k_D)^x = k_H/k_T$. Theoretical calculations put x in the range 1.33–1.58, and a simple zero-point energy formulation of isotope effects gives $x = 1.44$. Deviations from this range have recently been used as evidence of tunneling in enzyme-catalyzed reactions, so the accuracy of experimental results in simple chemical systems is important. Such a simple system is the hydroxide ion-catalyzed enolization of acetone (a prototype ketone ionization reaction), but the originally published measurements give an anomalously low exponent, $x = 1.08$.[25] This exponent was based upon rates of detritiation of labeled acetone, which were determined by a technique that involved a difficult separation of acetone from water.

This reaction was reinvestigated[26] using tritium NMR to monitor the exchange of tritium between acetone and water in situ,[22] thus avoiding the troublesome acetone–water separation. In NMR experiments using a range of base concentrations, the tritium signal from the acetone at $\delta = 2.1$ ppm was seen to decay while another resonance at $\delta = 4.7$ ppm, attributable to HTO, appeared (Figure 9). Least-squares fitting of the observed exponential rise and decay gave first-order rate constants that agreed well with each other.

When the NMR result was combined with k_H and a literature value of k_D determined by mass spectrometric measurements, it yielded $k_H/k_D = 7.2$ and $k_H/k_T = 19.2$. These isotope effects provide the Swain–Schaad exponent $x = 1.49 \pm 0.07$, which is in complete agreement with theory. Thus, tritium NMR provided a simple method for the measurement of

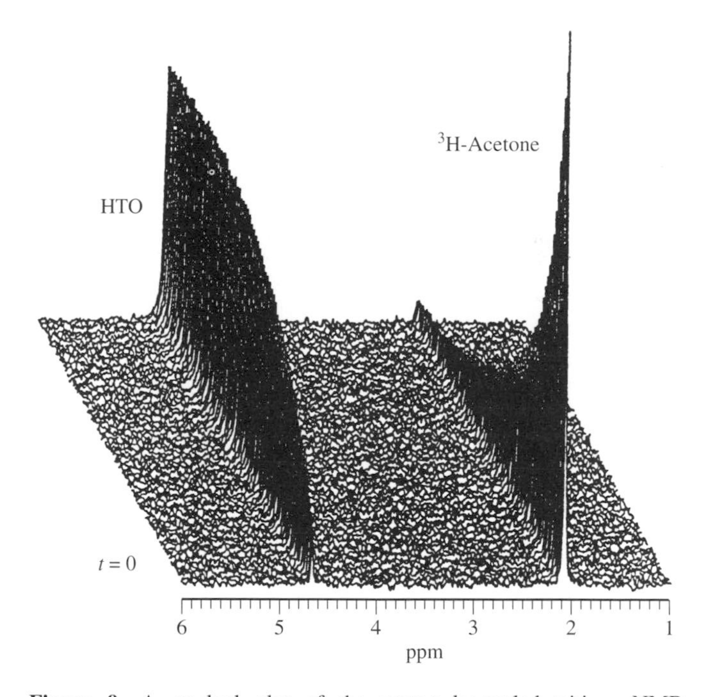

Figure 9 A stacked plot of the proton-decoupled tritium NMR spectra of the base-catalyzed detritiation of acetone ($\delta = 2.1$ ppm) with exchange into the solvent (HTO, 4.7 ppm). The experimental parameters were: 500 μL, 0.04 M, NaOH; 5 μL, ^{3}H-acetone; $T = 298$ K; 120 transients; 100 experiments over 10.5 h (ca. 4-min interval for the first 50, then 8-min interval); spectrometer unlocked; $t = 0$ when the acetone was added, and the first spectrum was finished at $t = 20$ min. Seventy-seven spectra are plotted

For References see p. 4830

an important exchange rate which was essentially inaccessible by chemical techniques.

4.6 Miscellaneous

Tritium NMR spectroscopy has been used in a number of other very specific chemical and physical investigations, most of which have been comprehensively reviewed elsewhere.[5] These applications have included: (i) the observation of hydrogen-bonding effects on the chemical shift of proton versus tritium NMR signals; (ii) determination of solvent isotope effects and fractionation factors; (iii) aspects of radiation chemistry including sample radiolysis, studies of DNA hydration, and in situ production of labeled carbocations; (iv) investigations of labeled molecules in a liquid crystal environment; (v) solids tritium NMR and characterization of trapped gases in irradiated lithium hydride (tritide) samples;[27] and (vi) study of the origin of an anomalously large proton–proton coupling in transition metal polyhydrides.[28]

5 CONCLUSION

We have discussed a wide range of applications where the use of tritium NMR has been useful, and some instances where it was essential. The similarity in NMR properties between protons and tritons means that many systems may be studied using either nucleus. Moreover, when proton systems become too complex, the spectral simplification of tritium spectroscopy suggests that tritium NMR should be considered. Radioactivity is the only issue hindering the universal application of tritium in many more general NMR studies.

6 RELATED ARTICLES

Enzymatic Transformations: Isotope Probes; Multidimensional Spectroscopy: Concepts; Tritium NMR in Biology.

7 REFERENCES

1. H. L. Anderson and A. Novick, *Phys. Rev.*, 1947, **71**, 372.
2. F. Bloch, A. C. Graves, M. Packard, and R. W. Spence, *Phys. Rev.*, 1947, **71**, 373.
3. F. Bloch, A. C. Graves, M. Packard, and R. W. Spence, *Phys. Rev.*, 1947, **71**, 551.
4. G. V. D. Tiers, C. A. Brown, R. A. Jackson, and T. N. Lahr, *J. Am. Chem. Soc.*, 1964, **86**, 2526.
5. E. A. Evans, D. C. Warrell, J. A. Elvidge, and J. R. Jones, 'Handbook of Tritium NMR Spectroscopy and Applications', Wiley, Chichester, 1985, pp. 1–249.
6. J. P. Bloxsidge, J. A. Elvidge, J. R. Jones, R. B. Mane, and M. Saljoughian, *Org. Magn. Reson.*, 1979, **12**, 574.
7. J. M. A. Al-Rawi, J. P. Bloxsidge, C. O'Brien, D. E. Caddy, J. A. Elvidge, J. R. Jones, and E. A. Evans, *J. Chem. Soc., Perkin Trans. 2*, 1974, 1635.
8. J. M. A. Al-Rawi, J. A. Elvidge, J. R. Jones, and E. A. Evans, *J. Chem. Soc., Perkin Trans. 2*, 1975, 449.
9. J. P. Bloxsidge, J. A. Elvidge, J. R. Jones, E. A. Evans, J. P. Kitcher, and D. C. Warrell, *Org. Magn. Reson.*, 1981, **15**, 214.
10. P. G. Williams, H. Morimoto, and D. E. Wemmer, *J. Am. Chem. Soc.*, 1988, **110**, 8038.
11. J. M. A. Al-Rawi, J. P. Bloxsidge, J. A. Elvidge, J. R. Jones, V. E. M. Chambers, V. M. A. Chambers, and E. A. Evans, *Steroids*, 1976, **28**, 359.
12. F. M. Kaspersen, C. W. Funke, E. M. G. Sperling, and G. N. Wagenaars, *J. Labelled Compds. Radiopharm.*, 1987, **24**, 219.
13. T. L. Ceckler and R. S. Balaban, *J. Magn. Reson.*, 1991, **93**, 572.
14. P. G. Williams, in 'Isotopes in the Physical and Biomedical Sciences', ed. E. Buncel and J. R. Jones, Elsevier, Amsterdam, 1991, Vol. 2, p. 55.
15. J. A. Elvidge, J. R. Jones, R. M. Lenk, Y. S. Tang, E. A. Evans, G. L. Guilford, and D. C. Warrell, *J. Chem. Res. (S)*, 1982, 82.
16. A. S. Culf, Ph.D. Thesis, University of Surrey, UK, 1994.
17. L. J. Altman and L. Thomas, *Anal. Chem.*, 1980, **52**, 992.
18. F. A. L. Anet, D. J. O'Leary, and P. G. Williams, *J. Chem. Soc., Chem. Commun.*, 1990, 1427.
19. N. D. Priestley, H. G. Floss, W. A. Froland, J. D. Lipscomb, P. G. Williams, and H. Morimoto, *J. Am. Chem. Soc.*, 1992, **114**, 7561.
20. J. A. Elvidge, E. A. Evans, J. R. Jones, and L. M. Zhang, 'Synth. Appl. Isot. Labeled Compd., Proc. 2nd Int. Symp.', ed. R. R. Muccino, Elsevier, Amsterdam, 1986, p. 401.
21. F. M. Kaspersen, C. W. Funke, E. M. G. Sperling, F. A. M. Van Rooy, and G. N. Wagenaars, *J. Chem. Soc., Perkin Trans. 2*, 1986, 585.
22. R. E. Dixon, P. G. Williams, M. Saljoughian, M. A. Long, and A. Streitwieser, *Magn. Reson. Chem.*, 1991, **29**, 509.
23. E. Buncel, J. P. Davey, G. J. Buist, J. R. Jones, and K. D. Perring, *J. Chem. Soc., Perkin Trans. 2*, 1990, 169.
24. D. W. Boerth, and A. Streitwieser, Jr., *J. Am. Chem. Soc.*, 1981, **103**, 6443.
25. J. R. Jones, *Trans. Faraday Soc.*, 1969, **65**, 2138.
26. Y. Chiang, A. J. Kresge, H. Morimoto, and P. G. Williams, *J. Am. Chem. Soc.*, 1992, **114**, 3981.
27. R. C. Bowman, Jr., A. Attalla, P. C. Souers, C. L. Folkers, T. McCreary, G. D. Snider, F. Vanderhoofven, and R. T. Tsugawa, *J. Nucl. Mater.*, 1988, **154**, 318.
28. K. W. Zilm, D. M. Heinekey, J. M. Millar, N. G. Payne, and P. Demou, *J. Am. Chem. Soc.*, 1989, **111**, 3088.

Acknowledgements

M.G.K. is supported by the Office of Energy Research, Office of Health and Environmental Research, Health Effects Research Division of the US Department of Energy under Contract DE-AC03-76SF00098, and through Instrumentation Grants from the US Department of Energy, DE FG05-86ER75281, and the National Science Foundation, DMB 86-09035. P.G.W. is supported by the Biomedical Research Technology Program, National Center for Research Resources, US National Institutes of Health, under Grant P41 RR01237, through Contract DE-AC03-76SF00098 with the US Department of Energy.

Biographical Sketches

Mark G. Kubinec. *b* 1964. B.S., 1986, biochemistry, B.S., 1987, chemistry, Michigan State University, USA, Ph.D., 1994, Chemistry, University of California, Berkeley, USA. Postdoctoral fellow, Lawrence Berkeley National Laboratory, 1995–present.

Philip G. Williams. *b* 1956. B.Sc., 1980, chemistry, Ph.D., 1984, chemistry, University of New South Wales, Australia. Research Fellow, Ludwig Institute for Cancer Research, Sydney, Australia, 1984–85; staff scientist, Lawrence Berkeley National Laboratory, USA, 1986–present. Approx. 70 publications Current research specialties: tritium labeling, tritium NMR.

Tritium NMR in Biology

Mark G. Kubinec & Philip G. Williams

Lawrence Berkeley National Laboratory, Berkeley, CA, USA

1 INTRODUCTION

In the study of biologically active molecules, the effectiveness of NMR spectroscopy is limited by proton resolution. Even using modern techniques with ^{13}C- and ^{15}N-labeled molecules of uniform enrichment the upper molecular weight limit for NMR applications is ca. 40 kDa. Site specific or ligand labeling methods are the only alternative for systems over this 40 kDa limit, and are most useful in the study of catalytic or dynamic regions of macromolecules. These latter approaches eliminate the resolution and assignment limitations of uniform labeling while bringing to bear the power of NMR for the study of structural and dynamic problems.

Tritium is the most attractive NMR active nucleus available for site specific or ligand labeling. We have included some chemical and radiochemical properties of tritium in Table 1 and compared the NMR characteristics of ^{3}H with ^{1}H, ^{13}C, and ^{15}N in Table 2. Among the most important NMR properties of tritium are the high gyromagnetic ratio and low spin number ($\frac{1}{2}$), yielding the highest molar sensitivity of any nucleus. This is important in biological systems where sensitivity is limited by low solubility and/or catalytic yield of biological molecules. The low natural abundance of tritium ensures that the 'tritium window' on biological systems is not clouded by spurious signals, and few published ^{3}H NMR spectra have more than a half dozen signals to assign and monitor. Assignment is aided by the fact that tritons and protons have essentially identical chemical shifts.[1] Replacing one isotope of hydrogen with another allows basically native species to be studied by tritium NMR spectroscopy. In contrast, labeling with ^{19}F or some different *element* can completely change the substrate under investigation. In general, labeling with tritium leads to minor isotope effects on the labeled system, and the most serious, kinetic isotope effects, may be helpful in unraveling the mechanism of reactions at the labeled site (as discussed below).

Despite its many advantages, tritium NMR development has lagged behind that of other heteronuclei. Initially, hardware development and concern over highly radioactive samples hindered progress. Indeed, the first recorded tritium NMR spectra required many TBqs (curies) of material to obtain an adequate signal. Modern high-field spectrometers can readily detect 3.7 MBq (0.1 mCi) of tritium (i.e. ca. 2.5×10^{15} nuclei). This makes radiation safety a less dominant issue and yields a detection limit more suitable for biological samples. To contain the radiation, samples are often doubly sealed in Teflon tubes inside glass tubes. Since tritium is a weak β emitter (E_{max} = 18.6 keV), this approach totally blocks radiation while providing insurance against leakage or spilling of samples.[1] To understand the amounts of radioactivity typically encountered in tritium NMR experiments one must be familiar with the commonly used units, which are included in Table 1 along with several conversion factors. The most important equivalence is: 100% substitution of a tritium atom for a hydrogen atom in a molecule leads to a specific activity of 1064 GBq mmol^{-1} (28.76 Ci mmol^{-1}). For example, 250 μL of a 1 mM solution (0.25 μmol) of a substrate having 100% abundance of tritium at a single site contains ca. 266 MBq (7.2 mCi). For perspective, safety lights in aircraft have up to 370 GBq (10 Ci) of tritium, a marine compass may contain as much as 28 GBq (0.75 Ci), and commonly available night sights for rifles contain 2 GBq (50 mCi) of tritium.

The sensitivity of the tritium nucleus in modern spectrometers, coupled with advances in synthetic labeling techniques, have paved the way for the use of tritium NMR in biology. Today, most small biologically active molecules can be readily tritium-labeled, and the chemistry is often more straightforward than labeling the same molecule with either ^{13}C or ^{15}N. The tritiated molecules can be studied in conjunction with macromolecules in virtually any biological system accessible by NMR spectroscopy, including solutions of proteins and nucleic acids, cellular suspensions, and whole animals. Tritium NMR spectra can provide a remarkably simple and revealing window into the structure, chemistry, and dynamics of a biological system.

Three general applications of ^{3}H NMR are discussed below: Section 2 will cover enzyme mechanistic studies; Section 3 will review enzyme ligand interactions; and Section 4 will discuss macromolecular labeling.

Table 1 General Physical and Chemical Data for Tritium, and Common Radioactivity Units[a]

Production	^{6}Li(n, α)^{3}H
Radiation	β (100%); range = 4.5–6 mm in air, 6 μm in H_2O
β Energy	E_{max} = 18.6 keV, E_{mean} = 5.7 keV
Halflife	12.43 y (4540 d)
Maximum specific activity	1064 GBq milliatom^{-1} (28.76 Ci milliatom^{-1})

[a] 37 gigabecquerels (GBq) = 1 curie (Ci) = 3.7×10^{10} disintegrations per second; 1 gray (Gy) = 10 000 erg g^{-1} = 100 rad (units of absorbed dose); 1 sievert (Sv) = 100 rem (units of radiation exposure).

For References see p. 4838

Table 2 NMR Properties of ^3H and Common NMR Nuclei[a]

Nucleus	Natural abundance	Spin	γ	Resonant frequency	Relative sensitivity[b]
^1H	99.984	$\frac{1}{2}$	26.510	300.13	1.00
^3H	**$<10^{-16}$**	$\frac{1}{2}$	**28.5335**	**320.13**	**1.21**
^{13}C	1.10	$\frac{1}{2}$	6.7263	75.46	1.59×10^{-2}
^{15}N	0.37	$\frac{1}{2}$	−2.7116	30.40	1.04×10^{-3}

[a]7.1 T field.
[b]Sensitivity given for equal numbers of nuclei in the same field.

2 ELUCIDATION OF ENZYME STEREOCHEMISTRY

For decades, radioactive tracer nuclei such as ^{14}C, ^3H, ^{32}P, and ^{35}S have been used to study metabolic pathways in cells by determination of the radiolabel position in a biological product or the stereochemistry of its addition. This often requires extensive chemical or enzymatic manipulation of the biosynthetic product. The complexity of these methods makes them difficult and error prone. Tritium NMR is the simple, rapid, accurate, and nondestructive product analysis technique researchers have sought: an NMR spectrum is acquired as opposed to chemical degradation or analysis. In addition to its relative simplicity, tritium NMR provides more detailed structural and stereochemical information than any other technique.

The group at Surrey University which pioneered ^3H NMR spectroscopy[1] first demonstrated the power of ^3H NMR in biology by following up ^{14}C work on the biosynthesis of penicillin with tritium labeling and tritium NMR. By growing bacteria on media supplemented with tritium-labeled acetate, they were able to determine unambiguously the fate of the methyl protons of acetate in the biosynthesis of the penicillin molecule.[2] This, and other early work, underscored many of the advantages of using tritium labeling coupled with ^3H NMR analysis. Since 1974 this approach has been applied to much more subtle and complicated problems of enzyme stereochemistry. It has proven especially useful in instances of so-called 'cryptic' stereochemistry, where the fate of an individual H atom must be traced, as illustrated in Figure 1(a). Tritium NMR and a fascinating species known as a 'chiral methyl group' have emerged as the most powerful method for studying these reactions.

Chiral methyl groups are the smallest chiral entity ever synthesized.[3] They are created by attaching each of the isotopes of hydrogen, a proton, deuteron, and triton, to a single carbon atom. If a reaction occurs at this chiral center as in Figure 1(a), ^3H NMR analysis of the resulting methylene will give the stereochemical course of the reaction provided that the diastereotopic positions ['a' and 'b' in Figure 1(a)] are resolvable in the spectrum.

Figures 1(b)–(e) show a breakdown of an ideal substitution reaction to demonstrate how the NMR pattern of products from a chiral methyl reaction may be used to determine enzyme stereospecificity. Reaction of an (*S*) chiral methyl with retention of configuration is illustrated. Three isotopomeric species can result from the reaction, depending on which isotope of hydrogen is removed. Two of these species contain a tritium atom and may be observed by tritium NMR spectroscopy. One of these products, with a triton at the pro-*R* position, will give an NMR signal at chemical shift 'a', as shown in Figure 1(b). Here the proton was replaced leaving an (*R*) R–CDT–X species, and the ^3H signal is a triplet due to the triton–deuteron coupling (since ^2H is spin 1). The chemical shift of this species will also be affected by a deuterium isotope effect. A doublet will also occur in the tritium NMR spectrum from the other product, containing a triton in the pro-*S* position [at chemical shift 'b', Figure 1(c)]. In this species, the deuterium atom was replaced, giving (*S*) R–CHT–X. If the addition of X had occurred with inversion, the opposite pattern (a doublet at chemical shift 'a' and a triplet at 'b') would be observed. There is often a primary kinetic isotope effect in the substitution of X, and this will alter the intensity of the lines in the spectrum in a way that also reflects the stereochemistry of the reaction. If configuration is retained and there is a 'normal' kinetic isotope effect, the signal at 'a' will have a larger integral since it is kinetically more favorable to remove the proton [cf. Figures 1(d) and (e)]. The ratio of the intensities of the lines will directly yield the magnitude of the isotope effect. Thus both the intensity and coupling pattern of the lines in the ^3H NMR spectrum are indicative of the stereochemistry of the substi-

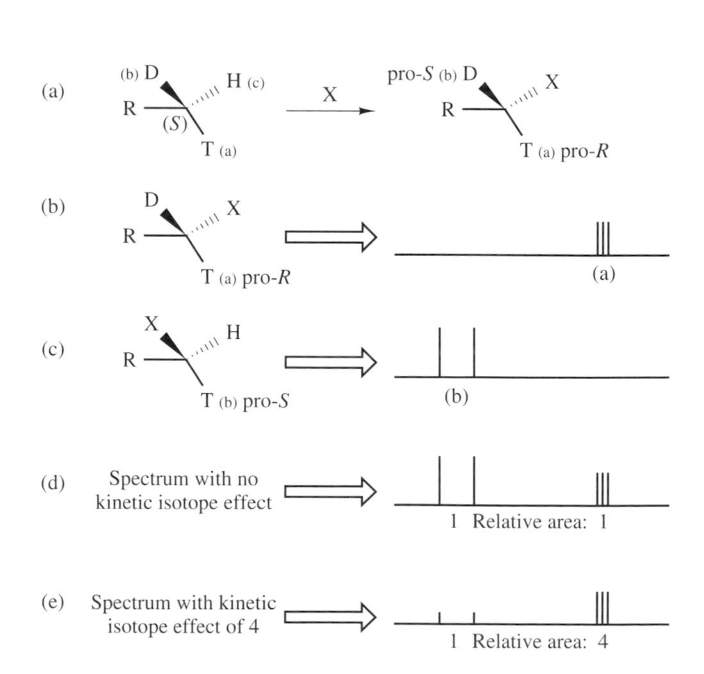

Figure 1 (a) A chiral methyl group undergoes a substitution reaction: D = deuterium, T = tritium. For a substitution with retention of configuration, the products detectable by ^3H NMR spectroscopy are shown in (b) and (c), along with idealized spectra. The effect of a kinetic isotope effect is shown in (d) and (e)

tution. If racemization occurs the effect will be obvious, since the mixing of the product species leads to equal intensities at all labeled positions.

Analysis of enzyme stereochemistry using chiral methyl groups is becoming common.[3–8] Aside from the ability to resolve pro-*R* and pro-*S* chemical shifts, there are other technical requirements. It must be possible to synthesize the chiral tritiated substrate at sufficient specific activity. For probing reactions that are occurring in vivo or in cell-free extracts, a high level (5–50 GBq) is usually necessary to ensure sufficient uptake and yield for processing. Biosynthetic yield must also be high enough to provide product in the 4–400 MBq range for NMR analysis. This can be as much as 10^6-times more radioactivity than commonly used in a simple tracer experiment. In ^{13}C, ^{15}N, and ^{2}H spectroscopy of biomolecules, natural abundance signals may interfere with observation of the added label. In contrast, any tritium signals observed are directly related to the tritium originally introduced as labeled substrate. This feature facilitates the use of tritium NMR for in vivo studies, where tritium may be detected in several metabolites, including solvent.[9]

Another crucial and often difficult requirement of this technique is knowledge of the stereochemistry of the methyl group prior to reaction with the enzyme. Currently the most popular method involves the conversion of chiral acetate to acetyl coenzyme A, followed by digestion with malate synthase and fumarase. (This technology is very well described by Floss and Lee,[3] which also contains numerous primary references.) The percentage of tritium retention in the fumarase reaction is determined by liquid scintillation counting (LSC), and the enantiomeric excess in the starting acetate can then be calculated. The method depends on the ability to convert the subject chiral methyl group by a sequence of reactions of known stereochemistry into the methyl group of acetate. A more direct approach uses ^{3}H NMR to analyze *N*-methylated 2-methylpiperidine,[8] where the ability to transfer the chiral methyl species to 2-methylpiperidine builds on much of the same chemistry as the chiral acetate approach. The hindered rotation of the labeled *N*-methyl group makes the chemical shift of the triton dependent on the specific chirality of the methyl carbon. The only disadvantage of this approach compared with the enzymatic determination is the much larger quantities of tritium required for NMR analysis.

2.1 The Berberine Bridge Enzyme

A classic example of the application of chiral methyl groups for understanding cryptic stereochemistry is the study of the berberine bridge enzyme (BBE).[7] BBE catalyzes the ring-closure reaction shown in Figure 2 to yield scoulerine, which has axial (a) and equatorial (e) methylene hydrogens at the 8-position. These are well resolved ($\Delta\delta = 0.5$ ppm) and their proton chemical shifts are known. The four possible product configurations, $^{3}H_a$–$^{1}H_e$, $^{3}H_a$–$^{2}H_e$, $^{3}H_e$–$^{1}H_a$, and $^{3}H_e$–$^{2}H_a$, should all be distinguishable in the ^{3}H NMR spectrum by a combination of their chemical shifts and splitting patterns. Labeled scoulerine was enzymatically synthesized by incubation of *S*-adenosylmethionine with nor-reticuline in the presence of nor-reticuline-*N*-methyltransferase, followed by incubation with BBE. Independent incubations were performed using *S*-adenosylmethionine having exclusively an (*S*)-methyl or (*R*)-methyl, and the ^{3}H NMR spectrum of the purified scoulerine from the (*S*) reaction is reproduced in Figure 3. Spectra were acquired with and without proton decoupling to help assign tritons coupled to protons, and the deuterium couplings were too small to be resolved at this field [J(T,D) = 2.3 Hz]. The most striking feature of these spectra is the abundance of tritium at the 8-axial position (pro-*R*). This peak is unchanged by proton decoupling, indicating that its geminal partner is a deuteron. The equatorial (pro-*S*) triton is coupled to a proton [J(T,H) = 15.3 Hz]. Since these are the only two tritiated species observed, the berberine bridge enzyme must be acting stereospecifically with exclusive *inversion* during the ring closure reaction. The 4:1 axial to equatorial intensity ratio is explained by a kinetic isotope effect, where $k_H/k_D = 4$.

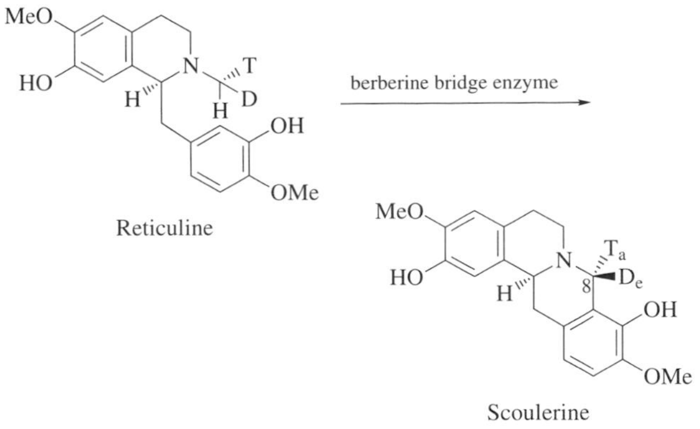

Figure 2 The biosynthesis of scoulerine by ring closure of a methyl group in reticuline, catalyzed by the berberine bridge enzyme

In this study, 29.6 MBq (800 μCi) of tritium was introduced into the in vitro enzymatic system via tritiated *S*-adenosylmethionine, and 5.7 MBq (153 μCi) of scoulerine was purified. One tritium per molecule represents a specific activity of 1064 GBq mmol^{-1}, so 5.7 MBq of scoulerine is 5.3 nmol (ca. 3.2×10^{15} tritiated molecules). At this level of tritium abundance, 24 h of signal averaging (around 50 000 transients) were required to achieve each of the NMR spectra in Figure 3. This is comparable to the length of a standard 2D NMR experiment on a biological sample, and is therefore a small instrumental time commitment relative to the information obtained.

2.2 Cephalosporin C

In the BBE study described above, the amount of radioactivity employed was relatively small since enzymatic reactions were carried out in a cell-free extract. In contrast, studies with a true in vivo step may require tritium levels in the multiple curie range. The use of high levels of radioactivity is readily justified in cell systems where the enzymes involved may be unknown or difficult to purify, and elucidation of the biosynthetic route to Cephalosporin C is a good example (Figure 4).[6] In particular, both the pro-*R* and pro-*S* methyl groups of valine are stereospecifically derivatized in this synthesis, whereby the pro-*S* methyl (*) becomes the exocyclic ester (position 3′) and the pro-*R* methyl (¶) participates in ring expansion (position 2). The two proton abstractions are catalyzed by different enzymes with very similar properties. Both are 2-oxoglutarate dependent dioxygenases that require Fe(II) and a reducing agent. The

For References see p. 4838

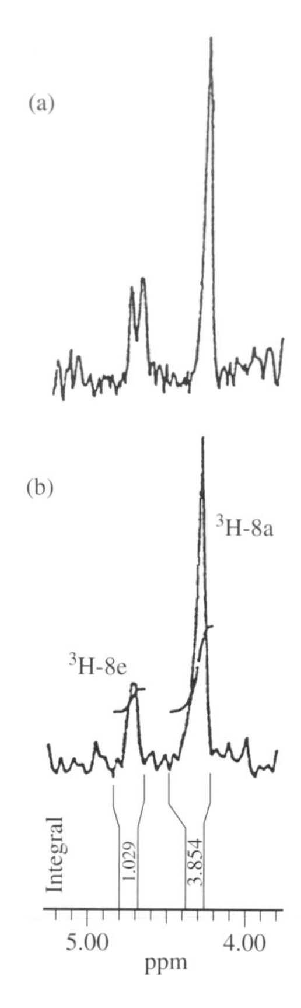

Figure 3 Tritium NMR spectra of scoulerine. The sample contained 5.7 MBq (153 μCi) of tritium in CD_3OD. (a) ^{1}H gated decoupled, WALTZ-16 ^{2}H broadband-decoupled, 47 347 transients. (b) Composite pulse broadband ^{1}H-decoupled, 76 472 transients. (Reprinted with permission from *J. Am. Chem. Soc.*, 1988, **110**, 7878. Copyright 1988 American Chemical Society)

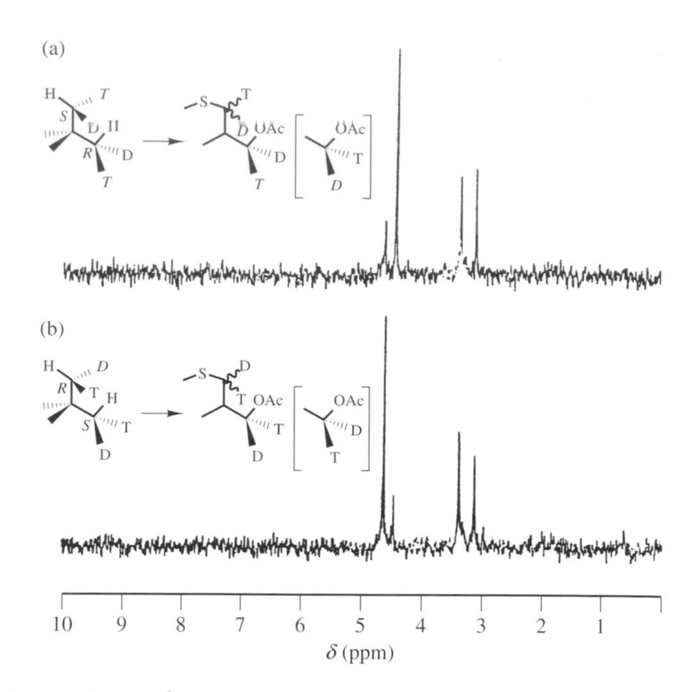

Figure 5 The ^{3}H NMR spectrum of Cephalosporin C derived from the starting species shown on each figure section. (Reprinted with permission from *Biochem. J.*, 1984, **222**, 777. Copyright 1984 Biochemical Society, London)

stereochemistry of the proton abstractions in these reactions was studied in a very elegant set of experiments using valine molecules with either (*R*) or (*S*) chiral methyl at either the pro-*R* (¶) or pro-*S* (*) positions. In this way, both enzyme mechanisms can be studied simultaneously in an in vivo synthesis. Each NMR study required ca. 90 GBq (2.5 Ci) of labeled valine substrates added in the cellular media in order to isolate 110 MBq (3 mCi) of cephalosporin (ca. 0.1% yield).

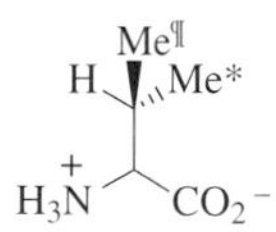

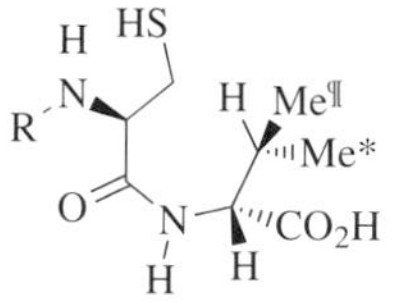

Figure 4 The biosynthesis of Cephalosporin C. The methyl groups of valine are stereospecifically incorporated into the product. This allows analysis of both methyl reactions in the same tritium NMR experiment (see text)

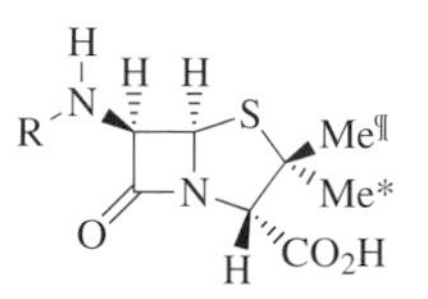

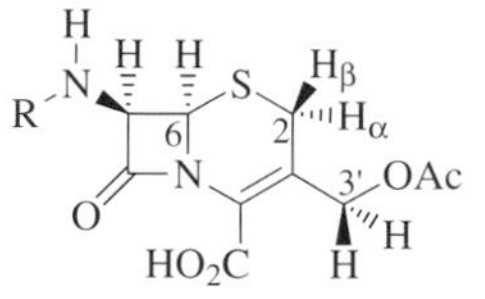

The ^{3}H NMR spectra of the isolated products are shown in Figure 5, and the combination of starting valine species for each sample is shown on the figure. The signals at 3.2 ppm and 3.4 ppm are the C-2 tritons, and clearly the ring expansion at the 2-position shows little stereochemical preference, since an equal mixture of the possible products is formed. This is consistent with a radical mechanism at this step in the biosynthesis.

The resonances at 4.6 ppm and 4.8 ppm arise from the C-3′ tritons, but were not stereospecifically assigned. In contrast to the ring expansion reaction, there is a strong stereochemical preference in reactions at the 3′-position, as shown by the dependence of the 4.8 ppm and 4.6 ppm signal intensities on the configuration of the pro-*S* methyl in the starting valine. The patterns are very similar to those observed in the BBE study described above (see Figure 3), but in this case the minor signal does not represent an R–CHT–X species since no proton coupling is observed. Since impurities have been ruled out, the remaining possibility is that the enzyme is only about 90% stereospecific, and a 'partial' preference is not uncommon in enzyme-catalyzed reactions.[10] When signals from the R–CHT–X species are not detected, the isotope effect (k_H/k_D) can only be estimated by the size of the spectral noise. In this case, k_H/k_D must be at least 5 for the R–CHT–X species to be unobservable.

The two similar enzymes studied here have clearly different modes of action in this biosynthetic pathway. The radioactivity and excellent NMR properties of tritium were essential in the synthesis of precursors, isolation of products, and final product analysis in this study.[6]

3 PROTEIN–LIGAND INTERACTIONS

In the study of protein–ligand interactions, the goal is both detailed structural information on binding sites and dynamic information on binding modes and kinetics.[11] There are two major techniques for acquiring detailed structural information on protein structure; X-ray crystallography and NMR spectroscopy. Crystallography can give high-resolution structures but only limited information on dynamics. NMR is an excellent tool for dynamic studies but the size of the system that can be studied is limited, usually by proton resolution, to around 35–40 kDa. Tritium NMR can complement both these approaches. In cases where a crystal- or NMR-derived structural model exists, tritium NMR provides a method for probing dynamics that far exceeds data from crystallography and surpasses NMR with more conventional isotopes in its simplicity and application to larger protein systems. When NMR or crystal structural data are absent, useful information from ^{3}H chemical shifts, relaxation studies, and ^{3}H–^{1}H NOEs can be used to characterize the ligand binding site.

Here we review two works which are not only exemplary in their demonstration of the power of ^{3}H NMR in ligand binding investigations, but also show two distinct approaches to the problem which are used repeatedly in the literature.[12–16] We highly recommend these two studies for those interested in these techniques. The study of maltose binding protein shows how a ligand undergoing exchange at an intermediate rate can be analyzed by standard techniques, e.g. one-dimensional spectra, saturation transfer, and 2D EXSY spectroscopy.[13] Next we describe how the relaxation behavior of ^{3}H augmented extensive computer simulation in studying the dynamics of a small inhibitor peptide binding to bacterial collagenase.[14]

3.1 Maltose Binding Protein

Maltose binding protein (MBP) is involved in the transport of maltodextrins (maltose and its oligomers) across the periplasmic space in *E. coli*. MBP is a soluble, monomeric protein of 40 kDa, with known primary sequence, which has a dissociation constant for maltose of about 10^{-6} M, increasing somewhat for the longer maltose oligomers. Although there has been a crystal structure determined for MBP, there were questions remaining about the binding kinetics and possible differences in binding of the α and β anomers.

Maltose [Figure 6(d)] and longer maltodextrins were labeled, and tritium was incorporated to a level of 5–20% at the C-1 position. Hence, a limited percentage of ligand molecules were visible to tritium NMR detection. The labeled ligands were characterized by tritium and proton NMR techniques prior to binding studies. The proton-decoupled ^{3}H NMR spectrum of tritiated maltose in Figure 6(a) is representative of all the oligosaccharides, and shows lines from the α and β anomers at 5.15 ppm and 4.55 ppm (relative to DSS) respectively, with the expected intensity ratio of 40:60.

Initial investigations of the interaction of a labeled ligand and macromolecule invariably involve simple 1D NMR observations of a number of ligand/macromolecule mole ratios. When MBP (ca. 2.5 mM solution, in phosphate buffer at pH 7.0) is titrated with labeled maltose two lines appear, at 3.5 ppm and 3.4 ppm [Figure 6(b), 0.25 equiv.]. The lines are broadened relative to free maltose. This is a common effect in ligand binding studies, which results from the *bound* sugar molecules tumbling with the longer correlation time of the protein. As further additions are made exceeding a 1:1 mole ratio of maltose to MBP, additional peaks due to free maltose are observed at the same chemical shifts as in the absence of protein [Figure 6(b), 2 equiv.].

Very often several deductions can be drawn even from simple 1D ^{3}H NMR experiments. In this case, the observations are: (a) the large shifts of the bound resonances to lower frequency suggests that the labeled end of the maltose molecule makes direct contact with an aromatic residue in the protein binding site, probably tryptophan 230; (b) the degree of shift of the two anomers is different, which suggests an asymmetric environment around the binding site; and (c) the relative intensity of the bound peaks differs below and above stoichiometric addition. With excess protein, the intensities of the bound forms of the two anomers initially remain at their equilibrium solution value. If the intensities are followed over time, a gradual conversion of the β form into α is observed. When the sugar is present in excess, the *bound* α and *free* β peaks both exceed their equilibrium levels. Both of these observations indicate that there is significant anomeric specificity in binding.

The absolute identity of the bound resonances was determined by saturation transfer and 2D EXSY [Figure 6(c)] techniques, which confirmed that the α anomer is preferentially bound by the protein. These experiments were enhanced by the large chemical shift differences for bound and free sugar resonances, and the saturation transfer studies allowed independent determinations of the dissociation rate constants for α and β anomers. Both the identity of the bound species and the rate constant information is accessible from 2D EXSY experiments, where transfer of magnetization by chemical exchange between bound and free maltose gives rise to the observed cross peaks, and confirms the bound resonance at low frequency as the bound α anomer [Figure 6(c)].

Spectra of longer maltodextrins, containing three to six glucose units, in the presence of excess MBP contained resonances with varying linewidths and average chemical shifts, as a direct result of the range of binding affinities and chemical exchange rates prevalent for the various bound α and β species. Such an array of cases would have been impossible to understand without the simplification conferred by tritium labeling of the ligand and tritium NMR analysis.

In these studies tritium proved to be an excellent direct detection probe with which (a) clear anomeric specificity was observed in a situation where there were few alternative methods for unequivocally demonstrating it, (b) two types of binding (reducing end versus middle or remote) were deduced, this difference being undetectable by other means, and (c) the binding of longer oligomers was characterized with the same or **greater** ease than the maltose case. An added convenience was that the labeling chemistry was straightforward for all the ligands.

3.2 Bacterial Collagenase

As we have mentioned, the other common way to study protein–ligand interactions is by an analysis of the relaxation properties of tritium. An example of using ^{3}H NMR to study the dynamic and conformational aspects of an inhibitor inter-

For References see p. 4838

action with bacterial collagenase is the most thorough example of this technique. The protein system was bacterial collagenase from *Achromobacter iophagus* (ca. 71 kDa), and the labeled inhibitor was succinyl-L-prolyl-L-alanine ($K_i = 2.1 \times 10^{-4}$ M), one of a range of competitive inhibitors under study.

The studies involved rigorous analysis of the relaxation behavior of selected spin systems in labeled inhibitor molecules in the absence and presence of bacterial collagenase. In contrast to the MBP/maltodextrin experiments described above, in this example the experimental data are only the starting point for extensive calculation and interpretation of physical models. Hence sample preparation included rigorous deoxygenation to facilitate accurate relaxation rate measurements, and repetitive determination of these parameters.

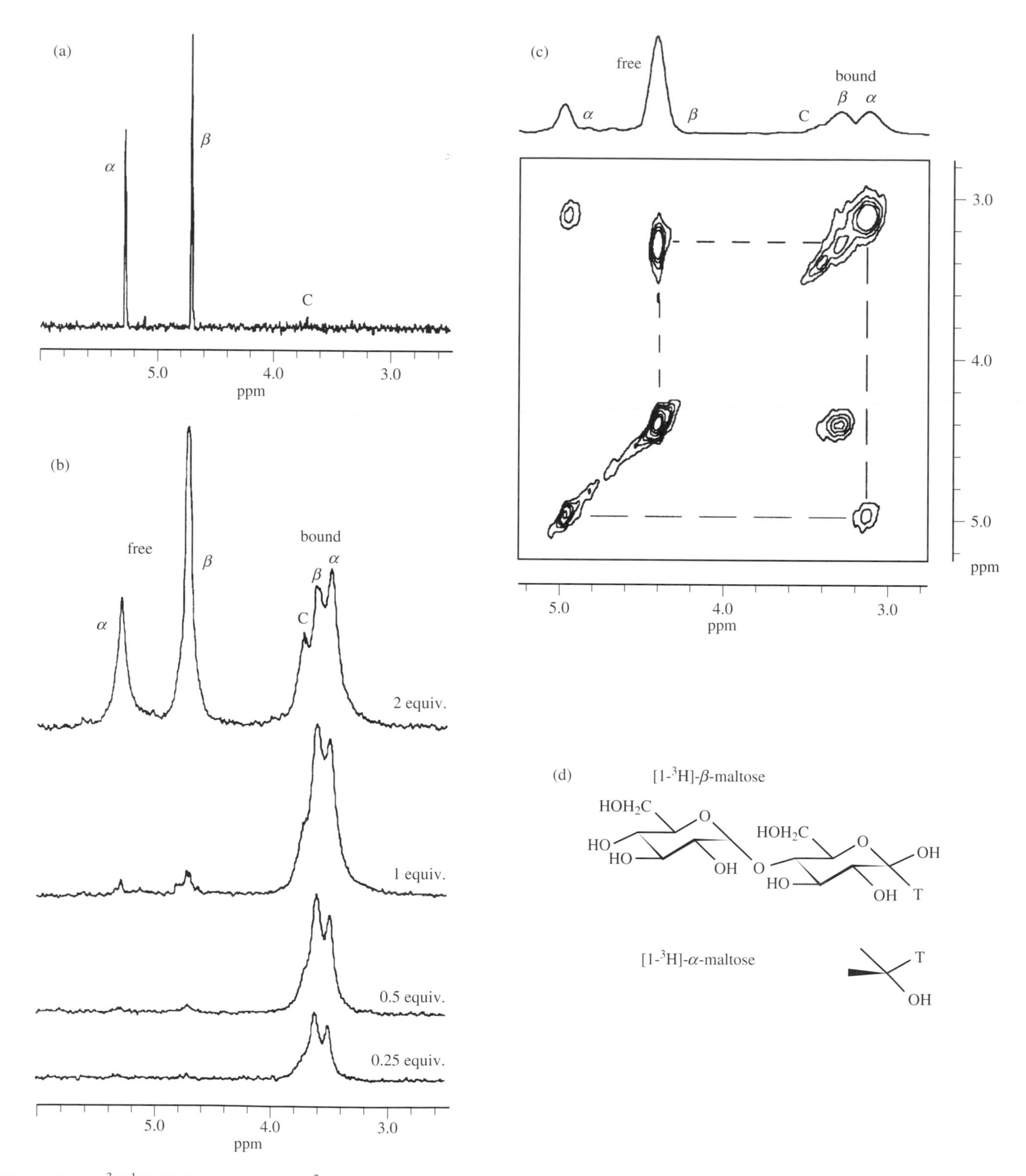

Figure 6 (a) $^3H[^1H]$ NMR spectrum of [1-^{3}H]maltose, (b) the titration of maltose binding protein with tritium-labeled maltose, (c) the 2D EXSY spectrum of excess maltose in the presence of maltose binding protein, (d) the structure of [1-^{3}H]maltose

The solution behavior of the free ligand was first characterized, before proceeding to inhibition studies. Measurement of the relaxation properties of the tritium pair on the contiguous methylenes in the succinyl residue, and comparison with the relaxation behavior of the complementary protons of the analogously deuterated molecule, allowed the calculation of overall and internal correlation times. From these values, intertriton and interproton distances were estimated, rotamers proposed, and the fractional populations and coupling constants so derived were compared with the experimentally determined parameters. In these two ways, both the motional (i.e. correlation times) and conformational (rotamer) populations of the free inhibitor were deduced.

Comparable relaxation measurements in the inhibitor/collagenase samples were conducted with a 20–30-fold excess of inhibitor (e.g. 97% free, 3% bound in many experiments), and the spectra showed the weighted average of the NMR parameters of these subspectra, i.e. the chemical shifts and linewidths, etc. We calculate that the tritium samples contained approximately 1000 MBq (27 mCi); [protein] = 40 μM, therefore [inhibitor] = 1.2 mM, assume 0.5 mL sample (= 0.6 μmol), specific activity = 45 mCi μmol^{-1}. By analysis of the cross relaxation of the Ala signals of the proton spectra of the inhibitor/collagenase sample, an indication of the motional freedom of this residue in the presence of protein could be gained, and this indicated that the α-proton had restricted motion. In a similar manner to the free inhibitor, relaxation measurements on the succinyl (Suc) methylenes in the presence of protein permitted the deduction of overall correlation times, internuclear distances, and a qualitative estimate of the order parameter (a measure of the internal mobility of the Suc) for the bound inhibitor. The similarity of tritium and proton cross-relaxation terms for the complementary methylene pairs suggests a similar geometry, and this is satisfied by the carbonyl and carboxy groups adopting a *trans* conformation when bound to the protein.

The natural substrate for this enzyme has the general sequence X–Gly–Pro–Y, and is cleaved at the peptide bond between X and Gly. The studies described above clearly showed that the carboxylate group of the inhibitor Suc residue has the strongest interaction with the enzyme, and that the carbonyl of the Ala residue appeared to be another anchoring site. Using these clues and detailed knowledge of the active site of other zinc proteinases, the structure of the inhibitor may be aligned with that of the natural substrate to identify important structural features. This approach was used to guide the design of improved inhibitors.

4 MACROMOLECULES LABELED WITH TRITIUM

Biological molecules are produced for NMR studies in a variety of ways. The most common are: solid phase synthesis for DNA, peptides, and smaller proteins; in vitro synthesis for RNA production; and in vivo or cellular expression systems, commonly used to prepare ^{13}C and ^{15}N uniformly enriched protein samples. Tritium labeling may be used in conjunction with each of these techniques. The cephalosporin example above demonstrates the in vivo method, while both DNA and RNA have been labeled at several sites demonstrating the use of solid phase and in vitro methods.

The self-complementary DNA dodecamer with sequence 5′ dCGCGAATTCGCG 3′ has been extensively studied. The proton spectrum is fully assigned and detailed X-ray crystal and NMR structures have been published. Both the crystal structure and recently published NMR data show interesting DNA–water interactions. This molecule has been labeled at the 2- and 8-positions of adenosine-5 (5′ dCGCGA*ATTCGCG 3′) by addition of a tritiated phosphoramidite precursor at the appropriate step of the solid phase synthesis. Tritium NMR was expected to be helpful in the further investigation of water interactions since a major difficulty with the proton NMR studies is water suppression.

The one-dimensional spectrum of the tritiated DNA has two lines at 8.1 ppm (8-position) and 7.2 ppm (2-position). Two-dimensional tritium-detected ^3H–^1H NOESY spectra of the molecule in H_2O solution showed all the expected intramolecular DNA proton cross peaks, plus cross peaks at the water chemical shift (Figure 7). No water suppression was necessary. In many respects the spectrum in Figure 7 represents a milestone in tritium NMR studies. It is the first and essentially the only heteronuclear ^3H–^1H NOESY collected on a macromolecule and fully assigned. Because it directly mirrors the proton spectrum, it demonstrates the fidelity of tritium NMR to proton NMR for the analysis of 'close in space' interactions.

The signal at the water chemical shift for both tritons is antiphase, while analogous proton data show purely in-phase cross peaks for this interaction. The major difference between the homo- and heteronuclear spectra is that the water is off-resonance in the tritium-detected experiments. This anomaly is explained by recent observations of multiple quantum inter-

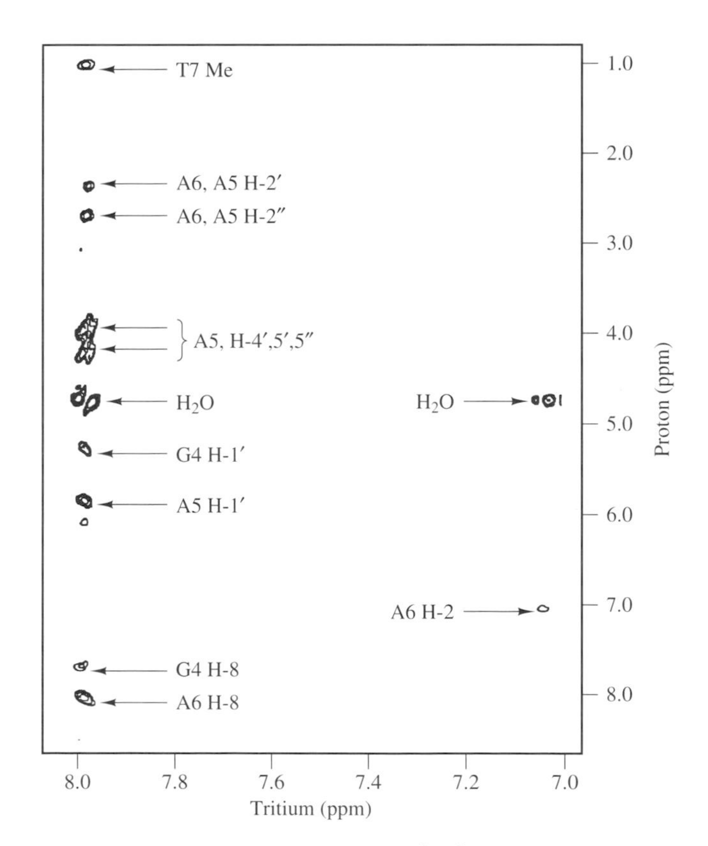

Figure 7 The 640-MHz tritium-detected ^3H–^1H NOESY spectrum of a ^3H-labeled DNA oligomer

For References see p. 4838

actions between solvent and solute when the solvent is off-resonance (see '***Concentrated Solution Effects***'). Although the multiple quantum nature of these cross peaks limits their usefulness in studying DNA–water interactions, their intensity and their acquisition without water suppression indicate that tritium NMR is an excellent method for further investigation of this multiple quantum phenomenon.

RNA has been a particularly difficult molecule to study by NMR methods. The poor dispersion of spectral lines in the ^{1}H, ^{13}C, and ^{15}N spectra has plagued the structural spectroscopist. Tritium NMR is an effective method for increasing resolution in RNA spectra. Figure 8 compares the ^{3}H NMR spectrum of a 64-base RNA molecule dissolved in H_2O and the corresponding proton spectrum of an analogous sample dissolved in D_2O (spectral conditions were not identical). The RNA molecule was labeled at the 2-position of each adenosine, giving tritium at 18 of the 75 aromatic sites in the molecule. The spectral simplification afforded by ^{3}H NMR is actually much more dramatic if it is compared with the proton spectrum in H_2O solution, where dozens of additional exchangeable protons also resonate in the region shown.

With these simpler spectra, it is clear from Figure 8 that the 18 labeled sites on the molecule are not equally represented. Temperature studies confirmed that a significant amount of conformational exchange is occurring in this molecule. Various assignment techniques, including one-dimensional tritium–proton NOE experiments, indicated that the highest degree of conformational exchange is in the central region of the molecule, which includes the bases most important for the self-cleaving properties of this RNA molecule. This conformational exchange is totally undetectable in the proton spectrum and would be difficult to analyze by a technique other than tritium NMR.

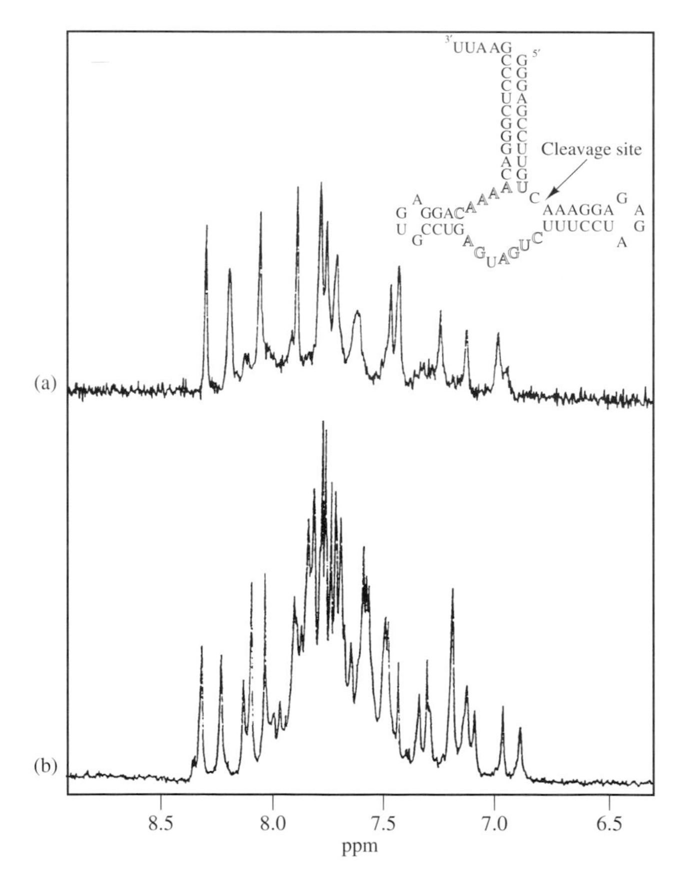

Figure 8 A comparison of the aromatic regions of (a) the ^{3}H (533 MHz) and (b) ^{1}H (600 MHz) NMR spectra of the 65 base RNA molecule shown in (a)

The DNA and RNA samples were labeled at high specific activity. The DNA was approximately 1500 GBq mmol^{-1} (40 Ci mmol^{-1}) at two sites and the RNA was more than 15 000 GBq mmol^{-1} (400 Ci mmol^{-1}) at 18 sites. These high levels of activity are beneficial for the rapid acquisition of spectra, but also accelerate radiation decomposition of the substrates. As a result, the samples discussed here had a useful lifetime of only 3 to 5 weeks. In cases where samples are concentration limited, high specific activity is the only alternative and a short sample lifetime should be expected. If the working concentration is not strictly limited by the properties of the sample, it is better to label at lower abundance and store the sample in dilute form, concentrating it only for NMR experiments. DNA and RNA labeling with tritium requires about the same amount of effort as the more common ^{13}C and ^{15}N techniques. In contrast to these nuclei, tritium is much more sensitive and can be directly detected without the need for water suppression.

5 CONCLUSION

The NMR properties of tritium are slightly better than those of ^{1}H, with essentially zero natural abundance. Hence it is possible to detect tritium with more sensitivity than the hydrogen it replaces, and spectral simplification is assured. Radioactivity is the only issue hindering the universal application of tritium in biological NMR studies.

6 RELATED ARTICLES

Biological Macromolecules: Structure Determination in Solution; Biological Macromolecules; Biological Macromolecules: NMR Parameters; Biosynthesis & Metabolic Pathways: Carbon-13 and Nitrogen-15 NMR; Biosynthesis & Metabolic Pathways: Deuterium NMR; Concentrated Solution Effects; DNA: A-, B-, & Z-; Enzymatic Transformations: Isotope Probes; Multidimensional Spectroscopy: Concepts; NOESY; Nucleic Acid Structures in Solution: Sequence Dependence; Nucleic Acids: Chemical Shifts; Nucleic Acids: Phosphorus-31 NMR; Nucleic Acids: Spectra, Structures, & Dynamics; Selective Relaxation Techniques in Biological NMR; Structures of Larger Proteins, Protein–Ligand, & Protein–DNA Complexes by Multi-Dimensional Heteronuclear NMR; Tritium NMR.

7 REFERENCES

1. E. A. Evans, D. C. Warrell, J. A. Elvidge, and J. R. Jones, 'Handbook of Tritium NMR Spectroscopy and Applications', Wiley, Chichester, 1985, pp. 1–249.
2. J. M. A. Al-Rawi, J. A. Elvidge, D. K. Jaiswal, J. R. Jones, and R. Thomas, *J. Chem. Soc., Chem. Commun.*, 1974, 220.
3. H. G. Floss and S. Lee, *Acc. Chem. Res.*, 1993, **26**, 116.
4. L. J. Altman, C. Y. Han, A. Bertolino, G. Handy, D. Laungani, W. Muller, S. Schwartz, D. Shanker, W. H. de Wolf, and F. Yang, *J. Am. Chem. Soc.*, 1978, **100**, 3235.

5. D. J. Aberhart, 'Synth. Appl. Isot. Labeled Compd., Proc. Int. Symp.', ed. W. P. Duncan and A. B. Susan, Elsevier, Amsterdam, 1983, p. 309.
6. C.-P. Pang, R. L. White, E. P. Abraham, D. H. G. Crout, M. Lutstorf, P. J. Morgan, and A. E. Derome, *Biochem. J.*, 1984, **222**, 777.
7. T. Frenzel, J. M. Beale, M. Kobayashi, M. H. Zenk, and H. G. Floss, *J. Am. Chem. Soc.*, 1988, **110**, 7878.
8. F. A. L. Anet, D. J. O'Leary, J. M. Beale, and H. G. Floss, *J. Am. Chem. Soc.*, 1989, **111**, 8935.
9. R. D. Newmark, S. Un, P. G. Williams, P. J. Carson, H. Morimoto, and M. P. Klein, *Proc. Natl. Acad. Sci. USA*, 1990, **87**, 583.
10. N. D. Priestley, H. G. Floss, W. A. Froland, J. D. Lipscomb, P. G. Williams, and H. Morimoto, *J. Am. Chem. Soc.*, 1992, **114**, 7561.
11. D. E. Wemmer and P. G. Williams, in 'Methods in Enzymology', ed. T. L. James and N. J. Oppenheimer, Academic Press, Orlando, FL, 1994, Vol. 239, p. 739.
12. J. N. S. Evans, G. Burton, P. E. Fagerness, N. E. Mackenzie, and A.I. Scott, *Biochemistry*, 1986, **25**, 905.
13. K. Gehring, P. G. Williams, J. G. Pelton, H. Morimoto, and D. E. Wemmer, *Biochemistry*, 1991, **30**, 5524.
14. V. Dive, A. Lai, G. Valensin, G. Saba, A. Yiotakis, and F. Toma, *Biopolymers*, 1991, **31**, 305.
15. A. Bergerat, W. Guschlbauer, and V. Fazakerley, *Proc. Natl. Acad. Sci. USA*, 1991, **88**, 6394.
16. T. M. O'Connell, J. T. Gerig, and P. G. Williams, *J. Am. Chem. Soc.*, 1993, **115**, 3048.

Acknowledgements

M.G.K. is supported by the Office of Energy Research, Office of Health and Environmental Research, Health Effects Research Division of the US Department of Energy under Contract DE-AC03-76SF00098, and through Instrumentation Grants from the US Department of Energy, DE FG05-86ER75281, and the National Science Foundation, DMB 86-09035. P.G.W. is supported by the Biomedical Research Technology Program, National Center for Research Resources, US National Institutes of Health, under Grant P41 RR01237, through Contract DE-AC03-76SF00098 with the US Department of Energy.

Biographical Sketches

Mark G. Kubinec. *b* 1964. B.S., 1986, biochemistry, B.S., 1987, chemistry, Michigan State University, USA, Ph.D., 1994, Chemistry, University of California, Berkeley, USA. Postdoctoral fellow, Lawrence Berkeley National Laboratory, 1995–present.

Philip G. Williams. *b* 1956. B.Sc., 1980, chemistry, Ph.D., 1984, Chemistry, University of New South Wales, Australia. Research Fellow, Ludwig Institute for Cancer Research, Sydney, Australia, 1984–85; staff scientist, Lawrence Berkeley Laboratory, USA, 1986–present. Approx. 70 publications Current research specialties: tritium labeling, tritium NMR.

Two-Dimensional Carbon–Heteroelement Correlation

Stefan Berger
Philipps University Marburg, Germany

1 INTRODUCTION

Two-dimensional correlation spectroscopy, originally proposed by Jeener[1] in 1971, was developed by Ernst et al.[2] and abbreviated with the term COSY. Since then, this method has played the most important role in the structure elucidation of organic compounds. In 1978, Freeman and Morris[3] created the polarization transfer experiment to correlate chemical shifts of carbons and protons. This type of H,C COSY experiment dominated the following decade because it could be easily implemented on FT spectrometers which were constructed for proton-decoupled carbon NMR measurements. Although the more attractive 'inverse' (i.e. proton) detection was invented in the early 1980s,[4] these new methods struggled because the existing spectrometer generation did not support the necessary phase coherence between the proton decoupler and the receiver. Since 1989, inverse detection has been a standard feature of modern commercial instruments. For the inverse H,C COSY experiment, two pulse sequences were used: HMQC[5] and HSQC.[6] The main difference between them is the order of the coherence which is relevant to the information transfer between the nuclei. The abbreviations HMQC and HSQC refer to Heteronuclear *Multiple* and *Single* Quantum Coherence, respectively.

Heteronuclear correlation spectroscopy is not restricted to the spin pair ^{1}H,^{13}C. ^{1}H,^{n}X correlations are also used for structure elucidation, e.g. ^{1}H,^{31}P for organophosphorus compounds. In fact, the inverse experiments were introduced for the spin pair ^{1}H,^{15}N. Today, indirect detection is the method of choice for rare nuclei.

The molecular skeleton, however, is usually made from carbon. In organometallic, bioorganic, and heterocyclic chemistry we find, in addition, one or more heteroelements embedded in the carbon framework, whereas the protons only cover the outer surface of the molecule. It is therefore straightforward to create 2D NMR methods to detect directly the connectivity between the carbon atoms and magnetically active heteroelements. In this article we review the hardware requirements

For References see p. 4844

needed for this purpose, discuss a variety of pulse methods used, and discuss the chemical applications achieved so far with these techniques.

2 SELECTION OF NUCLEI

Possible correlation partners of ^{13}C are all spin $\frac{1}{2}$ nuclei, such as ^{19}F, ^{15}N, ^{31}P, ^{29}Si, ^{77}Se, ^{119}Sn, ^{125}Te, ^{195}Pt, ^{199}Hg, or ^{207}Pb. For special applications quadrupolar nuclei such as ^{2}H, ^{6}Li, or ^{11}B may also be of interest. Of course, the relative importance of these correlations is dictated by the chemistry of these classes of compounds. In a $^{13}C,^{n}X$ correlation one is faced with the question of which nucleus one should detect and which should develop chemical shift during t_1. There are complex arguments based on the pulse sequences used, the T_1 and T_2 relaxation times of the two nuclear species involved, the spectral width required in both the F_1 and F_2 dimensions, and many more. At first sight, perhaps surprisingly, the natural abundance of a nucleus gives no direct argument, since for correlation NMR it is always the product of both natural abundances that is decisive. There is, however, a practical reason why the relative natural abundances have to be considered. In the case of, for example, a $^{13}C,^{31}P$ correlation, each rare ^{13}C spin 'sees' a phosphorus, whereas only 1% of the phosphorus atoms 'see' a ^{13}C spin. The suppression of unwanted signals must therefore be rather efficient if one is to detect the nucleus with the higher abundance.

Next, one should consider the linewidth of the 1D spectra of both nuclei. The detected nucleus should be the one with the smaller linewidth, since a short T_2 leads to signal losses during the pulse sequence. In the case of $^{13}C,^{n}X$ correlations (^{n}X = ^{77}Se, ^{125}Te, ^{195}Pt, ^{199}Hg, and ^{207}Pb), carbon has the smaller linewidth. Because of the chemical shift anisotropy and the temperature dependence of the chemical shift of heavy nuclei, the measurement of ^{13}C would therefore be favored in these cases.

For carbon detection, in addition, the large NOE enhancement caused by proton irradiation has to be taken into account. Heteronuclei usually have no directly bound proton, their NOE factors are therefore usually less than the maximum value. To facilitate further discussion, the γ values and the natural abundances of the nuclei used for $^{13}C,^{n}X$ correlation are given in Table 1.

By inspection of Table 1, it is possible to group the possible correlation partners of carbon into three classes. One group has lower γ values (^{2}H, ^{6}Li, ^{15}N), one group quite comparable γ values (^{29}Si, ^{195}Pt, ^{199}Hg, ^{207}Pb), and a third group definitively higher γ values (^{11}B, ^{19}F, ^{31}P, ^{119}Sn, ^{125}Te). For this last group, therefore, using the HMQC method, ^{n}X detection should be preferable if arguments based on γ values were the only factors. However, even in this group polarization transfer with carbon detection might be competitive, since the higher γ value of the X nucleus is transferred to the carbon signal, and this itself is enhanced by NOE.

The repetition time of these experiments is governed by the T_1 relaxation of nucleus 1 in HMQC and HSQC, but of nucleus 2 in the polarization transfer experiment. At high magnetic fields, heteronuclei relax not only by dipolar but also, to a considerable extent, by other relaxation mechanisms such as chemical shift anisotropy. Hence, the relaxation times of metal nuclei are very often in the range of 1 s. This again would favor the HMQC method with ^{n}X detection or the polarization transfer method with ^{13}C detection.

A final argument stems from the required spectral width in both dimensions. It is very inconvenient to have a large spectral width in F_1 because this would be very time-consuming; furthermore, signal folding in F_1 cannot be filtered out. Thus, it is often better to detect the nucleus with the larger spectral width. In addition, this can be more easily separated into different parts if the spectral width is too large to allow reasonably uniform 180° pulses.

Table 1 Gyromagnetic Ratios and Natural Abundances of Selected Nuclei

Nucleus	γ (10^7rad $T^{-1}s^{-1}$)	Natural abundance
^{2}H	4.106	0.015
^{6}Li	3.937	7.42
^{11}B	8.583	80.42
^{13}C	6.726	1.108
^{15}N	−2.711	0.37
^{19}F	25.166	100
^{29}Si	−5.314	4.7
^{31}P	10.829	100
^{77}Se	5.101	7.58
^{119}Sn	−9.971	8.58
^{125}Te	−8.452	6.99
^{195}Pt	5.75	33.8
^{199}Hg	4.769	16.84
^{207}Pb	5.597	22.6

3 HARDWARE CONSIDERATIONS

The experiments described in this article are usually performed under complete proton decoupling. Thus, in addition to the ^{1}H coil and the standard ^{2}H lock signal, one needs two more frequencies to be delivered to the sample. The simplest approach would be a fixed probehead, with an additional coil double-tuned to, for example, ^{13}C and ^{15}N. This is very popular in protein research, because proteins doubly labeled with ^{13}C and ^{15}N are now fairly readily available. The drawback of such a probehead is obvious; it is restricted to the special spin triple $^{1}H,^{13}C,^{15}N$, and no other experiment is possible. A second approach would be to have one multinuclear tuneable coil, and double tune the second X nucleus to the ^{1}H coil. One has to decide which X nucleus to choose. For the experiments described here, ^{13}C was chosen since in organoelement chemistry carbon is the common denominator. The spectrometer needs an independent third radiofrequency channel, which can deliver its pulses to the appropiate coil of the probehead. Such a third radiofrequency channel usually requires a second frequency synthesizer, some gating electronics, and a power amplifier. The feature of a third radiofrequency channel is standard in research instruments built later than 1990; however, it has been adapted in many NMR laboratories to older instruments.

4 SELECTION OF PULSE METHODS

The basic pulse sequences known from $^{13}C,^{1}H$ correlation have all been adapted for $^{13}X,^{n}X$ correlation under proton decoupling. In Figure 1 the three common pulse sequences are shown.

In the HMQC technique [Figure 1(a)], transverse magnetization of nucleus 1 is first generated. Antiphase magnetization which evolves during the first delay is transferred to nucleus 2 which evolves under the chemical shift during t_1. This is transferred back to nucleus 1. During the final acquisition time, t_2, the chemical shift of nucleus 1 and heteronuclear coupling to nucleus 2 evolve; thus, at the end of this pulse sequence we measure the magnetization which originates from the equilibrium magnetization of the detected nucleus. Neglecting, at this stage, relaxation arguments, the overall sensitivity[7] of the HMQC sequence is then proportional to $\gamma_1^{5/2}$. Thus, for a $^{13}C,^{n}X$ correlation with this technique, theory strongly suggests the choice of the nucleus with the higher γ value as the detected nucleus.

The HSQC sequence [Figure 1(b)] consists of two INEPT sandwiches, which are separated by the evolution time t_1. Here we again start with the equilibrium magnetization of nucleus 1. If we compare the magnetization transfer ways with the HMQC method, we realize that in both experiments the same equilibrium magnetization is finally used for detection; hence, the sensitivity arguments stay the same. For the application discussed here, homonuclear spin couplings can be neglected and thus the HSQC method should yield comparable results to HMQC. The HMQC experiment, however, does need more than double the number of radiofrequency pulses. For practical reasons of error propagation, it might therefore be less favorable. Another point of discussion would be the different relaxation properties of single- and multiple-quantum coherences which, in the case of heteronuclear correlation, are not predominant.

In contrast to the HMQC and the HSQC experiment, transverse magnetization of nucleus 2, which is **not** detected later, will be created in the beginning of the polarization transfer experiment [Figure 1(c)]. In the following evolution time this magnetization is modulated with the chemical shift of nucleus 2. Since nucleus 1 is detected, the overall sensitivity in this experiment[7] is related to the factors $\gamma_1^{3/2}\gamma_2$; hence, the choice of the detected nucleus is not as obvious as in the two former pulse sequences. There is a further point to be considered in a comparison of the three pulse sequences. In both HMQC and HSQC magnetization is transferred twice, forwards and back-

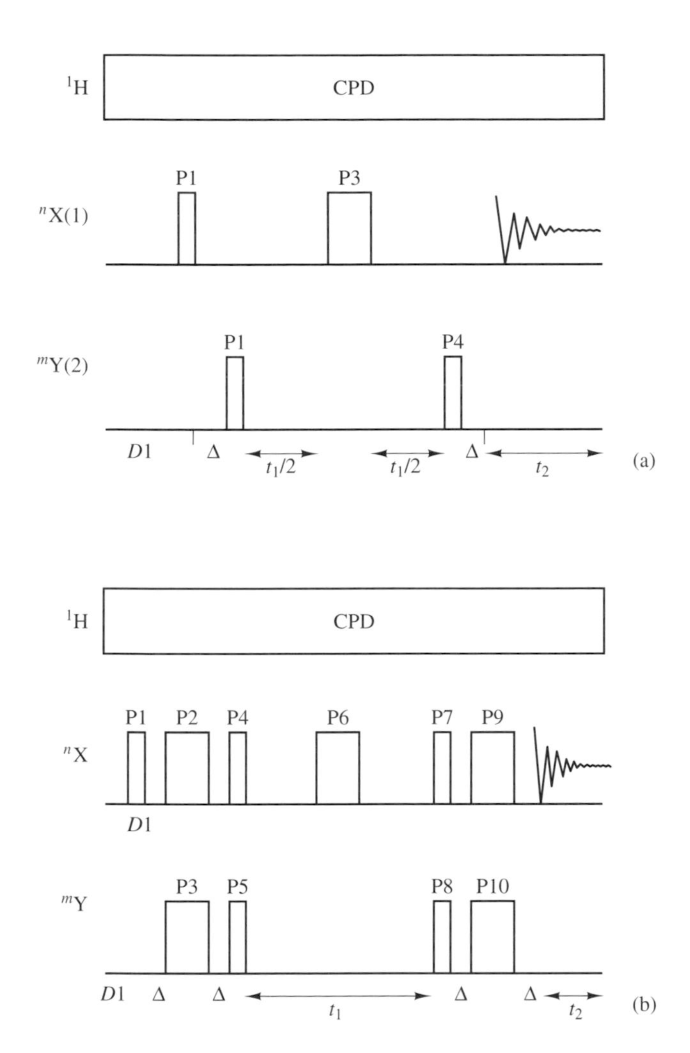

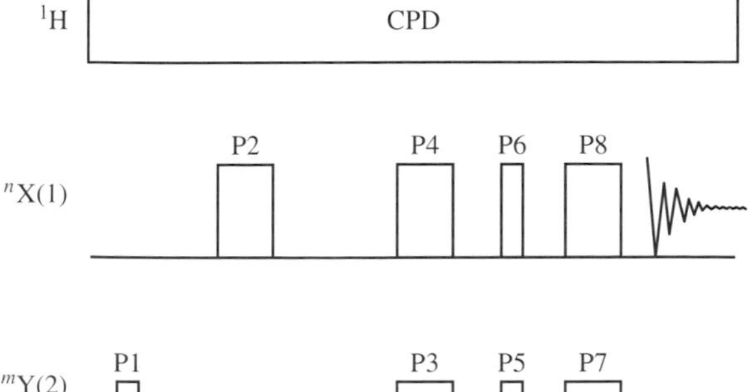

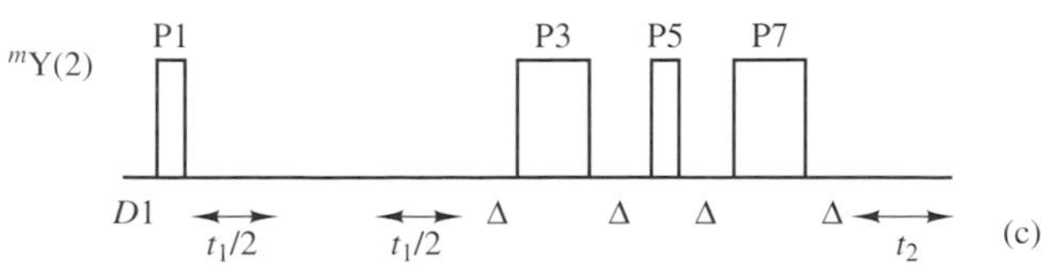

Figure 1 Pulse sequences for X,Y correlation under proton decoupling. (a) HMQC pulse sequence adapted for X,Y correlation; P1, P2, and P4 represent 90° pulses, P3 a 180° pulse, $D1$ = relaxation delay, $\Delta = \frac{1}{2}J(\text{X,Y})$. The pulse and receiver phases are: P1, P3: x, P2: x, $-x$, incremented by time proportional phase incrementation (TPPI), P4: x, x, $-x$, $-x$, rec: x, $-x$, $-x$, x. (b) HSQC pulse sequence adapted for X,Y correlation under ^{1}H decoupling; P1, P4, P5, P7, and P8 represent 90° pulses, P2, P3, P6, P9, and P10 180° pulses, $D1$ = relaxation delay, $\Delta = \frac{1}{4}J(\text{X,Y})$. The pulse and receiver phases are: P1, P2, P6, P9: x, P3, P10: $(x)_4$, $(-x)_4$, P4, P7: y, P5: x, $-x$, incremented by TPPI, P8: x, x, $-x$, $-x$, rec: x, $-x$, $-x$, x. (c) Polarization transfer sequence adapted for X,Y correlation under ^{1}H decoupling; P1, P5, and P6 represent 90° pulses, P2, P3, P4, P7, and P8 180° pulses, $D1$ = relaxation delay, $\Delta = \frac{1}{4}J(\text{X,Y})$. The pulse and receiver phases are: P1: x, P2, P3, P4, P7: x, x, $-x$, $-x$, P5: y, $-y$, incremented by TPPI, P6: $(x)_4$, $(y)_4$, $(-x)_4$, $(-y)_4$, P8: $(x)_2$, $(-x)_2$, $(y)_2$, $(-y)_2$, rec: $(x, -x)_2$, $(y, -y)_2$, $(-x, x)_2$, $(-y, y)_2$

For References see p. 4844

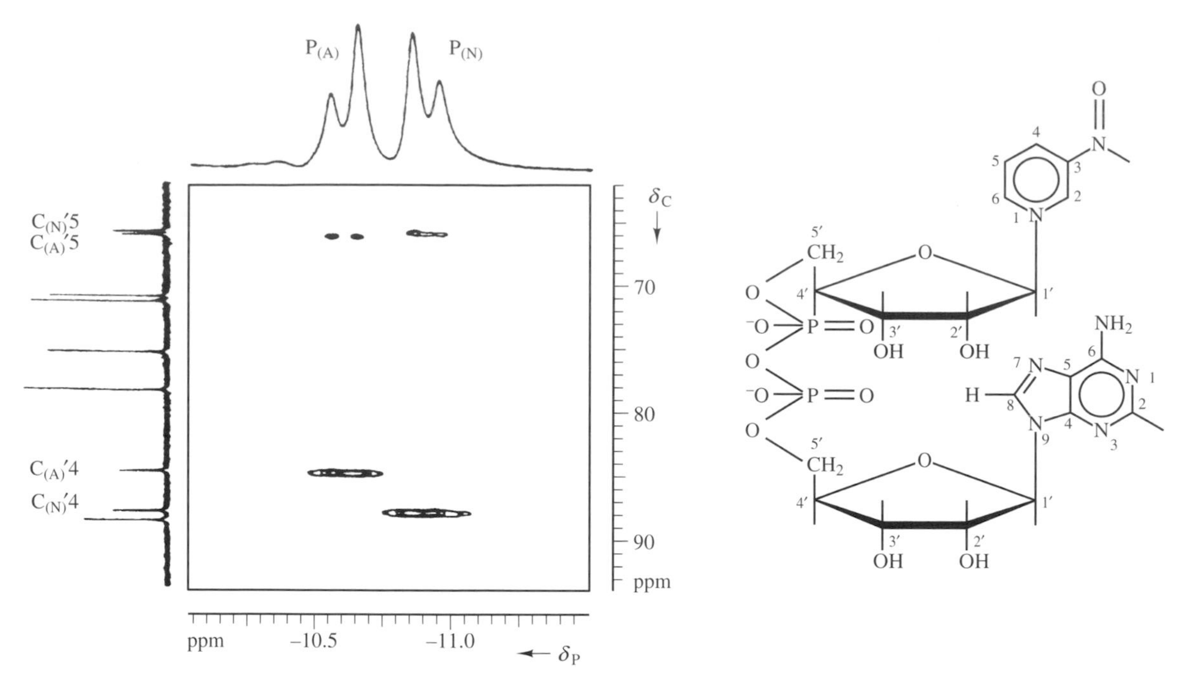

Figure 2 2D $^{13}C,^{31}P$ correlation of NAD^+ using ^{31}P detection (taken from Bast et al.[11])

wards, between the two nuclei. In the polarization transfer method this occurs only once. If this transfer is not optimal due to a range of different spin coupling constants, there are inevitable losses of signal in each transfer; thus, the method with only one transfer might be favored. Modifications of the above pulse sequences employ an INEPT transfer from protons to one heteronucleus prior to the start of the actual correlation.[8,9]

5 CHEMICAL APPLICATIONS

5.1 $^{13}C,^{31}P$ Correlation

Most of the published work has been performed on the spin pair $^{13}C,^{31}P$; hence, it seems appropriate to review this nucleus first. Martin and co-workers[10] showed the feasibility of this technique with a homebuilt probe and the HMQC method. They used triphenylphosphane oxide as an example. Since the spectrometer used did not have three rf channels, proton decoupling could not be employed. Of course, 2D correlation on this simple compound does not give new chemical insights, it merely served as a model compound. A more systematic study compared the three pulse methods taking triphenylphosphane as an example using ^{31}P and ^{13}C detection.[11] The merits of the different methods were discussed and it was found that for ^{31}P detection the polarization transfer method was the most suitable; for ^{13}C detection the HMQC method was the most suitable. As a first real example, the $^{13}C,^{31}P$ correlation of nicotinamide adenine dinucleotide (NAD^+) was measured. This example is given in Figure 2 and shows very nicely the ease of assignment with such methods. Further examples given by Bast et al.[11] were a mixture of mononucleotides.

Examples from the chemistry of phosphonates have been published by Grossmann and co-workers.[12] For a mixture of stereoisomeric diphosphonates they could, with the help of $^{13}C,^{31}P$ correlation, assign the three different sets of signals and determine the P,P′ spin coupling constant in these compounds. Other examples from the same group include bis(dialkoxythiophosphoryl) sulfides and their polysulfides.[13] An example from the field of ligands used in organometallic chemistry is provided by the molecule prophos (see Figure 3) in which we find all the phenyl groups to be inequivalent due to chirality, and one might question the relative assignments of these carbon atoms.[14] The ^{31}P NMR spectrum is easily assigned via H,P correlation. Phosphorus A ($\delta = -25$ ppm) couples with all aliphatic protons while phosphorus B ($\delta = -4.7$ ppm) only couples with the methylene and methine group. The phosphorus-substituted *ipso*-carbon atoms of the four phenyl rings are particularly difficult to assign by any other method. In Figure 3 the extension of a 2D $^{13}C,^{31}P$ correlation is shown, which immediately relates the connectivity of the pairs of diastereotopic *ipso*-carbon atoms with their corresponding phosphorus atoms. Close inspection of Figure 3 reveals, from the slant of the cross peaks, that the signs of the ^1J(P,C) and ^3J(P,P) spin coupling constants are the same; ^1J(P,C) spin coupling constants for threefold coordinated phosphorus are known to be negative.[15] The possibility of evaluating the relative sign of coupling constants can be viewed as a further asset of $^{13}C,^{n}X$ correlations.

Very recently, $^{19}F,^{13}C$ correlations have been published.[27–29]

5.2 $^{13}C,^{15}N$ Correlation

The $^{13}C,^{15}N$ correlation is especially important in the protein field and was therefore, up to now, only applied in this class of compounds. Doubly-labeled material has to be used, but this is readily available by biological methods. Using the HMQC method, the first paper in this field[16] correlated the signals of a protein of molecular weight of 20 000 Da (see Figure 4). A huge number of cross peaks describing the amide links

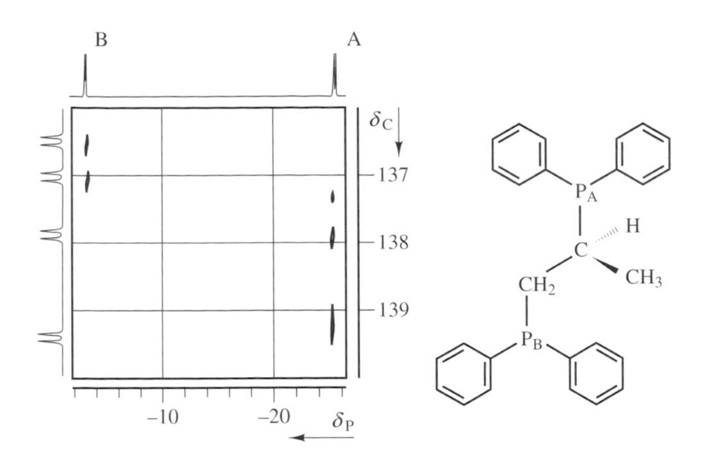

Figure 3 2D $^{13}C,^{31}P$ correlation of the chiral ligand prophos using ^{31}P detection and the HMQC method with extension for the *ipso*-carbon atoms being shown

of the protein can be seen. After methodical research on labeled acetamide[8] to find the best possible pulse sequence, another group used this method to elucidate the structure of a cyclic decapeptide.[17]

5.3 Correlation with Metal Nuclei

This application of correlation spectroscopy is very recent. For ^{29}Si, a new pulse sequence, a combination of INEPT and HMQC, was proposed and applied to decamethyltetrasiloxane as a model compound and to a polymer silicon oil as an example.[9] This technique has also been used in organoselenium chemistry, where, for example, a full assignment of the ^{13}C and ^{77}Se chemical shifts could be achieved for olefins bearing two different phenylseleno groups. Furthermore, incorrect literature structures could be corrected.[18] Another olefin with two different trimethyltin groups served as the first example for $^{13}C,^{119}Sn$ correlation.[19] It was shown that long range coupling constants can also be used to correlate carbon and tin chemical shifts. Conformational assignment can therefore be made by

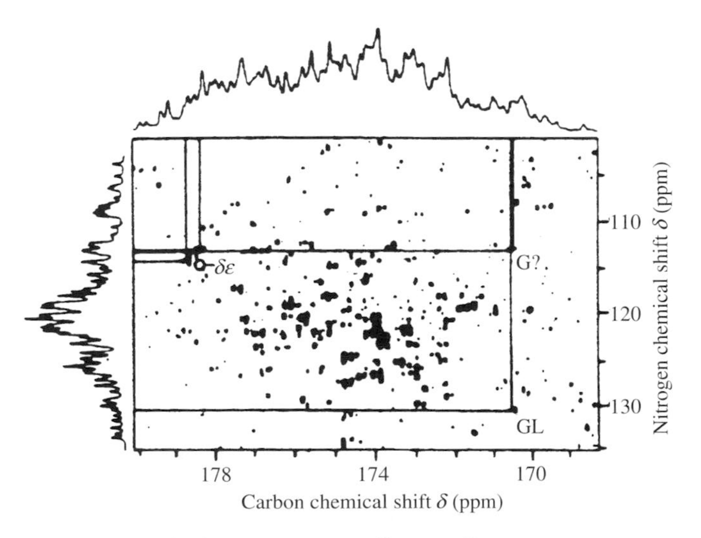

Figure 4 2D $^{13}C,^{15}N$ correlation of ^{13}C and ^{15}N labeled flavodoxin from *Anabaena* 7120 (taken from Westler et al.[16])

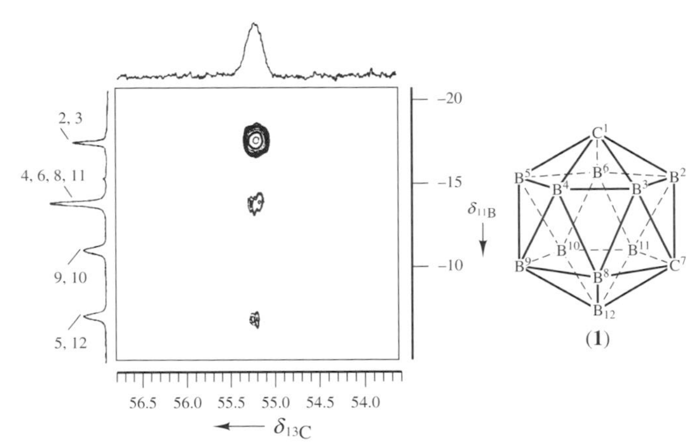

Figure 5 2D $^{13}C,^{11}B$ correlation of 1,7-dicarba-*closo*-dodecaborane(12) (taken from Fäcke and Berger[26])

these means. Recently, two examples from organomercury[20] and organolead[21] chemistry have been tested.

5.4 Correlation with Quadrupolar Nuclei

Although, at first sight, 2D correlation spectroscopy of ^{13}C with quadrupolar nuclei does not seems feasible due to relaxation problems, there are two nuclei where, due to the relatively small quadrupolar moment, such a correlation has been performed, i.e. the ^2H and the ^6Li nuclei. Günther and coworkers were the first to employ these methods for both nuclei. For ^2H they chose 2,4,5,6-pyridine-d_4 as an example to test the correlation technique[22] and, later on, applied this to the analysis of isotopomeric mixtures of bicyclic compounds in which the deuterium label was distributed during a chemical reaction.[23] By means of the $^{13}C,^2H$ correlation the signals in this mixture could be easily assigned. For this technique, the usual ^2H lock of modern instruments cannot be used and must be replaced by a ^{19}F lock system. In organolithium compounds one is often faced with dynamic equilibria between aggregates of different sizes, depending on temperature and solvent. The structures obtained by X-ray crystallography do not necessarily correspond to those predominant in solution. In two reports[24,25] it was shown that a 2D $^{13}C,^6Li$ correlation can contribute to the analysis of organolithium compounds. Carboranes and metallacarboranes provide a large class of compounds where sometimes, due to symmetry reasons, the ^{11}B signals display reasonably small linewidths on the order of the $^{13}C,^{11}B$ spin coupling constant. A $^{13}C,^{11}B$ correlation was therefore attempted for selected carboranes. In Figure 5\ the result for the *meta*-dodecaborane (**1**) is shown,[26] displaying nicely the different connections within this cage compound.

6 CONCLUSIONS

We have shown in this article the feasibility and possibilities of 2D $^{13}C,^nX$ correlation, and demonstrated that this technique often provides a very direct and straightforward signal assignment. The further development and application of

For References see p. 4844

these techniques will certainly give new possibilities in structural elucidation in the field of organoelement chemistry.

7 RELATED ARTICLES

COSY Two-Dimensional Experiments; Heteronuclear Assignment Techniques; Heteronuclear Shift Correlation Spectroscopy; Organometallic Compounds.

8 REFERENCES

1. J. Jeener, lecture given at *AMPÈRE International Summer School*, Basko Polje, Yugoslavia, 1971.
2. W. P. Aue, E. Bartholdi, and R. R. Ernst, *J. Chem. Phys.*, 1976, **64**, 2229.
3. R. Freeman and G. A. Morris, *J. Chem. Soc., Chem. Commun.*, 1978, 684.
4. L. Müller, *J. Am. Chem. Soc.*, 1979, **101**, 4481.
5. A. Bax, R. H. Griffey, and B. L. Hawkins, *J. Magn. Reson.*, 1983, **55**, 301.
6. G. Bodenhausen and D. J. Ruben, *Chem. Phys. Lett.*, 1980, **69**, 185.
7. R. R. Ernst, G. Bodenhausen, and A. Wokaun, 'Principles of Nuclear Magnetic Resonance in One and Two Dimensions', Clarendon Press, Oxford, 1987, p. 468.
8. W. P. Niemczura, G. L. Helms, A. S. Chesnick, R. E. Moore, and V. Bornemann, *J. Magn. Reson.*, 1989, **81**, 635.
9. S. Berger, *J. Magn. Reson., Ser. A*, 1993, **101**, 329.
10. L. D. Sims, L. R. Soltero, and G. E. Martin, *Magn. Reson. Chem.* 1989, **27**, 599.
11. P. Bast, S. Berger, and H. Günther, *Magn. Reson. Chem.*, 1992, **30**, 587.
12. H. Beckmann, G. Grossmann, and G. Ohms, *Magn. Reson. Chem.*, 1992, **30**, 860.
13. G. Grossmann and H. Komber, *Phosphorus, Sulfur, Silicon, Relat. Elem.*, 1991, **61**, 269.
14. T. Fäcke, R. Wagner, and S. Berger, *Concepts Magn. Reson.*, 1994, **6**, 293.
15. S. Berger, S. Braun, and H. O. Kalinowski, 'NMR-Spektroskopie von Nichtmetallen', Thieme Verlag, Stuttgart, 1993, Vol. 3, p. 135.
16. W. M. Westler, B. J. Stockman, and J. L. Markley, *J. Am. Chem. Soc.*, 1988, **110**, 6256.
17. R. E. Moore, V. Bornemann, W. P. Niemczura, J. M. Gregson, J. L. Chen, T. R. Norton, G. M. L. Patterson, and G. L. Helms, *J. Am. Chem. Soc.*, 1989, **111**, 6128.
18. T. Fäcke, R. Wagner, and S. Berger, *J. Org. Chem.*, 1993, **58**, 5475.
19. S. Berger and T. N. Mitchell, *Organometallics*, 1992, **11**, 3481.
20. T. Fäcke and S. Berger, *J. Organomet. Chem.*, in press.
21. T. Fäcke and S. Berger, *Main Group Metal Chem.*, 1994, **17**, 463.
22. J. R. Wesener, P. Schmitt, and H. Günther, *Magn. Reson. Chem.*, 1984, **22**, 468.
23. J. R. Wesener and H. Günther, *J. Am. Chem. Soc.*, 1985, **107**, 1537.
24. D. Moskau, F. Brauers, H. Günther, and A. Maercker, *J. Am. Chem. Soc.*, 1987, **109**, 5532.
25. H. J. Gais, J. Vollhardt, H. Günther, D. Moskau, H. J. Lindner, and S. Braun, *J. Am. Chem. Soc*, 1988, **110**, 978.
26. T. Fäcke and S. Berger, *Magn. Reson. Chem.*, 1994, **32**, 436.
27. A. A. Ribeiro and M. J. Geen, *J. Magn. Reson., Ser. A*, 1994, **107**, 158.
28. S. Berger, *J. Fluorine Chem.*, 1995, **72**, 117.
29. R. Eujeu and A. Putorru, *J. Organomet. Chem.*, 1992, **438**, 57.

For list of General Abbreviations see end-papers

Biographical Sketch

Stefan Berger. *b* 1946. Ph.D., 1973, University Tübingen, under supervision of Prof. Dr. A. Rieker 1973. Introduced to NMR by H. Suhr and A. Rieker. Postdoctoral fellow with Prof. J. D. Roberts, California Institute of Technology, 1973–74. Habilitation, 1981. Faculty, University Marburg, 1974–present. Approx. 100 publications. Current research interests: development of selective pulse methods and applications of NMR in the field of organometallic and bioorganic chemistry.

Two-Dimensional *J*-Resolved Spectroscopy

Gareth A. Morris
University of Manchester, Manchester, UK

1 INTRODUCTION

J-resolved 2D spectroscopy (also commonly known as 2D *J* spectroscopy) is a class of two-dimensional NMR methods that separate chemical shifts from multiplet structure. In pulse sequences for *J*-resolved 2D spectroscopy, a 180° pulse is applied at the centre of the evolution period t_1, generating a spin echo. The modulation of spin echoes by scalar coupling causes signals to be dispersed in the f_1 domain of the resultant 2D spectrum according to their position within scalar coupling multiplets. This allows the separation of multiplet structure from chemical shifts, improving the resolution of individual signals in spectra with overlapping multiplets. There is a further gain in resolution in the f_1 domain of the 2D spectrum, because the spin echo suppresses the effects of $\boldsymbol{B}_0$ inhomogeneity, giving f_1 linewidths approaching the natural limit $1/\pi T_2$. *J*-resolved 2D spectroscopy is now much less widely used than correlated 2D methods, but still finds application, for example in the analysis of complex spin systems and in the separation of homonuclear from heteronuclear coupling structure.

Both homonuclear and heteronuclear multiplet structure can be investigated by *J*-resolved spectroscopy. Since broadband suppression of homonuclear couplings during measurement of the free induction decay is not possible, homonuclear *J*-resolved 2D spectroscopy[1] produces spectra with multiplet structure in f_1, and both multiplet structure and chemical shifts in f_2. In heteronuclear *J*-resolved 2D spectroscopy,[2,3] on the other hand, it is normal to apply broadband decoupling during data acquisition, so that multiplet structure is suppressed in f_2 and a complete separation is achieved between multiplet structure in f_1 and chemical shifts in f_2. For weakly coupled spin systems, a similar separation can be achieved in homonuclear *J*-resolved 2D spectroscopy by 'shearing' or 'tilting' the 2D data matrix[1] to give a 45° tilt in frequency space:

$$S(f_1,f_2') = S(f_1,f_2 - f_1) \qquad (1)$$

A further difference between homonuclear and heteronuclear methods lies in the origin of the echo modulation. In the homonuclear experiment, the modulation arises naturally from the action of the 180° pulse used to generate the spin echo, since it affects both the partners in any scalar coupling. In the heteronuclear experiment, couplings to weakly coupled heteronuclei do not give rise to any modulation of the spin echo unless a suitable perturbation is applied to the heteronuclei.

The simplest form of *J*-resolved spectroscopy is the homonuclear experiment of the following sequence, in which the basic Carr–Purcell method A spin echo sequence[4] forms the evolution period of a 2D experiment:

$$90^\circ \quad \tfrac{1}{2}t_1 \quad 180^\circ \quad \tfrac{1}{2}t_1 \quad \text{Acquire} \qquad (2)$$

This experiment is normally used in proton NMR, but is applicable to any nuclei with homonuclear couplings. Because the resulting signals are phase-modulated with respect to t_1, it is not normally practicable to use phase-sensitive display. In order to secure acceptable absolute value mode lineshapes, it is usual to apply either pseudoecho[5] or sine-bell[6] weighting in both frequency dimensions, despite the signal-to-noise ratio penalty. This forces the time domain signals into an approximately symmetric envelope, so that dispersion-mode contributions (which have odd symmetry) are absent from the resultant frequency domain lineshapes.

Sequence (2) will not normally generate any echo modulation as a result of heteronuclear couplings; this is the basis of the use of homonuclear *J*-resolved spectroscopy to distinguish between homonuclear and heteronuclear multiplet structure. (Echo modulation due to heteronuclear couplings can arise, however, if there is coupling to a group of heteronuclei that are strongly coupled amongst themselves.[7]) For simplicity, it will be assumed here that the nucleus being observed is ^{13}C and that the coupled nuclei are protons. Although the great majority of experimental applications of heteronuclear *J*-resolved spectroscopy are to this pair of nuclei, there are of course many other possibilities.

Two methods are commonly used to ensure that heteronuclear couplings generate echo modulation. In the 'gated decoupler' method,[2,3,8,9] broadband decoupling is applied during one half only of the evolution period t_1; decoupling is normally also used during the preparation period (to provide the NOE) and during measurement of the free induction decay (to remove multiplet structure from f_2):

$$\begin{array}{llllll} ^1\text{H} & \leftarrow \text{Decouple} & & \rightarrow & & \leftarrow \text{Decouple} \rightarrow \\ ^{13}\text{C} & 90^\circ & \tfrac{1}{2}t_1 & 180^\circ & \tfrac{1}{2}t_1 & \text{Acquire} \end{array} \qquad (3)$$

This gives a 2D spectrum in which the multiplet structure is scaled down to half its normal width in f_1, but otherwise faithfully reflects the conventional spectrum, even where there is strong proton–proton coupling.

The second, 'proton flip', technique[3,8,10] uses a 180° pulse to invert the coupled heteronuclei at the midpoint of the evolution period; again, decoupling is normally used during the preparation and detection periods:

$$\begin{array}{llllll} ^1\text{H} & \leftarrow \text{Decouple} \rightarrow & & 180^\circ & & \leftarrow \text{Decouple} \rightarrow \\ ^{13}\text{C} & 90^\circ & \tfrac{1}{2}t_1 & 180^\circ & \tfrac{1}{2}t_1 & \text{Acquire} \end{array} \qquad (4)$$

Here the multiplet structure appears without any scaling, but only matches that in the 1D spectrum if the heteronuclei (here protons) are weakly coupled. The analysis of strongly coupled proton flip *J*-resolved spectra is discussed in Section 3.2, and closely parallels the analysis for the homonuclear case.

2 HISTORICAL REVIEW

The earliest use of Fourier transformation of the dependence of echo amplitude on the spin echo delay t_1 was the '*J* spectroscopy' experiment of Freeman and Hill.[11] This involved measuring the amplitude of the signal at the peak of the echo for a series of equally spaced values of t_1, and Fourier transformation with respect to t_1 to generate a '*J* spectrum' in which chemical shifts and line broadening due to field inhomogeneity were suppressed. This experiment was of little practical utility, because all the multiplets in a spectrum were superimposed, so that only the simplest of spin systems could be studied. The effect of measuring just the single data point at the maximum of the echo is equivalent to carrying out an integral projection of a 2D *J*-resolved spectrum onto the f_1 axis.

The first two-dimensional experiments to use spin echoes were reported in Aue, Bartholdi, and Ernst's classic introductory paper[12] on 2D NMR; these were effectively COSY experiments with 180° mixing pulses. Homonuclear *J*-resolved spectroscopy proper was introduced a few months later,[1] in a communication that pointed out the possibility of obtaining a 'decoupled' proton spectrum from a 45° projection. The first heteronuclear experiments were proposed independently by Freeman and co-workers shortly afterwards.[2] Heteronuclear *J*-resolved spectroscopy proved a useful vehicle for the technical development of 2D NMR,[13] since it required only modest data storage.

The observation of artifacts caused by imperfect 180° pulses prompted the development of EXORCYCLE,[14] one of the first and most widely used phase cycling techniques; the acronym derived from the use of the terms 'ghost' and 'phantom' for the spurious signals. The differences in multiplet structure between gated decoupler and proton flip heteronuclear *J*-resolved spectra seen in the presence of strong proton–proton coupling[8–10] led to the development of analytical solutions and numerical software for calculating strongly coupled homonuclear and proton flip heteronuclear *J*-resolved spectra.[10,15,16] Heteronuclear *J*-resolved spectroscopy was also the vehicle for early investigations of the problem of lineshapes in phase-sensitive 2D spectra. Phase modulation leads to 'phasetwist'

For References see p. 4849

lineshapes, which are an inseparable mixture of 2D absorption and dispersion mode lineshapes.[13] This problem can be cured in heteronuclear J-resolved spectra by combining results from two experiments that have opposite senses of precession in t_1; the apparent sense of precession can be reversed either by adding a second 180° pulse at the end of the evolution period[17] or, in the gated decoupler experiment, by changing from decoupling during one half of t_1 to decoupling in the other.[18] Such expedients are actually seldom used, since in f_2-decoupled heteronuclear J-resolved spectroscopy, phase-sensitive f_1 cross-sections (which can be phased to absorption mode, since they pass through the midpoints of the phasetwisted lineshapes) can usually be obtained free of overlap from neighboring signals in f_2.

With the introduction of better spectrometer computers, J-resolved spectroscopy was quickly eclipsed by the more powerful and general class of correlated 2D experiments such as COSY and NOESY, but, nevertheless, a number of useful extensions have been demonstrated. Making the 180° proton pulse of sequence (4) selective restricts the modulation of the spin echo to couplings to protons with a particular chemical shift.[19] If the carbon and proton 180° pulses of a proton flip experiment are replaced by a BIRD (bilinear rotation decoupling) sequence,[20] the refocusing can be made selective for either long-range or one-bond heteronuclear couplings. This is known as semiselective J spectroscopy,[21] giving J-resolved spectra in which, depending on the relative phases of the pulses in the BIRD sequence, either the one-bond or the longer range couplings are removed from the f_1 dimension, simplifying analysis. Again, strong proton–proton coupling can complicate spectra very considerably, but numerical simulations can be performed;[22] the large number of lines produced can actually be an advantage, because it increases the accuracy with which spin system parameters may be derived.

Another extension is known as indirect J spectroscopy;[23,24] this adds a polarization transfer step to a proton spin echo, transferring the proton J modulation to ^{13}C signals. This works well where the ^{13}C satellites in the proton spectrum are weakly coupled, allowing clean resolution of individual proton multiplets, even in systems where the proton spectrum is completely unresolved. It has been little used, however, and for most purposes a high-f_2-resolution HMQC experiment[25] is preferable; this does not give any proton linewidth improvement, but is quicker, and doubles the chances of avoiding strong coupling if run without ^{13}C decoupling in t_2.

3 THEORY

3.1 Weak Coupling

Consider first a weakly coupled homonuclear system of two spins $\frac{1}{2}$, I and S; the energy level diagram and spectrum are shown in Figure 1. There are two I transitions: one for the half of the spin systems in which S is in the α state and one for those in which it is β, and a similar pair of S transitions. Initially, the magnetizations for the four transitions will be at equilibrium; the state of the spin system may represented in the product operator formalism[26] as $\hat{I}_z + \hat{S}_z$, and the four magnetizations will point along the z direction in the rotating frame of reference. The initial 90° pulse of sequence (2) will rotate the

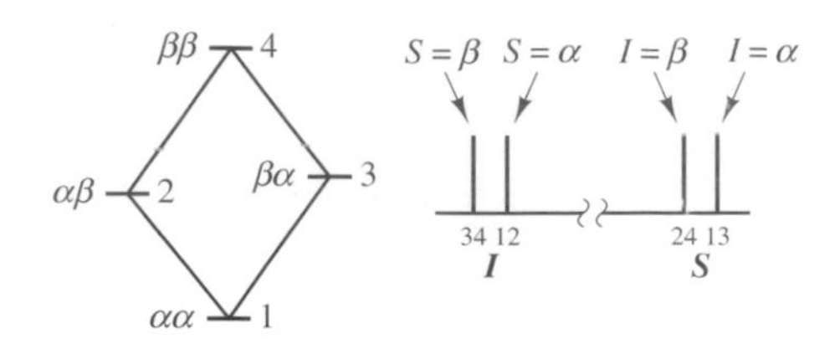

Figure 1 Energy level diagram and spectrum for a weakly coupled IS system of two spins $\frac{1}{2}$

four magnetizations down to the $-y$ axis of the rotating frame to give $-\hat{I}_y - \hat{S}_y$; the I-spin magnetizations are shown diagrammatically in Figure 2(a). During the first half of the evolution period, the chemical shifts of spins I and S will cause the magnetizations to precess through angles $\phi_I = \pi\delta_I t_1$ and $\phi_S = \pi\delta_S t_1$ (in rad), respectively, where δ_I and δ_S are the offsets from resonance (in Hz) of the two spins. The scalar coupling J_{IS} will cause the two components of each multiplet to diverge by an angle $\theta_1 = \pi J_{IS} t_1$, giving the state

$$\begin{aligned} &-\hat{I}_y \cos\phi_I \cos\tfrac{1}{2}\theta_1 + \hat{I}_x \sin\phi_I \cos\tfrac{1}{2}\theta_1 + 2\hat{I}_x\hat{S}_z \cos\phi_I \sin\tfrac{1}{2}\theta_1 \\ &+ 2\hat{I}_y\hat{S}_z \sin\phi_I \sin\tfrac{1}{2}\theta_1 - \hat{S}_y \cos\phi_S \cos\tfrac{1}{2}\theta_1 + \hat{S}_x \sin\phi_S \cos\tfrac{1}{2}\theta_1 \\ &+ 2\hat{S}_x\hat{I}_z \cos\phi_S \sin\tfrac{1}{2}\theta_1 + 2\hat{S}_y\hat{I}_z \sin\phi_S \sin\tfrac{1}{2}\theta_1 \end{aligned} \quad (5)$$

for which the I magnetizations are shown in Figure 2(b); for the time being, relaxation will be neglected. The 180° pulse, applied about the x axis in the rotating frame, will have two effects. It will change the signs of all the terms in $\hat{I}_y$ and $\hat{S}_y$, reflecting the corresponding magnetization vectors about the (x,z) plane of the rotating frame. It will also invert all terms in $\hat{I}_z$ and $\hat{S}_z$; this corresponds to interchanging α and β spins, so that, for example, the I magnetization associated with transition 12 before the 180° pulse (for which S is in the α state) is transferred to transition 34 ($S = \beta$), giving the situation shown in Figure 2(c).

During the second half of the evolution period, the I and S magnetizations will again precess through angles ϕ_I and ϕ_S respectively, and the two components of the I and S pairs will again diverge by $\frac{1}{2}\theta_1$ radians. The effect of the chemical shift

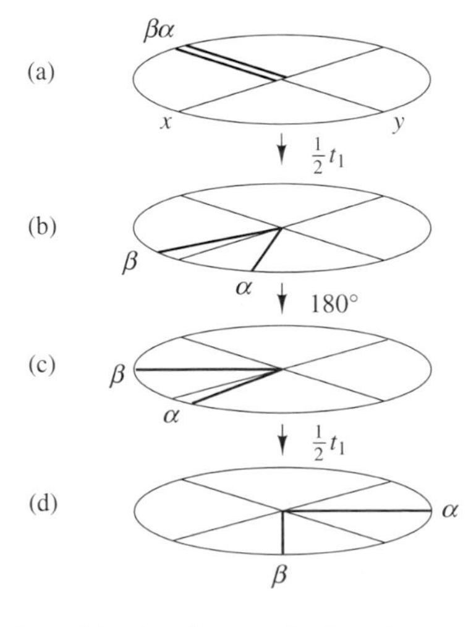

Figure 2 Motion of the two I magnetizations in a weakly coupled IS system of two spins $\frac{1}{2}$ during sequence (2)

precession during the second half of t_1 exactly matches the shift precession during the first half, so that in the absence of any coupling J_{IS}, all four magnetizations would refocus along the $\hat{I}_y$ axis. In the presence of a coupling, the two components of the I and S multiplets end up disposed symmetrically either side of the $\hat{I}_y$ axis, an angle θ apart (Figure 2d), giving the state $+\hat{I}_y\cos\theta_1 - 2\hat{I}_x\hat{S}_z\sin\theta_1 + \hat{S}_y\cos\theta_1 - 2\hat{S}_x\hat{I}_z\sin\theta_1$. The signal at time t_1 is thus unaffected by chemical shifts (or inhomogeneity of the static magnetic field): a spin echo is produced, which is modulated only by the scalar coupling J_{IS}. Each multiplet is a mixture of in-phase absorption mode signals ($\hat{I}_y$) and antiphase dispersion signals ($2\hat{I}_x\hat{S}_z$); the net effect of the modulated spin echo is to multiply one multiplet component by a phase factor $e^{-i\pi J_{IS}t_1}$ and the other by $e^{i\pi J_{IS}t_1}$.

At time t_1, the second half of the modulated echo is recorded as a free induction decay $S(t_2)$. Neglecting relaxation, the complex signal $S(t_1,t_2) = M_y - iM_x$ recorded can then be written as

$$S(t_1,t_2)=\tfrac{1}{2}(e^{i\pi J_{IS}t_1}e^{i(2\pi\delta_I+\pi J_{IS})t_2}+e^{-i\pi J_{IS}t_1}e^{i(2\pi\delta_I-\pi J_{IS})t_2} + e^{i\pi J_{IS}t_1}e^{i(2\pi\delta_S+\pi J_{IS})t_2}+e^{-i\pi J_{IS}t_1}e^{i(2\pi\delta_S-\pi J_{IS})t_2}) \quad (6)$$

which consists of four signals with frequency coordinates (f_1,f_2) of $(\frac{1}{2}J_{IS}, \delta_I + \frac{1}{2}J_{IS})$, $(-\frac{1}{2}J_{IS}, \delta_I - \frac{1}{2}J_{IS})$, $(\frac{1}{2}J_{IS}, \delta_S + \frac{1}{2}J_{IS})$ and $(-\frac{1}{2}J_{IS}, \delta_S - \frac{1}{2}J_{IS})$: the signals are dispersed according to their normal frequencies $\delta \pm \frac{1}{2}J_{IS}$ in f_2, and with just their multiplet structure $\pm\frac{1}{2}J_{IS}$ in f_1. The extension to larger spin systems is straightforward.

The same logic can be applied to heteronuclear J-resolved experiments. The analysis of the proton flip method parallels that for the homonuclear case, except that the initial 90° pulse affects only (say) spin I, so that all terms in $\hat{S}_x$ and $\hat{S}_y$ disappear. In the gated decoupler method, the use of broadband S-spin decoupling during the first half of the evolution period suppresses the effects of J_{IS} in the first half of t_1, so that the frequencies in f_1 are halved to $\pm\frac{1}{4}J_{IS}$. In both methods, broadband decoupling is normally used during t_2 to remove multiplet structure from f_2.

3.2 Strong Coupling

In the presence of strong coupling, the simple analysis given in Section 3.1 breaks down, and it is generally necessary to use density matrix theory. Although analytical results have been presented for some simple spin systems,[16] it is usual to employ numerical methods; a modified version of the 1D spin simulation program LAOCN3,[27] SONOFLAOCOON,[10] may be used both for simulation and for iterative analysis of strongly coupled J-resolved spectra.

Consider a system containing two weakly coupled groups of spins I and S, either of which may be strongly coupled, where only the spins I are to be observed. In the heteronuclear case [sequence (4)], group I will generally contain a single ^{13}C spin and S the coupled protons; in the homonuclear case [sequence(2)], all the spins will belong to group I. If the matrix representation of the q component of spin angular momentum for group I is denoted by $\mathbf{I}^q$, the initial reduced density matrix ρ_0^I for group I will be equal to some constant m_0 times $\mathbf{I}^z$, and the effect of the 90° I pulse will be to rotate ρ_0 into $-m_0\mathbf{I}^y$. If calculations are performed in the eigenbasis, the ijth element of ρ^I at time $\frac{1}{2}t_1$ will be $-m_0\exp[i\pi t_1(\nu_i - \nu_j] I^y_{ij}$, where ν_i is the ith energy eigenvalue of the spin system (in Hz). If the matrix representation for 180° rotation of spins I and S is $\mathbf{T}^{IS}$ then the effect of the 180° I and S pulses will be to transfer coherences ρ_{ij} into other coherences ρ_{kl}:

$$\rho_{kl} = \sum_i\sum_j \rho_{ij}T^{IS*}_{ik}T^{IS}_{jl} \quad (7)$$

Thus, at the end of the evolution period, the total y signal will be

$$S = -m_0\sum_i\sum_j\sum_k\sum_l I^y_{ij}T^{IS*}_{ik}T^{IS}_{jl}I^y_{lk} \times \exp[i\pi t_1(\nu_i - \nu_j + \nu_l - \nu_k)] \quad (8)$$

The 2D J-resolved spectrum thus consists of a series of responses S_{ijkl} of f_1 frequencies $\frac{1}{2}(\nu_i - \nu_j + \nu_l - \nu_k)$, with intensities given by the preexponential terms in equation (8). Only when all spins are weakly coupled are the f_1 frequencies determined solely by coupling constants; where there is strong coupling, they will also contain terms in chemical shifts (often leading to folding in f_1), and the signals may have negative intensity.

The calculation of a 2D J-resolved spectrum thus reduces to the diagonalization of the nuclear spin Hamiltonian, as in conventional spin simulation programs such as LAOCN3,[27] followed by construction of the eigenbasis matrix representations of the y component of angular momentum and of the 180° rotation operator. By suitable choice of the groups of spins I and S, the same program may be used to calculate both homonuclear and proton flip heteronuclear J-resolved spectra. More complex experiments such as semiselective J spectroscopy[21,22] can be simulated by replacing $\mathbf{T}^{IS}$ with the matrix representation of a BIRD operator.[20]

4 APPLICATIONS

4.1 Practical Implementation

J-resolved spectroscopy is one of the easiest 2D methods to implement. Sequences (2) and (3) require only the usual calibration of the 90° observe pulse width, which is not critical; if α is the actual flip angle of a nominal 90° pulse, the signal obtained is proportional to $\sin\alpha\cos 2\alpha$. Calibration of the proton 180° pulse in sequence (3) can be carried out by setting t_1 equal to $1/^1J(C,H)$ and varying the proton pulse width; the signal is nulled when the flip angle is 90°, and inverted when it is 180°. For all three basic sequences the phases of the 90° pulse, 180° refocusing pulse, and receiver should follow a scheme such as that in Table 1, which combines EXORCYCLE[14] and CYCLOPS.[28]

The two principal advantages of J-resolved spectroscopy are the decrease in f_1 linewidth afforded by suppressing $\boldsymbol{B}_0$ inhomogeneity effects, and the improved ability to distinguish signals that results from separating multiplet structure from chemical shifts. The former advantage is generally restricted to small molecules with long T_2s, and is most commonly exploited in $^1H-^{13}C$ heteronuclear J-resolved spectroscopy, while the latter is most useful in ^1H homonuclear J-resolved spectroscopy.

For References see p. 4849

Table 1 Phase Cycling for Sequences (2)–(4); Phase Shifts are Indicated as Multiples of 90° for the 90° Pulse (ϕ_1), 180° Pulse (ϕ_2), and Receiver (ϕ_R); the Phase of the Proton 180° Pulse in Sequence (4) is Immaterial

ϕ_1	0321	1032	2103	3210
ϕ_2	0000	1111	2222	3333
ϕ_R	0123	1230	2301	3012

4.2 Homonuclear *J*-Resolved 2D spectroscopy

Figure 3 shows the sugar ring region of the 300 MHz proton spectra of a solution of sucrose octaacetate in deuterochloroform. At the right is the normal 1D spectrum, together with a contour plot of the *J*-resolved spectrum. Pseudoecho weighting was used, with absolute value display; good results are also achievable with sine-bell weighting. Strong coupling signals are visible, for example in the region around 5.4 ppm, in addition to the expected multiplets. The effects of 'tilting' the data matrix at 45° in frequency space are illustrated in the left-hand contour plot of Figure 3, which shows the *J*-resolved spectrum after applying equation (1). At the far left of Figure 3, the results of integrating or 'projecting' the tilted spectrum along the f_1 direction to give a 'proton-decoupled proton' spectrum are shown; again, strong coupling artifacts are visible, for example around 5.4 ppm.

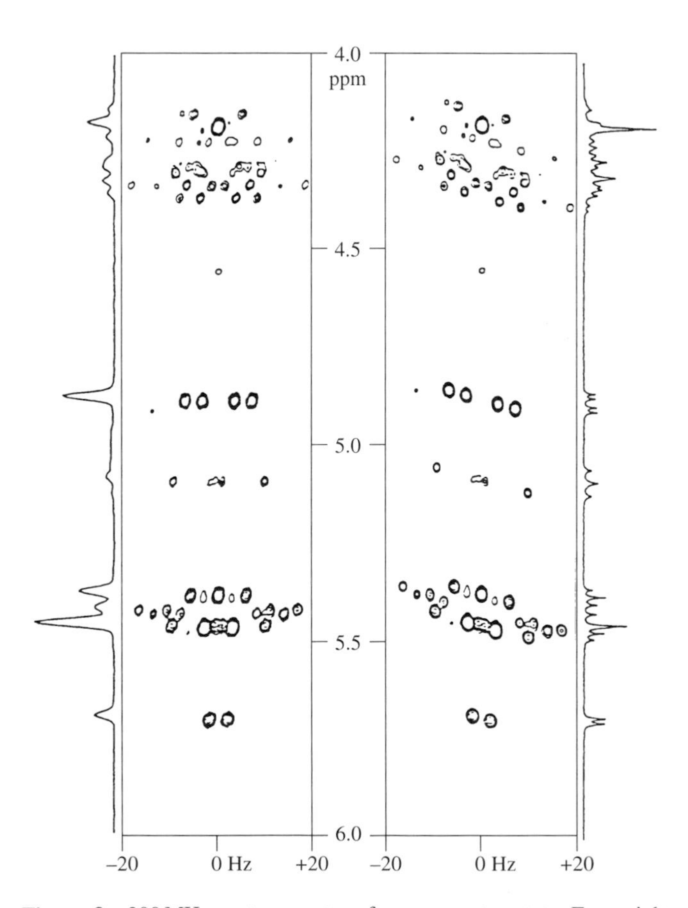

Figure 3 300 MHz proton spectra of sucrose octaacetate. From right to left: normal 1D spectrum; *J*-resolved spectrum; tilted *J*-resolved spectrum; and f_2 projection of the tilted *J*-resolved spectrum

4.3 Heteronuclear *J*-Resolved 2D spectroscopy

The two classical uses of heteronuclear *J*-resolved spectroscopy are with low f_1 resolution for the determination of multiplicity, and with high f_1 resolution for the measurement of long-range ^1H–^{13}C couplings in small molecules. DEPT is now generally preferred for multiplicity determination, but *J*-resolved spectroscopy remains the method of choice for high-resolution measurements of ^nJ(C,H). One advantage of heteronuclear *J*-resolved spectroscopy for multiplicity determination is that it gives clear results in the presence of ^{13}C shift degeneracy: Figure 4 shows a contour plot of a gated decoupler spectrum of a solution of cholesteryl acetate in deuterochloroform, together with an expanded stacked trace plot of an overlapping region, showing the coincidence of a methine and a methylene signal at about 29 ppm.

4.4 Semiselective *J*-Resolved 2D spectroscopy

Figure 5 shows experimental and theoretical cross sections through a high-resolution heteronuclear semiselective 2D *J* spectrum of thiophene in acetone-d_6. One-bond couplings have been suppressed by choosing phases 0, 0, and 180° for the three successive proton pulses of the BIRD sequence.[20] The cross sections illustrate both the high resolution achievable and

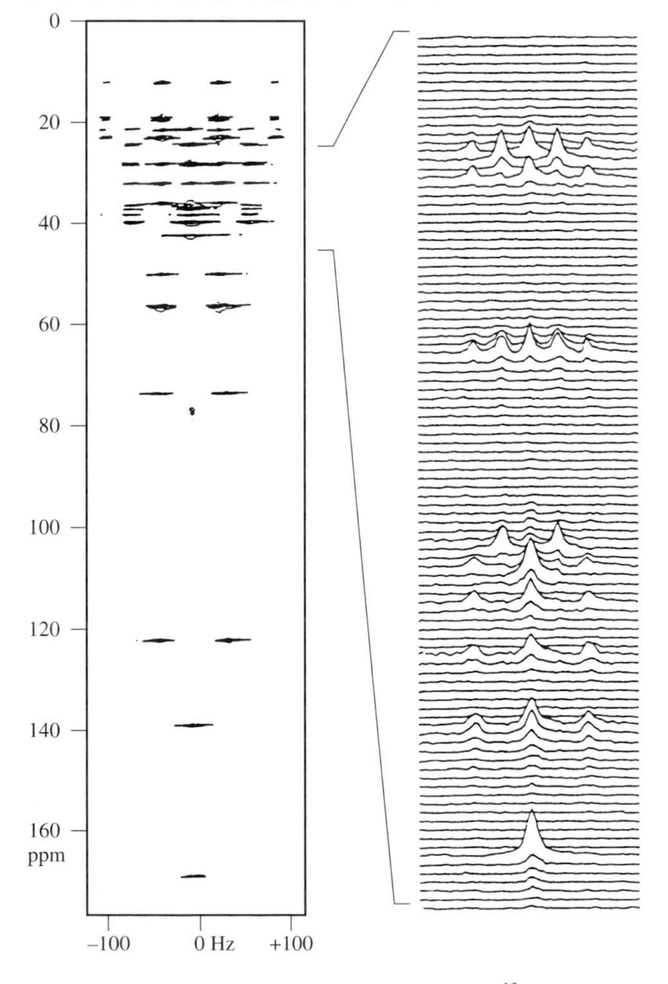

Figure 4 Gated decoupler heteronuclear *J*-resolved ^{13}C spectrum of cholesteryl acetate, showing a stacked trace plot of an expanded region

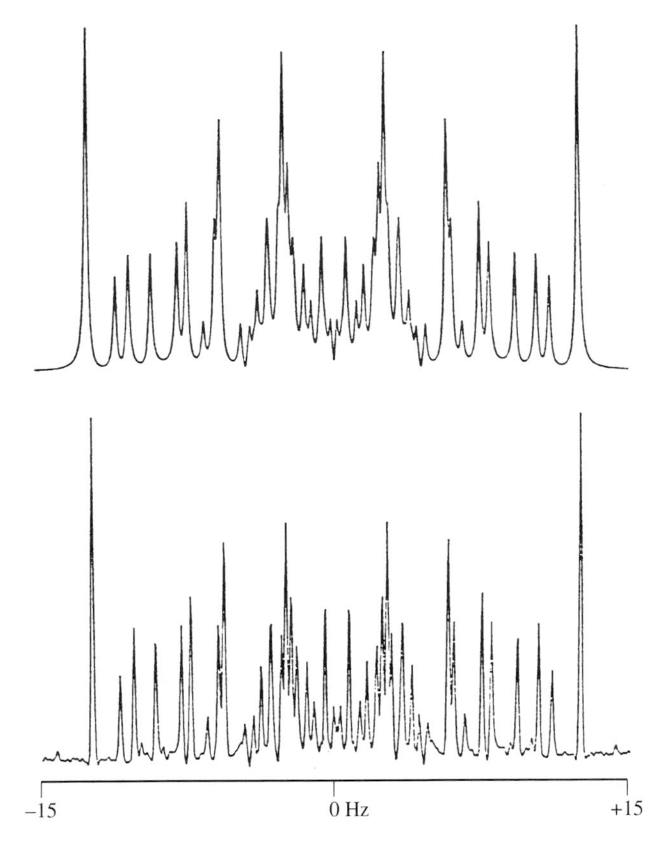

Figure 5 Semiselective one-bond suppressed heteronuclear *J*-resolved ^{13}C spectra for C-2 of thiophene (60% v/v solution in deuteroacetone): (top) calculated and (bottom) experimental. The sequence of Bax[21] was used, on a 400 MHz spectrometer

the complexity of the spectra that can result even from relatively small strongly coupled spin systems. The theoretical trace was calculated using the approach outlined in Section 3.2;[22] the delays during the BIRD sequence allow some of the strong coupling responses to appear in dispersion mode.

5 RELATED ARTICLES

Analysis of High Resolution Solution State Spectra; Indirect Coupling: Theory & Applications in Organic Chemistry; Liouville Equation of Motion; Multidimensional Spectroscopy: Concepts; Phase Cycling; Spin Echo Spectroscopy of Liquid Samples; Two-Dimensional NMR of Molecules Oriented in Liquid Crystalline Phases.

6 REFERENCES

1. W. P. Aue, J. Karhan, and R. R. Ernst, *J. Chem. Phys.*, 1976, **64**, 4226.
2. G. Bodenhausen, R. Freeman, and D. L. Turner, *J. Chem. Phys.*, 1976, **65**, 839.
3. L. Müller, A. Kumar, and R. R. Ernst, *J. Magn. Reson.*, 1977, **25**, 383.
4. H. Y. Carr and E. M. Purcell, *Phys. Rev.*, 1954, **94**, 630.
5. A. Bax, R. Freeman, and G. A. Morris, *J. Magn. Reson.*, 1981, **43**, 333.
6. J. C. Lindon and A. G. Ferrige, in 'Progress in Nuclear Magnetic Resonance Spectroscopy', ed. J. W. Emsley, J. Feeney, and L. H. Sutcliffe, Pergamon Press, Oxford, 1980, Vol. 14, p. 27.
7. A. Kumar and R. R. Ernst, *Chem. Phys. Lett.*, 1976, **37**, 162.
8. G. Bodenhausen, R. Freeman, R. Niedermeyer, and D. L. Turner, *J. Magn. Reson.*, 1976, **24**, 291.
9. R. Freeman, G. A. Morris, and D. L. Turner, *J. Magn. Reson.*, 1977, **26**, 373.
10. G. Bodenhausen, R. Freeman, G. A. Morris, and D. L. Turner, *J. Magn. Reson.*, 1977, **28**, 17.
11. R. Freeman and H. D. W. Hill, *J. Chem. Phys.*, 1971, **54**, 301.
12. W. P. Aue, E. Bartholdi, and R. R. Ernst, *J. Chem. Phys.*, 1976, **64**, 2229.
13. G. Bodenhausen, R. Freeman, R. Niedermayer, and D. L. Turner, *J. Magn. Reson.*, 1977, **26**, 133.
14. G. Bodenhausen, R. Freeman, and D. L. Turner, *J. Magn. Reson.*, 1977, **27**, 511.
15. A. Kumar, *J. Magn. Reson.*, 1978, **30**, 227; *J. Magn. Reson.*, 1980, **40**, 413.
16. G. Bodenhausen, R. Freeman, G. A. Morris, and D. L. Turner, *J. Magn. Reson.*, 1978, **31**, 75.
17. P. Bachmann, W. P. Aue, L. Müller, and R. R. Ernst, *J. Magn. Reson.*, 1977, **28**, 29.
18. R. Freeman, S. P. Kempsell, and M. H. Levitt, *J. Magn. Reson.*, 1979, **34**, 663.
19. A. Bax and R. Freeman, *J. Am. Chem. Soc.*, 1982, **104**, 1099.
20. J. R. Garbow, D. P. Weitekamp, and A. Pines, *Chem. Phys. Lett.*, 1982, **93**, 504.
21. A. Bax, *J. Magn. Reson.*, 1983, **52**, 330.
22. P. Sándor, G. A. Morris, and A. Gibbs, *J. Magn. Reson.*, 1989, **81**, 255.
23. G. A. Morris, *J. Magn. Reson.*, 1981, **44**, 277.
24. V. Rutar, *J. Magn. Reson.*, 1984, **56**, 413.
25. A. Bax, and S. Subramaniam, *J. Magn. Reson.*, 1986, **67**, 565.
26. O. W. Sørensen, G. W. Eich, M. H. Levitt, G. Bodenhausen, and R. R. Ernst, in 'Progress in Nuclear Magnetic Resonance Spectroscopy', ed. J. W. Emsley, J. Feeney, and L. H. Sutcliffe, Pergamon Press, Oxford, 1983, Vol. 16, p. 163.
27. A. A. Bothner-By and S. M. Castellano, in 'Computer Programs for Chemistry', ed. D. F. DeTar, Benjamin, New York, 1968, Vol. 1.
28. D. I. Hoult and R. E. Richards, *Proc. R. Soc. Lond. Ser. A*, 1975, **344**, 311.

Biographical Sketch

Gareth A. Morris. *b* 1954. M.A. (Nat.Sci.), Ph.D., 1978, (supervisor Ray Freeman) Oxford University, UK. Fellow by Examination, Magdalen College, Oxford, 1978–81; Izaak Walton Killam Postdoctoral fellow, University of British Columbia (with Laurie Hall) 1978–79. Lecturer (1982–89) and Reader (1989 to date) in Chemistry, University of Manchester. Approx. 100 publications. Current research interests: development and application of novel NMR techniques to problems in chemistry, biochemistry, and medicine.

Two-Dimensional Methods of Monitoring Exchange

Keith G. Orrell

University of Exeter, Exeter, UK

1 INTRODUCTION

The suitability of NMR spectroscopy for monitoring exchange processes of molecules has been realized since the earliest days of the technique. In those formative years, rate processes were studied primarily by their effects on internal resonance frequencies of nuclei (bandshape analysis)[1] or on signal intensities (magnetization transfer experiments).[2,3] The first method relies on exchange processes occurring at frequencies comparable to differences in nuclear resonance frequencies $\Delta\nu_i$ (see ***Chemical Exchange Effects on Spectra***), whereas the second approach requires exchange rates to be comparable to spin–lattice relaxation rates R_{1i}. The consequences of such criteria are that bandshape analysis methods are applicable to rather faster rate processes, with pseudo-first-order rate constants k in the range $1–10^4\,s^{-1}$, than magnetization transfer methods, where typically k values are in the range $10^{-2}–10^2\,s^{-1}$. Such ranges, however, are very imprecise, and depend on the nucleus chosen to monitor the process and the nature of the chemical species itself—for example, whether it is diamagnetic or paramagnetic.

In magnetization transfer experiments, based on the classic work of Forsén and Hoffman,[2,3] nuclei in one of the exchanging sites are labeled by disturbing their energy population distribution, either by selective decoupling, which equalizes their populations (saturation), or by applying a selective 180° pulse, which inverts their populations (inversion). Chemical exchange causes these nuclei to move to one or more different sites, thus introducing a further disturbance to the populations in that site. This leads to a time dependence of populations of exchanging sites, which is reflected in the changing intensities of signals following the perturbing pulse. Measurements of this intensity variation can provide accurate exchange rate data. The method can be made very reliable for two-site exchange,[4] and has been very widely applied. The method can also be generalized for multisite exchange, where it has been shown that for a system of N exchanging sites $\frac{1}{2}N(N-1)$ exchange rates and N spin–lattice relaxation rates need to be evaluated.[5] However, this requires many experiments, and this type of problem is far more efficiently handled by two-dimensional experiments[6–8] (see ***Multidimensional Spectroscopy: Concepts***), which is the subject of this article.

Rate processes that cause exchange of magnetization between nuclear sites are many and varied. They range from examples of intermolecular exchange (e.g., ligand or solvent exchange processes) to subtle intramolecular rearrangements between chemically identical species (e.g., fluxional processes) (see ***Fluxional Motion***). This article presents the basic theory of two-dimensional exchange spectroscopy (EXSY), illustrates the scope and limitations of the technique, and provides examples of its application, employing a variety of different sensor nuclei, to different areas of chemistry.

The focus will be on chemical exchange arising from structural rearrangements of molecules, but the reader is reminded that essentially the same technique is used for studying motional processes as revealed by cross relaxation rather than chemical exchange effects. In this context, the technique is known as two-dimensional NOE spectroscopy, and has been widely exploited[9] (see ***NOESY***).

2 THEORY AND METHODOLOGY OF 2D EXCHANGE SPECTROSCOPY

The application of 2D NMR for chemical exchange studies was first described by Jeener, Meier, Bachmann, and Ernst.[10] Its theoretical basis was elaborated by Macura and Ernst,[11] and has been the subject of a number of reviews.[12–15]

2.1 Basic EXSY Experiment

The fundamental principle of the EXSY experiment is the frequency labeling of the spin–lattice (longitudinal) magnetization of various sites before exchange occurs, in order that, after exchange, the pathways of magnetization can be traced back to their origins.

The basic pulse sequence is shown in Figure 1. A two-spin system of spins A and B, mutually interacting purely through chemical exchange with no cross relaxation, will be considered. Equal concentrations of spins (i.e., $k_{AB} = k_{BA} = k$), equal spin–lattice relaxation rates ($R_1^A = R_1^B = R_1$), and equal transverse relaxation times ($T_2^A = T_2^B = T_2$) will be assumed. After the first 90°_x pulse, the magnetization vectors will evolve in the (x', y') plane in the usual manner of free induction decay (FID) for a period t_1 (evolution time). The total magnetization M^+ is given by

$$M^+ = M_y(t) + iM_x(t) \quad (1)$$

If exchange is slow, the effect of lineshape may be neglected, and the two magnetization components are given by

$$M_A^+(t_1) = M_{A0}\exp(i\omega_A t_1 - t_1/T_2) \quad (2)$$

$$M_B^+(t_1) = M_{B0}\exp(i\omega_B t_1 - t_1/T_2) \quad (3)$$

where $\omega_{A,B}$ are the effective precession frequences in the rotating frame. When a second 90°_x pulse, the mixing pulse, is applied, the magnetization vectors are tipped into the $-z$ direction, the real components of the transverse magnetization being converted into longitudinal magnetization:

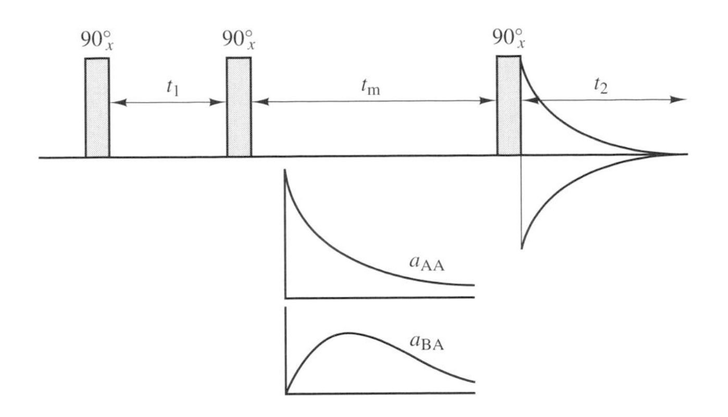

Figure 1 Basic EXSY pulse sequence, showing typical variations in the mixing coefficients for diagonal and cross peaks during the mixing time t_m

$$M_{AZ}(t_m = 0) = -M_{A0} \cos\omega_A t_1 \exp(-t_1/T_2) \quad (4)$$

$$M_{BZ}(t_m = 0) = -M_{B0} \cos\omega_B t_1 \exp(-t_1/T_2) \quad (5)$$

The imaginary x components are unaffected by this pulse, and may be destroyed by a field gradient pulse during the mixing time t_m. The t_1-modulated longitudinal components in equations (4) and (5) migrate from one site to another by chemical exchange, with a rate constant k, while the spin–lattice relaxation attenuates the memory of the initial labeling:

$$\begin{aligned} M_{AZ} = {} & M_{AZ}(t_m = 0)\tfrac{1}{2}[1 + \exp(-2kt_m)] \exp(-t_m R_1) \\ & + M_{BZ}(t_m = 0)\tfrac{1}{2}[1 - \exp(-2kt_m)] \exp(-t_m R_1) \end{aligned} \quad (6)$$

$$\begin{aligned} M_{BZ} = {} & M_{AZ}(t_m = 0)\tfrac{1}{2}[1 - \exp(-2kt_m)] \exp(-t_m R_1) \\ & + M_{BZ}(t_m = 0)\tfrac{1}{2}[1 + \exp(-2kt_m)] \exp(-t_m R_1) \end{aligned} \quad (7)$$

The final 90°_x pulse converts these longitudinal components into observable magnetization. Double Fourier transformation will produce signals at (ω_1, ω_2) = (ω_A, ω_A) and (ω_B, ω_B) for nuclei that have not exchanged during the time t_m. Nuclei that have exchanged sites will produce signals at (ω_A, ω_B) and (ω_B, ω_A). The amplitudes $I_{kl}(t_m)$ of the diagonal and off-diagonal signals depend on the equilibrium magnetizations M_{l0} and on mixing coefficients $a_{kl}(t_m)$, which correspond to the factors in equations (6) and (7):

$$a_{AA} = a_{BB} = \tfrac{1}{2}[1 + \exp(-2kt_m)] \exp(-t_m R_1) \quad (8)$$

$$a_{AB} = a_{BA} = \tfrac{1}{2}[1 - \exp(-2kt_m)] \exp(-t_m R_1) \quad (9)$$

In this simple two-site exchange system, the equilibrium magnetizations M_{A0} and M_{B0} are equal, and the exchange rate can be determined from the ratio of diagonal and off-diagonal (cross peak) intensities:

$$\frac{I_{AA}}{I_{AB}} = \frac{a_{AA}}{a_{AB}} = \frac{1 + \exp(-2kt_m)}{1 - \exp(-2kt_m)} \approx \frac{1 - kt_m}{kt_m} \quad (10)$$

This intensity ratio is a sensitive function of the mixing time t_m, and care must be taken in the length chosen in order to obtain optimum accuracy of the rate constant. The amplitudes of the diagonal signals decay in the biexponential manner with increasing mixing time, whereas the off-diagonal signals first increase owing to exchange and then decrease because of spin–lattice relaxation (see Figure 1). The optimal value of t_m corresponds to the region on the graph where the intensity of the cross peak is almost linearly proportional to t_m and well before it reaches a level where its intensity is more dependent on T_1 than on t_m. It has been shown theoretically[13] that an approximate expression for the optimal value of t_m for an equally populated two-site exchange system is

$$t_m(\text{opt}) \approx (R_1 + 2k)^{-1} \quad (11)$$

One of the first reported EXSY spectra is shown in Figure 2.[10] It depicts the diagonal and cross-peak signals of the *N*-methyl hydrogens of *N,N*-dimethylacetamide, which exhibit exchange due to the slow rotation about the C–N bond in the molecule.

This theory can be extended to general N-site exchange when $N \geqslant 3$ using a matrix formulation. This has been described fully elsewhere,[12–15] and here space permits only a summary. For nuclei in site j, the initial 90°_x pulse causes their magnetization to evolve according to

$$M_j = M_{j0} \exp(\mathrm{i}\omega_j t_1 - t_1/T_{2j}) \quad (12)$$

The second 90°_x pulse, plus a gradient pulse, converts this transverse magnetization to z magnetization according to

$$M_{jz}(t_m = 0) = -M_{j0} \exp(\mathrm{i}\omega_j t_1 - t_1/T_{2j}) \quad (13)$$

During the mixing period, chemical exchange and longitudinal relaxation affect this magnetization. We define m_j to be the deviation of M_{jz} from its equilibrium value by

$$m_j = M_{jz} - M_{j0} \quad (14)$$

The time dependences of these deviations are given by equations of the type

$$\frac{\mathrm{d}m_1}{\mathrm{d}t} = -\left(R_{11} + \sum_l k_{1l}\right) m_1 + k_{21} m_2 + \cdots + k_{N1} m_N \quad (15)$$

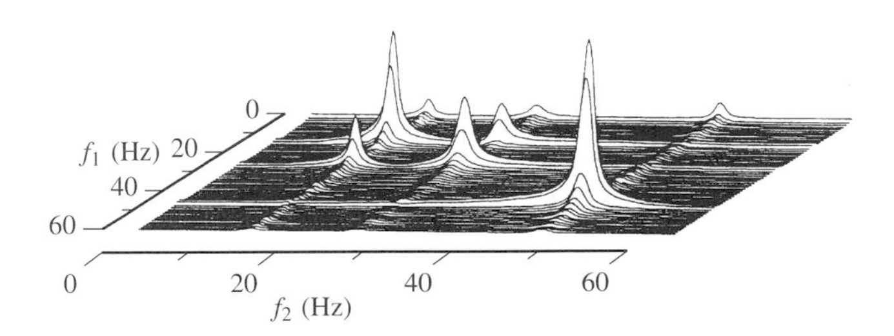

Figure 2 Hydrogen-1 EXSY spectrum of *N,N*-dimethylacetamide, showing the *N*-methyl exchange cross peaks. Experimental conditions: 128 equidistant t_1 values from 0 to 1016 ms, each FID represented by 256 samples, and a mixing time t_m = 0.5 s. Note that diagonal signals go from top left to bottom right

For References see p. 4856

$$\frac{dm_2}{dt} = k_{12}m_1 - \left(R_{12} + \sum_l k_{2l}\right)m_2 + \cdots + k_{N2}m_N \tag{16}$$

$$\vdots$$

$$\frac{dm_N}{dt} = k_{1N}m_N + k_{2N}m_2 + \cdots - \left(R_{1N} + \sum_l k_{N1}\right)m_N \tag{17}$$

where k_{ij} are the rate constants for exchange from site i to site j, and R_{1i} are the spin–lattice relaxation rates of nuclei in site i. These equations may be written in matrix form as

$$\dot{\mathbf{m}} = -\mathbf{L}\mathbf{m} \tag{18}$$

where $\mathbf{m}$ is a column matrix whose components are $m_1, m_2, \ldots, m_N$, $\dot{\mathbf{m}}$ is the time derivative of $\mathbf{m}$, and $\mathbf{L}$ is a square $N \times N$ matrix whose off-diagonal elements are $L_{ij} = -k_{ji}$ and whose diagonal elements are $L_{ii} = \sum_l k_{il} + R_{1i}$. This matrix $\mathbf{L}$ contains all site-to-site rate constants, and the problem is to relate the elements of $\mathbf{L}$ to the signal intensities in an EXSY spectrum. The formal solution is given by

$$\mathbf{m}(t_m) = \exp(-\mathbf{L}t_m)\mathbf{m}(0) \tag{19}$$

where $\mathbf{m}(0)$ is the column matrix of the deviations $m_j(0)$ at the start of the mixing period.

The explicit expression for the intensity of any EXSY signal (diagonal or off-diagonal) at ω_i along the frequency coordinate ω_2 and at ω_j along ω_1 is given by[13]

$$\begin{aligned} I_{ij}(t_m) &= M_{j0}\exp(-\mathbf{L}t_m)_{ij} \\ &= \left(\delta_{ij} - t_m L_{ij} + \tfrac{1}{2}t_m^2 \sum_k L_{ik}L_{kj} - \cdots\right)M_{j0} \end{aligned} \tag{20}$$

where $\delta_{ij} = 1$ $(i = j)$ or 0 $(i \neq j)$. Equation (20) shows that an EXSY spectrum represents a graphical display of the exchange pathways, with cross peaks corresponding to nuclei that exchange sites. Intensities of signals correspond to the exponential of the matrix $\mathbf{L}$, allowance being made for the fact that the mathematical convention of a matrix is with the diagonal running from top left to bottom right whereas an EXSY spectral map consists of a diagonal running from bottom left to top right.

2.2 Extraction of Rate Data from EXSY Spectra

The complexity of equation (20) arises because cross peaks for exchange between sites i and j can arise from both direct exchange $i \rightleftharpoons j$ and indirect exchange, $i \rightleftharpoons k \rightleftharpoons j$. For very short t_m values, the indirect cross peaks vanish, and equation (20) simplifies. However, short t_m values cause all cross-peak intensities to be low, so this approach is of limited applicability. Three strategies have been employed to extract reliable rate data from EXSY spectra.

2.2.1 *Initial rate approximation*

This involves measuring cross-peak intensities as a function of small values of t_m.[16] Rate constants k_{ij} are determined from the slope of the graph of $I_{ij}(t_m)$ versus $M_{j0}(t_m)$, but the method involves repeating the experiment for a range of mixing times and extrapolating to $t_m = 0$, which is a time-consuming approach.

2.2.2 *Iterative Analysis*

In this method,[17] trial values of k_{ij} and R_{1i} are chosen, and a total magnetization transfer matrix $\mathbf{L}$ is constructed. Diagonalization of this yields a signal intensity matrix, which is compared with the experimental intensity array. An iterative fitting is then performed by varying exchange and relaxation rates until a 'best fit' is obtained.

2.2.3 *Direct Matrix Transformation*

Unlike the previous method, this approach does not require any prior knowledge of k_{ij} and R_{1i} values, and uses a matrix method for inverting equation (20) first presented by Perrin and Gipe.[18] It is based on expressing an EXSY spectrum as an intensity matrix $\mathbf{I}$ with elements

$$I_{ij} = p_i \sum_{k=1}^{N} A_{ijk}\exp(\lambda_k t_m) \tag{21}$$

where p_i is the relative population of site i, and A_{ijk} and λ_k are constants depending on the experimental conditions. In matrix form,

$$\mathbf{I} = \mathbf{PJ} \tag{22}$$

where $\mathbf{P}$ is a column matrix of site populations and $\mathbf{J}$ is a matrix with elements $\exp(\lambda_k t_m)$. The kinetic matrix $\mathbf{K}$ consisting purely of rate constants k_{ij} can be related to the EXSY spectral intensity matrix $\mathbf{I}$ by

$$\mathbf{I} = \mathbf{P}\exp(\mathbf{K}t_m) \tag{23}$$

Now, $\mathbf{J} = \mathbf{IP}^{-1}$, and diagonalization of $\mathbf{J}$ will generate the array $\exp(\mathbf{\Lambda}t_m)$:

$$\mathbf{J} = \mathbf{X}\exp(\mathbf{\Lambda}t_m)\mathbf{X}^{-1} \tag{24}$$

Taking the natural logarithm of the eigenvalues and dividing by t_m, i.e.,

$$\frac{\ln \mathbf{J}}{t_m} = \frac{\mathbf{X}\ln\exp(\mathbf{\Lambda}t_m)\mathbf{X}^{-1}}{t_m} \tag{25}$$

generates the matrix $\mathbf{\Lambda}$.

Finally, matrix multiplication

$$\mathbf{X\Lambda X}^{-1} = \mathbf{K} \tag{26}$$

yields the kinetic matrix $\mathbf{K}$, consisting of all rate constants for the multisite process. It should be noted that in this formulation,[19] all off-diagonal elements are positive and diagonal elements negative, in contrast to the kinetic part of the $\mathbf{L}$ matrix defined earlier [equation (18)].

It is also necessary to include spin–lattice relaxation of nuclei. This can take the form of a relaxation matrix $\mathbf{R}$, which when added to $\mathbf{K}$ gives the total magnetization transfer matrix $\mathbf{L}$:

$$\mathbf{L} = \mathbf{K} + \mathbf{R} \tag{27}$$

The matrix $\mathbf{R}$ consists of diagonal elements $-R_{1i}$ and off-diagonal elements $-\sigma_{ij} = -R_{ij}$, where these refer to cross-relaxation rates.

In cases where cross relaxation between nuclei can be neglected, the above procedures have been incorporated into a general computer program.[19] Its input consists of the number of exchanging sites (2–8), the relative populations of each site, spin–lattice relaxation rates, and experimental signal intensities, the latter being measured by integration of appropriate rows of the 2D spectral map, or, preferably, by volume integration. The accuracy of the final rate data depends on the choice of mixing time and on the accuracy of measured signal intensities. Pure absorption mode spectra provide better resolution of closely separated signals than magnitude mode spectra. This can be achieved using time proportional phase incrementation (TPPI), and so this is the recommended procedure for accurate quantitative EXSY studies. The ^{1}H EXSY spectrum of [^{15}N]acetamide using this procedure is shown in Figure 3.[20] Both NH signals are doublets [^{1}J(NH) $\approx$ 90 Hz], and exhibit cross peaks due to slow C–N bond rotation. Intensity measurements of all eight signals provided accurate rate and R_1 values for the NH hydrogens. The much weaker cross peaks are due to ^{15}N spin–lattice relaxation occurring during the mixing period.

The assessment of errors in EXSY-based rate constants is an important matter which has received careful consideration.[19,21,22] The choice of mixing time has already been shown to be crucial. In multisite exchanges, it is usually impossible to find a single value of t_m that is optimal for all exchange pathways. Additional cross peaks due to indirect exchanges [equation (20)] may then arise if the mixing time is considerably greater than optimal, and so qualitative interpretations of exchange pathways from EXSY spectra must be made with caution.

2.3 Separation of Chemical Exchange and Cross-Relaxation Effects

The theoretical treatment above has neglected cross relaxation between nuclei; i.e., off-diagonal elements of the matrix **R** have been assumed to be negligibly small. This is generally valid for intermolecular exchanges and studies of intramolecular exchange using low-abundance sensor nuclei (e.g., ^{13}C). Also, it is often valid for studies of abundant nuclei (e.g., ^{1}H, ^{19}F, and ^{31}P) in small molecules, which can undergo rapid tumbling in solution. For large, slowly tumbling molecules, however, separation of chemical exchange and cross-relaxation rates is essential.

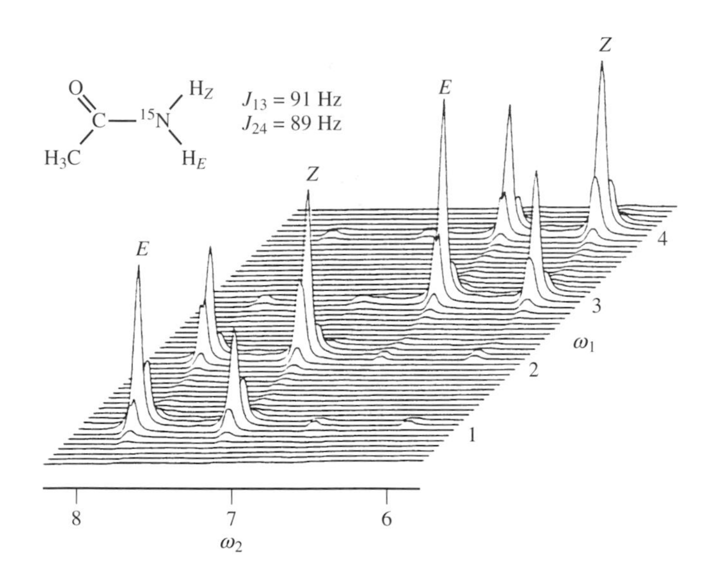

Figure 3 Partial phase-sensitive ^{1}H EXSY spectrum of ^{15}N-enriched-acetamide obtained with the TPPI method for t_m = 400 ms. A 1K data block and spectral width in ω_2 of 614 Hz were used. A total of 256 t_1 values was used and zero filled to 1K, resulting in a digital resolution of 1.2 Hz point^{-1} in both frequency domains

Cross-relaxation rates are given by

$$\sigma_{ij} = \frac{\gamma^4 \hbar^2 \tau_c}{10 r_{ij}^6} \left(\frac{6}{1 + 4\omega^2 \tau_c^2} - 1 \right) \tag{28}$$

where r_{ij} is the distance between nuclei i and j, and τ_c is the rotational correlation time of the molecule. The r^{-6} dependence provides proximity information, which has been widely exploited in conformational studies of complex molecules.[9] For small molecules, where τ_c is short, cross-relaxation cross peaks are of opposite sign to diagonal signals, whereas for macromolecules with relatively large τ_c values, cross-relaxation signals are of the same sign as the diagonals. Pure absorption mode (TPPI) spectra, therefore, can in theory distinguish the two cases, but in either case cross-peak intensities may not reflect chemical exchange alone.

Separation of cross relaxation and chemical exchange is most easily achieved by temperature dependence studies, since k_{ij} values increase greatly with temperature whereas σ_{ij} values do not. More recently, new pulse sequences have been suggested for achieving this separation. These include zz-EXSY,[23] an interlayed NOESY/ROESY method,[24] and gradient-enhanced exchange spectroscopy (GEXSY).[25]

2.4 One-Dimensional Variants of EXSY

The time required to obtain kinetic data from EXSY experiments can often be considerable, especially if long relaxation delays, long mixing times, and high digital resolution in both dimensions are required. Much time is incurred scanning through the whole range of incremental evolution periods, t_1, yet the majority of rows in the final spectrum do not contain useful information. However, certain t_1 values cannot simply be omitted, since each value contributes to all frequencies through the FID. One method for shortening the experimental time would be to dispense with the t_1 period completely and rely on selective excitation (see ***Selective Pulses***). The feasibility of this approach has been demonstrated for COSY experiments and is known as pseudo-COSY (ψ-COSY).[26] A similar technique has recently been proposed for exchange spectroscopy. It is described as multiplet-selective excitation or MUSEX,[27] and is based on selective excitation of only one nucleus of a coupled system using a typical DANTE pulse[28] (see ***Double Resonance***).

Another variant uses the normal nonselective pulse sequence and difference spectroscopy.[29] For N-site exchange, data for N evolution times must be collected, measuring each at zero and nonzero mixing times. Difference spectra obtained by subtracting the equilibrium spectrum from spectra recorded both without and with mixing yields exchange information. For a system of three spins A, B, and C, six one-dimensional spectra are measured for three t_1 periods, each without and with mixing. It follows that

For References see p. 4856

$$\begin{bmatrix} A(t_m) - A_0 \\ B(t_m) - B_0 \\ C(t_m) - C_0 \end{bmatrix} = \exp(-\mathbf{L}t_m) \begin{bmatrix} A(t_m = 0) - A_0 \\ B(t_m = 0) - B_0 \\ C(t_m = 0) - C_0 \end{bmatrix} \qquad (29)$$

The matrix **L** can then be computed by a version of the back transformation method (Section 2.2.3). Greater accuracy can be achieved if the experiment is performed for more t_1 values than are strictly necessary.[30]

2.5 Other Variants of EXSY

In 1989, Jeener, one of the pioneers of 2D NMR and its application to chemical kinetics, returned to this subject by examining the theoretical basis of the bandshapes of EXSY signals.[31] He has shown that the shapes, particularly of off-diagonal signals, provide a clear signature of the chemical exchange process even for a single temperature. Superoperator theory was applied to one- and two-spin exchanging signals. Computation, however, is not trivial, and no obvious advantages of this approach over conventional 1D bandshape analysis or 2D EXSY experiments are apparent at this stage.

In its most general form, the EXSY or NOESY pulse sequence involves three time variables, t_1, t_m, and t_2. Fourier transformation of all of which would lead to three-dimensional spectra. In practice, however, the mixing time t_m is not taken as a time variable, and 2D spectra based on a very limited range of t_m values are usually obtained. An alternative approach is to make t_m proportional to t_1 (i.e., $t_m = \kappa t_1$). The resulting 'accordion' technique[32] leads to EXSY-type spectra in which the signals are distorted from their usual shapes, and rate constants need to be extracted from signal widths rather than intensities. This restricts the versatility of the method.

Three-dimensional NMR experiments are now rapidly gaining ground[33] (see ***Multidimensional Spectroscopy: Concepts***). In 3D exchange spectroscopy, two successive exchange processes may be mapped. These may be chemical exchange (EXSY) or cross relaxation in either the laboratory frame (NOESY) or in the rotating frame (ROESY), resulting in such experiments as EXSY–EXSY and NOESY–ROESY. In the context of this article, it should be mentioned that a 3D EXSY–EXSY spectrum of heptamethylbenzenonium sulfate (in sulfuric acid) has been obtained,[33] showing the 1,2-commutation of the methyl group round the ring. It is too early to predict whether such experiments will provide more insight into dynamic processes than can be achieved by 1D and 2D experiments. Dynamic studies of highly complex molecules would appear to be most likely to gain by the extra frequency dimension.

3 APPLICATIONS OF EXSY

A few representative examples of EXSY experiments using different sensor nuclei will now be given.

3.1 Carbon-13 and Hydrogen-1

In Section 2.3, the separation of chemical exchange and cross-relaxation contributions to cross peak intensities was discussed. This problem does not arise in EXSY spectra of low-abundance nuclei (e.g., ^{13}C), where cross-relaxation effects are negligible, but is a potential problem with abundant nuclei such as ^{1}H. A specific case has been described where the effects of ^{1}H–^{1}H cross relaxation have been quantified.[34] In the trinuclear rhenium complex $[Re_3(\mu\text{-}H)_4(CO)_{10}]$, carbonyl scrambling occurs by a concerted mechanism (see ***Organometallic Compounds***), the kinetics of which can be measured accurately by ^{13}C EXSY. Lower accuracy would be attached to ^{1}H-EXSY-based rates, since one-dimensional selective perturbation data showed that the cross-relaxation rate of a pair of bridging hydrogens was $0.04\,s^{-1}$, a small but not negligible rate. On the other hand, a combined ^{13}C and ^{1}H EXSY study of the metal migration in the cyclooctatetraene (COT) complexes $[M(CO)_3(\eta^6\text{-COT})]$ (M = Cr or W) produced kinetic data in excellent agreement with each other, indicating that ^{1}H–^{1}H cross relaxation in this case was negligible.[35]

3.2 Lithium-7

The first ^{7}Li EXSY spectra have recently been produced in connection with a kinetic study of the complex [lithium monobenzo-15-crown-5]$^{+}$ in nitromethane.[36] Mixing times varied between 5 and 100 ms, nine experiments being performed, each requiring 1 h of spectrometer time. Rate data were in close agreement with those measured by bandshape analysis, showing that the EXSY method can be applied with confidence to systems such as cryptands, where exchange is too slow for bandshape treatment.

3.3 Boron-11

The well-known redistribution reactions of boron trihalide mixtures have been examined by ^{11}B EXSY.[37] The particular mixture studied was a nearly 1:1 M mixture of BCl_3 and BBr_3 at 400 K. The spectrum for a mixing time of 50 ms is shown in Figure 4. Rate data were calculated by the three methods described in Section 2.2, with the direct matrix diagonalization method being preferred.

3.4 Silicon-29

Much insight into the silicate anion exchange pathways in ^{29}Si-enriched potassium silicate solutions has been gained by ^{29}Si EXSY.[38] Four intermolecular exchanges were discovered, involving monomeric anion, dimer, linear trimer, linear tetramer, and substituted cyclic trimer species, plus two intramolecular exchanges. Although this was only a qualitative study, it demonstrates the power of 2D EXSY over its 1D counterpart.

3.5 Phosphorus-31

This nucleus is well suited to EXSY studies in view of its high receptivity, wide chemical shift range, and common absence of ^{31}P–^{31}P cross relaxation. Numerous studies have been reported, one example being the octahedral organochromium(0) complexes $[Cr(CO)_2(CX)\{(MeO)_3P\}_3]$ (X = S, Se).[39] These stereochemically nonrigid complexes undergo rearrangements via trigonal-prismatic (Bailar) intermediates. The ^{31}P EXSY experiments were of the 'accordion' type[32] (Section 2.5), with the factor κ having a magnitude of 30.

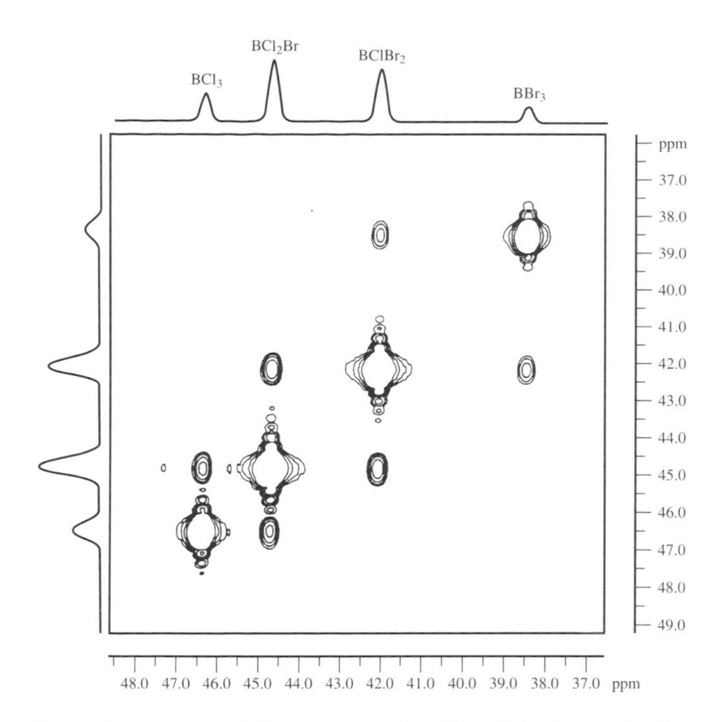

Figure 4 Boron-11 EXSY spectrum of a 1.05:1.0 M mixture of BCl_3 and BBr_3 at 400 K. The mixing time was 50 ms. The F_1 dimension contained 128 points, zero filled to 256 points. The F_2 dimension contained 512 real plus complex points. The number of scans per FID was 96

3.6 Vanadium-51

Quantitative kinetic data on oligomerization reactions of vanadate in aqueous solutions has been provided by ^{51}V EXSY.[40] At pH 8.6, the EXSY spectrum shows all major oligomers, namely monomer, dimer, tetramer, and pentamer, exchanging with each other. Rates of several exchange pathways were identified. Mixing times were chosen in the range 8–10 ms.

3.7 Tin-119

The first example of ^{119}Sn EXSY was in connection with the dynamic stereochemistry of a ditin compound $CH_2[PhSn(SCH_2CH_2)_2NMe]_2$.[41] The spectrum, with a short mixing time of 5 ms, showed pairwise exchange of three isomers associated with uncorrelated isomerization of each tin centre. The same research group has also investigated $[MeSn(CH_2CH_2CH_2)_2NMe]_2$, which in solution exists as at least three isomers, two of which interconvert at rates measurable by ^{119}Sn EXSY.[42]

3.8 Tellurium-125

Platinum(IV) complexes of the type $[PtXMe_3\{MeE(CH_2)_n\text{-}EMe\}]$ (E = S, Se, Te; X = Cl, Br, I; n = 2, 3) undergo a variety of internal rearrangements, the most facile being pyramidal inversion of the coordinated chalcogen atoms. When the chalcogen, E, is Se or Te, the inversion process can in theory be followed by ^{1}H, ^{13}C, ^{77}Se, ^{125}Te, and ^{195}Pt spectra. To avoid any complications from cross-relaxation effects, the rare nuclei are the preferred probes. The inversion process causes exchange between four isomers according to Scheme 1.

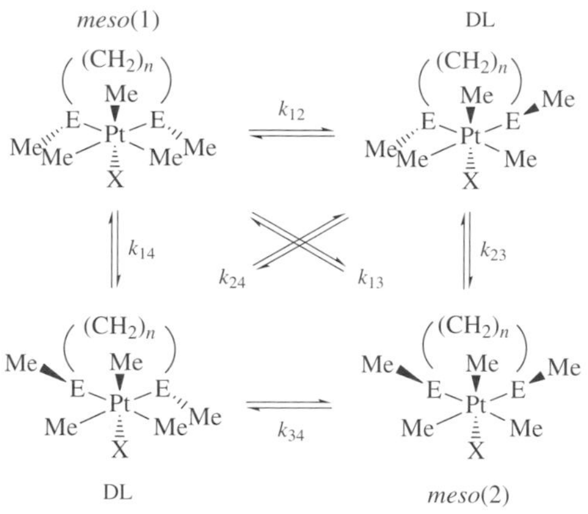

Scheme 1

For the complex $[PtIMe_3\{MeTe(CH_2)_3TeMe\}]$, the Te inversion rates were measured by ^{125}Te EXSY.[43] Cross peaks were detected between the DL and *meso* signals only, indicating that rates of synchronous inversion of both Te atoms are negligible.

3.9 Platinum-195

^{195}Pt EXSY experiments are particularly powerful in monitoring exchange rates in the above Pt(IV) complexes. The EXSY spectrum of $[PtIMe_3(MeSCH_2CH_2SMe)]$ is illustrated in Figure 5. The three isomers are identified as shown, the signals due to DL and *meso*(1) also possessing ^{13}C satellites. For the chosen mixing time of 0.3 s, cross peaks were detected between all three species. The inversion process is strictly represented by a four-site exchange problem (Scheme 1), but, when monitored by ^{195}Pt nuclei, it reduces to a three-site exchange since the DL pair are indistinguishable. Signal intensity data from the spectra were used to set up the intensity matrix **I** as defined earlier,[19] and the relative populations of the isomers enabled the matrix **P** to be constructed. The matrix conversion program[19] then generated the kinetic-plus-relaxation matrix **L**, the off-diagonal elements of which provided rate constant magnitudes, namely, $k_{12} = 0.52 \pm 0.03\,s^{-1}$, $k_{23} = 0.50 \pm 0.03\,s^{-1}$, and $k_{13} = 0.004 \pm 0.04\,s^{-1}$. This latter rate constant for synchronous double inversion of both sulfur atoms is clearly zero within experimental uncertainty, despite the appearance of weak cross peaks I_{13} and I_{31}. This illustrates the point made earlier (Section 2.2) that the existence of a cross peak does not necessarily imply direct exchange between the two sites in question, and so qualitative interpretations of EXSY spectra must be made with considerable caution. The same research group has studied the sulfur inversion dynamics of $[PtIMe_3\text{-}(MeSCH_2CH_2SEt)]$.[44] This constitutes a four-site exchange problem, a case too complex to be analyzed by total bandshape analysis. In contrast, the ^{195}Pt EXSY spectrum provided definitive rate data for the six distinct inversion pathways.

For References see p. 4856

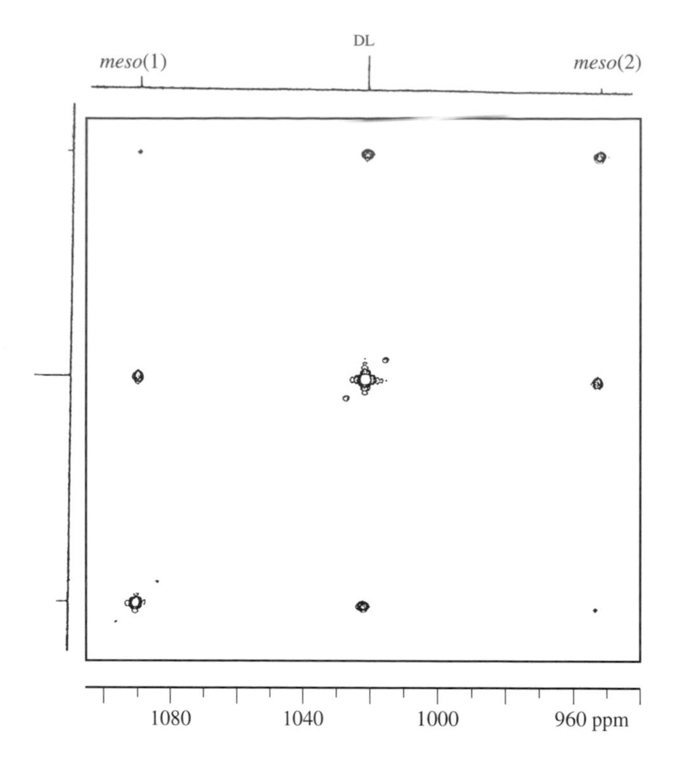

Figure 5 Platinum-195 EXSY spectrum of $[PtIMe_3(MeSCH_2CH_2SMe)]$ at 253 K. The mixing time was 300 ms. The F_1 dimension contained 256 points, zero filled to 512 points. The F_2 dimension contained 1024 points

4 RELATED ARTICLES

Chemical Exchange Effects on Spectra; Fluxional Motion; Homonuclear Three-Dimensional NMR of Biomolecules; Inorganic Chemistry Applications; Multidimensional Spectroscopy: Concepts; NOESY; Relaxation Processes: Cross Correlation & Interference Terms.

5 REFERENCES

1. H. S. Gutowsky and C. H. Holm, *J. Chem. Phys.*, 1956, **25**, 1228.
2. S. Forsén and R. A. Hoffman, *J. Chem. Phys.*, 1963, **39**, 2892; *J. Chem. Phys.*, 1964, **40**, 1189.
3. R. A. Hoffman and S. Forsén, *J. Chem. Phys.*, 1966, **45**, 2049.
4. J. J. Led and H. Gesmar, *J. Magn. Reson.*, 1982, **49**, 444.
5. H. Gesmar and J. J. Led, *J. Magn. Reson.*, 1986, **68**, 95.
6. R. Benn and H. Gunther, *Angew. Chem., Int. Ed. Engl.*, 1983, **22**, 350.
7. H. Kessler, M. Gehrke, and C. Griesinger, *Angew. Chem., Int. Ed. Engl.*, 1988, **27**, 490.
8. R. R. Ernst, G. Bodenhausen, and A. Wokaun, 'Principles of Nuclear Magnetic Resonance in One and Two Dimensions', Clarendon Press, Oxford, 1987, Chaps. 6–9.
9. D. Neuhaus and M. Williamson, 'The Nuclear Overhauser Effect in Structural and Conformational Analysis', VCH, New York, 1989.
10. J. Jeener, B. H. Meier, P. Bachmann, and R. R. Ernst, *J. Chem. Phys.*, 1979, **71**, 4546.
11. S. Macura and R. R. Ernst, *Mol. Phys.*, 1980, **41**, 95.
12. R. Willem, in 'Progress in Nuclear Magnetic Resonance Spectroscopy', ed. J. W. Emsley, J. Feeney, and L. H. Sutcliffe, Pergamon Press, Oxford, 1987, Vol. 20, p. 1.
13. C. L. Perrin and T. J. Dwyer, *Chem. Rev.*, 1990, **90**, 935.
14. K. G. Orrell, V. Šik, and D. Stephenson, in 'Progress in Nuclear Magnetic Resonance Spectroscopy', ed. J. W. Emsley, J. Feeney, and L. H. Sutcliffe, Pergamon Press, Oxford, 1990, Vol. 22, p. 141.
15. K. G. Orrell and V. Šik, in 'Annual Reports on NMR Spectroscopy', ed. G. A. Webb, Academic Press, London, 1993, Vol. 27, p. 103.
16. A. Kumar, G. Wagner, R. R. Ernst, and K. Wüthrich, *J. Am. Chem. Soc.*, 1981, **103**, 3654.
17. G. E. Hawkes, L. Y. Lian, E. W. Randall, K. D. Sales, and S. Aime, *J. Magn. Reson.*, 1985, **65**, 173.
18. C. L. Perrin and R. K. Gipe, *J. Am. Chem. Soc.*, 1984, **106**, 4036.
19. E. W. Abel, T. P. J. Coston, K. G. Orrell, V. Šik, and D. Stephenson, *J. Magn. Reson.*, 1986, **70**, 34.
20. E. R. Johnston, M. J. Dellwo, and J. Hendrix, *J. Magn. Reson.*, 1986, **66**, 399.
21. P. W. Kuchel, B. T. Bulliman, B. E. Chapman and G. L. Mendz, *J. Magn. Reson.*, 1988, **76**, 136.
22. C. L. Perrin and R. E. Engler, *J. Magn. Reson.*, 1990, **90**, 363.
23. G. Wagner, G. Bodenhausen, N. Muller, M. Rance, O. W. Sorensen, R. R. Ernst, and K. Wuthrich, *J. Am. Chem. Soc.*, 1985, **107**, 6440.
24. J. Fejzo, W. M. Westler, S. Macura, and J. L. Markley, *J. Magn. Reson.*, 1991, **92**, 20.
25. C. T. W. Moonen, P. Van Gelderen, G. W. Vuister, and P. C. M. Van Zijl, *J. Magn. Reson.*, 1992, **97**, 419.
26. S. Davies, J. Friedrich, and R. Freeman, *J. Magn. Reson.*, 1987, **75**, 540.
27. Yu E. Chernysh, G. S. Borodkin, E. V. Sukholenko, and L. E. Nivorozhkin, *J. Magn. Reson.*, 1992, **96**, 131.
28. G. A. Morris and R. A. Freeman, *J. Magn. Reson.*, 1978, **29**, 433.
29. R. E. Engler, E. R. Johnston, and C. G. Wade, *J. Magn. Reson.*, 1988, **77**, 377.
30. B. T. Bulliman, P. W. Kuchel, and B. E. Chapman, *J. Magn. Reson.*, 1989, **82**, 131.
31. J. Jeener, *J. Chem. Phys.*, 1989, **90**, 2959.
32. G. Bodenhausen and R. R. Ernst, *J. Am. Chem. Soc.*, 1982, **104**, 1304.
33. C. Griesinger, O. W. Sorensen, and R. R. Ernst, *J. Magn. Reson.*, 1989, **84**, 14.
34. T. Beringhelli, G. D'Alfonso, H. Molinari, G. E. Hawkes, and K. D. Sales, *J. Magn. Reson.*, 1988, **80**, 45.
35. E. W. Abel, K. G. Orrell, K. B. Qureshi, V. Šik, and D. Stephenson, *J. Organomet. Chem.*, 1988, **353**, 337.
36. K. M. Briere, H. D. Dettman, and C. Detellier, *J. Magn. Reson.*, 1991, **94**, 600.
37. E. F. Derose, J. Castillo, D. Saulys, and J. Morrison, *J. Magn. Reson.*, 1991, **93**, 347.
38. C. T. G. Knight, R. J. Kirkpatrick, and E. Oldfield, *J. Magn. Reson.*, 1988, **78**, 31.
39. A. A. Ismail, F. Sauriol, and I. S. Butler, *Inorg. Chem.*, 1989, **28**, 1007.
40. D. C. Crans, C. D. Rithner, and L. A. Theisen, *J. Am. Chem. Soc.*, 1990, **112**, 2901.
41. C. Wynants, G. Van Binst, C. Muegge, K. Jurkschat, A. Tzschach, H. Pepermans, M. Gielen, and R. Willem, *Organometallics*, 1985, **4**, 1906.
42. K. Jurkschat, A. Tzschach, C. Muegge, J. Piret-Meunier, M. Van Meerssche, G. Van Binst, C. Wynants, M. Gielen, and R. Willem, *Organometallics*, 1988, **7**, 593.
43. E. W. Abel, K. G. Orrell, S. P. Scanlon, D. Stephenson, T. Kemmitt, and W. Levason, *J. Chem. Soc., Dalton Trans.*, 1991, 591.
44. E. W. Abel, I. Moss, K. G. Orrell, V. Šik, and D. Stephenson, *J. Chem. Soc., Dalton Trans.*, 1987, 2695.

Biographical Sketch

Keith G. Orrell. *b* 1940. B.Sc.Tech., 1960, Ph.D., 1964, D.Sc., 1989, University of Manchester, UK. Successively lecturer, senior lecturer and (currently) reader, University of Exeter, UK. Approx. 150 publications on dynamic NMR, NMR of paramagnetic species and studies in liquid crystalline phases. Current research specialties: NMR studies of intramolecular motions, particularly in inorganic and organometallic compounds.

Two-Dimensional NMR of Molecules Oriented in Liquid Crystalline Phases

Anil Kumar

Indian Institute of Science, Bangalore, India

1 INTRODUCTION

The direct dipole–dipole interaction between nuclear spins plays a dominant role in magnetic resonance. Several kilohertz in magnitude, it causes severe line broadening of NMR lines in solids owing to the presence of a large number of coupled spins. Anisotropic reorientation coupled with translational diffusion of molecules embedded in ordered matrices, such as a liquid crystalline medium, yields extremely interesting NMR spectra.[1] In such cases, the intermolecular dipolar interactions are averaged to zero, and the intramolecular interactions, scaled by the order parameter, are reduced to a few hundreds of hertz. As a result, only a finite number of spins are dipolar-coupled, yielding a large number of sharp and often well-resolved spectral lines. These dipolar-coupled spectra are extremely useful, and contain detailed information on the conformation, symmetry and anisotropy of reorientation of the molecules. However, since the dipolar couplings are often large compared with the chemical shifts, the spectra exhibit strong coupling features, making the analysis of such spectra nontrivial.[2–5] Only in some simple spin systems is the Hamiltonian analytically diagonalizable, allowing straightforward interpretation of the spectra. In most cases, one has to perform a numerical analysis using iterative computer programs such as LAOCOON[6] and its various modifications.[4,7,8] The input to these programs comprises the line frequencies, intensities, and initial parameters of the Hamiltonian. Trapping of the algorithm into local minima is a frequently encountered problem, and any additional information about the spectrum, the spin system, and the molecular reorientations is extremely useful.

Two-dimensional (2D) NMR spectroscopy has played a significant role in simplifying spectra and in often providing the key information. The various 2D NMR studies of oriented molecules discussed in this article are classified into the following categories:

1. resolved spectroscopy;
2. experiments utilizing transverse spin orders using single and multiple quantum coherences;
3. experiments utilizing longitudinal spin orders such as NOESY and Z-COSY.

The emphasis in this article is on proton spectra of small molecules oriented in liquid crystal matrices. Reference may be made to a review on this subject,[9] and periodic reviews of NMR of oriented molecules.[10]

2 TWO-DIMENSIONAL SPIN ECHO RESOLVED SPECTROSCOPY

The spin echo 2D resolved experiment utilizes a pulse sequence (90°, $\frac{1}{2}t_1$, 180°, $\frac{1}{2}t_1$, t_2). The 180° pulse reverses the sign of all terms linear in spin operators over which the 180° pulse acts, and leaves all bilinear terms invariant. For homonuclear (spin-$\frac{1}{2}$) spins, the various terms contributing to the spectra of oriented molecules are

$$\hat{\mathcal{H}} = \hat{\mathcal{H}}_Z + \hat{\mathcal{H}}_J + \hat{\mathcal{H}}_D \tag{1}$$

where $\hat{\mathcal{H}}_Z$ involves chemical shift terms linear in spin operators, and $\hat{\mathcal{H}}_J$ and $\hat{\mathcal{H}}_D$ are the indirect spin–spin and direct dipole–dipole couplings, respectively, which are both bilinear in spin operators. The 180° pulse reverses the sign of $\hat{\mathcal{H}}_Z$ and leaves $\hat{\mathcal{H}}_J$ and $\hat{\mathcal{H}}_D$ invariant. If all the terms commute, as happens only in weakly coupled isotropic cases where $\hat{\mathcal{H}}_D$ is absent and $\hat{\mathcal{H}}_Z >> \hat{\mathcal{H}}_J$, then $\hat{\mathcal{H}}_Z$ refocuses, and the effective Hamiltonian is simply $\hat{\mathcal{H}}_J$. In such cases, the 2D spin echo spectrum consists of a simple separation of chemical shifts and couplings, with the multiplets of each chemical site appearing along a 45° line from the $\omega_1 = 0$ axis[11,12] (see ***Two-Dimensional J-Resolved Spectroscopy***). When the various terms in equation (1) do not commute, the 180° pulse causes coherence transfer between various transitions, and calculations performed for strongly coupled spins show that the line intensities and frequencies along the ω_1 direction are modified and that there are additional lines in the 2D spectra.[13,14]

The first application of this experiment in oriented systems was on an AB spin system formed by the protons of 1-thio-3-selenole-2-thione oriented in MBBA.[15] The 1D spectrum of the two spins consists of four lines symmetrically displaced from the average chemical shift. The difference between the outer

For References see p. 4865

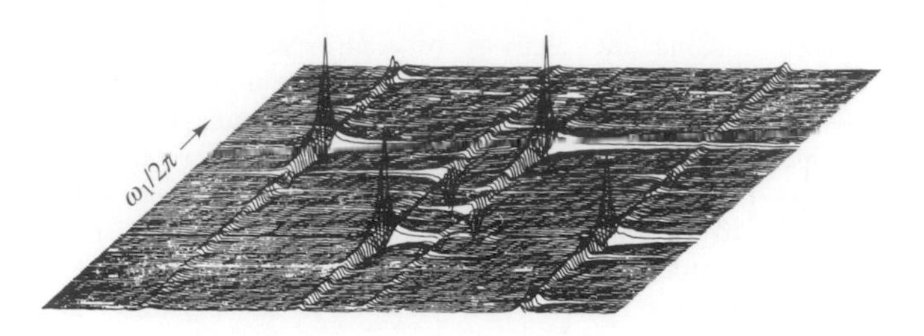

Figure 1 Phase-sensitive spin echo 2D resolved spectrum of two strongly coupled protons of 1-thio-3-selenole-2-thione oriented in MBBA [obtained using the pulse sequence (90°, $\frac{1}{2}t_1$, 180°, $\frac{1}{2}t_1$, t_2)], recorded at 270 MHz. While the 1D spectrum consists of four lines, the 2D spectrum has eight, two of which appear with negative intensity. (Reproduced by permission of Academic Press from Khetrapal et al.[15])

and inner lines is the sum $|J_{AB} + 2D_{AB}|$. To assign the lines, one needs additional information, which can be provided by conventional double resonance experiments or by the above 2D resolved experiment. The 2D resolved experiment has eight lines, two of which appear with negative intensity (Figure 1). There are four lines of equal intensity along the ω_1 dimension, uniquely identifying the parameter $|J_{AB} + 2D_{AB}|$.[15]

Another 2D spin echo study has been carried out on the AB_2 spin system of 1-bromo-2,6-dinitrobenzene oriented in the room temperature nematic, Merck Phase V.[16] Significant rf inhomogeneity gave rise to artifacts in the spectra, which were simulated as errors in the refocusing 180° pulse. Errors in the 180° pulse cause additional coherence transfer pathways of the type obtained in spin echo correlated spectroscopy (SECSY),[17] and such spectra thus have features of resolved spectroscopy and SECSY, both affected by strong couplings.

Accurate values for homonuclear dipolar couplings have been extracted from 1D spin echo spectroscopy of oriented molecules containing AA′BB′-type spin systems, formed by the protons in *para*-substituted pyridines, which are coupled to an increasingly complex group of heteronuclei and to fluorines in tetrafluorinated benzenes.[18] All the echo spectra were analyzed with a modified LAOCOON-type computer program.[16,19] The echo spectra differed from the normal spectra principally in the presence of extra lines, which were interpreted as being due to $\boldsymbol{B}_1$ inhomogeneity and analyzed iteratively assuming a refocusing pulse of angle 160° ± 10°.[18]

3 TWO-DIMENSIONAL CORRELATION SPECTROSCOPY

3.1 Single Quantum Correlation (COSY)

Two-dimensional COSY utilizes a pulse sequence (90°, t_1, 90°, t_2), and gives cross peaks between transitions that are coupled to each other in a coupled network. The first application of COSY for the spectra of oriented molecules was to the AA′BB′ spin system formed by the protons of benzoselenodiazole oriented in the liquid crystal EBBA.[20] The COSY spectrum separated the symmetric and antisymmetric transitions by the absence of cross peaks between them. This separation further facilitated complete identification and analysis of the antisymmetric part of the spectrum, virtually by inspection. A more detailed analysis of the intensities of the cross peaks, by changing the flip angle of the coherence transfer pulse (the second pulse of COSY), further facilitated identification of connected transitions and the assignment of resonances.[20]

The absence of cross peaks between symmetric and antisymmetric domains in COSY has been exploited further for systems with higher symmetry. The proton spectra of oriented acetone and oriented benzene have been studied.[21] Acetone has two rotating methyl groups, each having C_{3v} symmetry, yielding a spin system of the type A_3A_3'. The eigenstates of the Hamiltonian can be grouped into four irreducible representations, namely, A_g, A_u, G_1, and G_{2g}, with analytical expressions available for the energies. The 2D technique separated the spectrum into four parts, with cross peaks within each group and no cross peaks between the groups. This yielded a complete symmetry filtering of all the transitions, with straightforward identification of each group and the values of the dipolar couplings. Benzene has D_{6h} symmetry, and the spin states are divided into six irreducible representations (A_1, A_2, B_1, B_2, E_1, and E_2). The A_2 representation has only one spin state and no transition. The remaining five representations are filtered by COSY (Figure 2) in a straightforward manner, and separate iteration of each of the domains using a modified LACOONOR program yielded all the parameters in an efficient manner.[21]

Flip-angle-dependent COSY (COSY-30; 90°, t_1, 30°, t_2) experiments have also been performed on the protons of *p*-dimethoxy[$^2H_6$]benzene oriented in the liquid crystal PCH 1132, yielding information on the directly connected transitions.[8] This system of four ring protons coupled to six methyl deuteriums forms one of the largest spin systems giving a well-resolved spectrum. The ^1H–^2H dipolar couplings were focused by a spin echo experiment yielding a 10-line spectrum typical of an AA′A″A‴ spin system. The connectivity information allowed a complete analysis of this spectrum, and yielded the various spectral parameters.[8] The 2D COSY spectrum of ethyl iodide oriented in a nematic phase has also been reported.[22]

3.2 Multiple Quantum Correlation

In the previous section, the use of experiments that correlate single quantum coherences with each other were described. In this section, the use of the correlation of multiple quantum coherences to single quantum coherences by 2D spectroscopy (MQ spectroscopy) and simplification of COSY spectra by multiple quantum filtering is discussed.

3.2.1 MQ Spectroscopy

Two-dimensional multiple quantum spectra are obtained by the use of the pulse sequence (90°, τ, 90°, t_1, 90°, t_2), where the (90°, τ, 90°) pulse sandwich excites all orders. These multiple quantum coherences cannot be detected directly, and therefore are monitored indirectly by the use of 2D NMR. These coherences, allowed to evolve during the period t_1, are frequency-labeled, and at the end of t_1, a 90° pulse is used to transform them into single quantum (SQ) coherences, which are detected during the period t_2. Thus, the 2D spectrum contains MQ frequencies in the ω_1 dimension correlated with various SQ frequencies in the ω_2 dimension.[12,23,24] There is no net transfer of magnetization from one order to another, but only a differential transfer; therefore, many correlations appear with negative intensity. The projection of the absolute mode of the 2D spec-

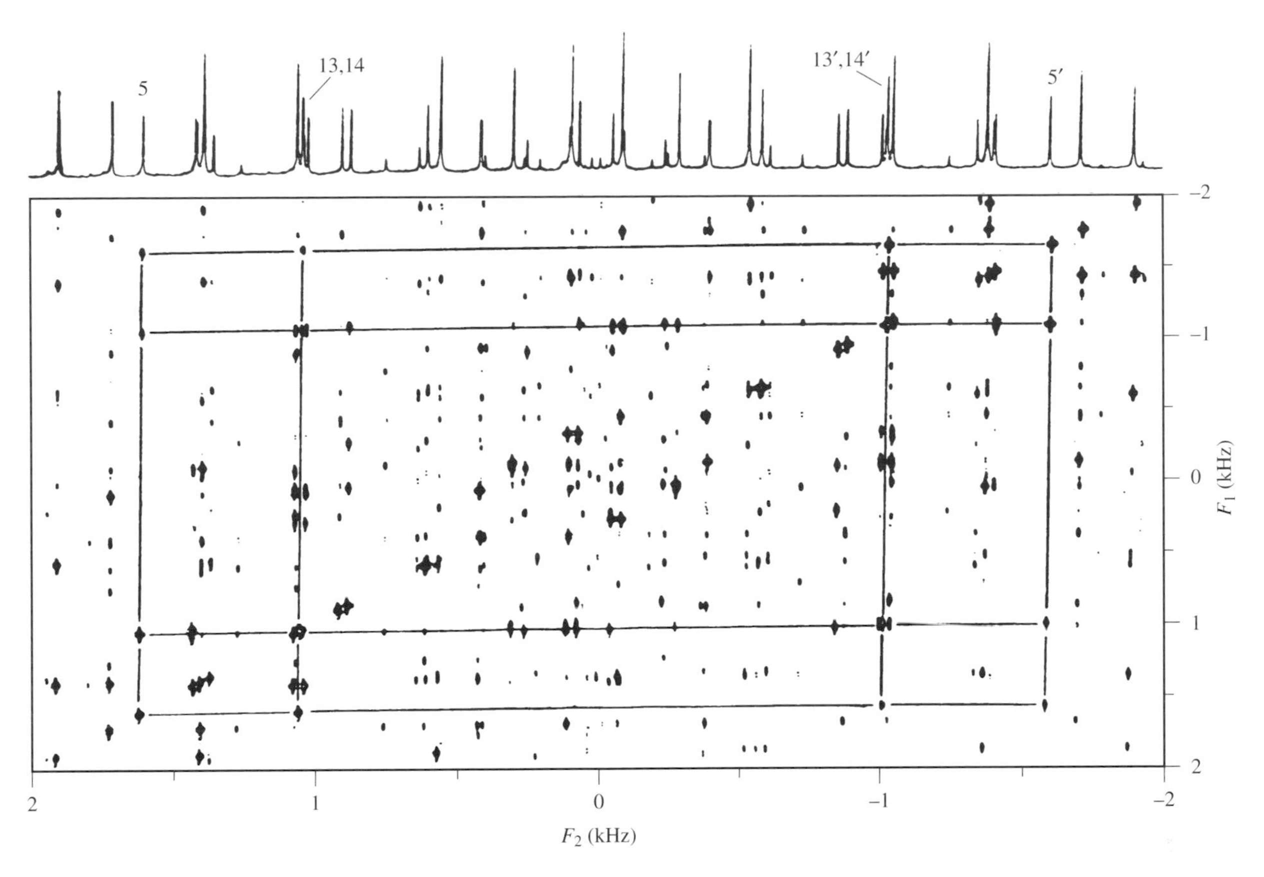

Figure 2 COSY spectrum of benzene oriented in ZLI-1167, recorded at 270 MHz. The connectivity in the B_1 symmetry domain containing lines 5, 13, 14, 14′, 13′, and 5′ is marked in the spectrum. (Reproduced by permission of Elsevier Science Publishers BV from Rukmani and Kumar[21])

trum on the ω_1 axis yields a 1D MQ spectrum, which can also be obtained by single point detection of the signal in t_2.[22,25–40] While most of the applications have been demonstrated using such one-dimensional spectra, 2D spectra have also been recorded (see ***Multiple Quantum Spectroscopy of Liquid Samples***). Figure 3 shows the 2D MQ spectrum of oriented benzene.[22,39] The spectrum becomes progressively simpler as one goes to higher multiple quantum orders, there being seven four-quantum frequencies, two five-quantum frequencies, and one six-quantum frequency present in the six-proton system of oriented benzene.

The various multiple quantum orders in Figure 3 have been separated by offsetting the carrier from the middle of the single quantum spectrum. If $\Delta\omega$ is the offset parameter then various orders are separated out by $n\,\Delta\omega$. The magnetic field inhomogeneity contribution to the linewidth increases proportionally for various MQ orders. In order to focus the field inhomogeneity, a 180° pulse has been introduced in the middle of the period t_1, which also focuses the offset dependence of various MQ orders. The various MQ orders in such cases are separated out by the use of time proportional phase incrementation (TPPI),[25] in which the phases of all the pulses in the excitation sequence are incremented in sequential t_1 experiments as $\phi = \Delta\omega\, t_1$. This results in a separation of various MQ orders by an effective offset parameter of $n\,\Delta\omega$. The 180° pulse in the middle of the period t_1 leaves all homonuclear couplings invariant, and, if there are no chemical shifts, yields an undistorted MQ spectrum, for example, in benzene.[25,26,32] However, if there are chemical shifts present along with dipolar and/or J couplings and the spectra are strongly coupled then the 180° pulse causes coherence transfer within each order, giving rise to new transitions and changed frequencies.[13,41,42]

Using the above pulse sequences, the 1D MQ spectrum has been obtained for the partially deuterated nematic liquid crystal 4-cyano-4′-(n-pentyl-d_{11})biphenyl (5CB-d_{11}), which has eight ring protons coupled to each other and yields a rather broad deuterium-decoupled single-quantum spectrum, but well-resolved five-, six-, and seven-quantum spectra (Figure 4).[33] Detailed analyses of these spectra have been carried out to obtain the dipolar couplings between various protons. The two biphenyl rings were either considered equivalent (D_4 symmetry) or inequivalent (D_2 symmetry). While an unequivocal fit was not obtained for either model, the D_4 symmetry resulted in the most physically reasonable molecular parameters. However, several splittings could not be explained by either of these models or by the effect of strong couplings, and it was suggested that rf inhomogeneity might provide an explanation for the extra splittings not predicted by the D_4 symmetry.[33]

3.2.2 *Selectivity of Excitation and Detection*

Attempts have been made to excite/retain only a few selected orders instead of all orders of MQ, thereby increasing the intensity of the particular order and the resolution of the detected order. Retaining a particular order is rather straightforward by phase cycling the detection pulse (the third 90° pulse)

For References see p. 4865

and the receiver or the excitation sequence and the receiver, in a concerted manner.[12] An example is shown in Figure 5.[34] The selection of a particular MQ order can also be achieved by using the technique of coherence transfer echo filtering (CTEF). A field gradient pulse of length τ in the MQ evolution period followed by a field gradient pulse of length $n\tau$ during the detection period refocuses only the nth order. This has been demonstrated on hexachloroethane oriented in EBBA.[43]

A large amount of effort has been made by Pines and co-workers in attempting to excite selectively all the transitions of a particular order using dipolar refocusing.[26,29,38–40] The introduction of a 180° pulse in the middle of the excitation

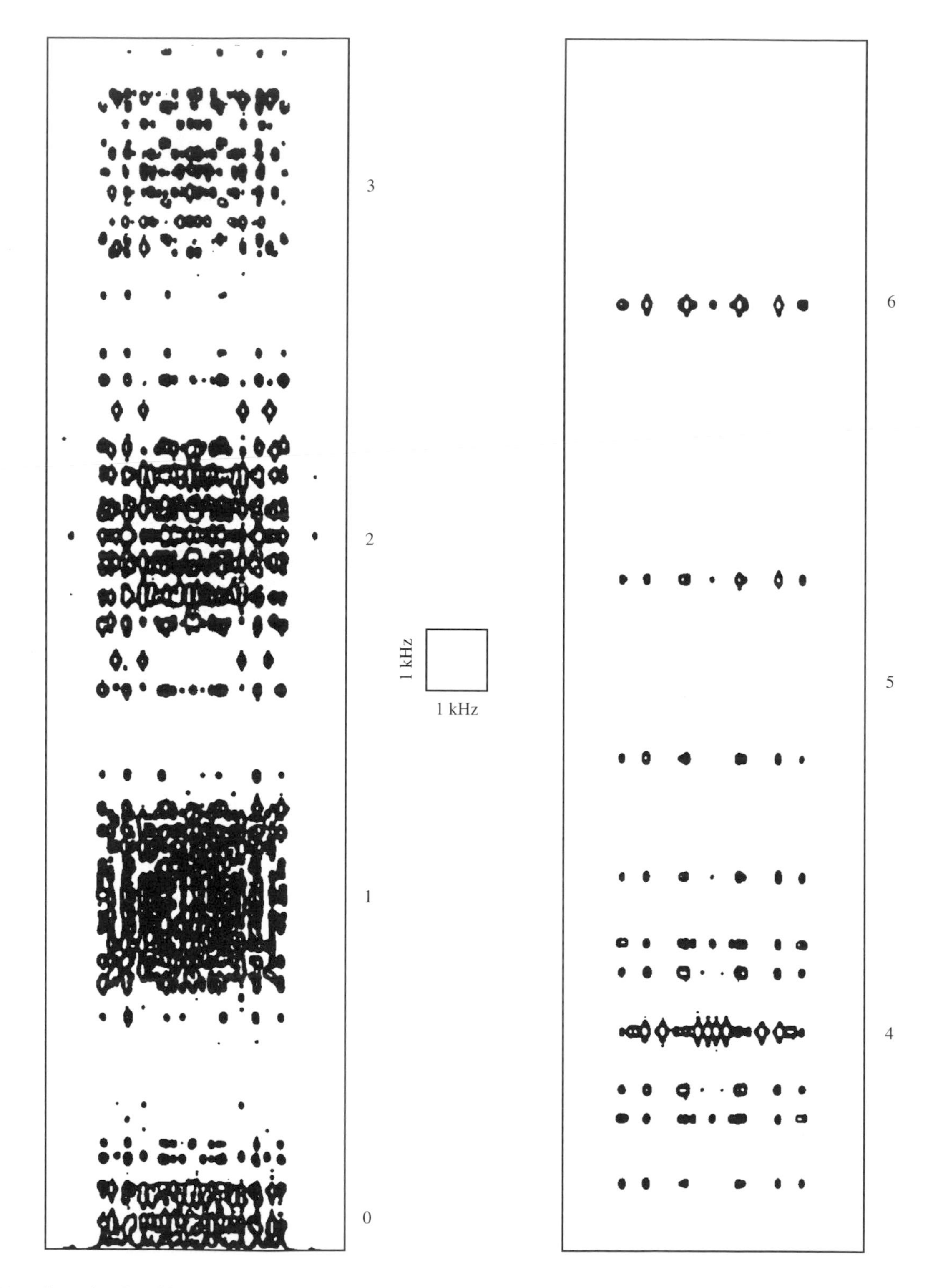

Figure 3 Two-dimensional multiple quantum spectrum of oriented benzene obtained using the pulse sequence (90°, τ, 90°, t_1, 90°, t_2), recorded at 200 MHz with a spectral offset of 6 kHz. (Reproduced by permission of Elsevier Science Publishers BV from Drobny[22])

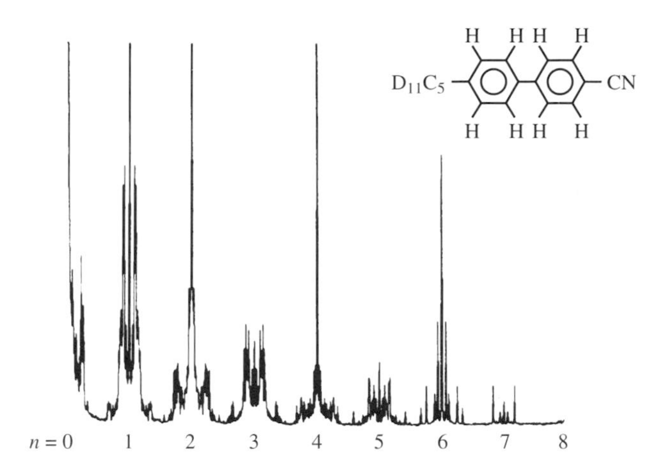

Figure 4 Multiple quantum proton spectrum of 5CB-d_{11} obtained using the pulse sequence (90°, τ, 90°, $\frac{1}{2}t_1$, 180°, $\frac{1}{2}t_1$, 90°, τ), at 182 MHz. The various multiple quantum orders are separated by TPPI, and the spectrum is the sum of magnitude mode spectra recorded for several values of τ ranging from 0.4 to 1.4 ms. (Reproduced by permission of Elsevier Science Publishers BV from Sinton and Pines[27])

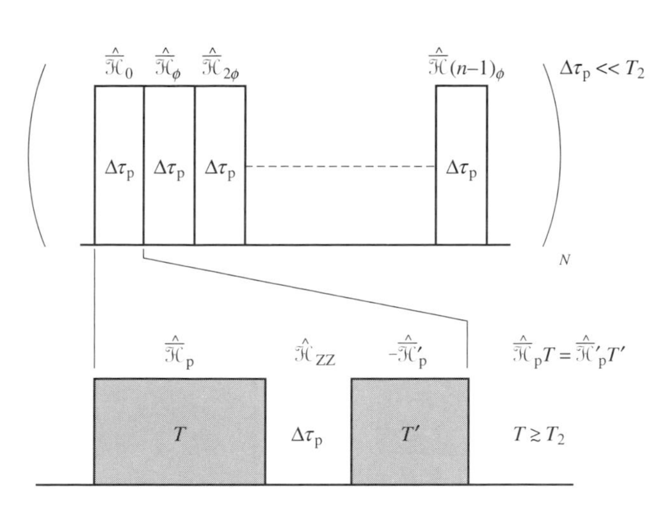

Figure 6 Pulse scheme for selective excitation of the n-quantum spectrum of dipolar-coupled spins. The order n is selected by incrementing ϕ in steps of $2\pi/n$. (Reproduced by permission of the American Physical Society from Warren et al.[26])

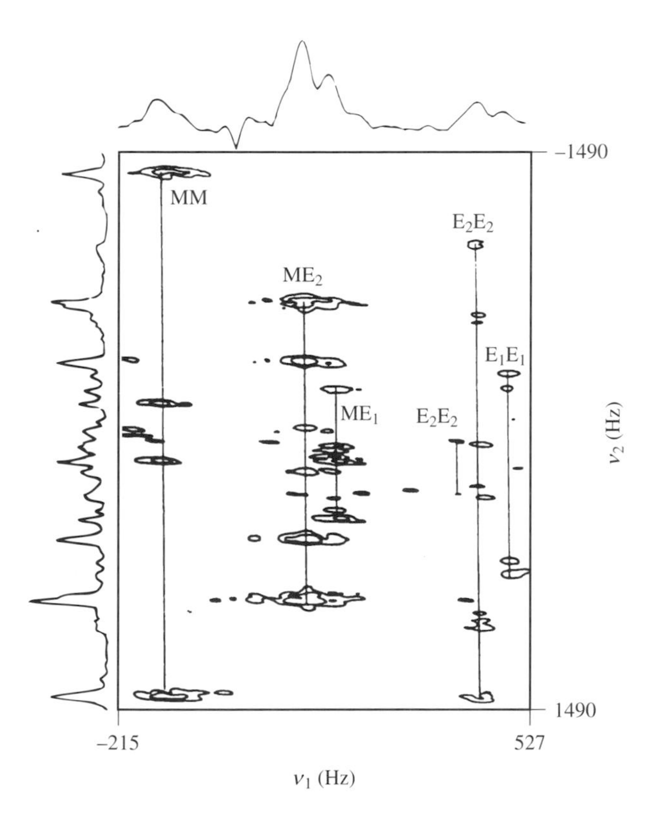

Figure 5 Double quantum versus single quantum correlation proton spectrum of 81% randomly deuterated n-hexane recorded by the sequence ($90°_\phi$, $\frac{1}{2}\tau_1$, $180°_\phi$, $\frac{1}{2}\tau_1$, $90°_\phi$, t_1, $90°_\alpha$, $\frac{1}{2}\tau_2$, $180°_\alpha$, $\frac{1}{2}\tau_2$, t_2), recorded at 360 MHz, under deuterium decoupling. The phase ϕ was incremented by 90° while the receiver phase was incremented by 180°. This retains double quantum coherences. Six vertical lines have been drawn at ν_1 frequencies corresponding to double quantum frequencies $2\nu_M$, $\nu_M + \nu_E$, $\nu_M + \nu_E$, $2\nu_E$, $2\nu_E$, and $\nu_E + \nu_E$. (Reproduced by permission of the American Chemical Society from Gochin et al.[34])

sequence, along with appropriate phases of the two 90° pulses, yields an even- or odd-order excitation sequence for weakly coupled spins.[12] However, for dipolar-coupled spins, it is possible to reverse the sign of the effective dipolar Hamiltonian by using multiple pulses, allowing one to design pulse sequences for exciting a particular order. The basic building block of the pulse sequence is shown in Figure 6. This block replaces the nonselective excitation sequence 90°–τ–90°. Each sub-block consists of three parts. The first has a sequence of 90° pulses with interval τ, yielding a first-order average Hamiltonian $\hat{\bar{\mathcal{H}}}_p$, and is applied for a time T. The second part consists of a delay $\Delta\tau_p$, and the third consists of a sequence of 90° pulses with delays of τ and 2τ, having an average Hamiltonian $-\hat{\bar{\mathcal{H}}}'_p$ applied for a time T' such that $\hat{\bar{\mathcal{H}}}_pT = \hat{\bar{\mathcal{H}}}'_pT'$. Together, these yield the first sub-block, having an average Hamiltonian $\hat{\bar{\mathcal{H}}}_0$, where the subscript refers to the phase of this sub-block. These sub-blocks are repeated n times, where n is the order to be selected, and the phase is incremented each time for the entire subcycle by $\phi = 2\pi/n$. For better selectivity, the whole sequence may be repeated N times, where N may be 4 or 5. The results of such selective excitation are shown in Figure 7. More details about such excitations are given by Warren et al.[26,29] and Drobny et al.[40] The highest order spectrum of N coupled spins $\frac{1}{2}$ is always a single line, whose position is determined by the sum of the chemical shifts of the coupled spins and is independent of spin–spin and dipolar couplings. Pines and co-workers have demonstrated that the $(N-1)$th- and $(N-2)$th-order spectra contain enough information to allow a spectral fit by iteration on the dipolar couplings, if there are no chemical shift differences, and, in general, it is possible to derive all the information from high-order MQ spectra that have a finite number of lines to be iterated upon.[25,28,30,35–40]

Most of the spectra using the above experimental procedures have been recorded using 2D pulse sequences, but recorded in a 1D manner using a single point detection in t_2. We shall restrict further discussion to 2D spectra.

For References see p. 4865

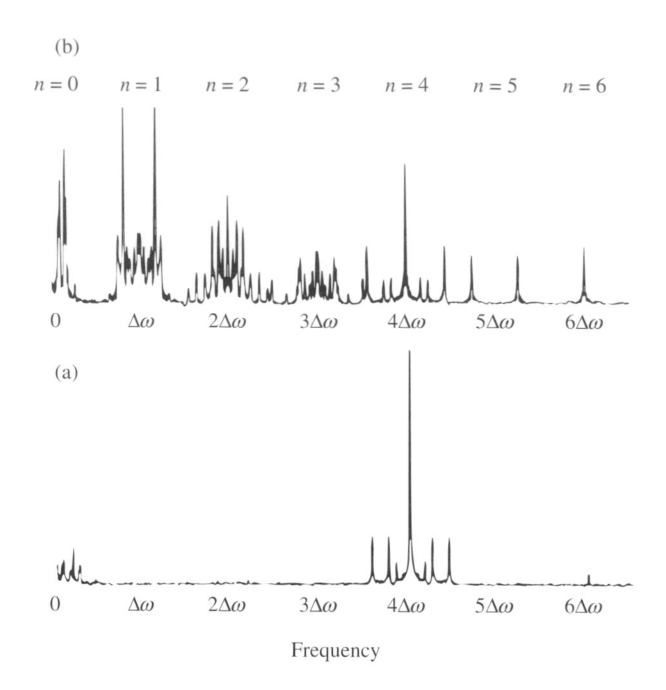

Figure 7 Four-quantum selective (a) and nonselective (b) MQ spectra of oriented benzene. (Reproduced by permission of the American Institute of Physics from Warren et al.[29])

3.2.3 *Simplification of Spectra by Specific and Random Deuteration*

As the number of dipolar-coupled spins increases beyond six or seven, the spectra become too complex for analysis by conventional procedures. Specific deuteration, coupled with deuterium decoupling—a straightforward technique for reducing the number of coupled protons—has been used in several cases. The main studies have been carried out on *n*-hexane-d_6, deuterated at methyl positions, and on 4-cyano-4′-(*n*-pentyl-d_{11})biphenyl (5CB-d_{11}), in which all the aliphatic protons have been deuterated, reducing both to systems of eight coupled protons.[27,33,35,39] The six- and seven-quantum 2D MQ spectra of *n*-hexane-d_6 (Figure 8) have been used to obtain the various dipolar couplings and their signs to estimate the dynamic molecular structure in the form of conformational probabilities.[35,39]

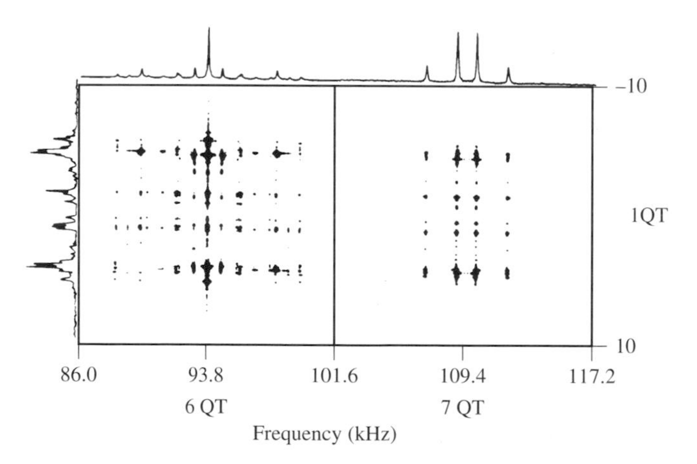

Figure 8 Absolute value six- and seven-quantum versus single-quantum correlation 2D spectrum of *n*-hexane-d_6 deuterated at methyl positions obtained by using the same pulse sequence as in Figure 5. The various orders were separated by TPPI using $\Delta\phi$ = 22.5°. (Reproduced by permission of Elsevier Science Publishers BV from Gochin et al.[35])

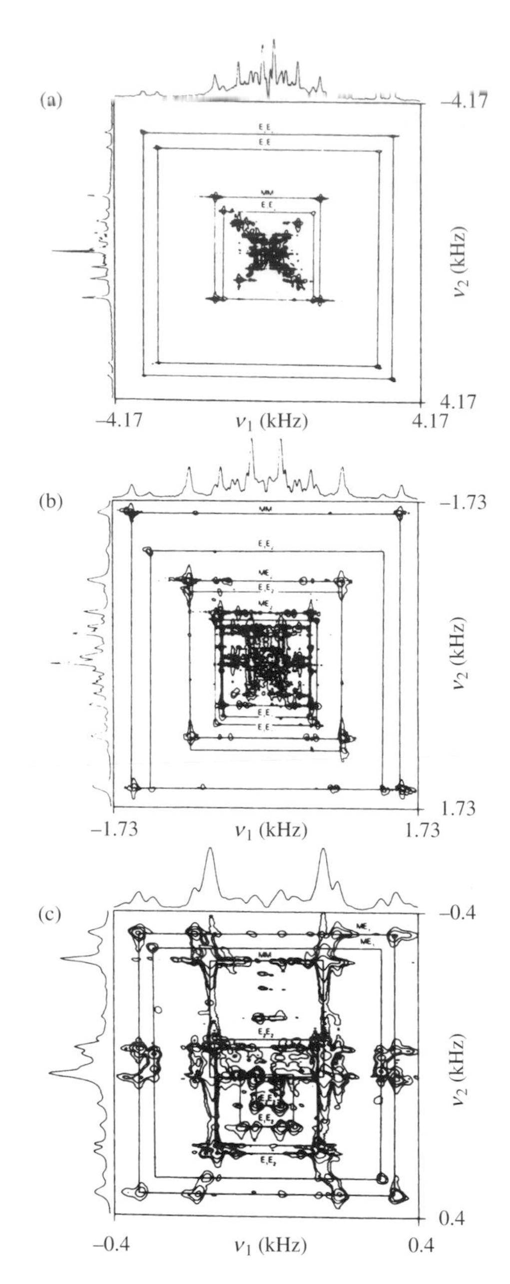

Figure 9 Double quantum filtered COSY (DQFC) spectrum of 81% randomly deuterated *n*-hexane-$d_{0.81}$ obtained using the pulse sequence ($90°_\phi$, $\frac{1}{2}t_1$, $180°_\phi$, $\frac{1}{2}t_1$, $90°_\phi$, $\frac{1}{2}\tau_1$, $180°_x$, $\frac{1}{2}\tau_1$, $90°_x$, $\frac{1}{2}\tau_2$, $180°$, $\frac{1}{2}\tau_2$, t_2), with phase cycling to retain double quantum coherence during the period τ_1. (a), (b), and (c) show successively expanded regions of the spectrum. The various two-spin systems are identified with square patterns. (Reproduced by permission of Taylor and Francis Ltd. from Gochin et al.[44])

Another method of reducing the size of the spin system, but retaining information on dipolar couplings between all the protons of a molecule, is the technique of random deuteration combined with multiple quantum 2D spectroscopy or multiple quantum filtered COSY.[24,36,44,45] For example, an 81% random deuteration of *n*-hexane, which has 14 protons, yields approximately 6%, 18%, 25%, 25%, 14%, 5%, 3%, 2%, and 1% molecules having 0, 1, 2, 3, 4, 5, 6, 7, and 8 protons, respectively.[36] Of these, only the one-, two-, and three-proton cases give significant intensity. The one-proton spectra can be easily filtered out using double quantum spectroscopy/filtering. The three- and high-spin spectra are quite low in intensity owing to the large number of lines and large number of possible spin systems. The double quantum filtered COSY (DQFC) spectrum thus contains prominently only the spectra of two coupled spins, and yields square patterns for each of the spin systems.[34,36] Numbering the carbons as M, E_1, E_2, E_2', E_1', and M′ in *n*-hexane, the various two-spin systems will be of the A_2 type for the protons in MM, E_1E_1, E_2E_2, E_2E_2'(*cis*), E_2E_2'(*trans*), E_1E_1'(*cis*), E_1E_1'(*trans*), and MM′ positions, and of AB type for ME_1, ME_2, ME_1', E_1E_2(*cis*), E_1E_2(*trans*), E_1E_2'(*cis*), and E_1E_2'(*trans*) positions. All such spin systems have been readily identified in the DQFC spectrum of 81% deuterated *n*-hexane recorded under deuterium decoupling (Figure 9), and several in the double-quantum 2D spectrum (Figure 5).[34,36] This yields all the dipolar couplings in *n*-hexane. Detailed analysis of the measured dipolar couplings has been carried out in terms of conformation and orientations of *n*-hexane in the liquid crystal solvent.

Recently, this idea has been extended to a series of *n*-alkanes, from hexane to decane (random deuteration level of 80–90%), oriented in a nematic liquid crystal.[44] The DQFC spectrum in each case yields a large number of square patterns, each of which belongs to a two-spin system, yielding the totality of dipolar couplings between pairwise protons in the above molecules. The interpretation of the measured dipolar couplings is complicated, since they are averaged over molecular conformations and orientations sampled by the molecules. Nevertheless, they provide constraints on the time-averaged conformations and orientations of the molecules, which in turn allow an examination of various possible models for the solute–liquid crystal interactions.[44]

DQFC spectroscopy has also been utilized for a study of 78.7% randomly deuterated benzene oriented in a nematic phase.[36] Subspectra of three diprotonated and three triprotonated species could be identified in the spectrum. The subspectra were compared with the simulated spectra for various assignments, and the spectral parameters were obtained. A five-quantum versus single-quantum correlation 2D spectrum was also recorded for 34% randomly deuterated benzene. The five-quantum spectrum can be obtained either from fully protonated species or singly deuterated species. The singly deuterated species has only one five-quantum transition, while the fully protonated species has two five-quantum transitions, the correlation of which with various single quantum transitions yields various spectral parameters.[36]

3.2.4 *Application to Chemically Exchanging Systems*

Multiple quantum 2D spectroscopy has also been applied to chemically rearranging systems of oriented *s*-trioxane, which undergoes ring inversion, and oriented cyclooctatetracene, which undergoes a bond shift rearrangement.[46] Two-dimensional MQ–SQ correlation spectra were recorded at various temperatures, from which projected MQ spectra were obtained. The high-order (four and six) MQ spectra are much simpler than the single-quantum spectra, and exhibit characteristic exchange broadening and narrowing effects very similar to the single-quantum spectra. It is much easier to compare the high-order MQ spectra with the simulated spectra, calculated using exact as well as approximate expressions for the dynamic lineshapes, which then yield information on the rates of the dynamic processes. The uses, advantages, and limitations of the technique have been discussed in detail.[46]

4 EXPERIMENTS UTILIZING LONGITUDINAL SPIN ORDER

4.1 Effect of Strong Coupling in NOESY

It is known that each cross-section of the NOESY experiment using the (90°, t_1, 90°, τ_m, 90°, t_2) sequence is equivalent to a 1D difference NOE experiment in which the spin corresponding to the diagonal peak is selectively inverted at $\tau_m = 0$. It has been shown recently that this equivalence breaks down for strongly coupled spins, resulting in cross peaks in NOESY experiments, even without relaxation.[47,48] Explicit theoretical calculations have been carried out for an AB spin system, and the results are given in Figure 10.

Calculations have also been carried out for an ABX spin system, showing the existence of cross peaks between all X and AB transitions, even when X is only weakly coupled to AB.[48] The existence of such cross peaks has been experimentally demonstrated in oriented acetone (Figure 11), where a quantitative fit to the experimental cross sections, after removal of zero quantum interference, has been demonstrated by detailed calculations.[47] These cross peaks arise owing to the combined effect of strong coupling and the nonlinearity of the rf pulses. In order to examine whether the cross peaks are due to nonlinearity of the second or the third pulse in the NOESY experiment, calculations for the ABX spin system have been carried out for a small-angle second or third pulse. In an ABX spin system, there are two X transitions between pure states. In

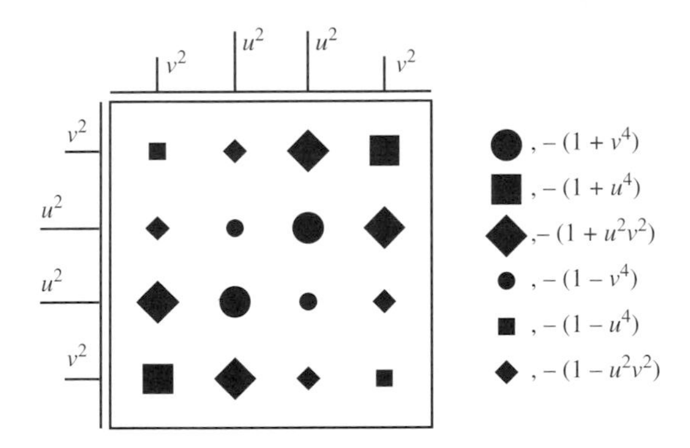

Figure 10 Schematic spectrum of an AB spin system calculated for the NOESY experiment (90°, t_1, 90°, τ_m, 90°, t_2), with phase cycling to retain longitudinal magnetization during the short period τ_m. The intensities of the 2D peaks are obtained by multiplying the symbols with the 1D intensities indicated on the ω_1 and ω_2 axes. For example, the intensity of the bottom left diagonal peak is given by $-v^4(1+u^4)$, where $u = \cos\theta + \sin\theta$, $v = \cos\theta - \sin\theta$, and $\tan 2\theta = J_{AB}/(\nu_A - \nu_B)$. All cross peaks disappear for weakly coupled spins ($\theta = 0$, $u = v = 1$). (Reproduced by permission of the *Bulletin of Magnetic Resonance* from Christy Rani Grace and Kumar[48])

For References see p. 4865

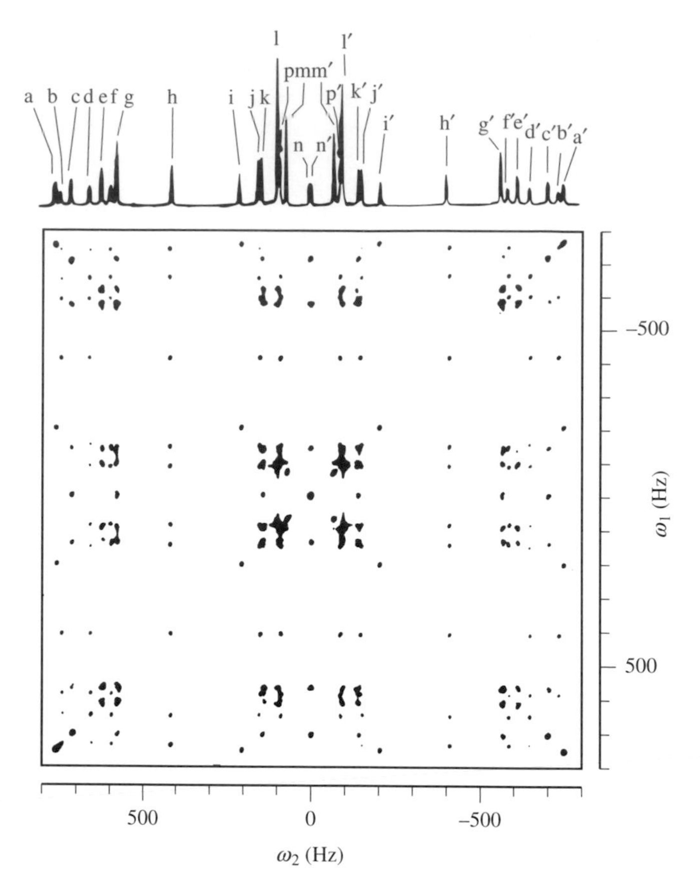

Figure 11 Phase-sensitive NOESY spectrum of oriented acetone obtained using the pulse sequence (90°, t_1, 90°, τ_m, 90°, t_2), with phase cycling to retain longitudinal magnetization during the short period τ_m (= 20 μs). Filled contours represent negative intensity. The various cross peaks here arise mainly owing to the strong coupling effect. (Reproduced by permission of Academic Press from Christy Rani Grace and Kumar[47])

the experiments with the second pulse being of small angle and the third being 90°, there are no cross peaks from these X transitions to the AB transitions. Similarly, in the experiment in which the second pulse is 90° and the third is of small angle, there are no cross peaks from AB transitions to these X transitions. These observations indicate that the nonlinearity of a pulse is restricted to transitions within a spin or within groups of strongly coupled spins.[48]

4.2 Construction of Energy Level Diagrams

4.2.1 *Using NOESY-α,β*

In the linear regime, the NOESY-α,β experiment corresponds to the sequence (90°, t_1, α, τ_m, β, t_2), where α is a small-angle pulse and only longitudinal magnetization is retained during the period τ_m. It has been shown recently that, irrespective of the strength of coupling, each cross section of this experiment is equivalent to a 1D difference experiment in which the peak corresponding to the diagonal is selectively inverted at $\tau_m = 0$ and the state of spin system is monitored by the β pulse. If β is also a small-angle pulse then the state of the spin system is measured faithfully. In a coupled spin system, selective inversion of a transition gives rise to a direct population effect on all the transitions that share energy levels with the inverted transition. The transitions that increase or decrease in intensity are called progressively or regressively connected to the inverted transition, respectively. Use has been made of this information, without bringing in relaxation (for small values of τ_m), to construct energy level diagrams of oriented molecules.[49] This experiment is similar to a Z-COSY experiment,[50] as well as to a large number of difference 1D experiments in which one peak is selectively inverted at a time.

Figure 12 shows the 2D spectrum of oriented acetone recorded using NOESY-α,β for τ_m = 10 μs and $\alpha = \beta = 14°$. As in COSY-β, the cross peaks here are also only between directly connected transitions within each irreducible representation, with the advantage that the 2D spectrum has pure absorptive lines.[49] From various cross sections of such spectra, a connectivity matrix is built up by entering (+1), (−1), and 0 for progressively, regressively, and unconnected transitions, respectively. This connectivity matrix is supposed to be symmetric, and its systematic analysis leads to a complete identification of the connected network of transitions, and in turn the complete energy level of a given symmetry.[49] The well-known problems of zero quantum coherences, which are not cancelled by NOESY phase cycling, also interfere in this experiment, and therefore have to be removed by known procedures.[12] The connectivity information has also been utilized for the analysis of the spectrum of *cis,cis*-mucanonitrile oriented in ZL1-1167, a four-spin system having C_{2h} symmetry.[51]

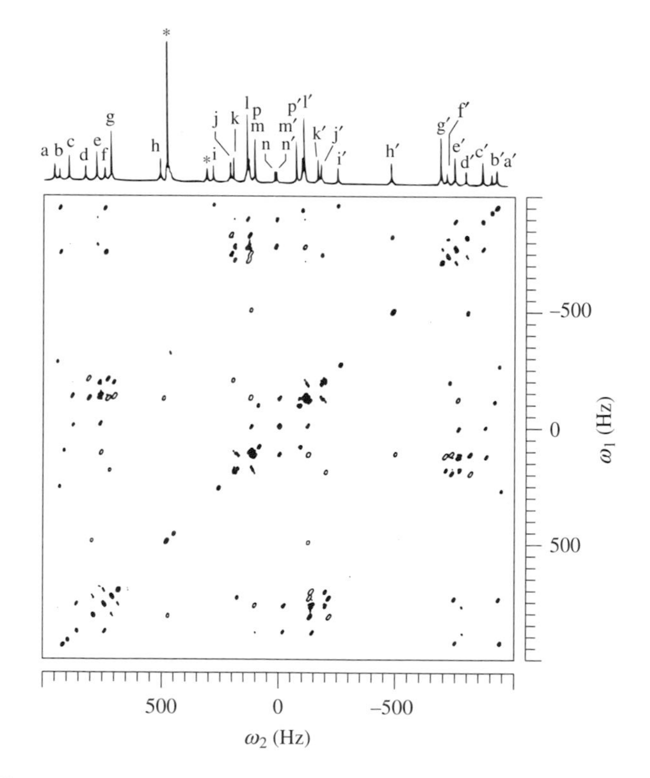

Figure 12 Phase-sensitive small-flip-angle NOESY (NOESY-α,β) spectrum of oriented acetone for $\alpha = \beta = 14°$ and τ_m = 20 μs, using the sequence (90°, t_1, α, τ_m, β, t_2) while retaining only the longitudinal magnetization during the short interval τ_m. Each diagonal peak has cross peaks only to directly connected transitions within the irreducible representation. Filled contours represent negative intensity. (Reproduced by permission of Academic Press from Christy Rani Grace and Kumar[49])

4.2.2 *Using Z-COSY*

Similar experiments (Z-COSY: 90°, t_1, α, τ_m, α) have also been performed on oriented methanol (an A_3B type of spin system) and oriented 1,3-dichlorobenzene (A_2BC). By systematically reducing the flip angle α, the direct progressive and regressive connectivities were identified. Patterns of connectivity were identified with the help of such spectra, which were then utilized for the construction of energy level diagrams. This information was in turn utilized for iteratively obtaining the complete Hamiltonian and its parameters.[50]

5 OTHER STUDIES

There has been an interesting application of near-magic-angle sample spinning combined with 2D NMR (COSY and 2D *J*-resolved experiments) of substituted benzenes dissolved in lyotropic liquid crystals. The normal spectra of these molecules were too complex to be analyzed. The near-magic-angle spinning simplified the spectrum considerably, and, in addition, the 2D *J*-resolved spectrum showed well-resolved multiplet patterns for various protons, while COSY helped in identifying the peaks of these protons.[52]

Another interesting study has been a 2D correlation of the liquid crystal spectrum with the isotropic spectrum by the use of a temperature jump during the mixing period. This allows a straightforward separation of the liquid crystal spectrum into various chemical sites.[53] There are several detailed and informative 2D NMR studies of both the spin echo and the correlated type of $I \geqslant 1$ spins in oriented systems, as well as many 2D NMR studies of liquid crystal molecules with and without magic angle sample spinning. Further, there have been several studies of zero field NMR of molecules oriented in liquid crystal matrices. These studies are discussed in other articles in this Encyclopedia.

6 RELATED ARTICLES

Analysis of High Resolution Solution State Spectra; Analysis of Spectra: Automatic Methods; COSY Spectra: Quantitative Analysis; COSY Two-Dimensional Experiments; Deuterium NMR in Solids; INADEQUATE Experiment; Liquid Crystalline Samples: Carbon-13 NMR; Liquid Crystalline Samples: Deuterium NMR; Liquid Crystalline Samples: Diffusion; Liquid Crystalline Samples: Relaxation Mechanisms; Liquid Crystalline Samples: Spectral Analysis; Liquid Crystalline Samples: Structure of Nonrigid Molecules; Multidimensional Spectroscopy: Concepts; Multiple Quantum NMR in Solids; NOESY; Spin Echo Spectroscopy of Liquid Samples; Spinning Liquid Crystalline Samples; Two-Dimensional *J*-Resolved Spectroscopy; Zero Field NMR.

7 REFERENCES

1. A. Saupe and G. Englert, *Phys. Rev. Lett.*, 1963, **11**, 462.
2. A. D. Buckingham and K. A. McLauchlan, in 'Progress in Nuclear Magnetic Resonance Spectroscopy', ed. J. W. Emsley, J. Feeney, and L. H. Sutcliffe, Pergamon Press, Oxford, 1967, Vol. 2, p. 63.
3. P. Diehl and C. L. Khetrapal in 'NMR Basic Principles and Progress', ed. P. Diehl, E. Fluck, and R. Kosfeld, Springer-Verlag, Berlin, 1969, Vol. 1, p. 1.
4. P. Diehl, C. L. Khetrapal, and H. P. Kellerhals, *Mol. Phys.*, 1968, **15**, 333.
5. J. W. Emsley and J. C. Lindon, 'NMR Spectroscopy Using Liquid Crystal Solvents', Pergamon Press, Oxford, 1975.
6. S. Castellano and A. A. Bothner-By, *J. Chem. Phys.*, 1964, **41**, 3863.
7. P. Diehl, H. P. Kellerhals, and W. Niederberger, *J. Magn. Reson.*, 1971, **4**, 352.
8. J. W. Emsley, D. L. Turner, A. M. Giroud, and M. Longeri, *J. Chem. Soc., Faraday Trans. 2*, 1985, **81**, 603.
9. A. Kumar, *Current Sci.*, 1988, **57**, 109.
10. C. L. Khetrapal and K. V. Ramanathan, in 'Specialist Periodical Reports on NMR', ed. G. A. Webb, The Royal Society of Chemistry, London, 1994, Vol. 23, p. 439 and references therein.
11. W. P. Aue, J. Karhan, and R. R. Ernst, *J. Chem. Phys.*, 1976, **64**, 4226.
12. R. R. Ernst, G. Bodenhausen, and A. Wokaun, 'Principles of Nuclear Magnetic Resonance in One and Two Dimensions', Clarendon Press, Oxford, 1987.
13. A. Kumar, *J. Magn. Reson.*, 1978, **30**, 227 (erratum, 1980, **40**, 413).
14. A. Kumar and C. L. Khetrapal, *J. Magn. Reson.*, 1978, **30**, 137.
15. C. L. Khetrapal, A. Kumar, A. C. Kunwar, P. C. Mathias, and K. V. Ramanathan, *J. Magn. Reson.*, 1980, **37**, 349.
16. D. L. Turner, *J. Magn. Reson.*, 1982, **46**, 213.
17. K. Nagayama, A. Kumar, K. Wüthrich, and R. R. Ernst, *J. Magn. Reson.*, 1980, **40**, 321.
18. A. G. Avent, J. W. Emsley, and D. L. Turner, *J. Magn. Reson.*, 1983, **52**, 57.
19. J. W. Emsley and D. L. Turner, *J. Chem. Soc., Faraday Trans. 2*, 1981, **77**, 1493.
20. M. Albert Thomas, K. V. Ramanathan, and A. Kumar, *J. Magn. Reson.*, 1983, **55**, 386.
21. K. Rukmani and A. Kumar, *Chem. Phys. Lett.*, 1987, **133**, 485.
22. G. Drobny, *Chem. Phys. Lett.*, 1984, **109**, 132.
23. W. P. Aue, E. Bartholdi, and R. R. Ernst, *J. Chem. Phys.*, 1976, **64**, 2229.
24. A. Wokaun and R. R. Ernst, *Chem. Phys. Lett.*, 1977, **52**, 407.
25. G. Drobny, A. Pines, S. Sinton, D. P. Weitekamp, and D. Wemmer, *Faraday Symp. Chem. Soc.*, 1978, **13**, 49.
26. W. S. Warren, S. Sinton, D. P. Weitekamp, and A. Pines, *Phys. Rev. Lett.*, 1979, **43**, 1791.
27. S. Sinton and A. Pines, *Chem. Phys. Lett.*, 1980, **76**, 263.
28. J. Tang and A. Pines, *J. Chem. Phys.*, 1980, **72**, 3290; 1980, **73**, 2512.
29. W. S. Warren, D. P. Weitekamp, and A. Pines, *J. Chem. Phys.*, 1980, **73**, 2084.
30. W. S. Warren and A. Pines, *J. Chem. Phys.*, 1981, **74**, 2808; *J. Am. Chem. Soc.*, 1981, **103**, 1613.
31. D. P. Weitekamp, J. R. Garbow, and A. Pines, *J. Magn. Reson.*, 1982, **46**, 529; *J. Chem. Phys.*, 1982, **77**, 2870.
32. J. B. Murdoch, W. S. Warren, D. P. Weitekamp, and A. Pines, *J. Magn. Reson.*, 1984, **60**, 205.
33. S. W. Sinton, D. B. Zax, J. B. Murdoch, and A. Pines, *Mol. Phys.*, 1984, **53**, 333.
34. M. Gochin, K. V. Schenker, H. Zimmermann, and A. Pines, *J. Am. Chem. Soc.*, 1986, **108**, 6813.
35. M. Gochin, H. Zimmermann, and A. Pines, *Chem. Phys. Lett.*, 1987, **137**, 51.
36. M. Gochin, D. Hugi-Cleary, H. Zimmermann, and A. Pines, *Mol. Phys.*, 1987, **60**, 205.
37. M. Munowitz and A. Pines, *Science (Washington, DC)*, 1986, **233**, 525.

38. W. S. Warren, J. B. Murdoch, and A. Pines, *J. Magn. Reson.*, 1984, **60**, 236.
39. G. P. Drobny, *Annu. Rev. Phys. Chem.*, 1985, **36**, 451.
40. G. P. Drobny, A. Pines, S. W. Sinton, W. S. Warren, and D. P. Weitekamp, *Phil. Trans. R. Soc. Lond., Ser. A*, 1981, **299**, 585.
41. M. Albert Thomas and A. Kumar, *J. Magn. Reson.*, 1982, **47**, 535.
42. M. Albert Thomas and A. Kumar, *J. Magn. Reson.*, 1984, **56**, 479.
43. H. Fujiwara, N. Shimizu, T. Takagi, Y. Sasaki, and K. Takahashi *J. Magn. Reson.*, 1985, **64**, 325.
44. M. Gochin, A. Pines, M. E. Rosen, S. P. Rucker, and C. Schmidt, *Mol. Phys.*, 1990, **69**, 671.
45. M. E. Rosen, S. P. Rucker, C. Schmidt, and A. Pines, *J. Phys. Chem.*, 1993, **97**, 3858.
46. D. Gamliel, A. Maliniak, Z. Luz, R. Poupko, and A. J. Vega, *J. Chem. Phys.*, 1990, **93**, 5379.
47. R. Christy Rani Grace and A. Kumar, *J. Magn. Reson.*, 1992, **97**, 184.
48. R. Christy Rani Grace and A. Kumar, *Bull. Magn. Reson.*, 1992, **14**, 42.
49. R. Christy Rani Grace and A. Kumar, *J. Magn. Reson.*, 1992, **99**, 81.
50. P. Pfländler and G. Bodenhausen, *J. Magn. Reson.*, 1991, **91**, 65.
51. R. Christy Rani Grace, N. Suryaprakash, A. Kumar and C. L. Khetrapal, *J. Magn. Reson., Ser. A*, 1994, **106**, 79.
52. H. Fujiwara, in 'Proceedings of 11th Conference of the International Society of Magnetic Resonance, Vancouver, 1992', Abstract 0–69, p. 114; in 'Proceedings of 32nd Annual NMR Symposium, Japan, 1993', L35 p. 141.
53. K. Akasaka, A. Naito, and M. Kimura, in 'Proceedings of 11th Symposium of the International Society of Magnetic Resonance, Vancouver, 1992', Abstract 0–36, p. 81; A. Naito, M. Imanari, and K. Akasaka, *J. Magn. Reson.*, 1991, **92**, 85; K. Akasaka, A. Naito, and M. Imanari, *J. Am. Chem. Soc.*, 1991, **113**, 4688.

Biographical Sketch

Anil Kumar. *b* 1941. M.Sc., 1961, Agra, Ph.D., 1969, Indian Institute of Technology, Kanpur (supervisor, B. D. N. Rao). Postdoctoral research at Georgia Tech., Atlanta, GA (Sidney Gordon) 1969–70, University of North Carolina, Chapel Hill, NC (Charles Johnson) 1970–72, and E.T.H. Zürich, Switzerland (Richard Ernst) 1973–76. Faculty of Physics, Indian Institute of Science, Bangalore, 1977–present. Sabbatical leave at E.T.H. Zürich with Kurt Wüthrich, 1979–80. Approx 100 publications. Research interests: methodology of NMR, two-dimensional techniques, relaxation studies, applications to structure determination of biomolecules and characterization of molecular motions.

Two-Dimensional Powder Correlation Methods

Craig D. Hughes

Los Alamos National Laboratory, Los Alamos, NM, USA

1 INTRODUCTION

There are two general motivations to measure the ^{13}C chemical shift tensor in a solid. The first is that it is sensitive to the electronic structure of the molecule in the vicinity of the nucleus, and therefore provides useful experimental data for testing theoretical models of molecular electronic structure. The second, more practical, motivation is analytic in nature; the chemical shift tensor is a sensitive and often unique tool for characterizing heterogeneous materials, and its orientational dependence can be exploited to help characterize partially oriented materials.

The measurement of the ^{13}C chemical shift tensor's principal values and its orientation with respect to the molecular coordinate system is often difficult. A common method is to rotate a single crystal of the sample in a magnetic field and map the relationship between the field direction and the resonance frequency for each magnetically equivalent carbon nucleus in the sample.[1] As the number of magnetically unique carbon nuclei in the crystal increases, correlated 2D exchange NMR spectroscopy is a powerful tool in resolving and correlating the peaks resulting from similar chemical shift tensors.[1,2]

When a good single crystal of the sample is not available, powder techniques must be used. In simple cases, the principal values may be measured from the breakpoints of a 1D powder pattern. The powder spectrum is then simulated with a chemical shift powder pattern for each unique carbon in the sample in order to measure the principal values. This technique is limited to relatively simple spectra, owing to the complexity of fitting multiple, overlapping 1D powder patterns. There are a number of alternate experimental approaches to solving this problem, many of which are reviewed elsewhere in this Encyclopedia.

A relatively simple 2D technique for measuring principal values of overlapping ^{13}C chemical shift tensor patterns in powders is the chemical shift–chemical shift correlation experiment developed for single crystal measurements.[2–4] The sample is rotated relative to the external magnetic field during the mixing time of a 2D exchange experiment, which correlates

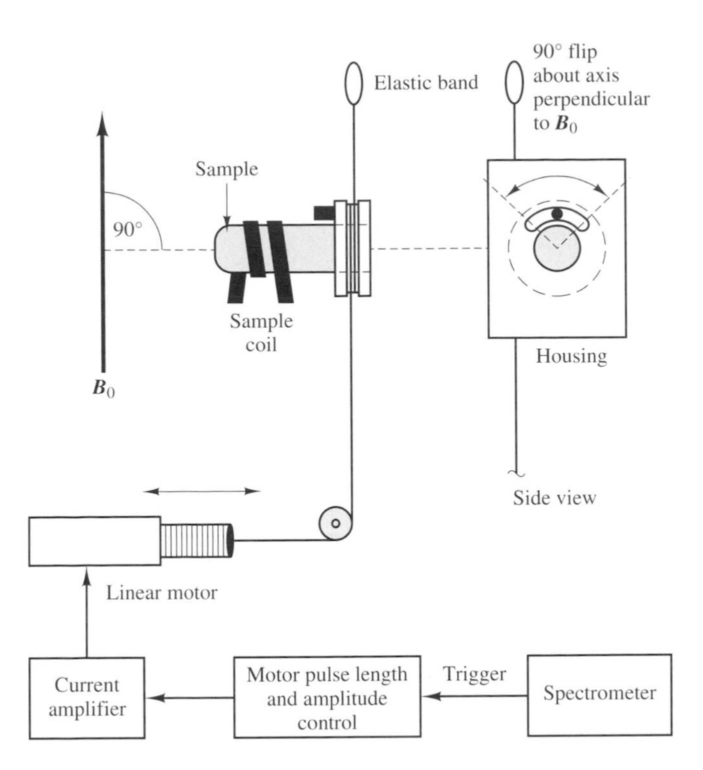

Figure 1 90° sample reorientation apparatus. The powdered sample is placed in a shortened 12 mm NMR tube, which is then epoxied into the sample pulley. The pulley is rotated inside a circular recess in the housing by pulling the attached string

the powder patterns in the two different orientations. The resulting 2D powder patterns have a characteristic shape that depends on the angle of the rotation and the principal values of the chemical shift tensors in the sample. This method utilizes the additional resolution available in 2D spectroscopy, and is especially useful for sorting the principal values of chemical shift tensors in badly overlapped 1D powder patterns.

2 EXPERIMENTAL

The probe appropriate for the 2D powder correlation experiment is a double tuned, large-sample-volume system designed to allow ^{1}H decoupling and rapid 90° rotation of the sample about an axis perpendicular to the external magnetic field, as shown in Figure 1. The results reported here were obtained using a Bruker CXP-200 NMR spectrometer, with a resonance frequency of 50.3 MHz for ^{13}C.

The pulse sequence used in the experiment is shown in Figure 2. The first part of the pulse sequence is the common cross-polarization technique with Hartmann–Hahn matching and spin locking between the ^{1}H and ^{13}C spins. The remainder of the sequence is a chemical exchange experiment that correlates two different ^{13}C chemical shift evolutions. Here, v and w are the phases of the ^{1}H rf pulses, and x and Φ_i those of the ^{13}C pulses. In this experiment, the measured chemical shift is determined solely by the orientation of the crystallite in the magnetic field. The sample is rotated by 90° during the mixing time, so each crystallite in the powdered sample 'sees' two different magnetic field directions during t_1 and t_2, which results in two different evolution frequencies ω_1 and ω_2. The resulting 2D powder patterns contain information about all of the ^{13}C chemical shift tensors in the sample, and are characteristic of the 90° sample rotation.

The 2D data are gathered according to the prescription of States, Haberkorn, and Ruben,[5] which gives amplitude-modulated hypercomplex multidimensional data sets.

3 RESULTS

The spectrum in Figure 3 is a contour plot of the 2D chemical shift correlation spectrum of ferrocene, with the more familiar 1D projections of the spectrum shown at the sides. Ferrocene has an axially symmetric ^{13}C chemical shift tensor whose orientation is determined by its motion about the molecular rotational axis; $\delta_{\parallel}$ is the principal value of the chemical shift tensor when the field is parallel to ferrocene's rotational

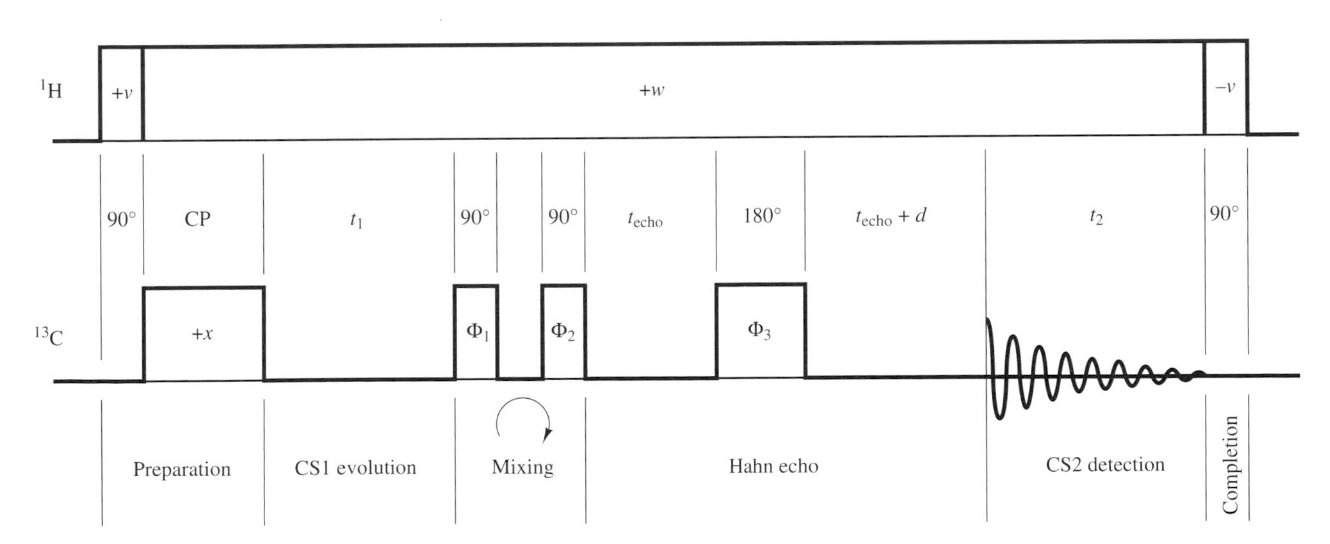

Figure 2 Chemical shift–chemical shift correlation pulse sequence. Phase cycling corrects for artifacts resulting from ^{13}C T_1 relaxation during the mixing time and cross polarization during the 180° echo pulse, and allows for quadrature detection during the evolution dimension. The mixing period is 35 ms long

For References see p. 4872

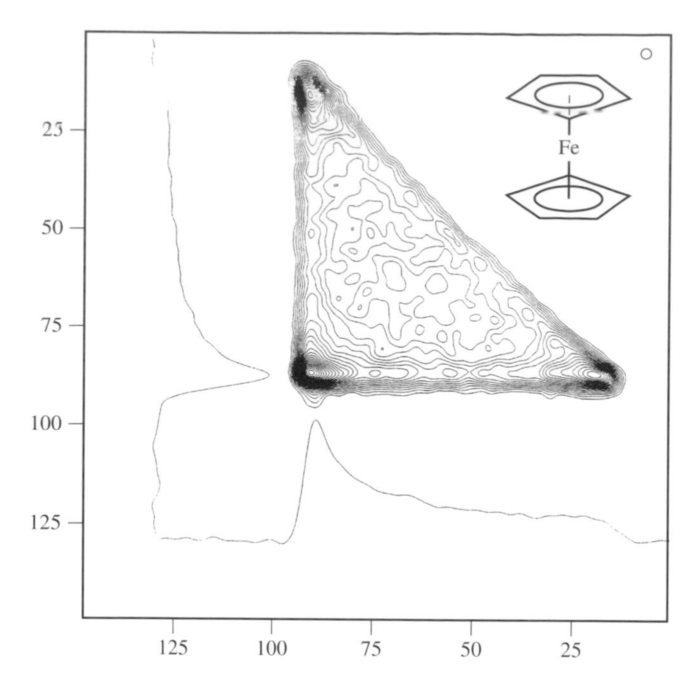

Figure 3 Equal-amplitude contour plot of the 2D ^{13}C chemical shift–chemical shift correlation experimental spectrum of ferrocene (dicyclopentadienyliron). The scales are in ppm from liquid TMS (positive values are to high frequency). ω_1 is the vertical dimension and ω_2 is the horizontal dimension. The diameter of the small circle plotted in the upper right-hand corner is the full width at halfheight of the Gaussian broadening ($\Delta\omega_G$) function that is convolved with the data. In this figure, $\Delta\omega_G$ = 3 ppm. The 1D powder projections are plotted on the vertical and horizontal axes

axis, and $\delta_\perp$ is the degenerate principal value when the field lies parallel to the ring plane. An axially symmetric 2D powder pattern has three large peaks connected by ridges forming a right isosceles triangular pattern with a deep basin in its interior. Comparing the features of the 2D spectrum with its 1D projections indicates that the peaks or cusps define the principal values of the chemical shift tensor.

The ferrocene spectrum is a simple example of a 2D chemical shift correlation pattern. The principal values could have been obtained equally well from a 1D powder spectrum. The 2D technique is significantly more useful when multiple overlapping bands in a 1D powder spectrum leave unresolvable ambiguities. The spectrum of coronene shown in Figure 4 illustrates how powerful the 2D correlation technique is. Coronene has three sets of unique carbons, and the molecule rapidly rotates about an axis perpendicular to its plane, averaging all of its chemical shift tensors to axial symmetry. The foremost challenge in measuring the principal values of the three coronene tensors from the 1D powder spectrum was to determine which pairs of the six breakpoints are grouped together as the principal values for each of the three different ^{13}C nuclei in the molecule. The spectrum in Figure 4 exhibits how these groupings were readily determined with the 2D chemical shift correlation experiment. The correlative properties of the powder pattern allows immediate connection of principal value pairs from each chemically distinct nucleus.

The 2D chemical shift correlation spectrum of isotactic polypropylene is shown in Figure 5. This spectrum indicates

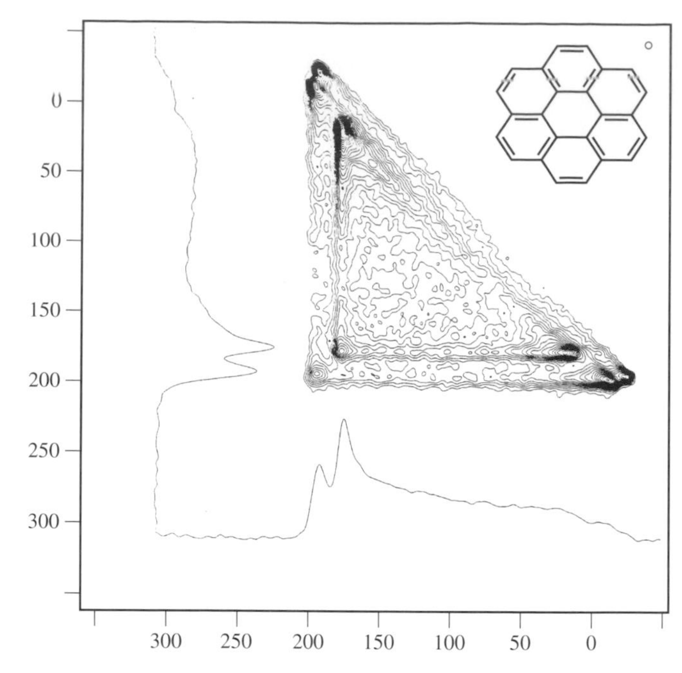

Figure 4 Contour plot of the spectrum of coronene (hexabenzobenzene). The Gaussian broadening in this figure is 3 ppm

that the added correlation and resolution of two dimensions resolves three different chemical shift tensor patterns in a situation where the 1D powder pattern is intractable.

In an effort to determine the limits of resolution of this technique, a spectrum of methyl α-D-glucopyranoside was acquired (Figure 6). Methyl α-D-glucopyranoside has seven carbon types, whose severely overlapped ^{13}C chemical shift tensors have been measured in a single crystal, and therefore is an excellent candidate for comparison of techniques. The 1D pow-

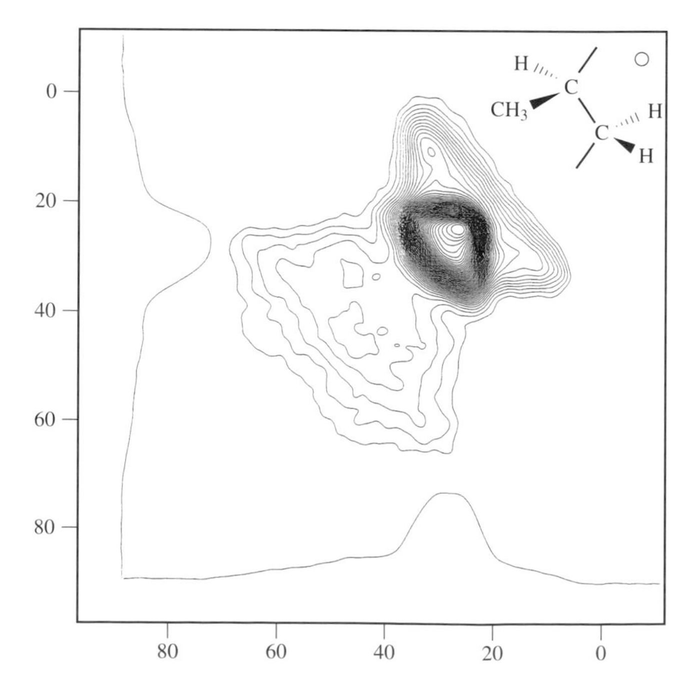

Figure 5 Contour plot of the spectrum of isotactic polypropylene. The Gaussian broadening in this figure is approximately 2.5 ppm

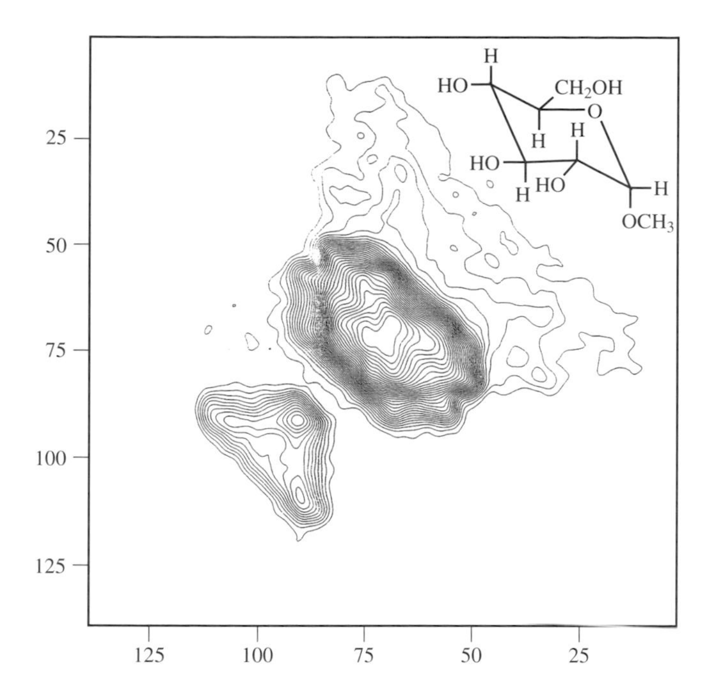

Figure 6 Contour plot of the spectrum of methyl α-D-glucopyranoside

der spectrum is severely overlapped, with almost no distinguishable features to facilitate conventional fitting techniques. However, the 2D powder pattern does show a considerable improvement in resolution. At least four distinct powder patterns may be immediately recognized by the outlines of the cliff-like features that form the boundaries of the patterns. This spectrum was simulated by successively subtracting out simulated components from the pattern and finding new component patterns in the residual spectrum. Prior knowledge of the tensor principal values was not used. The results of this fitting agree well with previous results,[6] as shown in Figure 7. Although this fitting technique left two possible shift tensors unresolved, one of the simulated component patterns has roughly twice the integrated amplitude of the remaining five patterns, indicating that the two unresolved tensor patterns have very similar principal values, which could not be reliably distinguished with this analysis technique.

The measurement of six of the seven ^{13}C chemical shift tensor principal values in methyl α-D-glucopyranoside represents a remarkable increase in resolution over 1D powder spectroscopy, and perhaps an upper limit on the number of overlapping powder patterns that can be successfully resolved with the 2D chemical shift correlation technique. The agreement between the single crystal data and the powder fitting data is very good.

The increased resolving power of the chemical shift correlation experiment has been applied to complex samples such as coals in order to measure the aromaticity of the sample. Figure 8 shows the spectrum of the Argonne premium coal, Pocahontas #3. This spectrum has a clear segregation of amplitude between two different regions. The small volume in the upper right-hand corner of the spectrum has a relatively narrow chemical shift anisotropy and an average chemical shift of approximately 25 ppm. These chemical shifts fall into a range expected from aliphatic carbon nuclei; therefore, this volume represents the aliphatic fraction of the sample. The triangular shaped volume in the lower left-hand corner of this spectrum has a broader chemical shift anisotropy, and an average chemical shift of approximately 130 ppm, and represents the aromatic

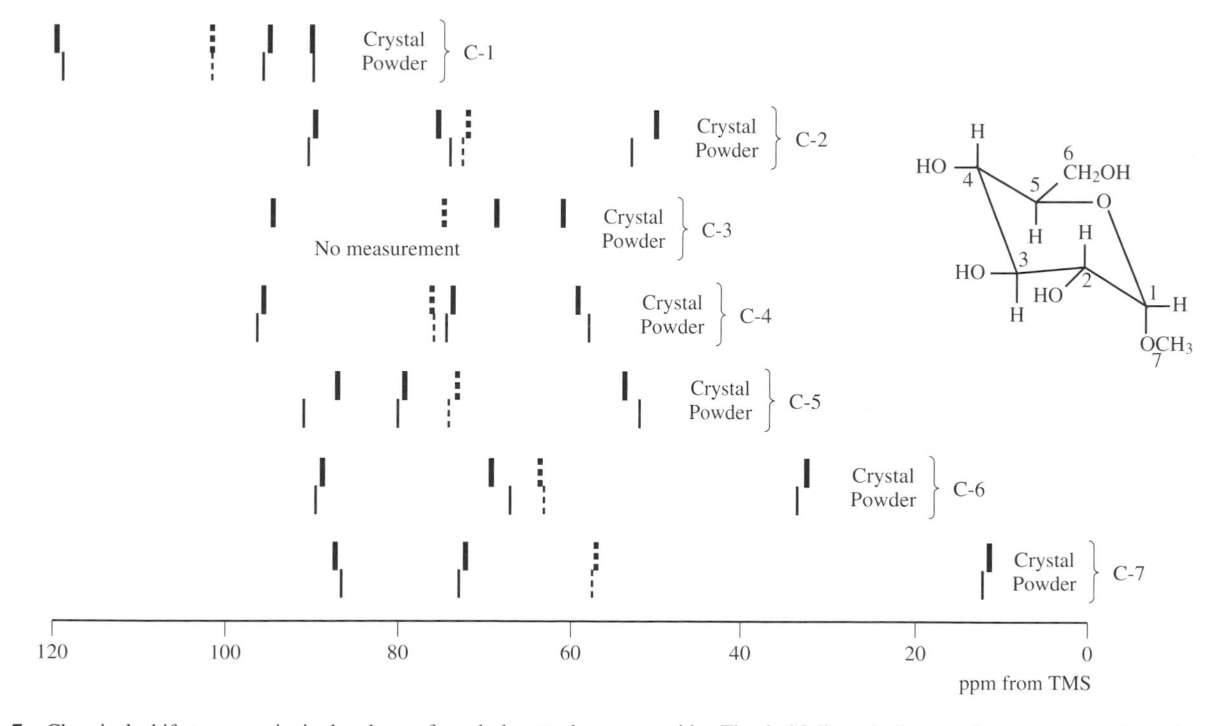

Figure 7 Chemical shift tensor principal values of methyl α-D-glucopyranoside. The bold lines indicate values measured in a single crystal experiment.[6] The lighter faced lines indicate values measured from fitting the 2D powder spectrum. The dashed lines indicate the values of the average chemical shift, i.e., $\delta_{\text{average}} = \frac{1}{3}(\delta_{11} + \delta_{22} + \delta_{33})$

For References see p. 4872

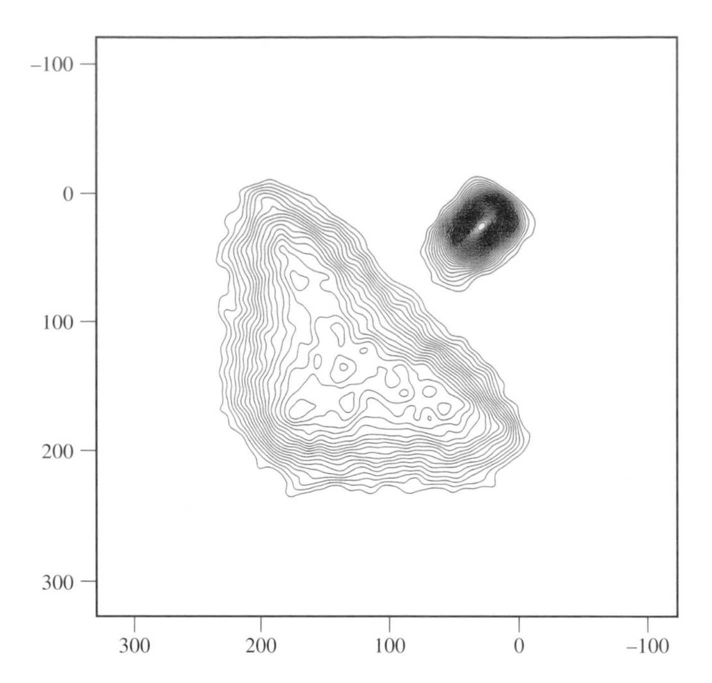

Figure 8 Contour plot of the spectrum of the Argonne premium coal, Pocahontas #3. The Gaussian broadening in this figure is approximately 20 ppm

fraction of the sample. The ratio of these two volumes gives an estimate of the sample's aromaticity:

$$\%\ \text{Aromaticity} = \frac{\text{Volume}_{\text{aromatic}}}{\text{Volume}_{\text{aromatic}} + \text{Volume}_{\text{aliphatic}}} \times 100\% \qquad (1)$$

This quantity has been measured from the spectrum in Figure 8, and the result is 87%. There is good agreement between the aromaticity measured from these data and the more conventional CP MAS result of 86%. An advantage of the chemical shift correlation technique is that it is less sensitive to the Hartmann–Hahn matching condition than is the CP MAS technique. This data also has the potential for testing theoretical models that attempt to correlate the components of the chemical structure of coals and related materials with the distributions of chemical shift anisotropy.

4 SIMULATIONS

In order briefly to establish the correspondence between the principal values of the chemical shift tensor and the features of 2D chemical shift correlation spectra, a series of computer simulations for the 90° sample rotation experiment representing the different classes of chemical shift tensors are shown in Figure 9. The general triangular or hexagonal shape of the simulations is a characteristic of the 90° sample rotation experiment. These simulations show the relationships between the ridges and peaks of this new type of 2D spectrum and the breakpoints in the 1D powder spectra, which are plotted as projections below the 2D simulations.

Often, the measurement of principal values from 1D powder patterns is seriously hindered by nonideal distribution of amplitude across the spectrum, especially when cross-polarization techniques are used. The ability to simulate this amplitude distortion in a 2D powder pattern would increase the accuracy of the measurement of principal values. The frequency and amplitude of the NMR signal from a particular nucleus in a given crystallite in a powdered sample is determined by the principal values of its chemical shift tensor and by that crystallite's orientation relative to the external magnetic field. Therefore, the simulation model is parameterized in terms of chemical shift tensor principal values and the three Euler orientation angles α, β, and γ that describe the relative orientation of the nuclear environment with the external magnetic field in a randomly oriented powder.

In this experiment, each nucleus in the sample has two resonance frequencies, which correspond to the two different orientations that the nucleus may have with the external field during t_1 and t_2. Assuming a random distribution of crystallite orientations, the simulated hypercomplex 2D FID from a single nucleus in an experiment where the external field is along the x sample axis during t_1 and along the y sample axis during t_2 can be written as

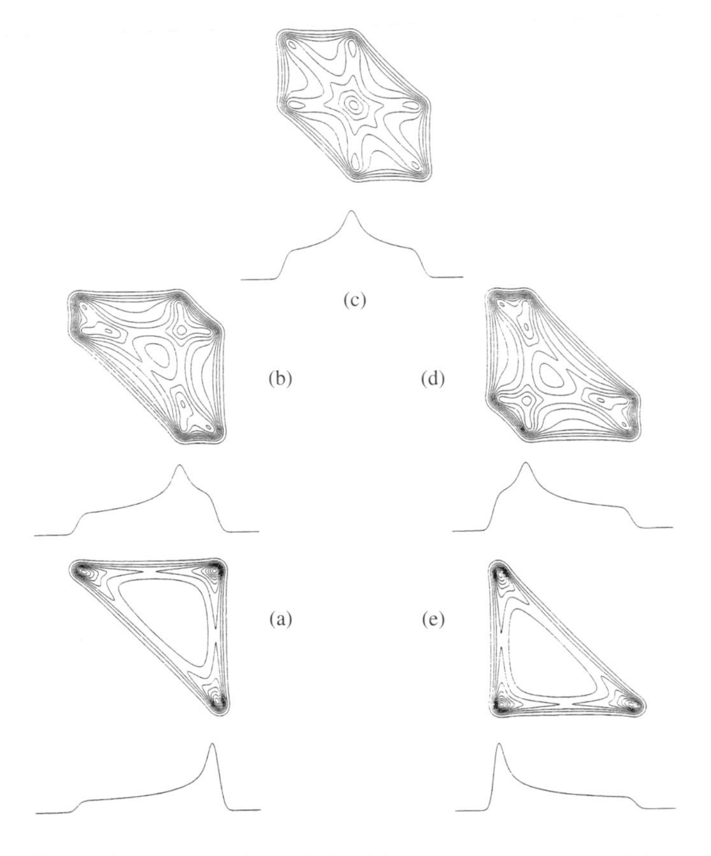

Figure 9 Contour plots of 2D 90° sample rotation simulations resulting from five different chemical shift tensors, showing the correspondence between the features in the 2D powder patterns and conventional 1D powder patterns: (a) axially symmetric chemical shift tensor with low-frequency (upfield) degenerate principal values; (b) general chemical shift tensor with $\delta_{11} - \delta_{22} = 3(\delta_{22} - \delta_{33})$; (c) general chemical shift tensor with $\delta_{11} - \delta_{22} = \delta_{22} - \delta_{33}$; (d) general chemical shift tensor with $3(\delta_{11} - \delta_{22}) = \delta_{22} - \delta_{33}$; (e) axially symmetric chemical shift tensor with high-frequency degenerate principal values

$$s_{\text{sim}}(t_1,t_2) = \frac{1}{8\pi^2}\int \mathrm{d}\alpha \sin\beta\, \mathrm{d}\beta\, \mathrm{d}\gamma\, A(\alpha,\beta,\gamma) \times \left[\exp\{\mathrm{i}_1[\omega_x(\alpha,\beta,\gamma)-\omega_0]t_1\}\exp\left(-\frac{t_1^2\Delta\omega_{\text{G1}}^2}{G^2}\right)\right] \times \left[\exp\{\mathrm{i}_2[\omega_y(\alpha,\beta,\gamma)-\omega_0]t_2\}\exp\left(-\frac{t_2^2\Delta\omega_{\text{G2}}^2}{G^2}\right)\right] \quad (2)$$

where $A(\alpha,\beta,\gamma)$ is an amplitude weighting function that depends upon the crystallite orientation in the field. The frequency ω_0 is the spectrometer carrier frequency, and $\Delta\omega_{\text{G1}}$ and $\Delta\omega_{\text{G2}}$ are Gaussian broadenings in the t_1 and t_2 dimensions, respectively. G^2 is a constant equal to $4\ln(2)/\pi^2$. The imaginary units i_1 and i_2 (with $\mathrm{i}_1^2 = \mathrm{i}_2^2 = -1$) distinguish the independent imaginary parts of the hypercomplex t_1 and t_2 data.[7] The resonance frequencies for the x and y field directions of a given crystallite with an α, β, γ orientation relative to the principal axis frame of the tensor are given by the matrix products

$$\left.\begin{aligned}\omega_x(\alpha,\beta,\gamma) &= \boldsymbol{x}^{-1}\mathbf{R}^{-1}(\alpha,\beta,\gamma)\boldsymbol{\delta}\mathbf{R}(\alpha,\beta,\gamma)\boldsymbol{x}\\ \omega_y(\alpha,\beta,\gamma) &= \boldsymbol{y}^{-1}\mathbf{R}^{-1}(\alpha,\beta,\gamma)\boldsymbol{\delta}\mathbf{R}(\alpha,\beta,\gamma)\boldsymbol{y}\end{aligned}\right\} \quad (3)$$

where $\boldsymbol{x}$ and $\boldsymbol{y}$ are unit column vectors in the x and y sample frame directions, respectively. The unitary rotation matrix $\mathbf{R}(\alpha,\beta,\gamma)$ rotates the chemical shift tensor $\boldsymbol{\delta}$ from its principal axis frame to the sample frame, and this tensor in the principal axis frame is

$$\boldsymbol{\delta} = \begin{bmatrix}\delta_{11} & 0 & 0\\ 0 & \delta_{22} & 0\\ 0 & 0 & \delta_{33}\end{bmatrix} \quad (4)$$

Nonideal amplitude distributions in powder spectra are primarily artifacts resulting from the orientational dependence of both cross polarization and spin relaxation. Functions of orientation described by the Euler angles α, β, and γ are conveniently expanded in the Wigner rotation matrices $\mathcal{D}^{(j)}_{m',m}(\alpha,\beta,\gamma)$, which form a complete, orthonormal basis set.[8,9] The three-dimensional amplitude weighting function that accounts for the polarization efficiency can be expanded as

$$A(\alpha,\beta,\gamma) = \sum_{j,m',m} a^{(j)}_{m',m}\mathcal{D}^{(j)}_{m',m}(\alpha,\beta,\gamma) \quad (5)$$

where the j sum is from 0 to ∞ and the m and m' sums run from $-j$ to $+j$. The expansion coefficients $a^{(j)}_{m',m}$ thus provide a complete description of the different amplitudes resulting from all possible crystallite orientations. The mathematical construct of equation (5) can also be used to characterize ordering in partially oriented samples.

It can be shown that the simulated 2D FID may be expanded in a complete set of Wigner 2D subFIDs, where the linear coefficients $a^{(j)}_{m',m}$ are the weighting factors for each of the orthonormal Wigner subFIDs $s^{(j)}_{m',m}$ (t_1, t_2) in the basis set

$$s_{\text{sim}}(t_1,t_2) = \sum_{j,m',m} a^{(j)}_{m'm}s^{(j)}_{m',m}(t_1,t_2) \quad (6)$$

These functions can be used to determine any general amplitude distribution in the spectrum using standard linear fitting techniques.

Although all of the fitting is done in the time domain, the 2D Fourier transform of the 2D FID produces the more familiar simulated 2D spectra:[7]

$$S(\omega_1,\omega_2) = \int_{-\infty}^{+\infty}\mathrm{d}t_1\exp(-\mathrm{i}_1\omega t_1) \times \int_{-\infty}^{+\infty}\mathrm{d}t_2\exp(-\mathrm{i}_2\omega t_2)s(t_1,t_2) \quad (7)$$

Likewise, the Fourier transforms of the subFIDs give the subspectra:[10]

$$S^{(j)}_{m',m}(\omega_1,\omega_2) = \int_{-\infty}^{+\infty}\mathrm{d}t_1\exp(-\mathrm{i}_1\omega t_1) \times \int_{-\infty}^{+\infty}\mathrm{d}t_2\exp(-\mathrm{i}_2\omega t_2)s^{(j)}_{m',m}(t_1,t_2) \quad (8)$$

The first subspectrum in the series, $S^0_{0,0}$, is the 'ideal' powder spectrum. This subspectrum represents the case where each spin in the sample contributes the same amplitude to the spectrum, and where every crystallite orientation is equally probable. This term is also the only member of the expansion whose integrated spectrum has a nonzero volume. The higher order subspectra in the series are positive in some areas and negative in others, so that they integrate to zero volume over the domain of the spectrum. These higher order subFIDs or subspectra describe modifications in the spectral amplitude that deviate from the 'ideal' powder response, and may be used to characterize the orientational dependence of the amplitude that results either from experimental artifacts or from macroscopically ordered samples. It has been found for most powder spectra that the number of terms necessary to simulate effectively all of the deficiencies in the spectrum is relatively small, indicating that $A(\alpha,\beta,\gamma)$ is a fairly smooth function in α, β, and γ. It has also been found for the 2D powder spectra presented here that the coefficient $a^0_{0,0}$ is much larger than the higher order coefficients, supporting the general observation that the amplitude distribution across a powder pattern is nearly ideal, with only relatively small nonideal contributions.

5 DISCUSSION

The chemical shift tensor principal values reported here result from fitting simulations calculated from equation (6) to the data. The fitting is done in the time domain, where the fractional residual is taken as the figure of merit for any given simulation, and is minimized to find the 'best fit'. The fractional residual is defined as

$$\text{fractional residual} = \frac{\sum_{t_1,t_2}|s_{\text{data}}(t_1,t_2) - ks_{\text{sim}}(t_1,t_2)|^2}{\sum_{t_1,t_2}|s_{\text{data}}(t_1,t_2)|^2} \quad (9)$$

where $s_{\text{data}}(t_1, t_2)$ is the 2D hypercomplex data FID, and k is a scaling constant that adjusts for experimental factors in the acquisition and processing of the data.

The 'best fit' results for ferrocene, coronene, polyethylene, and polypropylene given in Table 1 are in excellent agreement with previous results.[11–14] The chemical shift principal values differ by 3 ppm or less from the values measured with other techniques, while the average chemical shifts i.e., δ_{average} =

For References see p. 4872

Table 1 Chemical Shift Tensor Principal Values from Fits (ppm from External Liquid TMS)

	$\delta_{\perp}$		$\delta_{\parallel}$	$\delta_{average}$	δ_{liquid}
Ferrocene:					
This work	94		17	68	
Previous results[11]	94		18	69	69.4
	$\delta_{\perp}$		$\delta_{\parallel}$	$\delta_{average}$	$\delta_{CP\ MAS}$
Coronene:					
(Protonated)					
This work	178		15	124	
Previous results[12]	175		15	122	123.5
(Bridging)					
This work	197		−15	126	
Previous results[12]	196		−16	125	126.0
(Inner)					
This work	193		−26	120	
Previous results[12]	191		−26	119	119.8
	δ_{11}	δ_{22}	δ_{33}	$\delta_{average}$	$\delta_{CP\ MAS}$
Polypropylene:					
(Methine)					
This work	37	22	22	27	26.5
Previous results	38[14]	21[14]	21[14]	27[14]	26.5[13]
(Methylene)					
This work	65	45	25	45	44.5
Previous results	65[14]	44[14]	26[14]	45[14]	44.5[13]
(Methyl)					
This work	32	32	5	23	22.5
Previous results	32[14]	32[14]	3[14]	22[14]	22.5[13]

$\frac{1}{3}(\delta_{11} + \delta_{22} + \delta_{33})$ differ from published CP MAS isotropic chemical shifts by 1 ppm or less.

This 2D technique has some distinct advantages over other methods for measuring ^{13}C shift tensor principal values in complex powdered solids. The inherent correlation and resolution advantage of 2D powder spectroscopy over conventional 1D techniques is evident in the spectra of coronene, polypropylene, methyl α-D-glucopyranoside, and coal. Powder patterns that overlap severely in 1D spectra can often be untangled with this 2D spectroscopy. Another powerful advantage of the technique is that the data can be simulated and fit with a high degree of fidelity using a straightforward physical model.

Another significant advantage of the technique is its simplicity. The sample does not spin at high speeds at the magic angle; therefore, the probe can be designed to accept a larger volume sample inside a transverse solenoidal coil in order to optimize sensitivity. The 90° rotation of the sample is not difficult to accomplish accurately, both in terms of constructing and driving the mechanism. The sample is stationary during the cross-polarization time, which results in more effective Hartmann–Hahn matching. The principal values of the chemical shift are represented in frequencies measured from the spectrum, not from information encoded in spinning sideband amplitudes, as in slow spinning techniques. Furthermore, the frequencies of the principal values of the chemical shift tensors are measured directly from the spectra; there is no scaling factor relating the frequencies measured in the spectrum to the principal values of the chemical shift tensor.

6 RELATED ARTICLES

Chemical Shift Tensor Measurement in Solids; Chemical Shift Tensors; Chemical Shift Tensors in Single Crystals; Coal Structure from Solid State NMR; Double Resonance; Fossil Fuels; Magic Angle Turning & Hopping; Multidimensional Spectroscopy: Concepts; Polymer Dynamics & Order from Multidimensional Solid State NMR; Polymer Physics; Sideband Analysis in Magic Angle Spinning NMR of Solids; Variable Angle Sample Spinning.

7 REFERENCES

1. J. Jeener, B. H. Meier, P. Bachmann, and R. R. Ernst, *J. Chem Phys.*, 1979, **71**, 4546.
2. C. M. Carter, D. W. Alderman, and D. M. Grant, *J. Magn. Reson.*, 1985, **65**, 183.
3. P. M. Henrichs, *Macromolecules*, 1987, **20**, 2099.
4. C. D. Hughes, M. H. Sherwood, D. W. Alderman, and D. M. Grant, *J. Magn. Reson., Ser. A*, 1993, **102**, 58.

5. D. J. States, R. A. Haberkorn, and D. J. Ruben, *J. Magn. Reson.*, 1982, **36**, 286.
6. D. L. Sastry, K. Takegoshi, and C. A. McDowell, *Carbohydr. Res.*, 1987, **165**, 161.
7. R. R. Ernst, G. Bodenhausen, and A. Wokaun 'Principles of Nuclear Magnetic Resonance in One and Two Dimensions', Clarendon Press, Oxford, 1987, Chap. 6.
8. A. R. Edmonds, 'Angular Momentum in Quantum Mechanics', Princeton University Press, Princeton, NJ, 1960, Chaps. 1 and 4.
9. M. E. Rose, 'Elementary Theory of Angular Momentum', Wiley, New York, 1957.
10. R. Hentschel, J. Schlitter, H. Sillescu, and H. W. Spiess, *J. Chem. Phys.*, 1978, **68**, 56.
11. D. E. Wemmer and A. Pines, *J. Am. Chem. Soc.*, 1980, **103**, 34.
12. H. A. Resing and D. L. VanderHart, *Z. Phys. Chem.*, 1987, **151**, 137.
13. A. Bunn, M. E. A. Cudby, R. K. Harris, K. J. Packer, and B. J. Say, *J. Chem. Soc., Chem. Commun.*, 1981, 15.
14. T. Nakai, J. Ashida and T. Terao, *Magn. Reson. Chem.*, 1989, **27**, 666.

Biographical Sketch

Craig D. Hughes. *b* 1963. B.S., 1985, New Mexico Institute of Mining and Technology; Ph.D., University of Utah, 1992. Introduced to NMR by L. G. Werbelow and D. M. Grant. Research specialties; characterization of powdered, heterogeneous and oriented materials, catalytic zeolites, and imaging.

Ultraslow Motions in Solids

David C. Ailion

University of Utah, Salt Lake City, UT, USA

1 INTRODUCTION

The use of conventional NMR relaxation measurement techniques for studying moderately rapid atomic motions may result in observation of the following phenomena:

1. a decrease (or a minimum)[1] of the spin–lattice relation time T_1 in a plot of T_1 versus temperature;
2. motional narrowing of the resonance linewidth[2] (or motional lengthening of the free induction decay or the spin–spin relaxation time T_2);
3. a decrease[2] in the Hahn echo decay time compared to the Carr–Purcell–Meiboom–Gill (CPMG)[3] time constant (see ***Diffusion Measurements by Magnetic Field Gradient Methods***).

The term ultraslow motions[4] refers to motions that are observable by NMR but are sufficiently infrequent that they produce no motional narrowing of the resonance line and make no measurable contribution to the spin–lattice relaxation rate. Typical T_1 minima correspond to motions for which the mean atomic jump rate $1/\tau_c$ is of the order of the nuclear precession frequency Ω_0 ($\approx 10^8\,s^{-1}$, typically).[1] Motions having jump frequencies one or two orders of magnitude higher or lower than that for the minimum usually still give measurable contributions to the relaxation rate. Since the condition for motional narrowing is that the atomic jump time τ_c be less than T_2, ultraslow motions are typically those for which $T_2 < \tau_c < T_1$.[4] A very wide range of atomic motions falls under this classification, since $T_2 \approx 10^{-4}$–10^{-5} s in solids, while T_1 can be as long as several seconds to hours in length.

It is clearly desirable to extend to lower frequencies the frequency range (and thus the temperature range) of motions that can be studied. One advantage of studying a wider range of atomic motions is that motions having slightly different activation energies may be distinguished if they are observed over a sufficiently wide temperature range. Also, motions associated with a particular phenomenon, like a phase transition, may occur only in a particular temperature region.

The basic idea of the ultraslow motion techniques can be understood by considering what happens to a T_1 minimum if the field B_0, and thus the resonant frequency, is greatly reduced. According to the Bloembergen, Purcell, and Pound (BPP) theory,[1] atomic diffusion or molecular reorientations will cause fluctuations in the dipolar interactions that can be treated as perturbations on the Zeeman levels, and, furthermore, these fluctuations can be described by an exponential correlation function. BPP derived for T_1 an expression of the form

$$\frac{1}{T_1} \approx \frac{\Omega_d^2 \tau_c}{1 + \Omega_0^2 \tau_c^2} \tag{1}$$

where Ω_0 is the Larmor angular frequency and Ω_d is the angular frequency corresponding to a typical dipolar splitting. (Actually, there are additional terms of similar form. For interactions involving identical spins, there is an extra term proportional to $1/(1 + 4\Omega_0^2\tau_0^2)$; for heteronuclear systems, there are also terms having $[1 + (\Omega_I - \Omega_S)^2\tau_c^2]$ and $[1 + (\Omega_I + \Omega_S)^2\tau_c^2]$ in the denominator.) Equation (1) predicts that, in a temperature plot of T_1, there will be a minimum value for T_1 at the temperature for which $\tau_c \approx \Omega_0^{-1}$, in agreement with the idea that the transition probability $1/T_1$ arises largely from frequency components in the perturbation that equal the energy level spacing divided by $\hbar$. If Ω_0 is reduced, the minimum will correspond to a longer value of τ_c, and thus to lower frequency motions. For a typical diffusion mechanism, atomic jumps become less frequent as the temperature is lowered. Thus the T_1 minimum in a low field will occur at a lower temperature than would the T_1 minimum at a higher field.

There are experimental problems associated with attempting to study relaxation in very weak fields where the NMR signal might be very small (see ***Zero Field NMR***). There are also theoretical difficulties, since the theory underlying equation (1) is a perturbation theory that treats the fluctuating dipolar Hamiltonian due to diffusion as a perturbation on the Zeeman Hamiltonian; such an assumption may not be valid in very weak applied fields. Solutions to both these problems will now be discussed. Ultraslow motion techniques have been reviewed by Ailion.[5–7]

2 SPIN–LATTICE RELAXATION IN THE ROTATING FRAME ($T_{1\rho}$)

One approach to studying relaxation in weak applied fields and yet having the strong signal strength for a large field is to use the field cycling method[8,9] (see ***Field Cycling Experiments***), in which the external magnetic field is reduced adiabatically to a small value and subsequently increased adiabatically back to the original value, where the signal is measured. If there are no irreversible processes, the entropy will be constant, and the full signal will be measured after the remagnetization. However, atomic diffusion constitutes an irreversible process, which will cause a loss of dipolar order. Thus, by plotting the signal strength versus time in the demagnetized state, the weak field relaxation can in principle be measured. One difficulty with this approach is that it may be difficult to satisfy the requirement that the demagnetization and remagnetization processes be sufficiently slow that the adiabatic condition is satisfied and yet be rapid compared with T_1.

An alternate approach is to measure $T_{1\rho}$, the spin–lattice relaxation time in the rotating frame of the rf field $\boldsymbol{B}_1$ (see ***Magnetization Transfer between Water & Macromolecules in Proton MRI***). According to Redfield,[10] a spin system in an rf field strong enough to saturate the resonance line (i.e., $\gamma^2 B_1^2 T_1 T_2 >> 1$) will form a canonical distribution among energy levels separated by γB_{eff} rather than γB_0, where the effective field $\boldsymbol{B}_{\text{eff}}$ in a frame rotating at angular frequency ω is given by

$$\boldsymbol{B}_{\text{eff}} = B_1 \boldsymbol{i} + (B_0 - \omega/\gamma)\boldsymbol{k} \tag{2}$$

In this case, the spins will, after equilibrium has been achieved, be characterized by a 'spin temperature in the rotating frame', which means that Curie's law will hold, but with $\boldsymbol{M}_{\text{eq}} = C\boldsymbol{H}_{\text{eff}}/T$. The significance is that, at exact resonance ($B_0 = \omega/\gamma$), the equilibrium magnetization $\boldsymbol{M}_{\text{eq}}$ will be parallel to $\boldsymbol{B}_1$ rather than to $\boldsymbol{B}_0$. The $T_{1\rho}$ minimum will then occur when $\gamma B_1 \tau_c \approx 1$, for B_0 at exact resonance. Since B_1 is typically of order 1–50 G, this minimum arises from much lower frequency motions than does the normal T_1 minimum. If B_1 is much larger than the magnetic dipolar or electric quadrupolar local fields, the dipolar and quadrupolar fluctuations can still be treated by perturbation theory (this approach is called the 'weak collision' theory).[11] If, on the other hand, B_1 is comparable to or less than the dipolar or quadrupolar local fields, perturbation theory fails, and a different theory, called the 'strong collision' theory,[12] must be used. In this article, we shall consider the case for a spin-$\frac{1}{2}$ nucleus or one in which quadrupolar interactions can be neglected; the inclusion of quadrupolar interactions in the strong collision theory has been described by Ailion.[5]

2.1 Weak Collision Theory

All of the weak collision calculations are applications of the BPP theory in which the fluctuating dipolar Hamiltonian is treated as a perturbation on the Zeeman system. The only important difference is that the basic laboratory frame Hamiltonian is replaced by a rotating frame Hamiltonian in which $\boldsymbol{B}_0$ is replaced by $\boldsymbol{B}_{\text{eff}}$, as given by equation (2). At exact resonance, this theory predicts that $T_{1\rho}$ should exhibit a low-frequency minimum of the form

$$\frac{1}{T_{1\rho}} = K \frac{\tau_c}{1 + 4\Omega_1^2 \tau_c^2} \tag{3}$$

where Ω_1 ($= \gamma B_1$) is proportional to the energy level splitting in the rotating frame. Thus the $T_{1\rho}$ minimum occurs when $\tau_c \approx (2\Omega_1)^{-1}$, which may be several orders of magnitude longer than the value of τ_c at a typical T_1 minimum. Accordingly, a $T_{1\rho}$ minimum is deeper than a T_1 minimum, and corresponds to slower atomic motions occurring at lower temperatures, as shown in Figure 1.

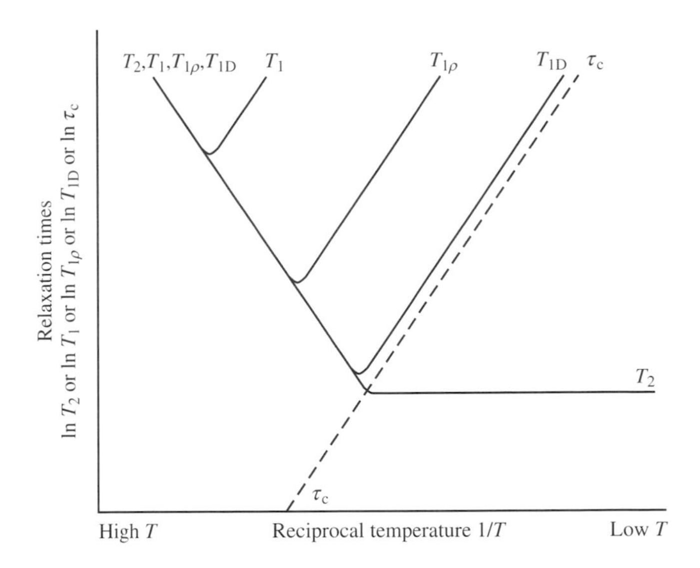

Figure 1 Temperature plot of relaxation times (T_2, T_1, $T_{1\rho}$, and T_{1D}) and atomic jump time τ_c. (Reproduced by permission of Academic Press from D. C. Ailion, in 'Methods of Experimental Physics', ed. J. N. Mundy, S. J. Rothman, M. J. Fluss, and L. C. Smedskjaer, Academic Press, Orlando, 1983, Vol. 21, Chap. 6)

2.2 Strong Collision Theory

In order to get even more effective relaxation so as to observe even lower frequency motions, B_1, and thus Ω_1 in equation (3), can be reduced to a value comparable to the local dipolar field, or even to zero (e.g., by adiabatic demagnetization in the rotating frame (ADRF),[13] which is discussed in Section 3.3). In this case, equation (3) would predict that the minimum would not occur unless τ_c were infinite. However, perturbation theories like BPP and the weak collision theory will not be valid, since the perturbation (the dipolar Hamiltonian) in this case is comparable to or larger than the unperturbed Hamiltonian (the Zeeman Hamiltonian). Nevertheless, the condition for the relaxation time minimum in zero magnetic field can still be easily understood. The underlying physical basis for all relaxation time minima is that the most effective relaxation (and thus the minimum) occurs when the principal frequency components of the fluctuating Hamiltonian are comparable to the energy level splitting (in units of $\hbar$). We have seen, for Zeeman relaxation in the laboratory or rotating frames, that this requirement is equivalent to requiring that $1/\tau_c \approx \Omega_0$ or $1/\tau_c \approx \Omega_1$, respectively. For relaxation due to a fluctuating dipolar Hamiltonian in zero field, the energy level splittings $\hbar\Omega_d$ are determined by dipolar interactions. Thus the minimum in zero field occurs when $1/\tau_c \approx \Omega_d$ or $\tau_c \approx T_2$, which is at the onset of motional narrowing. The zero field relaxation time for the case in which quadrupolar interactions can be neglected is commonly called the dipolar relaxation time T_{1D},[14] and is also shown in Figure 1.

In order to study the dynamics of the atomic motions, it is desirable to obtain theoretical expressions that relate the measured relaxation time $T_{1\rho}$ or T_{1D} to the atomic jump time τ_c throughout the temperature range of observations. Experimental procedures for measuring these relaxation times are described in Sections 2.3 and 3.3, respectively. The strong collision theory, originally developed by Slichter and Ailion[12] for

For References see p. 4881

low-field relaxation, is not a perturbation theory, but is based on the assumption that sufficient time elapses between diffusion jumps to allow the dipolar and rotating frame Zeeman interactions to cross relax to a common spin temperature between each jump. This temperature is then disturbed by the sudden change in dipolar energy resulting from a jump, but reequilibrates in a time of order T_2 to a new value prior to the next jump. This spin temperature assumption is thus equivalent to requiring that $\tau_c >> T_2$; hence the strong collision theory will be valid in the 'rigid lattice' below both the $T_{1\rho}$ minimum and the temperature for motional narrowing. If the ADRF process is complete so that $\boldsymbol{B}_1$ is reduced to zero in the demagnetized state, the ADRF process will have transferred long-range Zeeman order due to spins aligned preferentially along $\boldsymbol{B}_{\text{eff}}$ to short-range dipolar order from spins aligned along their individual local fields. A single diffusion jump per spin will then have a major effect on the relaxation so that, in zero B_1, we should expect $T_{1\rho}$ (= $T_{1\text{D}}$ for zero field) to be of order τ_c. In the weak collision regime, much less of the order is dipolar and much more is Zeeman; accordingly, it will take many jumps to relax the magnetization, and $T_{1\rho}$ will be much larger than τ_c. Both these effects are shown in Figure 1.

The actual strong collision result[12] for $T_{1\rho}$ at exact resonance due to dipolar fluctuations of a homonuclear (i.e., having only one nuclear species) system in a field B_1 is

$$\frac{1}{T_{1\rho}} = \frac{1}{T_{1\text{D}}} \frac{B_\text{D}^2}{B_1^2 + B_\text{D}^2} \tag{4}$$

where B_D is the dipolar local field. We see that $T_{1\rho}$ reduces to $T_{1\text{D}}$ in the limit as B_1 approaches zero. Note that the local field B_D can be determined experimentally by measuring $T_{1\rho}$ versus B_1^2 for fixed $T_{1\text{D}}$ (i.e., at a fixed temperature).

2.3 Experimental Techniques for Measuring $T_{1\rho}$

Essentially all methods for measuring $T_{1\rho}$ start by tipping the magnetization from its original direction (the z direction) to a direction parallel to $\boldsymbol{B}_{\text{eff}}$ in the rotating frame, a procedure called spin locking. For exact resonance, the magnetization would then be aligned parallel to $\boldsymbol{B}_1$ (in contrast to the situation immediately after a $\frac{1}{2}\pi$ pulse, when the magnetization would be perpendicular to $\boldsymbol{B}_1$). $T_{1\rho}$ can then be determined by measuring the magnetization as a function of time τ in the aligned state, since $T_{1\rho}$ is given by

$$M_x = M_0 \exp(-\tau/T_{1\rho}) \tag{5}$$

where M_x is the height of the FID following the turn-off of the variable length B_1 pulse. Figure 2 shows several different pulse sequences for measuring $T_{1\rho}$.

3 DIPOLAR RELAXATION IN THE LABORATORY FRAME ($T_{1\text{D}}$)

3.1 Homonuclear Systems

For homonuclear systems, $T_{1\text{D}}$ can be calculated by considering the fractional change in dipolar energy per jump for N jumping atoms with a time τ_c between jumps

$$\frac{1}{T_{1\text{D}}} = \frac{N}{\tau_c}\frac{E_{\text{Df}} - E_{\text{Di}}}{E_{\text{Di}}} = \frac{N}{\tau_c}\frac{\langle\hat{\mathcal{H}}_{\text{Df}}^{(0)}\rangle - \langle\hat{\mathcal{H}}_{\text{Di}}^{(0)}\rangle}{\langle\hat{\mathcal{H}}_{\text{Di}}^{(0)}\rangle} \tag{6}$$

where E_{Di} and E_{Df} are, respectively, the dipolar energies before and after a jump. $\hat{\mathcal{H}}_{\text{Di}}^{(0)}$ and $\hat{\mathcal{H}}_{\text{Df}}^{(0)}$ are, similarly, the secular parts of the dipolar Hamiltonian (i.e., those parts that commute with the Zeeman Hamiltonian), and the angle brackets indicate expectation values. If the system can be characterized by a spin temperature before each jump, the expectation values can be expressed as traces of the quantum mechanical operators, and the following expression is obtained in the high-temperature approximation:[2]

$$\frac{1}{T_{1\text{D}}} = \frac{N}{\tau_c}\frac{\text{Tr}(\hat{\mathcal{H}}_{\text{Di}}^{(0)})^2 - \text{Tr}(\hat{\mathcal{H}}_{\text{Di}}^{(0)}\hat{\mathcal{H}}_{\text{Df}}^{(0)})}{\text{Tr}(\hat{\mathcal{H}}_{\text{Di}}^{(0)})^2} \tag{7}$$

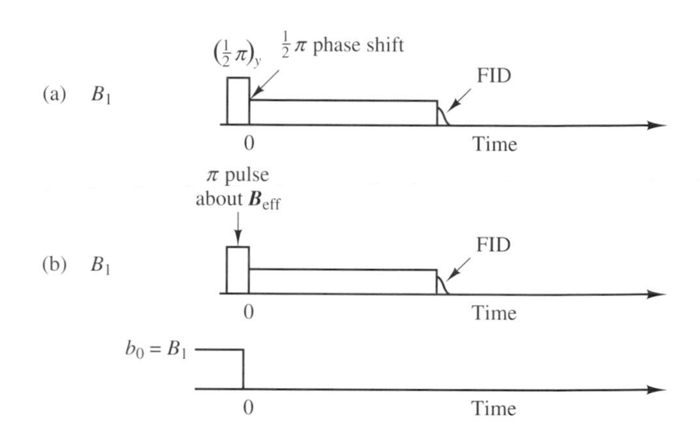

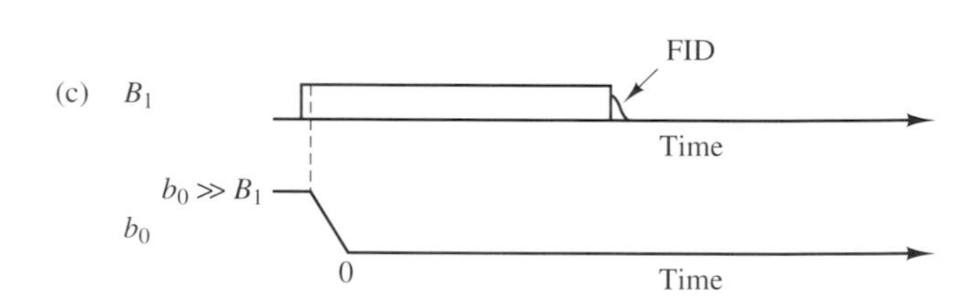

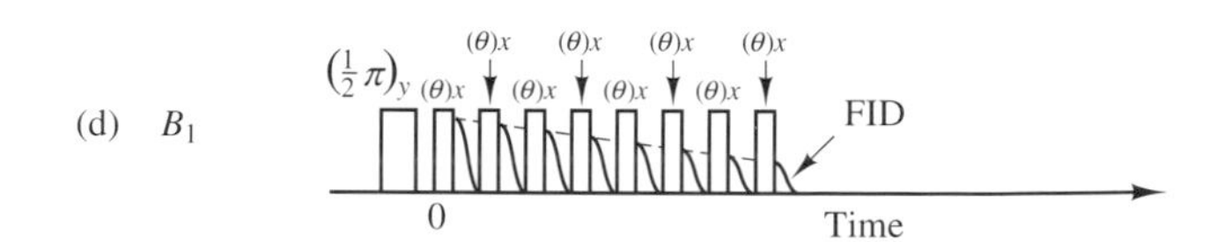

Figure 2 Pulse sequences for measuring $T_{1\rho}$. (a), (b), and (c) all have a spin locking sequence followed by a long rf pulse. The spin locking in (a) consists of a $\frac{1}{2}\pi$ pulse followed by a $\frac{1}{2}\pi$ phase shift. In (b), spin locking is achieved by a π pulse about $\boldsymbol{B}_{\text{eff}}$ (at an angle of 45° with respect to the z direction). In (c), $\boldsymbol{B}_1$ is turned on when the z field (or frequency) is far off resonance, and spin locking is achieved as a result of an adiabatic return to resonance. In (d), spin locking is achieved, as in (a), by a $\frac{1}{2}\pi$ pulse followed by a $\frac{1}{2}\pi$ phase shift; $T_{1\rho}$ is measured by monitoring the FID following short pulses of duration θ. (Reproduced with modifications by permission of Academic Press from D. C. Ailion, in 'Methods of Experimental Physics', ed. J. N. Mundy, S. J. Rothman, M. J. Fluss, and L. C. Smedskjaer, Academic Press, Orlando, 1983, Vol. 21, Chap. 6)

This equation can be rewritten as

$$\frac{1}{T_{1D}} = \frac{2(1-p)}{\tau_c} \tag{8}$$

where $1 - p$ is a calculable geometrical factor of order unity that characterizes the fractional change in energy resulting from a diffusion jump. Because of this factor, T_{1D} shows a 10–20% dependence on the orientation of the field $\boldsymbol{B}_0$ relative to the crystal axes, and, furthermore, this anisotropy depends on diffusion mechanism.[15] We also note that T_{1D} is of order τ_c, as we have seen intuitively. Diffusion can be studied by measuring either T_{1D} or $T_{1\rho}$, provided the relaxation rate due to diffusion is greater than the relaxation rate due to other T_1 processes. The condition for observability of diffusion effects is then that $\tau_c < T_1$. Since T_1 values are typically 10^{-3}–10^3 s, this condition is much less stringent than the condition for motional narrowing ($\tau_c < T_2$).

We should note that the strong collision formula, equation (8), can be used to obtain τ_c from the measured relaxation time, assuming $\tau_c >> T_2$. BPP-type theories can be used to determine τ_c in the motionally narrowed region where $\tau_c << T_2$. In the vicinity of the T_{1D} minimum, neither theory is valid.

3.2 Heteronuclear Systems

A problem arises in the interpretation of T_{1D} measurements for heteronuclear systems in which one nuclear species is diffusing and the others are stationary. For simplicity, consider a system containing two species of comparable abundance and gyromagnetic ratio, whose spin quantum numbers are labeled *I* and *S*. Assume further that the *I* and *S* spins are in strong contact. (A good example would be the alkali halide LiF, in which every lithium atom is surrounded by fluorines, and vice versa.) The problem is that if the *I* spins are irradiated and T_{1D} is measured, the result will be the same as when the *S* spins are irradiated and their T_{1D} measured, independent of which spin is actually diffusing.[16] Hence, from a T_{1D} measurement alone, it would not be possible in this case to determine which species is diffusing. This feature arises because there is normally strong coupling between parts of the dipolar interaction involving different species (i.e., the *I–I* and *S–S* dipolar terms are strongly coupled through the *I–S* interaction). So, if an *S* spin undergoes a jump, there will be rapid cross relaxation between *I–I* and *S–S* terms, which will result in the local heating being transferred rapidly to the entire spin system. Thus, in the strong collision limit, the entire dipolar reservoir is describable by a single temperature prior to a diffusion jump, but either spin species alone is not.

Fortunately, there is a way to determine in a T_{1D} experiment which spin species is diffusing for the case where the *S* spins are weakly magnetic (small γ_S) compared with the more strongly magnetic *I* spins (large γ_I). If the relaxation is due to the motion of *S* spins, T_{1D} will have an enormous anisotropy, whereas for diffusion of *I* spins the anisotropy will be very small [arising only from the $1 - p$ factor in equation (8)]. A more general strong collision theory,[17] appropriate for heteronuclear systems, shows that, for the case in which the strongly magnetic *I* spins are diffusing, only the *I–I* interaction terms need be kept. The resulting expression for $1/T_{1D}$ is similar to equation (8), only with p replaced by p_{II} and τ_c by τ_I:

$$\frac{1}{T_{1D}} = \frac{2(1-p_{II})}{\tau_I} \tag{9}$$

However, if T_{1D} is due to the diffusion of weakly magnetic *S* spins then the diffusion produces significant changes in the *I–S* interaction, with the result that

$$\frac{1}{T_{1D}} = \frac{1-p_{SI}}{\tau_S}\frac{B_{DIS}^2}{B_{DII}^2} \tag{10}$$

where B_{DIS} and B_{DII} are the *I–S* and *I–I* contributions to the dipolar local fields and τ_S is the time between jumps of the *S* spins. The precise definitions of these local fields and the parameters p_{II} and p_{SI} are given by Stokes and Ailion.[17] Equation (10) predicts a strong anisotropy, arising from the factor B_{DIS}^2/B_{DII}^2, which is absent for motion by *I* atoms. The observation of this anisotropy would confirm that the diffusion arises from the motion of weakly magnetic spins. The experimental verification of this idea is shown in Figure 7 in Section 5 below.

3.3 Experimental Techniques for Measuring T_{1D}

There are basically two commonly used methods for measuring T_{1D}, and each involves transferring order to the dipolar system and then monitoring the decrease in dipolar order arising from dipolar relaxation. The first technique consists in following a spin locking sequence by an adiabatic demagnetization–remagnetization cycle of the rf field.[5] The second consists in creating a state of dipolar order by a pair of phase-shifted pulses and then using a third pulse to monitor the decrease in dipolar order due to the relaxation.[14] Both are shown in Figure 3.

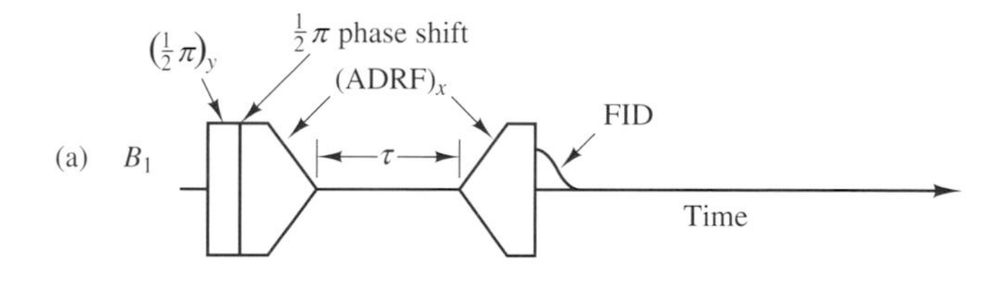

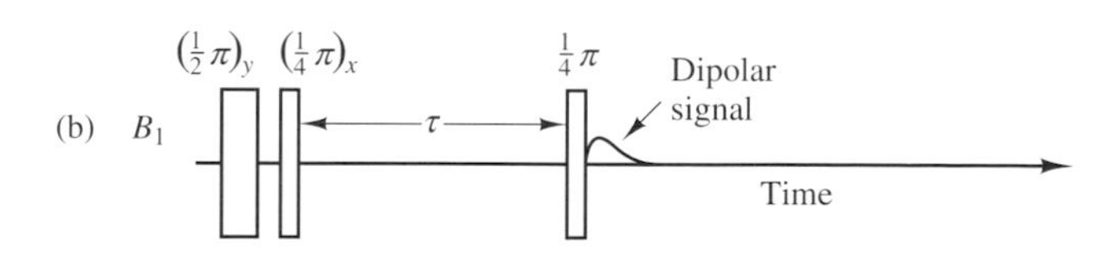

Figure 3 Pulse sequences for measuring T_{1D}. (a) The ADRF sequence showing demagnetization of B_1 following spin locking. (b) The phase-shifted pulse pair sequence. The first two pulses are separated by a time that is approximately T_2 for maximum signal. (Reproduced with modifications by permission of Academic Press from D. C. Ailion, in 'Methods of Experimental Physics', ed. J. N. Mundy, S. J. Rothman, M. J. Fluss, and L. C. Smedskjaer, Academic Press, Orlando, 1983, Vol. 21, Chap. 6)

For References see p. 4881

3.3.1 Adiabatic Demagnetization in the Rotating Frame (ADRF)

This method, illustrated in Figure 3(a), is similar to the methods for measuring $T_{1\rho}$, shown in Figures 2(a)–(c), except that spin locking is now followed by an adiabatic demagnetization of B_1 to zero followed a variable time later by an adiabatic remagnetization, so that the NMR signal can be measured. The adiabatic demagnetization transfers order from Zeeman order in the rotating frame to dipolar order (in alignment of the spins along the local fields). The subsequent remagnetization converts the remaining dipolar order (not destroyed by the dipolar relaxation) back to Zeeman order, where it is observed in the form of an FID following the turn-off of the rf pulse. Dipolar relaxation results in a reduction in this FID.

One limitation of this technique arises from the conflicting requirements that the B_1 sweep be adiabatic yet that it also be completed in a time short compared with T_{1D}. This limitation makes it difficult to measure a T_{1D} less than about 0.1–1 ms.

3.3.2 Phase-Shifted Pulse Pair

This method, shown in Figure 3(b), avoids the difficulty in measuring a short T_{1D} inherent in ADRF. In this technique, a state of dipolar order is created by applying two closely spaced pulses that are 90° out of phase. The dipolar order is observed by applying a third pulse of arbitrary phase in the (x,y) plane. The dipolar signal that appears after the third pulse is proportional to the time derivative of the FID.

A major disadvantage of this technique is that the maximum efficiency for transfer of order is only about 56%, in contrast to the 100% efficiency theoretically obtainable with ADRF.

In summary, the ADRF technique is superior when measuring longer relaxation times or when $T_{1\rho}$ measurements are also desired, while the phase-shifted pulse pair method is better for very short T_{1D} measurement.

4 DIPOLAR RELAXATION IN THE ROTATING FRAME ($T_{1D\rho}$)

The last paragraph of Section 3.2 described the possibility of using a measurement of the anisotropy of T_{1D} in a single crystal to verify that a weakly magnetic spin species is diffusing. There are two disadvantages of this approach:

1. single crystals are not always available;
2. even though S spins are diffusing, they may not be dominating T_{1D} if they are sufficiently weakly magnetic or low in abundance.

This second feature can be seen in equation (10), since, for $B_{DIS} << B_{DII}$, we have $T_{1D} >> \tau_S$ for weak (S) spin motion. In contrast, for strong (I) spin motion, $T_{1D} \approx \tau_I$. These results arise from the fact that the heat capacities of the terms involving S spins are much smaller than the I–I heat capacities. Accordingly, it may be very difficult, or even impossible, in many cases to study the diffusion of sufficiently weak S spins in a T_{1D} experiment.

In order to determine the diffusing species in a slow motion experiment, it is necessary to identify a parameter that is sensitive to which species is diffusing. If this parameter can be varied by the experimenter then the diffusing species can be identified and studied. Furthermore, for the case of diffusion by spins that are so weakly magnetic as to preclude their study through direct T_{1D} measurement, it is desirable to vary the parameter so as to enhance the temperature change of the entire system resulting from the weak spins' motions. A way to perform such experiments is to observe the dipolar relaxation in the rotating frame of one of the species, say, the I spins. In this frame, only part of the original dipolar Hamiltonian will be secular. The symbol $T_{1D\rho}$ is used for the relaxation time of this secular part of the rotating-frame Hamiltonian, in analogy to T_{1D}, which is the relaxation time for the secular part of the laboratory-frame dipolar Hamiltonian. $T_{1D\rho}$ may be considered to be the dipolar relaxation time in the rotating frame, whereas T_{1D} is the dipolar relaxation time in the laboratory frame. In contrast to T_{1D}, $T_{1D\rho}$ may depend on the angle θ_I between $\boldsymbol{B}_{\text{eff}}$ and $\boldsymbol{B}_0$ in the rotating reference frame. For diffusion by the strong I spins, $T_{1D\rho} = T_{1D}$ and is described by equation (9) for all θ_I except near the magic angle θ_m (defined by $\cos^2\theta_m = \frac{1}{3}$), where there is a small change.[17] However, for diffusion by weak S spins, the following expression is obtained for $T_{1D\rho}$:[17]

$$\frac{1}{T_{1D\rho}(\theta_I)} = \frac{1}{\tau_S}\,\frac{\cos^2\theta_I B_{DIS}^2(1-p_{SI}) + 2B_{DSS}^2(1-p_{SS})}{\left[\frac{1}{2}(3\cos^2\theta_I - 1)^2\right]^2 B_{DII}^2 + \cos^2\theta_I B_{DIS}^2 + B_{DSS}^2} \tag{11}$$

where $1 - p_{SI}$ and $1 - p_{SS}$ are geometrical factors of order unity (analogous to p of the conventional strong collision theory). In contrast to the result obtained in the case of diffusion by the strong I spins, this expression predicts a very strong dependence of $T_{1D\rho}$ on θ_I, which can be observed experimentally, using the pulse sequence described in the next section.

4.1 Experimental Technique for Measuring $T_{1D\rho}$

Figure 4 shows a pulse sequence that can be used to measure $T_{1D\rho}$.[17] A spin locking sequence followed by an ADRF is applied on resonance to the I spins. This step transfers Zeeman order, which was originally along $\boldsymbol{B}_0$, to dipolar order. A large off-resonance pulse is then applied, which does not change the dipolar order but rather changes its reference frame from the laboratory frame to the rotating frame. After establishing a spin temperature in the rotating frame, the frequency of the rf pulse is then reduced adiabatically toward resonance, thereby reducing B_{eff} and changing θ_I. By reversing this sequence, the remaining order is transferred back to laboratory-frame Zeeman order, where it appears as an FID following

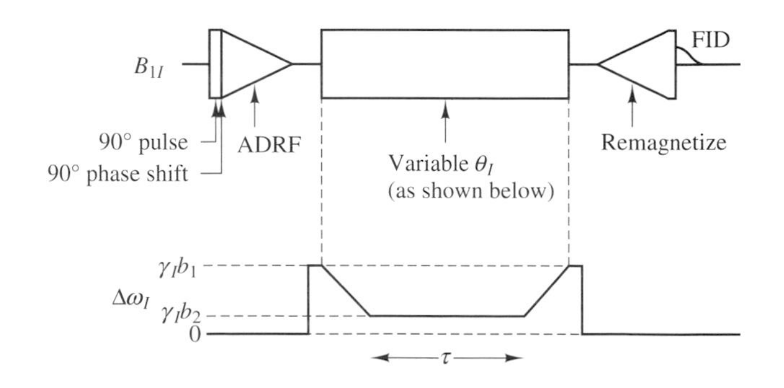

Figure 4 Pulse sequence for measuring $T_{1D\rho}$. The lower part of the figure shows the variation of $\Delta\omega_I = \gamma_I B_0 - \omega_I$, the off-resonance frequency of B_{1I}. (Reproduced with modifications by permission of the American Physical Society from H. T. Stokes and D. C. Ailion, *Phys. Rev. B*, 1978, **18**, 141)

the turn-off of the rf pulse. If one increases the time τ in the state at θ_I, the magnitude of the resulting FID will be reduced, owing to $T_{1D\rho}$ relaxation. A plot of magnetization versus τ on a semilogarithmic plot will yield a straight line of slope $-1/T_{1D\rho}(\theta_I)$. To vary θ_I, one merely needs to vary the frequency of the pulse (and thus b_2) in the variable time interval τ.

5 EXPERIMENTAL RESULTS USING ULTRASLOW MOTION TECHNIQUES

Ultraslow motions have been widely studied in the last 30 years, generally via $T_{1\rho}$ and T_{1D} measurements. It would be impossible in this article to summarize all the applications. Instead, just a few examples will be given to confirm various features of the previous discussion and illustrate the kinds of additional information that may be obtained from such measurements.

Figure 5 shows the original $T_{1\rho}$ data of Ailion and Slichter[18] in metallic lithium. The principal feature is that the $T_{1\rho}$ minimum is symmetric and occurs at the onset of motional narrowing. Furthermore, the minimum value of $T_{1\rho}$ is within an order of magnitude of T_2, as expected from the strong collision theory. Figure 6 shows a plot of the atomic jump time obtained from the experimental $T_{1\rho}$ values, using the strong collision theory. As can be seen, the data exhibit Arrhenius behavior over eight orders of magnitude, and correspond to extremely slow diffusion (of the order of one atomic jump per second) at the lowest temperatures.

Figure 7 shows the verification of the prediction of a very strong anisotropy in T_{1D} described in Section 3.2 for a system having diffusion dominated by weakly magnetic spins. These measurements were performed on a single crystal of KF doped with 0.1% CaF_2. In this case, mobile potassium vacancies are created to compensate for the extra charge of the Ca^{2+}. We thus have a case in which the diffusion of weak spins (potassium) dominates T_{1D} (of fluorine). The solid curve is the calculated anisotropy of the local field factor of equation (10).

Figure 8 shows experimental verification of the strong collision theory for $T_{1D\rho}$. The sample used in this experiment is the same KF:0.1% CaF_2 single crystal used for the data of Figure 7. The curve is calculated from equation (11). All the strong dependence of $T_{1D\rho}$ on θ_I (described earlier) appears, thereby indicating that the weak S spins (in this case, ^{39}K) are diffusing rather than the strong I spins (^{19}F here).

A possibly important potential application of this $T_{1D\rho}$ technique is the study of impurity diffusion. In this case, the S spins are weak because their concentration N_S is small (but γ_S may be large). As in the case of small γ_S, diffusion by the

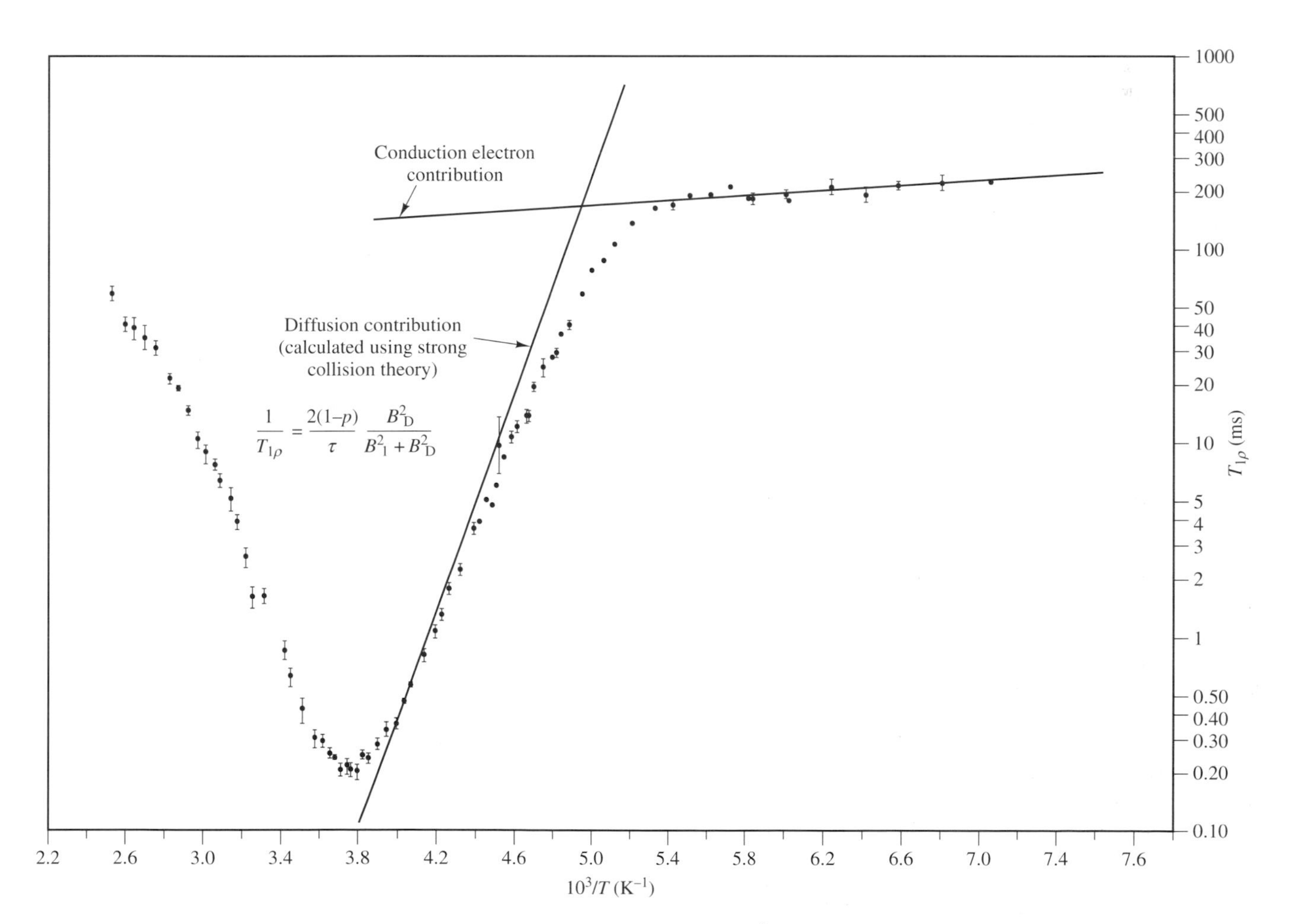

Figure 5 $T_{1\rho}$ versus reciprocal temperature in metallic lithium, with $B_1 = 1.3 \times 10^{-4}$ T. (Reproduced with modifications by permission of the American Physical Society from D. C. Ailion and, C. P. Slichter, *Phys. Rev.*, 1965, **137**, A235)

For References see p. 4881

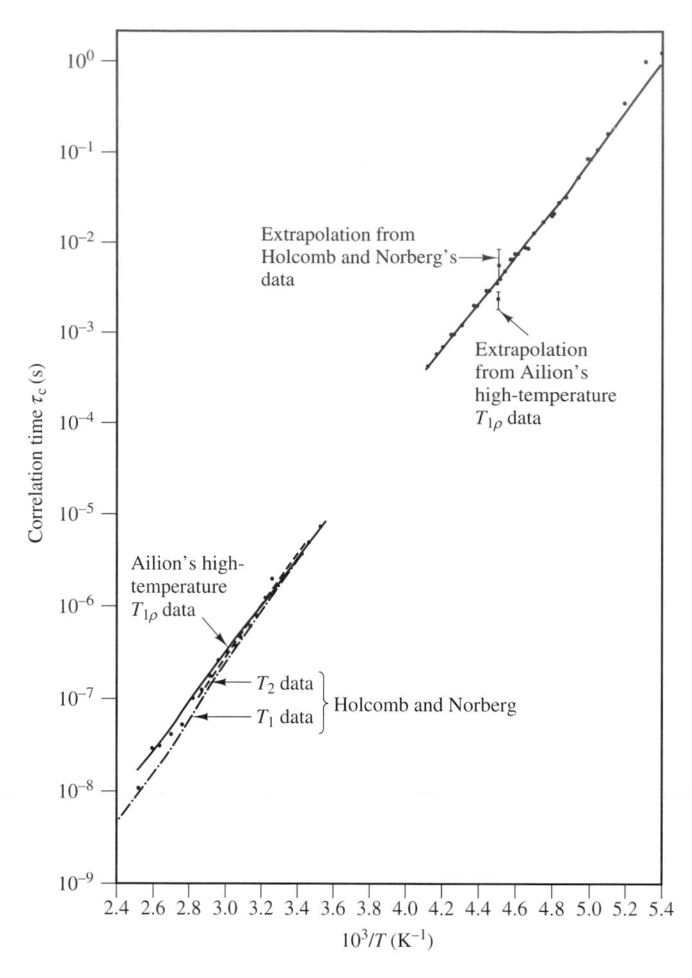

Figure 6 τ_c versus reciprocal temperature in metallic lithium. The experimental points were obtained from the data of Figure 5 using the strong collision theory in the low-temperature region and the weak collision theory in the high-temperature region. The 'gap' in the middle corresponds to the region around the $T_{1\rho}$ minimum where neither theory is valid. In the high temperature region, values of τ_c obtained from Holcomb and Norberg[21] were also plotted. (Reproduced with modifications by permission of the American Physical Society from D. C. Ailion and C. P. Slichter, *Phys. Rev.*, 1965, **137**, A235)

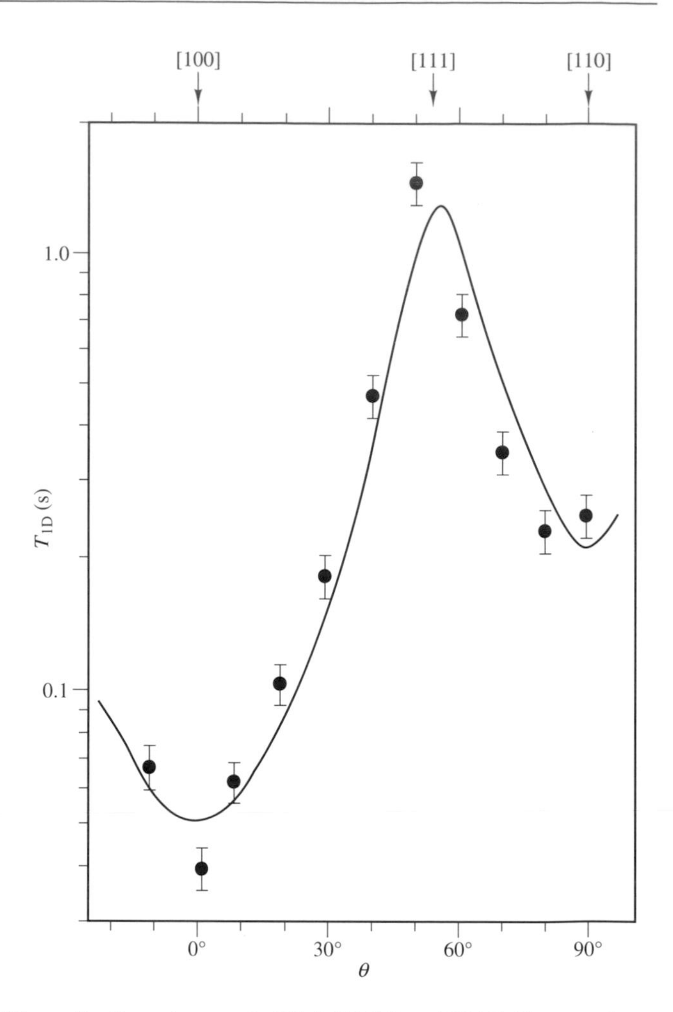

Figure 7 T_{1D} anisotropy in KF:0.1% CaF_2 at 227 °C. The crystal was rotated about its [110] axis. The curve is the calculated anisotropy of $B^2_{D//}/B^2_{DIS}$. (Reproduced by permission of the American Physical Society from H. T. Stokes and D. C. Ailion, *Phys. Rev. B*, 1978, **18**, 141)

weak S spins should again be characterized by a sharp dip in $T_{1D\rho}$ at the magic angle. For this application to be fruitful, an abundant, easily observable I spin species must exist and be in good thermal contact with the S spins. However, it has been shown[19] that nonadiabatic effects will also give rise to a dip in $T_{1D\rho}$ at the magic angle, which could mask the effect of the S spins' diffusion. Nevertheless, by appropriately modifying the 'adiabatic' sweeps of Figure 4, these effects can be minimized.[19]

An interesting new application of ultraslow motion NMR has been developed[20] to study the ultraslow hopping of selenium atoms between two sites in crystalline Se having well-resolved separate NMR lines. By selective excitation, one of the lines is saturated. Then, as the ^{77}Se atoms jump between the two sites, the magnetization of the previously saturated site builds up and the magnetization of the previously unsaturated site decreases. By measuring the time constant for these decays, the jump time τ_c and the diffusion constant D can both be directly determined. Excellent agreement was found between D measured by this NMR approach and D measured by radioactive tracers. As with the T_{1D} measurements, this technique will be able to observe jump times limited by T_1 for these specialized systems.

6 CONCLUSIONS

The most obvious advantage of the slow motion methods is to extend downward in temperature the range over which atomic diffusion can be studied. As a result, diffusion mechanisms occurring only at lower temperatures can be observed. Moreover, since the effect of diffusion on the dipolar relaxation time will be much greater than its effect on the spin–lattice relaxation time, going to zero field magnifies the relaxation effects, thereby making possible the observation of diffusion whose relaxation effects might otherwise be too weak to be observable. A measurement of the frequency dependence of the relaxation time can determine the relaxation mechanism and can check the validity of the theory used, such as the assumption of an exponential correlation function, underlying equation

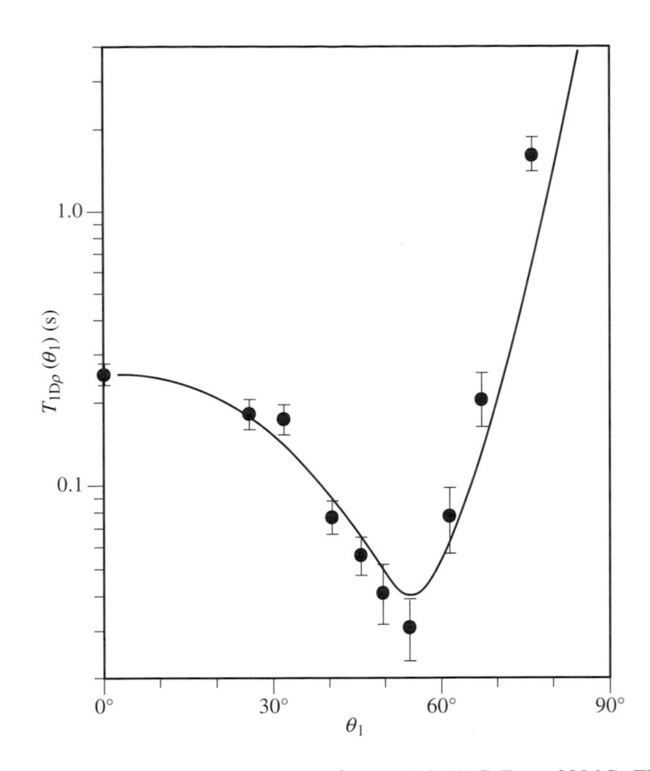

Figure 8 $T_{1D\rho}$ as a function of θ_l in KF:0.1% CaF_2 at 200 °C. The curve was calculated from equation (11). Note that the data point at $\theta_l = 0$ is T_{1D}. (Reproduced by permission of the American Physical Society from H. T. Stokes and D. C. Ailion, *Phys. Rev. B*, 1978, **18**, 141)

(1). $T_{1\rho}$ and T_{1D} measurements greatly increase the frequency range available, since they are effectively very low frequency measurements (of order 10^3 Hz), in contrast to typical T_1 measurements (of order 10^7 Hz). Finally, by having a wider temperature range available, it is possible to observe the onset of other diffusion mechanisms having different activation energies. $T_{1D\rho}$ measurements can be used to enhance the NMR effects arising from the ultraslow motions of weakly magnetic nuclei and, possibly, impurities.

7 RELATED ARTICLES

Brownian Motion & Correlation Times; Diffusion Measurements by Magnetic Field Gradient Methods; Diffusion in Rare Gas Solids; Diffusion in Solids; Double Resonance; Field Cycling Experiments; Low Spin Temperature NMR; Magnetization Transfer between Water & Macromolecules in Proton MRI; Relaxation Theory: Density Matrix Formulation; Rotating Frame Spin–Lattice Relaxation Off-Resonance; Slow & Ultraslow Motions in Biology; Zero Field NMR.

8 REFERENCES

1. N. Bloembergen. E. M. Purcell, and R. V. Pound, *Phys. Rev.*, 1948, **73**, 679.
2. C. P. Slichter, 'Principles of Magnetic Resonance', 3rd edn., Springer-Verlag, Berlin, 1990.
3. H. Y. Carr and E. M. Purcell, *Phys. Rev.*, 1954, **94**, 630; S. Meiboom and D. Gill, *Rev. Sci. Instrum.*, 1958, **29**, 6881.
4. D. C. Ailion and C. P. Slichter, *Phys. Rev. Lett.*, 1964, **12**, 168.
5. D. C. Ailion, in 'Advances in Magnetic Resonance', ed. J. S. Waugh, Academic Press, New York, 1971, Vol. 5, Chap. 4.
6. D. C. Ailion, in 'Methods of Experimental Physics', ed. J. N. Mundy, S. J. Rothman, M. J. Fluss, and L. C. Smedskjaer, Academic Press, Orlando, 1983, Vol. 21, Chap. 6.
7. D. C. Ailion, in 'Proceedings of 10th Ampère Summer School and Symposium', ed. R. Blinc, M. Vilfan, and J. Slak, J. Stefan Institute, Ljubljana, 1988, p. 47.
8. L. C. Hebel and C. P. Slichter, *Phys. Rev.*, 1959, **113**, 1504; A. G. Anderson and A. G. Redfield, *Phys. Rev.*, 1959, **116**, 583.
9. F. Noack, in 'Progress in Nuclear Magnetic Resonance Spectroscopy', ed. J. W. Emsley, J. Feeney, and L. H. Sutcliffe, Pergamon Press, Oxford, 1986, Vol. 18, p. 171.
10. A. G. Redfield, *Phys. Rev.*, 1955, **98**, 1787.
11. D. C. Look and I. J. Lowe, *J. Chem. Phys.*, 1966, **46**, 2995.
12. C. P. Slichter and D. C. Ailion, *Phys. Rev.*, 1964, **135**, A1099.
13. C. P. Slichter and W. C. Holton, *Phys. Rev.*, 1961, **122**, 1701.
14. J. Jeener and P. Broekaert, *Phys. Rev.*, 1967, **157**, 232.
15. D. C. Ailion and P. Ho, *Phys. Rev.*, 1968, **168**, 662.
16. H. T. Stokes and D. C. Ailion, *Phys. Rev. B*, 1977, **16**, 4746.
17. H. T. Stokes and D. C. Ailion, *Phys. Rev. B*, 1978, **18**, 141.
18. D. C. Ailion and C. P. Slichter, *Phys. Rev.*, 1965, **137**, A235.
19. B. Günther and D. C. Ailion, *J. Magn. Reson.*, 1985, **61**, 482.
20. B. Günther and O. Kanert, *Phys. Rev. B*, 1985, **31**, 20.
21. D. F. Holcomb and R. F. Norburg, *Phys. Rev.*, 1955, **98**, 1074.

Biographical Sketch

David C. Ailion. *b* 1937. A.B., 1956, M.S., 1958, Ph.D., 1964, Physics, University of Illinois, under supervision of C. P. Slichter. Faculty member (currently Professor of Physics) University of Utah, 1964–present. Fellow of American Physical Society, Humboldt Senior Research Award (1994–6), University of Utah Distinguished Research Award (1995), NIH Special Fellow (1971–2), Guest Member of AMPERE Committee. Approx. 130 publications. Current research specialties: magnetic resonance imaging of lungs, NMR of partially disordered systems (incommensurate insulators and glasses).

Ultrasonic Irradiation & NMR

John Homer

Aston University, Birmingham, UK

1 INTRODUCTION

Although the relationship might not be apparent immediately, NMR spectroscopy and sound are inextricably linked. Their marriage is secured, most evidently in solids, by spin–lattice relaxation, which itself controls the utility, and limitations, of the NMR technique. While we direct quanta of electromagnetic energy at nuclear spin systems to make them do our bidding, Nature cleverly generates its own missiles, phonons. By bouncing them back and forth between the nuclear spins and their surrounding lattice, Nature can facilitate the necessary initial polarization of spins in our magnetic fields, and then undo our experimental meddling by using them to return perturbed systems to natural equilibrium.

The state of a spin system may be characterized by its so-called spin temperature, and the magnitude of this in relation to the temperature of the bulk lattice can be imagined to govern the flow of thermal energy between the spins and lattice. To conceptualize the interchange mechanism, it is convenient to refer to Debye's theory of the specific heats of solids.[1] This relies on the existence of a large number of standing elastic waves (of which sound is an example) of high frequencies that are associated with thermal lattice vibrations. Each of the quantized elastic waves has a characteristic frequency, ν (= $\omega/2\pi$), and is called a phonon which carries an energy $h\nu$. It behaves as though it has momentum, and the creation or destruction of phonons corresponds with the increase or decrease of the energy of the elastic wave. The sum total of the phonon energies is the internal energy of the solid.

By treating a solid as a continuum, Debye provided a simple theory that enables the prediction of the dependence of phonon density on frequency, as exemplified in Figure 1.

Although not revealing all of the detailed features of more precise lattice theory, it serves to illustrate two important points. First, there is a cut-off frequency above which the phonon density vanishes. Second, below the cut-off frequency there is appreciable phonon density at all frequencies. The existence of an effective continuum of phonon frequencies, and energies, creates the possibility of energy transfer between spins and the lattice either at the nuclear spin Larmor frequency, when in a magnetic field, or indirectly at other frequencies.

As liquids possess thermal energy that involves the motion of molecules or clusters of molecules, they can also be considered to have phonons associated with them. The greater motion in liquids than in solids results in the creation and destruction of phonons at any particular location in the liquid.

Phonons, of course, are not limited to Nature's domain and control. In addition to thermal phonons, an acoustic wave can be considered to be a beam of phonons over which the experimenter has control. Accordingly, phonons of varying energy are at our disposal by generating sound (manifest as sinusoidally varying pressure variations) at a variety of frequencies. Sound that has a frequency above about 18 kHz is referred to as ultrasound. In the terms of the rapidly increasing body of sonochemists, ultrasound with frequencies above 1 MHz is referred to as diagnostic ultrasound, while that with lower frequencies is referred to as power ultrasound.[2a] It is ultrasound with the higher frequencies that is of particular interest to those concerned with experiments that rely on, for example, ultrasonic phonon–thermal phonon and –spin interactions.

The combination of ultrasound and NMR experiments is not new, but such experiments have been confined largely to acoustic nuclear magnetic resonance (ANMR).[3] However, experiments have been reported recently that, in general terms, use ultrasound at frequencies away from the nuclear Larmor frequency in order to modify normal NMR spectral parameters.[4–6] For example, ultrasound has been used to reduce spin–lattice relaxation times in both liquids[4] and solids,[6] and separately to effect the sonically-induced narrowing of the NMR spectra of solids (SINNMR).[5,6] In the interests of both coverage and the benefits of highlighting relevant theoretical features, some attention will be paid to ANMR before covering aspects of the more recent studies. While the former subject has been treated in theoretical detail, the understanding of the principles underlying the latter studies remains largely at the empirical level and their treatment may be deemed speculative.

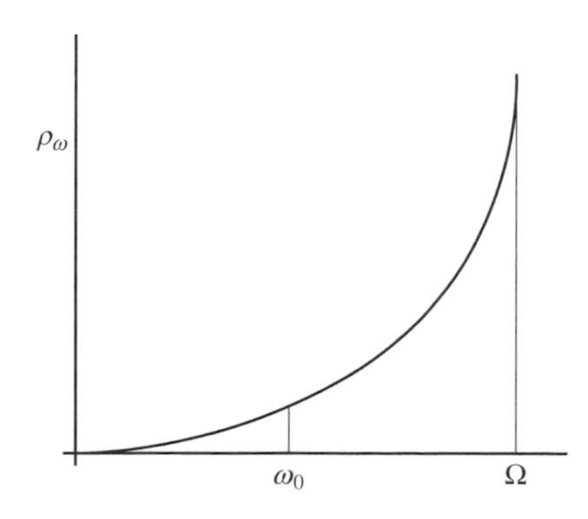

Figure 1 Variation in phonon spectral density (ρ_ω) as a function of frequency. (Adapted by permission of Academic Press (London) Ltd, from R. A. Levy, 'Principles of Solid State Physics', 1968, p. 133)

2 LATTICE PHONONS IN SOLIDS

Although there are a variety of mechanisms than can be considered to contribute to spin–lattice relaxation in solids, and different ways of treating these, attention will be focused here on the consequences of the thermal motions of constituents of the lattice.

The atomic constituents of a solid (the lattice) are in continual thermal agitation and the approximately elastic interatomic forces cause the disturbances to be propagated as elastic waves. Lattice oscillators with particular energies can be equated to there being appropriate numbers of lattice quanta or phonons in the lattice. As mentioned earlier, the principal features of lattice phonons can be deduced from the Debye treatment. Inevitably Abragam[7] has summarized the salient features of this and shows that the density of phonon states (ρ_ω) in a solid continuum that contains N atoms in a sample of volume V depends on the frequency (ω) according to

$$\rho_\omega = \frac{3V\omega^2}{2\pi^2\nu^3} \quad (1)$$

where ν is the velocity of propagation. Obviously the number of modes must be limited to the $3N$ degrees of freedom of the lattice and it is necessary to have an upper cut-off frequency (Ω) that is related to the Debye temperature (θ) by

$$k\theta = h\Omega/2\pi \quad (2)$$

The dependence of phonon density on frequency and the cut-off (typically at about 10^{13} Hz for many materials) is illustrated in Figure 1: note the approximate position corresponding to a typical nuclear spin Larmor frequency, ω_0.

3 RELAXATION TRANSITION PROBABILITIES

3.1 Solids

Bearing in mind the typical value of the cut-off frequency, Figure 1 reveals that there will be a finite phonon spectral density at the nuclear Larmor frequency (ω_0) that can, in principle, facilitate spin–relaxation transitions. These can occur by a *direct* process but phonons at other frequencies can facilitate *indirect* processes.

3.1.1 Direct

In the direct relaxation process, a spin makes a transition from one energy level to another by the emission (or absorption) of a single phonon with the resonant frequency, as illustrated in Figure 2(a). The transition probability (P_1) and the relaxation time for this process may be represented in a simplified form by

$$P_1 \sim T_1^{-1} = CB_0^2T \quad (3)$$

where C is a constant, B_0 the applied magnetic field, and T the absolute temperature. It emerges that C, being proportional to $k\Omega^{-3}$, is exceedingly small in relation to the product of the other two terms in equation (3) so that the transition probability is very low. The reason for this is that only phonons

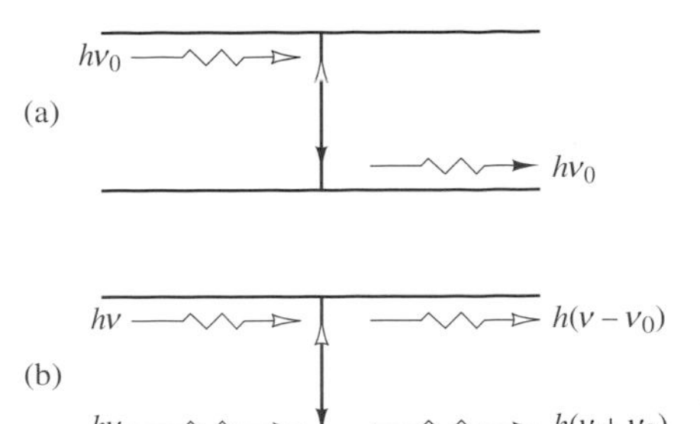

Figure 2 (a) Direct and (b) indirect relaxation transition mechanisms involving phonons. (Adapted by permission of Academic Press (Orlando), from R. T. Beyer and S. V. Letcher, 'Physical Ultrasonics', 1969, p. 290)

essentially at ω_0 contribute to the process and the spectral density of thermal energy, proportional to the number of phonons at this frequency, is very low.

3.1.2 Indirect

Of the indirect relaxation processes, those sequentially involving two phonons, or the absorption or emission of two phonons, are highly improbable[7,8] because the number of pairs of phonons whose energies combine to match $h\nu_0$ is very small. Accordingly, attention is focused on the Raman process, shown in Figure 2(b), in which the spin may change levels with the simultaneous absorption and emission of phonons. If the absorbed phonon has a frequency ν, the frequency of the emitted phonon may be either $\nu + \nu_0$ (spin emission) or $\nu - \nu_0$ (spin absorption). Obviously, by reference to Figure 1, absorption can only occur when ω lies between ω_0 and Ω. However, emission can occur for any value of ω up to Ω, and has high probability. The Raman relaxation rate is independent of ω_0 and varies as T^2 for $T \geqslant \theta$ and as T^7 for $T << \theta$. Abragam[7] shows that the ratio of the probability of this type of Raman relaxation (P_2) to that of the direct process is

$$\frac{P_2}{P_1} \approx \frac{kT\Omega^2}{m\nu^2\omega_0^2} \quad (4)$$

and accordingly $P_2 > P_1$.

3.1.3 Other Processes

Lattice vibrations can result in the modulation of the separations of nuclear spins and in the electric field gradients at nuclei. In principle, therefore, both dipolar and quadrupolar relaxation processes may be influenced by lattice vibrations. Whereas the magnetic nuclear spin–phonon coupling relaxation mechanism turns out to be negligible, the coupling of lattice vibrations with nuclear quadrupole moments can be important. In the case of quadrupole spin–lattice relaxation, both direct and Raman processes are possible, but the probability of the latter process is some 10^{10} times more probable than the former.

3.2 Liquids

Many mechanisms may be attributed to nuclear spin relaxation in liquids. Inevitably the increased molecular motion, particularly translation and rotation, is important to the charac-

For References see p. 4890

terization of the relaxation mechanisms. Although the coupling of a spin system with the lattice can in principle be described quantum mechanically, an alternative is to describe the lattice classically and consider the effects of fluctuating magnetic and electric fields. Accordingly, the enormous complexity of the subject is frequently camouflaged by simplistically representing the spin–lattice relaxation time in the form

$$1/T_1 = 1/T_1^{\text{trans}} + 1/T_1^{\text{rot}} \quad (5)$$

where T_1^{trans} and T_1^{rot} encompass contributions arising respectively from molecular translational and rotational motions that are assigned appropriate correlation times, τ. For the purposes of the following discussion this oversimplified treatment of relaxation in liquids will suffice.

The above deliberately biased treatment of selected aspects of spin–lattice relaxation leads to several interesting questions. For example: (a) if lattice phonons induce transitions of nuclear spins between allowed levels can we use sound to mimic this experimentally and detect a net NMR effect?; (b) can we use externally applied sound to modify effectively the natural phonon spectral density in a solid sample and change T_1 from its normal value?; (c) can sound be used to modulate normal molecular motion and modify T_1 values relative to their normal values in liquids?; (d) if the answer to (c) is yes, can we use sound to induce motion of small solid particles suspended in a liquid in such a way that the resulting motions resemble those of rather large molecules so that the usual dipolar and quadrupolar line broadenings from solids are reduced? Answers to these questions will now be embraced, and although not all are definitive they are included in the hope that they may stimulate work that will further enhance the already immense diagnostic capacity of NMR.

4 ACOUSTIC NUCLEAR MAGNETIC RESONANCE IN SOLIDS

If thermal phonons facilitate the net transfer of energy from a nonequilibrium spin system to the lattice, it is not unreasonable to speculate that ultrasonic phonons may be used to cause a net absorption of energy by a spin system, as proposed by Kastler[9] and Al'tshuler.[10] It is indeed possible to use ultrasound to excite vibrations in a solid lattice and concentrate energy in a narrow frequency range. If the ultrasound has the frequency ν_0 it can stimulate nuclear spin transitions by a mechanism that is directly analogous to the direct thermal relaxation process. The important difference between the natural and experimental processes stems from the fact that in the latter the spectral density at the Larmor frequency is much greater than in the former case. While this increases the rate of stimulated transitions by a factor of around 10^{11} for dipoles, it is unlikely that this is enough to make the direct process experimentally valuable in such cases. However, in the case of nuclear quadrupoles, the ultrasonic transition rates can be increased by a further factor of about 10^4, so that they can compete successfully with relaxation transitions and a net absorption of acoustic energy is observed in ANMR.

There have been many experimental demonstrations of ANMR for quadrupolar nuclei in solids. As the selection rules for allowed transitions due to quadrupole coupling are $\Delta m =$ ± 1 and ± 2 (but not $m = -\frac{1}{2} \leftrightarrow +\frac{1}{2}$), ANMR can be detected at both ν_0 and $2\nu_0$. Usually, ANMR is detected by monitoring the ultrasonic saturation of resonances by conventional NMR techniques. Indeed the first demonstration of ANMR by Proctor et al.[3] relied on this approach. Essentially, after applying a conventional 90° rf pulse, the sample was irradiated with a relatively long pulse of ultrasound which prevented complete restoration of M_z to $\boldsymbol{M}_0$ due to saturation, and the resulting magnetization, read by a subsequent 90° pulse, was found to be less than that obtained in the absence of ultrasound.

ANMR can also be observed without the use of rf irradiation, through the direct detection of the loss of acoustic energy to the spin system,[11] although this is by no means easy due to the low acoustic absorption coefficient.

In view of the experimental difficulties associated with ANMR [e.g. the sample may have to be specially prepared and a transducer (piezoelectric crystal[8]) bonded to it in order to enable coupling of the ultrasound with it], doubt may be cast on its value as being of other than academic interest. However, it should be pointed out that the penetration of ultrasound into metals is not restricted by skin depth problems[7,11] and the direct, non-rf, detection of ANMR can have value in this area of study.

5 ACOUSTIC NUCLEAR MAGNETIC RESONANCE IN LIQUIDS

The theoretical arguments for and against the possibility of ANMR in liquids are plentiful and diverse. For example, Kessel[12] argues that ANMR should not be observable for either dipolar or quadrupolar nuclei. On the other hand, Iolin[13] suggests that the reorientation of the axis of an anisotropic molecule by an acoustic wave could lead to narrow ANMR resonances in low viscosity liquids. Alekseev and Kopvillem[14] essentially reason that the variation in sound pressure should induce molecular motion and, if the spectral density at the resonant frequency due to this is greater than that corresponding to Brownian motion, ANMR should be possible even if the applied acoustic frequency is not at the Larmor frequency. An experiment on ^{1}H in an aqueous copper sulfate solution using ν_0 = 16 MHz with sound at 2.1 MHz and 72 W cm^{-2} showed a small signal amplitude modulation that may be interpreted to indicate that ANMR is possible in liquids with an acoustic wave frequency lower than ν_0.

The increasing evidence in support of the possibility of conducting ANMR in liquids prompted Homer and Patel[4] to attempt ^{14}N ANMR (ν_0 = 6.42 MHz) saturation experiments on acetonitrile and *N*,*N*-dimethylformamide. The results as presented in Figures 3(a) and (b) are similar. As the acoustic power (more precisely, reference should be made to the intensity of an acoustic beam which is the average rate of energy flux across unit area perpendicular to the direction of propagation, and here is derived from the number of watts delivered to immersed 5-mm diameter piezoelectric crystals) is increased at 1.12 MHz little change in the S/N ratio is observed. At 6 MHz there is some evidence of decreasing S/N ratio with increasing applied power. At 6.42 MHz only a very small residual ^{14}N signal is seen, even at low applied acoustic power. At 10 MHz the signal is again detected: similar results were obtained from corresponding studies of *N*,*N*-dimethylaceta-

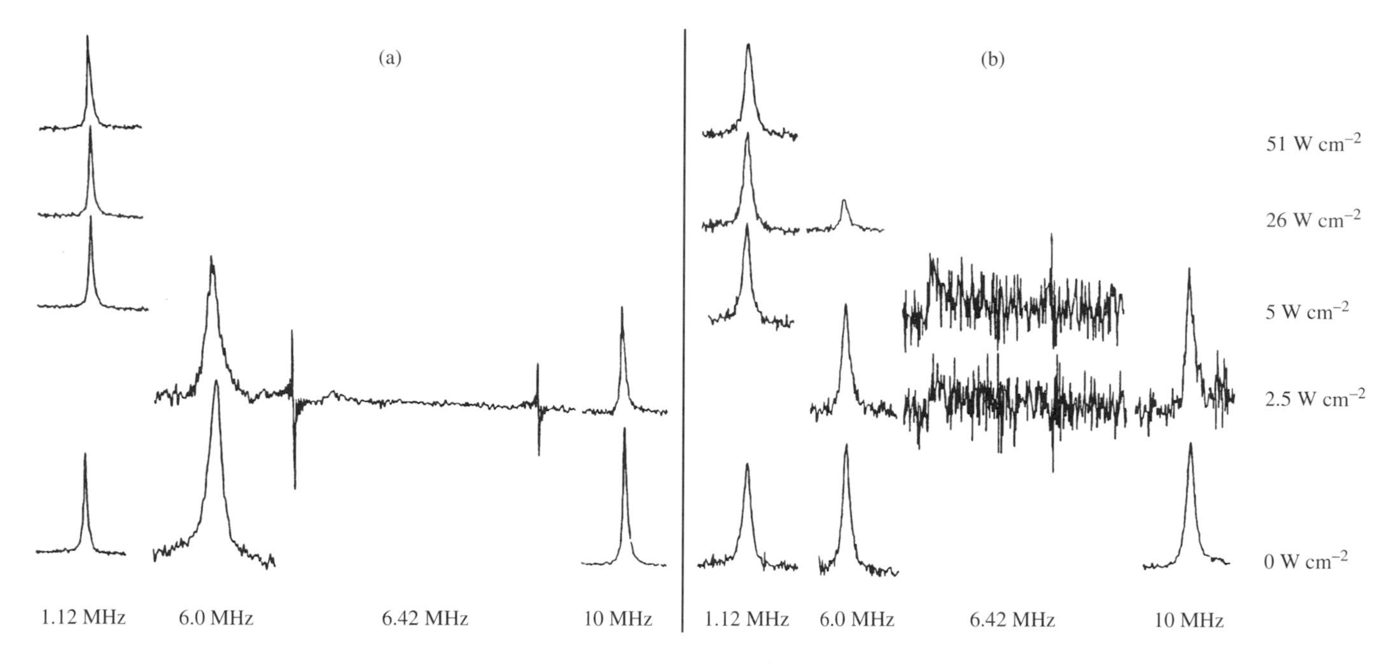

Figure 3 The effect of ultrasound at various frequencies and powers on the ^{14}N spectra of (a) acetonitrile and (b) *N,N*-dimethylformamide in equimolar mixtures with $CDCl_3$. (Adapted from S. U. Patel, Ph.D. Thesis, Aston University, 1989)

mide.[15] The implications of these experiments are that ANMR is possible for liquids. However, it must be noted that when the acoustic frequency matches the Larmor frequency, interference beats are observed that, in principle, could reduce the S/N of the ^{14}N signal.

6 ULTRASONICALLY INDUCED CHANGES TO T_1 IN LIQUIDS AND SOLIDS

Almost three decades ago, Bowen[16] noted a reduction of T_1 during ANMR studies on an aqueous colloidal sol of As_2S_3. As discussed earlier the irradiation of solids with ultrasound may lead to reductions in T_1: if this can be achieved routinely it could have a major impact on the investigation of a range of solid materials for which magnetic resonance studies normally rely on very slowly relaxing nuclei. The indications are that such reductions might be best achieved through the indirect Raman process using acoustic frequencies less than the Larmor frequency.

In the case of liquids, there is no reason why ultrasound should not be used to reduce T_1 values if sound can be used to cause unusual relative molecular motions. However, in this case predictions become far more problematical in the absence of a detailed characterization of the structures of liquids and of acoustically induced molecular motions. A pragmatic approach may prove to be the best way forward.

Homer and Patel[4,15] have conducted extensive investigations on the effects of ultrasound at various frequencies on the T_1 values of ^1H, ^{13}C, and ^{14}N in a variety of liquids and liquid mixtures. As a detailed discussion of the data is beyond the scope of the present work, only the main conclusions will be mentioned. It is important to note at the outset that the studies were conducted using immersed piezoelectric transducers, but each component of the assemblies used was checked to ensure that it had no effect on the measured relaxation times. The T_1 values were measured using the rapid DESPOT[17] (driven equilibrium single pulse observation of T_1) technique so that the effects of direct heating of the sample were minimized. No effect was observed using low-frequency ultrasound at 20 kHz, but changes to T_1 were observed in thc MHz region. Changes to T_1 were observed only for liquid mixtures, suggesting that ultrasound causes the relative motion of different molecular species and that, at least, it modifies the translational contribution to T_1. A possible reason for this observation may stem from the fact that liquid mixtures are normally inhomogeneous at the molecular level due to different interaction energies between like and unlike molecular pairs. If ultrasound modifies this local inhomogeneity, the time-average environment of a particular molecule may be changed and, therefore, the value of T_1 for a nucleus within it may also be changed. Further support for the suggestion that ultrasonically induced changes to T_1 largely originate from a translational mechanism derives from the fact that no acoustically stimulated changes to T_1 were observed[15] for ^{14}N in a conformationally rigid molecule for which intramolecular mechanisms are responsible for the relaxation characteristics of the quadrupolar nucleus. The fact that little or no changes were observed for pure liquids indicates that the effects of acoustic streaming,[8] which could enhance the effects on T_1 of molecular diffusion into and out of the NMR detector region, are unimportant. The effects of MHz ultrasound on T_1 are greater for deoxygenated than oxygenated liquids: the implication is that the ultrasonically induced changes to T_1 that were observed largely for oxygenated liquids are in fact less than the absolute effect of ultrasound on the T_1 values of the major molecular components of the mixtures. Detailed examination of the extensive results reveals that they are not totally reproducible in an absolute sense, probably due to the differing initial state of oxygenation of the samples. Nevertheless, the overall trends in the results

For References see p. 4890

Table 1 The Effects of 1.115 MHz Ultrasound on ^1H (89.56 MHz) and ^{13}C (22.5 MHz) T_1 Values for a 1:1:2 Molar Ratio of Cyclohexane/Benzene/Chloroform-*d*. (Adapted from S. U. Patel, Ph.D. Thesis, Aston University, 1989)

	T_1 Values (s)			
Power (W cm^{-2})	Aryl ^1H	Aryl ^{13}C	CH_2 ^1H	CH_2 ^{13}C
0	4.9	12.0	3.9	10.1
2.6	4.1	10.6	3.6	9.9
5.1	3.9	10.8	2.8	9.8
10.2	2.8 (−43%)	9.8 (−18%)	2.2 (−44%)	9.6 (−5%)
25.5	4.1	10.9	3.1	9.3
51.0		11.6		11.2

are reproducible. Some selected results for ^1H and ^{13}C for cyclohexane/benzene/chloroform-*d* mixtures are presented in Table 1: little change in the data is observed when the ultrasound frequency is increased to 6 MHz. These data serve to illustrate several important general features of the experiments.

As the power of the applied ultrasound is increased progressively, the T_1 values first decrease from their normal values to a minimum, whereafter they increase. The maximum decreases (approximate percentages are shown in parentheses) in T_1 are always larger for ^1H than for ^{13}C, consistent with the well-known fact that translational relative intermolecular motion influences the former more than the latter.

Whenever a process removes energy from a sound wave and returns it at an appreciable time later in the wave cycle, the acoustic energy is dissipated. Simplistically, therefore, the acoustic wave will modify normal molecular translation but processes such as molecular collisions can transfer translational energy to, for example, molecular rotational and vibrational energies. The changes to T_1, as illustrated in Table 1, may therefore also accommodate changes arising from the rotational contribution to T_1.

If resonances are normally in the extreme narrowing region with $\omega_0\tau << 1$, the only way that dipolar governed T_1 values can be artificially reduced at the lower ultrasound powers is for the spectral density at ω_0 to increase. This necessitates an increase in molecular correlation times to intermediate values where $\omega_0\tau \approx 1$ (see Figure 4). If, pro temp, applied ultrasound is assumed mainly to modify molecular translational motion, and the propagation of the acoustic wave is considered to be effected by almost instantaneous and sequential molecule to molecule interactions, it is not unreasonable to assume that the acoustically imposed translational correlation time is of the order of the inverse of the acoustic frequency (~10^{-6} s). Consequently, if nuclear relaxation rates are considered to be influenced by a normal correlation time of the order of say 10^{-10} s and an ultrasonically, and translationally, dominated correlation time of the order of 10^{-6} s is imposed in an approximately additive fashion, it is evident that the latter will have the principal effect. In this case $\omega_0\tau \approx 1$, and there will be a significant increase in the spectral density at ω_0 and result in a reduction in T_1, as observed. The fact that 20 kHz ultrasound does not change T_1 is consistent with the fact that in this case $\omega_0\tau >> 1$ and there may be little change from the normal spectral density at ω_0. The fact that the values of T_1 ultimately increase with increasing ultrasonic power is interesting and lends support to the above speculations. As the relaxation times are dipolar controlled, the only way they can increase in the extreme narrowing limit is by a reduction in the correlation time (see Figure 4) with an accompanying decrease in the spectral density at ω_0. The implication is that now a much higher proportion of the increased translational energy available on molecular collisions is transferred to rotational energy by acoustic thermal relaxation, with a net decrease in correlation time and an increase in T_1. It must, of course, be noted that direct heating of the sample by the immersed transducer, when operating at the higher powers, could also account for a reduction in correlation times and an increase in T_1. An additional point that must be recognised is that ultrasonic effects are highly nonlinear and threshold effects are observed, such as that due to the onset of cavitation. It is possible, therefore, that minima in acoustically controlled T_1 values, such as

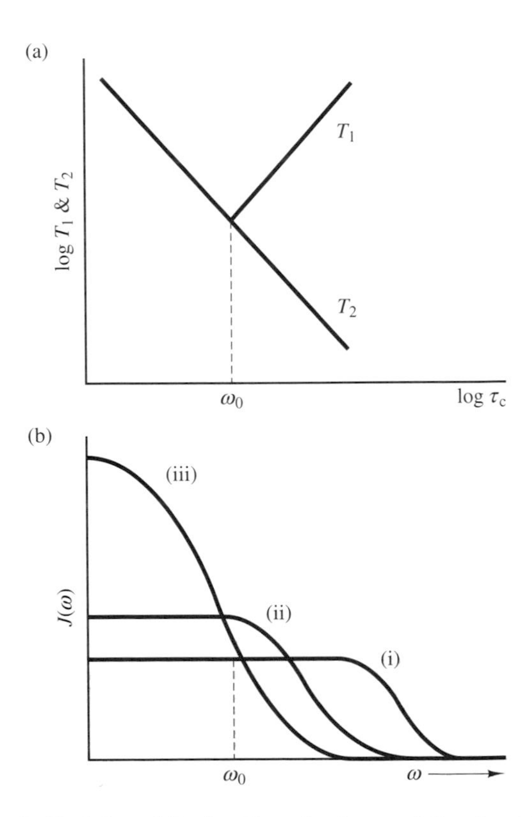

Figure 4 Variation of (a) T_1 with molecular correlation times (τ) and (b) spectral density (J_ω) with frequency (ω) [(i) short correlation times, $\omega\tau_c << 1$; (ii) intermediate correlation times, $\omega\tau_c \approx 1$; (iii) long correlation times, $\omega\tau_c >> 1$]

is evident in Table 1, could simply arise because of, for example, the onset of cavitation.

It must be emphasized that the above speculative explanations regarding the effect of ultrasound ignore the various other acoustic relaxation processes that may occur. Of interest among these is the transfer of energy from the translational mode of motion to the potential energy of the rearrangement between energetically different structures. Apparently, only one successful NMR detection of such a process has been recorded.[2b,15] This was for the classical example of *N,N*-dimethylacetamide. Progressively increasing the power of 20 kHz ultrasound was reported to result in the onset of rotation about the central N–C bond and cause the coalescence of the two ^{1}H methyl resonances that are observed at room temperature. To have produced the same effect by electrically raising the temperature would have necessitated the sample reaching about 100 °C, and normally the direct heating of a sample during ultrasonic irradiation does not result in the sample temperature exceeding about 55 °C.

7 SONICALLY INDUCED NARROWING OF THE NMR SPECTRA OF SOLIDS (SINNMR)

It is well known that the NMR spectra of mobile liquids can provide more structural information directly than can the corresponding spectra of solids. The basic reason for this is that rapid incoherent molecular motion in liquids averages out the anisotropic line-broadening effects that are present in solids. However, various experimental techniques are available to reduce line broadening for solids. Among these is the MAS NMR technique, originating from the pioneering work of Andrew et al.,[18] that involves the rapid spinning of a sample about its axis at the so-called magic angle (54.74°) to the polarizing field direction in order to remove dipolar broadening effects. Quadrupolar broadening interactions can also be minimized by rapid sample rotation about other appropriate angles, e.g., 30.56°. The simultaneous minimization of both dipolar and quadrupolar broadenings was first achieved through the elegant methods of DOR and DAS.[19a,b,20] Other relevant techniques involve rapid pulsed manipulation of magnetization vectors to place them on time average at appropriate angles to minimize the line-broadening effects, e.g., M-REV8.[21] The combination of rapid sample spinning and pulse sequences is possible through CRAMPS.[22] Essentially, all of these techniques depend on coherent motions of, or within, a sample that are induced by the experimenter. An alternative is to make solid samples undergo incoherent motion, similar to molecular motion in liquids, to effect resonance line narrowing.

Although various ways of inducing incoherent motion of solids are possible, probably the most obvious methods depend on suspending solid particles in a suitable support liquid. If the particles are extremely small, Brownian motion can be used to effect line narrowing. This was first proposed by Yesinowski,[23] and again, a decade later, by Kimura and Satoh.[24] Naturally, the preparation of sufficiently small particles capable of experiencing effective Brownian motion is not without difficulties. In order to induce incoherent motion of much larger particles, some external perturbation of them is necessary. Homer et al.[5,6] have demonstrated that 20 kHz ultrasound can be used to induce adequate incoherent motion to enable line narrowing for particles having dimensions up to 2 mm.

In principle, ultrasound can be used to induce motion of suspended particles through several mechanisms. Of these, three are particularly important and are now outlined.

The theoretical work of Dysthe[25] reveals that a single suspended anisotropic particle subject to a standing acoustic wave, with a wavelength much greater than the particle dimensions, has three possible equilibrium orientations that are not necessarily energetically equivalent. If the particle is transiently perturbed in some way from an equilibrium orientation it will rapidly reorientate to the most stable orientation, whereafter its motion will be purely translational. In a many-particle system, interparticle collisions may provide the necessary perturbation to cause incoherent particle motion.

One consequence of the passage of ultrasound through a liquid that has not been detailed so far is the production of cavities[8] in the liquid. If ultrasonic irradiation of a liquid is considered to result in the subjection of a region of the liquid to periodic compression and decompression, the latter can, in the presence of a suitable nucleation center, result in the formation of a small bubble or cavity. Subsequent decompression cycles of the acoustic wave can cause the cavity to grow in size to result in one of two types. First, a stable cavity may be formed that either resonates synchronously with the wave when the frequency of the latter is the same as the frequency of the oscillation of the cavity, or, when the two frequencies are not equal, has its oscillation frequency modulated by that of the acoustic wave. Second, an unstable cavity may be formed that can collapse violently (it is this mechanism that lies at the heart of sonochemistry and facilitates the enhancement of the rates of chemical reactions). The details of the mechanism have not been definitively characterized, but two possibilities are favored. One, the so-called 'hot spot' theory,[26] requires violent collapse of a cavity with the generation of local temperatures in excess of 2000 °C and pressures exceeding 2000 atm. The other, the electric field theory,[27] suggests that fragmentation of a cavity produces sufficient molecular ordering of the double layer at the constriction to generate intense electric fields that are capable of producing an electrical discharge (this might be expected to be modulated by magnetic fields, but preliminary studies of the effect of these on the sonochemically accelerated maleate–fumarate isomerization revealed no effect on the reaction rate in the presence and absence of a 2.3 T magnetic field[28]). Whatever the detailed mechanism of cavity collapse may be, there are two incontrovertible consequences of it. First, shock waves are generated that can cause interparticle collisions of such violence that, for certain metallic particles, fusion may occur on impact.[29] Second, if cavity collapse occurs near a solid surface, a microjet of liquid passes through the cavity and strikes the surface with speeds up to several hundreds of meters per second.[30]

If any of the above-mentioned consequences of ultrasonic irradiation of liquids causes energy dissipation on a solid along an axis tangential, or near tangential, to the surface, the particle will rotate in a fashion that will become incoherent after sufficient perturbations. Even when this is achieved it is necessary to arrange for the particles to be brought to spatial equilibrium in the NMR detector coil region. The pioneering work of

For References see p. 4890

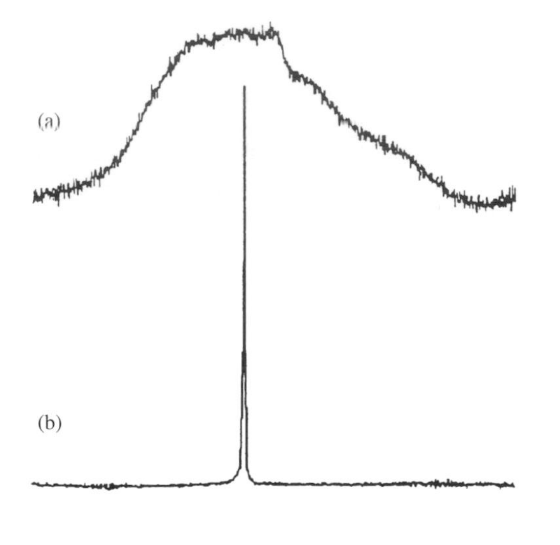

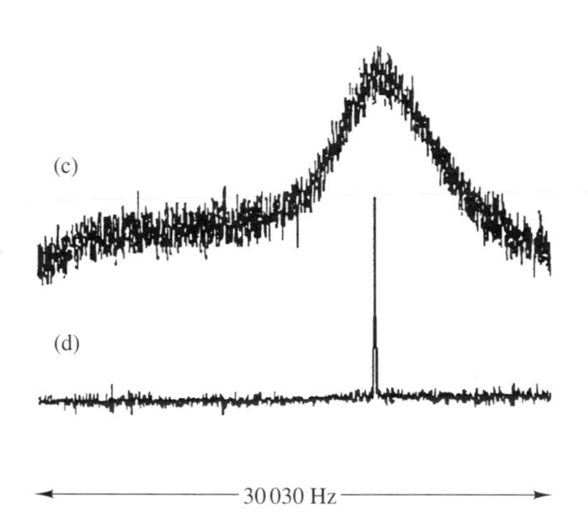

Figure 5 Sodium-23 spectra (23.64 MHz) of trisodium phosphate dodecahydrate (a) static and (b) solid suspended in a $CHCl_3/CHBr_3$ mixture and irradiated with ultrasound at 20 kHz. Plots (c) and (d) are the corresponding ^{31}P spectra (36.26 MHz). (Reproduced by permission of the Royal Society of Chemistry, from J. Homer and M. J. Howard, *J. Chem. Soc., Faraday Trans.*, 1993, **89**, 3029)

Homer et al.[5,6] was restricted by access only to an iron magnet NMR spectrometer (^{1}H frequency at 89.56 MHz) in which ultrasonic irradiation of the sample was only readily possible vertically downwards from the top of the sample. Nevertheless, by using a pseudoprogressive wave, particles were appropriately positioned by balancing the forces due to the acoustic wave, gravity, buoyancy, and viscous drag. Although the original expectation was that it would be necessary to use ultrasound with frequencies in the megahertz region, the first reproducible SINNMR spectra for ^{23}Na and ^{31}P in trisodium phosphate dodecahydrate (TSP) were obtained at 20 kHz. These are compared in Figure 5 with the corresponding static powder spectra. The full width at half-maximum height (FWHM) of the SINNMR resonances are more than two orders of magnitude smaller than the corresponding bandwidths of the static sample resonances for both the quadrupolar and dipolar nuclei, and the spectra show no spinning sidebands.

Detailed investigations on TSP revealed the following important findings. SINNMR spectra can be obtained from samples with different particle sizes, but for each size the density of the support liquid (in this case mixtures of chloroform and bromoform in order to achieve sufficiently high densities relevant to the experimental configuration used) has to be appropriately adjusted to optimize the SINNMR resonance signal-to-noise ratio (S/N). If a range of particle sizes is used, the smaller particles pass through and the larger do not reach the detector coil region, and are not detected. Table 2 presents ^{23}Na SINNMR S/N and FWHM data obtained for TSP with optimized support liquid densities and the same acoustic power and NMR pulse and acquisition parameters. The interesting feature here is that for particles with dimensions less than about 100 μm, the S/N decreases dramatically while the FWHM increases. This dimension corresponds well with a resonant cavity radius $R_r \approx 95 \mu m$ calculated using[31]

$$R_r = \frac{1}{2\pi\nu}\left(\frac{3\kappa P_0}{\rho}\right)^{1/2} \tag{6}$$

with values for the acoustic frequency (ν), a polytropic constant (κ) (varying between 1 and the ratio of the specific heats), the ambient liquid pressure (P_0), and the liquid density (ρ) appropriate to the support media used. As indicated earlier, microjet action is only effective when the particle dimensions are greater than that of the resonant cavity. Taken together with the fact that SINNMR spectra were only obtained using power levels above the estimated cavity threshold of about 16 $W\,cm^{-2}$, the implications are that cavitation is important to the efficient production of SINNMR spectra at low acoustic frequencies.

Measurements of T_1 for ^{31}P in TSP were made to elucidate the motional characteristics of particles producing SINNMR spectra. A reference value of T_1 was obtained first by MAS NMR at 121.441 MHz and adjusted to give 7.1 s (incorrectly stated elsewhere[6] to be 46 s) at the SINNMR frequency of

Table 2 ^{23}Na SINNMR S/N and FWHM for Different Particle Size Ranges of Trisodium Phosphate Dodecahydrate[a]. (Reproduced by permission of the Royal Society of Chemistry, from J. Homer and M. J. Howard, *J. Chem. Soc., Faraday Trans.*, 1993, **89**, 3029)

Particle size range (μm)	^{23}Na signal-to-noise ratio	^{23}Na FWHM (Hz)
1000–500	110:1	88
500–210	110:1	103
210–105	102:1	121
105–90	20:1	161

[a]For each particle size range, the support medium density was optimized to ensure that particle displacement was between the limits of the NMR probe region. All acquisition parameters were identical.

36.26 MHz. Subsequently, T_1 was found to be 0.52 ±0.05 s for the SINNMR resonance, whereas a value of 2.1 ± 0.2 s was obtained under identical conditions for a sonicated sample constrained in an open mesh nylon bag. These data reveal two important results. First, the reduction of T_1 under conditions of MAS NMR to that obtained by acoustically irradiating a static sample is consistent with the possibility of using Raman-type processes to reduce T_1 values, as speculated on earlier. Second, the reduction in T_1 from the value obtained for the irradiated static sample to that measured for the freely moving particles must be attributed to the motion of the particles in the latter case. The two relevant values were used with the theoretical dependence of T_1 on correlation times to deduce that in the SINNMR experiment the particles had a correlation time of the order of 4×10^{-7} s. The validity of this value was substantiated by using it to predict a ^{31}P SINNMR FWHM of 32 Hz that was in close agreement with the measured value of 31 Hz. Further detailed studies of particle correlation times[32] have revealed, surprisingly, that the larger particles move more rapidly than the smaller particles. This unexpected finding has been rationalized in terms of the relative probabilities of microjet/shock wave energy dissipation at the particle surface resulting in either particle translation or rotation; the probability of induced particle rotation being higher than translation for the larger particles with dimensions greater than that of the resonant cavity. While the investigations outlined above were specifically directed to establishing that ultrasound does in fact induce the incoherent motion of suspended solid particles that is essential to the production of SINNMR spectra, the finding that ultrasound reduces T_1 in solids may have additional important benefits in the study of materials via very slowly relaxing nuclei.

It is important to observe that throughout the SINNMR studies outlined above Homer and Howard[6] paid particular attention to the possibility that sample melting, dissolution, chemical reaction, phase changes, etc. could have been responsible for the line narrowings observed. While each of these were in fact shown to be unimportant, it was found that ultrasound did stimulate effectively free rotation of species in the solid TSP lattice, but the resulting narrowed line contributed only about $\frac{1}{80}$th to the signal intensity in an optimized SINNMR experiment.

Encouraged by the implications of the investigations of TSP that SINNMR is a genuine phenomenon, further SINNMR investigations using 20 kHz ultrasound were conducted. For example, the major resonance in the ^{19}F spectrum of PTFE (polytetrafluoroethylene) was narrowed by a factor of about two. Normally, the narrowing of such strongly dipolar broadened resonances using MAS NMR is inefficient and pulse techniques such as M-REV8 have to be used. Greater success has been achieved with the ion exchange resin Amberlite, for which ^{1}H SINNMR studies resolved three separated major resonances at shifts corresponding to inflections in the otherwise broad resonance obtained from the static solid. A further particularly important set of SINNMR observations has also been made on aluminum and its alloys: the particles were protected by a resin matrix to prevent sonochemical reactions with the haloform support liquids or more probably with Cl_2, Br_2, HCl, or HBr generated by the sonication of the haloforms. Some results on ^{27}Al are presented in Figure 6. These reveal FWHMs in the range 350–500 Hz that are significantly less than the

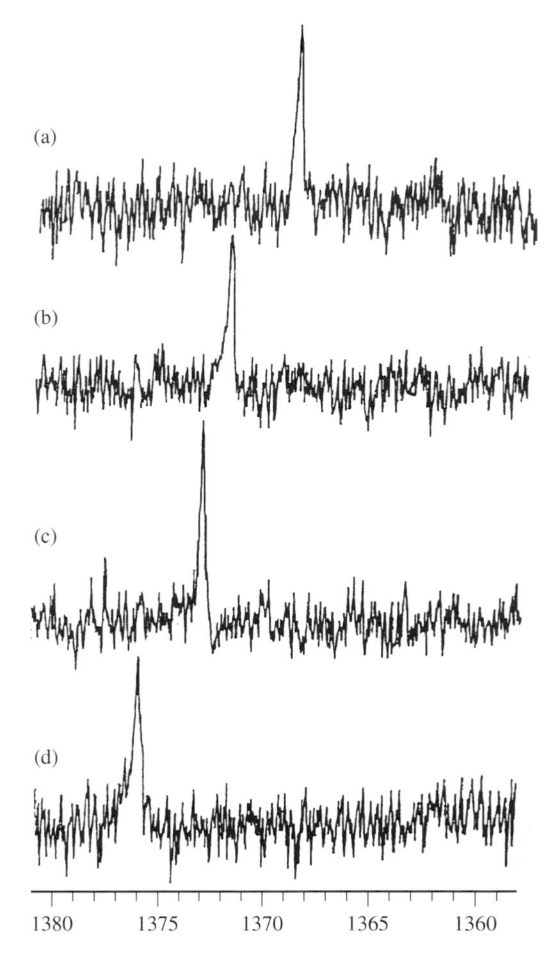

Figure 6 Aluminum-27 SINNMR spectra (23.3 MHz) of (a) Li(3%)/Al alloy, (b) Dural, (c) 99% Al, and (d) Al foil, relative to aqueous 0.05 M $AlCl_3$. (Reproduced from M. J. Howard, Ph.D. Thesis, Aston University, 1994)

value of 700 Hz obtained by Andrew et al.[33] for aluminum using MAS NMR. Of particular importance is the fact that the SINNMR chemical shifts relative to aqueous $AlCl_3$ are typically Knight shifted,[34] and could not arise from reaction products but only from the metallic state. These particular results may be viewed as definitive evidence for SINNMR being a genuine phenomenon. The small shift differences for ^{27}Al between the spectra shown in Figure 6 have not yet been characterized. While they could be due to the effect of differing alloying components, it is possible that, as Knight shifts are pressure-dependent,[34] the shift differences arise from differing phonon–Fermi electron interactions.

Superficially, the foregoing discussion might be interpreted to indicate that the utility of SINNMR is limited to the study of fairly large solid particles. However, it must be appreciated that this implication stems from the use of ultrasound at the low frequency of 20 kHz. Even in the event that SINNMR relies solely on cavitational processes, reference to equation (6) reveals that the use of higher acoustic frequencies will facilitate the study of small particles. For example, if water was used as the support liquid, the use of ultrasound at 280 kHz should permit the study of 10 μm size particles. Following this prediction, studies using acoustic frequencies in the megahertz region have very recently been initiated. Unfortunately, these

For References see p. 4890

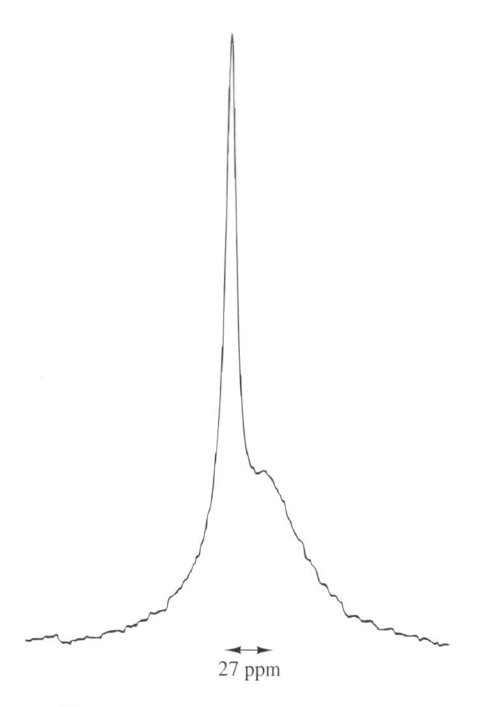

Figure 7 The ^{11}B SINNMR spectrum (28.75 MHz) of a $Na_2O/B_2O_3/Al_2O_3$ (20:70:10 mol%) glass (provided by Dr. C. Jaeger, Max Planck Institute, Mainz) from a suspension in $CHBr_3$ irradiated with 2 MHz ultrasound. (Reproduced from M. J. Howard, Ph.D. Thesis, Aston University, 1994)

have so far been restricted to the acoustic irradiation of samples from the top in a conventional high-resolution iron magnet spectrometer. Nevertheless, encouraging results have been obtained, and Figure 7 provides an example of the ^{11}B SINNMR spectrum obtained by 2 MHz ultrasonic irradiation of a finely ground boron-rich glass in bromoform. The spectrum appears to reveal the chemically-shifted isotropic BO_3 and BO_4 resonances that are known from MAS NMR studies to have respective abundances of 80% and 20%.

An interesting finding that has been obtained during the preliminary high acoustic frequency small particle size work is that nonirradiated suspensions of particles that have been produced simply by grinding in a pestle and mortar are quite capable of yielding narrowed resonances relative to those of the static powder: subsequent high acoustic frequency irradiation of the samples further reduces the linewidths. Concurrent particle size measurements have revealed no detectable submicron particles, so that, strangely, it would appear that there are no sufficiently small particles present, as required by the principles implicit in the work of Yesinowski[23] and of Kimura and Satoh,[24] to be influenced by Brownian motion.

Although SINNMR shows promise as a new tool in the armory of NMR techniques, it is still in its infancy with many problems to be overcome before it becomes a readily usable and versatile method of studying structural problems. Major advances should become possible by its development to enable the levitation of particles in cryomagnet spectrometers: work is currently under way on this aspect of the technique.

8 CONCLUSIONS

NMR studies of samples simultaneously irradiated with ultrasound appear to have promise and be worthy of further development. While ANMR of solids and possibly liquids may hold little more than academic interest, it may be that its use for the study of metals, where phonons do not experience the skin depth problems of photons, may be worthy of increased attention. The fact that ultrasound can be used to modify T_1 values in liquids could be put to use in conjunction with magnetic resonance imaging, because of the ability to direct ultrasound reasonably precisely. If the strong indications that ultrasound can be used to reduce T_1 values in solids are definitively substantiated, it will undoubtedly prove beneficial to develop the approach to facilitate studies of nuclei that have naturally long relaxation times. Probably the major use of combined ultrasound and NMR investigations will lie with SINNMR. Not only does it result in the reduction of T_1 values in solids, but it also enables significant reductions in resonance line broadenings of both dipolar and quadrupolar origins to be achieved. Additionally, it has the ability to yield isotropic chemical shifts directly from spectra of solids that show resonances, often with better resolution than those obtained by MAS techniques, without spinning sidebands.

9 RELATED ARTICLES

Aluminum-27 NMR of Solutions; Boron NMR; Brownian Motion & Correlation Times; Carbon-13 Relaxation Measurements: Organic Chemistry Applications; Chemical Shift Tensors; Colloidal Systems; CRAMPS; Double Rotation; Dynamic Angle Spinning; Fluorine-19 NMR; High Speed MAS of Half-Integer Quadrupolar Nuclei in Solids; Line Narrowing Methods in Solids; Low Spin Temperature NMR; Phosphorus-31 NMR; Quadrupolar Interactions; Quadrupolar Nuclei in Glasses; Quadrupolar Nuclei in Solids; Relaxation of Quadrupolar Nuclei Measured via Multiple Quantum Filtration; Shielding: Overview of Theoretical Methods; Sideband Analysis in Magic Angle Spinning NMR of Solids; Sodium-23 NMR.

10 REFERENCES

1. See e.g. R. A. Levy, 'Principles of Solid State Physics', Academic Press, New York, 1968.
2. (a) See, for example, G. J. Price, p. 1; (b) J. Homer, S. U. Patel, and M. J. Howard, p. 136, in 'Current Trends in Sonochemistry', ed. G. J. Price, Royal Society of Chemistry, Special Publication No. 116, Cambridge, 1992.
3. W. G. Proctor and W. H. Tanttila, *Phys. Rev.*, 1956, **101**, 1757; W. G. Proctor and W. A. Robinson, *Phys. Rev.*, 1956, **104**, 1344.
4. J. Homer and S. U. Patel, *J. Chem. Soc., Faraday Trans.*, 1990, **86**, 215.
5. J. Homer, P. McKeown, W. R. McWhinnie, S. U. Patel, and G. J. Tilstone, *J. Chem. Soc., Faraday Trans.*, 1991, **87**, 2253.
6. J. Homer and M. J. Howard, *J. Chem. Soc., Faraday Trans.*, 1993, **89**, 3029.
7. A. Abragam, 'Principles of Nuclear Magnetism', Oxford University Press, New York, 1989.
8. See, for example, R. T. Beyer and S. V. Letcher, 'Physical Ultrasonics', Academic Press, London, 1969.

9. A. Kastler, *Experientia*, 1952, **8**, 1.
10. S. A. Al'tshuler, *Zh. Eksp. Teor. Fiz.*, 1953, **24**, 681; 1954, **28**, 49 (translated in *Sov. Phys. JETP*, 1955, **1**, 29; 1955, **1**, 37); S. A. Al'tshuler, B. I. Kochelaev, and A. M. Leuslin, *Usp. Fiz. Nauk*, 1961, **75**, 459 (translated in *Sov. Phys. Usp.*, 1962, **4**, 880).
11. D. I. Bolef and M. Menes, *Phys. Rev.*, 1958, **109**, 218; 1959, **114**, 1441.
12. A. R. Kessel, *Sov. Phys. Acoust.*, 1971, **16**, 425, 'Effect of Molecular Motion on Acoustic Resonance', (translated from *Akust. Zh.*, 1970, **16**, 497).
13. E. M. Iolin, *J. Phys. C. Solid State Phys.*, 1973, **6**, 3469; *Sov. Phys. Dokl.*, 1974, **18**, 527; E. M. Iolin and V. V. Kozlov, *Mol. Phys.*, 1978, **35**, 419.
14. A. V. Alekseev and U. K. Kopvillem, *Ultrasonics*, 1980, 76.
15. S. U. Patel, Ph.D. Thesis, Aston University, UK, 1989.
16. L. O. Bowen, *Br. J. Appl. Phys.*, 1964, **15**, 1451; *Proc. Phys. Soc. London*, 1966, **87**, 717.
17. J. Homer and M.S. Beevers, *J. Magn. Reson.*, 1985, **63**, 287.
18. E. R. Andrew and R. G. Eades, *Proc. R. Soc. London, Ser. A*, 1953, **216**, 398; E. R. Andrew, A. Bradbury, and R. G. Eades, *Nature (London)*, 1958, **182**, 1659.
19. (a) A. Samoson, E. Lipmaa, and A. Pines, *Mol. Phys.*, 1988, **65**, 1013; (b) A. Llor and J. Virlet, *Chem. Phys. Lett.*, 1988, **152**, 248.
20. B. F. Chmelka, K. T. Mueller, A. Pines, J. Stebbings, Y. Wu, and J. W. Zwanziger, *Nature (London)*, 1989, **339**, 42.
21. P. Mansfield, *J. Phys. C: Solid State Phys.*, 1971, **4**, 1444; P. Mansfield, M. J. Orchard, D. C. Stalker, and K. H. B. Richard, *Phys. Rev. B*, 1973, **7**, 90; W.-K. Rhim, D. D. Ellerman, and R. W. Vaughan, *J. Chem. Phys.*, 1973, **58**, 1772; 1973, **59**, 3740.
22. B. C. Gerstein, R. G. Pembleton, R. C. Wilson, and L. M. Ryan, *J. Chem. Phys.*, 1977, **66**, 361.
23. J. P. Yesinowski, *20th Exp. NMR Conf.*, Asilomar, CA, 1979; J. P. Yesinowski, *J. Am. Chem. Soc.*, 1981, **103**, 6266.
24. K. Kimura and N. Satoh, *Chem. Lett.*, 1989, 271.
25. K. B. Dysthe, *J. Sound Vib.*, 1969, **10**, 331.
26. V. L. Levshin and S. N. Rzhevkin, *Dokl. Akad. Nauk SSSR*, 1937, **16**, 407: B. E. Noltingk and E. A. Neppiras, *Proc. Phys. Soc. London*, 1950, **B63**, 674; 1951, **B64**, 1032.
27. V. Griffing, *J. Chem. Phys.*, 1950, **18**, 997: M. A. Margulis, *Russ. J. Phys. Chem.*, 1985, **59**, 882.
28. J. Homer, J. Reisse, and S. A. Palfreyman, unpublished results.
29. K. S. Suslick and S. J. Doktycz, in 'Advances in Sonochemistry', ed. T. J. Mason, JAI, New York, 1990, Vol. 1, pp. 197–230.
30. W. Lauterborn and H. J. Bolle, *J. Fluid Mech.*, 1975, **72**, 391.
31. See, for example, R. E. Apfel, 'Methods of Experimental Physics', ed. P. D. Edmonds, Academic Press, New York, 1981, Vol. 19, Chap. 7.
32. J. Homer and M. J. Howard, unpublished results.
33. E. R. Andrew, W. S. Hinshaw, and R. S. Tiffen, *Phys. Lett. A*, 1973, **46**, 57.
34. W. D. Knight, *Solid State Phys.*, 1956, **2**, 93.

Biographical Sketch

John Homer. *b*, 1938. B.Sc., 1959, Leeds, Ph.D., 1962, Birmingham, D.Sc., 1980 Leeds. Introduced to NMR by J. A. S. Smith and interest nurtured by L. F. Thomas (Ph.D. supervisor). Albright and Wilson (MFG) Ltd., 1962–64; Aston University, 1964–present, Reader in Physical Chemistry. Approximately 100 publications. Current research interests: new techniques, either theoretical or experimental; NMR techniques including DESPOT, TIRADES, SINNMR, and PENDANT; generalization of London's dispersion theory to polyatomic molecules; generalized approach to phase equilibria (AGAPE).

Vanadium Catalysts: Solid State NMR

Vjatcheslav M. Mastikhin & Olga B. Lapina

Boreskov Institute of Catalysis, Siberian Branch of Russian Academy of Sciences, Novosibirsk, Russia

1 INTRODUCTION

Measurements of ^{51}V NMR spectra of solids (see ***Quadrupolar Nuclei in Glasses***; ***Quadrupolar Nuclei in Solids***) with the use of modern NMR spectrometers (see ***Spectrometers: A General Overview***) equipped with line-narrowing techniques for obtaining high-resolution spectra of solids (see ***Double Rotation***; ***High Speed MAS of Half-Integer Quadrupolar Nuclei in Solids***; ***Magic Angle Spinning***; ***Variable Angle Sample Spinning***) provide important information on the local environment of V nuclei.[1] The latter is known to be a key factor determining the activity and selectivity of heterogeneous catalysts. Due to this fact ^{51}V NMR solid state spectroscopy is successfully used for the studies of vanadia-based catalysts. The latter are used for the production of sulfuric acid, the selective oxidation of hydrocarbons, and also for reduction of atmospheric pollution.[2]

The results presented below show that ^{51}V NMR provides important information concerning the structure and reactivity of surface V sites. Herein will be discussed the main features of ^{51}V NMR in solids and its application for the studies of vanadium catalysts.

The vanadium-51 nucleus (natural abundance 99.76%) has a spin $I = \frac{7}{2}$, its electric quadrupole moment being $-4 \times 10^{-2} \times 10^{-24}$ cm^2; the relative intensity of ^{51}V NMR is 0.38 compared with an equal number of protons.

In general, three different types of interactions influence the ^{51}V NMR spectra of solid diamagnetic samples (see ***Internal Spin Interactions & Rotations in Solids***): (1) the dipole interaction (see ***Dipolar & Indirect Coupling Tensors in Solids***) of the magnetic moment of the ^{51}V nucleus with the magnetic moments of other nuclei that broadens the lines; (2) the quadrupole interaction (see ***Quadrupolar Interactions***) of the ^{51}V nucleus with the electric field gradient that splits the lines and contributes to the shift of the central ($m_I = \frac{1}{2} \leftrightarrow m_I = -\frac{1}{2}$) line; and (3) the chemical shielding interaction (see ***Chemical Shift Tensors***) that changes the position of the lines and makes them asymmetric.

In general, the lineshape might be rather complicated due to the simultaneous action of all types of interactions. For powdered samples (the case typically met in heterogeneous catalysts) only the central transition ($m_I = \frac{1}{2} \leftrightarrow m_I = -\frac{1}{2}$) can most commonly be observed, while other transitions are too broad to be recorded. The line from the central transition is typically anisotropic, i.e. it has a rather complicated shape. This complicated shape reflects the fact that the observed line is actually a superposition of individual lines from vanadium sites having various orientations with respect to the external magnetic field. When these individual lines are narrow enough, the positions of the so-called discontinuity points can be readily identified in the overall ^{51}V NMR spectrum. From the position of the discontinuity points one can easily obtain the quadrupole coupling constant, χ, the asymmetry parameter, η, and the shielding anisotropy parameters σ_{11}, σ_{22}, and σ_{33}.[3]

Unfortunately, in the ^{51}V NMR spectra of many solid catalysts the effects from the discontinuity points are often completely or partially obscured due to the large broadening of individual lines. In this case computer analysis of the spectra recorded at various frequencies becomes necessary if one wishes to obtain the above-mentioned values. Note, that interactions affecting ^{51}V NMR spectra exhibit different frequency dependences. Indeed, the dipole interaction and the first-order quadrupole interaction do not depend on the spectrometer frequency ν_0, while the second-order quadrupole effects are inversely proportional to ν_0. The effects of the shielding anisotropy are directly proportional to ν_0. Thus, at a sufficiently high ν_0, the second-order quadrupole effects are suppressed and can be neglected, while the effects of the chemical shielding anisotropy become more pronounced and can be measured more precisely (***Chemical Shift Tensor Measurement in Solids***). A comparison of ^{51}V NMR spectra recorded at various ν_0 with computer simulated spectra has been demonstrated to allow sufficiently precise measurements of the spectral parameters characterizing vanadium sites in the solid catalysts.[1]

Typical examples of computer simulated ^{51}V NMR spectra of V(V) sites in solid V_2O_5 are presented in Figure 1. Calculations were performed using the program for high magnetic fields ($B_0 > 7$ T) when the perturbation theory approach can be applied, i.e. the values of the quadrupole and nuclear shielding terms are small when compared with the nuclear Zeeman energy.[1,4]

Figures 1(a) and (b) show the influence of the first-order quadrupole effects on the calculated spectra of solid V_2O_5 at $\nu_0 = 105.15$ MHz. Figure 1(c) demonstrates the spectral broadening in powdered or polycrystalline samples due to the distribution of quadrupole coupling constant.

2 VANADIUM-51 NMR SPECTRA OF SOLIDS WITH A WELL-CHARACTERIZED STRUCTURE. INTERRELATION BETWEEN ^{51}V NMR PARAMETERS AND THE LOCAL ENVIRONMENT OF V NUCLEI

To analyze the spectra of real catalytic systems one needs information on the spectra of vanadium compounds that might be present in the catalysts, as well as data on the interrelation between ^{51}V NMR parameters and the characteristics of the local environment of the V nuclei.

The ^{51}V NMR spectra of various solid vanadium compounds[1,3,5–8] with well-characterized structures[9] and of the vanadates of alkali metals, which are often used for the preparation of catalysts, have been studied.

In Table 1 the ^{51}V NMR chemical shielding parameters[1,3,5,7,8] for vanadates are listed.

The ^{51}V NMR parameters of the large number of V compounds with a well-characterized crystal structure presented in Table 1 prove the high sensitivity of NMR spectra toward details of the local environment of V nuclei.

In most cases the shape of ^{51}V NMR spectra measured in high magnetic fields ($B_0 > 7$ T) depends on the anisotropy of the chemical shielding tensor. The influence of the second-order quadrupole effects is typically smaller and often is quite negligible.[10]

The following conclusions on V coordination (tetrahedral or octahedral, regular or distorted) and the extent of association of vanadium–oxygen polyhedra can be drawn based on the type and magnitude of the chemical shielding anisotropy.[1,7,10]

(a) For vanadium in regular tetrahedral sites (sites of the Q^0 type), the isotropic spectra with close values of the chemical shielding tensor components ($\sigma_{11} \approx \sigma_{22} \approx \sigma_{33}$) typically have $\Delta\sigma < 100$ ppm.
(b) For vanadium in slightly distorted tetrahedral sites with adjacent VO_4 tetrahedra sharing one common oxygen atom (sites of the Q^1 type), a fully anisotropic shielding tensor ($\sigma_{11} \neq \sigma_{22} \neq \sigma_{33}$) is typical, but with a larger $\Delta\sigma$ value ($\Delta\sigma$ = 70–300 ppm).
(c) For vanadium in strongly distorted tetrahedral sites with adjacent VO_4 tetrahedra sharing two common oxygen atoms (sites of the Q^2 type), a fully anisotropic shielding tensor ($\sigma_{11} \neq \sigma_{22} \neq \sigma_{33}$) is typical with $\Delta\sigma$ = 300–375 ppm).
(d) For vanadium in distorted octahedral sites (i.e. in a distorted octahedral environment of oxygen atoms), a nearly axial shielding anisotropy is typical with $\sigma_{11} \approx \sigma_{22} < \sigma_{33}$ ($\Delta\sigma$ = 450–900 ppm for structures 4 in Figure 2 and $\Delta\sigma$ = 900–1300 ppm for structures 5 in Figure 2).

Conclusions (a) to (d) are illustrated in Figure 2 for a family of Tl vanadates and V_2O_5.

Note, that ^{51}V NMR spectra with approximate axial symmetry of the σ tensor should also be expected for VO_4 tetrahedra with symmetry close to C_{3v}, i.e. when one V–O bond differs significantly from the other three (which have about the same length) (sites of the Q^1 type). In this case the direction of $\sigma_{\parallel}$ will coincide with the direction of the V–O bond which differs from the three others, with $|\sigma_{\perp}| < |\sigma_{\parallel}|$ if this bond is shorter than the three others and $|\sigma_{\perp}| > |\sigma_{\parallel}|$ if this bond is the longest one.

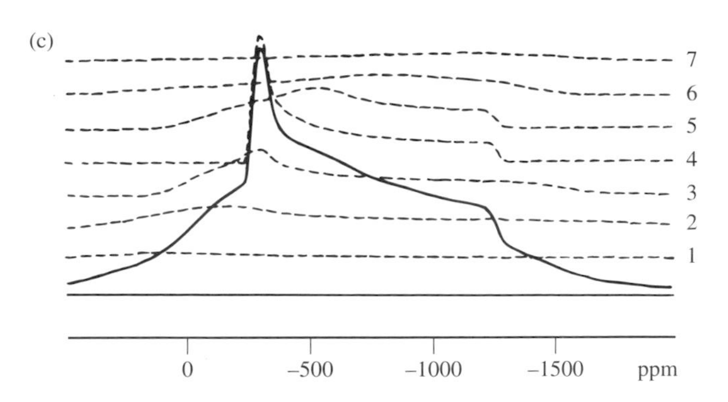

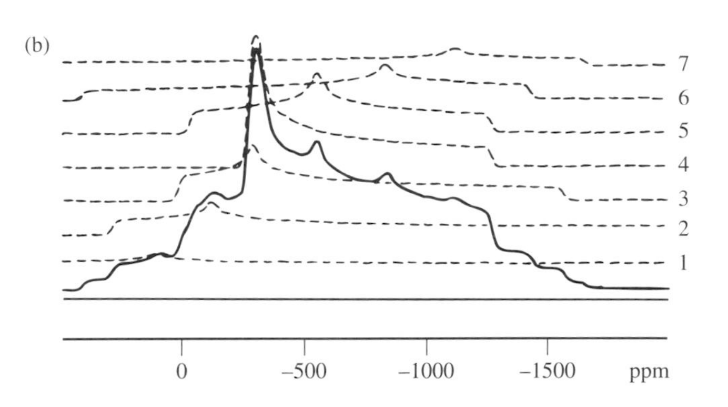

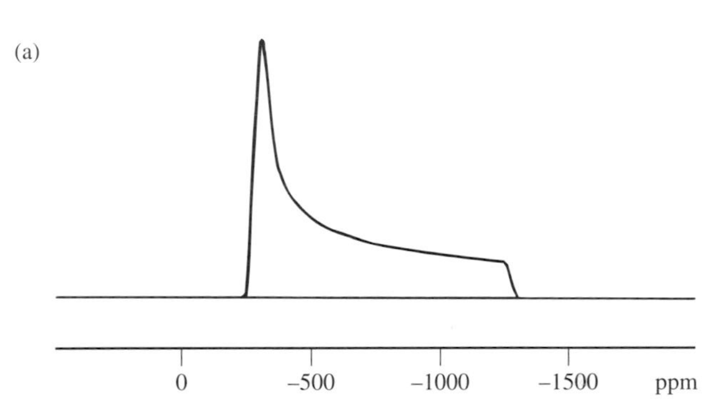

Figure 1 Computer simulated ^{51}V NMR spectra demonstrating the influence of the first-order quadrupole effects and line broadening on the spectra of solid V_2O_5 (at the frequency 105.15 MHz). The following parameters were used: χ = 0.805 MHz, η = 0.04;[5] σ_{11} = 310, σ_{22} = 1280, σ_{33} = 270 ppm. (a): The lineshape for the central transition ($\frac{1}{2}$, $-\frac{1}{2}$). (b) (Dotted lines): the lineshapes for all possible transitions: 1 ($-\frac{5}{2}$, $-\frac{7}{2}$), 2 ($-\frac{3}{2}$, $-\frac{5}{2}$), 3 ($-\frac{1}{2}$, $-\frac{3}{2}$), 4 ($\frac{1}{2}$, $-\frac{1}{2}$), the central transition, 5 ($\frac{3}{2}$, $\frac{1}{2}$), 6 ($\frac{5}{2}$, $\frac{3}{2}$), 7 ($\frac{7}{2}$, $\frac{5}{2}$). (b) (Solid line): the overall spectrum from all transitions. (c): Spectrum demonstrating the broadening of the lines from all transitions [dotted lines, numbered exactly as in (b)] and the overall spectrum from all transitions (solid line) due to the Gaussian distribution of quadrupole coupling constant χ. Dipolar linewidth DF = 1.5 kHz

A comparison of the shielding anisotropy $\Delta\sigma = |\sigma_{33} - \frac{1}{2}(\sigma_{11} + \sigma_{22})|$ for metavanadates whose structural data are known shows a good correlation between $\Delta\sigma$, on the one hand, and the angle O_3–V–$O_3{}'$ on the other hand [Figure 3(a)]. According to the structures of the metavanadates schematically presented in Figure 3(a) this angle characterizes the distortion of the tetrahedral environment of V. The larger the deviation of

For References see p. 4903

Table 1 Components of ^{51}V Chemical Shieldings (Measured in ppm with Respect to $VOCl_3$ with an Accuracy of ±10 ppm) and Quadrupole Constants χ and η for Vanadates

Compound	Site	σ_{11}	σ_{22}	σ_{33}	σ_{iso}	$\Delta\sigma$	χ(MHz)	η
Virtually regular tetrahedra VO_4, Q^0 type								
Li_3VO_4[1]					544	50	1.52	
							1.51[3]	
Na_3VO_4[7]					545			
							1.05[3]	
K_3VO_4[1]					560	100		
Cs_3VO_4[1]		520	580	626	576	70		
Tl_3VO_4[1]					480	30		
$Mg_3(VO_4)_2$[1]					557	40		
$Ca_3(VO_4)_2$[1]					615	100		
							2.05[3]	
$Sr_3(VO_4)_2$[1]					610	20		
							0.53[3]	
$Ba_3(VO_4)_2$[1]					605	20		
							0.75[3]	
$Zn_3(VO_4)_2$[7]					522			
$Pb_3(VO_4)_2$[22]		521	508	467	486	48		0.41
$AlVO_4$[1]	V1	630	640	730	668	95		
	V2	605	745	800	747	55		
	V3	710	800	830	780	75		
$BiVO_4$[1]		355	405	500	420	120		
YVO_4[1]					664	30		
							4.75[3]	0.00[3]
$LaVO_4$[1]		555	616	657	609	72		
							5.21[3]	0.69[3]
$LuVO_4$[1]					663	30		
							4.23[3]	0.00[3]
Slightly distorted tetrahedra, Q^1 type								
$Na_4V_2O_7$[1]	V1				560			
	V2				575			
$K_4V_2O_7$[1]		500	582	642	578	100		
$Cs_4V_2O_7$[1]	V1				543			
	V2				567			
$Tl_4V_2O_7$[1]		443	512	556	504	89		
α-$Mg_2V_2O_7$[1]	V1	510	570	585	555	45		
	V2	570	580	700	617	125		
β-$Mg_2V_2O_7$[1]	V1	460	560	680	560	170		
	V2	590	660	700	650	75		
$Ca_2V_2O_7$[1]	V1	528	564	630	574	84		
	V2	530	564	640	578	93		
$Sr_2V_2O_7$[1]	V1	523	548	600	557	65		
	V2	480	620	650	582	155		
	V3	480	632	652	588	160		
	V4	480	638	658	592	168		
$Ba_2V_2O_7$[1]	V1	530	555	652	579	109		
	V2	575	574	615	588	140		
	V3	535	625	640	600	100		
$Zn_2V_2O_7$[7]		500	640	720	625	150		
$Cd_2V_2O_7$[7]		370	660	660	579	290		
$Pb_2V_2O_7$[7]		430	480	620	522	165		
ZrV_2O_7[1]		710	802	824	774	110		
Distorted tetrahedra, Q^2 type								
$LiVO_3$[8]		385	540	794	573	332	3.18	0.87
NH_4VO_3[1]		380	530	807	572	352		
		366[8]	534[8]	810[8]	570[8]	360[8]		
							2.95[8]	0.30[8]
							2.88[3]	0.30[3]
							2.95[3]	0.19[3]
							2.76[3]	0.37[3]
β-$AgVO_3$[22]		202	285	609	364	366		0.34
α-$NaVO_3$[1]		368	530	820	582	371		
		355[8]	531[8]	833[8]	573[8]	390[8]		
							3.80[8]	0.46[8]
							3.70[3]	0.52[3]

						3.65[3]	0.60[3]
						3.15[3]	0.64[3]
						3.94[3]	0.64[3]
$TlVO_3$[1]	300	490	793	528	398		
	296[8]	497[8]	794[8]	529[8]	398[8]	3.67[8]	0.71[8]
$CsVO_3$[1]	330	522	863	583	437		
						3.92[3]	0.62[3]
						3.84[3]	0.63[3]
$RbVO_3$[1]	313	508	863	570	453		
						4.33[3]	0.72[3]
KVO_3[1]	294	490	856	548	464		
	313[8]	501[8]	842[8]	552[8]	435[8]	4.20[8]	0.80[8]
						4.21[3]	0.65[3]
						4.35[3]	0.75[3]
						4.06[3]	0.76[3]
						4.34[3]	0.77[3]
Distorted octahedra							
$K_2V_6O_{16}$[1]	290	290	935	503	630		
$Rb_2V_6O_{16}$[1]	290	290	935	503	630		
$Cs_2V_6O_{16}$[1]	296	296	953	508	645		
$Tl_2V_6O_{16}$[1]	485	485	1165	700	680		
α-$Mg(VO_3)_2$[1]	310	470	950	576	560		
						6.79[3]	0.63[3]
$Ca(VO_3)_2$[1]	278	355	1080	575	764		
						3.30[3]	0.8[3]
						3.16[3]	0.60[3]
						2.81[3]	0.60[3]
$Sr(VO_3)_2$[22]	533	577	833	643	283		0.29
$Ba(VO_3)_2$[1]	540	614	950	660	373		
$Zn(VO_3)_2$[7]	270	410	920	517	580		
$Pb(VO_3)_2$[7]	310	320	1000	533	685		
$Cd(VO_3)_2$[1]	305	365	830	500	495		
V_2O_5[1]	310	310	1270	610	960		
						0.8[5]	0.04[5]
$VOAsO_4$[1]	218	255	1370	617	1134		
$VOPO_4$[1]	285	285	1547	734	1260		

angle O_3–V–O_3' from that in the regular tetrahedron (109°28′), the larger is $\Delta\sigma$.

Distortion of the VO_4 tetrahedra also results in an increase in the electric field gradient on the V nucleus. It seems natural to expect a correlation between $\Delta\sigma$ and the quadrupole interaction parameters, χ and η. Indeed, such a correlation has been found for metavanadates [Figure 3(b)].

In contrast to ^{27}Al[11] and ^{29}Si[12] (see ***Aluminum-27 NMR of Solutions***; ***Molecular Sieves: Crystalline Systems***; ***Silicon-29 NMR***), the isotropic chemical shifts σ_{iso} for ^{51}V nuclei in vanadates are not too sensitive to V coordination (octahedral or tetrahedral). Compounds with rather different coordination may have very close σ_{iso} values. Thus, for vanadates, the anisotropy of the ^{51}V chemical shift is much more sensitive to the character of the local arrangement of oxygen atoms around V than the isotropic chemical shift σ_{iso}. At the same time, for compounds with the same type of first coordination sphere, σ_{iso} depends notably on the type of atoms in the second coordination sphere of V, as also found for ^{27}Al and ^{29}Si atoms.[11,12]

3 STUDY OF VANADIUM CATALYSTS

Typically, other techniques and approaches are used for identification of individual surface V species, in addition to ^{51}V NMR. These include studies with magic angle spinning (MAS), NMR of nuclei of elements other than V and other spectroscopic methods, as well as chemical approaches. A combination of spectroscopic data with the measurements of catalytic activity allows one to identify the catalytically active sites.

3.1 V_2O_5/SiO_2 Catalysts

3.1.1 Prepared by Impregnation

Spectra of V_2O_5/SiO_2 samples prepared via impregnation of SiO_2 with NH_4VO_3 solution are presented in Figure 4. The spectrum of the sample dried at 120 °C indicates only a slight change in the vanadium surroundings in comparison to that in NH_4VO_3 (spectrum 1a) and can be ascribed to (VO_4) species adsorbed on the SiO_2 surface. Calcination in vacuum provides a species that may differ both in the number of bonds with the surface and with neighboring vanadium species (spectra 2, 3, and 4 in Figure 4). Thus, lines present in spectra 2, 3, and 4 (calcination at 200 °C, 500 °C, and 700 °C) may be assigned to V species bonded to the surface via one, two, or three V–O bonds, respectively, as well as to V species bonded to neighboring vanadium species via one or two bonds.

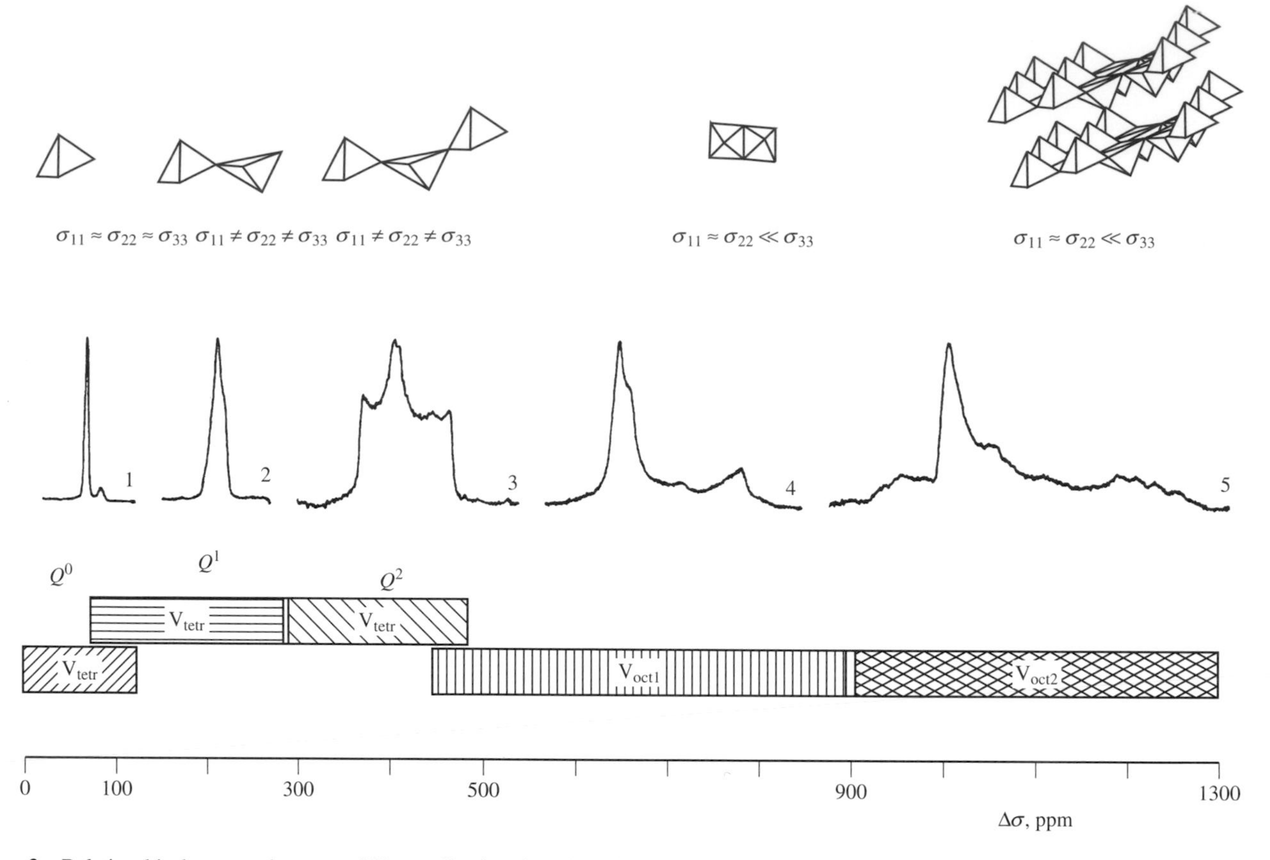

Figure 2 Relationship between the type of V coordination (tetrahedral or octahedral), the extent of polyhedra association (Q), and the shielding anisotropy $\Delta\sigma$. [As an example, the ^{51}V NMR spectra of Tl vanadates and V_2O_5 are presented: 1, Tl_3VO_4 (regular tetrahedra, Q^0); 2, $Tl_4V_2O_7$ (slightly distorted tetrahedra, Q^1); 3, $TlVO_3$ (strongly distorted tetrahedra, Q^2); 4, $Tl_2V_6O_{16}$ (distorted octahedra); 5, V_2O_5 (strongly distorted octahedron or square-pyramid)]

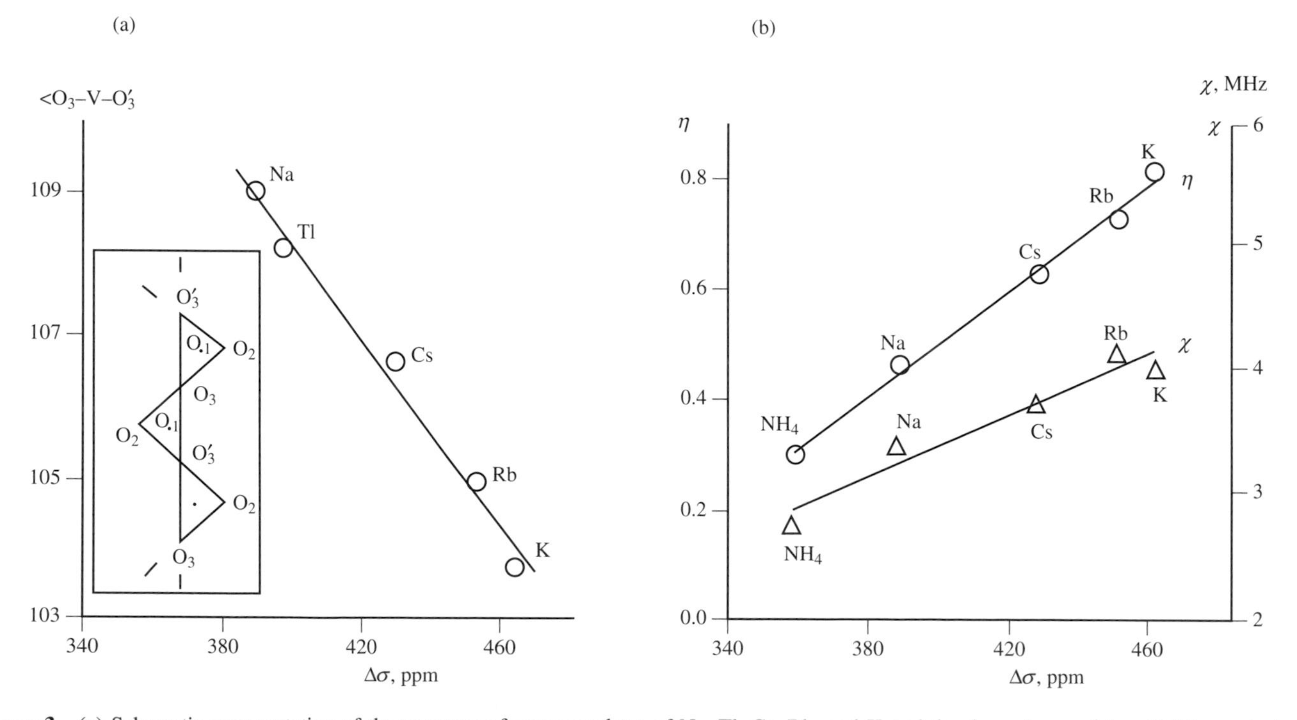

Figure 3 (a) Schematic representation of the structure of metavanadates of Na, Tl, Cs, Rb, and K and the dependence of the shielding anisotropy ($\Delta\sigma$) of their ^{51}V NMR spectra upon the O_3–V–O_3' angle. This angle characterizes the distortion of the VO_4 tetrahedron from the regular form where O_3–V–O_3' = 109°28′. (b) Correlation between the shielding anisotropy $\Delta\sigma$ (data of Table 1) and the quadrupole coupling parameters χ and η (the latter taken from Pletnev et al.[3] and Eckert and Wachs[7]) for metavanadates

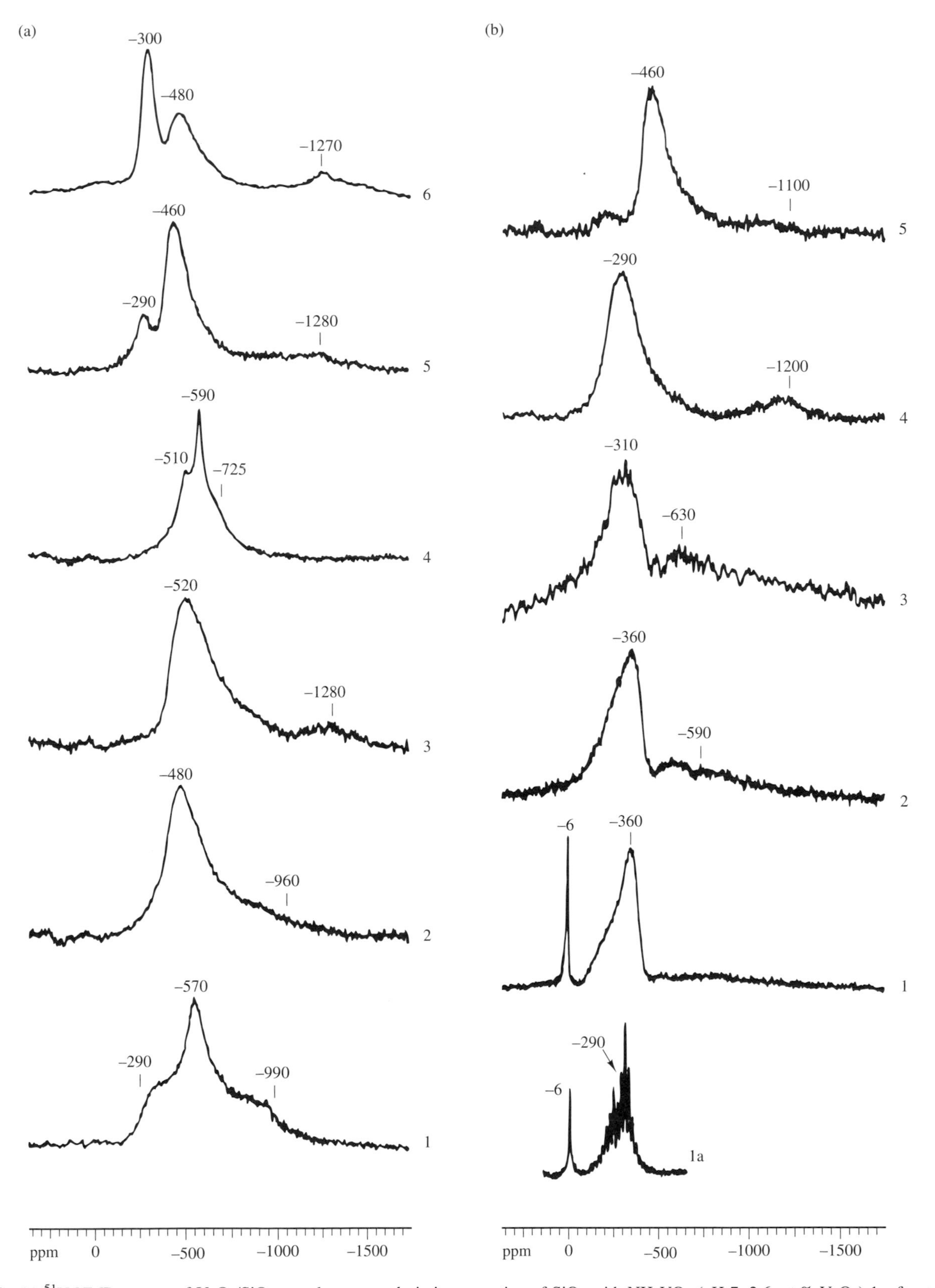

Figure 4 (a) ^{51}V NMR spectra of V_2O_5/SiO_2 samples prepared via impregnation of SiO_2 with NH_4VO_3 (pH 7, 2.6 wt.% V_2O_5) 1, after treatment in vacuum at 120 °C; 2, at 200 °C; 3, at 500 °C; 4, at 700 °C. Spectra 5 and 6 correspond to samples with a V content of 5.5 and 6.4 wt.% V_2O_5, respectively, calcined at 500 °C. (b) ^{51}V NMR spectra of $VOCl_3/SiO_2$ samples prepared by interaction of $VOCl_3$ from the gas phase with SiO_2: 1, after $VOCl_3$ deposition; 1a, MAS spectrum of the same sample [line at 6 ppm corresponds to physically adsorbed $VOCl_3$, anisotropic line with σ_{iso} = 290 ppm corresponds to $VOCl_2(OSi)$ surface species]; 2, 3, 4, and 5, increase of H_2O content (replacement of Cl atoms to OH groups takes place during the first stage of hydration, then interaction of V species with the next H_2O molecules occurs); 6, spectrum 5 after evacuation at 100 °C. All the spectra here and in the remaining figures were recorded at 105.15 MHz, pulse duration = 0.5 μs, pulse repetition frequency = 1 Hz. Chemical shifts are presented with respect to $VOCl_3$ as external reference

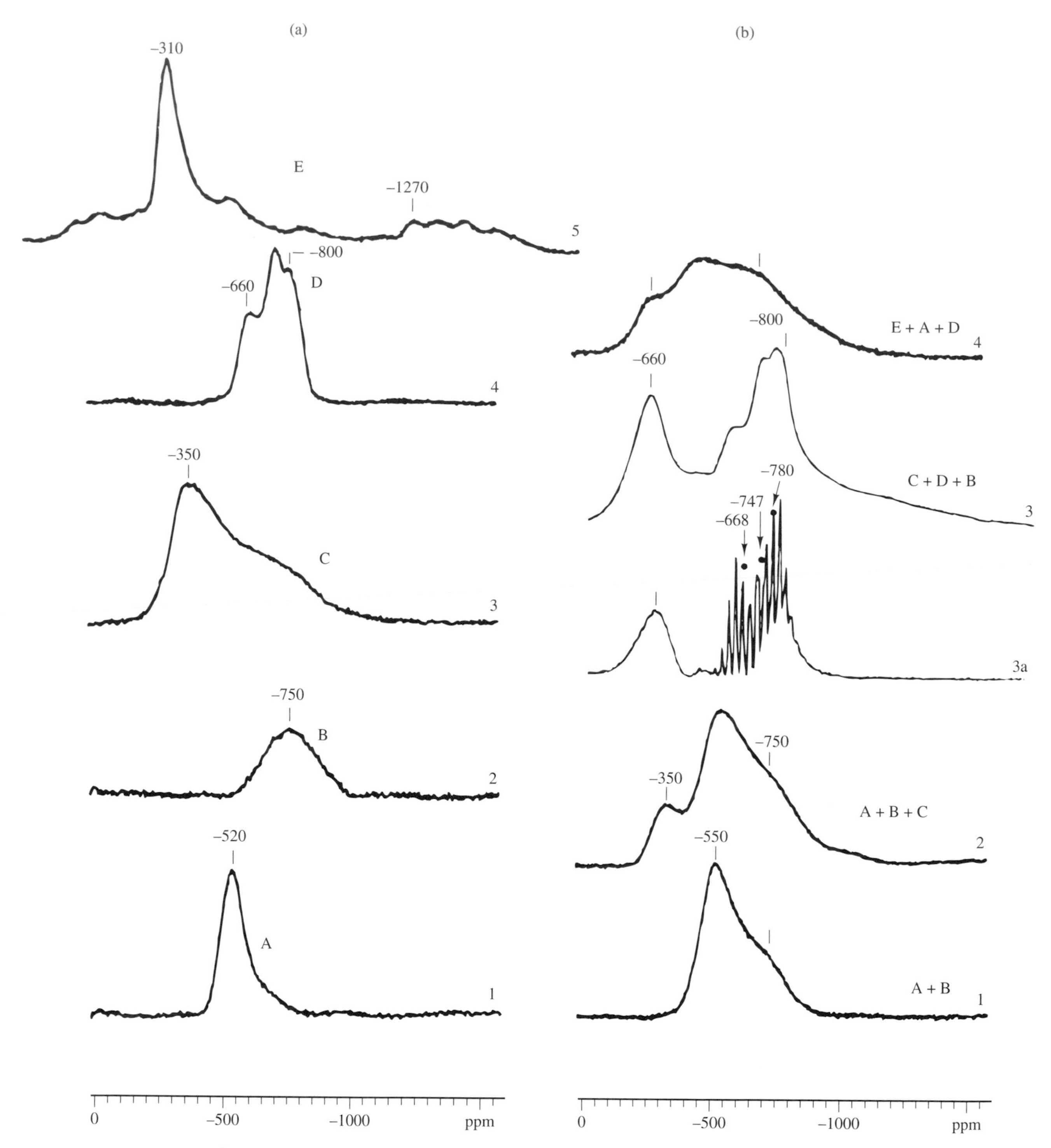

Figure 5 Classification of ^{51}V NMR lines observed in V_2O_5/Al_2O_3 catalysts prepared by Al_2O_3 impregnation with NH_4VO_3 solution. (a) Individual lines: 1, line A, σ_{iso} = 520–590 ppm, observed in wet samples at a V content from 1–2 wt.%; 2, line B, σ_{iso} = 750 ppm observed at a V content of 1 wt.% for samples calcined at 500 °C; 3, line C, $\sigma_{\perp}$ = 350 ppm observed for wet samples at high V content; 4, line D from polycrystalline $AlVO_4$; 5, line E from polycrystalline V_2O_5. (b) Spectra of various catalysts: 1, prepared by impregnation with NH_4VO_3 (pH 12, 2.2 wt.% V) dried at 110 °C, superposition of lines A and B; 2, the same sample calcined at 550 °C, superposition of lines A, B, and C; 3a, MAS spectrum of sample 3; 3, catalyst prepared via impregnation with VOC_2O_4 (pH 0.4, 14 wt.% V) calcined at 500 °C, superposition of lines B, C, and D (line D gives well-resolved sidebands with MAS); 4, catalyst prepared via impregnation with NH_4VO_3 (pH 7, 14.8 wt.% V), the superposition of lines A, D, and E

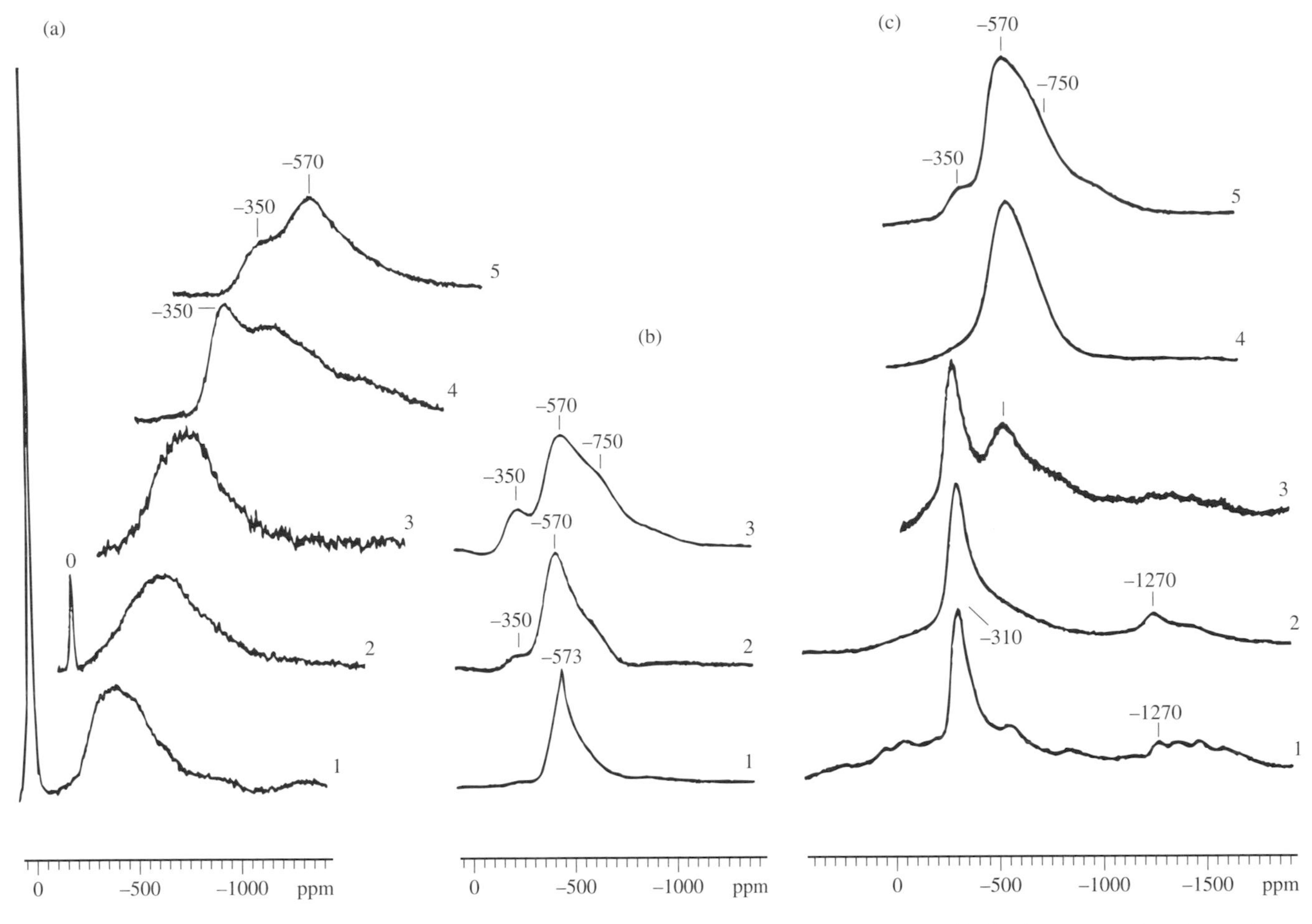

Figure 6 Vanadium-51 NMR spectra of V_2O_5/Al_2O_3 catalysts at different stages in their preparation: (a) Catalysts prepared by interaction with gaseous $VOCl_3$: 1, spectrum after $VOCl_3$ deposition; 2, 3, and 4, increase of H_2O content; 5, sample 4 after calcination at 500 °C. (b) Catalysts prepared by impregnation with NH_4VO_3 (pH 12, 3.8 wt.% V): 1, wet sample; 2, sample dried at 120 °C; 3, sample calcined at 500 °C. (c) Catalysts prepared by ultra-high-intensity grinding (UHIG): 1, V_2O_5; 2, V_2O_5 after UHIG; 3, V_2O_5-Al_2O_3 mixture (5 wt.% V_2O_5) calcined at 500 °C for 14 h; 4, after UHIG treatment at room temperature; 5, UHIG with subsequent calcination of the samples at 500 °C for 2 h

Spectra of samples with high V concentrations (spectra 5 and 6) suggest the formation of the amorphous (nonregular) precursor of V_2O_5, since no sidebands are observed in the MAS spectra of these samples. Increasing the treatment temperature up to 700 °C results in the appearance of a line with highly resolved sidebands, due to rearrangement of the amorphous V_2O_5 structure to the crystalline one.

3.1.2 *Prepared by $VOCl_3$ Interaction with SiO_2*

The spectrum of the sample after $VOCl_3$ deposition suggests the presence of physically adsorbed $VOCl_3$ (narrow line with $\delta = -6$ ppm) and of $VOCl_{3-n}(OSi)_n$ species bonded with the SiO_2 surface [a line with an axial anisotropy, Figure 4(b), spectrum 1]. Deposition of $VOCl_3$ at room temperature yields complexes with $n = 1$ (a line with $\sigma_{iso} = 290$ ppm, spectrum 1a). Increasing the deposition temperature to 400 °C leads to complexes with $n = 2$ or 3. The first additions of water cause the disappearance of physically adsorbed $VOCl_3$ [Figure 4(b), spectrum 2], then water interacts with the $VOCl_{3-n}(OSi)_n$ surface species producing partially hydrated $VO(OH)_y$-$Cl_{3-n-y}(OSi)_n$ ($y \leqslant n$, $n = 1$, 2, or 3) species (spectra 2 and 3) and fully hydrated $VO(OH)_{3-n}\cdot 2H_2O(OSi)_n$ species (for these species the sign of the anisotropy is opposite to that for species containing Cl in their coordination sphere, $\sigma_\parallel > \sigma_\perp$). Dehydration under mild conditions (100 °C) results in the removal of water from the first coordination sphere and in the formation of associated tetrahedral complexes (spectrum 5). The latter was ascribed to $VO(OSi)_3$ by Das et al.[13]

Application of magic angle spinning does not improve the resolution, unlike the situation when complexes contain Cl in the first coordination sphere. This indicates a nonregularity in the structure of the surface oxide complexes.

These data along with results presented by Lapina et al.[14] show that after calcination at 400–500 °C similar complexes are formed on the SiO_2 surface by the different deposition procedures.

3.2 V_2O_5/Al_2O_3 Catalysts

Vanadium-51 NMR spectra of V_2O_5/Al_2O_3 catalysts prepared by interaction of Al_2O_3 with $VOCl_3$, impregnation of Al_2O_3 with NH_4VO_3, and ultra-high-intensity grinding (UHIG) of V_2O_5-Al_2O_3 mixtures are shown in Figures 5 and 6.

For References see p. 4903

3.2.1 *Prepared by Impregnation*

The various ^{51}V NMR spectra of these catalysts can be classified in terms of the five different types of lines contained[15,16] (Figure 5). Line A [spectrum (a)-1] observed for wet samples of low V content (1 wt.%) has been attributed to tetrahedral VO_4 species relatively loosely bound to the Al_2O_3 surface via one or two oxygen atoms. Line B, which was detected for calcined samples of low V content (1 wt.%) [spectrum (a)-2] was attributed to VO_4 tetrahedra bound more firmly to the Al_2O_3 surface via two or more oxygen atoms. Line C observed for samples with a large V content either dried or calcined [spectrum (a)-3] has been attributed to polynuclear V species of the decavanadate type on the Al_2O_3 surface. Line D is attributed to the $AlVO_4$ phase. It has been detected in calcined samples with a large V concentration [spectrum (a)-4]. From the MAS spectra, the σ_{iso} values for each of the three nonequivalent V sites in the structure of this compound were as follows: $\sigma_{iso\,1} = 668 \pm 5$; $\sigma_{iso\,2} = 747 \pm 5$; $\sigma_{iso\,3} = 780 \pm 5$ ppm. Line E ($\sigma_{\perp} = 310$ ppm, $\sigma_{\parallel} = 1270$ ppm) relates to V_2O_5 [spectrum (a)-5]. In the spectra of real catalysts, superposition of lines from different V sites occurs as shown in Figure 5(b).

3.2.2 *Prepared by Interaction of Al_2O_3 with $VOCl_3$*

Figure 6(a) shows that at least four types of particles are formed on the Al_2O_3 surface after $VOCl_3$ deposition: physically adsorbed $VOCl_3$ (narrow line with $\delta \approx 0$ ppm) and particles containing Cl atoms in the first coordination sphere, i.e. $VOCl_{3-n}(OAl)_n$, where $n = 1$, 2, or 3 (broad lines in the range from 0 to -1000 ppm, spectrum 1). The process of hydration occurs initially by hydration of the physically adsorbed $VOCl_3$ (disappearance of the signal at $\delta \approx 0$ ppm) and then exchange of Cl atoms in the first coordination sphere of V by OH groups takes place; after sample treatment at 500 °C, the formation of V sites typical for catalysts prepared by impregnation are also formed [cf. Figure 6(a), spectrum 5 and Figure 6(b), spectrum 3].

3.2.3 *Prepared by UHIG Treatment*

The ultra-high-intensity grinding (UHIG) of V_2O_5-Al_2O_3 mixtures was found to induce a heterogeneous chemical reaction between these oxides producing highly dispersed V species on the support surface.[17] The ^{51}V NMR spectrum of V_2O_5 after UHIG treatment demonstrates the broadening of the

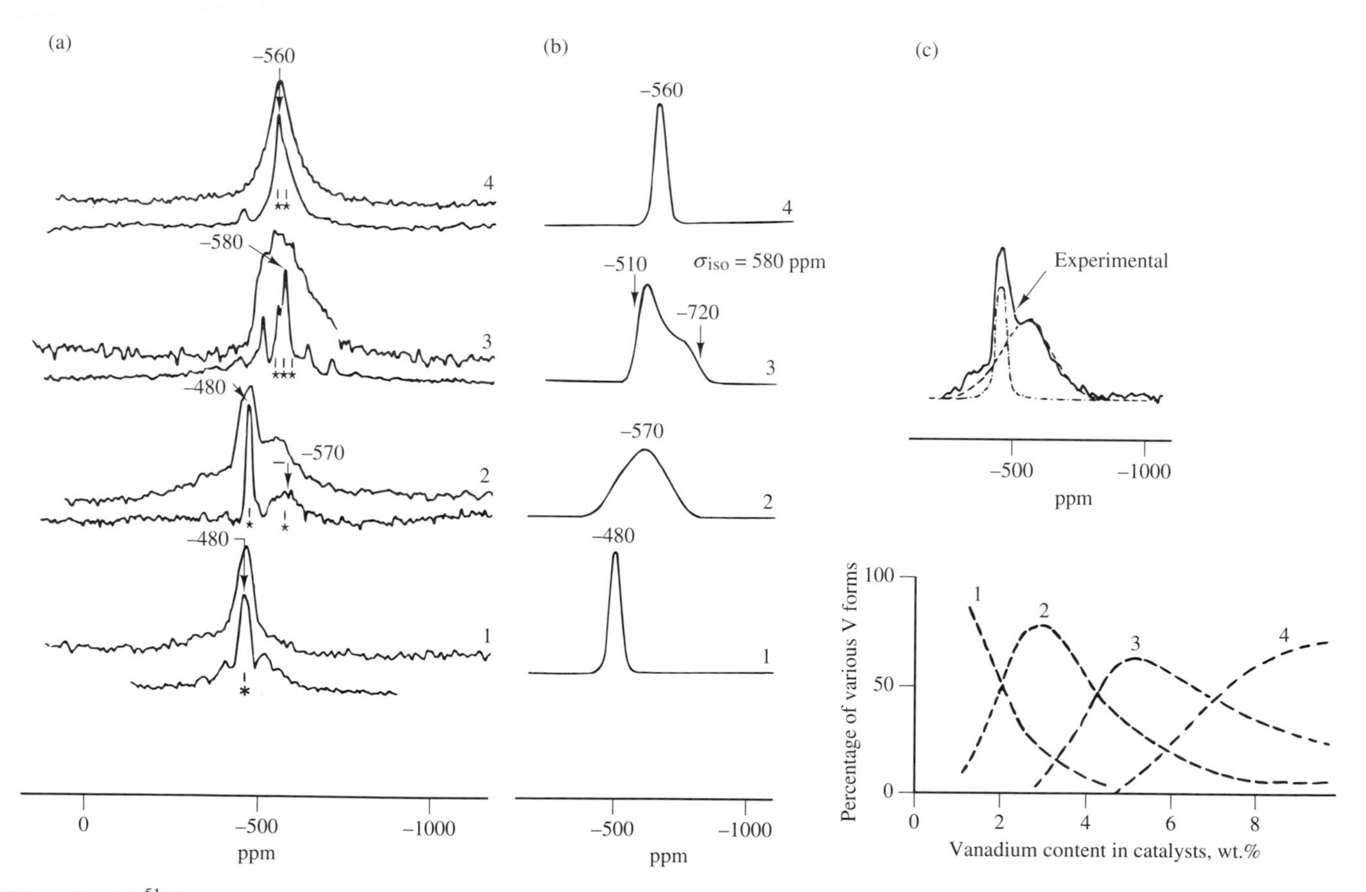

Figure 7 (a) ^{51}V NMR spectra of V_2O_5/MgO catalysts of various vanadium content: 1, 1.6 wt.%; 2, 2.6 wt.%; 3, 5.0 wt.%; 4, 9.6 wt.%. The upper spectra were recorded without MAS; the MAS spectra are shown below. The isotropic chemical shifts are indicated by asterisks (*). (b) The simulated lineshapes of ^{51}V NMR spectra for various V species found in V_2O_5/MgO catalysts: 1, isolated tetrahedral surface species; 2, tetrahedral surface clusters; 3, tetrahedral surface clusters of another type; 4, $Mg_3(VO_4)_2$. (c) Deconvolution of an experimental spectrum (2.6 wt.% V) into individual lines and the percentage of various V species on the MgO surface as a function of the total V content in V_2O_5/MgO catalysts: 1, isolated tetrahedral surface species; 2, tetrahedral surface clusters; 3, tetrahedral surface clusters of another type; 4, $Mg_3(VO_4)_2$

peaks from first-order quadrupole effects due to distortion of the local vanadium environment while the general crystal structure of V_2O_5 is retained [spectra 1 and 2 in Figure 6(c)]. Only a small part of the V_2O_5 reacts with Al_2O_3 when a V_2O_5-Al_2O_3 mixture is calcined for 14 h at 500 °C (peak at −570 ppm) (spectrum 3). A more complete interaction between V_2O_5 and Al_2O_3 takes place in the samples after UHIG treatment at room temperature (spectrum 4). As the concentration of V increases, the other tetrahedral V state (peak near −750 ppm) appears. Subsequent calcination at 500 °C produces V in octahedral coordination. In general, ^{51}V NMR spectra of the samples treated at 500 °C after UHIG treatment show the formation of surface structures that are similar to those for V_2O_5/Al_2O_3 catalysts prepared by the conventional impregnation/calcination method.

Thus, the types of surface V species on Al_2O_3 are the same for all three preparation procedures discussed above. Their relative amounts depend in a similar way on the total V concentration on the Al_2O_3 surface and on the treatment temperature. At low V concentration species A prevail, while at the increased V concentration species B and C appear on the catalyst surface. At still larger amounts of supported V (>10 wt.%), V_2O_5 and $AlVO_4$ also form.

3.3 V_2O_5/MgO Catalysts

Vanadium-51 NMR spectra of V_2O_5/MgO catalysts, recorded at various vanadium content, show four different V sites on the MgO surface (Figure 7).[18] Only one of them can be attributed to magnesium vanadate, $Mg_3(VO_4)_2$. Other V sites have been identified on the basis of correlations between the structure of the local environment of the V nuclei and the shielding anisotropy.

The narrow isotropic line with σ_{iso} = 480 ppm observed at low vanadium concentration (from 1.6–2.6 wt.% V) [spectra 1 and 2 in Figure 7(a)] has been attributed to isolated surface V atoms which have a virtually regular tetrahedral oxygen environment. Two broadened lines, i.e. an almost isotropic line with σ_{iso} = 570 ppm and a line with $\sigma_{\perp}$ = 510 ppm, σ = 720 ppm have been assigned to surface clusters with V atoms in a tetrahedral environment (spectra 2, 3, and 4). At large V content, orthovanadate $Mg_3(VO_4)_2$ (σ_{iso} = 560 ppm) is formed in the catalysts (spectrum 4). The calculated shapes of the individual lines for various V sites in V_2O_5/MgO catalysts and deconvolution of the experimental spectrum into individual lines are also shown in Figures 7(b) and (c). Figure 7(c) also presents the relative amounts of the various V species as measured by deconvolution of the experimental spectra.

Hence, the ^{51}V NMR spectra of V_2O_5/MgO catalysts allow one to form certain conclusions on the structure of surface V species and their concentration as a function of the total vanadium content.

3.4 V_2O_5/TiO_2 Catalysts

The influence of the admixtures on the structure of the supported species has been demonstrated for V_2O_5/TiO_2 supported catalysts[1,19] (Figure 8). Spectra of catalysts prepared via the gas-phase reaction of $VOCl_3$ with TiO_2 followed by subsequent hydration and H_2O removal at 350 °C are shown in Figure 8(a). The catalyst prepared from the most pure anatase (impurities less than 0.1 wt.%) contains two tetrahedral (peaks at δ = −440 and −560 ppm) and one octahedral species (peak at δ = −320 ppm) as follows from spectrum 1 in Figure 8(a). An admixture containing sulfate anions (3.9 wt.%) increases the content of the octahedral vanadium species (spectrum 2). The presence of 3 wt.% Si leads to tetrahedral vanadium complexes typical for V_2O_5/SiO_2 samples (peak at δ = −490 ppm, spectrum 3) and with anatase containing Na (2.3 wt.%) the formation of Na_3VO_4 was observed (line with δ = −535 ppm, spectrum 4). Similar surface species have also been identified for the same catalysts prepared by impregnation of the same anatase samples with a solution of vanadyl oxalate [cf. Figure 8(a) and (b)].

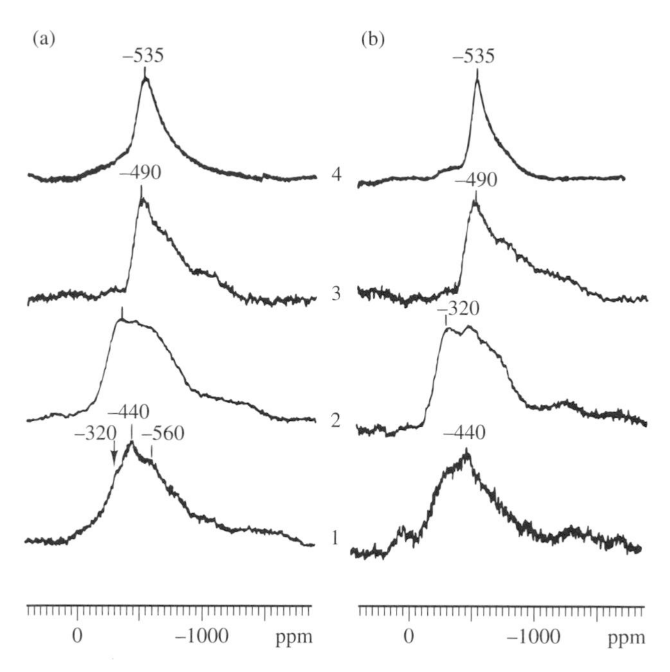

Figure 8 Vanadium-51 NMR spectra of V_2O_5/TiO_2 catalysts (a) prepared by gas phase reaction of $VOCl_3$ with TiO_2 followed by subsequent hydration and removal of H_2O with a stream of He at 350 °C and (b) prepared via impregnation with a solution of VOC_2O_4: 1, anatase (all impurities less 0.1 wt.%); 2, TiO_2 (3.9 wt.% SO_4); 3, TiO_2 (3 wt.% Si); 4, TiO_2 (2.3 wt.% Na)

Thus, the type of surface V species on TiO_2 depends mainly on the presence of the impurities in anatase with the preparation procedure having a considerably smaller effect on the type of V sites on the catalyst surface.

Data analysis presented above and by Lapina et al.[1] shows that the structure of surface V species depends mainly on the type of support. The preparation procedure (impregnation, grafting, UHIG treatment) appears to have much less of an effect on the type of V sites on the catalyst surface provided that the concentration of vanadium and the treatment temperatures are the same.

At a V concentration less than 0.5 monolayer, surface tetrahedral species with different extents of distortion of the oxygen environment are formed, with the distortion depending on the type of the support used.

(a) For MgO and $AlPO_4$, the surface vanadium species has an almost regular tetrahedral environment with σ_{iso} depending on the type of support employed (σ_{iso} = 480 and 880 ppm for MgO and $AlPO_4$, respectively) and $\Delta\sigma$ < 100 ppm.

For References see p. 4903

(b) For Al_2O_3, ZrO_2, ZrO_2/TiO_2, and SnO_2, slightly distorted tetrahedral V species exist with σ_{iso} values ranging from 500–600 ppm and $\Delta\sigma < 200$ ppm.
(c) For SiO_2 and TiO_2, distorted tetrahedral V species with one short V–O bond are formed. They exhibit axial anisotropy of chemical shielding with $\Delta\sigma < 600$ ppm.

An increase in V concentration (up to monolayer coverage) results in association of surface species. The following features associated with their ^{51}V NMR spectra can be noted.

(a) On MgO, $AlPO_4$ associated tetrahedral surface V species exhibit spectra with a fully anisotropic σ tensor, which points to a distortion of the VO_4 tetrahedra in these species (σ_{iso} = 560–570 ppm). The lines for such associated species are usually broader when compared with those for isolated species.
(b) For associated V species on Al_2O_3, ZrO_2, SnO_2, and TiO_2/ZrO_2, fully anisotropic lines from distorted VO_4 surface tetrahedra are also observed (σ = 700–800 ppm). The widths of the lines are close to those for isolated V species on this surface.
(c) For associated V species on Al_2O_3, ZrO_2, TiO_2/ZrO_2, and SnO_2 of another type, lines with an axial anisotropy of the shielding tensor arising from distorted octahedral V sites are observed ($\sigma_\perp$ = 320–330 ppm, $\Delta\sigma$ = 600 ppm).
(d) For associated V species on SiO_2 and TiO_2, lines with an axial anisotropy of the shielding tensor arising from distorted octahedral V sites are observed ($\sigma_\perp$ = 290–300 ppm, $\Delta\sigma > 900$ ppm).

A further increase of V concentration above monolayer coverage provides crystalline compounds [V_2O_5 on SiO_2, TiO_2, $AlPO_4$, TiO_2/ZrO_2, and SnO_2; $AlVO_4$ on Al_2O_3; and $Mg_3(VO_4)_2$ on MgO].

Water adsorption changes the V coordination from tetrahedral to octahedral on SiO_2 and TiO_2. In these cases, water molecules insert into the first coordination sphere of V. For V sites on ZrO_2, $AlPO_4$, TiO_2/ZrO_2, and SnO_2, water molecules adsorb in the second coordination sphere, providing slight shifts of the ^{51}V NMR lines upon adsorption.

Addition of promoters, such as rhodium, titania, or zirconia (or impurities), results in notable changes in both the structure of the surface V species and their redistribution on the surface.

3.5 Vanadium Catalysts for SO_2 Oxidation

The active component of vanadium catalysts for SO_2 oxidation is known to consist of vanadium oxide and sulfates or pyrosulfates of alkali metals (K, Na, Cs are most commonly used) supported on porous materials such as silica or silica alumina. At ambient temperature the active component forms a thin vitreous film dispersed over the support. Under the reaction conditions (400–500 °C), the active component exists as a melt forming a very thin liquid layer on the support surface.

We present here the results of ^{51}V NMR studies of oxosulfatovanadates(V), chemical compounds formed between V_2O_5 and alkali pyrosulfates that are assumed to be present in the active component of these catalysts, as well as of V_2O_5-$K_2S_2O_7$ alloys and of commercial catalysts. Combined with other spectroscopic methods and kinetic studies, these results have helped to reveal the active sites in SO_2 oxidation and the mechanism of this catalytic reaction.[1,20,21]

The spectra of oxosulfatovanadate(V) $K_3VO_2(SO_4)_2$, the alloy $V_2O_5\cdot 3K_2S_2O_7$, and a commercial catalyst are presented in Figure 9. All oxosulfatovanadates(V) exhibit axial anisotropy of the chemical shielding tensor with parameters close to those for V_2O_5. Thus the local environment of the vanadium atom in oxosulfatovanadates(V) is similar to that in V_2O_5, where vanadium has a distorted octahedral (bipyramidal) oxygen atom environment with one V–O bond being considerably shorter than the others. Comparison of the spectrum of $V_2O_5\cdot 3K_2S_2O_7$ alloy (Figure 9) with that for oxosulfatovanadate(V) $K_3VO_2(SO_4)_2$ shows that the former exhibits σ values close to those for oxosulfatovanadate(V). This indicates the same local environment for the V atoms in the glassy alloy and in crystalline oxosulfatovanadate(V).

The ^{51}V NMR spectra for various catalysts after treatment with a reaction gas mixture become quite similar and exhibit the same two lines, namely an almost isotropic line with σ_{iso} = 520–560 ppm and a line with axial anisotropy of the shielding tensor. Only the relative intensities of the two lines but not their character vary from one catalyst to another. In particular, the average chemical shifts for both the isotropic and anisotropic lines ($\sigma_\perp$ = 320–350 ppm, $\sigma_\parallel$ = 1200–1300 ppm) are almost the same for all the catalysts studied. This indicates that the active component in these catalysts is the same and is actually formed during the course of catalytic reaction.[2] Initially, catalysts arising from different preparations contain a variety of V sites. However, on interaction with the components of the reaction media only the two sites mentioned above are formed. The nearly isotropic line belongs to V atoms in a slightly distorted tetrahedral environment and can be attributed to vanadium bonded to the support. This line exhibits an increase in relative intensity with a decrease in the overall V content of the sample. Thus, the isotropic line belongs to V complexes on the SiO_2 surface.

Measurement of the catalytic activity for a series of samples with different contents of surface tetrahedral V has showed the latter to be inactive in SO_2 oxidation.[20]

To elucidate the nature of the vanadium complexes which are active in SO_2 oxidation, ^{51}V, ^{17}O, ^{23}Na, ^{39}K, and ^{133}Cs NMR were combined with catalytic activity measurements of

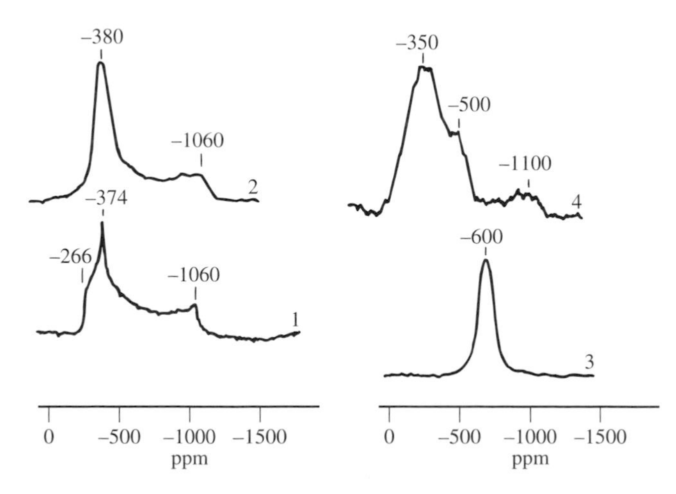

Figure 9 Vanadium-51 NMR spectra of (1) oxosulfatovanadate(V) $K_3VO_2(SO_4)_2$; (2) the solid alloy $V_2O_5\cdot 3K_2S_2O_7$; (3) the industrial catalyst BAV before treatment under the reaction conditions; (4) the same after treatment

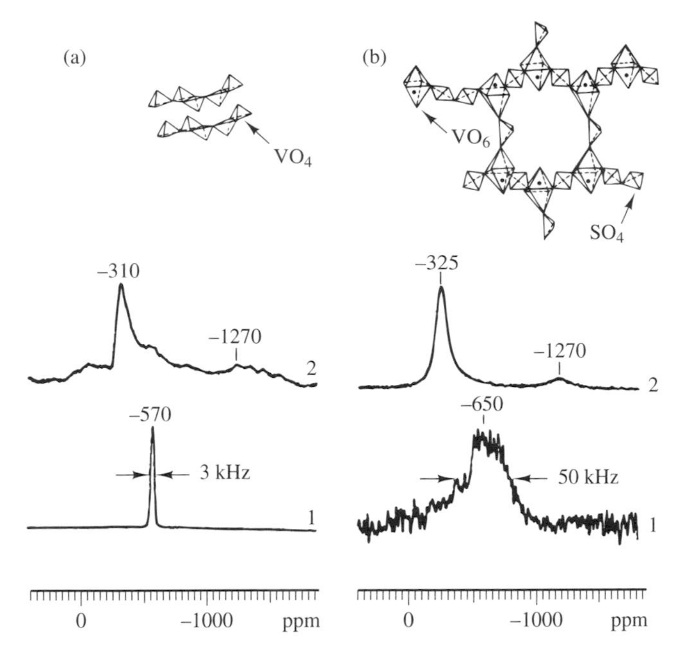

Figure 10 Vanadium-51 NMR spectra of (a) V_2O_5 and (b) of the alloy $V_2O_5 \cdot 3Cs_2S_2O_7$ (1) below the melting point (i.e. at 20 °C) and (2) above the melting point (670 °C and 370 °C for (a) and (b), respectively). Above: the structures formed in molten V_2O_5 and $V_2O_5 \cdot 3Cs_2S_2O_7$ alloy

V_2O_5-$K_2S_2O_7$ melts and of catalysts with different amounts of the active component on SiO_2.[1,20,21]

Vanadium-51 NMR spectra of V_2O_5 and $V_2O_5 \cdot 3Cs_2S_2O_7$ alloy at temperatures below and above their melting points (Figure 10) demonstrate the spectrum of molten $V_2O_5 \cdot 3Cs_2S_2O_7$ (see ***Molten Salts***) to be substantially broader than that of V_2O_5. This means, that unlike V_2O_5, where melting leads to a separation of the vanadium–oxygen layers and breaks them into relatively short fragments, in molten $V_2O_5 \cdot 3Cs_2S_2O_7$ substantially larger particles are formed.

Oxygen-17 NMR (see ***Oxygen-17 NMR***) has also been used to characterize the melt of the active component in SO_2 oxidation at 500 °C in the presence of an $SO_3 + SO_2 + O_2$ mixture, i.e. under typical conditions for the catalytic process in industry.[1,20,21] Addition of V_2O_5 to $K_2S_2O_7$ leads to a shift and a broadening of the ^{17}O line, indicating V coordination with pyrosulfate anions. A more sophisticated analysis using a thermodynamic model of the ^{17}O linewidths has shown that at small vanadium concentrations approximately two to three pyrosulfate anions are coordinated to one V atom.

Oxygen-17 NMR data also suggest a fast exchange between terminal and bridging oxygen atoms in the pyrosulfate anion which may occur via the reaction $S_2O_7^{2-} \rightleftarrows SO_4^{2-} + SO_3$. The characteristic time τ during which this equilibrium is established is less than 10^{-3} s. Oxygen-17 and ^{51}V NMR data show that on addition of V_2O_5 to $K_2S_2O_7$, complexes are formed according to the reaction: $V_2O_5 + 3K_2S_2O_7 \rightleftarrows 2K_3VO(SO_4)_3$. The increase in the ^{51}V and ^{17}O linewidths on increasing the V concentration suggests a further association of the V species leading to larger oligomers of the type shown in Figure 10. The large size of these oligomers makes their rotational diffusion very slow. Internal rotation of their fragments can also be hindered because of branching and linking of the oligomeric chains. Because of these factors, the ^{17}O and ^{51}V NMR lines of these species are too broad to be detected.

When supported on SiO_2, the dimeric or oligomeric vanadium species can be more stabilized on the surface due to their interaction with Si–OH groups. These results agree well with studies of the catalytic activity of thin films of active melts on Pyrex glass and on SiO_2, as well as with kinetic studies of SO_2 oxidation.[20]

4 RELATED ARTICLES

Aluminum-27 NMR of Solutions; Chemical Shift Tensor Measurement in Solids; Anisotropy of Shielding & Coupling in Liquid Crystalline Solutions; Chemical Shift Tensors; Dipolar & Indirect Coupling Tensors in Solids; Internal Spin Interactions & Rotations in Solids; High Speed MAS of Half-Integer Quadrupolar Nuclei in Solids; Internal Spin Interactions & Rotations in Solids; Magic Angle Spinning; Molecular Sieves: Crystalline Systems; Molten Salts; Oxygen-17 NMR; Quadrupolar Interactions; Quadrupolar Nuclei in Glasses; Quadrupolar Nuclei in Solids; Silicon-29 NMR; Variable Angle Sample Spinning.

5 REFERENCES

1. O. B. Lapina, V. M. Mastikhin, A. A. Shubin, V. N. Krasilnikov, and K. I. Zamaraev, *Prog. Nucl. Magn. Reson. Spectrosc.*, 1992, **24**, 457.
2. B. E. Leach (ed.), 'Applied Industrial Catalysis', Academic Press, New York, 1983.
3. R. N. Pletnev, V. A. Gubanov, and A. A. Fotiev, 'NMR in Oxide Vanadium Compounds', Nauka, Moscow, 1979 (in Russian); [*Chem Abs.*, 1980, **92**, 159 005x].
4. M. H. Cohen and F. Reif, *Solid State Phys.*, 1957, **5**, 321.
5. S. D. Gornostansky and G. V. Stager, *J. Chem. Phys.*, 1967, **46**, 4959.
6. D. Rehder, *Bull. Magn. Reson.*, 1982, **4**, 33.
7. H. Eckert and I. E. Wachs, *J. Phys. Chem.*, 1989, **93**, 6796.
8. J. Skibsted, N. C. Nielsen, H. Bildsoe, and H. J. Jakobsen, *J. Am. Chem. Soc.*, 1993, **115**, 7351.
9. A. A. Fotiev, B. V. Slobodin, and M. Ya. Hodos, 'Vanadates, their Synthesis, Composition and Properties', Nauka, Moscow, 1988 (in Russian); [*Chem. Abs.*, 1989, **110**, 68 625c].
10. V. M. Mastikhin, O. B. Lapina, V. N. Krasilnikov, and A. A. Ivakin, *React. Kinet. Catal. Lett.*, 1984, **24**, 119.
11. D. Muller, W. Gessner, and A. R. Grimmer, *Z. Chem.*, 1977, **B12**, 453.
12. J. Klinowsky, *Prog. Nucl. Magn. Reson. Spectrosc.*, 1984, **16**, 237.
13. N. Das, H. Eckert, H. Hu, I. E. Wachs, J. F. Walzer, and F. J. Feher, *J. Phys. Chem.*, 1993, **97**, 8240.
14. J. B. Lapina, V. M. Mastikhin, A. V. Nosov, T. Beutel, and H. Knozinger, *Catal. Lett.*, 1992, **13**, 203.
15. J. B. Lapina, V. M. Mastikhin, L. G. Simonova, and Yu. O. Bulgakova, *J. Mol. Catal.*, 1991, **69**, 61.
16. L. R. Le Costumer, B. Taouk, M. Le Meur, E. Payen, M. Guelton, and J. Grimblot, *J. Phys. Chem.*, 1988, **92**, 1230.
17. Z. Sobalik, O. B. Lapina, O. N. Novgorodova, and V. M. Mastikhin, *Appl. Catal.*, 1990, **63**, 191.
18. O. B. Lapina, A. V. Simakov, V. M. Mastikhin, S. A. Veniaminov, and A. A. Shubin, *J. Mol. Catal.*, 1989, **50**, 55.
19. H. Eckert, G. Deo, I. E. Wachs, and A. M. Hirt, *Colloids Surf.*, 1990, **45**, 347.

20. V. M. Mastikhin, O. B. Lapina, B. S. Balzhinimaev, L. G. Simonova, L. M. Karnatovskaya, and A. A. Ivanov, *J. Catal.*, 1987, **103**, 160.
21. B. S. Balzhinimaev, A. A. Ivanov, O. B. Lapina, V. M. Mastikhin, and K. I. Zamaraev, *Faraday Discuss. Chem. Soc.*, 1989, **87/88**, 133.
22. S. Hayakawa, T. Yoko, and S. Sakka, *J. Solid State Chem.*, 1994, **112**, 329.

Biographical Sketches

Vjatcheslav M. Mastikhin. *b* 1937. Graduated 1959, Kharkov University, Ph.D. 1969, Dr.S., 1986, Boreskov Institute of Catalysis. Leading Scientist, Boreskov Institute of Catalysis, Novosibirsk, 1986–1995. Approx. 200 publications. Research interests: application of solid state NMR to problems of heterogeneous catalysis.

Olga B. Lapina. *b* 1953. Graduated 1976, from Novosibirsk University, Ph.D., 1984, Dr.S., 1995, Boreskov Institute of Catalysis. Senior Scientist, Boreskov Institute of Catalysis, Novosibirsk, 1990–present. Approx. 100 publications. Research interests: application of solid state NMR to problems of heterogeneous catalysis.

Variable Angle Sample Spinning

Naresh K. Sethi

Amoco Research Center, Naperville, IL, USA

1 INTRODUCTION

In polycrystalline solids, where the molecules have restricted mobility, tensorial NMR interactions such as chemical shift anisotropy (CSA) and dipolar couplings are manifested in broadening of resonance lines. Complete measure of these interactions provides valuable information on molecular geometry, bonding, and electronic structure. However, analysis of a solid state NMR spectrum to extract this information is often made very difficult by extensive overlap of broad lines from chemically different spins. Variable angle sample spinning (VASS) techniques are designed to unscramble overlapping solid state NMR spectral features from chemically different spins.

The resonance line-broadening effects of CSA and dipolar couplings can be removed by acquiring the NMR spectrum while spinning the sample at high speed along an axis inclined at the magic angle (54.74°) from the direction of the external magnetic field $\boldsymbol{B}_0$.[1,2] This technique of magic angle spinning (MAS) permits obtaining a 'liquid-like' spectrum for polycrystalline solids, which facilitates chemical analysis. However, this is accomplished at the expense of losing all the important information regarding the anisotropy of the interaction tensor. Of the various approaches that have been proposed to reintroduce the anisotropy information back into the NMR spectrum of spinning solids, VASS techniques are, conceptually, by far the simplest. In VASS, the anisotropy information is reintroduced in the NMR spectrum by simply spinning the sample at angles other than the magic angle.[3–5]

This article will discuss the basic principles behind VASS techniques, spectral features of the NMR signal when the sample spinning speed is slow or fast compared with the strength of the anisotropic interaction, and the use of two-dimensional (2D) techniques to overcome the limitations of one-dimensional (1D) VASS.

2 BACKGROUND

To illustrate VASS, we shall assume that CSA is the only NMR interaction affecting the spectral shape. The extension of these concepts to the case of dipolar couplings is fairly straightforward. Other articles in this Encyclopedia discuss the concept of chemical shift anisotropy. (Cross references to related articles are provided at the end of this one.) Also, a rigorous mathematical description of the NMR of spinning solids is beyond the scope of this article. Mathematical expressions relevant to the present discussion are used here in their final forms. Details may be found in the literature.[6–9]

The resonance frequency of a spin in the solid state depends upon the relative orientation of $\boldsymbol{B}_0$ and an axis system fixed to the crystallite geometry called the principal axis system (PAS) of the shielding tensor. In the case of polycrystalline or powdered solids, with no preferred crystallite orientation, all relative orientations of PAS and $\boldsymbol{B}_0$ are equally probable. The solid state NMR spectrum is therefore a superposition of resonance lines from all orientations. This results in a broad but very characteristic spectral pattern, called a powder pattern. The significant feature of the powder pattern is that it shows discontinuities or breakpoints at the three principal frequencies of the shielding tensor. These principal frequencies are precisely what we wish to be able to measure to characterize the tensor.

For a sample undergoing spinning motion, the relative orientation of the PAS and $\boldsymbol{B}_0$ acquires a periodic time dependence. The instantaneous resonance frequency of any spin therefore also changes periodically in time. The instantaneous resonance frequency of any given spin during the rotation period depends upon the initial orientation of the PAS and $\boldsymbol{B}_0$, the angle of spinning, and the strength of the anisotropic interaction. The dependence of the time evolution of the resonance frequency Ω of a given spin on the various variables that affect it is given by[6,9]

$$\Omega(\Theta, \Phi, \beta, t) = \Omega_0(\Theta, \Phi, \beta) + \Omega_a(\Theta, \Phi, \beta, t) \quad (1)$$

where

$$\Omega_0(\Theta, \Phi, \beta) = \sigma_i + \tfrac{1}{4}(3\cos^2\beta - 1)[(3\cos^2\Theta - 1)(\sigma_{33} - \sigma_i) + \sin^2\Theta \cos 2\Phi(\sigma_{11} - \sigma_{22})] \quad (2)$$

Here, Θ and Φ are the polar angles that define the relative orientation of the spinning axis direction in the PAS, and β is the angle between the sample spinning axis and $\boldsymbol{B}_0$. The three principal frequencies of the shielding tensor are given by σ_{11}, σ_{22}, and σ_{33}, with σ_i the isotropic frequency, which is the average of the three principal frequencies. The principal frequencies are defined such that $\sigma_{33} - \sigma_i \geqslant \sigma_{22} - \sigma_i \geqslant \sigma_{11} - \sigma_i$.

For the purposes of this article, equation (1) is written as the sum of two parts: a time-independent part Ω_0 and a time-dependent part Ω_a. This separation into two parts of the parametric dependence of the resonance frequency illustrates two essential features of the spectral response from spinning solids. The time-independent part determines the average frequency during a spinning period, and specifies the position of the main resonance line or the central line in the spectrum from a crystallite with given initial PAS orientation. The time-dependent part incorporates the explicit effect of periodic changes in the instantaneous resonance frequency on the spectral response. The exact form of Ω_a and methods adopted for solving it to calculate its effect on spectral response can be found in the literature.[6–9] For the purpose of discussion here, two significant properties of Ω_a are as follows:

$$\Omega_a(\Theta, \Phi, \beta, t) = \Omega_a(\Theta, \Phi, \beta, t \pm nt_r) \quad (3)$$

where t_r is the time of one rotation period and $n = 1,2,3,\ldots$, and

$$\Omega_a(\Theta, \Phi, \beta, t) \propto \frac{\sigma_{33} - \sigma_i}{\omega_r} \quad (4)$$

where $\omega_r = 1/t_r$ is the spinning speed. Equation (3) illustrates the periodicity of Ω_a, i.e., after every rotation period, Ω_a returns to its original value. Since Ω_a from all crystallites return to their original values at the same instant, an echo is formed in the time domain NMR signal—the free induction decay (FID). The FID may contain multiple echoes spaced in time by t_r. The Fourier transform of the FID containing such an echo envelope shows the appearance of extra lines, called spinning sidebands, in the frequency domain NMR spectrum.[6] The spinning sidebands appear at intervals equal to ω_r from the centerband frequency, which is given by Ω_0. Equation (4) states that the effect of Ω_a diminishes with increasing ω_r or decreasing $\sigma_{33} - \sigma_i$. In reality, the effect of Ω_a on spectral response is negligible under the condition $\sigma_{33} - \sigma_i < \omega_r$, in which case the entire spectral shape is determined by Ω_0.

Figure 1 shows computer-calculated VASS powder patterns from a single spin under fast spinning conditions, i.e., $\sigma_{33} - \sigma_i < \omega_r$. The bottom trace is the nonspinning powder pattern. The three discontinuities in the powder pattern correspond to the three principal values of the shielding tensor. Under fast spinning VASS conditions, the shape of the powder pattern remains identical to that of the nonspinning pattern, except that the entire pattern is scaled by a factor of $\frac{1}{2}(3\cos^2\beta - 1)$ about σ_i. This scaling factor is 1 when $\beta = 0$, i.e., the same as the nonspinning case. As β is increased, the powder pattern is scaled by this factor, which at the magic angle ($\beta = 54.74°$) is zero, and the entire powder pattern collapses into a single peak at σ_i. As β is moved past the magic angle, the scaling factor becomes negative, causing the entire powder pattern to be reflected about σ_i. The scaling factor has the maximum negative value of 0.5 at $\beta = 90°$. As the powder pattern is scaled with changing β, so are the discontinuities in the powder pattern at the three principal values of the shielding tensor. The appearance of a discontinuity in the scaled powder pattern corresponding to any σ value is given by

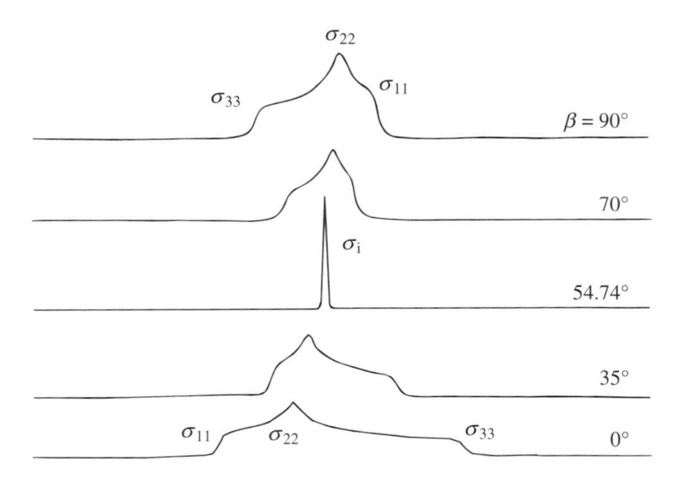

Figure 1 Effect of changing the angle of spinning β on powder patterns under fast spinning VASS conditions. Note that the positioning of all σs is reflected about σ_i for $\beta > 54.74°$

$$\sigma = \sigma_i + \tfrac{1}{2}(3\cos^2\beta - 1)(\sigma - \sigma_i) \quad (5)$$

Figure 1 shows how the positions of discontinuities in the powder pattern change under fast spinning VASS conditions.

Figure 2 shows calculated VASS powder spectra under slow spinning conditions, i.e., $\sigma_{33} - \sigma_i > \omega_r$. For samples spinning at the magic angle, the entire pattern is broken up into a series of sharp peaks. The intensity distribution of spinning sidebands under slow magic angle spinning conditions is strongly dependent upon the principal frequencies of the shielding tensor.[7] Under VASS conditions, however, the sidebands take on a fascinating variety of shapes. Under slow spinning VASS conditions, the shapes of the centerband as well as the sidebands deviate significantly from the ideal powder pattern shown in Figure 1.[9,10]

The shapes of slow spinning VASS spectral bands in Figure 2 can be understood by considering how the signal amplitude from any given crystallite is distributed between the centerband and the spinning sidebands. This distribution varies greatly amongst crystallites, depending upon the initial orientation of their PAS, i.e., two crystallites with different initial PAS orientations may not contribute the same signal amplitude to the centerband under slow spinning VASS conditions.[9] This results in the appearance of a distorted powder pattern, because the ideal powder pattern, shown in Figure 1, is predicated upon the

For References see p. 4908

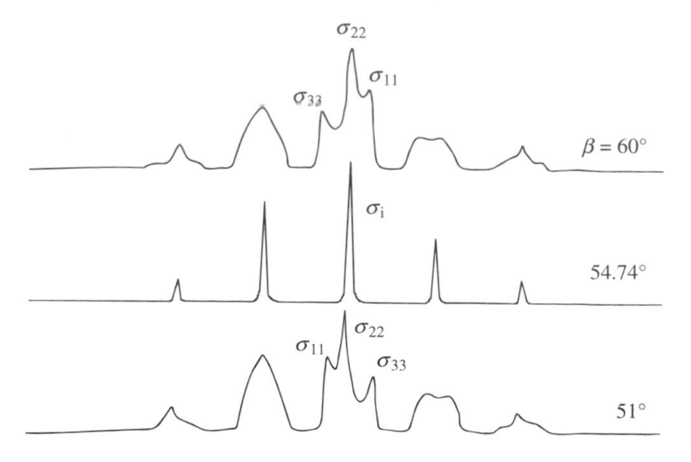

Figure 2 Distorted patterns under slow spinning VASS conditions. Note the cusps in the center band. (Reproduced with the permission of Taylor & Francis Ltd)

fact that each crystallite, irrespective of its initial orientation, contributes the same amplitude to the pattern. This can only occur under nonspinning or fast spinning VASS conditions.

An interesting consequence of such a nonuniform distribution of signal amplitudes is that the centerband develops cusps at frequencies where the discontinuities from the principal frequencies of the shielding tensor are expected in the scaled powder pattern. This happens because those particular orientations of the crystallites with Ω_0 close to a scaled principal frequency, equation (5), contribute significantly more amplitude to the centerband than they do to the sidebands, while those having Ω_0 values in between any two principal frequencies contribute significantly more amplitude to the sidebands.[9] Consequently, the shapes of the bands become highly distorted, and the centerband develops cusps at the principal frequencies.

3 EXAMPLES

From a practical viewpoint, VASS exploits the fact that when spinning at angles other that the magic angle, NMR spectral patterns are influenced by the relationship between the three principal frequencies and the isotropic frequency in a well-defined manner. This permits the design of experiments to measure the three principal frequencies in situations where the nonspinning spectrum has an overlap of different powder patterns to the extent that individual patterns cannot be identified.

We illustrate the applications of fast spinning VASS with an example. We wish to determine the principal values of CSA tensors for the three unique aromatic carbons in hexahydropyrene (HHP). Figure 3 shows the series of experimental and computer-calculated VASS spectra for HHP.[11] A broad pattern with few distinctive features is obtained in the nonspinning case (bottom trace, Figure 3), from which it is not possible to determine nine independent parameters, i.e., three σ values for three different types of aromatic carbons. Moreover, the σ_{33} values for all three aromatic carbons happen to be hidden under the uninteresting peaks from the methylene carbons, and therefore cannot even be detected. Figure 3 shows four fast spinning VASS spectra obtained at different β. Two immediate benefits of VASS are clear.

1. Since the entire aromatic band has been scaled, the uninteresting methylene peaks can now be completely ignored for analysis.
2. More importantly, the aromatic bands at different β are not merely scaled versions of the nonspinning spectrum or of each other. This is a direct consequence of the fact that, although all three patterns are scaled by the same factor, the relative positioning of the three overlapping patterns changes as β is changed, because of the differences in their isotropic frequency.

The level of spectral information available to determine principal frequencies of shielding tensors is therefore augmented by obtaining VASS spectra at different β. The discontinuity in a pattern corresponding to some principal frequency may be clearly identified at a particular choice of β, which otherwise was indistinguishable because of overlap.[5,11]

All VASS spectra contain complementary information. Analysis of these spectra to determine principal frequencies of shielding tensors is best done by simultaneous computer fitting

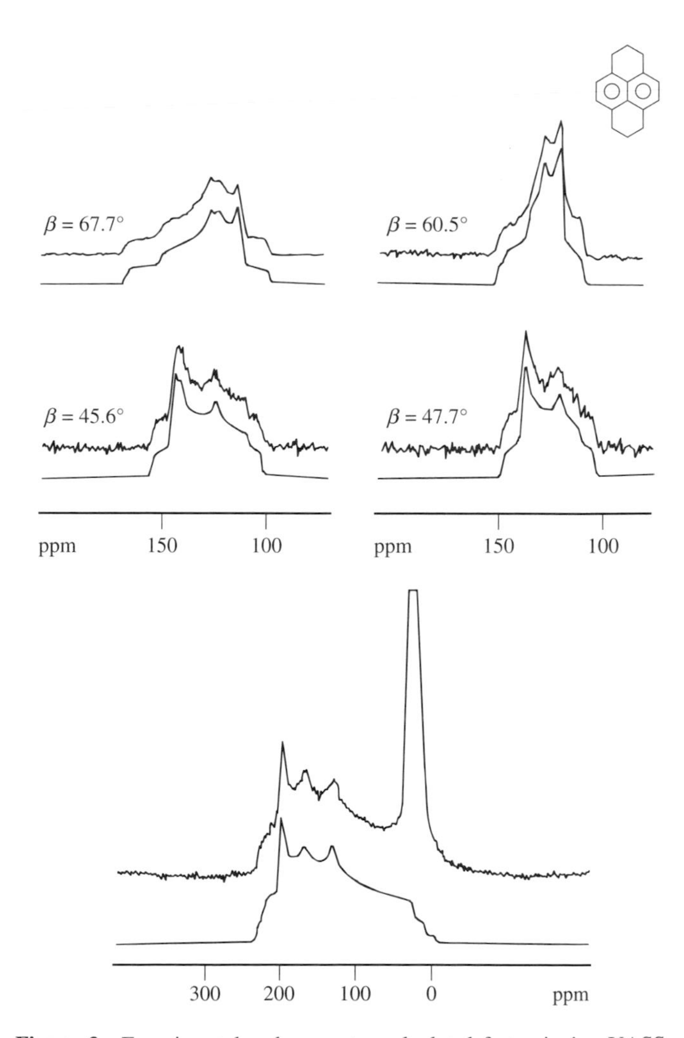

Figure 3 Experimental and computer calculated fast spinning VASS patterns from hexahydropyrene (HHP). The bottom trace shows the nonspinning spectrum and that calculated using parameters from VASS analysis. (Reproduced with the permission of the American Chemical Society)

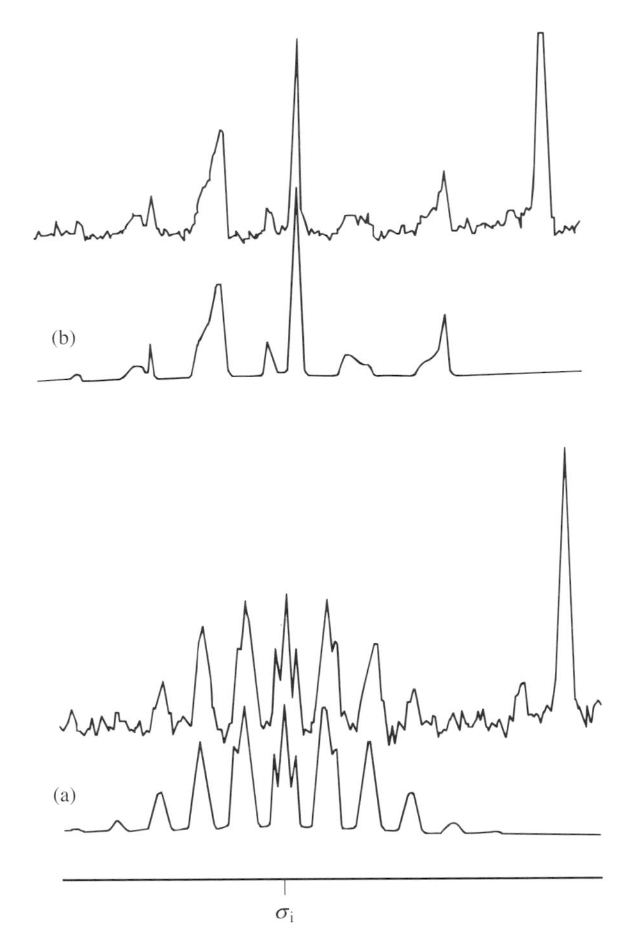

Figure 4 Experimental examples of slow spinning VASS, showing distorted band patterns from (a) malonic acid (with $\beta = 58.2°$ and $\omega_r = 832\,Hz$) and (b) hexamethylbenzene (with $\beta = 51.6°$ and $\omega_r = 510\,Hz$). In both cases, the intense peak on the right-hand side of the spectra is due to aliphatic carbons. (Reproduced with the permission of Taylor & Francis Ltd)

of all the spectra. The best values of principal frequencies are obtained when a single set of parameters is found to fit optimally all the VASS spectra. The results of such an analysis are plotted under each VASS spectrum in Figure 3. The parameters thus obtained can now be used to calculate the composite nonspinning spectra. This is shown at the bottom of Figure 3. Note that, while it may not have been possible to analyze unambiguously the nonspinning spectrum by itself, the data obtained from the analysis of multiple VASS spectra permitted characterizing it satisfactorily.

Figure 4 shows two experimental examples of slow spinning VASS. The bottom two traces are the experimental and computer-calculated spectra of carboxylic carbons of malonic acid. The shielding tensor for these carbons is known to be highly asymmetric.[12] The centerband, as expected, shows three spikes, corresponding to the three σ values, at positions given by equation (5). The top two traces in Figure 4 similarly show the experimental and calculated spectra of aromatic carbons of hexamethylbenzene (HMB). It is known that $\sigma_{11} = \sigma_{22}$ for HMB at room temperature.[13] Thus, only two spikes are seen in the centerband, indicating that two principal frequencies are sufficient to characterize this tensor. The simulations for both these examples are used to explain quantitatively the distorted lineshapes.[9]

Under optimum conditions, one could simply determine the principal frequencies by noting the frequencies of the cusps in the centerband for a given β, and using equation (5). For the examples shown in Figure 4, this procedure indeed yields satisfactory results for the principal frequencies of the shielding tensors.[9] However, it may not always be possible to resolve all the cusps from multiple patterns. Nevertheless, one should be aware of the phenomenon to avoid fortuitous misinterpretation of multiplicities of lines in solid state NMR spectrum when using MAS for chemical analysis, which can occur if β is not set precisely to the magic angle and the spinning speed is less than the anisotropy of the interaction tensor.

4 TWO-DIMENSIONAL (2D) METHODS

The VASS technique outlined above relies on obtaining a few VASS spectra individually, followed by computer analysis to determine principal frequencies of shielding tensors. All such 1D techniques are eventually limited by the resolution that they can afford. Two 2D methods are mentioned here that utilize the principles of VASS outlined above, but permit isolation of a VASS spectrum from each individual spin, thus allowing considerable simplification of analysis. The reader is also referred to other related NMR techniques: magic angle hopping (MAH),[14] stop and go (STAG),[15] shift scaling with rotation synchronized pulses,[16–19] and slow magic angle turning (MAT).[20]

In the switching angle sample spinning (SASS) (also called dynamic angle spinning, DAS) 2D technique, the isotropic chemical shift information is correlated with the VASS scaled anisotropic information.[21–23] During the preparation time of the SASS experiment, the scaled anisotropic information is encoded on the time domain NMR signal for a variable time t_1, while the sample spins at an angle far from the magic angle, say β_{t_1}. During the evolution time of a few milliseconds, β is mechanically switched to the magic angle. This is followed by the detection period, t_2, where a normal FID is collected. Since the FID is acquired under the MAS conditions, it only contains the isotropic frequency information, but each isotropic frequency had been modulated by the evolution during t_1 owing to scaled anisotropic information. A double Fourier transform of the data matrix generated with data acquired by systematically incrementing t_1 shows correlations between the isotropic peaks for each spin and its VASS spectrum for β_{t_1}. This allows VASS spectra for all spins to be isolated from each other, assuming their isotropic frequencies are sufficiently different, and analyzed individually.

The variable angle correlation spectroscopy (VACSY) 2D technique is a clever way to synthesize spectral information that is similar to SASS but without the need for any sudden mechanical switching of the spinning axis.[24,25] Instead, a series of 1D VASS spectra are obtained, one at a time, by systematically incrementing β. A 2D data matrix is then constructed, post hoc, by interpolating among all the 1D VASS data while exploiting the relationship between time evolution of the VASS NMR signal during data acquisition and the scaling factor $\frac{1}{2}(3\cos^2\beta - 1)$. Fourier analysis of these data permits

For References see p. 4908

isotropic versus full, i.e., unscaled, anisotropic information to be correlated.

The VACSY technique provides advantages over SASS/DAS in experimental simplicity, because it does not require switching β rapidly while the sample spins at high speed—a very demanding technical requirement. Also, the VACSY technique yields an unscaled powder pattern, thus providing somewhat superior resolution. However, a potential problem for the VACSY technique is that the scaling factor is bound within the limits 1.0 to −0.5. This causes incomplete data sampling in one dimension of the 2D experiment. Fourier analysis of incompletely sampled data results in baseline distortions that can complicate data analysis. In practice, though, this has not been found to be a serious drawback in the applicability of the VACSY technique.

The 2D techniques remove the limitations of the 1D VASS analysis, particularly for samples containing a large number of chemically different spins for which the 1D spectra are perhaps too complicated to permit an unambiguous analysis. This advantage, however, must be balanced against the technical difficulties and the time required to perform a full 2D experiment. Eventually, the experimenter must decide between the simplicity of the analysis from a complex 2D experiment and the complexity of analysis from a simple 1D spectrum—a ubiquitous dilemma in modern day NMR.

5 RELATED ARTICLES

Chemical Shift Tensor Measurement in Solids; Dynamic Angle Spinning; Line Narrowing Methods in Solids; Magic Angle Spinning; Magic Angle Turning & Hopping; Rotating Solids; Sideband Analysis in Magic Angle Spinning NMR of Solids.

6 REFERENCES

1. E. R. Andrew, A. Bradbury, and R. G. Eades, *Nature (London)*, 1959, **183**, 1802.
2. I. J. Lowe, *Phys. Rev. Lett.*, 1959, **2**, 285.
3. E. O. Stejskal, J. Schaefer, and R. A. McKay, *J. Magn. Reson.*, 1977, **25**, 569.
4. P. D. Murphy, and B. C. Gerstein, *J. Am. Chem. Soc.*, 1981, **103**, 3282.
5. N. K. Sethi, D. M. Grant, and R. J. Pugmire, *J. Magn. Reson.*, 1987, **71**, 476.
6. M. M. Maricq and J. S. Waugh, *J. Chem. Phys.*, 1979, **70**, 3300.
7. J. Herzfeld, and A. E. Burger, *J. Chem. Phys.*, 1980, **73**, 6021.
8. T. Tereo, H. Miura, and A. Saika, *J. Chem. Phys.*, 1986, **85**, 3816.
9. N. K. Sethi, D. W. Alderman, and D. M. Grant, *Mol. Phys.*, 1990, **71**, 217.
10. E. M. Merger, D. P. Raleigh, and R. G. Griffin, *J. Magn. Reson.*, 1985, **63**, 579.
11. A. M. Orendt, M. S. Solum, N. K. Sethi, C. D. Hughes, R. J. Pugmire, and D. M. Grant, in 'Magnetic Resonance of Carbonaceous Solids', ed. R. E. Botto and Y. Snada, Advances in Chemistry Series 229, American Chemical Society, Washington, DC, Chap. 22.
12. J. Tegenfeldt, H. Feucht, G. Ruschitzka, and U. Haeberlen, *J. Magn. Reson.*, 1980, **39**, 509.
13. D. E. Wemmer, D. J. Ruben, and A. Pines, *J. Am. Chem. Soc.*, 1981, **103**, 28.
14. A. Bax, N. M. Szeverenyi, and G. E. Maciel, *J. Magn. Reson.*, 1983, **52**, 147.
15. R. C. Zeigler, R. A. Wind, and G. E. Maciel, *J. Magn. Reson.*, 1988, **79**, 299.
16. E. Lippma, M. Alla, and T. Tuherm, in 'Proccedings of 19th Congrès Ampère', Heidelberg, 1976, ed. H. Brunner, K. H. Hausser, and D. Schweiter, p. 113.
17. Y. Yarim-Agaev, P. N. Tutunjian, and J. S. Waugh, *J. Magn. Reson.*, 1982, **47**, 51.
18. A. Bax, N. M. Szeverenyi, and G. E. Maciel, *J. Magn. Reson.*, 1983, **51**, 400.
19. R. Tycho, G. Dabbagh, and P. A. Miura, *J. Magn. Reson.*, 1989, **85**, 265.
20. Z. Gan, *J. Am. Chem. Soc.*, 1992, **114**, 8307.
21. A. Bax, N. M. Szeverenyi, and G. E. Maciel, *J. Magn. Reson.*, 1983, **55**, 494.
22. G. E. Maciel, N. M. Szeverenyi, and M. Sardarshti, *J. Magn. Reson.*, 1985, **64**, 365.
23. T. Tereo, T. Fujji, T. Onodera, and A. Saika, *Chem. Phys. Lett.*, 1984, **107**, 145.
24. L. Frydman, G. C. Chingas, Y. K. Lee, P. J. Grandinetti, M. A. Eastman, G. A. Barrall, and A. Pines, *J. Chem. Phys.*, 1992, **97**, 4800.
25. L. Frydman, G. C. Chingas, Y. K. Lee, P. J. Grandinetti, M. A. Eastman, G. A. Barrall, and A. Pines, *Isr. J. Chem.*, 1992, **32**, 161.

Biographical Sketch

Naresh K. Sethi. *b* 1961. B.S., 1982, University of Delhi, India; Ph.D., 1988, University of Utah. Introduced to NMR by D. M. Grant. Analytical research scientist at Amoco Research Center, 1989–present. Research interests include: application of solid state and high resolution NMR for characterizing petroleum oils, polymers and catalysts, and the use of electron spectroscopy (XPS and Auger) for surface analysis.

Vicinal Coupling Constants & Conformation of Biomolecules

Cornelis Altona

Leiden University, Leiden, The Netherlands

1 INTRODUCTION

The discovery[1] that the vicinal (three-bond) proton–proton coupling constant, ^3J(H,H), varies smoothly with the associated H–C–C–H torsion angle ϕ(HH) in ethane had a profound impact on the acceptance of NMR as an indispensable tool in stereochemistry and conformational analysis. The well-known Karplus equation,[1,2] also called the cosine-square rule [equation (1)], was derived from valence bond calculations on the unperturbed ethane molecule.

$$^3J(\mathrm{H,H}) = C_0 + C_1\cos(\phi_{\mathrm{HH}}) + C_2\cos(2\phi_{\mathrm{HH}}) \tag{1a}$$

The Fourier series embodied in equation (1a) is more often written as:

$$^3J(\mathrm{H,H}) = A\cos^2(\phi_{\mathrm{HH}}) + B\cos(\phi_{\mathrm{HH}}) + C \tag{1b}$$

Of course, simple relationships exist between C_0, C_1, C_2 and A, B, C, respectively: $C_0 = 0.5A + C$, $C_1 = B$, $C_2 = 0.5A$. Karplus[2] predicted that ^3J(H,H) depends on various other molecular parameters, such as the electronegativity of the substituents, bond lengths, and bond angles. In the same paper[2] he issued a caveat against the indiscriminate application of his equation and the theoretically derived parameters [coefficients in equation (1)] to 'estimate torsion angles to an accuracy of one or two degrees'.

The general shape of the Karplus curve, large ^3J at $\phi = 0°$ (eclipsed, sp) and at or near $\phi = 180°$ (*trans*, *anti* or *ap*), and decreasing to a minimum value ($^3J \approx 0$ Hz) near $\phi = 90°$, in the course of time was shown to be applicable to many molecular fragments, including ^3J(H,H) with heteroatoms in the coupling path and also ^3J(X,H) and ^3J(X,Y) with X,Y NMR-active heteronuclei like ^{13}C, ^{15}N, and ^{31}P, for example ^3J(H,C,O,H), ^3J(H,N,C,H), ^3J(N,C,C,H), ^3J(C,C,C,H), ^3J(H,C,O,P), ^3J(C,C,O,P). Several of these more exotic vicinal couplings are useful in the conformational analysis of biomolecules such as peptides, proteins, nucleic acids, and polysaccharides.

In suitable cases direct (^1J) and geminal (^2J) coupling constants provide additional structural information, albeit of a more qualitative nature. Long-range couplings (^nJ, $n > 3$) are often highly useful, especially ^4J couplings (planar W path). However, the present contribution focuses entirely on vicinal (^3J) couplings.

The coupling constant ^3J(H,C,C,H) with C saturated carbon is by far the best investigated and will be treated in depth in the present article. A selection of more simple Karplus equations involving heteroatoms will be presented and briefly discussed. Familiarity with the basic tenets and methods of conformational analysis is assumed. The multitude of techniques that are currently available to measure the coupling constants of interest remain outside the scope of the present contribution (see Section 6 for related articles). Clearly, the approaches that are found useful in the conformational analysis of biomolecules are applicable to many other molecular fragments in organic chemistry. The article concludes with some applications in the field of (poly)peptides and nucleic acids, and caveats are issued. The reader is advised to consult the selection of references given for further details and for references therein.

2 VICINAL PROTON–PROTON COUPLING CONSTANTS ^3J(H,C,C,H)

2.1 General Background

In the late 1950s experiments already showed[3] that the *time-averaged* ^3J(H,H) [equation (2)] in monosubstituted ethanes CH_3CH_2X decreases more or less linearly with increasing electronegativity of the substituent X.

$$\langle ^3J(\mathrm{H,H})\rangle = [J(60) + J(180) + J(300)]/3 \tag{2}$$

Studies carried out in the early 1960s[4,5] confirmed these findings. Perhaps unfortunately, the (erroneous) notion that ^3J(H,H) in all situations decreases with increasing electronegativity became firmly lodged in the minds of NMR spectroscopists. A second, also erroneous, assumption was generally accepted and that is the linear additivity of (relative) electronegativities $\Delta\chi_i = \chi_i - \chi_{\mathrm{H}}$. In other words, the general validity of equation (3) was accepted without question, although proper analysis of the experimental data on monosubstituted ethanes and 2-substituted propanes, already available in the early 1960s, could have shown differently.

$$^3J(\mathrm{H,H}) = \mathrm{f}(\phi) + a\Sigma_i\Delta\chi_i \tag{3a}$$

or alternatively:

$$^3J(\mathrm{H,H}) = \mathrm{f}(\phi)[1 - a(\Sigma_i\Delta\chi_i)/\mathrm{f}(\phi)] \tag{3b}$$

where f(ϕ) denotes a Karplus function of ϕ_{HH}; a/f(ϕ) is usually written as k.

In the case of coupling to the fast-rotating methyl protons, for example in substituted ethanes and 2-substituted propanes [equation (2)], the ϕ-dependent terms in equation (1a) vanish and f(ϕ) reduces to the 'constant' C_0. The following compari-

For References see p. 4922

son is obtained (Hz):

$$CH_3CH_2X:^3\ C_0 = 7.5, \quad a = -0.4$$

$$CH_3CH_2X:^4\ C_0 = 7.9, \quad a = -0.7$$

$$CH_3CHXCH_3:^3\ C_0 = 7.0, \quad a = -0.55$$

$$\text{Combined:}^5\ C_0 = 7.7, \quad a = -0.80$$

In the remainder of this article the slope (a) will be denoted C_{01}. The 'goodness of fit' of these least-squares regressions (±0.3 Hz) is disappointing; moreover, several workers[4,5] noted a breakdown of linearity in cases where X and Y in CH_3CHXY represent highly electronegative substituents like fluorine or oxygen. It should be noted that the measured coupling to methyl cannot yield the electronegativity dependence of the 'constants' C_1 and C_2 in equation (1a); it was simply assumed that these dependences were all equal.

In 1965 Booth[6] noted that an electronegative substituent located on the C–C fragment constituting the coupling path and in antiparallel (*ap*) position relative to one of the two coupling *gauche* (*sc*) protons causes an 'extra' diminution of their mutual coupling constant (Booth's rule). In hindsight it is easy to see that this observation implies a breakdown of the symmetry inherent to a Fourier series with cosine terms only, e.g., equation (1a), but it was not so interpreted at the time. On the contrary, one might say, and for a long time the Durette–Horton equation[7] [equation (4)], was popular in the field.

$$^3J(\mathrm{H,H}) = [C_0 + C_1\cos(\phi) + C_2\cos(2\phi)](1 - k\sum_1^4 \Delta\chi_i) \quad (4)$$

Equation (4) is simply a combination of equations (1a) and (3b), rewritten here in terms of cosine Fourier coefficients C; ϕ denotes the torsion angle between the coupling nuclei indicated in parentheses on the left-hand side of equation (4), $\Delta\chi_i$ are relative electronegativities of substituents according to either the Pauling[8] or the Huggins[9] scale.

Unfortunately, equation (4) embodies at least four questionable or even erroneous assumptions:

1. The orientation of any electronegative substituent with respect to the coupling protons is unimportant (in negation of Booth's rule).
2. Coefficients C_0, C_1, C_2 have the same electronegativity dependence.
3. Substituent effects are linearly additive.
4. The standard atomic electronegativity scales[8,9] can be applied to polyvalent substituents, regardless of the remaining atoms or groups attached.

At this point it should be mentioned that many workers in the field of stereochemistry in a way circumvented assumptions 2, 3, and 4 by the calibration of Karplus parameters for given classes of molecules of interest with the aid of model compounds with known or estimated torsion angles. Even today this approach often remains the only sensible course of action available in the case of couplings to heteronuclei, e.g., ^3J(C,H), ^3J(N,H), ^3J(P,H), ^3J(P,C), vide infra. More refined approaches are now possible in the case of ^3J(H,H) in sp^3–sp^3 H–C–C–H fragments.

2.2 Substituent-Induced Asymmetry in the Karplus Curve

The first systematic investigation into the orientational effects of substituents on ^3J(H,H) was carried out by Abraham and Gatti[10] with the aid of a series of 1,2-disubstituted ethanes CH_2X–CH_2Y. Their results are summarized in Figure 1.

For most practical purposes one needs to distinguish three different situations:

(a) *Gauche* (*sc*) protons, J_g, with X in an *anti* orientation X_t: J_g decreases strongly with increasing electronegativity of X (Booth's rule);
(b) *Gauche* (*sc*) protons, J_g, and a substituent X in a *gauche* orientation with respect to its vicinal coupling proton, X_g: J_g increases with increasing electronegativity of X (Abraham's rule);
(c) *Anti* or *trans* (*ap*) protons, J_t, with X necessarily in a *gauche* position (not shown in Figure 1): J_t decreases with increasing electronegativity of X.

For X = oxygen in conformationally rigid pyranose sugars, a statistical analysis[11] of a large set of couplings indicated the following changes of ^3J from X = H: situation (a) −1.8 Hz; (b) +0.5 Hz; (c) −1.45 Hz. This example clearly illustrates the fact that *gauche* (*sc*) couplings do not obey the simple cosine-square rule. The various effects are additive to a good approximation, 24 additivity constants (eight different substituent types) gave an rms error of only 0.29 Hz for a test set consisting of 305 couplings.

Extended Hückel molecular orbital theory (EHT) calculations of ^3J(H,H), carried out by Pachler[12,13] on CH_3CH_2X (X = H, CH_3, NH_2, OH, F), CH_3CHF_2, and CH_2FCH_2F, indicated that an X substituent (or the C–X bond) induces an asymmetry in the Karplus curve. More refined EHT calculations[14,15] confirmed this important finding and showed[15] that mutual interaction between substituents cannot be neglected. Pachler[12,13] described the calculated influence of the X substituent in terms of the summation of two effects: (1) an attenuation of the Fourier coefficients, i.e., as in equation (4) but with different k values for C_0, C_1, and C_2; (2) a phase shift, assumed to be proportional to the electronegativity of the substituent (Figure 2).

A phase shift can be formulated in several ways, for example by the introduction of sine terms.[12,13] This leads to equation (5).

$$^3J(\mathrm{H,H}) = C_0 + C_1\cos(\phi) + C_2\cos(2\phi) + S_1\sin(\phi) + S_2\sin(2\phi) \quad (5a)$$

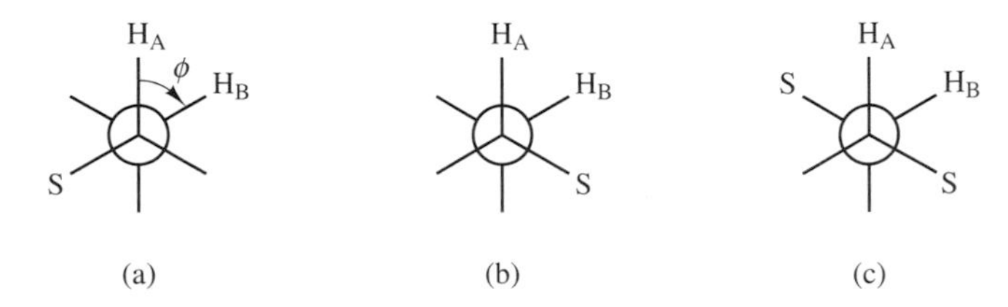

Figure 1 Newman projections along the central bond of a H$_A$–C–C–H$_B$ fragment, H$_A$ and H$_B$ are mutually coupled. (a) Situation pertaining to Booth's rule (strong decrease of ^3J(H,H) for an electronegative substituent S); (b) situation pertaining to Abraham's rule [weak increase of ^3J(H,H)]; (c) situation where Abraham's rule applies twice. This configuration occurs, for example, in cyclohexanes with H$_A$, H$_B$ diequatorial and the S substituents diaxial

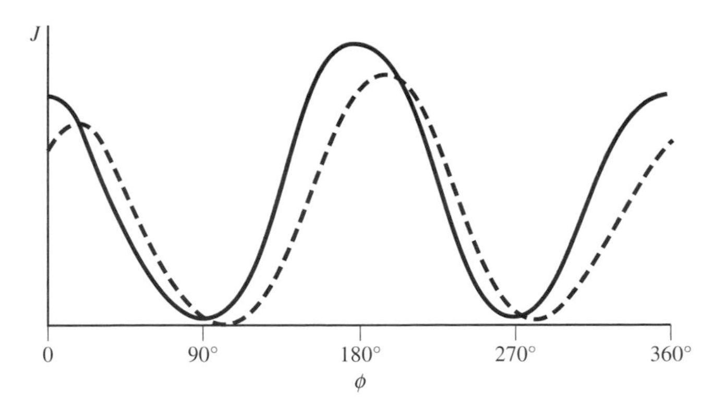

Figure 2 The calculated coupling constants (EHT method[14]) for ethane (solid line) and fluoroethane (dashed line) as a function of the HH torsion angle ϕ; the positive direction of ϕ is defined according to IUPAC,[17] the F atom occupies the S_1 substituent position shown in Figure 3 (sign factor s_i = +1). (Reproduced with permission from Elsevier Science Ltd., Kidlington, UK, from K. G. R. Pachler, *Tetrahedron*, 1971, **27**, 187)

with

$$C_m = C_{m0} + C_{m1}\Sigma_i\Delta\chi_i \quad \text{and} \quad S_m = S_{m1}\Sigma_i s_i\Delta\chi_i \tag{5b}$$

In equation (5) the original Pachler equation[12,13] is recast into a general Fourier expression.[15] Moreover, the use of a separate and cumbersome proton-substituent torsion angle Φ_{HX}[12,13] is avoided by the introduction of the 'sign' factor s_i for each substituent S_i, s_i is +1 or -1 according to the definition by Haasnoot et al.[16] (see Figure 3). (In this paper, as well as in others,[15] the sign factor S_i was denoted ξ_i.)

It is important to keep in mind that the sign factors s_i of the substituents S_1 and S_2 (*not* to be confused with the Fourier coefficients S_1 and S_2) depend exclusively upon their stereochemical position with respect to their 'own' geminal coupling proton H_A. The following recipe applies: choose a Newman projection along the coupling path in such a manner that the chosen coupling proton H_A is located on the front side (Figure 3): S_1 is found at 120° and S_2 at $-120°$ (positive angle defined as clockwise). Repeat the procedure with H_B up front, S_3 is at 120° and S_4 at $-120°$. The sign factors (s_i = +1 for substituents S_1 and S_3, $s_i = -1$ for S_2 and S_4) are thus uniquely defined and are *independent* of the actual value of the torsion angle ϕ_{AB}. It goes without saying that in the case of a CH_2 group ($CH_{A1}H_{A2}$) one has to take care to select the proper H_A (interchange of H_{A1} and H_{A2} means a change of sign factor for the remaining substituent). Equation (6) simplifies the computer programming:

$$\Sigma_i s_i\Delta\chi_i = \Delta\chi_1 - \Delta\chi_2 + \Delta\chi_3 - \Delta\chi_4 \tag{6}$$

It is easily appreciated that the sine terms vanish in all cases where equivalent substituents (same $\Delta\chi_i$) occupy a +1 and a -1 position at the same time, for example in $-CH_2-CHX_2$ and the +,− coupling in the $-CHX-CH_2X$ molecular fragment. The above sign factor definition is associated with the definition of the positive direction of ϕ_{AB} according to IUPAC rules:[17] positive when H_B rotates clockwise away from H_A, with the H_A, H_B eclipsed position taken as zero angle. Since the sine terms change sign at ϕ_{AB} = 0° a change of substituent position from +1 to -1 can be depicted as a mirroring of the ΔJ curve

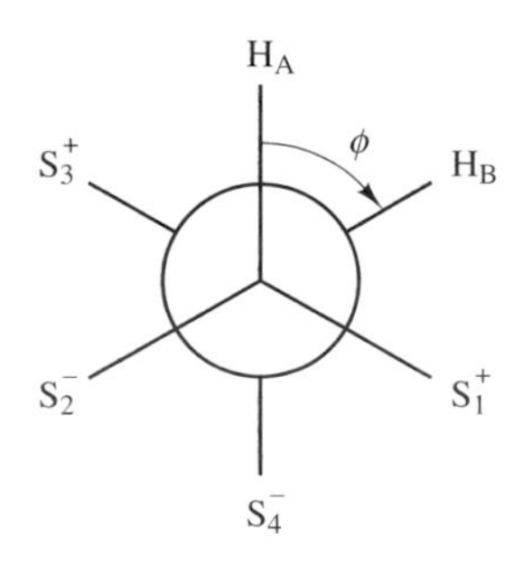

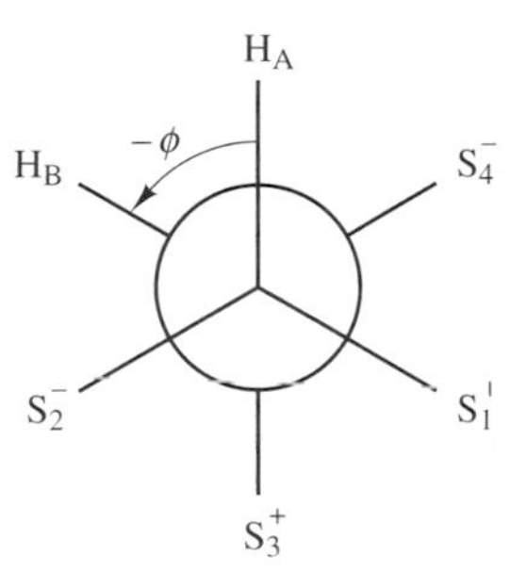

Figure 3 Definition of substituents with positive (S_1 and S_3) and negative (S_2 and S_4) sign factors s_i. Note that with each coupling nucleus in an sp^3–sp^3 fragment two geminal substituents with opposite sign factors are associated. The result of rotation of H_B with respect to H_A in the $-\phi$ direction is mathematically equivalent to the result of a $+\phi$ rotation with concomittant interchange of S_1 and S_2. (Reproduced with permission of Elsevier Science Ltd., Kidlington, UK, from C. A. G. Haasnoot, F. A. A. M. de Leeuw, and C. Altona, *Tetrahedron*, 1980, **36**, 2783)

(Figure 4). Equation (5) is invariant under electronegativity scale changes,[15] aside from changes in the values of the coefficients C_m and S_m. It is a valid procedure to choose the relative electronegativity of hydrogen as zero and to fix $\Delta\chi$ of one other substituent. For practical reasons, a value λ_{OR} = 1.40 was selected, vide infra.

Haasnoot et al.[16] also sought to describe the generalized Karplus equation by superimposing a phase shift upon the symmetrical ethane curve [equation (7) and Figure 4].

$$\begin{aligned} {}^3J(\mathrm{H,H}) = {} & P_1\cos^2(\phi) + P_2\cos(\phi) + P_3 \\ & + \Sigma_i\Delta\chi_i[P_4 + P_5\cos^2(s_i(\phi) + P_6|\Delta\chi_i|)] \end{aligned} \tag{7}$$

Comparison with equation (5) shows that C_{11} and S_{11} have been set to zero, and that in the cos(2ϕ) and sin(2ϕ) terms both a constraint (through a common parameter P_5) and a nonlinear $\Delta\chi_i$ dependence have been introduced. The invariance property (vide supra) does not hold for equation (7). The parameters P_1–P_6 were empirically determined with the aid of a test set containing 315 couplings and corresponding ϕ_{HH} values.[16] Each substituent was assigned a $\Delta\chi_i$ value according to Huggins,[9] but empirically modified to account for the moderation of the electronegativity of the α substituent by attached atoms or groups (β substituents), other than hydrogen. A least-squares regression (J_{calc} versus J_{obs}) showed an rms error of 0.48 Hz and a correlation coefficient R of 0.992, in contrast to a similar

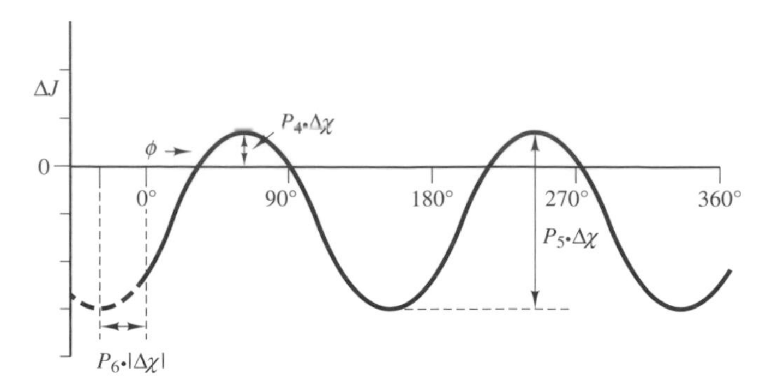

Figure 4 The calculated difference curve $\Delta J = J(CH_3CH_2S) - J(CH_3CH_3)$ for a substituent S with a positive sign factor, showing the rationale behind the phase-shift parameters P_4, P_5, and P_6 in the Haasnoot equation (7)[16]

regression using three parameters only, equation 1(a) (rms = 1.20 Hz, $R = 0.947$).[16]

2.3 Group Electronegativities and Nonadditivity Effects

Analysis[15] of the results of a systematic series of calculations of ^3J(H,H) in a variety of mono-, di-, tri-, and tetrasubstituted ethanes by means of the EHT method (1404 J values) suggested that: (a) the use of standard relative electronegativities $\Delta\chi_i$ should perhaps be abandoned in favor of 'best values' of substituent constants λ_i derived from the data set at hand; (b) the nonadditivity effect (interaction between substituents) can be well described by the introduction of cross terms involving products $\Delta\chi_i\Delta\chi_j$ or $\lambda_i\lambda_j$. Díez et al.[18] allowed the coefficients of equation (5b) to be quadratically dependent upon λ. It was shown that only certain combinations of cross terms can survive in such an expansion because of symmetry. Neglecting interactions between more than two substituents, one obtains equation (8), which is in fact an extension of equation (5).

$$C_m = C_{m0} + C_{m1}\Sigma_i\lambda_i + C_{m11}\Sigma_i\lambda_i^2 + C_{m12}(\lambda_1\lambda_2 + \lambda_3\lambda_4) + C_{m13}(\lambda_1\lambda_3 + \lambda_2\lambda_4) + C_{m14}(\lambda_1\lambda_4 + \lambda_2\lambda_3) \qquad (8a)$$

$$S_m = S_{m1}\Sigma_i s_i\lambda_i + S_{m11}\Sigma_i s_i\lambda_i^2 + S_{m13}(\lambda_1\lambda_3 + \lambda_2\lambda_4) \qquad (8b)$$

How can we determine the 'substituent constants' or 'group electronegativities' λ_i (which may contain other factors beside electronegativity)? Which of the cross terms in equation (8) are important and which can be neglected safely? In an attempt to answer these questions a combined empirical and theoretical approach was taken.

First, a large body of accurate (±0.02 Hz) coupling data from mono- and 1,1-disubstituted ethanes was measured and analyzed.[19] Because of torsional averaging, one isolates the 'constant' C_0 in equation (8a), $m = 0$. Excluding the self-quadratic term C_{011} from consideration,[15] we obtain equation (9).

$$\langle {}^3J(\mathrm{H,H})\rangle = C_{00} + C_{01}(\lambda_1 + \lambda_2) + C_{012}(\lambda_1\lambda_2) \qquad (9)$$

Not only were the coefficients C_{00}, C_{01}, and C_{012} optimized, but also the λ_i values. As explained above, parameter redundancies were removed by fixing the λ values of two different substituents present in the set: $\lambda_H = 0.00$ and $\lambda_{OR} = 1.40$ (measurements in solvent $CDCl_3$). The 'best' least-squares regression[19] yielded equation (10):

$$\langle {}^3J(\mathrm{H,H})\rangle = 7.836 - 0.594(\lambda_1 + \lambda_2) - 0.423(\lambda_1 \cdot \lambda_2) \qquad (10)$$

and in addition 50 λ_i values[19] from 84 couplings; rms deviation = 0.018 Hz, maximum deviation = 0.06 Hz. A similar regression with the use of Huggins's relative electronegativities yielded a 10-times larger rms deviation;[19] however, this regression also underlined the importance of the $\lambda_1\lambda_2$ cross term, at least for coupling to methyl protons.

Equation (10) proved to be highly useful for determining λ_i values of substituents not present in the original data set. Recently, special attention[20] was paid to solvent effects, in particular to the influence of the solvent D_2O in cases where the α substituent carries at least one nonconjugated lone pair of electrons that can readily act as a hydrogen bond acceptor. The vicinal coupling constants (and thus λ values) measured in common organic solvents (like acetone-d_6, CD_3CN, and DMSO)

Table 1 Substituent Parameters λ_i (Relative Group Electronegativities) Deduced from $\langle {}^3J\rangle$ of Various Mono- and 1,1-Disubstituted Ethanes in $CDCl_3$ and in D_2O, equation (10). More Detailed Data are given Elsewhere.[19,20,24] R Signifies Carbon, usually Alkyl; Z Signifies any Substituent, except Hydrogen and Halogens.[19] Fixed Points are Indicated by an Asterisk; 90% Confidence Limits (× 100) are Shown in Parentheses. Note that Several Values are not Applicable to Monosubstituted Ethanes[19,20]

Group	$\lambda_i(CDCl_3)$	$\lambda_i(D_2O)$
H	0.00*	0.00
CN	0.33(3)	0.33(3)
COOH	0.39(3)	0.33(3)
COO^-	–	0.41(4)
$COOR/CONH_2$	0.42(2)	–
Ph	0.45(4)	–
C(=O)R	0.50(3)	0.50
$C(Me)_3$	0.48	–
CHMeZ[a]	0.62	–
CH_2Z[b]	0.72	–
CH_3	0.80	–
SH	0.75(3)	0.75(3)
I	0.63(4)	–
Br	0.80(3)	–
Cl	0.94(3)	0.94(3)
F	1.37(5)	–
NRC(=O)R (*S-cis*)	0.52(4)	0.52(4)
9-Purines	0.56(2)	0.56(2)
1-Pyrimidines	0.56(2)	0.56(2)
NHC(=O)R	0.85(3)	0.81(3)
NH_3^+	–	0.82(4)
NR_2	1.12(5)	1.00(5)
NHR	1.15(5)	1.02(5)
NH_2	1.19(4)	1.09(4)
NO_2	0.77(3)	0.77(3)
OC(=O)R	1.16(3)	1.16(3)
OSO_2OR	1.19(9)	–
OPO_2OR	–	1.25(4)
$OPO(OR)_2$	1.29(4)	1.25(4)
OH	1.33(4)	1.25(4)
OR	1.40*	1.26(4)
OAr	1.42(5)	1.34(5)

[a] λ_i range 0.60–0.62; for Z = halogen: Cl 0.85, Br 0.92, I 1.02.
[b] λ_i range 0.65–0.78; for Z = halogen: F 0.65, Cl 0.82, Br 0.92, I 0.99.

hardly differ from those established in $CDCl_3$. Table 1 presents a selection of λ_i values of substituents that are of particular value in the analysis of couplings measured in biomolecules.

Second, the least-squares analysis of the set of 1404 J values indicated[15] that only two more cross terms (besides C_{012}) are probably important for use in equation (8a): C_{214} and either S_{211} or S_{213}.

2.4 Parameter Optimizations

In order to test the ideas developed in the previous sections, an empirical approach was again taken. The original data set (315 J set)[16] was first reduced to 299 J values (299 J set) by the deletion of norbornane- and norbornene-like structures [equations (5) and (7) do not contain correction terms for through-space effects,[21] nor for deviations of CCH bond angles from the tetrahedral value, see Section 5]. Group electronegativities λ_i were selected for the various substituents present in the data set. The corresponding torsion angles were recalculated with the aid of the MM2-85 force field.[22]

A series of least-squares parameter optimizations were carried out (see Table 2). It is of interest to watch the successive improvement, as judged by the drop in rms deviation, as we start (entry 1) from the original three-parameter Karplus equation [equation (1a)], and successively add (entry 2) an electronegativity dependence of the C_0 and C_2 coefficients, add (entry 3) the sine terms of the modified Pachler equation [equation (5)], use (entry 4) the Haasnoot approximation [equation (7)], and finally the Díez–Donders equation [equation (8)], the latter in three forms: (1) (entry 5), all cross terms zero, except the $S_{211}\sum_i S_i \lambda_i^2$; (2) (entry 6) as (1) but C_{012} $(\lambda_1\lambda_2 + \lambda_3\lambda_4)$ and C_{214} $(\lambda_1\lambda_4 + \lambda_2\lambda_3)\cos(2\phi)$ added; (3) (entry 7) as (2) plus a small threefold term $C_3\cos(3\phi)$ added. In the latter parameter optimization, the C_{01} parameter was kept fixed on the value obtained from the 'ethane set' [equation (10)]; C_3 was fixed at −0.10, a value close to that suggested by ab initio calculations.[23]

The current parameters are shown in equations (11)–(13), with reference to the entries in Table 2.

Entry 4, Haasnoot:

$$^3J(\mathrm{H,H}) = 14.64\cos^2(\phi) - 0.78\cos(\phi) + 0.58 + \Sigma_i\lambda_i[0.34 - 2.31\cos^2(s_i(\phi) + 18.40|\lambda_i|)] \quad (11)$$

Entry 5, Díez–Donders:

$$^3J(\mathrm{H,H}) = (7.82 - 0.79\Sigma_i\lambda_i) - 0.78\cos(\phi) + (6.54 - 0.64\Sigma_i\lambda_i)\cos(2\phi) + (0.70\Sigma_i s_i\lambda_i^2)\sin(2\phi) \quad (12)$$

Entry 7, Díez–Donders:

$$^3J(\mathrm{H,H}) = [7.41 - 0.59\Sigma_i\lambda_i - 0.27\Sigma_i(\lambda_1\lambda_2 + \lambda_3\lambda_4)] - 0.85\cos(\phi) + [6.84 - 0.87\Sigma_i\lambda_i + 0.29(\lambda_1\lambda_4 + \lambda_2\lambda_3)]\cos(2\phi) - 0.10\cos(3\phi) + (0.71\Sigma_i s_i\lambda_i^2)\sin(2\phi) \quad (13)$$

The coefficients of equation (13) differ from previous ones[24] mainly because a constraint is now laid upon the $\cos(3\phi)$ coefficient, vide supra.

Table 2 Final rms Deviations Obtained for ^3J(H,H) with the Aid of the 299 J Test Set in the Various Approximations[a]

Approximations	rms($\Delta\chi_i$)[b] (Hz)	rms(λ_i)[c] (Hz)	N[d]
1. Karplus[e]	1.18	1.18	3
2. Karplus (linear)[f]	0.73	0.70	5
3. Pachler[g]	0.38	0.39	6
4. Haasnoot[h]	0.37	0.36	6
5. Díez–Donders[i]	0.38	0.33	6
6. Díez–Donders[j]	0.37	0.32	8
7. Díez–Donders[k]	0.36	0.31	7

[a]The C_{11} coefficient in equation (5b) was set to zero in all parameter optimizations. It is strongly correlated to some other coefficients.
[b]With the use of Huggins's relative electronegativity scale[9] without β substituent correction.[16]
[c]With the use of experimental 'group electronegativity' values λ_i.[19,20]
[d]Number of optimized parameters.
[e]Equation (1a).
[f]Equation (5), C coefficients only.
[g]Equation (5), S_2 added (the rms error does not change when C_{11} and S_{11} are also added).
[h]Equation (7), the coefficients (based on λ_i) are shown in equation (11).
[i]Equation (8): C_{00}, C_{01}, C_{10}, C_{20}, C_{21}, S_{211} used, the coefficients (based on λ_i) are shown in equation (12).
[j]Equation (8): C_{00}, C_{01}, C_{012}, C_{10}, C_{20}, C_{21}, C_{214}, S_{211} used.
[k]Equation (8), as in footnote j, but C_{01} fixed on the 'ethane' value, and C_3 added (C_3 = 0.1 taken as the fixed value); the coefficients (based on λ_i) are shown in equation (13).

2.5 Summary of Results

1. In previous work the unsatisfactory rms deviation (0.48 Hz) obtained with the use of equation (7) and the full 315 J calibration set, prompted Haasnoot et al.[16] to propose a division of the set into three parts. In other words, separate parameters were presented for ethane-like fragments which contained two, three, and four substituents, i.e. for CH_2–CH_2, CH–CH_2, and CH–CH moieties, respectively. In contrast, all of the parameter optimizations reported in Table 2 were carried out on the full 299 J test set and it was found that a single optimization suffices to reach an acceptable rms deviation (0.37 Hz). Presumably, the improved agreement between J_{obs} and J_{calc} is at least, in part, the result of an improvement in calculated ϕ_{HH} values in the present calibration set.
2. The extended five-parameter Karplus equation [terms up to $\cos(2\phi)$ and linear electronegativity dependence of the Fourier coefficients C_0 and C_2] gives a 'best' rms deviation of only 0.70 Hz. Introduction of the asymmetry effect induced by a substituent diminishes the rms error in a dramatic way to less than 0.4 Hz. Systematic theoretical and experimental investigations into the effect of asymmetry on other types of vicinal couplings: ^3J(C,H), ^3J(N,H), ^3J(P,H), ^3J(P,C), unfortunately appear to be lacking, but see Section 3.2.
3. The Pachler Fourier series equation, equation (5), postulates a linear electronegativity dependence of the asymmetric part, expressed as $S_{21}\sum_i s_i\Delta\chi_i \sin(2\phi)$ or in the present work also as $S_{21}\sum S_i\lambda_i \sin(2\phi)$. The Haasnoot[16] formulation, equation (7), implicitly contains $\Delta\chi_i^2\cos(2\phi)$ and $\Delta\chi_i^2 \sin(2\phi)$ terms. The Díez–Donders equation, equation (8), explicitly uses an $S_{211}\sum_i S_i\lambda_i^2 \sin(2\phi)$ term. This term alone, in combination with the use of empirical group electronega-

For References see p. 4922

tivities λ_i, is responsible for a substantial decrease in the rms deviation compared with the linear Fourier: 0.33 Hz versus 0.39 Hz, compare entries 5 and 3 in Table 2.

4. The introduction of two more cross terms (C_{012} and C_{214}) in the Díez–Donders formulation is only slightly effective, compare entries 5 and 6. This result comes as a surprise in view of the importance of the C_{012} term in the ethane/2-propane series [equation (10)].
5. The use of a Pauling/Huggins-type electronegativity scale appears perfectly justified up to a certain level of approximation.

3 VICINAL COUPLING CONSTANTS INVOLVING HETEROATOMS

3.1 General Background

Measurement of heteronuclear coupling constants is an essential factor in the conformational analysis of biomolecules.[25–28] Over the past three decades much effort has been invested into the delineation of Karplus-type relations for couplings involving one or more heteronuclei, either as coupling partner or contained in the coupling path. Unfortunately, there exists no firmly established quantitative correlation between ^3J(H,C,C,H) and corresponding couplings to heteronuclei [^3J(X,C,C,H), X = ^{13}C, ^{15}N, ^{31}P], nor simple rules that predict the behavior of a vicinal coupling when a carbon atom in the coupling path is replaced by a heteroatom. In short, for each different combination of nuclei and coupling path, a Karplus-type equation must be parameterized from scratch, preferably on the basis of a large number of experimental couplings (calibration points). Semiempirical calculations of vicinal couplings cannot predict ^3J with an accuracy desired by NMR spectroscopists, but such calculations, especially when carried out for a series of related molecular fragments,[25] may reveal details that would otherwise not be considered.

Thus far, the proposed parameterizations, resulting in a set of Karplus coefficients, rarely appear to have gone beyond the level of the three-parameter equation (1b). The problem of substituent effects on the A, B, and C coefficients is usually circumvented by the use of model compounds that contain the same (or at least similar) substituents as the compounds that one is interested in. The torsion angles necessary for such a parameterization are either estimated from molecular models, calculated by means of force-field methods, or taken from X-ray studies. Thus, one may hope to work effectively on the level of the five-parameter Karplus expression. It should be

Table 3 Selection of Coefficients (Hz) for Use in the Standard $\cos^2$ Karplus Equation [equation (1b)] for Couplings Relevant to Amino Acids and Polypeptides; $^3J_{ap}$, $^3J_{sc}$ and the Rotationally Averaged Value $\langle ^3J \rangle$ are also given. In the Column Remarks MO stands for Semiempirical Calculations; (Exp(N) Denotes Experimental Calibration on N Data Points. See Figures 6 and 7 for Atom Numbering and Definition of Torsion Angles. In many Amino Acids X = C$^\gamma$

Main torsion angle	^3J monitored	A	B	C	$^3J_{ap}$	$^3J_{sc}$	$\langle ^3J \rangle$	Remarks
ϕ	HNC$^\alpha$H$^\alpha$	6.7	−1.3	1.5	9.5	2.5	4.8	Exp(112)[29]
		6.4	−1.4	1.9	9.7	2.8	5.1	Exp(37)[30]
		9.4	−1.1	0.4	10.9	2.2	5.1	Exp(22)[25]
ϕ	HNC$^\alpha$C$^\beta$	4.7	−1.5	−0.2	6.0	0.2	2.1	MO *trans*-amides[25]
ϕ	HNC$^\alpha$C′	5.7	−2.7	0.1	8.5	0.2	3.0	MO *trans*-amides[25]
ϕ	C′NC$^\alpha$H$^\alpha$	9.0	−4.4	−0.8	12.6	−0.8	3.7	Exp(3) tentative[25]
		4.5	−1.3	−1.2	4.6	−0.7	1.1	MO *trans*-amides[25]
		4.0	−1.8	0.8	6.6	0.9	2.8	Exp(12) *cis*-amides[31]
ϕ	C′NC$^\alpha$Cβ	1.8	−0.2	0.5	2.5	0.9	1.4	Exp(17) *cis*-amides[31]
		1.3	−0.6	−0.1	1.8	−0.1	0.5	MO *trans*-amides[25]
ϕ	C′NC$^\alpha$C′	1.4	−0.8	−0.3	1.9	−0.4	0.4	MO *trans*-amides[25]
χ	H$^\alpha$C$^\alpha$C$^\beta$H$^\beta$	see equation (12) and Table 5						Exp(299)
$\chi_1\chi_2$	H$^\alpha$C$^\alpha$C$^\beta$C$^\gamma$	10.2	−1.3	0.2	11.7	2.1	5.3	Exp(6) ornithines[32]
χ	C′C$^\alpha$C$^\beta$H$^\beta$	7.7	−0.9	0.0	8.6	1.5	3.9	Exp(10) norbornanes[33]
		–	–	–	10.0	1.0	4.0	Exp[a,34]
		–	–	–	11.9	0.4	4.2	Exp[a,34]
χ	NC$^\alpha$C$^\beta$H$^\beta$	–	–	–	−5.4	−1.7	−2.9	Exp[a,36]
χ	C′C$^\alpha$C$^\beta$C$^\gamma$	1.4	−0.8	−0.3	1.9	−0.4	0.4	MO *trans*-amides[25]
ψ	H$^\alpha$C$^\alpha$C′N	−5.1	2.2	0.9	−6.4	0.7	−1.7	MO diamide[25]
		−5.3	2.2	0.9	−6.6	0.7	−1.7	MO N–Me acetamide[37]

[a]Assumption made that populations about χ follow from ^3J(H,C,C,H): $^3J_{ap}$ = 13.6 Hz, $^3J_{sc}$ = 2.6 Hz.

Table 4 Selection of Coefficients (Hz) for Use in the Standard $\cos^2$ Karplus Equation [equation (1b)] for Couplings Relevant to Nucleic Acids. In the Column Remarks Exp(N) Denotes Experimental Calibration on N Data Points. See Figures 8, 10, 11 for Atom Numbering and Definition of Torsion Angles

Main torsion angle	^3J monitored	A	B	C	$^3J_{ap}$	$^3J_{sc}$	$\langle ^3J \rangle$	Remarks
χ	C(=C)NCH	6.2	−2.4	0.1	8.7	0.5	3.2	Exp(5)[47]
χ	C(=O)NCH	5.0	−2.1	0.1	7.2	0.3	2.6	Exp(4)[47]
β, ϵ	HCOP	15.3	−6.1	1.6	23.0	2.4	9.3	Exp(7)[45]
β, ϵ	CCOP	6.9	−3.4	0.7	11.0	0.7	4.1	Exp(10)[45]
γ Sugar	HCCH	see equation (12) and Table 7						Exp(299)

remembered that in the case of ^3J(H,C,C,H) this level yields an rms deviation of 0.7 Hz (299 experimental calibration points); inclusion of an asymmetry term causes the rms deviation to drop to less than 0.4 Hz, vide supra.

A selection of coefficients for use in equation (1b) that are currently used in structural and conformational studies of proteins (Table 3) and nucleic acids (Table 4) is presented here. Some parameterizations have been carried out solely on the basis of semiempirical MO calculations in various approximations and these should perhaps be taken only as a rough guide. The coefficients for ^3J(C,C,C,H) in Table 3 were determined from measurements[33] in norbornanes and are not necessarily representative for this coupling in open-chain compounds.

The most important asset that these coefficients presently offer is *not* their overall reliability (statistically tested in a small minority of cases) but the opportunities that are provided to remove ambiguities. For example, a single coupling constant ^3J(H,H), measured along a given bond in an open-chain molecule, in the general case (single conformer assumed) corresponds to any one out of four torsion angles ϕ_{HH}. The situation is even more complex in the case of a conformational mixture. Modern NMR techniques allow us to measure accurately one or more ^3J(X,H) and sometimes ^3J(X,Y) along the same bond. The various Karplus curves are phase-shifted with respect to one another, because the X and/or Y nuclei are shifted by ±120° (sp^3) or 180° (sp^2). Thus, comparison of the various couplings often solves an assignment problem, always acute when dealing with diastereotopic nuclei, and/or may result in the selection of only one conformation that fits all the data.[38] Other applications are discussed in Section 4.

3.2 Vicinal Carbon–Proton Couplings ^3J(C,C,C,H)

This particular coupling constant appears to have resisted[39] all attempts to formulate satisfyingly simple and general trigonometric relationships similar to those developed for ^3J(H,C,C,H). Although early MO calculations[40] on ^3J(C,H) in substituted propanes already indicated that these couplings indeed obey a Karplus-type equation, in practice severe complications arose and progress remains slow. The present situation concerning ^3J(C,H), Figure 5, can be summarized as follows:

(a) The ratio ^3J(C,H) / ^3J(H,H) in molecular fragments where C replaces a corresponding H varies from 0.50 to 0.85 (mean 0.61 ± 0.065).[27] Thus, this ratio cannot be used with confidence to scale ^3J(H,H) couplings or Karplus coefficients to fit ^3J(C,H) data.

(b) The influence of electronegative substituents on rotationally averaged $\langle ^3J$(C,H)$\rangle$ in *t*-butyl derivatives correlates well with ^3J(H,H), with the former slightly more sensitive to the electronegativity effect.[41] More recently, an analysis of ^3J(C,H) in carbohydrates[42] suggested that oxygen (and carbon) substituents require a different set of coefficients [including the sin(2ϕ) term] when moved from the β position to the γ position with respect to the coupling carbon nucleus, Figure 5. This implies that the minimum number of Karplus coefficients could be twice as large compared with the six that suffice to describe ^3J(H,H).

(c) When the coupling ^{13}C nucleus itself carries one or more substituents besides hydrogen, matters are even more complicated. It has long been known that the nature of the X substituent has a profound influence on the magnitude of the rotationally averaged ^3J(C,H) in $(CH_3)_3{}^{13}CX$ compounds.[43] Other factors besides electronegativity appear to be operative here. Moreover, MO calculations of ^3J(C,H) in 1-fluoropropane[40] (Figure 5, X = F) predicted that this coupling is substantially larger when ψ_{FC} is set at 180° than when it is set at 60°. In a recent MO study[44] on a series of 1-substituted propanes (Figure 5, X = H, CH_3, NH_2, OH, and F) both ϕ and ψ were varied from 0° to 360° in 30° steps. The rotation of ψ indeed appears to have a profound effect on ^3J(C,H). For example, for ϕ_{CH} = 180° the calculated ^3J(C,H) (X = F) varies from 8.0 Hz (ψ_{FC} = 0°) to 12.0 Hz (ψ_{FC} = 150°). These predictions were experimentally verified by measurements on *cis*- and *trans*-4-*t*-butylcyclohexanols, where the effect is even larger than predicted for X = OH.[44] In the same paper[44] the set of 340 calculated ^{13}C–H couplings was fitted (rms = 0.27 Hz) with the aid of a nine-coefficient Karplus-type expression. A 15-coefficient approximation was used to fit 39 experimental ^3J(C,H) values of cycloalkanols and pyranose sugars (rms = 0.30 Hz).[42]

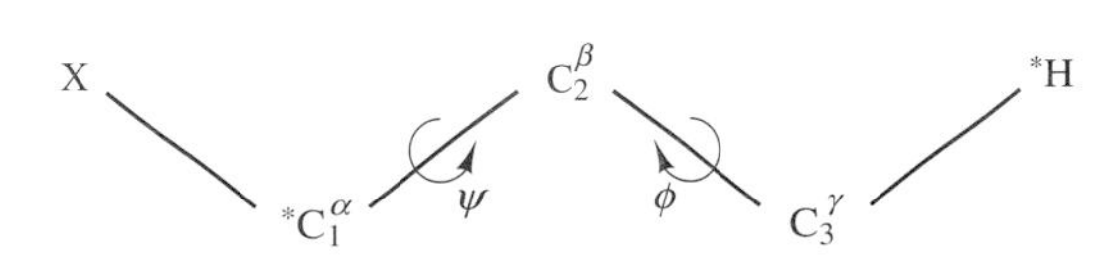

Figure 5 Torsion angles ϕ and ψ and atomic numbering for ^3J(C,C,C,H). The coupling nuclei are marked with an asterisk

For References see p. 4922

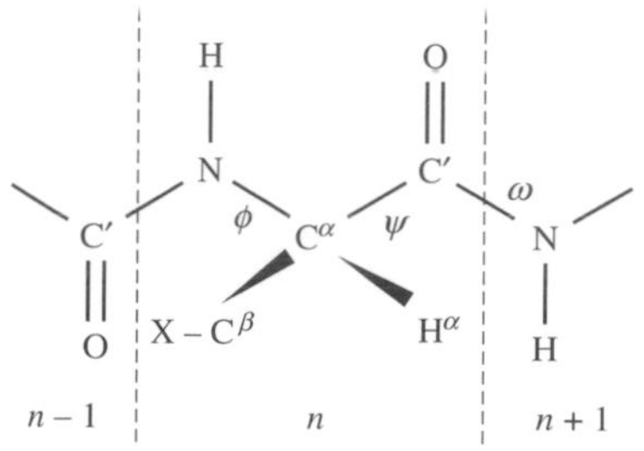

Figure 6 Schematic representation of the main-chain torsion angles ϕ, ψ, and ω in polypeptides and proteins. The χ_1 angle is defined as the torsion angle N–C$^\alpha$–C$^\beta$–X; in many amino acids X = C$^\gamma$

In summary, much work has to be done before ^3J(C,H) can be used to calculate torsion angles. In the meantime, it appears safe to apply the general rule: $^3J_{ap} >> {}^3J_{sc}$.

3.3 Vicinal Proton–Proton Couplings ^3J(H,N,C,H)

This coupling, usually denoted $^3J_{HN\alpha}$, deserves some attention because it can be measured with precision in proteins and affords important information concerning the local backbone geometry along the ϕ angle (Figures 6 and 7). It should be noted here that in protein chemistry the torsion angle between coupling nuclei is denoted as θ; for θ_{HNCH} one uses the relationship $\theta = \phi - 60°$ (Figure 7).

Independent calibrations in terms of *A*, *B*, and *C* [equation (1b)], were carried out by Pardi et al.[30] (37 data points, rms = 0.87 Hz) and by Ludvigsen et al.[29] (112 data points, rms = 1.01 Hz), Table 3, on the basis of known X-ray structures. Inspection of the plots given suggests that corrections for differences in electronegativity of the various side chains might improve the rms deviation, especially for θ_{HNCH} near 180° and near 0°. The Karplus coefficients proposed by Bystrov[25] predict $J(180)$ = 10.9 Hz and $J(0)$ = 8.7 Hz, the latter value is probably too large by about 1.5–2.0 Hz.

3.4 Vicinal ^{1}H and ^{13}C Couplings to ^{31}P in Phosphate Diesters

The vicinal couplings ^3J(P,O,C,H) and ^3J(P,O,C,C), taken in combination, yield important conformational information

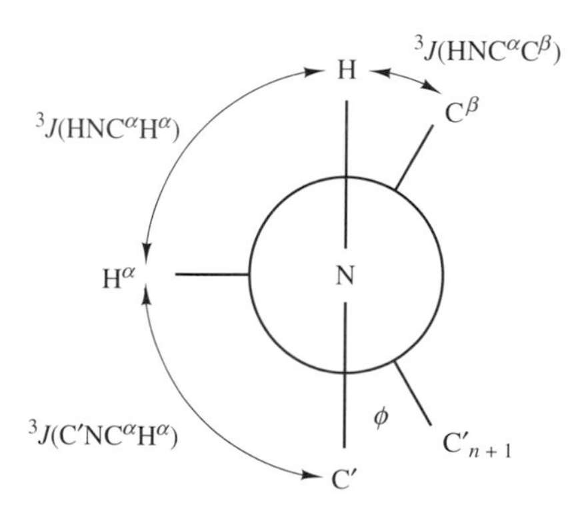

Figure 7 Newman projection along the N–C$^\alpha$ bond of a peptide backbone showing three important vicinal coupling constants that supply information about the ϕ torsion angle

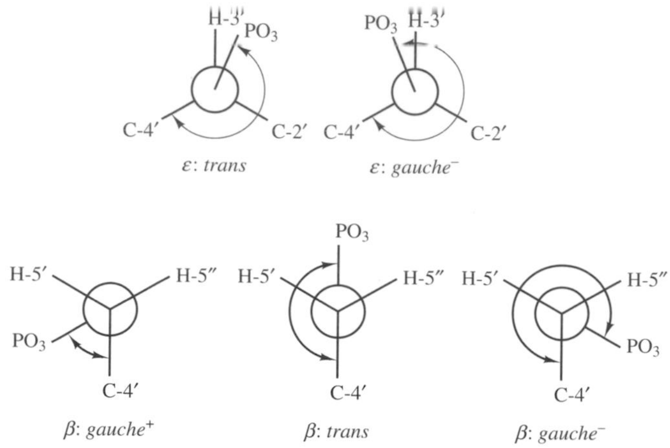

Figure 8 Newman projections along the main-chain torsion angle ϵ and β in nucleic acids. The negative charge on the phosphate group is omitted. These angles are monitored by ^3J(P,O,C,H) and ^3J(P,O,C,C). (Reproduced by permission of Adenine Press from P. P. Lankhorst, C. A. G. Haasnoot, C. Erkelens, and C. Altona, *J. Biomol. Struct. Dyn.*, 1984, **1**, 1387)

concerning two backbone torsion angles in nucleic acids: β (angle PO5′C5′C4′) and ϵ (angle C4′C3′O3′P), (Figure 8). The Karplus coefficients that describe the angular dependence of ^3J(P,H) and ^3J(P,C) are relatively large. This fact contributed to the success of a different approach taken in the parameterization,[45] necessitated by the knowledge that phosphates (and sulfates) often display 'nonclassical' torsion angles, i.e. large deviations from 'classical' 180° and 60° angles. Moreover, it is desirable to avoid the use of crystallographic information which in individual cases can be biased by crystal-packing effects. Instead, the unknown backbone angle (ϵ) can be determined simultaneously with the Karplus coefficients, provided that several different couplings along the same central bond are available and that conformational purity along this bond can be attained. In the case discussed here, six Karplus coefficients and four torsion angles were determined from 17 experimental couplings in RNA compounds (rms = 0.2 Hz). As it turned out, the *mean* NMR value of ϵ_{ap} in the RNA molecules in the stacked form (219°) corresponded well with the *mean* of crystallographic determinations in similar RNAs (218°). The same Karplus coefficients (Table 4) are transferable to DNA molecules.[46]

4 SELECTED APPLICATIONS OF ^3J IN THE FIELD OF BIOMOLECULES

4.1 General Remarks

For more than a decade the conformational situation ('structure') of biopolymers, such as proteins and nucleic acids, was judged from characteristic NOEs. More recently, interatomic distances or NOEs were introduced as constraints in MD (molecular dynamics calculations) trajectories. However, without additional information, it is very difficult to judge whether or not conformational homogeneity exists along each chemical bond of interest. This is especially true in cases where a given conformer predominates; minor forms may then escape detection altogether. Contradictory NOEs certainly point to the simultaneous existence of various rotamers, but these are not easily quantified from NOE information. With the advent of more powerful NMR spectrometers and techniques, investigators are becoming more and more aware that homo- and heteronuclear coupling constants offer a rich source of the additional information that is required.

An important recent application[48,49] of the Karplus coefficients involves the introduction of coupling constant restraints into molecular dynamics calculations. Defining $\Delta J = J_{obs} - J_{calc}$, one adds a penalty function in the form of a pseudo-energy term $\frac{1}{2}k_i\Delta J_i^2$ to the force field for each conformational constraint along given torsion angles, in addition to the usual distance constraints obtained from NOEs. One of the advantages of this approach, when applied to multiple couplings along a chosen bond, is that conformational equilibria, which occur on a short timescale, can be handled with more confidence. Of course, the k_i 'force constants' should be weighed according to the reliability of J_{obs} and J_{calc}, but to date such information on J_{calc} is limited to a few types of vicinal couplings, see Sections 3.2–3.5. It was also proposed to incorporate time-averaged J values[50] as constraints; this idea was succesfully tested on a cyclic peptide, antamanide ($^3J_{HN\alpha}$). Extension to multiple couplings about certain bonds undoubtedly will follow soon.

4.2 Amino Acids, Peptides, and Proteins

4.2.1 *Main Chain Torsion Angles in Polypeptides and Proteins*

The main chain in an amino acid sequence is characterized by three sequential torsion angles: ϕ, ψ, ω (Table 3 and Figure 6). For several reasons most attention was and still is focused on ϕ. This particular angle is correlated with the local shape of the sequence (α-helix, β-sheet, and various turns). In principle, ϕ can be monitored by the use of six different coupling constants (Table 3). Of these, $^3J(\mathrm{H,N,C^\alpha,H^\alpha})$ is the easiest one to measure, even in large proteins, followed by $^3J(\mathrm{H,N,C^\alpha,C^\beta})$.[38] Coupling to carbonyl carbon requires more effort and is difficult to obtain when the coupling is small. Nevertheless, a qualitative insight into the magnitude of two more ϕ couplings can be gained, at least in small cyclic peptides, from natural abundance NMR spectra; $^3J(\mathrm{C^\alpha,C',N,H})$ and $^3J(\mathrm{C',N,C^\alpha,H^\alpha})$.[38,51] Now that ^{13}C^{15}N isotopically-enriched proteins are becoming available, 3D and 4D NMR techniques offer new possibilities for measuring previously 'unmeasurable' couplings in large proteins, e.g., $^3J(\mathrm{H,N,C^\alpha,C'})$.[52] The ψ angle remains elusive. For $^3J(\mathrm{H^\alpha,C^\alpha,C',N})$ and $^3J(\mathrm{C^\beta,C^\alpha,C',N})$ very little reference material is available at present. A similar remark applies to $^3J(\mathrm{C^\alpha,C',N,H})$, which coupling monitors the ω angle. In rigid domains, both ψ and ω can be determined from NOEs, however.

4.2.2 *Side-Chain Conformations of Amino Acid Residues*

The heteronuclear $^3J(\mathrm{C',C^\alpha,C^\beta,H^\beta})$ and $^3J(\mathrm{N,C^\alpha,C^\beta,H^\beta})$ couplings along with the two homonuclear $^3J(\mathrm{H^\alpha,C^\alpha,C^\beta,H^\beta})$ couplings allow the unambiguous diastereotopic assignment of the two β-methylene protons of the side chain of amino acid residues.[51,53] Given the correct assignment of H$^\beta$(pro-*R*) and H$^\beta$(pro-*S*) (Figure 9), the conformational populations of the equilibrating three rotameric species can be calculated from the two $^3J(\mathrm{H^\alpha,C^\alpha,C^\beta,H^\beta})$ couplings when the limiting values ($^3J_{ap}$ and $^3J_{sc}$) are known. The Pachler[54] values $^3J_{ap}$ = 13.6 Hz and $^3J_{sc}$ = 2.6 Hz are still widely used in the literature, although Abraham et al.[55] in 1977 already pointed out that differentiation according to the electronegativity of the side chain as well as the inclusion of substituent orientation was called for. In that work[55] a simple additivity scheme of substituent constants for the prediction of $^3J_{sc}$ was proposed. For example, for X = carbon, the two $^3J_{sc}$ values predicted were 3.4 Hz and 4.2 Hz, for X = oxygen, values of 2.2 Hz and 4.9 Hz were proposed. A more refined, but similar, approach is now possible through the use of equations (11), (12), or (13) in conjunction with the experimental group electronegativities λ_i. It appears best to distinguish three different groups of side chains A through C according to the respective λ_i values given in Table 1.

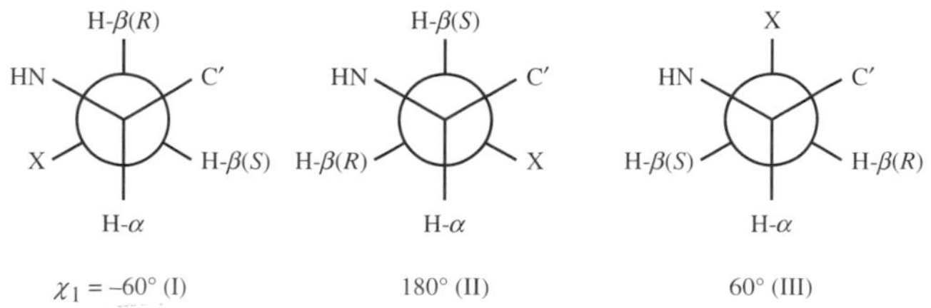

Figure 9 Newman projections of the three staggered χ_1 side-chain conformations in an amino acid with two H-β protons. The three rotamers are designated I, II, and III for χ_1 = −60°, 180°, and 60°, respectively. For the sake of clarity pro-*R* and pro-*S* are abbreviated as *R* and *S*, respectively

A: λ_i = 0.42 (Asn, Asp, His, Phe, Trp, Tyr)
B: λ_i = 0.60–0.75 (Arg, Cys, Gln, Glu, Leu, Lys, Met)
C: λ_i = 1.25 (Ser in D_2O), λ_i = 1.33 (Ser in $CDCl_3$)

The predicted H$^\alpha$H$^\beta$ limiting coupling constants for each of the three rotamers along χ_1 in peptides and proteins (idealized staggered geometry assumed), differentiated according to the electronegativity of the side chain, are shown in Table 5. The nature of the solvent plays a role only in the cases of free amino acids and N-terminal residues, because $\lambda(NH_2) > \lambda(NH_3^+)$ (Table 1).

Some examples of H$^\alpha$H$^\beta$ coupling constants and rotamer populations are displayed in Table 6. For purposes of comparison, the populations calculated with standard values[54] are also given. The differences between the new and the old population values are usually $\leqslant$ 7%, except for Ser ($\leqslant$10%). It is concluded that the trends recorded in the literature thus far remain valid. For future work requiring high precision, the new values are recommended. It is noted in passing that application of equations (11), (12), or (13) allows us to predict safely couplings pertaining to any deviation from ideal staggered geometries. A good example of nonstaggered situations occurs in proline residues. An analysis of the 10 coupling constants of proline and of 12 model compounds[59] with the aid of the Haasnoot equation [equation (7)][16] in conjunction with the pseudorotation model for five-membered rings[60] showed that the unsubstituted proline ring occurs in a practically 1 : 1 conformational equilibrium between N-type and S-type forms (see Section 4.3.1 for a discussion of N- and S-conformers). Recently, Karplus and co-workers[61] successfully used the Haasnoot equation [equation (7)] to probe the motional charac-

For References see p. 4922

Table 5 Predicted Limiting $H^{\alpha}H^{\beta}$ Coupling Constants (Hz) [equation (13)] for Each of the Three Rotamers along $C^{\alpha}C^{\beta}$ in Peptides and Proteins. Idealized Staggered Models Assumed. Three Classes of Amino Acid Residues A–C (See Text) are Differentiated According to the Group Electronegativity λ_i of the Side-Chain Substituent X, Figure 9. Because $\lambda(NH_2) > \lambda(NH_3^+) \approx \lambda[NHC(=O)R]$ (Table 1) N-Terminal Residues are Further Differentiated According to Solvent Conditions

	I(χ^-)		II(χ^t)		III(χ^+)	
ϕ_{HH}	pro-*R* 180°	pro-*S* −60°	pro-*R* 60°	pro-*S* 180°	pro-*R* −60°	pro-*S* 60°
D_2O (neutral or acidic: all residues) or nonaqueous solutions (except N-terminals)						
A	12.9	3.1	3.7	12.9	3.0	3.5
B	12.6	3.2	3.8	12.6	2.8	3.3
C	11.7	3.7	4.4	11.8	1.9	2.4
D_2O (basic) or nonaqueous solution: amino acids and N-terminal residues						
A	12.4	2.7	4.0	12.5	2.5	3.7
B	12.1	2.8	4.0	12.2	2.4	3.6
C	11.2	3.3	4.6	11.5	1.5	2.6

terisitics of the four proline residues in the cyclic decapeptide antamanide on a picosecond time scale.

4.3 Nucleic Acids

The structural properties and current nomenclature rules[17] of nucleic acid conformations have been treated by Wüthrich[28] and by Saenger.[62] The conformational analysis of oligonucleotides with the aid of coupling constants has been reviewed by Altona,[63] backbone torsion angles are shown in Figures 8, 10, and 12.

4.3.1 *The Furanose Ring*

The flexible five-membered sugar ring plays a pivotal role in nucleic acid structure and dynamic behavior. Structural differences between A-, B-, and Z-type duplexes are intimately correlated with specific conformational ranges of individual sugars. Within the B-DNA family each sugar responds to its surroundings (e.g., the base-stacking pattern) by an appropriate adaptation of its geometry. Careful analysis of sets of ^3J(H,H) values reveals intimate details on the 'atomic level' that cannot be obtained from NOEs.[64,65] Various reviews are available[24,63–65] and only a brief outline of the coupling constant approach is sketched here.

Two distinct D-ribose conformations (better: conformational ranges) occur in nucleic acids, N-type (North, roughly C-3′-*endo*) and S-type (South, roughly C-2′-*endo*), Figure 11, which occur in fast equilibrium unless the stereochemical requirements of the helix impose restraints. For example, an RNA double helix consists of an N–N–N sequence of sugars, Z DNA of an S–N–S–N alternation. In B DNA the duplex structure allows the sugars some mobility to flip from S to N and vice versa. Pyrimidine sequences are relatively more mobile (78 ± 13% S) than are purine sequences (90 ± 7% S).[24] One always measures time-averaged couplings and shifts under 'fast-exchange' conditions. Nevertheless, it remains possible to extract quantitative information on geometry and conformational population from the couplings thanks to the fact that the five

Table 6 Examples of $H^{\alpha}H^{\beta}$ Coupling Constants and χ Rotamer Populations in L-Amino Acids and Proteins.[a] The Appropriate Limiting Couplings are Given in Table 5. In Parentheses are the Populations Calculated with J_{ap} = 13.6 Hz and J_{sc} = 2.6 Hz[54]

		$^3J_{exp}$ (Hz)		Rotamer populations (%)		
Residue	pH or solvent	$\alpha\beta$(pro-*R*)	$\alpha\beta$(pro-*S*)	I	II	III
Phe[56]	9.8	8.4	4.9	57(53)	20(21)	23(26)
Phe[56]	2.2	8.3	4.5	53(52)	13(17)	34(31)
AcPSer[57]	8.0	5.6	3.3	37(27)	4(6)	59(67)
AcSer[57]	7.5	6.2	3.8	42(32)	9(11)	49(57)
Cyclosporin A[58] MeLeu[4]	$CDCl_3$	12.0	4.5	91(85)	12(17)	−3(−2)
Cyclosporin A[58] MeLeu[6]	$CDCl_3$	6.0	10.5	24(31)	77(72)	−1(−3)
Cyclohexapeptide[58] Phe[7]	DMSO	4.5	9.0	11(17)	59(58)	30(25)
Cyclohexapeptide[58] Trp[8]	DMSO	10.9	4.2	79(75)	11(15)	10(10)
Cyclohexapeptide[58] Phe[11]	DMSO	11.5	3.1	86(80)	2(5)	12(15)

[a]The rotamer populations are computed from three equations with three unknowns using standard matrix methods; for the sake of clarity pro-*R* and pro-*S* are abbreviated as *R* and *S*, respectively:

$J_{obs(R)} = p_I J_{I(R)} + p_{II} J_{II(R)} + p_{III} J_{III(R)}$

$J_{obs(S)} = p_I J_{I(S)} + p_{II} J_{II(S)} + p_{III} J_{III(S)}$

$p_I + p_{II} + p_{III} = 1$

The computation is simpler when it is assumed that all J_{sc} values are equal.[54]

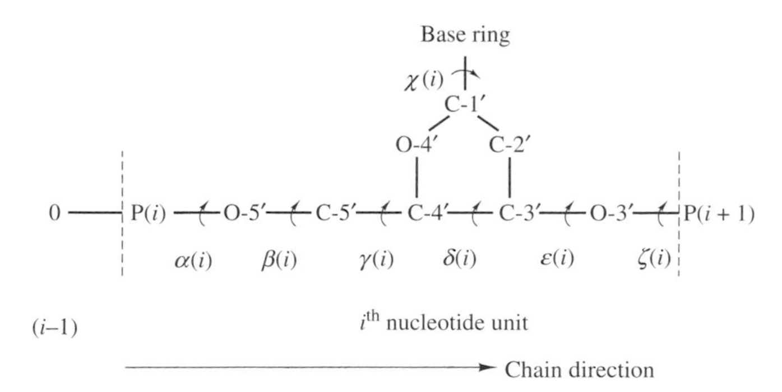

Figure 10 Conformational notation of the torsion angles of the backbone and χ of the base in nucleic acids[17]

endocyclic torsion angles ν_i are mutually dependent via the two-parameter equation (14).

$$\nu_j = \phi_m \cos\{P + 0.8\pi(j - 2)\}, \quad j = 0\text{–}4 \tag{14}$$

in which ϕ_m represents the amplitude of pucker and P the phase angle of pseudorotation.[60] The torsion angles are numbered clockwise, starting with ν_0 (C4′–O4′–C1′–C2′). In D-sugars, N-type conformers are characterized by a positive sign of ν_3, S-type forms by a negative sign of ν_3. The (exocyclic) proton–proton torsion angles ϕ_{HH} are related to the ν_j-values via the pseudorotation parameters ϕ_m and P [equation (15)]:[66]

$$\phi_{HH} = a + b\phi_m \cos(P + \text{phase shift}) \tag{15}$$

where, in the idealized approximation, $b = 1$ and $a = 0°$ for *cis* protons and ±120° for *trans* protons. Empirically established values for the parameters a and b in equation (15) for deoxyribose, ribose, arabinose, lyxose, and xylose are available.[66,67] Equations (14) and (15) and the extended Karplus equation [equation (13)] are employed in the iterative least-squares program PSEUROT.[68] Given a complete set of experimental proton–proton couplings $J(1',2')$, $J(2',3')$, and $J(3',4')$ for riboses and, in addition, $J(1',2'')$ and $J(2'',3')$ for deoxyriboses, the program minimizes the functional $R = \sum(J_{exp} - J_{calc})^2$ and yields the pseudorotation parameters of the two stable furanose conformers and the mole fractions. The accuracy of the procedure is greatly enhanced when the conformational equilibrium displays a strong shift with temperature. Version 4.0 of the PSEUROT program is in the public domain (*Quantum Chemistry Program Exchange No. 463*, Indiana State University,

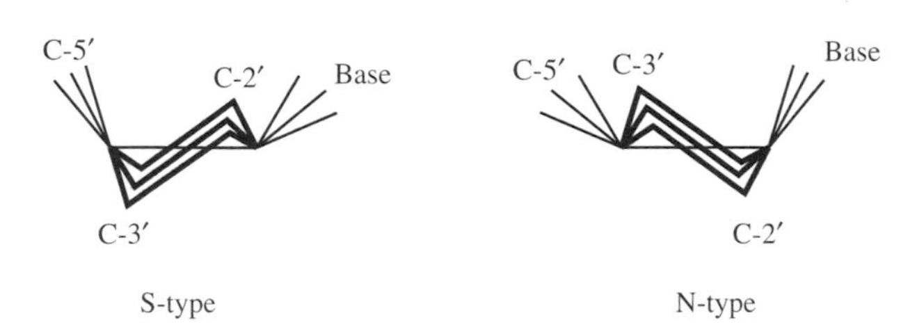

Figure 11 Schematic composite representation of the furanose conformations in nucleic acids: S-type (C-2′-*endo*/C-3′-*exo*) (left) and N-type (C-2′-*exo*/C-3′-*endo*) (right). (Reproduced by permission of Academic Press from J. van Wijk, B. D. Huckriede, J. H. Ippel, and C. Altona, *Methods Enzymol.*, 1992, **211**, 286)

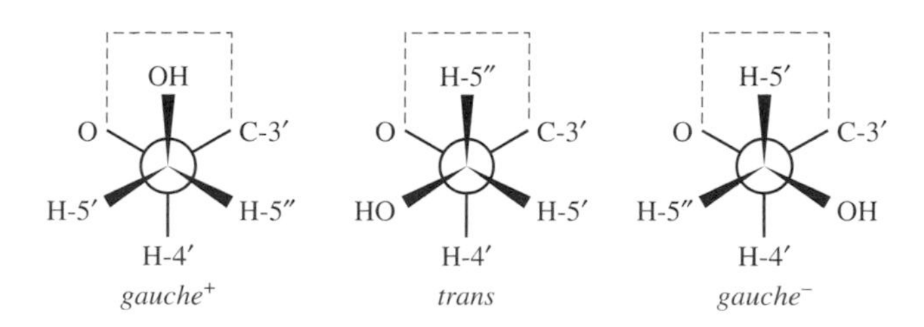

Figure 12 Newman projections of the three γ rotamers in nucleosides. In nucleotides the OH group is replaced by an $O\text{–}PO_3^-$ group

1983); this version employs equation (7) and the original coefficients.[16] Later versions (PC, IBM compatible) 5.4 and 6.0 (see text) are available under licence from the author. Note that the PSEUROT programs are designed to yield the conformational characteristics of any saturated five-membered ring, provided that the appropriate vicinal ^{1}H–^{1}H couplings are input.

In NMR practice it is not always feasible to carry out the laborious spectral simulations that are required to obtain the individual coupling constants. Often, the sums of couplings to given protons are easily available, however, and these contain much the same information.[24,69] In the case of DNA we define:

$$\Sigma 1' = J(1',2') + J(1',2'') \tag{16a}$$
$$\Sigma 2' = J(1',2') + J(2',3') + J(2',2'') \tag{16b}$$
$$\Sigma 2'' = J(1',2'') + J(2'',3') + J(2',2'') \tag{16c}$$
$$\Sigma 3'\{^{31}P\} = J(2',3') + J(2'',3') + J(3',4) \tag{16d}$$

The $\sum 2'$ and $\sum 2''$ terms include the absolute value of $^2J(2',2'')$ (approximately 14 Hz in deoxyribofuranose). A graphical method, based on the theory employed in PSEUROT, successfully utilizes graphs of $\sum J$ versus P and $\sum J$ versus populations to solve the conformational problems.[24,69] Recently, the graphical procedure was automated in program PSEUROT, Version 6.0.[68]

4.3.2 *Orientation of the Nucleic Acid Base*

The glycosyl torsion angle χ (not shown) defines the orientation of the base with respect to the sugar ring.[62] Two major orientations are known: *anti* and *syn*. In the more common *anti* form, the C-8–H-8 and C-6–H-6 vectors of the purine and the pyrimidine bases point roughly in the direction of the C-1′–O-4′ bond and are antiparallel to the C-1′–H-1′ bond. From pyrimidine model compounds the Karplus coefficients for ^3J(C-6,N-1,C-1′,H-1′) and ^3J(C-2,N-1,C-1′,H-1′) were determined[47] (Table 4) and from these values one predicts the following (very approximate) characteristics:

anti base: ^3J(C-6,H-1′) ≈ 9 Hz, ^3J(C-2,H-1′) ≈ 3 Hz
syn base: ^3J(C-6,H-1′) ≈ 4 Hz, ^3J(C-2,H-1′) ≈ 7 Hz

In a regular B-DNA structure with *anti* bases, the angle χ is well defined (±10°) from three NOE contacts: H-6/H-8–H-2′, H-6/H-8–H-2″, H-6/H-8–H-3′, but in exceptional cases where the base enjoys more conformational freedom measurement of ^3J(C,H) may well supply valuable additional information.

For References see p. 4922

4.3.3 *Backbone Conformations of RNA and DNA*

The six torsion angles along the nucleic acid backbone are denoted α–ζ, starting with P–O-5′ as the central bond for α, Figure 10. NMR spectroscopy provides no direct grip upon the important phosphate diester torsions ζ and α; these are usually determined from NOE-constrained MD trajectories. The remaining backbone angles β, γ, δ, and ϵ are monitored by various vicinal coupling constants. Because X-ray crystallography[62] has made it clear that important deviations from pure staggered geometries (180°, 60°) in double-helical structures commonly occur, it is the task of the NMR spectroscopist to deduce at the same time values for the backbone angles as well as conformational populations. This is possible in principle because Nature has conveniently provided a sufficient number of coupling nuclei along the backbone that can be put into use.

It is often helpful to study the behavior of sums of couplings versus rotation about a given torsion angle. As an aid to the discussion, Table 7 has been prepared. The Karplus coefficients used to calculate the ^3J(P,H) and ^3J(P,C) values in Table 7 are taken from Table 4.

Torsion angle β (P–O5′–C5′–C4′), Figure 8. In 5′-mononucleotides of RNA and DNA, the $\beta^t(ap)$ rotamer is invariably preferred (>70%). In stacked oligonucleotides, the β^t angle is confined to a small range of values (176 ± 4°). Under the additional assumption that in unstacked ('melted') states an equilibrium exists between two or three rotamers (β^t, 175°; β^+, 60°; β^-, 300°), a sum rule is easily derived [equation (17)]:

$$p(\beta^t) = (25.4 - \Sigma')/20.5 \qquad (17)$$

where p stands for the fractional population and $\sum' = J(5',\mathrm{P}) + J(5'',\mathrm{P})$. Stereospecific assignments of H-5′ and H-5″ are not necessary at this stage. Unfortunately, we have no information concerning the true values of the β^+ and β^- angles, but it can be verified that nonclassical β_{sc} rotamers (e.g. β_{sc}, ±90°) would introduce less than 2% error in the $p(\beta^t)$ value calculated from

Table 7 Backbone Angles β, γ, and ϵ(°) of Nucleic Acids in the Most Populated Ranges and Corresponding Calculated Vicinal Coupling Constants (Hz) (See Text)

β^t	170	175	180	185	190
J(P,H-5′)	4.0	3.1	2.4	1.8	1.3
J(P,H-5″)	1.3	1.8	2.4	3.1	4.0
$\sum'$	5.3	4.9	4.8	4.9	5.3
J(P,C-4′)	10.7	10.9	11.0	10.9	10.7
β^+	50	55	60	65	70
J(P,H-5′)	1.3	1.8	2.4	3.1	4.0
J(P,H-5″)	22.5	22.9	23.0	22.9	22.5
$\sum'$	23.8	24.7	25.4	26.0	26.5
J(P,C-4′)	1.4	1.0	0.7	0.5	0.3
β^-	−50	−55	−60	−65	−70
J(P,H-5′)	22.5	22.9	23.0	22.9	22.5
J(P,H-5″)	1.3	1.8	2.4	3.1	4.0
$\sum'$	23.8	24.7	25.4	26.0	26.5
J(P,C-4′)	1.4	1.0	0.7	0.5	0.3
γ^+	45	50	55	60	65
J(H-4′,H-5′)	1.1	1.5	2.0	2.5	3.2
J(H-4′,H-5″)	2.9	2.2	1.6	1.2	0.8
$\sum$	4.0	3.7	3.6	3.7	4.0
γ^t	165	170	175	180	185
J(H-4′,H-5′)	5.1	4.5	3.6	3.0	2.4
J(H-4′,H-5″)	11.0	11.0	10.9	10.6	10.2
$\sum$	16.1	15.5	14.5	13.6	12.6
γ^-	−80	−75	−70	−65	−60
J(H-4′,H-5′)	9.5	10.1	10.5	10.8	10.9
J(H-4′,H-5″)	1.9	2.4	3.1	3.8	4.6
$\sum$	11.4	12.5	13.6	14.6	15.4
ϵ^t	180	190	200	210	220
J(P,H-3′)	2.4	4.0	5.9	7.8	9.4
J(P,C-2′)	0.7	0.3	0.3	0.7	1.5
J(P,C-4′)	11.0	10.7	10.0	8.8	7.4
ϵ^-	−105	−100	−95	−90	−85
J(P,H-3′)	10.0	9.9	8.6	7.8	6.9
J(P,C-2′)	6.6	7.4	8.1	8.8	9.5
J(P,C-4′)	2.0	1.5	1.1	0.7	0.5

equation (17). The C-4′,P coupling can be used as an independent check on the β^t population [equation (18)]:

$$p(\beta^t) = [^3J(\text{C-4}',\text{P}) - 0.7]/10.2 \qquad (18)$$

Assuming that the H-5′ and H-5″ resonances are assigned stereospecifically from NOEs, approximate values of the minor populations $p\beta^+$ and $p\beta^-$ are derived from J(5′,P) and J(5″,P). Preference for β^+ was found to occur locally in a minihairpin loop.[70]

Torsion angle γ (O5′–C5′–C4′–C3′), Figure 12. Nucleosides and mononucleotides in the solid state display a preference for γ^+ (76%), followed by γ^t (18%) and γ^-(6%). This rotamer distribution in the crystalline state is mirrored by the distribution found in aqueous solution at ambient temperatures. Crystallographic data reveal trends toward nonclassical angles: γ^+, 53°; γ^t, 180°; γ^-, 290°. The right-handed helix requires a pure γ^+ rotamer; upon melting, conformational mobility is regained. The population γ^+ again follows from a sum rule [equation (19)]:

$$p(\gamma^+) = (13.6 - \Sigma)/10.0 \qquad (19)$$

where $\sum = J_{4'5'} + J_{4'5''}$. A pure or almost pure γ^t form is adopted by dG residues in Z DNA[62] and at the 5′–3′ loop stem junction in small hairpin loops in DNA, i.e. at the position of the local sharp turn of the backbone.[70] Such a turn is revealed by a large (>9 Hz) value of $^3J_{4'5''}$.

Torsion angle δ (C5′–C4′–C3′–O3′): This torsion angle is directly related to the pseudorotation parameters P and ϕ_m of the D-(deoxy)ribose sugar by equation (20).[63]

$$\delta = 120.6 + 1.1\phi_m \cos(P + 145.2) \qquad (20)$$

Torsion angle ϵ (C4′–C3′–O3′–P), Figure 8. The study of the conformational behavior of this backbone angle is virtually impossible without the aid of vicinal ^{13}C–^{13}P couplings because only a single ^3J(P,H) value is measured for each residue. The ϵ^+ rotamer is sterically ‘forbidden’[63] and need not be considered. The two allowed conformational ranges, ϵ^t and ϵ^-, are more or less symmetrically disposed with respect to H-3′ and, without further information, ^3J(P,H) can give few clues concerning the ϵ^t/ϵ^- rotamer population distribution. It is known, however, that ^3J(P,H-3′) of single-stranded RNA oligomers increases with increasing population of stacked (single-helical) conformer to a maximum of about 9 Hz, whereas ^3J(P,H-3′) in similar DNA compounds decreases to a minimum of about 4 Hz when fully stacked. This difference in behavior was quantitatively interpreted with the aid of ^{13}C NMR data.[45,46,71] It turned out that there exists a strong correlation between the conformation of the sugar ring and the ϵ^t angle. In RNA the stacked state strongly biases the sugar towards the N-type form [ϵ^t_N = 219°; ^3J(P,C-4′) $\approx$ 8 Hz; ^3J(P,H-3′) $\approx$ 9 Hz], whereas in DNA the sugar mainly occupies the S-type range [ϵ^t_S = 192°; ^3J(P,C-4′) $\approx$ 10.7 Hz; ^3J(P,H-3′) $\approx$ 4.5 Hz]. The ϵ^-_N combination appeared ‘forbidden’, whereas ϵ^-_S = 266°, corresponding with ^3J(P,C-2′) $\approx$ 8.3 Hz, remains allowed. The latter combination must be considered important, because X-ray crystallography has shown that it occurs in various residues throughout the double helix (BII form). We expect that the BII form will be studied in solution as soon as ^{13}C-enriched synthetic DNAs become available. Finally, it should be mentioned that ϵ^t_N in DNA (212°) is smaller than ϵ^t_N in RNA (219°).[71]

5 CAVEATS

The original optimism that arose among stereochemists when Karplus's first paper[1] on ^3J(H,H) in ethane appeared in 1959 soon gave way to grave scepticism in the course of the 1960s.[2] Nevertheless, the Karplus-type relationships are by now firmly rooted in stereochemical thinking. Perhaps unfortunately, not all the known ramifications are widely appreciated. Moreover, even in the situation of the best-known coupling, ^3J(H,C,C,H) along sp^3–sp^3 C–C bonds, some influences that affect couplings have not been taken into consideration in the present contribution. At least three of these (perhaps not unrelated) influences are known, but as yet not understood to the extent that quantitative corrections to calculated couplings can be brought into play: (a) the β effect, the orientation of a β substituent appears to affect ^3J(H,C,C,H), an increase of about 0.5 Hz is noted for $^3J_{ap}$ when a C–O bond occurs in the *ap* position with respect to one of the coupling protons;[11] (b) the through-space (Barfield) transmission effect,[72,73] first noted in norbornane-like structures but not necessarily limited to highly strained compounds; (c) bond angle deformations, semiempirical calculations[2,74] predict that vicinal couplings depend significantly on the bond angles HCC′ and CC′H′ of the HCC′H′ fragment. The effect would be maximal for $^3J_{ap}$ and $^3J_{sp}$. At this point a dilemma arises. Assuming that reliable empirical corrections for influences (a)–(c) can be established, it would be necessary to introduce much detailed structural information concerning the molecular fragment under study into play before a measured coupling can be interpreted quantitatively in terms of the same structure. Undoubtedly, in the future this will be done for selected fragments which are of vital interest to researchers. However, we may also expect that for times to come present-day knowledge will suffice to solve stereochemical problems routinely along the lines pioneered by Karplus in 1959.[1]

6 RELATED ARTICLES

Amino Acids, Peptides & Proteins: Chemical Shifts; Biological Macromolecules: Structure Determination in Solution; Biological Macromolecules: NMR Parameters; Carbohydrates & Glycoconjugates; Coupling Constants Determined by ECOSY; Coupling Through Space in Organic Chemistry; DNA: A-, B-, & Z-; DNA Triplexes, Quadruplexes, & Aptamers; Half-Filter Experiments: Proton–Carbon-13; Homonuclear Three-Dimensional NMR of Biomolecules; Indirect Coupling: Semiempirical Calculations; Nucleic Acid Structures in Solution: Sequence Dependence; Nucleic Acids: Phosphorus-31 NMR; Nucleic Acids: Spectra, Structures, & Dynamics; Quantitative Measurements; Structures of Larger Proteins, Protein–Ligand, & Protein–DNA Complexes by Multi-Dimensional Heteronuclear NMR; Three-Dimensional HMQC–NOESY, NOESY–HMQC, & NOESY–HSQC; Three- & Four-Dimensional Heteronuclear Magnetic Resonance; Two-Dimensional J-Resolved Spectroscopy; Vicinal ^{1}H,^{1}H Coupling Constants in Cyclic π-Systems.

For References see p. 4922

7 REFERENCES

1. M. Karplus, *J. Chem. Phys.*, 1959, **30**, 11.
2. M. Karplus, *J. Am. Chem. Soc.*, 1963, **85**, 2870.
3. R. E. Glick and A. A. Bothner-By, *J. Chem. Phys.*, 1956, **25**, 362.
4. C. N. Banwell and N. Sheppard, *Discuss. Faraday Soc.*, 1962, **34**, 115.
5. R. J. Abraham and K. G. R. Pachler, *Mol. Phys.*, 1964, **7**, 165.
6. H. Booth, *Tetrahedron Lett.*, 1965, 411.
7. P. L. Durette and D. Horton, *Org. Magn. Reson.*, 1971, **3**, 417.
8. L. Pauling, 'The Nature of the Chemical Bond', 3rd edn., Cornell University Press, Ithaca, NY, 1960.
9. M. L. Huggins, *J. Am. Chem. Soc.*, 1953, **75**, 4123.
10. R. J. Abraham and G. Gatti, *J. Chem. Soc. B*, 1969, 961.
11. C. Altona and C. A. G. Haasnoot, *Org. Magn. Reson.*, 1980, **13**, 417.
12. K. G. R. Pachler, *Tetrahedron*, 1971, **27**, 187.
13. K. G. R. Pachler, *J. Chem. Soc. Perkin Trans. 2*, 1972, 1935.
14. F. A. A. M. de Leeuw, C. A. G. Haasnoot, and C. Altona, *J. Am. Chem. Soc.*, 1984, **106**, 2299.
15. L. A. Donders, F. A. A. M. de Leeuw, and C. Altona, *Magn. Reson. Chem.*, 1989, **27**, 556.
16. C. A. G. Haasnoot, F. A. A. M. de Leeuw, and C. Altona, *Tetrahedron*, 1980, **36**, 2783.
17. IUPAC Tentative Rules for the Nomenclature of Organic Chemistry, *J. Org. Chem.*, 1970, **35**, 2849; for nucleic acids, see JCBN Recommendations, *Eur. J. Biochem.*, 1983, **131**, 9; *J. Biol. Chem.*, 1986, **261**, 13.
18. E. Díez, J. San-Fabián, J. Guilleme, C. Altona, and L. A. Donders, *Mol. Phys.*, 1989, **68**, 49.
19. C. Altona, J. H. Ippel, A. J. A. Westra Hoekzema, C. Erkelens, M. Groesbeek, and L. A. Donders, *Magn. Reson. Chem.*, 1989, **27**, 564.
20. C. Altona, R. Francke, R. de Haan, J. H. Ippel, G. J. Daalmans, and J. van Wijk, *Magn. Reson. Chem.*, 1994, **32**, 670.
21. F. A. A. M. de Leeuw, A. A. van Beuzekom, and C. Altona, *J. Comp. Chem.*, 1983, **4**, 438.
22. N. L. Allinger, 'Quantum Chemistry Program Exchange MMP(85)', Indiana State University, 1985.
23. J. San-Fabián, J. Guilleme, E. Díez, P. Lazzeretti, M. Malagoli, and R. Zanasi, *Chem. Phys. Lett.*, 1993, **206**, 253.
24. J. van Wijk, B. D. Huckriede, J. H. Ippel, and C. Altona, *Methods Enzymol.*, 1992, **211**, 286.
25. V. F. Bystrov, *Prog. Nucl. Magn. Reson. Spectrosc.*, 1976, **10**, 41.
26. P. E. Hansen, *Prog. Nucl. Magn. Reson. Spectrosc.*, 1981, **14**, 175.
27. J. L. Marshall, 'Carbon–Carbon and Carbon–Proton NMR Couplings', Verlag Chemie Int., Deerfield Beach, FL, 1983.
28. K. Wüthrich, 'NMR of Proteins and Nucleic Acids', Wiley, New York, 1986.
29. S. Ludvigsen, K. V. Anderson, and F. M. Poulsen, *J. Mol. Biol.*, 1991, **217**, 731.
30. A. Pardi, M. Billeter, and K. Wüthrich, *J. Mol. Biol.*, 1985, **180**, 741.
31. L.-F. Kao and M. Barfield, *J. Am. Chem. Soc.*, 1985, **107**, 2323.
32. A. de Marco and M. Llinas, *Biochemistry*, 1979, **18**, 3847.
33. R. Aydin, J. P. Loux, and H. Günther, *Angew. Chem., Int. Ed. Engl.*, 1982, **21**, 449.
34. M. C. Reddy, B. P. Nagy Reddy, K. R. Sridharan, and J. Ramakrishna, *Org. Magn. Reson.*, 1984, **22**, 464.
35. P. E. Hansen, J. Feeney, and G. C. K. Roberts, *J. Magn. Reson.*, 1975, **17**, 249.
36. D. Cowburn, D. H. Live, A. J. Fishman, and W. C. Agosta, *J. Am. Chem. Soc.*, 1983, **105**, 7435.
37. M. Barfield and H. L. Gearhart, *Mol. Phys.*, 1974, **27**, 899.
38. P. Schmieder and H. Kessler, *Biopolymers*, 1992, **32**, 435.
39. F. H. Cano, C. Foces-Foces, J. Jiménez-Barbero, A. Alemany, M. Bernabé, and M. Martin-Lomas, *J. Am. Chem. Soc.*, 1987, **52**, 3367.
40. R. Wasylishen and T. Schaefer, *Can. J. Chem.*, 1973, **51**, 961.
41. T. P. Forrest and S. Sukumar, *Can. J. Chem.*, 1977, **55**, 3686.
42. A. A. van Beuzekom, *Thesis*, Leiden University, 1989.
43. G. J. Karabatsos and C. E. Orzech, Jr., *J. Am. Chem. Soc.*, 1965, **87**, 560.
44. A. A. van Beuzekom, F. A. A. M. de Leeuw, and C. Altona, *Magn. Reson. Chem.*, 1990, **28**, 68.
45. P. P. Lankhorst, C. A. G. Haasnoot, C. Erkelens, and C. Altona, *J. Biomol. Struct. Dyn.*, 1984, **1**, 1387.
46. P. P. Lankhorst, C. A. G. Haasnoot, C. Erkelens, H. P. Westerink, G. A. van der Marel, J. H. van Boom, and C. Altona, *Nucleic Acids Res.*, 1985, **13**, 927.
47. D. B. Davies, P. Rajani, M. MacCoss, and S. Danyluk, *Magn. Reson. Chem.*, 1985, **23**, 72.
48. Y. Kim and J. H. Prestegard, *Proteins*, 1990, **8**, 377.
49. D. F. Mierke and H. Kessler, *Biopolymers*, 1993, **33**, 1003.
50. A. E. Torda, R. M. Brunne, T. Huber, H. Kessler, and W. F. van Gunsteren, *J. Biomol. NMR*, 1993, **3**, 55.
51. M. Eberstadt, D. F. Mierke, M. Köck, and H. Kessler, *Helv. Chim. Acta*, 1992, **75**, 2583.
52. G. W. Vuister and A. Bax, *J. Magn. Reson.*, 1992, **98**, 428.
53. P. Schmieder, M. Kurz, and H. Kessler, *J. Biolmol. NMR*, 1991, **1**, 403.
54. K. G. R. Pachler, *Spectrochim. Acta*, 1964, **20**, 581.
55. R. J. Abraham, P. Loftus, and W. A. Thomas, *Tetrahedron*, 1977, **33**, 1227.
56. J. Feeney, *J. Magn. Reson.*, 1976, **21**, 473.
57. L. Pogliani, D. Ziessow, and C. H. Krüger, *Tetrahedron*, 1979, **35**, 2867.
58. H. Kessler, C. Griesinger, and K. Wagner, *J. Am. Chem. Soc.*, 1987, **109**, 6927.
59. C. A. G. Haasnoot, F. A. A. M. de Leeuw, H. P. M. de Leeuw, and C. Altona, *Biopolymers*, 1981, **20**, 1211.
60. C. Altona and M. Sundaralingam, *J. Am. Chem. Soc.*, 1972, **94**, 8205.
61. J. M. Schmidt, R. Brüschweiler, R. R. Ernst, R. L. Dunbrack, Jr., D. Joseph, and M. Karplus, *J. Am. Chem. Soc.*, 1993, **125**, 8747.
62. W. Saenger, 'Principles of Nucleic Acid Structure', Springer, New York, 1984.
63. C. Altona, *Recl. Trav. Chim. Pays-Bas*, 1982, **101**, 413.
64. C. Altona, in 'Theoretical Biochemistry and Molecular Biophysics', ed. D. L. Beveridge and R. Lavery, Adenine Press, New York, 1990, p. 1.
65. F. J. M. van de Ven and C. W. Hilbers, *Eur. J. Biochem.*, 1988, **178**, 1.
66. C. A. G. Haasnoot, F. A. A. M. de Leeuw, H. P. M. de Leeuw, and C. Altona, *Org. Magn. Reson.*, 1981, **15**, 43.
67. F. A. A. M. de Leeuw and C. Altona, *J. Chem. Soc., Perkin Trans. 2*, 1982, 375.
68. F. A. A. M. de Leeuw and C. Altona, *J. Comp. Chem.*, 1983, **4**, 428.
69. L. J. Rinkel and C. Altona, *J. Biomol. Struct. Dyn.*, 1987, **4**, 621.
70. J. M. L. Pieters, E. de Vroom, G. A. van der Marel, J. H. van Boom, Th. M. G. Koning, R. Kaptein, and C. Altona, *Biochemistry*, 1990, **29**, 788.
71. P. P. Lankhorst, C. A. G. Haasnoot, C. Erkelens, and C. Altona, *Nucleic Acids Res.*, 1984, **12**, 5419.
72. M. Barfield, *J. Am. Chem. Soc.*, 1980, **102**, 1.
73. F. A. A. M. de Leeuw, A. A. van Beuzekom, and C. Altona, *J. Comp. Chem.*, 1983, **4**, 438.
74. M. Barfield and W. B. Smith, *J. Am. Chem. Soc.*, 1992, **114**, 1574.

Biographical Sketch

Cornelis Altona. *b* 1931. Ph.D., 1964, Leiden University. Introduced to NMR by E. Havinga and L. J. Oosterhoff. Visiting lecturer, Case Western University, 1967–69. Faculty of Chemistry, Leiden University, 1964–present. Approximately 185 publications. Research interests include: stereoelectronic effects (anomeric effect, *gauche* effect); conformational analysis (of five-membered rings in particular); generalized Karplus-type equations for vicinal spin–spin coupling constants; conformation, structure and dynamics of oligonucleotides (DNA, RNA) in aqueous solution, including hairpin loops and four-way junctions; force-field calculations and charge distributions; application of ab initio methods.

Vicinal ^{1}H,^{1}H Coupling Constants in Cyclic π-Systems

Harald Günther

University of Siegen, Germany

1 DISCUSSION

According to the Karplus theory of vicinal ^{1}H,^{1}H coupling constants, i.e. $^3J(^1\mathrm{H},^1\mathrm{H})$, such parameters depend on the structural features of the corresponding HC–CH fragment (dihedral angle Φ, bond length R_{cc}, and HCC valence angles) as well as on the electronic effects of substituents.[1,2] These results form the basis for numerous applications of ^{1}H NMR spectroscopy in organic, bio-organic, and organometallic chemistry.

While the dihedral angle dependence of $^3J(^1\mathrm{H},^1\mathrm{H})$—the well-known *Karplus curve*—is one of the major tools in conformational analysis,[3] structural research in the field of cyclic π-systems profits from the R_{cc} and valence angle dependence which dominates $^3J(^1\mathrm{H},^1\mathrm{H})$ values in planar unsaturated hydrocarbons, where $\Phi = 0°$ for a *cis*-HC$_\mu$$\overset{...}{-}C_\nu$H arrangement and where substituents are absent. This aspect was first highlighted as a linear dependence between $^3J(^1\mathrm{H},^1\mathrm{H})_{cis}$ and the Hückel (HMO) π-bond order, $P_{\mu\nu}$, of the C$_\mu$$\overset{...}{-}C_\nu$ bond was found for benzenoid aromatics[4] and later extended to olefinic systems.[5,6] At the same time, related correlations between $^3J(^1\mathrm{H},^1\mathrm{H})$ and the CC bond length $R_{\mu\nu}$ were established for aromatics[7] and polyenes.[8] An early application of $^3J(^1\mathrm{H},^1\mathrm{H})$ data for the analysis of the bonding situation in a cyclic π system was the detection of partial π-bond fixation in biphenylene (**1**),[9] which in terms of resonance theory demonstrated the preference for Kekulé structures with exocyclic double bonds at the central four-membered ring. Later, the olefinic nature of tropone (**2**) was established from its $^3J(^1\mathrm{H},^1\mathrm{H})$ values,[10] partial bond fixation in heterocyclic compounds was detected,[11,12] and the characteristic $^3J(^1\mathrm{H},^1\mathrm{H})$ data measured for cyclic dienes and trienes were recognized as important tools for the structural analysis of cyclic diene–triene isomers like norcaradienes and cycloheptatrienes.[13]

O

(**1**) (**2**)

A detailed investigation of the $^3J(^1\mathrm{H},^1\mathrm{H})_{cis}/R_{\mu\nu}$ and $P_{\mu\nu}$ relations in unsaturated six-membered rings of various benzenoid aromatics and other hydrocarbons,[14,15] where the magnitude of these constants ranges between 5 and 9 Hz (and—as for other $^3J(^1\mathrm{H},^1\mathrm{H})$ values—a positive sign is found[16]), finally led to equations (1) and (2) which can be used to derive bond length and π-bond order information from experimental $^3J(^1\mathrm{H},^1\mathrm{H})_{cis}$ data:

$$^3J(^1\mathrm{H},^1\mathrm{H})_{cis} = -351.0R_{\mu\nu}(\mathrm{nm}) + 56.65 \qquad (1)$$

$$^3J(^1\mathrm{H},^1\mathrm{H})_{cis} = 12.47P_{\mu\nu}(\mathrm{HMO}) - 0.71 \qquad (2)$$

The sensitivity of $^3J(^1\mathrm{H},^1\mathrm{H})_{cis}$ for bond length changes is high (Figure 1) and provides a valuable probe for the bonding situation in various cyclic π systems and annulenes. Steric compression and strain effects may cause deviations from equation (1) if changes of the HCC valence angles are induced.[17,18] This becomes apparent if one compares the data for 1,4-di-*t*-butylnaphthalene[18] with those of naphthalene[15] or data for *o*-di-*t*-butylbenzene[19] and the highly strained benzocyclopropene[17] with those of the unstrained 9,10-dihydroanthracene[17] (Figure 2).

Strain and steric compression are frequently encountered in fused ring systems where these factors cause a decrease in $^3J(1,2)$ and an increase in $^3J(2,3)$ if a small ring ($n < 6$) is fused to a large ring (**3**). The reverse trend—an increase of $^3J(1,2)$ and a decrease of $^3J(2,3)$—is found if a larger ring is fused to a smaller one (**4**).[20]

H-2 H-3 H-1 π π

H-1 H-2 H-3 π π

(**3**) (**4**)

More drastic changes of the HCC valence angles are induced in π systems of different ring size and modified correlations have been derived for these cases:

For References see p. 4925

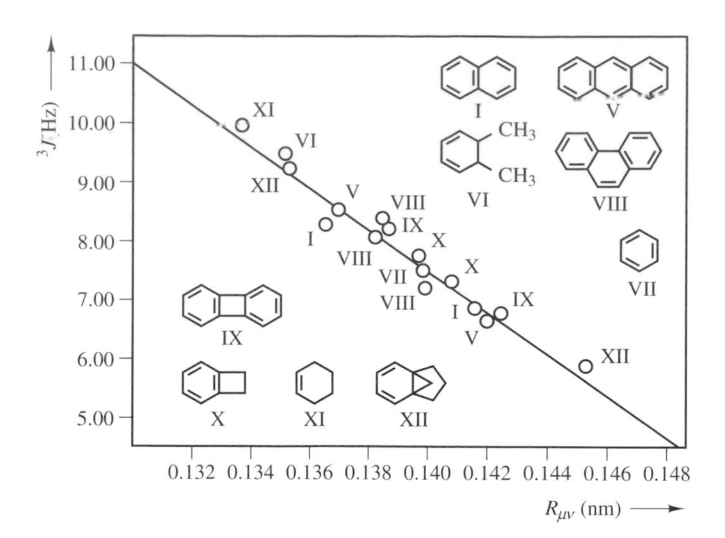

Figure 1 Dependence of $^3J(^1\mathrm{H},^1\mathrm{H})_{cis}$ on the CC bond length in the HC$\cdots$CH fragment of six-membered rings [equation (1)]. (Reproduced with permission from H. Günther, 'NMR Spectroscopy', 2nd edn., Wiley, Chichester, 1995)

Five-membered rings:[5,21]

$$^3J(^1\mathrm{H},^1\mathrm{H})_{cis} = -322.6R_{\mu\nu}(\mathrm{nm}) + 48.45 \quad (3)$$

$$^3J(^1\mathrm{H},^1\mathrm{H})_{cis} = 7.12P_{\mu\nu}(\mathrm{HMO}) - 1.18 \quad (4)$$

Seven-membered rings:[20,22]

$$^3J(^1\mathrm{H},^1\mathrm{H})_{cis} = -367.4R_{\mu\nu}(\mathrm{nm}) + 60.68 \quad (5)$$

$$^3J(^1\mathrm{H},^1\mathrm{H})_{cis} = 21.91P_{\mu\nu}(\mathrm{HMO}) - 3.85 \quad (6)$$

The ring size dependence of the HCC valence angles is a consequence of the changes in internal CCC valence angles for the respective perimeters and a linear correlation of ∢CCC with $^3J(^1\mathrm{H},^1\mathrm{H})_{cis}$ is thus not unexpected. Using five-, six-, seven-, and eight-membered ring systems, an increment of 0.29 Hz per degree was found after correcting for different π-bond orders.[23] Similar relations were established for the long-range ^{1}H,^{1}H couplings ^4J and ^5J.[23] For isolated double bonds, the effect of HCC valence angle changes is most clearly seen by the ring size dependence of $^3J(^1\mathrm{H},^1\mathrm{H})_{cis}$ in cycloalkenes,[24–27] as shown in Table 1.

	t-Bu/H compound	naphthalene
a	8.88	8.28
b	6.67	6.85

a	8.11	6.04	7.56
b	7.00	7.63	7.35

Figure 2 The effect of steric compression and strain on $^3J(^1\mathrm{H},^1\mathrm{H})_{cis}$ values (in Hz) of various HC$\cdots$CH fragments (a, b)

Table 1 The Ring Size Dependence of $^3J(^1\mathrm{H},^1\mathrm{H})_{cis}$ Values

Cycloalkene ring size	^3J(Hz)
3[28]	1.3
4[29]	2.85
5[27]	5.57
6[27]	10.11
7[27]	11.02
8[27]	10.41

A systematic study of $^3J(^1\mathrm{H},^1\mathrm{H})_{cis}$ for benzo-annelated annulenes[30] has shown that in addition to steric effects the electronic nature of the annulenes, i.e. the number of π electrons and the degree of delocalization, influences the $^3J(^1\mathrm{H},^1\mathrm{H})_{cis}$ data in the annelated benzene ring in a characteristic way. Thus, the π-electron structure of annulenes can be monitored by the $^3J(^1\mathrm{H},^1\mathrm{H})_{cis}$ data for the benzene ring in the corresponding benzoannulene. The π-bond orders P_{12} and P_{23} are determined from the measured coupling constants according to equation (7), which is based on PPP–SCF π-electron calculations,[31,32]

$$P_{\mu\nu}(\mathrm{SCF}) = 0.104\,^3J(^1\mathrm{H},^1\mathrm{H})_{cis} - 0.120 \quad (7)$$

a Q value being defined as $Q = P_{12}/P_{23}$. The magnitude of Q is related to the bonding situation in the annulene: $Q < 1.00$ for antiaromatic ($4n$) π systems, $Q > 1.10$ for aromatic ($4n + 2$) π systems, and $1.00 < Q > 1.10$ for olefinic systems. Typical examples are found with the benztropylium ion (**5**)[23] and benzocycloheptatriene-id (**6**)[33] or the two dehydroannulenes (**7**) and (**8**).[34]

6π, Q = 1.222 (**5**)
8π, Q = 0.899 (**6**)
14π, Q = 1.139 (**7**)
16π, Q = 0.986 (**8**)

Substituent effects have mostly been studied with substituted benzenes, where a careful determination of the $^3J(^1\mathrm{H},^1\mathrm{H})$ values[35] has established the effect of electronegativity: ^3J(a,b) increases in the (X)C–CH$_a$–CH$_b$–CH$_c$– fragment if X is more electronegative than hydrogen, while a small decrease is found for the coupling ^3J(b,c).

;Vicinal ^{1}H,^{1}H coupling via a *trans*-HC–CH fragment, $^3J(^1H,^1H)_{trans}$, is found only in larger annulene rings which tolerate *trans* double bonds. As a consequence of the different dihedral angle $\Phi = 180°$, such couplings are larger than the *cis* values, as the data reported for [18]annulene [$^3J(^1H,^1H)_{cis}$ = 8.0 and $^3J(^1H,^1H)_{trans}$ = 15.5 Hz][36] demonstrate. Other values have been collected in the literature,[34,37] but a detailed analysis of these data was not attempted. For bridged annulenes, various steric requirements imposed on the perimeter geometry by the bridging group usually lead to a loss of coplanarity of all HC–CH fragments and the magnitude of the vicinal ^{1}H,^{1}H coupling constants is subject to additional factors such as varying dihedral angles.[38–40] Early theoretical calculations for $^3J(^1H,^1H)_{cis}$ have been reviewed.[41]

In conclusion, vicinal ^{1}H,^{1}H coupling constants provide the basis for detailed insights into the electronic structure of annulenes and other cyclic π systems and complement the information obtained from shielding properties such as, for example, ring current effects.

2 RELATED ARTICLES

Heterocycles; Indirect Coupling: Theory & Applications in Organic Chemistry.

3 REFERENCES

1. M. Karplus, *J. Chem. Phys.*, 1959, **30**, 11.
2. M. Karplus, *J. Am. Chem. Soc.*, 1963, **85**, 2870.
3. H. Booth, *Prog. Nucl. Magn. Reson. Spectrosc.*, 1969, **5**, 149; C. Altona, J. H. Ippel, A. J. A. W. Hoekzema, C. Erkelenz, M. Groesbeek, and L. A. Donders, *Magn. Reson. Chem.*, 1989, **27**, 564.
4. N. Jonathan, S. Gordon, and B. P. Dailey, *J. Chem. Phys.*, 1962, **36**, 2443.
5. W. B. Smith, W. H. Watson, and S. Chiranjeevi, *J. Am. Chem. Soc.*, 1967, **89**, 1438.
6. H. Günther, *Tetrahedron Lett.*, 1967, 2967.
7. D. R. Eaton, A. D. Josey, W. D. Phillips, and R. E. Benson, *J. Chem. Phys.*, 1963, **39**, 3513.
8. G. Scheibe, W. Seiffert, G. Hohlneicher, C. Jutz, and H. J. Springer, *Tetrahedron Lett.*, 1966, 5053.
9. A. R. Katritzky and R. E. Reavill, *Recl. Trav. Chim. Pays-Bas*, 1964, **83**, 1230.
10. D. J. Bertelli, T. G. Andrews, Jr., and P. O. Crews, *J. Am. Chem. Soc.*, 1969, **91**, 5286.
11. N. M. D. Brown and P. Bladon, *Spectrochim. Acta*, 1968, **24A**, 1869.
12. A. J. Boulton, P. J. Halls, and A. R. Katritzky, *Org. Magn. Reson.*, 1969, **1**, 311.
13. H. Günther and H. H. Hinrichs, *Tetrahedron Lett.*, 1966, 787.
14. M. A. Cooper and S. L. Manatt, *J. Am. Chem. Soc.*, 1969, **91**, 6325.
15. J. B. Pawliczek and H. Günther, *Tetrahedron*, 1970, **26**, 1755.
16. P. C. Lauterbur and R. J. Kurland, *J. Am. Chem. Soc.*, 1962, **84**, 3405.
17. M. A. Cooper and S. L. Manatt, *J. Am. Chem. Soc.*, 1970, **92**, 1605.
18. M. A. Cooper and S. L. Manatt, *J. Am. Chem. Soc.*, 1970, **92**, 4646.
19. S. Castellano and R. Kostelnik, *Tetrahedron Lett.*, 1967, 5211.
20. H. Günther, H. Schmickler, M.-E. Günther, and D. Cremer, *Org. Magn. Reson.*, 1977, **9**, 420.
21. H. L. Ammon and G. L. Wheeler, *J. Am. Chem. Soc.*, 1975, **97**, 2326.
22. S. Braun and J. Kinkeldei, *Tetrahedron*, 1977, **33**, 3127.
23. H. Günther, A. Shyoukh, D. Cremer, and K.-H. Frisch, *Liebigs Ann. Chem.*, 1978, 150.
24. O. L. Chapman, *J. Am. Chem. Soc.*, 1963, **85**, 2014.
25. G. V. Smith and H. Kriloff, *J. Am. Chem. Soc.*, 1963, **85**, 2016.
26. P. Laszlo and P. v. R. Schleyer, *J. Am. Chem. Soc.*, 1963, **85**, 2017.
27. M. A. Cooper and S. L. Manatt, *Org. Magn. Reson.*, 1970, **2**, 511.
28. J. B. Lambert, A. P. Jovanovich, and W. L. Oliver, *J. Phys. Chem.* 1970, **74**, 2221.
29. E. A. Hill and J. D. Roberts, *J. Am. Chem. Soc.*, 1967, **89**, 2047.
30. H. Günther and D. Cremer, *Justus Liebigs Ann. Chem.*, 1972, **763**, 87.
31. R. Pariser and R. G. Parr, *J. Chem. Phys.*, 1953, **21**, 466.
32. J. A. Pople, *Trans. Faraday Soc.*, 1953, **49**, 1375.
33. S. W. Staley and A. W. Orvedal, *J. Am. Chem. Soc.*, 1973, **95**, 3384.
34. H. Günther, M.-E. Günther, D. Mondeshka, H. Schmickler, F. Sondheimer, N. Darby, and T. M. Cresp, *Chem. Ber.*, 1979, **112**, 71.
35. S. Castellano and C. Sun, *J. Am. Chem. Soc.*, 1966, **88**, 4741.
36. J. F. M. Oth, *Pure Appl. Chem.*, 1971, **25**, 573.
37. R. C. Haddon, V. R. Haddon, and L. M. Jackman, *Topics Curr. Chem.*, 1971, **16**, 103.
38. H. Günther, *Z. Naturforsch.*, 1969, **24B**, 680.
39. H. Günther and H.-H. Hinrichs, *Tetrahedron*, 1968, **24**, 7033.
40. A. Alscher, W. Bremser, D. Cremer, H. Günther, H. Schmickler, W. Sturm, and E. Vogel, *Chem. Ber.*, 1975, **108**, 640.
41. J. Kowalewski, *Prog. Nucl. Magn. Reson. Spectrosc.*, 1977, **11**, 1.

Biographical Sketch

Harald Günther, *b* 1935. Studied chemistry at the Universities of Stuttgart and Heidelberg, Ph. D., 1961, Heidelberg with G. Wittig. Research fellow at Mellon-Institute, Pittsburgh 1961–63; introduced to NMR by A. A. Bothner-By. 1963–77 Staff member at the Institute of Organic Chemistry, University of Cologne; Habilitation 1968 with E. Vogel; Associate Professor 1970; since 1977 Full Professor at the University of Siegen. 1973 Chemistry Award of the Academy of Sciences, Göttingen; 1973 Award of the Fonds der Chemischen Industrie, Frankfurt; ca. 200 publications; author of NMR textbook (German, English, Russian, Polish, French editions). Editor-in-Chief of *Magnetic Resonance in Chemistry* since 1992. Research interests include applications of high-resolution and solid state NMR in organic and organometallic chemistry.

Vitamin B_{12}

A. Ian Scott

Texas A&M University, College Station, TX, USA

1 INTRODUCTION

The story of how Nature synthesizes the corrin structure of vitamin B_{12} began about 4×10^9 years ago, when archaebacteria first grew on our planet in an anaerobic atmosphere. (The subject has been reviewed by Scott, by Battersby, and by Leeper.[1]) Largely through the power of NMR spectroscopy, in combination with molecular biology and enzymology, the last 25 years have witnessed the elucidation of the pathway to hydrogenobyrinic acid (Scheme 1), which serves as the precursor to the vitamin via side-chain amidation, cobalt insertion, and addition of the nucleotide loop.

In 1968, at Yale, armed with a home-made FT NMR spectrometer, and liters of *Propionibacterium shermanii* cells, we were able to obtain 2–5% incorporations of the ^{13}C-enriched substrates aminolevulinic acid (ALA), porphobilinogen (PBG), and, most importantly, uro'gen III (Scheme 1) to observe ^{13}C signals above natural abundance (1.1%) in vitamin B_{12}, thus linking the porphyrin and corrin pathways. Later, with L. Siegel (Duke) and G. Müller (Stuttgart), we found and identified the structure of the intermediates precorrins-2 and -3 in their oxidized forms,[1] although the spectroscopy involved 11 days of continuous acquisition of ^{13}C at natural abundance. After this initial burst of success, the next decade was spent in the search for further intermediates during which a great deal was learned about enzymes 2, 3, and 4 (Scheme 1), but, apart from finding the sequence of C-methylation by $^{13}CH_3$-*S*-adenosyl methionine (SAM) in a 'pulsed' $^{13}C/^{12}C$ time course, the steps between precorrin-3 and the corrin structure remained cryptic until 1987, when Roth showed that *Salmonella typhimurium* makes B_{12} anaerobically, and provided the necessary plasmids *and* the DNA sequences of the corrin operon—the *cbi* genes—for overexpression in *E. coli*.[1]

The corresponding *cob* genes of *Pseudomonas denitrificans* (Rhone-Poulenc, Paris) now became accessible,[2] so that, by mid-1990, the complete repertoire of corrin-synthesizing enzymes for both the anaerobic and aerobic pathways was in hand. Now came the fascinating problem of matching each gene product to its function. Collaborative work in Paris and Cambridge[1] (A. R. Battersby) had established the pathway from precorrin-6x to the cobalt-free corrin, hydrogenobyrinic acid (see Scheme 1). Our strategy for deducing the function of the enzymes connecting precorrin-2 and -6x was to enrich precorrin-2 biosynthetically with ^{13}C, then watch the growth of new signals in the NMR spectrum when incubated with each of the gene products in turn, in the presence of $^{13}CH_3$-SAM. Space does not permit other than a brief summary of this complex pathway, but the following examples taken from our own work covering the central section of Scheme 1 typify the NMR techniques used.

2 THE FIRST METHYLTRANSFERASES, THEIR SPECIFICITIES, AND THE SYNTHESIS OF PRECORRIN-2

Uro'gen III (enriched from [5-^{13}C]ALA at the positions (■) shown in Scheme 2) was prepared by mixing the enzymes 1, 2, and 3 (dehydratase, deaminase, and uro'gen III synthase) and then incubating with the first methyltransferase, M-1 (enzyme 4), and [$^{13}CH_3$]SAM. The resultant spectrum of precorrin-2 revealed only one sp^3 enriched carbon (■) assigned to C-15, thereby locating the reduced center. By using a different set of ^{13}C labels (● from [^{13}C-3]ALA) and [$^{13}CH_3$]SAM the sp^2 carbons at C-12 and C-18 were located as well as the sp^3 centers coupled to the pendant ^{13}C-methyl groups at (*) C-2 and C-7. This result[3] confirms an earlier NMR analysis[4] of precorrin-2 isolated by careful anaerobic purification of the methyl ester, and shows that no further tautomerism takes place during the latter procedure. The two sets of experiments mutually reinforce the postulate that precorrins-1, -2, and -3 all exist as hexahydroporphinoids, and recent labeling experiments[5] have provided good evidence that precorrin-1 is discharged from the methylating enzyme (SUMT) as the species with the structure shown in Scheme 1.

However, prolonged incubation (2 h) of uro'gen III with M-1 provided a surprising result, for the UV and NMR changed dramatically from that of precorrin-2 (a dipyrrocorphin) to the chromophore of a pyrrocorphin, hitherto known only as a synthetic tautomer of hexahydroporphyrin. At first sight, this event seemed to signal a further tautomerism of a dipyrrocorphin to a pyrrocorphin catalyzed by the enzyme, but when $^{13}CH_3$-SAM was added to the incubation, it was found that a *third* methyl group signal appeared in the 19–21 ppm region of the NMR spectrum. When uro'gen III was provided with appropriate ^{13}C labels (●) (Scheme 2), three pairs of doublets appeared in the sp^3 region (δ = 50–55 ppm) of the pyrrocorphin product. The necessary pulse labeling experiments together with appropriate FAB MS data finally led to the structural proposal (Scheme 2)[6] for the novel trimethylpyrrocorphin produced by 'overmethylation' of the normal substrate, uro'gen III, in the presence of high concentration of enzyme. Thus, M-1 has been recruited to insert a ring C methyl and synthesize the long-sought 'natural' chromophore corresponding to that of the postulated precorrin-4, although in this case the regiospecificity is altered from ring D to ring C. Under physiological conditions, however, the next step catalyzed by enzyme 5 (M-2; Scheme 1) inserts the next SAM-derived methyl group at C-20 to reach precorrin-3.

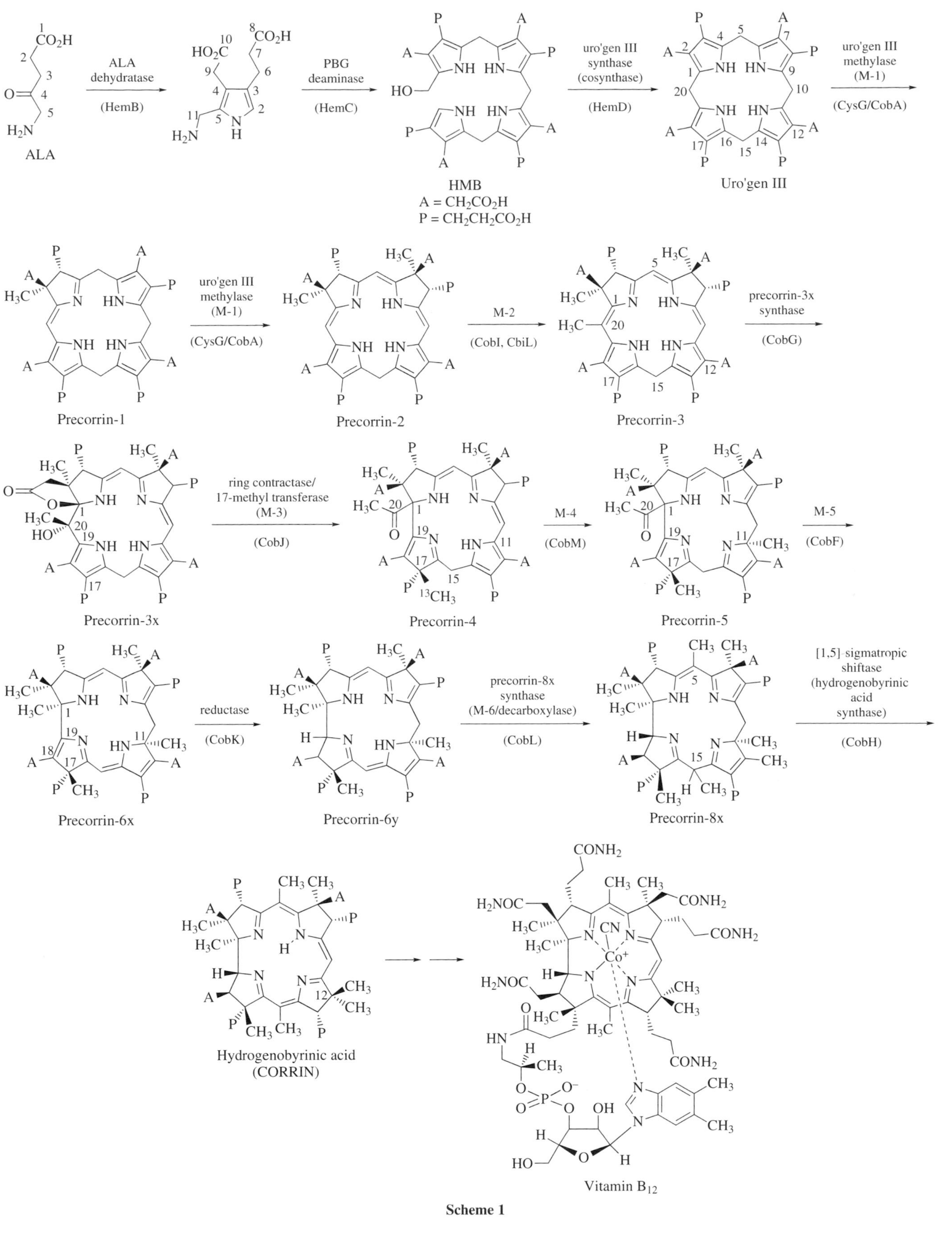

ALA

ALA dehydratase (HemB)

PBG deaminase (HemC)

HMB

$A = CH_2CO_2H$

$P = CH_2CH_2CO_2H$

uro'gen III synthase (cosynthase) (HemD)

Uro'gen III

uro'gen III methylase (M-1) (CysG/CobA)

Precorrin-1

uro'gen III methylase (M-1) (CysG/CobA)

Precorrin-2

M-2 (CobI, CbiL)

Precorrin-3

precorrin-3x synthase (CobG)

Precorrin-3x

ring contractase/ 17-methyl transferase (M-3) (CobJ)

Precorrin-4

M-4 (CobM)

Precorrin-5

M-5 (CobF)

Precorrin-6x

reductase (CobK)

Precorrin-6y

precorrin-8x synthase (M-6/decarboxylase) (CobL)

Precorrin-8x

[1,5] sigmatropic shiftase (hydrogenobyrinic acid synthase) (CobH)

Hydrogenobyrinic acid (CORRIN)

Vitamin B_{12}

Scheme 1

For References see p. 4929

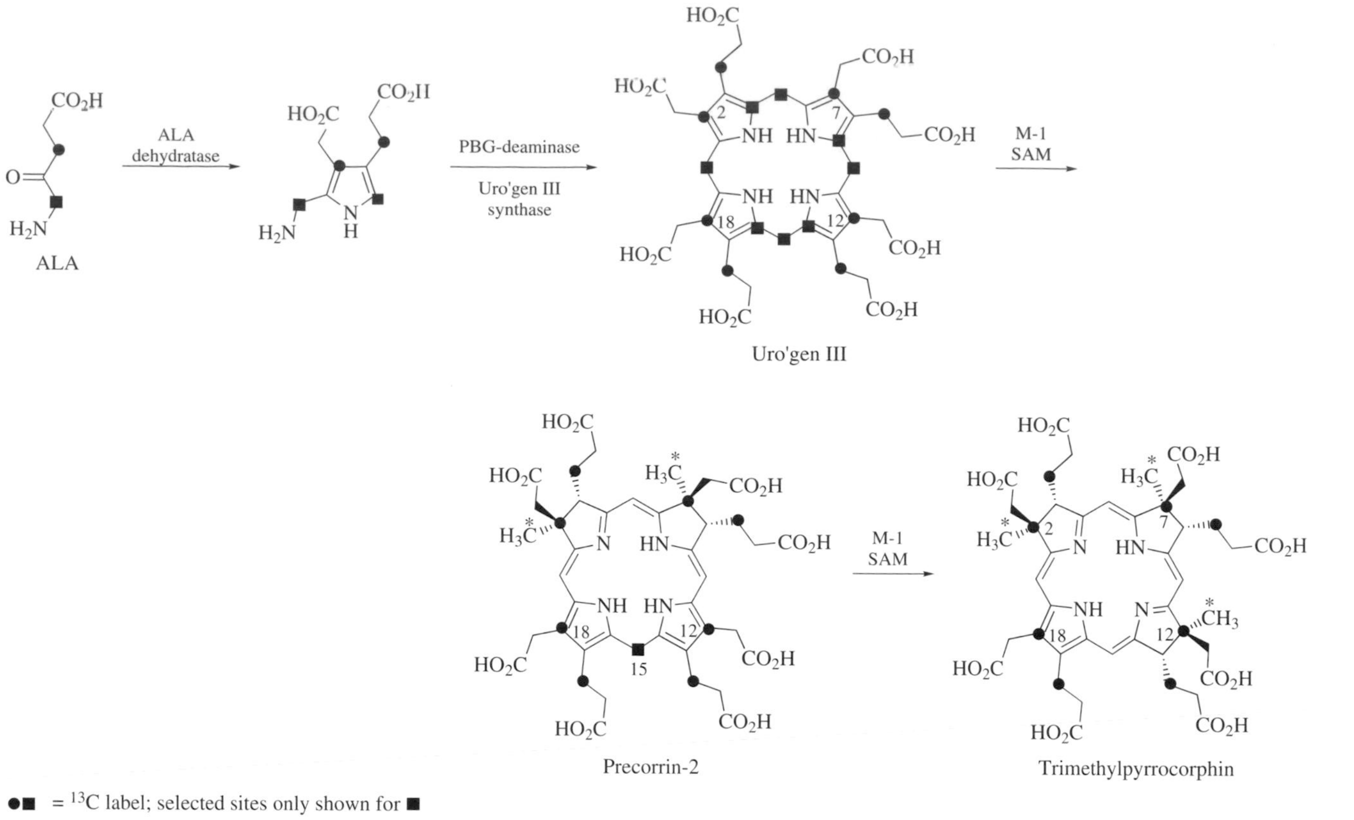

●■ = ^{13}C label; selected sites only shown for ■

Scheme 2

3 FROM PRECORRIN-3 TO HYDROGENOBYRINIC ACID

Using ^{13}C NMR spectroscopy as a probe for the activity of all the overexpressed enzymes of the B_{12} pathway in *P. denitrificans* each enzyme in turn was tested with precorrin-3 as substrate, and it was found that CobG[7] serves as an O_2-dependent enzyme whose role is to install oxygen-derived functionality at C-20, thus preparing the macrocycle for ring contraction. Remarkably, the resultant spring-loaded mechanism *does not operate until after the fourth C-methylation has occurred at C-17*, an event that is mediated by CobJ,[7] a SAM-requiring enzyme, thereby defining both the ring contraction and C-17 methylation sequences.

The NMR assays for the activities of CobG and J were developed as follows. First, precorrin-3 was prepared in two ^{13}C-isotopomeric versions—A, from [^{13}C-4](●)-, and B, from [^{13}C-5](■)-5-aminolevulinic acid, using the multienzyme synthesis described earlier,[8] as shown in Scheme 3. The reaction of isotopomer A (●) with CobG in the presence of O_2 and NADH resulted in a spectrum almost identical with that of the substrate except for the disappearance of the signal (●) for C-1 at δ = 146 ppm and the appearance of a peak (●) at δ = 106 ppm corresponding to sp^3 geometry and a new environment at C-1. When isotopomer B (■) served as substrate, the pattern of NMR signals remained constant, except for the resonance for C-20 (at δ = 103 ppm), which was replaced by a signal at δ = 78 ppm corresponding to oxygen insertion and resultant sp^3 hybridization at this center. In a separate experiment using precorrin-3 enriched at the C-20 methyl, the CH_3 signal (*) at C-20 (δ = 17.6 ppm) in precorrin-3 underwent a high-frequency shift to δ = 25 ppm during incubation with CobG. The above spectral changes are in accord with the addition of oxygen at both positions 1 and 20, and together with the observation by infrared spectroscopy of a γ-lactone (1799 cm^{-1}), lead to the structural proposal for the new product, precorrin-3x (Scheme 3), whose formation can be rationalized by epoxidation at C-1,20 by the O_2-dependent CobG enzyme, followed by participation of the ring A carboxylate in a lactonization–ring opening sequence. In the absence of O_2, no reaction of precorrin-3 with CobG was observed.[9]

Next, addition of SAM and, in turn, each of the remaining putative methyl transferases in the *P. denitrificans* repertoire (CobF, J, M) to precorrin-3x labeled from [4-^{13}C]ALA (●) resulted in a new signal for C-17 (δ = 66 ppm) corresponding to C-methylation at this center, *only when CobJ was present*. Confirmation that C-17 is indeed the site of the fourth methyl insertion came from double labeled incubation of isotopomer A (●) and $^{13}CH_3$-SAM (*) in the presence of *both* CobG and J. In this experiment, the new CH_3 signal (*) appearing at δ = 22.5 ppm was coupled (J = 37 Hz) to the C-17 resonance (●; δ = 66 ppm) (Scheme 3). Most significantly, the ^{13}C NMR spectrum of the new precorrin-4 also displays a pair of coupled carbons C-1, C-19 (δ = 82, 142 ppm, J = 52 Hz), showing that ring contraction occurs during incubation with CobJ. When the two-enzyme incubation was repeated using isotopomer B (■), the signal for C-20 in the product appeared at δ = 210 ppm, indicating that ring contraction is accompanied by the genesis of

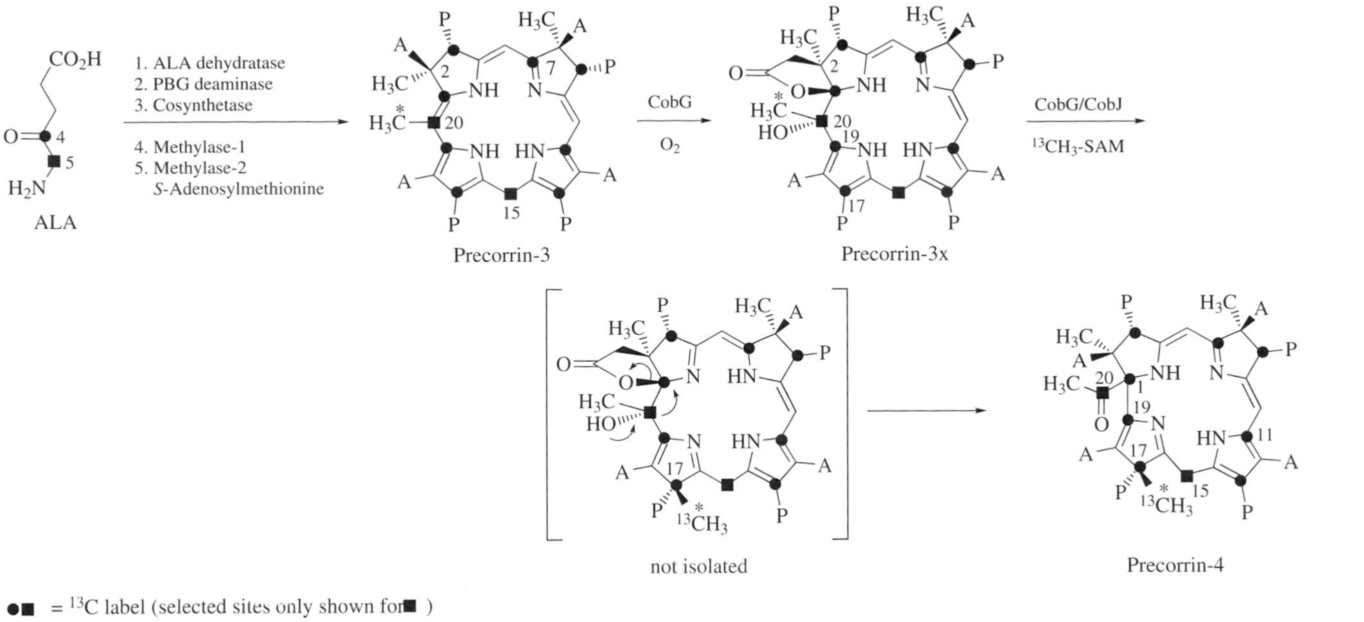

●■ = ^{13}C label (selected sites only shown for ■)

Scheme 3

a new methyl ketone function pendant from C-1, by a process that corresponds formally to the pinacol-type rearrangement illustrated in Scheme 3. Analysis of the full set of spectral data (NMR, FAB MS, and FT IR) leaves no doubt that the new isolate is precorrin-4,[7,10] the long-sought tetramethylated intermediate of corrin biosynthesis.

It is of considerable interest for the evolution of vitamin B_{12} synthesis that Nature should use both a 'modern' aerobic pathway (i.e., less than 2×10^9 years old) involving metal-free substrates for most of the way (Scheme 1), as well as the ancient, anaerobic sequence (dating to about 3.8×10^9 years) in which cobalt is inserted very early,[11] implying a corresponding dichotomy of the mechanism of ring contraction, in which O_2 is replaced in the metallo system by a two-electron valency change. Thus, the cobalt complex of precorrin-3x could be reached in the *Salmonella typhimurium* and *P. shermanii* series via internal redox of Co(III) → Co(I), which is formally equivalent to a two-electron oxidation of the ligand.

The discovery of the ring contractase system (CobG and CobJ) and the fourth methylating enzyme for C-17 (CobJ) (Scheme 3) reveals the way in which aerobic corrin biosynthesis uses an early oxidative ring contraction step concomitant with C-17 methylation on the β (upper) face of the substrate, precorrin-3x, leaving only the enzymes mediating the fifth (C-11) and sixth (C-1) C-methylation steps to be found. These have been identified recently[12] as CobM, which produces another new intermediate, precorrin-5 (step 8 in Scheme 1), and CobF (step 9), respectively. CobF was found to be the C-1 methyl transferase catalyzing both the insertion of the sixth methyl group in precorrin-5 and deacetylation at C-1 to reach precorrin-6x;. The route from precorrin-6x to (cobalt-free) hydrogenobyrinic acid (Scheme 1) has recently been described at the enzyme level,[13] so that almost every intermediate in the aerobic pathway has now been identified by NMR spectroscopy, and all of the functions necessary for corrin biosynthesis assigned in *P. denitrificans*. There now remains the experimentally challenging problem of analyzing the parallel, yet distinct, pathway in *S. typhimurium* and *P. shermanii*, for in these organisms the gene products are handling substrates in the form of cobalt complexes whose valency changes are believed to be at the heart of the mechanism of ring contraction in these anaerobic bacteria.[11]

It took an enormous international effort by Shemin, Scott (USA), Bykhovsky (Russia), Müller (Germany), Battersby (UK), and Arigoni (Switzerland) nearly 20 years to build the 'library' of the first seven intermediates,[1] yet, thanks to recombinant DNA technology, the remaining six structures (Scheme 1) have been discovered in the last three years—all of them determined by biosynthetic enrichment from ^{13}C-ALA and $^{13}CH_3$-SAM.

4 RELATED ARTICLES

Biosynthesis & Metabolic Pathways: Carbon-13 and Nitrogen-15 NMR; Enzymatic Transformations: Isotope Probes.

5 REFERENCES

1. A. I. Scott, *Acc. Chem. Res.*, 1990, **23**, 308; A. R. Battersby, *Acc. Chem. Res.*, 1993, **26**, 15; A. I. Scott, *Angew. Chem.*, 1993, **32**, 1223; F. J. Leeper, *Nat. Prod. Rep.*, 1989, **6**, 171.
2. J. Crouzet, B. Cameron, L. Cauchois, S. Rigault, M. Rouyez, F. Blanche, D. Thibaut, and L. Debussche, *J. Bacteriol.*, 1990, **172**, 5980.
3. M. J. Warren, N. J. Stolowich, P. J. Santander, C. A. Roessner, B. A. Sowa, and A. I. Scott, *FEBS Lett.*, 1990, **261**, 76.
4. W. M. Stork, M. G. Baker, P. R. Raithby, F. J. Leeper, and A. R. Battersby, *J. Chem. Soc., Chem. Commun.*, 1985, 1294.
5. R. D. Brunt, F. J. Leeper, I. Grgurina, and A. R. Battersby, *J. Chem. Soc., Chem. Commun.*, 1989, 428.
6. A. I. Scott, M. J. Warren, C. A. Roessner, N. J. Stolowich, and P. J. Santander, *J. Chem. Soc., Chem. Commun.*, 1990, 593.

7. A. I. Scott, C. A. Roessner, N. J. Stolowich, J. B. Spencer, C. Min, and S.-I. Ozaki, *FEBS Lett.*, 1993, **331**, 105.
8. M. J. Warren, C. A. Roessner, S.-I. Ozaki, N. J. Stolowich, P. J. Santander, and A. I. Scott, *Biochemistry*, 1992, **31**, 603.
9. J. B. Spencer, N. J. Stolowich, C. A. Roessner, C. Min, and A. I. Scott, *J. Am. Chem. Soc.*, 1993, **115**, 11 610.
10. D. Thibaut, L. Debussche, D. Frechet, F. Herman, M. Vuilhorgne, and F. Blanche, *J. Chem. Soc., Chem. Commun.*, 1993, 513.
11. G. Müller, F. Zipfel, K. Hlineny, E. Savvidis, R. Hertle, U. Traub-Eberhard, A. I. Scott, H. J. Williams, N. J. Stolowich, P. J. Santander, M. J. Warren, F. Blanche, and D. Thibaut, *J. Am. Chem. Soc.*, 1991, **113**, 9891.
12. C. Min, B. P. Atshaves, C. A. Roessner, N. J. Stolowich, J. B. Spencer, and A. I. Scott, *J. Am. Chem. Soc.*, 1993, **115**, 10 380.
13. D. Thibaut, F. Kiuchi, L. Debussche, F. J. Leeper, F. Blanche, and A. R. Battersby, *J. Chem. Soc., Chem. Commun.*, 1992, 139; D. Thibaut, M. Couder, A. Famechon, L. Debussche, B. Cameron, J. Crouzet, and F. Blanche, *J. Bacteriol*, 1992, **174**, 1043.

Biographical Sketch

A. Ian Scott. *b* 1928. B.Sc., 1949, Ph.D., 1952, D.Sc., 1963, M.A.(Hon.), Yale University, 1968. Lecturer, Glasgow University, 1957–62; Professor, University of British Columbia, 1962–65, University of Sussex, 1965–68, and Yale University, 1968–77. Davidson Professor of Science and Distinguished Professor of Chemistry and Biochemistry, Texas A&M University, 1981–present. Director of Center for Biological NMR, 1982–present. Three books, 1 patent, and approx. 320 publications. Research specialties: natural product biosynthesis and NMR spectroscopy.

Water Signal Suppression in NMR of Biomolecules

Maurice Guéron & Pierre Plateau
Ecole Polytechnique, Palaiseau, France

1 INTRODUCTION

1.1 Notation and Definition

An excitation sequence is made up of one or more pulses. It may be organized in subsequences of pulses. A selective pulse, subsequence, or sequence is one that excites spins in only a fraction of the spectral region of interest. Examples are a soft pulse and a DANTE subsequence. The prefix 'sub' may be omitted.

The term *saturation* is used to describe not only a state in which the longitudinal magnetization is zero throughout the sample, but any condition (e.g., that produced by forced nutation) that is operationally equivalent to uniform zero longitudinal magnetization.

1.2 Solvent Signal Suppression

It is often necessary to obtain proton NMR spectra of compounds whose concentration is in the millimolar range, dissolved in protonated solvents, such as H_2O rather than D_2O. This is the case for NMR of exchangeable protons, or in proton NMR in vivo. With nonselective pulse excitation, the high concentration of water protons, about 110 M, raises serious difficulties, such as saturation of the analog receiver or of the analog-to-digital converter, resulting in artifacts that could render the spectrum useless. It is therefore necessary to reduce the solvent signal.[1]

There are two stages in solvent signal suppression (SSS). The first is the reduction of the solvent contribution to the signal generated by the NMR probe, so as to prevent receiver saturation. This stage cannot be dispensed with. Methods for its implementation are of two types: those that perturb the solvent magnetization (e.g., by saturation), and those that leave it unchanged. The latter are preferred, or even indispensable, for the study of exchangeable protons.

The second stage consists in data postprocessing with the aim of further reduction of the residual solvent signal.

Suppression of the signal of the solvent, but not of the signals of the solutes, must use some difference between these. The most general difference is in the chemical shift, and most of this article is indeed devoted to SSS methods based on frequency differences between solvent and solutes.

However, any property distinguishing solvent from solute may be, and usually has been, recruited to aid in SSS. In the case of water, this includes

1. the small size of the molecule, leading to
   (a) fast rotational Brownian motion and hence long T_1,
   (b) fast translational Brownian motion and hence fast diffusion;
2. the equivalence of the two protons and consequent absence of J splitting;
3. the (obvious) lack of heteronuclear couplings to nitrogen or carbon.

An ideal SSS method would be universal, would suppress the solvent signal, and would leave all others unchanged. It would also be easy to implement, and would not be too demanding on the specifications of the spectrometer. In practice, the varieties of experimental situations and spectrometer limitations result in a large number of diverse cases, with different criteria for solvent suppression. Therefore, no SSS method is best for all cases at present. Despite this caveat, we shall risk general guidelines that apply to both 1D and nD NMR.

The first stage of SSS should reduce the solvent signal as required to avoid saturation of the receiver while keeping as small as possible any perturbations of the properties observable in the NMR spectrum, such as chemical exchange, NOE, signal-to-noise ratio and spectral (phase) distortions. Among frequency-selective SSS excitation sequences, the JR sequences described in Section 3.2.5 may be recommended on this basis.

All further reductions of the solvent signal should be left to data postprocessing. We recommend data processing in the frequency domain, including

1. the automatic subtraction of Lorentzians centered in a narrow window centered at the solvent frequency;
2. the automatic correction of spectral amplitude as required by the frequency-selective excitation procedure.

An elementary criterion for the quality of the SSS procedure is that NO linear phase correction or baseline correction should be needed, beyond those used on the same spectrometer in the case of excitation with a single 90° pulse.

1.3 A Single Long Pulse

In order to avoid excessive abstraction, we now describe a method of solvent signal suppression. It uses a single excitation

For References see p. 4942

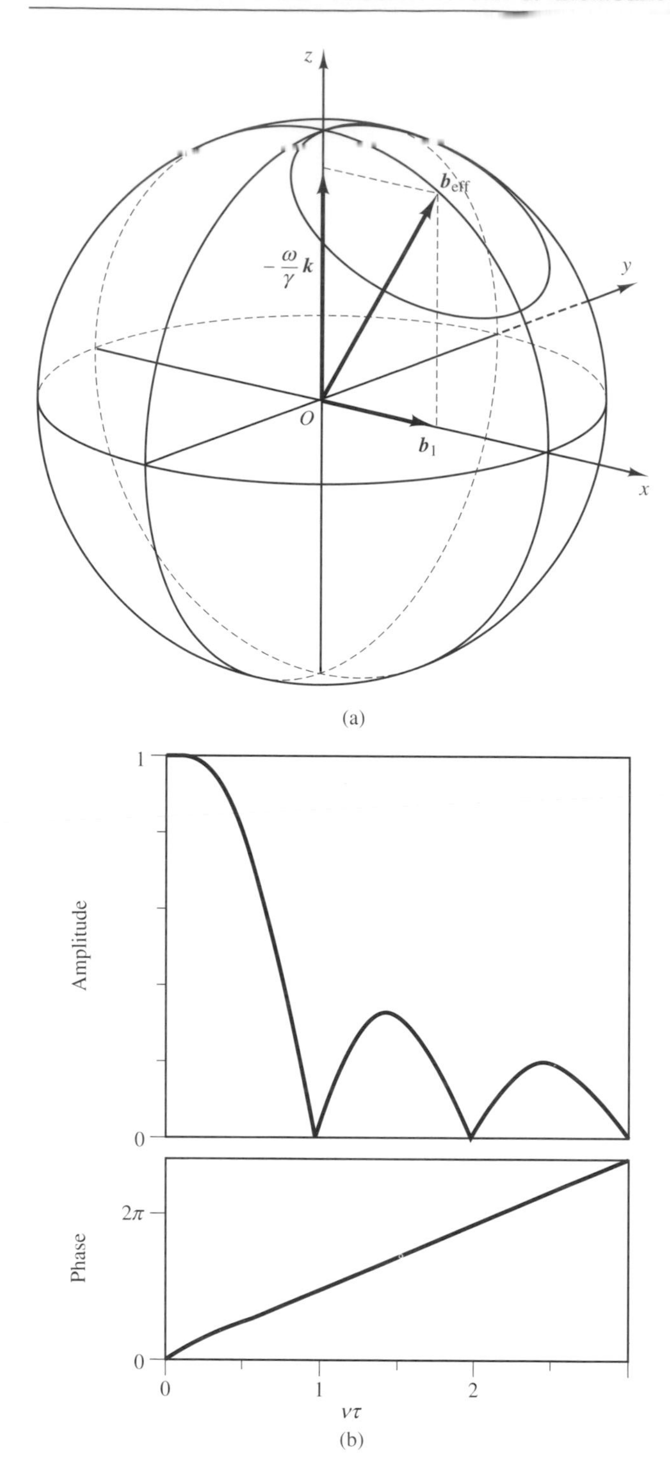

Figure 1 (a) The motion of the magnetic moment in the presence of a weak rotating field $\boldsymbol{b}_1$, as viewed in the frame rotating at the frequency of the field, aligned along Ox. The effective field $\boldsymbol{b}_{\text{eff}}$ is represented. Its components are $(-\omega/\gamma)\boldsymbol{k}$ and $\boldsymbol{b}_1$. The small circle is the trajectory of the magnetic moment, whose initial position is along the z axis. The scale is such that the lengths of the magnetic moment and of the effective field are equal. (b) Amplitude and phase of the transverse magnetic moment after a $\frac{1}{2}\pi$ pulse lasting τ, as a function of the frequency offset ν. The signal at $\nu = (\frac{15}{16})^{1/2}/\tau$ is suppressed (single long pulse technique). (Modified from M. Guéron, P. Plateau, and M. Decorps, in 'Progress in Nuclear Magnetic Resonance Spectroscopy', eds. J. W. Emsley, J. Feeney, and L. H. Sutcliffe, Pergamon Press, Oxford, 1991, Vol. 23, p. 138, Copyright 1991, with permission from Pergamon Press Ltd)

pulse at the frequency of the spins of interest, such that they rotate by an angle θ away from Oz, whereas the solvent spins end up along Oz after the pulse.

A good description of the single long pulse method can be made in a *rotating frame*—one that rotates around the z axis at the excitation frequency (Figure 1a). In this frame, the excitation field $\boldsymbol{b}_1$ is static, and this is the effective field for a spin whose resonance frequency is equal to the excitation frequency. The effective field $\boldsymbol{b}_{\text{eff}}$ for the solvent spins, offset by $\omega/2\pi$ from the excitation frequency, has, besides the transverse component $\boldsymbol{b}_1$, a longitudinal component $(-\omega/\gamma)\boldsymbol{k}$, where γ is the gyromagnetic ratio and $\boldsymbol{k}$ in the unit vector along the z axis. Each spin rotates around its effective field, at an angular frequency $\omega_{\text{eff}} = -\gamma\,|\boldsymbol{b}_{\text{eff}}|$.

At the end of a pulse lasting τ, the spins have rotated respectively by

$$\theta_{\text{resonant}} = -\gamma b_1 \tau \tag{1}$$

$$\theta_{\text{nonresonant}} = -\gamma \left[b_1^2 + \left(\frac{\omega}{\gamma}\right)^2 \right]^{1/2} \tau \tag{2}$$

Solvent signal suppression is obtained if $\theta_{\text{nonresonant}} = 2\pi$. Together with a choice of θ_{resonant}, this specifies the amplitude b_1 and duration τ of the pulse:

$$\tau = \frac{(4\pi^2 - \theta_{\text{resonant}}^2)^{1/2}}{|\omega|} \tag{3}$$

Equation (3) shows that the length of the pulse is a function of the frequency difference ω between the spins to be seen and those to be suppressed. More generally, for a strong variation of the excitation spectrum within a range ω_s, pulse lengths or intervals in the range of $2\pi/\omega_s$ are required. At 400 MHz, for a difference in chemical shift of 8 ppm, this corresponds to τ = 312 μs. To satisfy equation (1), a pulse of this length must be weak—or 'soft', as is commonly said.

The amplitude and phase of the transverse magnetization created by the soft pulse are shown in Figure 1(b) as functions of the frequency offset. The transverse magnetization is indeed zero at the solvent frequency, and maximum for the resonant spins. The phase of the transverse magnetization is approximately proportional to the frequency offset—the so-called linear phase shift.

This method has two qualities:

1. it is simple, and demands little of the spectrometer in terms of phase or amplitude modulation of the excitation pulse;
2. it restores the solvent magnetization to its original orientation.

It has two drawbacks:

1. the frequency of the excitation zero depends on the rf field amplitude [equation (2)], so that the degree of solvent signal suppression is susceptible to equipment drift;
2. compensation of the linear phase shift generates baseline distortions.

These features are important for SSS, and they will be considered anew as we examine principles and implementations of SSS.

1.4 Scope of This Article

Solvent signal suppression was originally developed in the framework of 1D NMR applications to chemistry and biochemistry. Nowadays, these applications often involve 2D and 3D NMR. SSS is also used in in vivo localized proton NMR spectroscopy, in which the rf field may be strongly inhomogeneous, and the water peak may have contributions not only from the volume of interest but also from large surrounding regions.

The emphasis of this article is on SSS fundamentals, which, despite differences in methodology, are the same in the three domains. Furthermore, the similarity in hardware and software leads to comparable limitations on SSS. The discussion is mostly in terms of 1D NMR, and specific problems of 2D and 3D NMR are treated in a separate section. See also ***Water Suppression in Proton MRS of Humans & Animals***.

The question of SSS also arises in the context of subtractive techniques, as used for instance in homo- or heteronuclear editing techniques like reverse INEPT, or multiquantum NMR. The subtractive character may help considerably with respect to SSS. This is mostly outside the scope of the present article, as are tailored excitation,[2] the use of notch filters,[3] rapid scan FT NMR,[4] and stochastic excitation.[5]

The reader in need of more information on SSS is referred to the extensive original literature, to recent reviews,[6,7] and to related articles in this Encyclopedia.

2 WHY SUPPRESS THE SOLVENT SIGNAL?

Solvent signal suppression is required because it is impossible to acquire the free induction decay of the large solvent signal while maintaining maximum sensitivity for other signals.

2.1 Acquisition of the Free Precession

For good sensitivity, the noise issuing from the probe should be amplified to a value no less than one least significant bit of the analog-to-digital converter (ADC). Signals that are buried in the noise can then be recovered without loss in signal-to-noise ratio (S/N). The RMS thermal noise of a resistor r in a bandwidth $\Delta\nu$, $(4rk_BT\Delta\nu)^{1/2}$, is 0.09 μV for 50 Ω and 10 kHz, and the appropriate receiver gain is then about 10^4 if the least significant bit of the ADC is 1 mV.

The maximum value of the FID must remain within the range of the ADC. It must therefore be less than 4096 times the noise for a 12-bit ADC, and 65 536 times for a 16-bit ADC. Since the noise increases with the bandwidth, the S/N of the FID, and together with it the required dynamic range of the receiver, may be reduced by increasing the bandwidth,[8] at no cost in the S/N of the spectrum.

The ratio of mean square signal to mean square noise $(\mathrm{S/N})^2$ at the beginning of the free precession following a $\frac{1}{2}\pi$ pulse is given by[7]

$$(\mathrm{S/N})^2 = \frac{mB_0/t_R}{4Fk_BT\Delta\nu} \quad (4)$$

where m is the magnetic moment within the probe, B_0 is the static field, and F is the noise factor of the receiver; t_R is the time for radiation damping[9] of a parallel-tuned circuit, given in SI units by

$$t_R^{-1} = \frac{M\gamma\eta Q}{2\epsilon_0 c^2} \quad (5)$$

Here, $4\pi\epsilon_0 c^2 = 10^7$, M is the magnetization (the magnetic moment per unit volume), Q is the quality factor of the tuned circuit, and η is the filling factor. For water protons, M at equilibrium in a field B at room temperature is equal to $3.24 \times 10^{-3}B$, and $t_R^{-1} = 168\eta QM$.

According to equation (4), the volt ratio ($\mathrm{S/N} \times 2\sqrt{2}$) of the *peak-to-peak* signal from 200 μL H_2O at 400 MHz to the RMS noise in a bandwidth of 10 kHz is about 3.8×10^5, or $2^{18.5}$, assuming that $Q\eta = 10$. Proper acquisition of the FID would thus require a 19-bit ADC together with a receiver having the same dynamic range. If a 12-bit ADC is used, the excitation procedure must reduce the water signal by a factor of at least 2^7, i.e., 128.

Note that the S/N of the FID is not equal to that of the spectrum. Indeed, the maximum value of the FID, right after a $\frac{1}{2}\pi$ pulse, corresponds to the integral of the spectrum, and is therefore independent of the linewidth, whereas the signal height in the spectrum is inversely proportional to it.

After attenuation by a factor of 128, sufficient for signal acquisition, the remaining water signal is still huge, but it may be reduced by postacquisition processing. Acquisition methods that provide a larger SSS factor than strictly required by the constraint of dynamic range are convenient, because they are more tolerant of spectrometer imperfections and may require less postacquisition processing. However, they have some problems. Typically, they use second- or third-order suppression, which creates a large region near the solvent frequency where the magnetic signals, but not the thermal noise of the probe, are reduced. The longer excitation sequence may affect measurements requiring multiple, closely spaced excitation pulses (spin echoes, inversion–recovery, 2D NMR, localized spectroscopy). It may also create spurious effects of cross polarization (NOE) and chemical exchange. Postacquisition processing suffers from none of these defects, and is therefore preferable in principle.

2.2 The 'Rolling Baseline'

Let us go back to the single long pulse described in Section 1. At the end of the pulse, the water magnetization is returned to the z axis, there is no transverse magnetization, and the amplitude of the water signal is zero.

Consider now a Bloch spin with central frequency ω_C close to, but distinct from that of the solvent. At the end of the excitation pulse, its magnetization will have a small transverse component. This will generate a FID, which will, as usual, decay exponentially according to T_2, and whose Fourier transform will therefore be the customary Lorentzian, or, in other words, the Bloch lineshape. The point is that this Lorentzian will have a finite, nonzero amplitude at the frequency of the solvent, that is, at the frequency of null excitation!

This shows that the solvent suppression scheme does not, strictly speaking, generate a frequency-dependent attenuation.[7] Rather, *it attenuates without distortion each Lorentzian line as a function of its central frequency: the suppression procedure acts on the entire Lorentzian.* In the same way, each Lorent-

For References see p. 4942

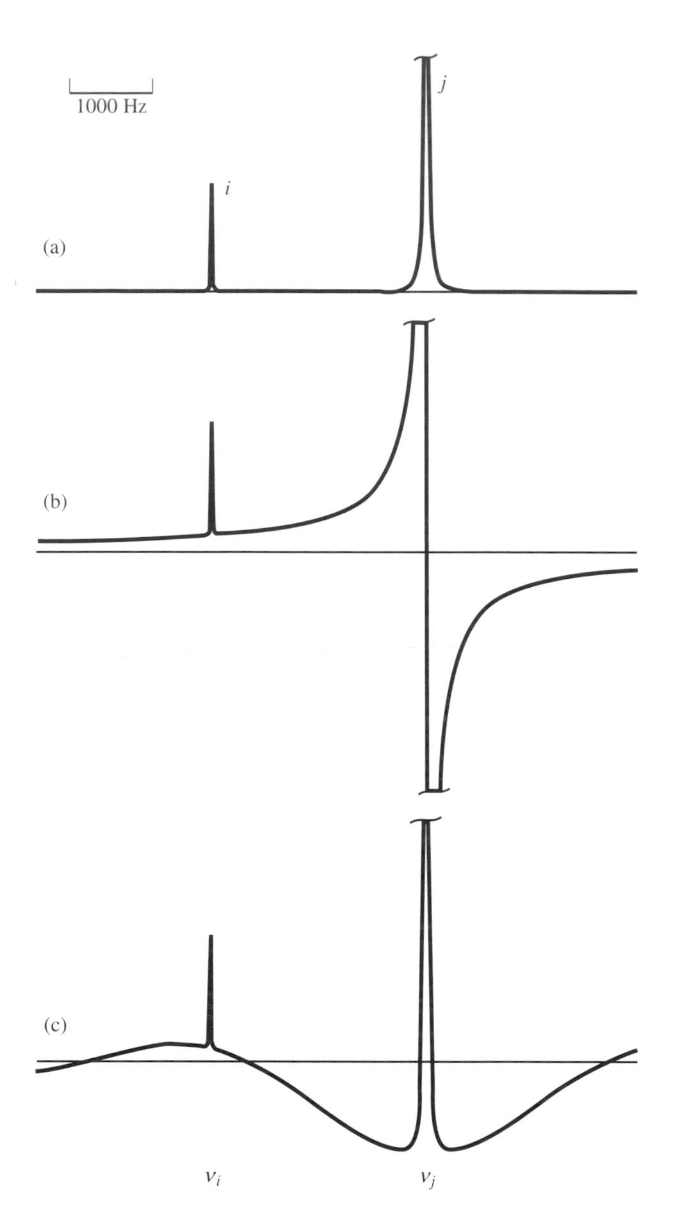

Figure 2 Simulated spectra of two Lorentzian lines (from species *i* and *j*), having the same width (5 Hz), and whose intensities are in the ratio of 1:100. The frequency separation is 2500 Hz. Only the absorption is shown. (a) The Fourier transform of the response to a hard pulse is the true spectrum, $S(\omega) = S_i(\omega) + S_j(\omega)$. (b) The Fourier transform $E(\omega)$ of the same response, except that acquisition begins 400 μs after the hard pulse. It is the sum of the two Lorentzians, with their phases offset by $\frac{1}{2}\pi$. The constant phase is chosen so that line *i* is phased in absorption. (c) The spectrum $D(\omega)$ that results from the application of a linear phase correction to $E(\omega)$, chosen so that each line is properly phased near its center. Since the phase varies with frequency, each line is distorted to a non-Lorentzian shape. The 'baseline roll' comes from the dispersion wing of the intense *j* line. (Reproduced from M. Guéron, P. Plateau, and M. Decorps, in 'Progress in Nuclear Magnetic Resonance Spectroscopy', eds. J. W. Emsley, J. Feeney, and L. H. Sutcliffe, Pergamon Press, Oxford, 1991, Vol. 23, p. 151, Copyright 1991, with permission from Pergamon Press Ltd)

zian is phase-shifted by a *constant* phase angle that is only a function of the central frequency.

This is not the same thing as a frequency-dependent phase shift,[1] and it cannot therefore be corrected by a true frequency-dependent phase shift of opposite value. Such a correction does not conserve Lorentzians, and this gives rise to the 'rolling baseline' defect that appears in Figure 2. This defect could be avoided by more sophisticated correction procedures.[7] Even better would be excitation procedures that do not produce a linear phase shift.

3 EXCITATION SCHEMES FOR SSS

The number and variety of SSS methods may seem overwhelming, with new ones being published monthly. One may nevertheless attempt to clarify the situation, as follows.

1. Many methods are useful only in particular conditions.
2. Interesting new methods are usually those that take advantage of the new possibilities of modern spectrometers, in particular fast and flexible control of rf phase and amplitude, and, most importantly, the generation of pulsed gradients.
3. Methods with modest hardware requirements are accessible to more users, and they are less susceptible to hardware imperfections. This argues in favor of symmetric sequences, which help to compensate spectrometer imperfections; short sequences, which minimize spurious evolutions (e.g., radiation damping and magnetization transfer) during the sequence; and strong pulses.
4. The linear phase shift is a source of distortions that are only corrected with difficulty: it can and should be avoided.
5. Many SSS excitation methods lead to saturation of the solvent resonance, and this may change the magnetization of some solute protons (e.g., imino protons of nucleic acids) by way of NOE or chemical exchange. Methods that leave the solvent magnetization at equilibrium do not produce such side effects. They are therefore more generally applicable.

In this section, we present a selection of SSS methods that provide examples of the characteristics just described. They are separated into two classes: those that alter the solvent magnetization and those that do not.

3.1 Methods That Alter the Solvent Magnetization

SSS may be obtained by canceling the solvent magnetization. Solvent/solute discrimination may be based on a difference in the chemical shift, or in the rates of longitudinal or transverse relaxation or even of diffusion. Methods based on relaxation require a time in the range of T_1 or T_2. Methods based on the inhomogeneity of the main field B_0 or of the rf field b_1 may operate much faster. The required time is in the range of $(\gamma\delta)^{-1}$, where δ is the variation of the field over the sample.

Let us start with relaxation-based methods. In method (a) (Figure 3), the solvent magnetization is selectively saturated by a long and weak pulse at the solvent resonance frequency, after which the strong observation pulse is applied. In method (b), the solvent magnetization is selectively inverted by a weak pulse or by a series of intense pulses, the so-called DANTE

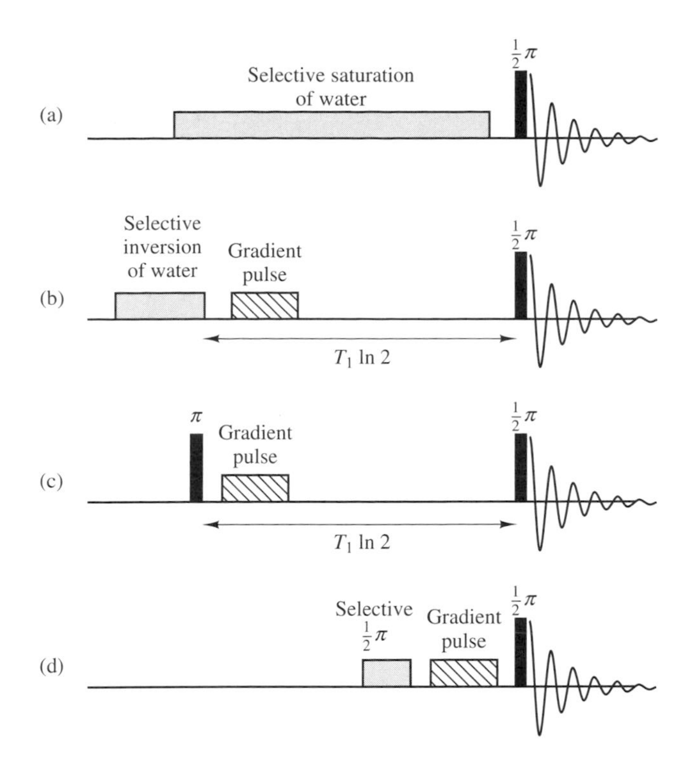

Figure 3 Four SSS sequences, based on saturation of the solvent signal. In methods (a) and (b), discrimination between solvent and solutes is achieved by frequency-selective excitation. In method (c), the water proton is selected by its long longitudinal relaxation time. Method (d), in contrast to (a)–(c), produces saturation quickly, independently of relaxation. (Modified from M. Guéron, P. Plateau, and M. Decorps, in 'Progress in Nuclear Magnetic Resonance Spectroscopy', eds. J. W. Emsley, J. Feeney, and L. H. Sutcliffe, Pergamon Press, Oxford, 1991, Vol. 23, p. 154, Copyright 1991, with permission from Pergamon Press Ltd)

sequence,[10] and residual transverse magnetization is destroyed with a gradient pulse. The observation pulse is applied when the longitudinal magnetization becomes null, after a time $T_1 \ln 2$, where T_1 is the longitudinal relaxation time of the solvent. Method (c) (WEFT)[11,12] is a variant of (b). It is simple to set up in that it uses only strong, nonselective pulses. Because all spins are inverted by the first pulse, it works only if the *longitudinal* relaxation of the solvent is slower than that of the protons of interest. This will often be the case, because of the relatively small size of the water molecule, which results in a large *rotational* diffusion coefficient.

When the *transverse* relaxation of the solvent is faster than that of the spins of interest, one can achieve SSS by the WATR method.[13] This is a spin echo method, in which the refocusing π pulse occurs after a time much longer than the transverse relaxation time of water protons, which are therefore not refocused. Relaxation may be enhanced by paramagnetic ions,[14] or by proton exchange with ammonia or urea.[15] The method is also applied to the elimination of the signal from water or fat in biological tissues. Both WEFT and WATR have a flat spectral response, including at the solvent frequency.

Spurious effects of solvent saturation are reduced in methods that achieve solvent saturation quickly, independently of relaxation. In method (d) of Figure 3, a selective $\frac{1}{2}\pi$ pulse is followed by a gradient pulse ('homospoil'), which destroys the transverse magnetization.[16] With forced nutation[17] in an inhomogeneous rf field $b_1 \pm \delta b_1$, saturation is obtained in a time $(\gamma\,\delta b_1)^{-1}$. Forced nutation of the solvent spins may be combined with *spin locking* of the solute spins, so that all signals, including the solvent residual, end up in-phase. The method applies to 1D and 2D NMR and to in vivo spectroscopy.[18–21] This is also the case for the 'WATERGATE' (water-suppression by gradient-tailored excitation) method when the observed signal is a spin echo, from which the water proton signal is absent because the refocusing π pulse is selective.[22]

In the 'DRYCLEAN' method,[23] one uses the large *translational* diffusion coefficient of water as compared with macromolecules: the transverse water magnetization is selectively destroyed by the application of a pulsed magnetic field gradient during a spin echo sequence.[24] A beautiful extension of this concept has been made in the case of proton NMR of perfused cells,[25] for which SSS is particularly important, since the cells occupy only a small fraction (about 1%) of the total volume. Furthermore, the signals of the extracellular solutes must also be suppressed. The diffusion method discriminates between intra- and extracellular products, because the range of diffusion of the latter is restricted by the cell membrane. For an appropriate duration of the gradient pulse, signals from extracellular water and solute are suppressed, while the intracellular ones are unaffected by diffusion. The signal of the intramolecular water is suppressed by other characteristics of the excitation sequence, such as frequency selectivity.

3.2 Methods That Avoid Solvent Saturation

3.2.1 Long Pulses

A weak or long or selective pulse is one for which b_1 is smaller than the chemical shift range of the spectrum, so that the pulse acts differently on different spin species. These differences are used in the 'long pulse' suppression methods, such as the 'single long pulse' described in Section 1.

The excitation spectrum of a long pulse may be usefully modified by modulating its phase, as in Redfield's '214' pulse,[1] whose time sequence is $2\underline{1}4\underline{1}2$, the underlined numbers corresponding to times during which the phase is shifted by π. Its excitation spectrum has two closely spaced zeros. Similar sequences have a second-order zero.[7] This protects against instrumental imperfections, such as inhomogeneity or drift in field or in temperature, all of which generate water signals at frequencies differing slightly from the nominal excitation zero. The sequence is also less sensitive to rf field inhomogeneity. The cost is reduced sensitivity close to the solvent frequency.

3.2.2 Short Pulses; Binomial Sequences

Binomial sequences consist of sequences of pulses whose amplitudes are proportional to the binomial coefficients, separated by equal interpulse delays. The simplest binomial sequence is 1, 1 (two pulses of equal amplitude and phase, separated by a delay), at the frequency of maximum signal. An exact, as opposed to linear, description may be given in the rotating frame. The first pulse tilts the spins by, for instance, $\frac{1}{6}\pi$ in the direction of the y axis. During the delay, the solvent spins rotate by $\omega\tau = \pi$ and the spins to be observed are immobile. The second and last pulse, at time 0, brings the water protons back along Oz, while the spins to be observed end up

For References see p. 4942

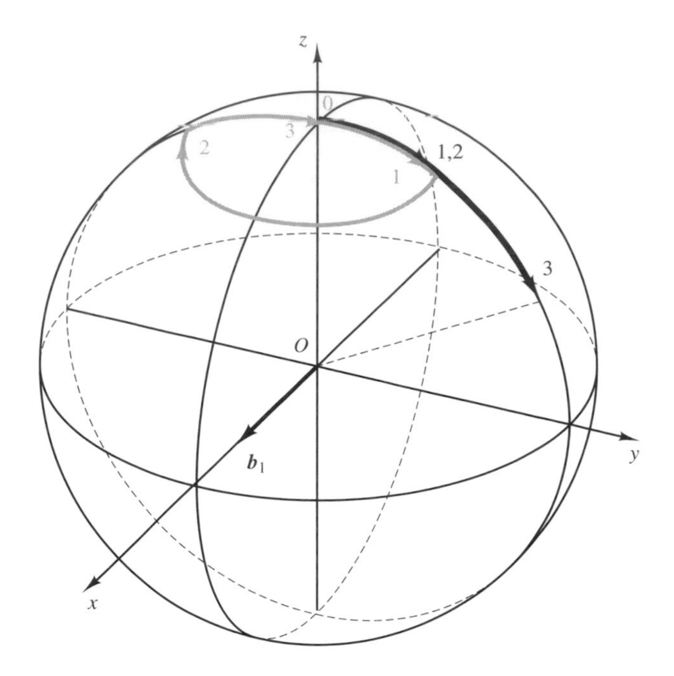

Figure 4 The simplest binomial SSS sequence: 1, 1. The gray line represents the motion of the solvent magnetization, ending in the *Oz* direction; the black line represents the motion of the magnetization of the resonant spins. (Modified from M. Guéron, P. Plateau, and M. Decorps, in 'Progress in Nuclear Magnetic Resonance Spectroscopy', eds. J. W. Emsley, J. Feeney, and L. H. Sutcliffe, Pergamon Press, Oxford, 1991, Vol. 23, p. 157, Copyright 1991, with permission from Pergamon Press Ltd)

tilted by $\frac{1}{3}\pi$ towards the *y* axis (Figure 4). Spins with intermediate chemical shifts are less tilted, and their transverse components have different directions, corresponding to a frequency-dependent phase shift.

The excitation may be described as two delta functions of opposite sign, at times $-\tau$ and 0. In the linear approximation, the spin response is determined by the Fourier transform of this sequence (the excitation spectrum), which is simply $e^{-i\omega\tau} + 1$, or $2e^{-i\omega\tau/2}\cos\frac{1}{2}\omega\tau$. The latter expression displays the frequency dependence of amplitude and phase. The amplitude is maximum for $\omega\tau = 0$, and zero for $\omega\tau = \pi$. The phase varies linearly from 0 to $\frac{1}{2}\pi$ in this frequency range.

This sequence has the following properties.

1. The evolution of the spins is easily described, because the excitation is a combination of hard pulses during which all spins rotate around the same pulse axis, and of waiting periods during which all spins rotate around the same *Oz* axis.
2. Hard pulses provide excitation over a large spectral region.
3. Owing to the symmetry of the sequence, spatial inhomogeneity or long-term drift of the pulse amplitude does not change the frequency of the excitation zero.

In the linear approximation, a binomial sequence of order *n* provides suppression to the same order. The 1, −2, 1 (i.e., 1, −1, immediately followed by −1, 1) and 1, 2, 1 excitation sequences provide second-order suppression for all pulse angles.

The main drawback of the binomial sequences is the linear phase shift of the excitation spectrum. It is considerably reduced, without change of the amplitude response, in a *modified binomial sequence*, which consists of a binomial inserted between a $-\pi$ and a $+\pi$ pulse.[7,26,27] More complex hard pulse SSS sequences can be built on the basis of extensive computer searches.[28,29]

3.2.3 Combinations of Selective and Nonselective Subsequences

As an example, consider a sequence consisting of a soft pulse of angle θ, followed by a hard pulse of angle $-\theta$, both at the solvent frequency.[11] The response to the soft pulse has strong phase shifts, but it is limited to a narrow range around the solvent frequency. The subsequent hard pulse provides nonselective excitation without phase shift, and cancels the effect of the soft pulse on the solvent magnetization. This sequence therefore provides a nearly flat and phased response. In this way, the two functions of the excitation sequence, signal suppression in a narrow range and observation in a broad range, are well separated. On the other hand, the inherent asymmetry of the method makes it sensitive to hardware imperfections.

3.2.4 Echoes, Gradients and EXORCYCLE

The linear phase may be avoided by generation of a spin echo. An echo excitation sequence consists of two parts: the excitation subsequence E1, which generates transverse magnetization, and a refocusing subsequence E2. E1 and/or E2 may be designed to provide SSS.

If E1 is a selective binomial, it produces a linear phase shift whose sign corresponds to a *delay*. This is changed to one corresponding to an *advance* in the echo. The latter is easily compensated by waiting for an equal time following the echo center time, prior to the acquisition. Another procedure is to acquire the complete echo. Its Fourier transform has an imaginary component, but nevertheless it has no contribution from the dispersion of any spin signal, so that a frequency-dependent phase correction will not produce much baseline roll. The properties of complete echoes have been applied to magnetic resonance imaging and 2D NMR, and recently to in vivo proton spectroscopy.[30] They can also be used for phase correction of the free precession itself, as in the 'pseudoecho' method.[28,31] The order of the zero in the spectral response to an echo sequence may be uncomfortably high. For instance if E1 and E2 have a first-order zero at the solvent frequency, the zero of the echo response is third order.

An echo may be generated by a 90° pulse for E1 and a 180° pulse for E2. With other subsequences, spurious free precession contributions combine with the echo. Such deviations occur when the rf field is inhomogeneous, for instance in in vivo spectroscopy with surface coils, and with frequency-selective excitation for SSS. To obtain a pure echo under such conditions, one may use sequences in which the transverse magnetization is suppressed at appropriate times with a pulsed gradient. Or one may add successive acquisitions in which the excitation phases are shifted according to the EXORCYCLE scheme.[7,32]

3.2.5 Jump and Return (JR)

The 'jump and return' (JR) excitation sequence includes two $\frac{1}{2}\pi$ pulses of opposite phases, at the solvent frequency.[33] The first pulse brings all spins along *Oy* in the rotating frame

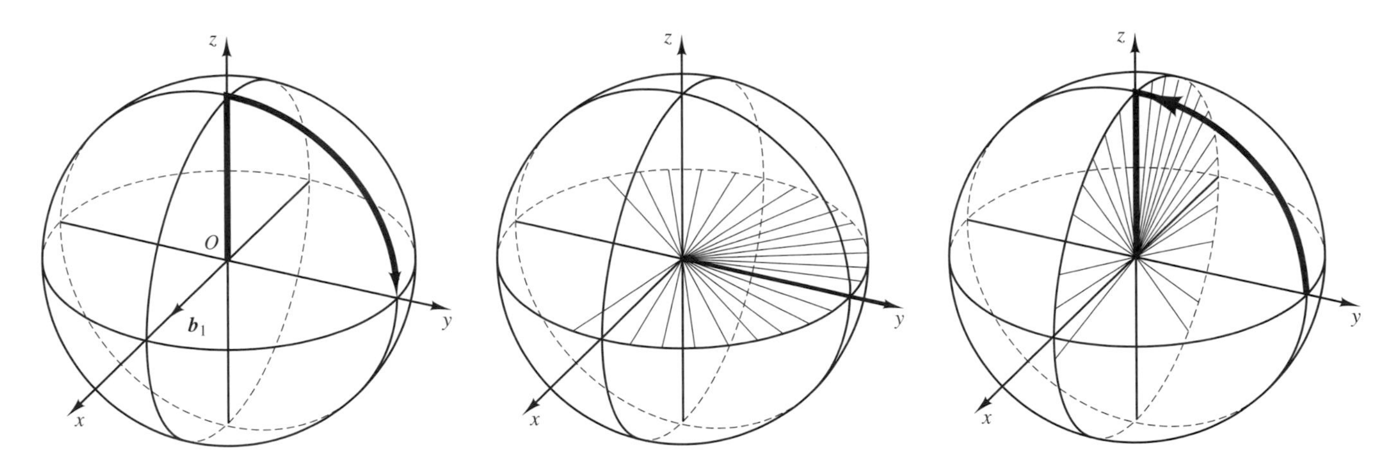

Figure 5 The 'jump and return' sequence. With the first $\frac{1}{2}\pi$ pulse, all spins jump to the Oy direction. During the waiting time τ, they rotate in the (x,y) plane by $\omega\tau$, ω being their frequency offset from the pulse frequency, which is the resonance frequency of the solvent. The second pulse, $-\frac{1}{2}\pi$, brings all spins in the (z,x) plane; hence, their transverse magnetizations are all along $\pm Ox$, so that there is no linear phase shift. The solvent spins return to the starting position along Oz. The effect of JR on each spin is a rotation by $\omega\tau$ around Oy. (Reproduced from M. Guéron, P. Plateau, and M. Decorps, in 'Progress in Nuclear Magnetic Resonance Spectroscopy', eds. J. W. Emsley, J. Feeney, and L. H. Sutcliffe, Pergamon Press, Oxford, 1991, Vol. 23, p. 160, Copyright 1991, with permission from Pergamon Press Ltd)

(Figure 5). During the interval between the pulses, the water protons remain along Oy and the spins of interest precess by $\frac{1}{2}\pi$ in the equatorial xOy plane. The second pulse rotates the xOy plane into the vertical xOz plane, so that all spin species end up with their transverse component along $\pm Ox$. Hence, the phase is constant across the spectrum, except for a step of π at the solvent frequency (Figure 6a). The solvent magnetization ends up along Oz, providing first-order SSS. Spins for which $\omega\tau = \frac{1}{2}\pi$ end up along Ox, corresponding to maximum signal.

The three successive rotations of the magnetization that are produced by JR have the same effect as a single rotation of angle $\omega\tau$ around Oy. This result is valid whatever the original orientation of the magnetization. For instance, two successive JR excitations rotate the ($\omega\tau = \frac{1}{2}\pi$) spins by π while preserving SSS.

The most interesting property of JR is the *lack of linear phase shift*, a property that obtains only for a specific pulse angle, $\frac{1}{2}\pi$ (or multiples), in the case of a very large b_1 field. If the angle differs slightly from this value, there will be a slight linear phase shift. But the SSS feature will not be affected. This makes the JR sequence robust with respect to spectrometer instability.

JR is the shortest known SSS sequence, being twice as short as the shortest binomial sequence, 1, -1: with JR, the spins of interest precess by $\frac{1}{2}\pi$ during the waiting period, whereas with 1, -1, the precession angle is π (Figure 4).

For spins whose frequency is close to that of maximum sensitivity, the rotation angle is close to $\frac{1}{2}\pi$. Thus, one can create an SSS version of any multipulse sequence of $\frac{1}{2}\pi$ pulses (or multiples) by simply replacing each pulse by a JR sequence. Application to 2D NMR is discussed in Section 6.

If the pulse angle is larger than $\frac{1}{2}\pi$, the JR sequence produces a phase shift corresponding to a negative delay, leading to a *spin echo*.[27]

The JR sequence is simple to adjust. The residual water signal should be observed on the free precession, whose phase and quadrature components should be available for display. The adjustment process is preferably carried out by analog controls in real time, rather than by computer-controlled procedures, whose response may be inconveniently slow. The Ox component is canceled by adjusting either the excitation frequency or the phase of one of the pulses. The Oy component is canceled by adjusting the length of one pulse. In this way, one corrects for hardware errors, as well as for movements of the water proton magnetization in the rotating frame due to frequency maladjustment, longitudinal relaxation, and radiation damping. JR has intrinsic compensation for transverse relaxation.

JR can be adapted to excitation frequencies distinct from the solvent frequency.

When the b_1 field is not very large, the JR response presents a quasilinear phase shift, and the frequency of maximum response is modified. These effects are avoided with the *power-adapted JR* (PJR) procedure,[27] whose frequency response is very close in phase and amplitude to that of JR with infinitely strong pulses. Let t_p be the length of a $\frac{1}{2}\pi$ pulse and let τ, as before, be equal to $\pi/2\omega_{max}$. Define a power parameter p by

$$p = t_p/\tau \tag{6}$$

In power-adapted JR, the pulse length is

$$t'_p = \frac{t_p}{1-(2/\pi)p} \tag{7}$$

and the interval τ' between the beginnings of the pulses is

$$\tau' = \tau(1 - \tfrac{1}{3}p) \tag{8}$$

The JR sequence shares with the 1, -1 sequence a linear amplitude response for Larmor frequencies close to that of the solvent. This is a blessing, since it enables one to observe signals close to the solvent. Although solvent suppression is not as good as with higher-order sequences, this may be corrected after the acquisition. Alternatively, one may use SJR, a second-order modification of JR.[33]

For References see p. 4942

4 EXPERIMENTAL ASPECTS

SSS is affected by instrumental imperfections of the excitation sequence, by relaxation and radiation damping during excitation, and by inhomogeneities in the static and rf fields. The practical analysis and adjustment of an SSS sequence require observation of the free induction decay in real time on an oscilloscope. A discussion of these matters may be found in Guéron et al.[7] Only a brief survey is given here.

Relaxation and radiation damping affect SSS by reorienting the magnetic moment of solvent spins during the excitation period. Radiation damping is the most significant, because its rate is usually the largest and because it may be inhomogeneous in the sample. Some SSS sequences (e.g., the binomial 1, −2, 1) automatically provide first-order compensation of radiation damping.

Static field inhomogeneities also affects SSS, particularly those such as Z^2 that give asymmetric lineshapes. It may also happen that small fractions of the sample suffer from large field offsets—either because they are rather far from the center of the sample or because of diamagnetic susceptibility anomalies. Examples are isolated drops on the sample tube wall, and the sample region located at the bottom of the sample tube. If they amount to 1% of the sample volume, they may limit the SSS factor to a few hundred. Even worse, they may generate artifactual lines in the spectrum. Appropriate probe designs help to reduce these problems. Sample tubes with a flat bottom and a long tail are recommended.[7]

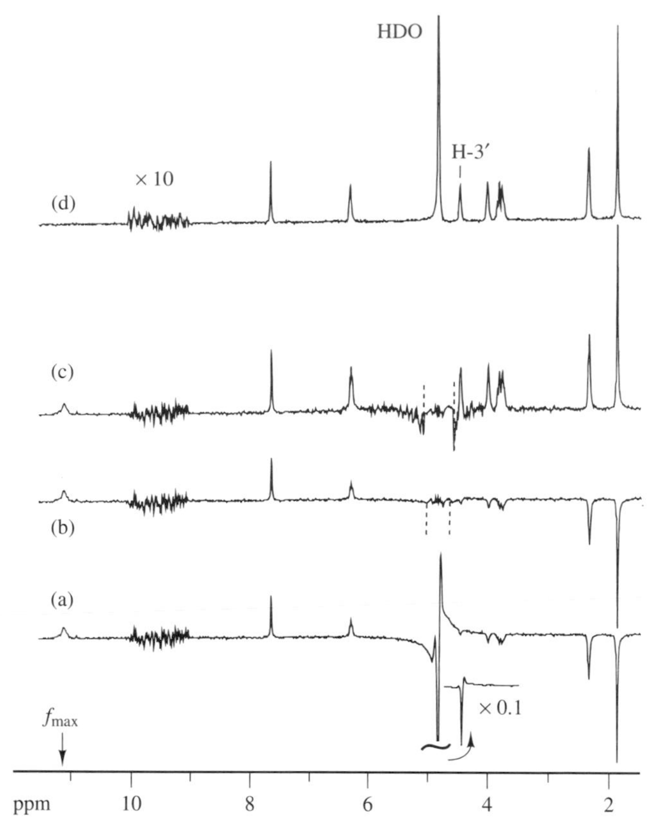

Figure 6 Proposed standard test for SSS. (a) JR spectrum of thymidine, 10 mM, in 90% H_2O, 10% D_2O. A single FID was collected. The length of a $\frac{1}{2}\pi$ pulse was 10 μs. The frequency of maximum sensitivity is at 11.1 ppm. No phase correction was applied, except for that used in the case of single pulse excitation. The imino proton signal (11.1 ppm) is broadened by exchange. Exponential multiplication provides 3 Hz broadening and insures that the peak heights are not sensitive to field inhomogeneity. By comparison with the H-5 signal at 7.7 ppm [spectra (a)–(d)], the suppression factor is about 1000. Proton frequency, 360 MHz; 5 mm outer-diameter sample tube; $T = 20\,°C$, pH 4. Acquisition and data processing were performed with networked 80386 PC-type computers, running MS-DOS, with 20 MHz clock. (b) Spectrum derived from (a) by repeated subtraction (10 times) of a Lorentzian simulating the tallest peak within ±75 Hz of water (dotted lines). Total processing time 4 s. (c) Spectrum derived from (b) by correction of the JR amplitude response. The correction, which is not applied to the range within ±95 Hz of water (dotted lines), does not change the signal-to-noise ratio. (d) Spectrum of thymidine, 10 mM, in D_2O, following a single $\frac{1}{2}\pi$ pulse. Other conditions are as in (a). This spectrum is provided as a reference for evaluation of spectrum (c), which is the result of JR excitation and processing. In spectrum (c), (i) the maximum sensitivity is the same as in (d), (ii) relative signal amplitudes are respected, and (iii) the H-3′ signal at 4.45 ppm demonstates good observability and lack of artifacts near the water position (4.82 ppm). Spectrum (c) was obtained from the FID in about 6 s processing time. (Reproduced from M. Guéron, P. Plateau, and M. Decorps, in 'Progress in Nuclear Magnetic Resonance Spectroscopy', eds. J. W. Emsley, J. Feeney, and L. H. Sutcliffe, Pergamon Press, Oxford, 1991, Vol. 23, p. 135, Copyright 1991, with permission from Pergamon Press Ltd)

4.1 Spectrometer Imperfections

Most SSS methods operate by compensation. For instance, one part of the excitation sequence moves the solvent magnetization away from Oz, another brings it back. Errors in *pulse amplitude*, *phase*, *timing* or *frequency* may ruin the compensation. Such errors may be static, or they may vary slowly (drift) or rapidly (fluctuations).

Static errors (e.g., of the phase) may be compensated for, if the spectrometer has provisions for appropriate adjustment. *Drift* of frequency, timing, or phase of the rf pulse generator should not be a problem in a spectrometer that is reasonably built and operated. The main sources of drift are the power amplifier, the temperature controller and the field/frequency lock. If the deuterated compound used for locking is not D_2O, the relative chemical shift of water will be changed by temperature drift, affecting SSS. Instrumental *fluctuations*, i.e., variations on the timescale of the SSS sequence, directly affect the compensation between pulses. They are easily observed as fluctuating differences between successive residual solvent signals following SSS excitation. For instance, we measured phase fluctuations of 1/1600 rad in the time between the two JR pulses. For a preliminary and easy measurement of the spectrometer fluctuations, one may look for fluctuations of the nominally null dispersion component of the free precession of a strong signal at the excitation frequency.[7]

4.2 Modern Spectrometers

Modern spectrometers have improved agility—meaning the capability of shaping rf excitation to high resolution in amplitude, phase, and frequency as a function of time. With the help of optimization techniques and/or on the basis of the Bloch equations, one may then impose desirable conditions on the excitation sequence. A recent example is a flexible procedure for the design of an excitation spectrum identical to any available

NMR spectrum.[34] New SSS sequences may now be endowed with many more parameters, and they may be designed, for instance, with the aim of obtaining constant phase and amplitude accross the spectrum, except for a 'notch' at the solvent frequency.[35] Earlier SSS sequences can be revised by changing the pulse shape (from square to Gaussian, half-Gaussian, etc.) and phase. SSS will certainly benefit from the enhanced agility. However, the benefits should be limited, because pulse sequences with improved selectivity properties tend to be longer.

Probably more important for SSS is the improvement in stability and quality of the rf pulse generators, which should result in better results with a given SSS sequence.

Another important feature of modern spectrometers is the generation of a pulsed gradient of the main field, resulting in the destruction, and possibly the later recall, of any transverse magnetization existing at that time. SSS sequences using a pulsed field gradient have been mentioned above, for example, the WATERGATE sequence (Section 3.1). Potential problems are the distortion of the pulsed field gradient by eddy currents and the perturbation of the field lock system during the pulse. An rf field gradient, which may provide the same benefits as a gradient of the main field, does not have these problems,[36] and could come into current use.

4.3 Evaluation Procedures and Test Samples

Detailed study of the performance of a method and of a spectrometer is usefully complemented by a global test, namely, the spectrum of a standard sample, designed to provide clear indications of the suppression factor, the sensitivity, and the type and amount of data processing required. We suggest a standard sample containing thymidine in H_2O, at a concentration of 10 mM. The excitation sequence should maximize the imino proton line, at -11.1 ppm. The spectrum should extend up to 1 ppm, exhibiting the aromatic, ribose, water, and methyl lines. A similar sample, but in D_2O, should be used for comparison. The spectrum should be examined, both before and after data processing, for reduction in sensitivity, artifacts, distortions of peaks and baseline, and reduction in visibility of lines close to the solvent line. Such defects are more important than the residual solvent signal per se. Figure 6 shows an application to the JR sequence.

5 DATA PROCESSING FOR SSS

The free precession resulting from SSS excitation, although it is within the dynamic range of the NMR receiver, may still contain a large solvent signal, which must be further reduced by postacquisition data processing. Data processing is invariably based on prior knowledge of the signal to be reduced (e.g., its frequency). It is carried out either in the time or frequency domain.

The elimination of a (solvent) signal on the basis of its frequency may be obtained with a frequency filter, which will strongly attenuate all of a Lorentzian centered at the filter frequency. For instance, subtracting from a FID its time-shifted copy filters out a signal at zero frequency.[37] Unfortunately, this convenient method introduces a linear phase shift.

A frequent defect of SSS spectra is the 'rolling baseline' resulting typically from linear phase correction acting on the solvent line. More or less empirical methods for baseline correction are usually provided on commercial equipment. The results are far from perfect. Worse, they may lead to systematic errors, for instance, in integration. To avoid this, the solvent signal should be eliminated before attempting linear phase correction. In our opinion, the best method is to treat the residual solvent signal as a collection of Lorentzians, which are simulated and subtracted recurrently, one at a time.[7] The procedure is fast and distortionless. It may be made entirely automatic (Figure 6). The general methods of maximum entropy[38] or linear prediction[39] may also be used.

In most SSS methods, the gain is a function of frequency, or, more precisely, the amplitude of a Lorentzian is a function of its center frequency. A frequency-dependent amplitude correction restores each Lorentzian peak to its correct height. This does produce distortions, but they are much less severe than those caused by phase correction.

Further spectral distortions in phase and amplitude arise owing to the noninfinite strength of the b_1 field (such effects motivate the use of PJR instead of JR). As a first step towards correcting these defects, it would be useful and possible to provide NMR spectrometers with software that would calculate amplitude and phase shift for an arbitrary excitation sequence.

6 SOLVENT SIGNAL SUPPRESSION IN 2D AND 3D NMR

This section, mainly contributed by A. G. Redfield of Brandeis University, is to a large extent based on limited personal experience. Therefore, references to recent excellent work may in some cases be inadvertently omitted.

6.1 General

We shall assume general familiarity with established methods and notations of 2D NMR.[40,41]

The solvent signal is not nearly so large a problem in 2D and 3D NMR as it is in 1D NMR, because, generally, one is interested only in cross peaks between two lines in the spectrum, and most often one of these is located far enough from the water peak to be observable with relatively simple means such as water presaturation. Cross peaks are commonly observable to protons of the amide and β-carbons, from protons of α-carbons that are within a few tens of hertz from the water resonance, and in some cases NOEs from peaks directly underneath water are observable. Furthermore, most 2D studies have been performed on extremely small biopolymers, i.e., proteins smaller than 15 kDa, or nucleic acid duplexes smaller than 15 base pairs. For these small molecules, the loss of signal by cross relaxation to saturated water[42] is relatively small. And, since the object of 2D NMR has often been primarily to determine structure, rather than study kinetics, the pH can often be chosen to minimize the other process of signal loss when water is saturated, namely, chemical exchange with water.

Furthermore, the nature of many 2D experiments is such as to cancel and filter out the direct effect of water signals and

For References see p. 4942

other large signals. Most heteronuclear 2D methods, and certain homonuclear methods such as double quantum filtered COSY,[41] are in this class. Especially for these experiments, the baseline roll far from water is not a problem. It is also often not a problem for many experiments, such as NOESY, where a large residual water self-peak does occur, because usually one of the two protons giving rise to interesting cross peaks is far from water, as mentioned above.

For these reasons, SSS methods have not been developed as intensively for 2D NMR as they have been, for example, by practitioners of in vivo NMR. Nevertheless, there is a need for continued development. Partly because each FID in a 2D data set results from the accumulation of only a few (typically 4–32) individual signals, averaging of random water signals due to random phase errors in selective sequences, or in water saturation, is not as effective as it is in some 1D experiments. These random signals result in random baseline variation in each 1D spectrum obtained from the 2D data set by Fourier transformation (FT) with respect to the real-time dimension t_2, before FT in the second, evolution or t_1, dimension. As a result, there is an ugly stripe in the final 2D spectrum along the line corresponding to the solvent frequency in the f_1 dimension, exhibiting streaks extending roughly 1 ppm in the f_2 direction. Similar noise emanates from residual HDO in D_2O solvent, if HDO is not saturated, and from other strong signals such as buffer signals.

These streaks are known as 'phase noise', and they can often be reduced by computer flattening of the baseline of each spectrum after the FT in the t_2 dimension, or by other operations on the frequency domain spectrum based on the known frequency of water,[43] by time domain removal of the low-frequency water signal,[44] or by simple convolution difference weighting. The latter weighting produces undesirable troughs near the diagonal ridge. We often work with two 2D FTs of the same spectrum, one with heavy convolution difference weighting for finding cross peaks close to the water phase noise line, and another with little or no such weighting, to find peaks close to the diagonal. It is desirable to have software conveniently set up for such treatment. If convolution difference weighting is used, it is important to have a versatile weighting function for which the convolution difference weighting, at the beginning of the FID, can be varied independently from the line broadening or enhancement functions (affecting the middle of the FID) and from apodization (at the end of the FID). The commonly used shifted sine-bell function does not have this quality.

Water presaturation in 2D NMR is often performed in much the same way as in 1D NMR. It is often desirable to saturate water with a sequence (such as a spin lock pulse) that acts for a short time just before the start of the sequence, rather than a saturation pulse during the entire recovery time. Doing so decreases the number of protons that will be cross saturated by water, at the cost of a wider spectral region of saturation. After selective presaturation of water and, unavoidably, of α-protons whose frequency is close to, or even at, the water frequency, one can take advantage of the short cross-relaxation time within the protein compared with the T_1 of water to restore quickly the polarization of the α-protons. One can thus observe NOESY cross peaks to these protons.[45] Water saturation is also used in the middle of 2D or 3D sequences.[46] In heteronuclear sequences, one can saturate water once heteronuclear correlation has been established,[47] because water is not involved in the correlation. A simple modification of the excitation sequence for generating heteronuclear magnetization transfer provides a time window when the heteronuclear correlation is along Oz. At that time, a gradient pulse[10] can be used to destroy selectively transverse water proton magnetization. In such a scheme, SSS is based on the lack of J coupling of the water protons to the heteronucleus. There is no need for any frequency-selective sequence.[48,49]

Three-dimensional NMR offers no problems beyond 2D NMR, as far as the solvent signal is concerned.

6.2 Selective Excitation in 2D NMR

Water saturation is less desirable for studies of larger proteins,[42] studies at a pH far from the minimum (pH = 4) for the rate of proton exchange with solvent, and the study of nucleic acids, whose water accessibility and smaller diameter promote cross relaxation to water. In these cases, selective sequences are useful. We discuss the approach in some detail, because we think it may be useful for 3D NMR in larger proteins, where good signal-to-noise ratio will become a serious consideration.

6.2.1 *Fully Selective Excitation*

Fully selective 2D heteronuclear multiple quantum coherence NMR, or HMQC, may be implemented using a JR-type selective pulse sequence in place of every $\frac{1}{2}\pi$ or π proton pulse in the sequence.[50,51] This has the advantage that water, being unperturbed, can act as a magnetization reservoir for the macromolecular spin system. Perturbation of the exchangeable protons is also avoided—a matter of importance for NMR of nucleic acids, and sometimes of proteins. The JR sequence has the advantage of being short and of not generating phase dispersion. A selective pulse sequence with a broader zero, such as 214 (Section 3.2.1), may be even better, because it can be set to excite predominantly the high-frequency amide region and not the region of the aliphatic protons, to low frequency of water. This is useful, because, to some extent, these protons can act as a magnetization reservoir, like those of water. We have obtained HMQC spectra using 214 in place of each pulse, but have not yet established whether or not 214 is more useful than JR in practice.

In HMQC, as in many other procedures, a proton π sequence is required in order to refocus unwanted proton evolution. Generally, for this purpose, we have used either back-to-back JR pulse sequences, or—what is nearly the same—a JR sequence with roughly twice the spacing of the 'traditional' JR sequence. An alternative, which requires a good magnet and rf field, is to use carefully adjusted π pulses. If these pulses occurred in pairs in the sequence (accomplished, if necessary, by adding an initial π pulse), the water resonance would remain relatively unsaturated. We have not yet attempted to use this approach.

There were two unexpected aspects of our use of JR for proton $\frac{1}{2}\pi$ and π sequences in HMQC. First, we eventually discovered that the primary amide ($^{15}NH_2$) region of the protein spectrum has complicated side-peak artifacts when selective JR sequences are used, compared with the expected two peaks at the same ^{15}N shift that are actually observed when HMQC is performed with water presaturation and hard $\frac{1}{2}\pi$ and π pulses. We do not have a quantitative explanation for this difference. But, once recognized, it is useful for distinguishing the primary amide peaks of glutamine and asparagine side chains from peptide NH peaks.

A second unexpected result of the use of a selective π sequence in HMQC is that, because the sequence is also relatively ineffective for perturbing the α-protons (whose chemical shift is close to that of water), it suppresses a splitting in the ^{15}N dimension due to the coupling of the α-proton to nitrogen, which would otherwise degrade resolution.[52] It is generally desirable to avoid this splitting.

6.2.2 *Mixed Excitation; Gradient Pulses*

In some cases, fully selective excitation is undesirable because cross peaks to water or to α-protons will be suppressed. In these cases, it is often—perhaps always—possible to use nonselective pulses as necessary within the first part of the sequence, and to use an SSS sequence in place of the last, read-out pulse. Before read-out, the large transverse water signal, which generally results from the first, 'hard' part of the sequence, must be destroyed—usually by application of gradient pulses placed so that they do not affect the evolution of the spins to be observed.

An example of this approach is the 'hard–soft NOESY' sequence,[53] in which the two encoding pulses are hard, the water signal is removed by relaxation or a gradient pulse during the mixing time, and SSS read-out follows. However, in high-field spectrometers equipped with efficient probes, there is a problem with this sequence when the first two pulses are phased so that they make the water magnetization negative: during acquisition, a residual water signal will then be amplified maser-like, by the phenomenon of radiation damping. For this phase combination, the SSS read-out subsequence should be replaced by one that, while not generating a transverse water signal, reverses the longitudinal water magnetization.[54]

A very useful example of mixed excitation is the use of a 'Z filter'[55] at the end of TOCSY (also called HOHAHA) or ROESY (also called CAMELSPIN) excitation. These are spin lock sequences, and it is difficult (but probably possible) and not very useful to make them water-selective. However, at the end of such a sequence, the magnetization can be flipped up to the z axis with a hard $\frac{1}{2}\pi$ pulse, and the residual transverse component of magnetization removed with a gradient pulse and/or appropriate phase cycling. Then, after a short homogeneity recovery delay, the final selective sequence is applied.[56] This is the method of choice to obtain amide–α-proton covalent correlations when water saturation is undesirable. The Z filter (as used here) functions exactly as the last two pulses of a Hahn stimulated echo.

The Z filter just described cannot generally be appended to a COSY or related sequence involving two-spin coherence. However, it seems likely that useful COSY and stimulated echo[57] homonuclear sequences could be assembled from combinations of selective pulses and gradient pulses.

Similar approaches have been applied to isotopically enriched proteins, in combination with heteronuclear selection, for the study of cross relaxation between amide and α-protons,[48] and also between amide protons and water.[58] In the latter work, frequency-selective subsequences are used to keep the solvent magnetization along the $+z$ axis except during the mixing period. Gradient pulses are used to destroy residual transverse magnetization, in order to avoid radiation damping artifacts and to enhance SSS. Applied in this way, they do not contribute to saturation of the solvent resonance. In particular, the final 'observe' subsequence is a WATERGATE-type sequence (Section 3.1) modified so as to avoid solvent saturation.

7 CONCLUSIONS

Proton NMR in protonated solvent used to be such a difficult proposition that it was carried out only if absolutely necessary—in particular for the study of exchangeable protons. Alternative approaches, such as NMR of nonexchangeable protons or of other nuclei, were much preferred. The situation has changed: SSS has become so simple that exchangeable protons are now a fruitful and relatively easy subject of investigation. It is now worthwhile to work in protonated water simply to avoid the H_2O-to-D_2O transfer of a precious protein. The mastering of NMR in protonated solvents leads to the highly sensitive reverse INEPT detection of ^{15}N, thus making possible studies of ^{15}N-labeled macromolecules that would have been impossible or very difficult previously. SSS is also applied to proton NMR spectroscopy of organs and organisms—a case where the solvent cannot be changed.

The value of signal solvent suppression stimulates methodological research, which is thriving if one gauges it by the number of publications. SSS methods also evolve as new equipment and processing software becomes available. They will benefit from enhanced stability of the rf phase and of the magnetic field; frequency-agile synthesizers would promote new variants; faster data processing would diminish the requirements on SSS excitation and acquisition.

Novel methods could emerge from two relatively recent developments, both stimulated by the needs of medical imaging and spectroscopy. One is the introduction of precise and flexible phase and amplitude control of the excitation pulses. This facilitates the design of excitation sequences that produce a spin response with complex, predesigned phase and amplitude properties.

The second—and in our opinion more important—development is the introduction of fast switching gradient coils. They are affecting all manners of NMR experiments, including those involving SSS. Switched gradients quickly eliminate transverse magnetization, so that a gradient pulse applied after a selective $\frac{1}{2}\pi$ sequence provides saturation of the solvent magnetic moment in a time short enough to avoid most difficulties due to spin exchange or chemical exchange. When used with echoes, gradients may eliminate the need for EXORCYCLE cycling. With these new possibilities, SSS methods using slow saturation, and those giving rise to a linear phase shift are, or should soon become, obsolete.

8 RELATED ARTICLES

Biological Macromolecules: Structure Determination in Solution; Bioreactors and Perfusion; Cell Suspensions; Complex Radiofrequency Pulses; ; Drug–Nucleic Acid Interactions; Field Gradients & Their Application; Hydrogen Exchange & Macromolecular Dynamics; Magnetization Transfer between Water & Macromolecules in Proton MRI; Nucleic Acids: Spectra, Structures, & Dynamics; Plant Physiology; Protein

For References see p. 4942

Hydration; Protein Structures: Relaxation Matrix Refinement; Shielding Theory: GIAO Method; Water Suppression in Proton MRS of Humans & Animals.

9 REFERENCES

1. A. G. Redfield, in 'NMR Basic Principles and Progress', eds. P. Diehl, E. Fluck, and R. Kosfeld, Springer-Verlag, Berlin, 1976, Vol. 13, p. 137.
2. B. L. Tomlinson and H. D. W. Hill, *J. Chem. Phys.*, 1973, **59**, 1775.
3. A. G. Marshall, T. Marcus, and J. Sallos, *J. Magn. Reson.*, 1979, **35**, 227.
4. R. K. Gupta, J. A. Ferretti, and E. D. Becker, *J. Magn. Reson.*, 1974, **13**, 275.
5. R. R. Ernst, *J. Magn. Reson.*, 1970, **3**, 10.
6. P. J. Hore, *Meth. Enzymol.*, 1989, **176**, 64.
7. M. Guéron, P. Plateau, and M. Decorps, in 'Progress in Nuclear Magnetic Resonance Spectroscopy', eds. J. W. Emsley, J. Feeney, and L. H. Sutcliffe, Pergamon Press, Oxford, 1991, Vol. 23, p. 135.
8. M. A. Delsuc and J. Y. Lallemand, *J. Magn. Reson.*, 1986, **69**, 504.
9. A. Abragam, 'The Principles of Nuclear Magnetism', Clarendon Press, Oxford, 1961, Chap. III.
10. G. A. Morris and R. Freeman, *J. Magn. Reson.*, 1978, **29**, 433.
11. S. L. Patt and B. D. Sykes, *J. Chem. Phys.*, 1972, **56**, 3182.
12. F. W. Benz, J. Feeney, and G. C. K. Roberts, *J. Magn. Reson.*, 1972, **8**, 114.
13. D. L. Rabenstein and A. A. Isab, *J. Magn. Reson.*, 1979, **36**, 281.
14. R. G. Bryant and T. M. Eads, *J. Magn. Reson.*, 1985, **64**, 312.
15. D. L. Rabenstein and S. Fan, *Anal. Chem.*, 1986, **58**, 3178.
16. R. L. Vold, J. S. Waugh, M. P. Klein, and D. E. Phelps, *J. Chem. Phys.*, 1968, **48**, 3831.
17. D. I. Hoult, *J. Magn. Reson.*, 1976, **21**, 337.
18. P. Blondet, M. Decorps, J. P. Albrand, A. L. Benabid, and C. Remy, *J. Magn. Reson.*, 1986, **69**, 403.
19. V. Sklenář and A. Bax, *J. Magn. Reson.*, 1987, **75**, 378.
20. D. Canet, D. Boudot, and J. Brondeau, *J. Magn. Reson.*, 1988, **79**, 377.
21. Z. Starčuk, L. Půček, R. Fiala, and Z. Starčuk, Jr., *J. Magn. Reson.*, 1988, **80**, 344.
22. M. Piotto, V. Saudek, and V. Sklenář, *J. Biomol. NMR*, 1992, **2**, 661.
23. P. C. M. Van Zijl and C. T. W. Moonen, *J. Magn. Reson.*, 1990, **87**, 18.
24. P. Stilbs, in 'Progress in Nuclear Magnetic Resonance Spectroscopy', eds. J. W. Emsley, J. Feeney, and L. H. Sutcliffe, Pergamon Press, Oxford, 1987, Vol. 19, p. 1.
25. P. C. Van Zijl, C. T. W. Moonen, P. Faustino, J. Pekar, O. Kaplan, and J. Cohen, *Proc. Natl. Acad. Sci. USA*, 1991, **88**, 3228.
26. G. J. Galloway, L. J. Haseler, M. F. Marshman, D. H. Williams, and D. M. Doddrell, *J. Magn. Reson.*, 1987, **74**, 184.
27. M. Guéron, P. Plateau, A. Kettani, and M. Decorps, *J. Magn. Reson.*, 1992, **96**, 541.
28. M. H. Levitt, *J. Chem. Phys.*, 1988, **88**, 3481.
29. A. L. Davis and S. Wimperis, *J. Magn. Reson.*, 1989, **84**, 620.
30. A. A. de Graaf, W. M. M. J. Bovée, N. E. P. Deutz, and R. A. F. M. Chamuleau, *Magn. Reson. Imaging*, 1988, **6**, 255.
31. R. R. Ernst, G. Bodenhausen, and A. Wokaun, 'Principles of Nuclear Magnetic Resonance in One and Two Dimensions', Clarendon Press, Oxford, 1987, p. 333.
32. G. Bodenhausen, R. Freeman, and D. L. Turner, *J. Magn. Reson.*, 1977, **27**, 511.
33. P. Plateau and M. Guéron, *J. Am. Chem. Soc.*, 1982, **104**, 7310.
34. Ē. Kupče and R. Freeman, *J. Magn. Reson., Ser. A*, 1994, **106**, 135.
35. H. Liu, K. Weisz, and T. L. James, *J. Magn. Reson., Ser. A*, 1993, **105**, 184.
36. W. E. Maas and D. G. Cory, *J. Magn. Reson., Ser. A*, 1994, **106**, 256.
37. K. Roth, B. J. Kimber, and J. Feeney, *J. Magn. Reson.*, 1980, **41**, 302.
38. D. S. Stephenson, in 'Progress in Nuclear Magnetic Resonance Spectrocopy', eds. J. W. Emsley, J. Feeney, and L. H. Sutcliffe, Pergamon Press, Oxford, 1988, Vol. 20, p. 515.
39. H. Barkhuijsen, R. de Beer, W. M. M. J. Bovée, and D. Van Ormondt, *J. Magn. Reson.*, 1985, **61**, 465.
40. K. Wüthrich, 'NMR of Proteins and Nucleic Acids', Wiley-Interscience, New York, 1986.
41. R. R. Ernst, G. Bodenhausen, and A. Wokaun, 'Principles of Nuclear Magnetic Resonance in One and Two Dimensions', Clarendon Press, Oxford, 1987.
42. J. D. Stoesz, A. G. Redfield, and D. Malinowski, *FEBS Lett.*, 1978, **91**, 320.
43. P. Tsang, P. E. Wright, and M. Rance, *J. Magn. Reson.*, 1990, **88**, 210.
44. D. Marion, M. Ikura, and A. Bax, *J. Magn. Reson.*, 1989, **84**, 425.
45. S. C. Brown, P. L. Weber, and L. Mueller, *J. Magn. Reson.*, 1988, **77**, 166.
46. M. Ikura, L. E. Kaye, and A. Bax, *Biochemistry*, 1990, **29**, 4659.
47. G. Otting and K. Wüthrich, *J. Magn. Reson.*, 1988, **76**, 569.
48. A. Bax and S. S. Pochapsky, *J. Magn. Reson.*, 1992, **99**, 638.
49. G. Wider and K. Wüthrich, *J. Magn. Reson., Ser. B*, 1993, **102**, 239.
50. S. Roy, M. Z. Papastavros, V. Sanchez, and A. G. Redfield, *Biochemistry*, 1984, **23**, 4395.
51. E. P. Nikonowicz and A. Pardi, *J. Mol. Biol.*, 1993, **232**, 1141.
52. A. Bax, M. Ikura, L. E. Kay, D. A. Torchia, and R. Tschudin, *J. Magn. Reson.*, 1990, **86**, 304.
53. J. D. Cutnell, *J. Am. Chem. Soc.*, 1982, **104**, 362.
54. P. R. Blake and M. F. Summers, *J. Magn. Reson.*, 1990, **86**, 622.
55. M. Rance, *J. Magn. Reson.*, 1987, **74**, 557.
56. A. Bax, V. Sklenár, G. M. Clore, and A. M. Gronenborn, *J. Am. Chem. Soc.*, 1987, **109**, 6511.
57. A. G. Redfield, *Chem. Phys. Lett.*, 1983, **96**, 537.
58. S. Grzesiek and A. Bax, *J. Biomol. NMR*, 1993, **3**, 627.

Acknowledgements

This article is partly based on M. Guéron, P. Plateau, and M. Decorps, in 'Progress in Nuclear Magnetic Resonance Spectroscopy', eds. J. W. Emsley, J. Feeney, and L. H. Sutcliffe, Pergamon Press, Oxford, 1991, Vol. 23, p. 135, Copyright 1991, with kind permission from Pergamon Press Ltd. We are grateful to A. G. Redfield for contributing most of Section 6, and for many suggestions. We thank J. L. Leroy for performing the experiments and draftwork for Figure 6.

Biographical Sketches

Pierre Plateau. *b*. 1956. Graduate, Ecole Polytechnique. Ph.D., Metabolism of diadenosine tetraphosphate and related nucleotides. Affiliated to the Centre National de la Recherche Scientifique as Directeur de Recherches. Introduced to NMR by M. Guéron. Recent research interests: structure–function relationships and the genetic expression of aminoacyl-tRNA synthetases.

Maurice Guéron. *b* 1935. Graduate, Ecole Polytechnique. Ph.D., NMR–EPR study of indium antimonide (adviser: I. Solomon). Postdoctoral position at Bell Labs with R. G. Shulman. Presently Director, Groupe de Biophysique de l'Ecole Polytechnique, also affiliated to the Centre Nationale de la Recherche Scientifique as Directeur de Recherches. Recent research interests: the applications of NMR to molecular biophysics, physiology and medicine.

Water Suppression in Proton MRS of Humans & Animals

Chrit T. W. Moonen
National Institutes of Health, Bethesda, MD, USA

&

Peter C. M. van Zijl
Johns Hopkins University Medical School, Baltimore, MD, USA

1 INTRODUCTION

In vivo NMR spectroscopy offers a noninvasive window on metabolism and physiology in healthy and diseased humans and animals. Most initial efforts in NMR spectroscopy of living systems were devoted to the phosphorus and carbon nuclei. Because proton NMR is more sensitive, and allows access to a range of different metabolites, attention to proton spectroscopy has been increasing. However, without specific efforts to suppress water and fat resonances, the latter compounds completely dominate the in vivo proton NMR spectrum. For example, the effective concentration of water hydrogen atoms may be as high as 110 M, whereas we wish to study metabolites in the millimolar range. Pioneering work on in vivo proton spectroscopy was performed at the laboratory of Shulman.[1,2] Recent advances are now permitting in vivo proton spectroscopy and spectroscopic imaging of the human brain in a routine examination. The topic of this review is to outline the possible strategies to suppress the water resonance, and other unwanted resonances such as those of fat (triglycerides).

The suppression of solvent resonances is an area of research that has long been in the domain of high-resolution (HR) NMR. Many techniques were well established before in vivo proton NMR was even contemplated. Recent developments such as shielded field gradient technology, and shaped rf excitation pulses have played an important role in NMR imaging. Now, these advanced tools are among the most powerful assets for water-suppressed proton spectroscopy in vivo, as well as in vitro.

When compared with HR NMR, in vivo spectroscopy has to overcome several typical problems related to $\boldsymbol{B}_0$ and $\boldsymbol{B}_1$ inhomogeneity, and motion, regardless of the nucleus being studied. These problems increase dramatically in proton spectroscopy because of the additional problem of water suppression and the narrow proton chemical shift range. For example, the range of water resonance frequencies over the human head amounts to several ppm (especially outside the brain area), comparable to the chemical shift range of most observable compounds. Therefore, chemical-shift-based water suppression cannot work unless a very high degree of localization is achieved. The use of surface coil receivers may help to avoid water signals from outside areas with a different resonance frequency. However, with the increasing attention being paid to spectroscopic imaging of large brain areas simultaneously, this solution is only of limited use. Motion effects further aggravate these problems.

Owing to limitations of space, in this article we emphasize general principles only, and highlight some important new developments. Because of the technical nature of this review, we do not restrict ourselves to human spectroscopy but also include important results in proton spectroscopy on animals. Detailed general reviews of water suppression in HR NMR and in vivo NMR have been published before.[3–6]

2 OVERVIEW OF SUPPRESSION METHODS BASED ON NMR AND OTHER BIOPHYSICAL PROPERTIES

2.1 Chemical Shift

Most suppression techniques are based on the frequency of the solvent resonance. Within this class of suppression strategies, we discriminate between methods that

1. accomplish a selective saturation of a frequency band before excitation of the spins of interest is started;[7–19] or
2. avoid altogether the excitation of a frequency band;[3,5,20,21] or
3. filter out a frequency band after signal reception.

In general, these methods have the disadvantage that all resonances of interest within the suppressed frequency band are also eliminated.

2.1.1 *Selective Saturation*

The general pulse sequence is

$$P_1 – d_1(G_1) – P_2 \quad \text{Acq} \qquad (1)$$

where P_1 is the soft rf pulse, d_1 is a short delay, preferably containing a gradient spoiling pulse G_1, and P_2 is the excitation pulse followed by acquisition time or the rest of a more complicated pulse sequence. This sequence, with a constant amplitude P_1 and a duration of the order of the solvent T_1 (the 'presaturation' method[7]) is one of the most common water suppression methods in HR NMR. It is now well established that it is more effective to employ a selective (shaped) P_1 pulse using a flip angle slightly larger than 90° to take T_1 relaxation during d_1 into account, thus nulling solvent magnetization at the time of rf pulse P_2.[8–14] The latter technique, including a spoiling gradient,[8,9] is often referred to as a CHESS sequence (chemical shift suppression). The method does not truly reflect

For References see p. 4952

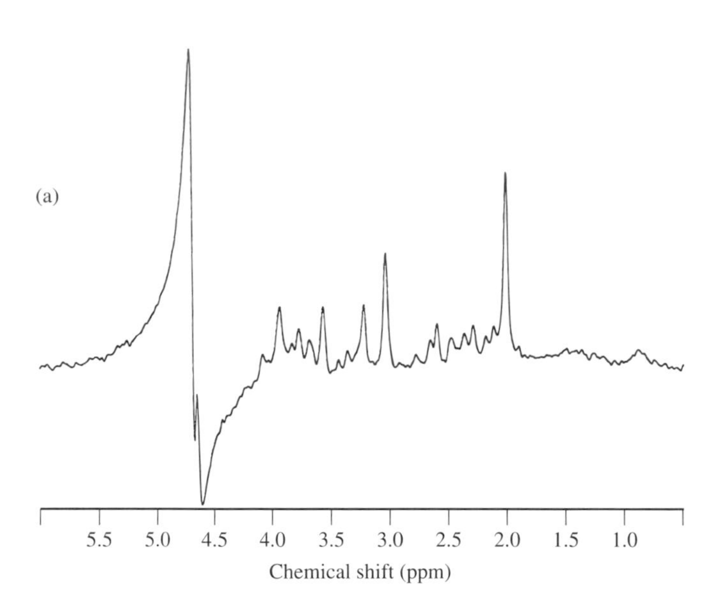

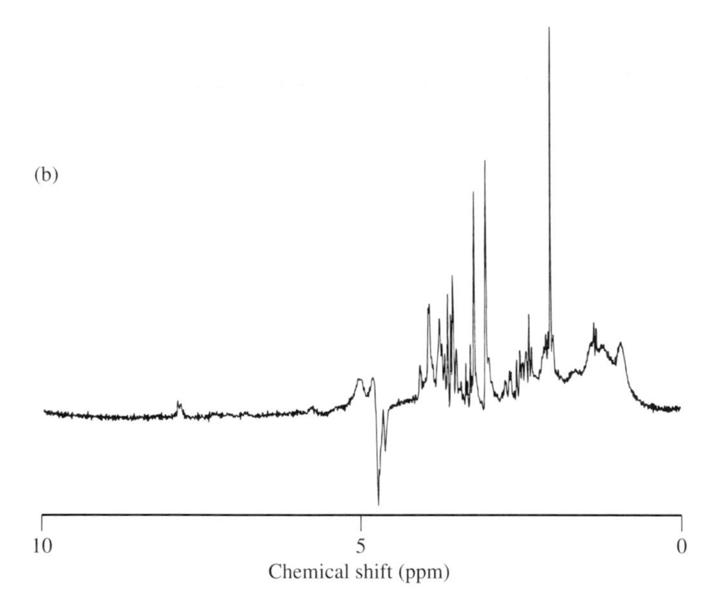

Figure 1 Examples of high quality water suppression in in vivo proton spectroscopy of (a) human and (b) animal brain. (a) In vivo proton NMR spectrum of an 18 ml volume obtained from normal gray matter of a healthy volunteer. The spectrum is obtained with STEAM localization (*TR*/*TE*/*TM* = 6000/20/30 ms), and three preceding CHESS water suppression procedures. The spectrum was obtained in approx. 6 min. A 2.0 T instrument was used, with a conventional quadrature head coil. Only exponential multiplication was used prior to FT, and no further data manipulation was applied. Courtesy of Dr. Jens Frahm, Biomedizinische NMR Forschungs GmbH, Max Planck Institut für Biophysikalische Chemie, Göttingen, Germany. (b) In vivo proton NMR spectrum from a 0.7 mL volume located in the superior part of the cat brain using STEAM localization (*TR*/*TE*/*TM* = 1000/18/80 ms). A 4.7 T instrument was used, equipped with 20 mT m^{-1} shielded gradients. A surface coil (outer diameter 48 mm) was used for transmission and reception. Three preceding CHESS water suppression sequences, and three CHESS water suppression sequences in the *TM* period were used. Lorentzian-to-Gaussian transformation was used prior to FT. For more details, see Moonen and van Zijl.[12]

a classical 'saturation' experiment. However, it leads to the condition of zero M_z magnetization and dephased transverse magnetization. In addition, selective magnetization inversion can be used followed by a waiting period to allow for zero M_z magnetization.[15]

The saturation procedure does not have to take place prior to the rest of the sequence. For example, a CHESS sequence can also be employed in the middle (*TM*) period of the stimulated echo sequence in in vivo NMR, when the magnetization due the resonances of interest is longitudinal. The basic sequence can be repeated for increased efficiency. However, care should be taken that different gradient directions and powers are used to avoid unwanted echoes of the solvent resonance. A detailed analysis has been presented previously.[12] Examples are given in Figure 1.

A $\boldsymbol{B}_1$ gradient may also be helpful in suppressing a frequency band.[16–19] A combination of hard pulses followed by spin locking procedures has been proposed for in vivo NMR[18,19] The basic idea is that all off-resonance magnetization will dephase in the inhomogeneous spin locking field. In addition, a fast pulse train of hard pulses (DANTE[20]) given with an inhomogeneous $\boldsymbol{B}_1$ field has also been shown to lead to rapid effective nulling of longitudinal magnetization of solvent.[16,17] These techniques are particularly useful for surface coils.

2.1.2 *Excitation Profiles with a Gap at the Water Frequency*

The difference in evolution frequency of transverse magnetization between solvent and solute can be employed to create a gap in the effective excitation profile at the solvent frequency.[3,5,21] The best-known and widely used is the so-called jump–return sequence of Plateau and Guéron.[21] The sequence starts with a $\frac{1}{2}\pi$ rf pulse with the transmitter on the solvent resonance. Following an evolution period t, a second $\frac{1}{2}\pi$ rf pulse is given with opposite phase. The magnetization of the solvent resonance is thus put back along the z axis. However, for a resonance with a relative frequency of $(4t)^{-1}$ with respect to the offset, the transverse magnetization will have rotated during t exactly 90°, and thus will remain unchanged by the second rf pulse. The frequency profile of such a sequence is approximately a sine function. Instead of two rf pulses, an increasing number of rf pulses can be used to improve the frequency profile. A disadvantage is the sometimes complicated phase roll over the spectrum. Shaped, self-refocusing rf pulses can improve the phase response dramatically.[22] Guéron et al.[5] have given a detailed review of this class of suppression techniques, along with further improvements.

Continuous wave (CW) and rapid scan techniques are based on a rapid sweep of the frequency band of interest and avoidance of the solvent resonances. However, these methods are now rarely used because of their low S/N per unit time.

2.1.3 *Suppression of Water Resonance Following the Complete Pulse Sequence*

When radiation damping is not a problem, the solvent resonance can be suppressed following the complete pulse sequence. The disadvantage is that even a very large dynamic range analog-to-digital converter ADC can often not handle the large voltage differences due to resonances of interest and solvent[3,6] (Section 3.3).

2.2 Relaxation

Differences in relaxation times T_1 and T_2 between solutes and solvent can be exploited to achieve a relative suppression of the water resonance. T_1 and T_2 values of some brain compounds have been given by Frahm et al.[23] If the solvent suppression mechanism is entirely based on relaxation differences, signal loss for the resonances of interest can be significant, because differences in relaxation times are not dramatic. The advantage of relaxation-based water suppression is that its mechanism is not frequency-selective, and therefore the method works regardless of the actual resonance frequency of water.

2.2.1 *T_1-Based Methods*

The general pulse sequence is as above. When P_1 in sequence 1 is a π pulse, and d_1 is a delay adjusted for zero longitudinal magnetization of water,[15,24,25] the sequence is often called a WEFT (water-eliminated Fourier transform). To avoid T_1 weighting of the resonances of interest, a selective π pulse can be used (see also Section 2.1.2). T_1-based methods require more time than CHESS-type methods to null M_z solvent magnetization (because of T_1), and are optimal only for a single T_1. In the case of more than two solvent populations with different T_1s, more inversion pulses can be used, with adjustable delays between π pulses.[25]

2.2.2 *T_2-Based Methods*

A general spin echo (or stimulated echo) sequence can be used, such as

$$P_1\text{–}\tau\text{–}P_2\text{–}\tau \quad \text{Acq} \tag{2}$$

where P_1 is the $\frac{1}{2}\pi$ excitation pulse, 2τ is the echo time, and P_2 is the refocusing (π) pulse. Acquisition is started at the top of the echo, or immediately following the P_2 pulse. Apart from water suppression, the spin echo sequence has the added advantage of the removal of broad resonances (owing to short T_2). The reduced overlap is often helpful for quantifying resonance intensities, for example in proton spectroscopic imaging. A disadvantage is the phase modulation due to J coupling. However, when several refocusing pulses are used, specific J modulation patterns of molecules of interest can be turned into an advantage, for example for selecting lactate and avoiding uncoupled resonances.[26]

2.3 Scalar Coupling

Many modern methods use nuclear coupling characteristics to select only coupled resonances and thus suppress the (singlet) solvent resonance automatically.[27–49] These methods are identical to many of the so-called 'editing' techniques. Some techniques can be understood in sufficient detail using a classical vector presentation, whereas some newer methods can best be treated with the modern coherence pathway formalism. We shall first explain some basic methods for heteronuclear applications (e.g. proton detection of ^{13}C-labeled compounds). For the sake of simplicity, we shall limit the discussion to weakly coupled AX systems.

2.3.1 *Indirect Detection by Spin Echo Methods*

The heteronuclear spin echo difference method (POCE, proton-detected carbon editing) is a two-scan experiment in which the heteronuclear π pulse is turned on/off in alternating experiments:[43,44,46,47]

$$\begin{array}{ll} ^1\mathrm{H}: \tfrac{1}{2}\pi\text{–}\tau\text{–}\pi\text{–}\tau & \mathrm{Acq} \\ ^{13}\mathrm{C}: \qquad\qquad \pi & (\mathrm{on/off}) \end{array} \tag{3}$$

The optimum delay τ is $(2J)^{-1}$. The evolution of the multiplet can be easily followed in the rotating frame. If the heteronuclear π pulse is off, the coupled spins change the direction of their evolution at the time of the π pulse. As a result, the components of the multiplet rephase at the top of the echo. If both π pulses are on, the chemical shift is rephased at the top of the echo, but evolution due to the coupling has continued for a period J^{-1}. As a consequence, the multiplet has opposite phase. Subtraction thus gives resonances only for coupled spins and eliminates those for all noncoupled spins.

2.3.2 *Indirect Detection by Polarization Transfer and Multiple Quantum Methods.*

So far, we have discussed spin systems in terms of longitudinal and transverse magnetization. A more general treatment involves the analysis of coherence orders during the pulse sequence, and is particularly useful for the understanding of polarization transfer and multiple quantum pulse sequences. Coherence orders are denoted by p, where $p = 0, \pm 1, \pm 2$ for zero, single, and double quantum order, respectively. For an introduction to the coherence formalism, the reader is referred to the literature.[27–29] We shall review two fundamental heteronuclear (^{1}H–^{13}C) sequences here. Proton spins will be indicated by I, carbon spins by S. The single quantum coherences, $\hat{I}_x$ and $\hat{I}_y$, refer to the classical transverse (observable) proton magnetization, $\hat{S}_x$ and $\hat{S}_y$ to single quantum carbon magnetization. $2\hat{I}_x\hat{S}_y$ consists of double quantum coherence and zero quantum coherence, and is not directly observable. $2\hat{I}_x\hat{S}_z$ and $2\hat{I}_z\hat{S}_x$ are so-called antiphase single quantum coherences. Radiofrequency pulses may generate transitions between coherence orders, but gradient pulses cannot. The complete pulse sequence can be seen as a coherence pathway vector $\boldsymbol{p}$ with as many elements as rf pulses in which the coherence order in period i is indicated with order p_i.

As an example, we shall follow an HMQC (heteronuclear multiple quantum coherence)[38,39,48,49] experiment (for the pulse sequence see Figure 2a). Following excitation of the protons ($\hat{I}_z \rightarrow -\hat{I}_y$) by the $\frac{1}{2}\pi$ rf pulse along the x axis in the rotating frame and neglecting the chemical shift, evolution of $\hat{I}_y$ will then lead to

$$-\hat{I}_y \xrightarrow{\text{evolution}} -\hat{I}_y\cos(\pi J_{IS}t) + 2\hat{I}_x\hat{S}_z\sin(\pi J_{IS}t) \tag{4}$$

The carbon pulse after an evolution period of $(2J)^{-1}$ will transfer the antiphase single quantum coherence into a multiple quantum coherence:

$$2\hat{I}_x\hat{S}_z \xrightarrow{\frac{1}{2}\pi_x^S} -2\hat{I}_x\hat{S}_y \tag{5}$$

The π proton pulse converts the zero-quantum into a double-quantum coherence, and the double-quantum into a zero-

For References see p. 4952

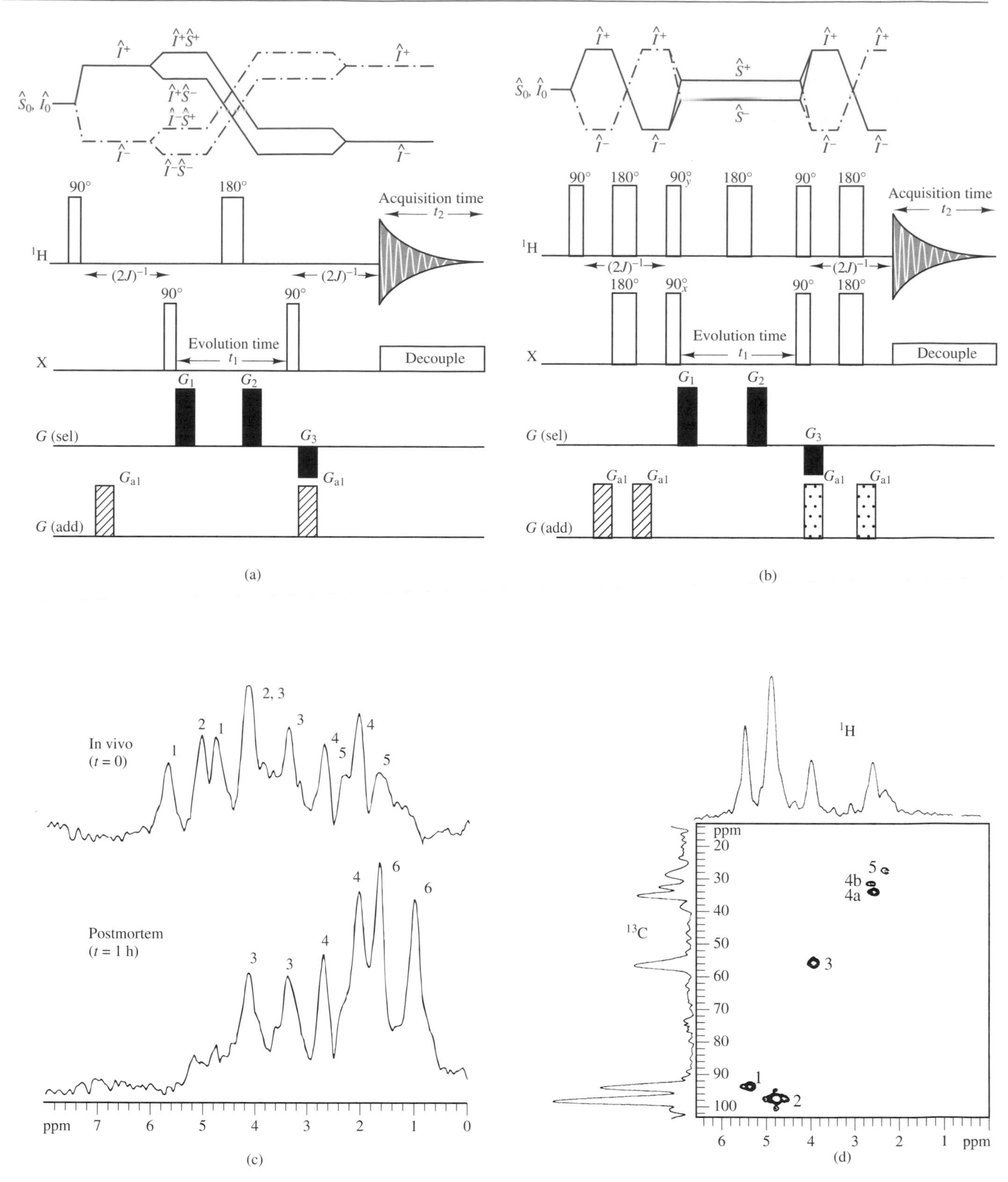

Figure 2 Coherence pathways and pulse sequences for (a) heteronuclear multiple quantum coherence (HMQC) and (b) heteronuclear single quantum coherence (HSQC) experiments. The phase of the rf pulses is irrelevant for the sequence when gradients G_{sel} and G_{add} are used for coherence selection. The phase of the rf pulses, and the multiscan phase cycle, is important when no gradients are used. Reproduced, with permission, from Ruiz-Cabello et al.[48] (c) ^{13}C–^{1}H HMQC spectra obtained from cat brain in vivo (upper spectrum), and 45 min following sacrifice (lower spectrum). Both spectra were obtained in 3 min 19 s from an 8 mm slice through the cat brain following infusion of ^{13}C-labeled glucose (final blood glucose level was 16.6 mM). Spectra are not ^{13}C-decoupled, and show ^{13}C multiplets. Note the complete absence of the water resonance. Assignments are: (1) α-[1-^{13}C]glucose; (2) β-[1-^{13}C]glucose; (3) [2-^{13}C]glutamate and glutamine; (4) [4-^{13}C]glutamate and glutamine; (5) [3-^{13}C]glutamate and glutamine; (6) [3-^{13}C]lactate. (d) Two-dimensional ^{13}C–^{1}H HMQC correlation spectrum obtained in a fully decoupled mode by incrementing the multiple quantum period. Total acquisition time was 30 min. The projections along the ^{1}H and ^{13}C axes are shown as well. Assignments are as in (c). Note the separation in peak 4 corresponding to a separation of the glutamate (peak a) and glutamine (peak b)

quantum coherence. The final $\pi/2$ carbon pulse transfers the $2\hat{I}_x\hat{S}_y$ coherences into observable antiphase proton magnetization ($2\hat{I}_x\hat{S}_z$ coherence), which is first allowed to refocus and then detected.

Figure 2(b) shows an HSQC heteronuclear polarization transfer experiment (heteronuclear single quantum coherence[50]). Following evolution as in the HMQC experiment [equation (4)] and refocusing of the chemical shift, the essential step is the simultaneous proton and carbon $\frac{1}{2}\pi$ pulses, which create antiphase carbon magnetization from antiphase proton magnetization ($2\hat{I}_x\hat{S}_z \rightarrow 2\hat{I}_z\hat{S}_y$). This is an effective polarization transfer from the proton spins to the coupled carbon spins. Now, the carbon nuclei can be detected with an enhanced polarization (the enhancement is γ_H/γ_C). The basic experiment is called the INEPT experiment. For maximum enhancement, a second (inverse) polarization transfer is carried out, and the sensitive hydrogen nucleus can be detected for optimum efficiency. Note the coherence pathway diagram in Figure 2(b).

When selecting a unique coherence pathway of Figure 2, noncoupled resonances are automatically eliminated. Therefore, coupled resonances under the water line can, in principle, be measured. However, note that the coherence pathway selection is generally a multiscan experiment (see Section 3.1 for important exceptions). The reason is that the rf pulses also lead to undesired coherences. This may be caused by the inhomogeneous $\boldsymbol{B}_1$ field, or by unwanted coherences that are generated even when ideal $\frac{1}{2}\pi$ pulses can be used (for example, owing to relaxation effects). In order to avoid the detection of unwanted coherences, the phase of the rf pulses and the receiver are varied using a phase cycling scheme.[27–29] For the purpose of water suppression, it is important to note that in a multiscan experiment, the ADC resolution may still be a limiting factor.

Limitations of space preclude a thorough analysis here of all hetero- and homonuclear editing pulse sequences with automatic water suppression. Generally, similar principles hold for homo- and heteronuclear coherence pathways.[27,28] However, note that single quantum evolution frequencies are similar in homonuclear sequences, whereas they are dependent on the gyromagnetic ratio of nucleus I or S in the heteronuclear case. Thus, a homonuclear editing sequence should involve either at least one rf pulse that avoids water excitation, or a multiple quantum order in part of the sequence in order to inherently eliminate the singlet solvent resonance.

In short, multipulse editing sequences, based on polarization transfer and multiple quantum coherences, can be used for 'automatic' water suppression. Multiscan phase cycling schemes are necessary for pure coherence pathway selection. In Section 3.1.3, an important exception to this rule will be reviewed, namely, the use of field gradients for single scan coherence pathway selection.

2.4 Diffusion

The diffusion constant of water is considerably higher than that of small metabolites and much higher than that of macromolecules. Pulse sequences can be sensitized to diffusion by the use of $\boldsymbol{B}_0$ magnetic field gradients without any effect on stationary molecules.[51–54] For two gradient pulses of strength G and duration δ, and with duration Δ between their starts, the attenuated signal S resulting from free diffusion can be expressed as

$$S/S_0 = \exp[-\gamma^2 G^2 \delta^2 (\Delta - \tfrac{1}{3}\delta) D] \tag{6}$$

where S_0 is the starting signal and D is the diffusion coefficient. For long diffusion times, restrictions by cell membranes and binding to macromolecules have to be taken into account. An additional advantage for water suppression is that the permeability of cell membranes to water is generally much higher than to most other compounds. The main disadvantage is that diffusion-sensitized pulse sequences are also very sensitive to motion. Therefore, such methods appear to have more potential for HR NMR than for in vivo NMR.[52,54]

2.5 Exchange

Exchange of protons between a dissolved compound and solvent water can result in different relaxation properties of the water protons. These exchange properties can be used to attenuate the solvent resonance.[55] For example, the saturation of exchangeable spins, whether by rf irradiation, T_2 relaxation, or dephasing in a magnetic field, can be transferred to the water resonance. This exchange mechanism is the basis of the so-called 'magnetization transfer contrast' in MRI,[56] but has not been used for the specific purpose of solvent suppression in in vivo NMR.

3 ADVANCED HARDWARE AND SOFTWARE TOOLS

3.1 Pulsed Field Gradients

The development of high-quality pulsed field gradients is probably the single most important tool that has made the routine applications of proton spectroscopy possible in humans. Historically, pulsed field gradients were used in HR NMR to dephase (spoil) the transverse magnetization—hence the often used name of a homospoil pulse.[57] Although the fundamental advantages of field gradients were known,[57–60] the limited quality with respect to amplitude, ramp times, current stability, and particularly residual gradients and vibration effects, prevented their general use. The extreme demands in MRI—in particular, with respect to the very fast echo planar imaging method[61]—have dramatically accelerated hardware developments. The advent of self-shielded gradient coils with minimal residual gradients should thus be viewed as a milestone.[62,63] An outer gradient coil is used that cancels the outside field of the inner coil, thus eliminating eddy currents in the cryostat. Pulsed field gradient technology is now also becoming a routine tool in HR NMR. Field gradients are used for the following purposes:

1. rapid dephasing of transverse magnetization (homospoil pulse);
2. spatial encoding and spatially selective excitation (in conjunction with shaped rf pulses);
3. single shot coherence pathway selection;
4. creating diffusion sensitivity (Section 2.1.4).

3.1.1 Rapid Dephasing of Transverse Magnetization

Following a gradient pulse of duration δ with magnitude G_α in direction α, the phase ϕ_α at location r_α as a function of coherence order p is

For References see p. 4952

$$\phi_\alpha(p) = p\gamma r_\alpha \delta G_\alpha \tag{7}$$

Therefore, a phase dispersion results that attenuates the transverse magnetization by a factor f:

$$\frac{1}{f} = \left| \frac{1}{r_2 - r_1} \int_{r_1}^{r_2} e^{-i\varphi_\alpha}\, dr_\alpha \right| \tag{8}$$

The limits r_2 and r_1 are the voxel dimensions in the gradient direction α. Equation (8) assumes equal spin density, and a homogeneous field. Note that the attenuation works as a result of the integration of the phase over the dimension of the volume.[12] In the case of spectroscopic imaging, the integration is over the voxel dimensions, not over the full object. The attenuation is higher with a higher gyromagnetic ratio, and with a higher coherence order. Homonuclear zero quantum order will thus not be dephased by the gradient pulse. Note also that the efficiency of the gradient crusher depends on the local static field homogeneity. For a detailed account, see Moonen et al.[64] Shimming is therefore important in the full region where high efficiency of gradient crushers is necessary, not only in the region of interest.

3.1.2 *Spatial Encoding and Spatially Selective Excitation*

The essence of field gradients in imaging is the encoding of spatial information,[65] whether used for phase encoding, frequency encoding, or spatially selective excitation. In proton spectroscopy of humans, field gradients are often employed for the same purpose.[66–82] For spectroscopic imaging,[83–89] phase encoding is generally used for spatial encoding in two or three directions. In conjunction with shaped rf pulses (leading to a desired frequency profile), field gradient pulses lead to spatially selective excitation or refocusing (see Section 4.2 for a more detailed account). The importance of spatial selection for water suppression purposes is further explained in Section 4.1. Spatial and spectral selection can be combined in a single rf pulse together with a special gradient waveform.[89–91]

3.1.3 *Single Shot Coherence Pathway Selection*

Conventionally (Section 2.3), a coherence pathway is selected using a multiscan approach using specific phase cycles of rf pulses and receiver. However, this ideal result is often not reached in vivo as a result of motion or of instrumental instabilities. The selection can also be achieved using pulsed field gradients, often completely eliminating the need for phase cycling. For example, a π refocusing pulse at the end of a pulse sequence may generate transverse magnetization as a result of an inhomogeneous $\boldsymbol{B}_1$ field. Using two scans with opposite phase of the π pulse, transverse magnetization is canceled. A pair of identical gradient pulses around the π pulse will lead to cancelation of the unwanted transverse magnetization in a single scan.

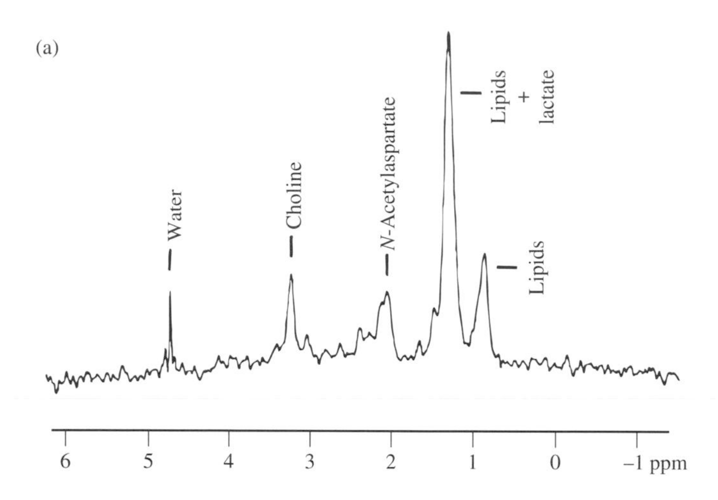

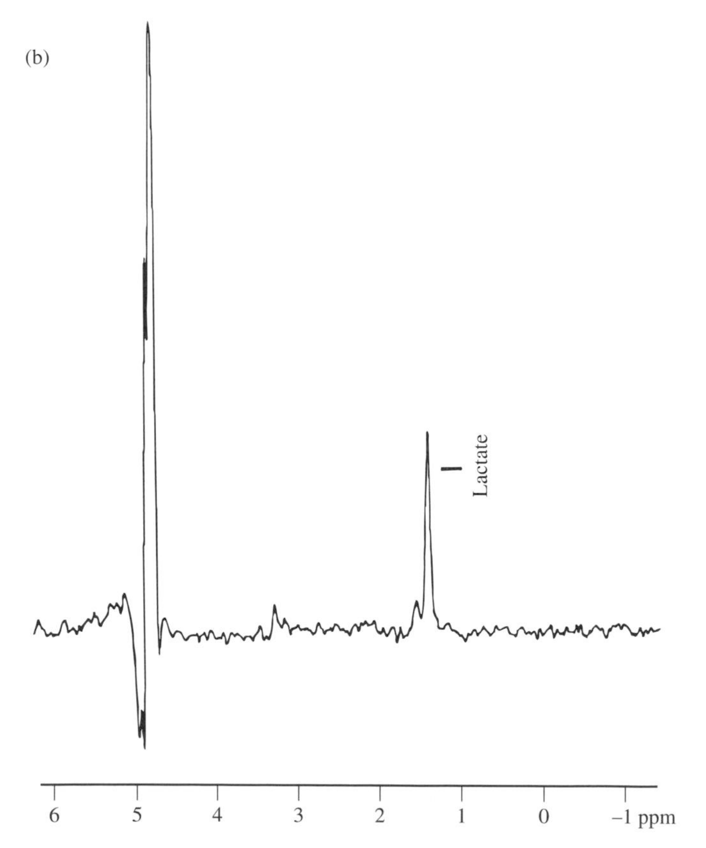

Figure 3 Unedited (a) and lactate-edited (b) localized proton spectra (volume 200 μl) of an intracerebral glioma in the rat obtained at 4.7 T with a surface coil for excitation in conjunction with adiabatic pulses (*TE*/*TR* = 272/2000 ms; 4 min acquisition time for each spectrum). The basic pulse sequence is a spin echo method with jump–return (Section 2.1.2) excitation and refocusing pulses for water suppression, combined [in (b)] with a frequency-selective inversion pulse at the resonance frequency of the C-α proton (4.1 ppm). The inversion pulse is alternated on/off, and the resulting signals are subtracted for the editing procedure.[47] All rf pulses have been replaced with adiabatic pulses. Instead of hard jump–return pulses, solvent-suppressive adiabatic pulses were used. The inversion pulse was an adiabatic DANTE pulse. To maintain better phase coherence in the editing procedure, a second (2π) adiabatic DANTE was inserted in the second $\frac{1}{2}TE$ period. Phase cycling using the EXORCYCLE procedure was performed. Localization was achieved with a modified 3D ISIS method using BIR-4 inversion pulses. Further details are given by Schapp et al.[103] and de Graaf et al.[124] Courtesy of Michael Garwood, Center for Magnetic Resonance Research, University of Minnesota, Minneapolis, MN

We can generalize equation (7) with respect to the phase of transverse magnetization at location r as a function of coherence order p during all periods i of the pulse sequence:

$$\phi_i(r) = \left(\sum_j p_j \gamma_j\right) r G_i t_i \tag{9}$$

where p_j is the coherence order of the individual nuclei (e.g. ^{13}C, ^{1}H) in period i and all other symbols have the same meaning as in equation (7). Thus, the phase evolves as a function of coherence order and gyromagnetic ratio, and equation (9) can be used to keep track of the phase evolution during the entire pulse sequence. A certain coherence order can thus be selectively rephased if the following condition is satisfied:

$$\sum_i \phi_i(r) = 0 \tag{10}$$

In the ideal case, spins of interest are refocused after the last rf pulse, and a gradient scheme is employed that avoids any rephasing of undesired coherences.[61,62] The required gradient power depends on the signal strength of the desired versus undesired coherences.

Localization, achieved by one or more spatially selective rf pulses, can be conveniently treated as a type of coherence pathway selection.[12,64] When more rf pulses are employed during a pulse sequence, it becomes harder to ensure sufficient dephasing of all undesired coherences. However, the use of gradients in all three principal axes with different amplitudes, and using different combinations at different time points in the sequence, offers great flexibility to optimize the single scan coherence selection.

Using equation (9), it can be easily seen that a coherence pathway selection involving a double quantum evolution ($p = \pm 2$) leads to dephasing of the solvent resonance without recourse to any frequency-dependent water suppression. The example shown in Figure 3 demonstrates that resonances under the water line may be observed, such as the resonances of the biologically important glucose molecule. Heteronuclear polarization transfer can also be achieved in a single scan with complete water suppression. However, homonuclear polarization transfer or any homonuclear sequence with $p = 0, \pm 1$, without chemical-shift-selective pulses, cannot achieve this, and additional water suppression methods must be employed. However, even in such cases, the use of field gradient pulses is beneficial because of the elimination of phase cycling (less motion sensitivity), and, in conjunction with other solvent suppression techniques, decreased water signal per scan and thus fewer demands on the receiver system.[92,93]

The use of field gradients for single shot coherence selection also has some disadvantages. Most important is the loss of a factor of two in the signal (except in the case of gradient pulses around a π refocusing pulse, and in special sequences using selective refocusing pulses[37] or sensitivity-enhanced gradient experiments[125]). The reason is that only one coherence pathway is selected, whereas phase cycling allows pathways of opposite signs to be selected simultaneously. For correlated multidimensional NMR, a second disadvantage (and a consequence of the first) is that phase-sensitive 2D spectra cannot be obtained using a single scan per t_1 increment. When using a second scan with selection of coherence path with opposite sign of coherence order in the t_1 period, phased resonances can be obtained. Of course, if these disadvantages are serious, pulsed field gradients may still be used only around refocusing pulses, and thus still eliminate part of the phase cycle.

3.2 Shaped rf Pulses

Shaped rf pulses play an important role in MRI, especially in slice-selective excitation and refocusing.[90,91,94–105] To a first approximation, the frequency response resulting from a particular rf pulse shape is given by the Fourier transform of the waveform. This is known as the linear response. For flip angles close to, and above 90°, linear response theory is no longer completely accurate, and the Bloch equations have to be used to determine the response.[5,106] Corrections to the basic sinc waveform are now commonly used in NMR imaging. In localized proton spectroscopy, shaped rf pulses are used for the same purposes. In addition, a frequency-selective excitation is often used to select a narrow frequency band at the water resonance. In contrast to slice-selective pulses, phase-coherent response is not desired for frequency-selective suppression. In other applications, it is useful to combine spatial and spectral selection in a single pulse. The different specifications in different pulse sequences clearly indicate the opportunities for a specific design of rf pulses. In modern applications, simple Fourier transform of a desired frequency profile is often used as a starting point for a shaped rf pulse. Then, an optimization routine is used to arrive at improved shapes using the complete Bloch equations. Recent advances use complex polynomials for mapping the rf pulse and then solving them analytically.[91,105]

The inhomogeneity in the $\boldsymbol{B}_1$ field (e.g., of a surface coil) has advantages and disadvantages in this respect. One can use the $\boldsymbol{B}_1$ gradient to advantage in order to arrive at a desired phase dispersal of the water magnetization following excitation (see Section 2.1.1). On the other hand, inhomogeneous $\boldsymbol{B}_1$ fields lead to loss of signal due to a spread in flip angle and to a spatially dependent population of desired and undesired coherence pathways, often necessitating the use of large gradient crushers for coherence path selection. The latter disadvantage can be overcome by using adiabatic pulses.

3.2.1 *Adiabatic rf Pulses*

Most water suppression techniques demand precise flip angles. Despite high $\boldsymbol{B}_1$ homogeneity in the modern volume coils that are now routinely available in whole body MR instruments, the range of the $\boldsymbol{B}_1$ field may still exceed the requirements—in particular, for applications of water-suppressed proton spectroscopic imaging. Accurate flip angles, independent of the local $\boldsymbol{B}_1$ field, can be achieved with adiabatic rf pulses.[94–104] The first such pulses accomplished excitation and spin inversion, and employed a continuous frequency ramp of the $\boldsymbol{B}_1$ field while maintaining a constant amplitude. The effect of this pulse can be visualized in the rotating frame by analyzing the direction of the effective field (a function of the $\boldsymbol{B}_1$ field and the rf frequency relative to the Larmor frequency). During the pulse, the effective field $\boldsymbol{B}_e$ rotates from the positive z axis to the negative z axis (in the case of inversion) or to the (x,y) plane (in the case of excitation). So long as the $\boldsymbol{B}_1$ field exceeds a threshold value (the adiabatic condition), the magnetization continues to rotate effectively with a small precession angle around the field $\boldsymbol{B}_e$

For References see p. 4952

during the entire pulse. In other words, the magnetization $\boldsymbol{M}$ remains collinear with $\boldsymbol{B}_e$ above a threshold $\boldsymbol{B}_1$. Therefore, if we let $\boldsymbol{B}_e$ rotate by π then $\boldsymbol{M}$ will also rotate by π, and spin inversion is accomplished. The flip angle is independent of the $\boldsymbol{B}_1$ field if the rate of change of the angle of the field $\boldsymbol{B}_e$ with the z axis remains much smaller than the rotation frequency of magnetization $\boldsymbol{M}$ around the effective field $\boldsymbol{B}_e$ (i.e., the adiabatic criterion). The frequency sweep does not need to be accomplished with a linear ramp. In fact, a whole range of frequency (or phase) modulation functions have been described. The effect of the different modulation functions affect the $\boldsymbol{B}_1$ and off-resonance range where the pulse remains adiabatic. The disadvantages of these original adiabatic pulses are that

1. they do not result in good slice profiles when used for slice selection;
2. they require increased rf power;
3. plane rotation is not possible.

The latter disadvantage, in particular, has prevented many applications, especially in areas related to coherence pathway selection and water suppression. However, recent advances have made it possible to perform adiabatic plane rotations and thus refocusing. In addition, slice-selective refocusing with adiabatic pulses has been made possible by using two consecutive adiabatic inversion pulses.[107]

The problem of plane rotation using the above adiabatic pulses is that, unlike magnetization that is initially collinear with the effective field $\boldsymbol{B}_e$, magnetization perpendicular to $\boldsymbol{B}_e$ at location r will precess through an angle β depending on the actual magnitude of $\boldsymbol{B}_e$ at location r. This angle therefore varies with location, and, as a result, the possibility of $\boldsymbol{B}_1$-independent rotation is apparently lost. The central idea behind the solution of Garwood, Ugurbil and colleagues[99–104] is that if the effective field $\boldsymbol{B}_e$ is suddenly inverted, and if $\boldsymbol{B}_e$ is properly rotated, leading to a precession of $-\beta$ at location r, then the magnetization $\boldsymbol{M}$ will undergo an effective plane rotation. One of the most versatile of this new class of adiabatic plane rotation pulses is the so-called BIR-4 pulse.[102] The pulse consists of three sections: an adiabatic half-passage in reverse, adiabatic inversion, and adiabatic half-passage. Two inversions of $\boldsymbol{B}_e$ occur, for example, between periods one and two, and between two and three, accompanied by two discontinuous phase jumps, $\Delta\phi_1$ and $\Delta\phi_2$. The BIR-4 pulse can accomplish uniform plane rotations through any angle, the magnitude being determined by the phase jumps $\Delta\phi_1$ and $\Delta\phi_2$.

The feature of BIR-4 that is uniquely suitable for water suppression purposes or coherence pathway selection purposes is the fact that the phase jumps can be accomplished not only using the phase of the $\boldsymbol{B}_1$ field, but also by the Larmor precession frequency or by J coupling.[101,104] The first possibility is achieved simply by inserting a delay at the time of the phase jumps, leading to frequency-selective adiabatic pulses. The second possibility leads to adiabatic editing based on spin–spin coupling and adiabatic polarization transfer.[104] These elegant features can be accomplished with a high insensitivity of more than 10-fold variations in the $\boldsymbol{B}_1$ field. Figure 3 shows an example.

3.3 Postacquisition Frequency Filters

Even when the pulse sequence has resulted in water suppression of such high quality that the preamplifier has not been overloaded and the dynamic range of the ADC is sufficient with respect to the S/N of the experiment, it may be advantageous further to reduce the water line using data processing algorithms. The reason lies in some remaining disadvantages of the (suppressed but still dominating) water resonance. First, the base of the water line extends far into other regions of the spectrum. Second, insufficient apodization and truncation of the time domain can cause 'ringing' over a large portion of the spectrum. The latter problem is especially relevant if the acquisition time is short, such as in fast spectroscopic imaging using multiple spin echoes.

Several data processing methods can be used to extract the information from the raw time domain signals. Baseline corrections and convolution difference methods have been used extensively.[3–6] Much progress has been made in fitting in the time and frequency domains.[108–110] The latter methods are preferred whenever additional a priori information is available about the spectrum. A simple approach—the frequency filter—is demonstrated in Figure 4. The method is robust, fast, and automatic.[111–113] First, with water on resonance, a moving average filter over n data points is applied to generate a time domain signal from the raw data, which thus contain only data around zero frequency. Generally, a Gaussian weighting function is used, together with some extrapolation to determine the first and last $\frac{1}{2}n$ points. Second, the filter output is subtracted from the raw data. It has been shown that hardware frequency filters can also be implemented before the preamplifier stage.[114]

4 EVALUATION OF EXPERIMENTAL PROBLEMS

4.1 Spatial Selection

Most human proton spectroscopy is performed with the standard quadrature resonator. Therefore, signal detection is about equally efficient over the whole head, including the neck region. Owing to susceptibility differences over the complete sensitive volume, and also in part to nonideal shimming or magnet imperfections, water in some regions will have similar resonance frequencies to the metabolites of interest in the region to be examined. Two important conclusions can be drawn. First, frequency-based water suppression may work in a well-shimmed area, but not in the complete sensitive coil volume. Second, frequency-based water suppression only makes sense in conjunction with high-quality localization.

Localization may be performed using a surface coil for reception and transmission purposes.[115] The relatively small coil limits the sensitive area and thus the source of artifactual signals. In addition, $\boldsymbol{B}_1$ gradients generated by the surface coil can be used for additional localization.[116] Far more common is the use of $\boldsymbol{B}_0$ gradients, which are, of course, routinely available on every clinical MRI instrument. One of the first methods used a field profiling method to spoil resonances originating outside the region of interest.[117] However, most methods employ switched $\boldsymbol{B}_0$ gradients for localization—either for phase encoding or for slice selection purposes.[66–82] Slice selection in one direction is achieved similarly to common methods in MRI. Volume selection is achieved using a combination of three frequency-selective rf pulses in the presence of three mutually orthogonal gradients. This can be done in a multiscan[66–70,73–76] or single scan[10,70,71,78–81] approach. Selection in two or three different directions can also be achieved

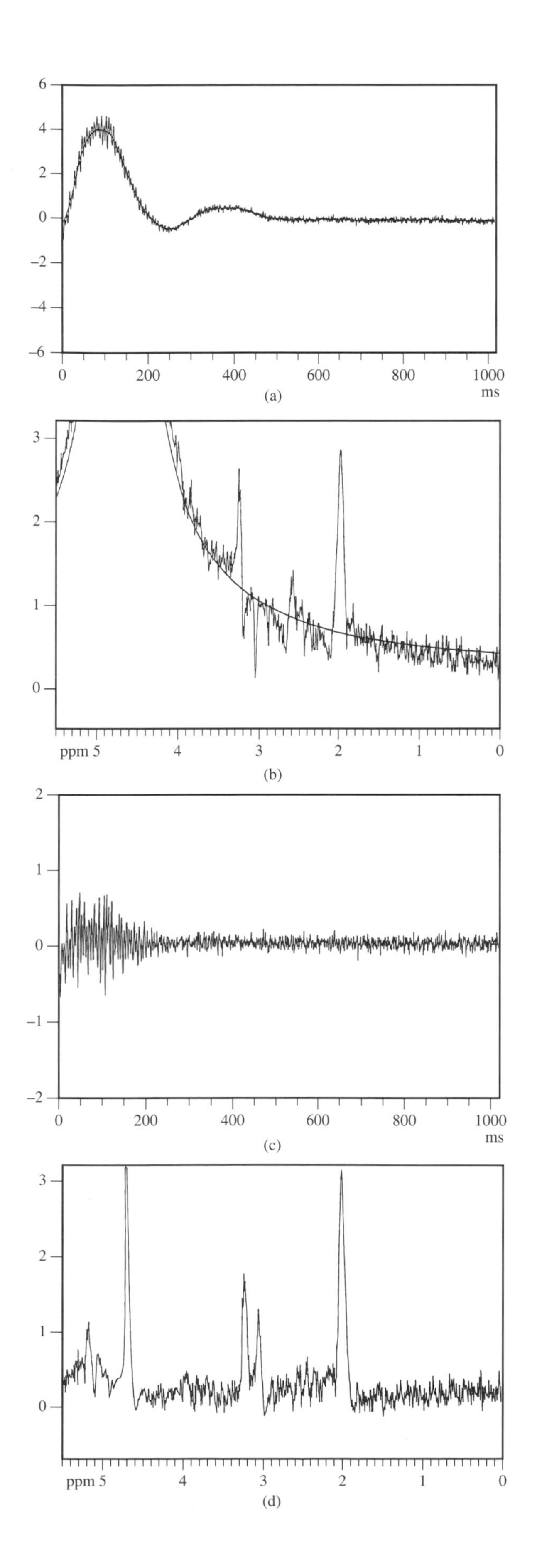

simultaneously.[90,118] Outer volume suppression may help the localization process. Because of the very high degree of localization, and the presence of motion, single shot methods are preferred. Note the important additional role of the $\boldsymbol{B}_0$ gradients in single shot localization (e.g., crushers used for coherence pathway selection without rf phase cycling). Phase encoding employing $\boldsymbol{B}_0$ gradients can be used for additional localization in spectroscopic imaging with Fourier transform[84] or alternative data processing methods using a priori information.[119,120] The quality of localization using phase encoding can be analyzed with the point spread function.

4.2 $\boldsymbol{B}_0$ Inhomogeneity

Perhaps the most significant problem in in vivo proton spectroscopy is the attainable homogeneity in the static magnetic field $\boldsymbol{B}_0$. Air–tissue and fat–tissue interfaces may lead to susceptibility gradients of the same magnitude as the generally used frequency range in in vivo proton spectroscopy. High-quality field homogeneity can be obtained in the brain, but is rather difficult in almost all other organs and tissues. This is the reason why proton spectroscopy is most commonly used in brain. Limited $\boldsymbol{B}_0$ homogeneity is not only detrimental to spectral appearance, but has also other disadvantages. Among these are

1. limitations in frequency-based water suppression methods;
2. errors in localization based on slice selection principles;
3. limitation in the efficiency of $\boldsymbol{B}_0$ gradients for coherence selection and for rapid dephasing;
4. echo shifting when the effect of an incompletely compensated $\boldsymbol{B}_0$ gradient is countered by the field inhomogeneity.

There are no complete solutions to the problem of limited $\boldsymbol{B}_0$ homogeneity. Partial remedies include optimal (possibly local) shimming, selecting small voxels, and using self-shielded gradients. Methods have been presented that result in sharp lines despite $\boldsymbol{B}_0$ inhomogeneities.[122] However, this result comes at the price of a severe S/N penalty.

4.3 $\boldsymbol{B}_1$ Inhomogeneity

The disadvantages of limited $\boldsymbol{B}_1$ homogeneity have been discussed in Section 2.1. In the context of water (or fat) suppression using the popular CHESS methods, the flip angle must be defined precisely. Recent advances in adiabatic pulse design may provide elegant improvements.

Figure 4 The use of a simple postprocessing frequency filter to limit detrimental effects of the suppressed but still dominant water resonance on spectral quality. The example is taken from a voxel of a water-suppressed spectroscopic image (for pulse sequence details, see Moonen et al.).[64] The raw time domain signals, and magnitude mode presentation following FT are shown in (a) and (b), respectively. The effect of the time domain frequency filter is shown on the filtered time domain (c) and the corresponding frequency spectrum (d) in magnitude mode presentation. Courtesy of Geoffrey Sobering, In Vivo NMR Research Center, NCRR, NIH, Bethesda, MD

For References see p. 4952

4.4 Motion

Motion is a common problem in MRI. Despite the much larger voxels in proton spectroscopy, it can be an even bigger problem in the latter technique—in particular, because of the extreme demands on water suppression and localization. In this regard, we may distinguish not only macroscopic motion of uncooperative patients, but also motion of arterial and venous blood, of CSF, and of brain tissue. The preferred method to minimize the detrimental effects of motion is to perform water suppression and localization in a single scan.[12] Multiscan phase cycling methods lead to incomplete suppression of unwanted signals in the presence of motion. However, single scan methods employing $\boldsymbol{B}_0$ gradients are not without problems. For example, motion during the scan may result in a phase shift of the echo. Tracking the phase using an additional echo[122] may help if the phase shift is identical for the complete voxel, but is of limited use if it is spatially dependent. In addition, if signal averaging is used, the motion sensitivity may be decreased when magnitude spectra are averaged, instead of pure phase spectra. In addition, shifts in the echo position may occur as a result of an imbalanced gradient in the presence of field inhomogeneities resulting from the motion.

4.5 Spectroscopic Imaging

The typical problems and solutions for solvent suppression in single voxels are similar to those in spectroscopic imaging methods. However, since modern spectroscopic imaging[86–91] is performed in a single session covering a large part of the brain, problems with $\boldsymbol{B}_0$ and $\boldsymbol{B}_1$ inhomogeneities are more severe. The problem of $\boldsymbol{B}_0$ homogeneity leads to the need for a larger frequency band affected by the water suppression procedure. Postprocessing water suppression (Section 3.3) is commonly used to minimize the effects of spatially dependent water suppression. Adiabatic pulses—now already routinely used for spin inversion purposes in spectroscopic imaging—may bring further improvements. Motion during spectroscopic imaging may lead to severe artifacts.

4.6 Fat Suppression

This article has dealt with suppression of the water resonance. Resonances of fat, or indeed, the family of triglycerides, can also dominate an in vivo proton NMR spectrum. Triglycerides lead to many resonances covering a large part of the useful proton spectrum. Therefore, typical frequency-selective suppression is not satisfactory. Three features of fat resonances are used for suppression purposes.

1. T_1 values are short (about 280 ms at 1.5 T), and can be used conveniently when preceding a pulse sequence with an inversion–recovery sequence.
2. In healthy brain tissue, no significant contributions originate from brain triglycerides.[2] The fat resonances arise predominantly from the skull tissue and can thus be suppressed by (a) outer volume suppression[67] (e.g., by spatial excitation of the entire skull area followed by rapid dephasing); or
   (b) avoiding excitation of skull area using accurate localization procedures (see Section 4.1).
3. Coupling patterns in trygliceride resonances, as distinct from those in the lactate doublet, are often used to separate the two compounds. Since the spatial origin of fat resonances is in the skull, phase encoding is helpful in defining the spatial origin of the fat signals, and thus suppressing the fat resonances in brain spectra.[123]

4.7 Safety Issues

The potential safety issues in water-suppressed proton spectroscopy are the same as for routine MRI. There are no specific safety issues with regard to water suppression, except in some cases heat deposition by rf irradiation when employing (semi)-continuous rf irradiation or $\boldsymbol{B}_1$ gradients to achieve scrambling of the water magnetization. However, most methods described in this article use less rf power per unit time than routinely used in multiple spin echo MRI methods.

5 SUMMARY

Solvent suppression has been, and remains, a rich area of research. Many advanced methods, developed over at least 30 years of high-resolution NMR, have been adapted for use in in vivo proton NMR spectroscopy. Water-suppressed human proton spectroscopy has now become almost a routine tool for brain NMR examination at 1.5 T. Although numerous developments have contributed to this extraordinary achievement, two fundamental technologies have been crucial:

1. the development of high-quality, fast switching, $\boldsymbol{B}_0$ gradient coils, in particular, the invention of the self-shielded gradient coil;
2. the achievement of excellent rf phase and amplitude stability and control.

The latter technology is especially relevant for adiabatic and other shaped pulses. Together with excellent progress in the field of single shot localization, these developments have made high-quality, water-suppressed proton spectroscopy of human and animal brain a reality. Solvent suppression in spectroscopic imaging is similar, but not identical, to the problem of single voxel solvent suppression. However, recent progress indicates that automated single slice and multislice spectroscopic imaging of the human brain will soon be available on standard 1.5 T instruments. The technological challenge that still remains largely unsolved is the routine use of proton spectroscopy of organs other than the brain.

6 RELATED ARTICLE

Single Voxel Proton NMR: Human Subjects.

7 REFERENCES

1. K. L. Behar, J. A. Den Hollander, M. E. Stromski, Y. Ogino, R. G. Shulman, O. A. C. Petroff, and J. W. Pritchard, *Proc. Natl. Acad. Sci. USA*, 1983, **80**, 4945.
2. C. C. Hanstock, D. L. Rothman, T. J. Jue, and R. G. Shulman, *Proc. Natl. Acad. Sci. USA.*, 1988, **85**, 1821.
3. P. J. Hore, *Methods Enzymol.*, 1989, **176**, 64.

4. J. E. Meier and A. G. Marshall, in 'Biological Magnetic Resonance', eds. L. J. Berliner and J. Reuben, Plenum Press, New York, 1990, Vol. 9, p. 199.
5. M. Guéron, P. Plateau, and D. Decorps, in 'Progress in Nuclear Magnetic Resonance Spectroscopy', eds. J. W. Emsley, J. Feeney and L. H. Sutcliffe, Pergamon Press, Oxford, 1991, Vol. 23, p. 135.
6. P. C. M. van Zijl, and C. T. W. Moonen, in 'NMR Basic Principles and Progress', eds. P. Diehl, E. Fluck, and R. Kosfeld, Springer-Verlag, Berlin, 1992, Vol. 26, p. 67.
7. D. I. Hoult, *J. Magn. Reson.*, 1976, **21**, 337.
8. A. Haase, J. Frahm, W. Hänicke, and D. Matthaei, *Phys. Med. Biol.*, 1985, **30**, 341.
9. D. M. Doddrell, G. J. Galloway, W. M. Brooks, J. Field, J. M. Bulsing, M. G. Irving, and H. Baddeley, *J. Magn. Reson.*, 1986, **70**, 176.
10. J. Frahm, K. D. Merboldt, and W. Hänicke, *J. Magn. Reson.*, 1987, **72**, 502.
11. I. M. Brereton, G. J. Galloway, J. Field, M. F. Marshman, and D. M. Doddrell, *J. Magn. Reson.*, 1989, **81**, 411.
12. C. T. W. Moonen and P. C. M. van Zijl, *J. Magn. Reson.*, 1990, **88**, 28.
13. R. H. Griffey and D. P. Flamig, *J. Magn. Reson.*, 1990, **88**, 161.
14. R. J. Ogg, P. B. Kingsley, and J. S. Taylor, *J. Magn. Reson.*, 1994, **104B**, 1.
15. C. A. G. Haasnoot, *J. Magn. Reson.*, 1983, **52**, 153.
16. D. Canet, J. Brondeau, E. Mischler, and F. Humbert, *J. Magn. Reson.*, 1993, **105A**, 239.
17. W. E. Maas and D. G. Gory, *J. Magn. Reson., Ser. A*, 1994, **106A**, 256.
18. P. Blondet, M. Decorps, and J. P. Albrand, *J. Magn. Reson.*, 1986, **69**, 403.
19. D. Bourgeois and P. Kozlowski, *Magn. Reson. Med.*, 1993, **29**, 402.
20. G. A. Morris and R. Freeman, *J. Magn. Reson.*, 1978, **29**, 433.
21. P. Plateau and M. Guéron, *J. Am. Chem. Soc.*, 1982, **104**, 7310.
22. H. Liu, K. Weisz, and T. L. James, *J. Magn. Reson., Ser. A*, 1993, **105**, 184.
23. J. Frahm, H. Bruhn, M. L. Gyngell, K.-D. Merboldt, W. Hänicke, and R. Sauter, *Magn. Reson. Med.*, 1989, **11**, 47.
24. T. Inubushi and E. D. Becker, *J. Magn. Reson.*, 1983, **51**, 128.
25. J. H. Duijn, G. B. Matson, A. A. Maudsley, J. W. Hugg, and M. W. Weiner, *Radiology*, 1992, **183**, 711.
26. J. H. Duijn and C. T. W. Moonen *Proc. Soc. Magn. Reson. Med. 1993 Meet.*, p. 316.
27. R. R. Ernst, G. Bodenhausen, and A. Wokaun, 'Principles of Nuclear Magnetic Resonance in One and Two Dimensions', Clarendon Press, Oxford, 1987.
28. N. Chandrakumar and S. Subramianan, 'Modern Techniques in High-Resolution FT-NMR', Springer-Verlag, New York, 1987.
29. O. W. Sørensen, in 'Progress in Nuclear Magnetic Resonance Spectroscopy', eds. J. W. Emsley, J. Feeney, and L. H. Sutcliffe, Pergamon Press, Oxford, 1989, Vol. 21, p. 503.
30. C. L. Dumoulin and D. Vatis, *Magn. Reson. Med.*, 1986, **3**, 282.
31. C. H. Sotak, D. M. Freeman, and R. E. Hurd, *J. Magn. Reson.*, 1988, **78**, 355.
32. R. E. Hurd and D. M. Freeman, *Proc. Natl. Acad. Sci. USA*, 1989, **86**, 4402.
33. D. M. Freeman, C. H. Sotak, H. H. Muller, S. W. Young, and R. E. Hurd, *Magn. Reson. Med.*, 1990, **14**, 321.
34. A. Knuttel and R. Kimmich, *Magn. Reson. Med.*, 1989, **10**, 404.
35. D. M. Doddrell, I. M. Brereton, L. N. Moxon, and G. J. Galloway, *Magn. Reson. Med.*, 1989, **9**, 132.
36. C. H. Sotak, *J. Magn. Reson.*, 1990, **90**, 198.
37. L. A. Trimble, J. F. Shen, A. H. Wilman, and P. S. Allen, *J. Magn. Reson.*, 1990, **86**, 191.
38. A. Knüttel, R. Kimmich, and K.-H. Spohn, *J. Magn. Reson.*, 1990, **86**, 526.
39. J. M. Bulsing and D. M. Doddrell, *J. Magn. Reson.*, 1986, **68**, 52.
40. C. J. Hardy and C. L. Dumoulin, *J. Magn. Reson.*, 1987, **5**, 75.
41. M. von Kienlin, J. P. Albrand, B. Authier, P. Blondet, S. Lotito, and M. Décorps, *J. Magn. Reson.*, 1987, **75**, 371.
42. A. Knüttel and R. Kimmich, *Magn. Reson. Med.*, 1989, **9**, 254.
43. D. L. Rothman, K. L. Behar, H. P. Hetherington, and R. G. Shulman, *Proc. Natl. Acad. Sci. USA*, 1984, **81**, 6330.
44. A. A. de Graaf, P. R. Luyten, J. A. den Hollander, W. Heindel, and W. M. M. J. Bovée, *Magn. Reson. Med.*, 1993, **30**, 231.
45. J. E. van Dijk, A. F. Mehlkopf, and W. M. M. J. Bovée, *NMR Biomed.*, 1992, **5**, 75.
46. D. L. Rothman, K. L. Behar, H. P. Hetherington, J. A. den Hollander, M. R. Bendall, O. A. C. Petroff, and R. G. Shulman, *Proc. Natl. Acad. Sci. USA*, 1985, **82**, 1633.
47. T. Jue, *J. Magn. Reson.*, 1987, **73**, 524.
48. J. Ruiz-Cabello, G. W. Vuister, C. T. W. Moonen, P. van Gelderen, J. S. Cohen, and P. C. M. van Zijl, *J. Magn. Reson.*, 1992, **100**, 282.
49. P. C. M. van Zijl, A. S. Chesnick, D. DesPres, C. T. W. Moonen, J. Ruiz-Cabello, and P. van Gelderen, *Magn. Reson. Med.*, 1993, **30**, 544.
50. G. Bodenhausen and D. J. Rubin, *Chem. Phys. Lett.*, 1980, **69**, 185.
51. E. O. Stejskal and J. E. Tanner, *J. Chem. Phys.*, 1965, **42**, 288.
52. P. C. M. van Zijl and C. T. W. Moonen, *J. Magn. Reson.*, 1990, **87**, 18.
53. J. Kärger, H. Pfeifer, and W. Heink, in 'Advances in Magnetic Resonance', ed. J. S. Waugh, Academic Press, New York, 1988, Vol. 12, p. 1.
54. P. C. M. van Zijl, C. T. W. Moonen, P. Faustino, J. Pekar, O. Kaplan, and J. S. Cohen, *Proc. Natl. Acad. Sci. USA*, 1991, **88**, 3228.
55. D. L. Rabenstein, S. Fan, and T. T. Nakashima, *J. Magn. Reson.*, 1985, **64**, 541.
56. S. D. Wolff and R. Balaban, *Magn. Reson. Med.*, 1989, **10**, 135.
57. R. L. Vold, J. S. Waugh, M. P. Klein, and D. E. Phelps, *J. Chem. Phys.*, 1968, **48**, 3831.
58. A. A. Maudsley, A. Wokaun, and R. R. Ernst, *Chem. Phys. Lett.*, 1978, **55**, 9.
59. A. Bax, P. G. de Jong, A. F. Mehlkopf, and J. Smidt, *Chem. Phys. Lett.*, 1980, **69**, 567.
60. P. Barker and R. Freeman, *J. Magn. Reson.*, 1985, **64**, 334.
61. P. Mansfield, *J. Phys. C*, 1977, **10**, L55.
62. P. Mansfield and B. Chapman, *J. Phys. E*, 1986, **19**, 540.
63. P. B. Roemer, W. A. Edelstein, and J. S. Hickey, *Proc. Soc. Magn. Reson. Med. 1986 Meet.*, p. 1067.
64. C. T. W. Moonen, G. S. Sobering, P. C. M. van Zijl, J. Gillen, M. von Kienlin, and A. Bizzi, *J. Magn. Reson.*, 1992, **98**, 556.
65. P. C. Lauterbur, *Nature (London)*, 1973, **242**, 190.
66. W. P. Aue, *Rev. Magn. Reson. Med.*, 1986, **1**, 21.
67. R. Sauter, S. Müller, and H. Weber, *J. Magn. Reson.*, 1987, **71**, 167.
68. P. R. Luyten, A. J. H. Mariën, B. Sijtsma, and J. A. den Hollander, *J. Magn. Reson.*, 1986, **67**, 148.
69. P. A. Bottomley, *Ann. NY Acad. Sci.*, 1987, **508**, 333.
70. R. J. Ordidge, P. Mansfield, J. A. B. Lohman, and S. B. Prime, *Ann. NY Acad. Sci.*, 1987, **508**, 376.
71. P. C. M. van Zijl, C. T. W. Moonen, J. R. Alger, J. S. Cohen, and A. S. Chesnick, *Magn. Reson. Med.*, 1989, **10**, 256.

72. C. T. W. Moonen, M. von Kienlin, P. C. M. van Zijl, J. Gillen, P. Daly, J. S. Cohen, and G. Wolf, *NMR Biomed.*, 1989, **2**, 201.
73. W. P. Aue, S. Müller, T. A. Cross, and J. Seelig, *J. Magn. Reson.*, 1984, **56**, 350.
74. P. A. Bottomley, T. H. Foster, and R. D. Darrow, *J. Magn. Reson.*, 1984, **59**, 338.
75. R. J. Ordidge, A. Connelly, and J. A. B. Lohman, *J. Magn. Reson.*, 1986, **66**, 283.
76. T. Mareci and H. R. Brooker, *J. Magn. Reson.*, 1985, **57**, 157.
77. A. Connelly, C. Counsell, J. A. B. Lohman, and R. J. Ordidge, *J. Magn. Reson.*, 1988, **78**, 519.
78. R. J. Ordidge, M. R. Bendall, R. E. Gordon, and A. Connelly, in 'Magnetic Resonance in Biology and Medicine', eds. Govil, Khetrapal, and Sran. McGraw-Hill, New Delhi, 1985, p. 387.
79. J. Granot, *J. Magn. Reson.*, 1986, **70**, 488.
80. G. McKinnon, *Proc. 5th Ann. Mtg. Soc. Magn. Reson. Med., Montreal*, 1986, p. 168.
81. R. Kimmich and D. Hoepfel, *J. Magn. Reson.*, 1987, **72**, 379.
82. J. Frahm, H. Bruhn, M. L. Gyngell, K.-D. Merboldt, W. Hänicke, and R. Sauter, *Magn. Reson. Med.*, 1989, **9**, 79.
83. T. R. Brown, B. M. Kincaid, and K. Ugurbil, *Proc. Natl. Acad. Sci. USA*, 1982, **79**, 3523.
84. A. A. Maudsley, S. K. Hilal, W. H. Perman, and H. E. Simon, *J. Magn. Reson.*, 1983, **51**, 147.
85. P. R. Luyten, A. J. H. Mariën, W. Heindel, P. H. J. van Gerwen, K. Herholz, J. A. den Hollander, G. Friedmann, and W. D. Heiss, *Radiology*, 1990, **176**, 791.
86. J. H. Duijn, G. B. Matson, A. A. Maudsley, and M. W. Weiner, *Magn. Reson. Med.*, 1992, **25**, 107.
87. J. H. Duijn, J. Gillen, G. Sobering, P. C. M. van Zijl and C. T. W. Moonen, *Radiology*, 1993, **188**, 277.
88. J. H. Duijn and C. T. W. Moonen, *Magn. Reson. Med.*, 1993, **30**, 409.
89. D. M. Spielman, J. M. Pauly, A. Macovski, G. Glover, and D. R. Enzmann, *J. Magn. Reson. Imaging*, 1992, **2**, 253.
90. P. G. Morris, in 'NMR Basic Principles and Progress', eds. P. Diehl, E. Fluck, and R. Kosfeld, Springer-Verlag, Berlin, 1992, Vol. 26, p. 149.
91. J. Pauly, P. Le Roux, D. Nishimura, and A. Macovski, *IEEE Trans. Med. Imaging*, 1991, **10**, 53.
92. R. E. Hurd and B. K. John, *J. Magn. Reson.*, 1991, **91**, 648.
93. M. von Kienlin, C. T. W. Moonen, A. van der Toorn, and P. C. M. van Zijl, *J. Magn. Reson.*, 1991, **93**, 423.
94. J. Baum, R. Tycko, and A. Pines, *J. Chem. Phys.*, 1983, **79**, 4643.
95. M. S. Silver, R. I. Joseph, and D. I. Hoult, *J. Magn. Reson.*, 1994, **59**, 347.
96. P. G. Morris, D. J. O. McIntyre, D. E. Rourke, and J. T. Ngo, *Magn. Reson. Med. Biol.*, 1989, **11**, 167.
97. S. Conolly, D. Nishimura, and A. Macovski, *J. Magn. Reson.*, 1989, **83**, 324.
98. C. J. Hardy, W. A. Edelstein, and D. Vatis, *J. Magn. Reson.*, 1986, **66**, 470.
99. K. Ugurbil, M. Garwood, and M. R. Bendall, *J. Magn. Reson.*, 1986, **72**, 177.
100. M. Garwood and K. Ugurbil, in 'NMR Basic Principles and Progress', eds. P. Diehl, E. Fluck, and R. Kosfeld, Springer-Verlag, Berlin, 1992, Vol. 26, p. 110.
101. B. D. Ross, H. Merkle, K. Hendrich, R. S. Staewen, and M. Garwood, *Magn. Reson. Med.*, 1992, **23**, 96.
102. M. Garwood and Y. Ke, *J. Magn. Reson.*, 1991, **94**, 511.
103. D. G. Schupp, H. Merkle, J. M. Ellermann, Y. Ke, and M. Garwood, *Magn. Reson. Med.*, 1993, **30**, 1.
104. M. Garwood and H. Merkle, *J. Magn. Reson.*, 1990, **94**, 180.
105. M. Shinnar, L. Bolinger, and J. S. Leigh, *Magn. Reson. Med.*, 1989, **12**, 88.
106. M. Goldman, 'Quantum Description of High-Resolution in Liquids', Clarendon Press, Oxford, 1988, Chap. 1.
107. S. Connolly, G. Glover, D. Nishimura, and A. Macovski, *Magn. Reson. Med.*, 1991, **18**, 28.
108. R. de Beer and D. van Ormondt, in 'NMR Basic Principles and Progress', eds. P. Diehl, E. Fluck, and R. Kosfeld, Springer-Verlag, Berlin, 1992, Vol. 26, p. 201.
109. A. A. de Graaf, J. E. van Dijk, and W. M. M. J. Bovée, *Magn. Reson. Med.*, 1989, **13**, 343.
110. J. S. Nelson and T. R. Brown, *J. Magn. Reson.*, 1989, **84**, 95.
111. H. Barkhuijsen, R. de Beer, W. M. M. J. Bovée, and D. van Ormondt, *J. Magn. Reson.*, 1985, **61**, 465.
112. D. S. Stephenson, in 'Progress in Nuclear Magnetic Resonance Spectroscopy', eds. J. W. Emsley, J. Feeney, and L. H. Sutcliffe, Pergamon Press, Oxford, 1988, Vol. 20, p. 515.
113. D. Marion, M. Ikura, and A. Bax, *J. Magn. Reson.*, 1989, **84**, 425.
114. O. Gonen and G. Johnson, *J. Magn. Reson., Ser. B*, 1993, **102**, 98.
115. J. J. H. Ackerman, T. H. Grove, G. G. Wong, D. G. Gadian, and G. K. Radda, *Nature (London)*, 1980, **283**, 167.
116. M. R. Bendall and R. E. Gordon, *J. Magn. Reson.*, 1983, **53**, 365.
117. R. E. Gordon, P. E. Hanley, D. Shaw, D. G. Gadian, G. K. Radda, P. Styles, P. J. Bore, and L. Chan, *Nature (London)*, 1980, **287**, 367.
118. C. J. Hardy, P. A. Bottomley, M. O'Donnell, and P. Roemer, *J. Magn. Reson.*, 1988, **77**, 233.
119. X. Hu, D. N. Levin, P. C. Lauterbur, and T. Spraggins, *Magn. Reson. Med.*, 1988, **8**, 314.
120. M. von Kienlin and R. Mejia, *J. Magn. Reson.*, 1991, **94**, 268.
121. R. L. Ehman and J. P. Felmlee, *Radiology*, 1989, **173**, 255.
122. L. D. Hall and T. J. Norwood, *J. Magn. Reson.*, 1986, **67**, 382.
123. S. Posse, B. Schuknecht, M. E. Smith, P. C. M. van Zijl, N. Herscovitch, and C. T. W. Moonen, *J. Comput. Assist. Tomogr.*, 1993, **17**, 1.
124. R. A. de Graaf, Y. Luo, M. Terpstra, H. Merkle, and M. Garwood, *J. Magn. Reson.*, in press.
125. J. Cavanagh, A. G. Palmer III, P. E. Wright, and M. J. Rance, *J. Magn. Reson.*, 1991, **429**, 91.

Acknowledgements

Peter C. M. van Zijl is supported by NIH Grant IR01 NS31490-01.

Biographical Sketches

Chrit Moonen. *b* 1955. M.S., 1980, Ph.D., 1983, Agricultural University, Wageningen, The Netherlands. Introduced to protein NMR by Kurt Wüthrich, Zürich, 1979. Faculty member, Department of Biochemistry at Wageningen 1984–86; sabbatical with George Radda, Department of Biochemistry, Oxford; one year with Morton Bradbury at University of California at Davis, 1987. Head of the NIH In Vivo NMR Research Center from 1987–present. Approx. 80 publications. Research interests include mapping of brain function, diffusion imaging and spectroscopy, proton spectroscopic imaging, physiological imaging, use of field gradients in HR NMR.

Peter C. M. van Zijl. *b* 1956. M.S., 1980, Ph.D., 1985, Free University, Amsterdam, The Netherlands. Postdoctoral fellow at Carnegie

Mellon University, Department of Chemistry, 1985–87; visiting associate at National Institutes of Health, National Cancer Institute, 1987–90. Research assistant professor at Georgetown University, Department of Pharmacology, 1990–92; associate professor at Johns Hopkins Medical School, Department of Radiology, 1992–present. Approx. 75 publications. Research interests: in vivo spectroscopy and imaging of brain function, including diffusion imaging and imaging of metabolite levels and active metabolism of magnetically labelled precursors. Mechanism of ischemia. Design of NMR technology for HR NMR.

Wavelet Encoding of MRI Images

Lawrence P. Panych & Ferenc A. Jolesz

Brigham and Women's Hospital and Harvard Medical School, Boston, MA, USA

1 INTRODUCTION

In Fourier transform MRI, the raw image data can be expressed mathematically as samples of the Fourier transform of a spatially dependent density function that is proportional to the magnitude distribution of the transverse magnetization. Thus, the raw Fourier image data represent a spatial frequency decomposition of the density function. In wavelet encoding of MRI images, one attempts to image in such a way that the raw image data can be expressed as samples of the wavelet transform of the density function along at least one spatial dimension. The raw wavelet image data have some characteristics of a spatial frequency decomposition, but they also retain features of the spatial domain representation.

More formally, let $s(x)$ represent the density function. The Fourier transform of $s(x)$ is a function $S(k)$ of the spatial frequency parameter k, and $|S(k)|^2$ gives a measure of the 'energy' content in $s(x)$ at the spatial frequency k (in cm^{-1}). The wavelet transform of $s(x)$ is a function $S_w(v,u)$ of two parameters: a 'scale' (spatial frequency band) parameter v and a location parameter u. $|S_w(v,u)|^2$ gives a measure of the 'energy' content in $s(x)$ at the scale v (in cm^{-1}) in the vicinity of the location u (in cm).

In Fourier transform MRI, the raw Fourier domain data are produced by manipulation of the phase angle of the transverse magnetization. In wavelet MRI, on the other hand, it is the angle of the magnetization vector between the longitudinal axis and the transverse plane that is manipulated. In other words, the amount of the magnetization tilted into the transverse plane is modulated according to wavelet-shaped excitation profiles. This is done by modifying the rf excitation pulses on each encoding step.

In this article, wavelet encoding in one of the dimensions of a two-dimensional imaging method is described. The second dimension is Fourier-encoded because, during the data acquisition period of an imaging sequence, it is not possible to manipulate the flip angle of the spins, nor is it possible (with linear gradients) to wavelet-encode the phase angle of the transverse magnetization. Thus, the so-called frequency-encoding dimension must be Fourier-encoded. Figure 1 shows a modified line scanning sequence that has been used[1] to encode two-dimensional images with wavelet encoding along one of the dimensions (y). The wavelet encoding method using this sequence applies ideas first proposed by Healy and Weaver.[2,3] There is a second wavelet encoding method also based on the use of specially designed excitation profiles where the raw data are expressed mathematically as samples of the wavelet transform of the Fourier space of the density function. This method will not be discussed here, and the reader is referred to the literature for details.[4]

2 WAVELET TRANSFORM THEORY

2.1 Continuous and Discrete Wavelet Transform

A wavelet transform[5] $F_\Psi(v,u)$ of the real-valued finite-energy function $f(x)$ is defined by

$$F_\Psi(v,u) = |v|^{-1/2} \int f(x)\Psi\left(\frac{x-u}{v}\right) \mathrm{d}x \quad v \neq 0 \tag{1}$$

$\Psi(x)$ is the real-valued 'basic' wavelet function, and is a window on $f(x)$ that is both scaled (v) and translated (u). For practical applications such as imaging, it is important to know if a function can be recovered from a discrete and finite sampling of its wavelet transform space. When the sampling is exponential as expressed by

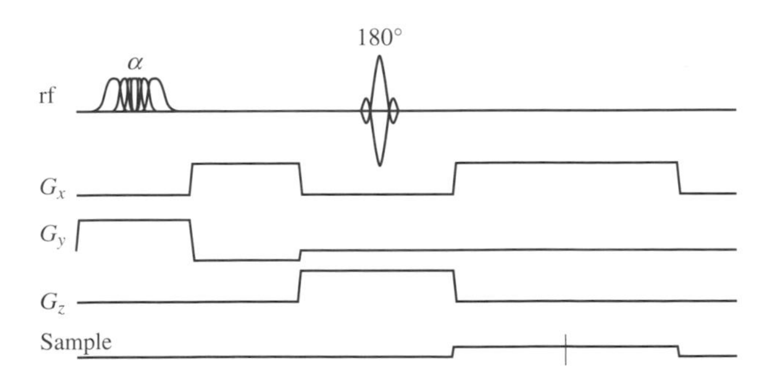

Figure 1 Pulse sequence for wavelet encoding in the y dimension

For References see p. 4959

$$v = 2^j, \quad u = 2^j n, \quad \text{where } j \text{ and } n \text{ are integers} \tag{2}$$

a family of 'dyadic' wavelets $\{\Psi_{j,n}(x)\}$ is produced as defined by

$$\Psi_{j,n}(x) \equiv 2^{-j/2}\Psi(2^{-j}x - n) \tag{3}$$

$\Psi(x)$ have been found such that the set of dyadic wavelets defined by equation (3) is orthonormal, and arbitrary functions can be represented by a series with the same number of terms used in a Fourier expansion. In fact, for functions with distinct edges (as are typical in MRI), even fewer terms are generally needed in the wavelet representation. Figure 2 demonstrates how a set of dyadic wavelets is constructed by scaling of the basic wavelet and then translating (in steps of $2^j\,\Delta x$) each of the scaled wavelets.

2.2 Wavelet Multiresolution Analysis

Multiresolution analysis[6] provides an elegant interpretation of the dyadic wavelet expansions. The central idea of multiresolution analysis is that the space of finite-energy functions can be decomposed into a hierarchy of subspaces $\{\mathbf{V}^{[j]}\}$ where $\mathbf{V}^{[j]}$ is the space of all approximations of the finite-energy functions at the resolution $2^j\,\Delta x$ (j integer). Resolution is defined such that, at the resolution $2^j\,\Delta x$, a function $f(x)$ is approximated by evenly spaced samples of width $2^j\,\Delta x$. Note that, by the convention adopted here, a higher index j corresponds to a poorer (coarser) resolution.

Approximation spaces are defined such that $\mathbf{V}^{[j+1]}$ is completely contained within the finer resolution subspace $\mathbf{V}^{[j]}$. The extra detail found in $\mathbf{V}^{[j]}$ that is not also found in $\mathbf{V}^{[j+1]}$ is included within the detail space $\mathbf{W}^{[j+1]}$ as formalized by

$$\mathbf{V}^{[j]} = \mathbf{W}^{[j+1]} \cup \mathbf{V}^{[j+1]} \tag{4}$$

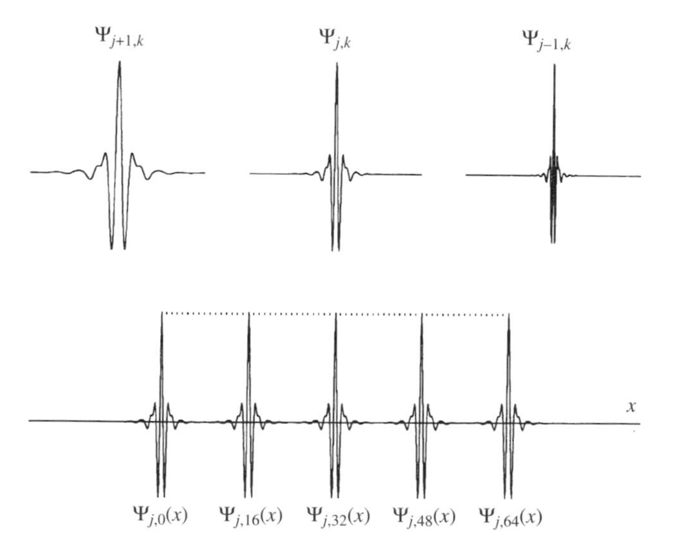

Figure 2 A wavelet basis is constructed by scaling the basic wavelet. The top row of plots shows Battle–Lemarie wavelets at three dyadic scales. At each scale, the wavelets are translated in increments of $2^j\,\Delta x$ as shown in the bottom row of plots for the scale j. In this figure, only every 16th translate is plotted, and the positions of all other translates are shown with tick marks

Each of the approximation subspaces $\mathbf{V}^{[j]}$ is spanned by a family of scaling functions, $\{\Phi_{j,n}(x)\}$, as defined by

$$\Phi_{j,n}(x) = 2^{-j/2}\Phi(2^{-j}x - n\,\Delta x), \quad \text{where } j \text{ and } n \text{ are integers} \tag{5}$$

and where Δx is the width of the basic scaling function[5] $\Phi(x)$. Each detail space $\mathbf{W}^{[j]}$ is spanned by a family of wavelets $\{\Psi_{j,n}(x)\}$ defined by

$$\Psi_{j,n}(x) = 2^{-j/2}\Psi(2^{-j}x - n\,\Delta x), \quad \text{where } j \text{ and } n \text{ are integers} \tag{6}$$

Figure 3 shows examples of basic scaling functions and wavelets used to build orthonormal bases.

A wavelet transform decomposes a function into an expansion in terms of functions built from the scaling functions and wavelets as follows. Assume that the function of interest is contained within the approximation space $\mathbf{V}^{[0]}$. Any function in $\mathbf{V}^{[0]}$ can be expressed as a weighted sum of the scaling functions $\{\Phi_{0,n}\}$ that span $\mathbf{V}^{[0]}$. However, since $\mathbf{V}^{[0]} = \mathbf{W}^{[1]} \cup \mathbf{V}^{[1]}$, it can also be expressed as a weighted sum of the scaling functions $\{\Phi_{1,n}\}$ and the wavelets $\{\Psi_{1,n}\}$ that span the subspaces. Further, since $\mathbf{V}^{[1]} = \mathbf{W}^{[2]} \cup \mathbf{V}^{[2]}$, any function in $\mathbf{V}^{[0]}$ can also be expressed as a weighted sum of the scaling functions $\{\Phi_{2,n}\}$ and the wavelets $\{\Psi_{2,n}\}$ and $\{\Psi_{1,n}\}$. The wavelet decomposition can be continued to an arbitrary scale J, so that the approximation of $f(x)$ in $\mathbf{V}^{[0]}$ is expressed as a weighted sum of wavelets $\{\Psi_{1,n}\}, \{\Psi_{2,n}\}, \ldots, \{\Psi_{J,n}\}$ and the scaling functions $\{\Phi_{J,n}\}$.

On a finite interval of length $N\,\Delta x$, J cannot be greater than $\log_2 N$. Each set of wavelets $\{\Psi_{j,n}\}$ or scaling functions $\{\Phi_{j,n}\}$ contains $N/2^j$ functions, since the translation step size for each scale j is $2^j\,\Delta x$. Problems at the edges of finite intervals are avoided if it is assumed that the basis functions wrap around at the edges.

The approximation of $f(x)$ in $\mathbf{V}^{[0]}$ on the finite interval of length $N\,\Delta x$ can be expressed by the wavelet series expansion

$$f^{[0]}(x) = \sum_{j=1}^{J}\sum_{k=0}^{N/2^j-1} D_{j,k}\Psi_{j,k}(x) + \sum_{l=0}^{N/2^J-1} A_{J,l}\Phi_{J,l}(x) \tag{7}$$

where the total number of terms in the expansion is N and the weights $A_{j,n}$ and $D_{j,n}$ are defined by

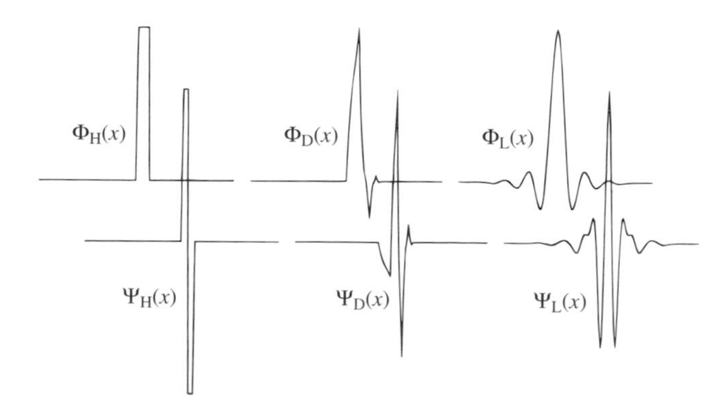

Figure 3 Basic scaling functions and wavelets for the Haar basis (Φ_H, Ψ_H), a Daubechies basis (Φ_D, Ψ_D), and the Battle–Lemarie basis (Φ_L, Ψ_L)

$$A_{j,n} = \int f(x)\Phi_{j,n}(x)\,\mathrm{d}x, \quad D_{j,n} = \int f(x)\Psi_{j,n}(x)\,\mathrm{d}x \tag{8}$$

2.3 Multiresolution Wavelet Reconstruction

The sets $\{D_{1,k}\},\ldots,\{D_{J,k}\}$ and $\{A_{J,l}\}$ in equation (7) are the coefficients of a discrete wavelet transform of $f(x)$. Reconstruction begins with the data sequences $\{A_{J,l}\}$ and $\{D_{J,l}\}$. The multiresolution interpretation says that the sequence $\{A_{J,l}\}$ represents the coarsest approximation of $f(x)$, and $\{D_{J,l}\}$ represents the detail between the coarsest approximation and an approximation one level finer.

To move to a finer level of approximation, zeros are inserted between each value in the two sequences. The zero-padded sequences are then individually convolved with the impulse response of a low- and a high-pass filter,[6] and the convolution results are added. In this way, the sequence $\{A_{J-1,k}\}$ is produced. It has twice as many values as $\{A_{J,l}\}$, and represents a discrete approximation of $f(x)$ at the second coarsest resolution. To move to another scale, $\{A_{J-1,k}\}$ and $\{D_{J-1,k}\}$ are padded with zeros and convolved with the same low- and high-pass filters, and $\{A_{J-2,k}\}$ is produced. The process is continued until all of the detail sequences are merged and the N-element sequence $\{A_{0,k}\}$ is the final result. $\{A_{0,k}\}$ represents a discrete approximation of $f(x)$ at the resolution Δx.

3 MR ENCODING BY SPATIALLY SELECTIVE EXCITATION

Assume that the modified line scan sequence shown in Figure 1 is used and, on each of N excitations, a set of profiles, $\{\phi_{0,n}(y)\}$, defined by

$$\phi_{0,n}(y) = \phi(y - n\Delta y), \quad n = 0, 1, \ldots, N-1 \tag{9}$$

is excited, where $\phi(y)$ is centered at $y = 0$, has a width of Δy, and is mostly positive. The general form of the echo data from the nth excitation, $S_n(k_x)$, is expressed by

$$S_n(k_x) = \int\int s(x,y)\phi_{0,n}(y)\mathrm{e}^{\mathrm{i}k_x x}\,\mathrm{d}x\,\mathrm{d}y \tag{10}$$

where $s(x,y)$ is the magnetization density function introduced in Section 1.

Assume that N (k-space) samples are taken from each echo. After performing an inverse fast Fourier transform (FFT) on each of the N sets of echo data with respect to k_x, the partially transformed data set $\{s_{m,n}\}$ can be expressed by

$$s_{m,n} = \int\int s(x,y)\mathrm{sinc}\left[\frac{\pi}{\Delta x}(x - m\,\Delta x)\right]\phi_{0,n}(y)\,\mathrm{d}x\,\mathrm{d}y$$
$$m, n = 0, 1, \ldots, N-1 \tag{11}$$

where the sinc function represents the point spreading due to finite k-space sampling. Clearly, $\{s_{m,n}\}$ is a discrete approximation of $s(x,y)$, since $\phi_{0,n}(y)$ in equation (11) also acts like a point-spread function. In fact, if $\phi(y)$ is a sinc function and $\Delta y = \Delta x$ then the N^2 values of $\{s_{m,n}\}$ should be identical to an image obtained using phase encoding.

Now, suppose that $\phi(y)$ is a scaling function $\Phi(y)$ such as one of those shown in Figure 3. In this case, $\{s_{m,n}\}$ is also a discrete approximation of $s(x,y)$, but with a different point-spread function in the y direction. If equation (11) is rewritten as

$$s_{m,n} = \int \bar{s}(m\,\Delta x, y)\Phi_{0,n}\,\mathrm{d}y \tag{12a}$$

where

$$\bar{s}(m\,\Delta x, y) \equiv \int s(x,y)\mathrm{sinc}\left[\frac{\pi}{\Delta x}(x - m\,\Delta x)\right]\mathrm{d}x \tag{12b}$$

and compared with the definition of $A_{j,n}$ in equation (8), it is clear that $s_{m,n}$ has the same form as $A_{0,n}$. For any x-location index m, $\{s_{m,0}, s_{m,1}, \ldots, s_{m,N-1}\}$ represents a set of coefficients $\{A_{0,n}\}$, and can be interpreted as a discrete approximation of $\bar{s}(m\,\Delta x, y)$.

At low flip angles, a scaling-function-shaped profile $\Phi(y)$ can be excited using an rf pulse whose shape is given by the Fourier transform of $\Phi(y)$. In order to translate the profiles across the field of view in steps of Δy, the rf offset frequency must be changed, on each excitation, in steps of $\Delta\omega = \gamma G_y \Delta y$, where G_y is the strength of the y gradient during the rf excitation. If $T_{3\,\mathrm{dB}}$ is the half-power (3 dB) duration of the rf pulse [see RFS(0) in Figure 4] then the width of the excitation profile is $(\gamma G_y T_{3\,\mathrm{dB}})^{-1}$, since the rf pulse excites a region proportional to its bandwidth. The translation step size Δy must be equal to the width of the excitation profile $\Phi(y)$; therefore, $\Delta\omega = 1/T_{3\mathrm{dB}}$.

For a fixed gradient strength and fixed total rf pulse duration T_{rf}, the achievable resolution Δy is strongly dependent on the type of scaling function. The best resolution is achieved when the ratio $r = T_{3\,\mathrm{dB}}/T_{\mathrm{rf}}$ approaches 1. To produce the sharp edges of the Haar scaling function, many sidelobes of a sinc-shaped rf pulse must be included, and r will be much less than 1. On the other hand, r for the Battle–Lemarie scaling function is $\frac{2}{3}$, and the resolution will be closer to the maximum, which is obtained when the rf pulse is box-shaped.

4 AN IMPLEMENTATION OF WAVELET-ENCODED MRI

4.1 Excitation Profiles and rf Pulses

From the multiresolution interpretation of the wavelet transform, we know that any function that can be represented by a weighted sum of scaling functions $\{\Phi_{0,n}(y)\}$ can also be represented by a weighted sum of the set of wavelets at the J scales and the set of scaling functions at scale J. For wavelet encoding, then, the pulse sequence shown in Figure 1 is used to excite specially shaped excitation profiles on each of N separate excitations in the same manner as in Section 3. In the wavelet case, however, the excitation profiles will be the $N/2$ wavelets $\{\Psi_{1,n}(y)\}$, the $N/4$ wavelets $\{\Psi_{2,n}(y)\},\ldots,$ the $N/2^J$ wavelet(s) $\{\Psi_{J,n}(y)\}$, and the $N/2^J$ scaling function(s) $\{\Phi_{J,n}(y)\}$.

To generate wavelets at different scales, there are two choices:

1. use a constant gradient strength while varying the duration of the rf pulse at each scale; or

For References see p. 4959

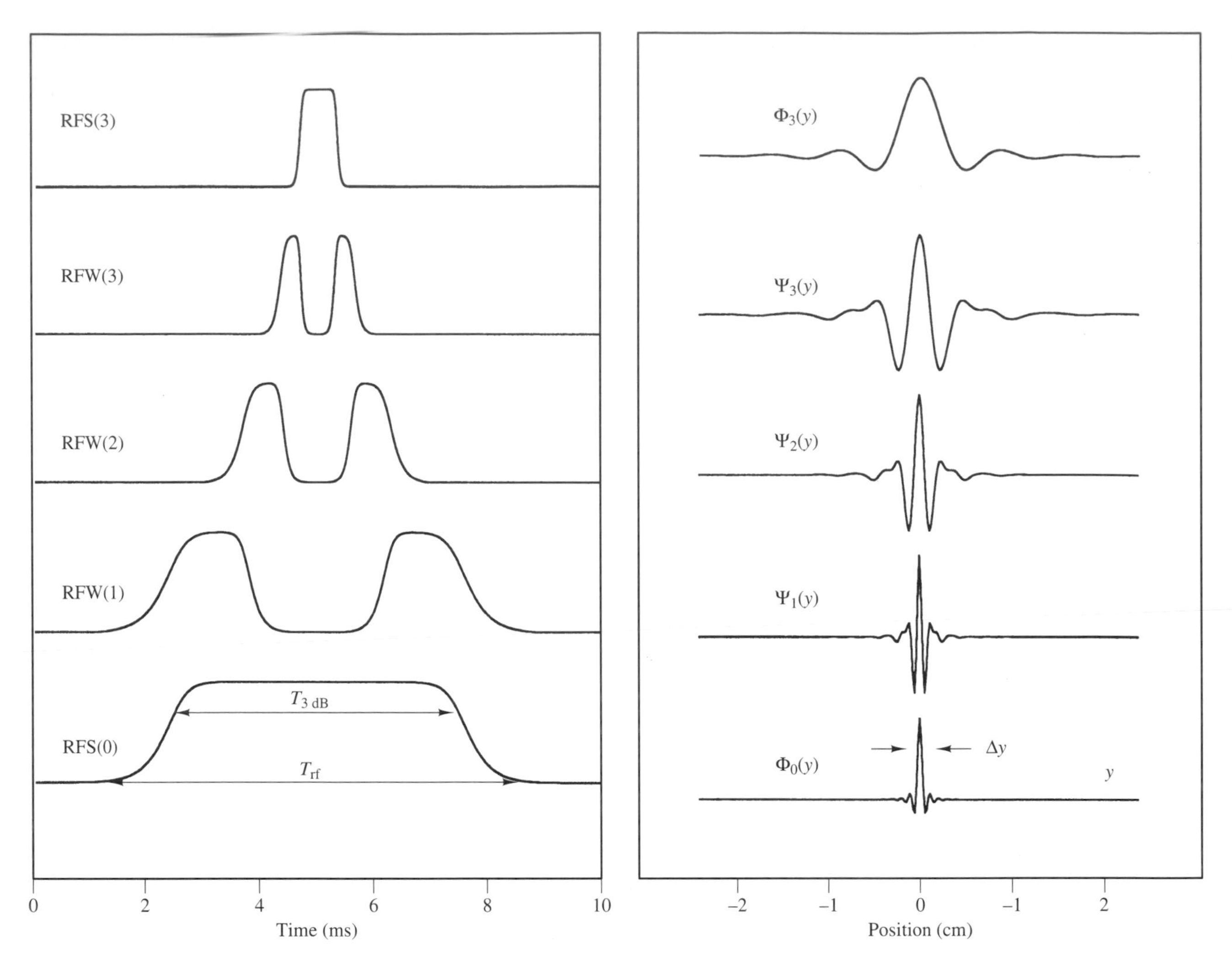

Figure 4 The top four pairs of plots are (amplitude-normalized) rf pulses and excitation profiles for a three-level wavelet encoding. See Section 3 for a discussion of RFS(0) and $\Phi_0(y)$. All profiles are computed assuming a gradient strength of $10\,\mathrm{mT\,m^{-1}}$

2. use an rf pulse of constant duration while varying the gradient strength at each scale.

Figure 4 shows a set of rf pulses to be used for wavelet encoding ($J = 3$) in the Battle–Lemarie basis using the constant gradient method. Note that RFS(0) is not needed for wavelet encoding. The rf pulses are shown as magnitude-normalized; however, if it is desired to maintain a constant flip angle at all scales, rf pulse amplitudes will be scale-dependent. Figure 4 also shows the (magnitude-normalized) profiles excited by each pulse. In the variable-gradient method, independent of the number of scales, only two different rf waveforms are needed (one for scaling functions and one for wavelets). If this method is used, however, gradient-strength-dependent imperfections such as eddy currents may result in image artifact.

4.2 Image Reconstruction

Assume that N wavelet-shaped profiles, as described in Section 4.1, are excited using the pulse sequence shown in Figure 1, and that N samples are collected on each echo. An inverse FFT is performed on each of the N sets of echo data, and the partially transformed data can be expressed as in equations (11) and (12a,b), with the exception that wavelet functions at the J scales and scaling function(s) at scale J replace the set of functions $\{\phi_{0,n}\}$ or $\{\Phi_{0,n}\}$. The expression for the partially transformed data has the same form as the definitions of $A_{j,n}$ and $D_{j,n}$ in equation (8). Thus, for any x-location index m, the N transformed data values represent the coefficients $\{D_{1,k}\},\ldots,\{D_{J,k}\}$ and $\{A_{J,k}\}$, which can be interpreted as a discrete wavelet transform of $\bar{s}(m\,\Delta x,y)$.

The reconstruction procedure outlined in Section 2.3 is used to transform the wavelet coefficient data for each x-location index m into a set $\{A_{0,n}\}$, which represents a discrete approximation of $\bar{s}(m\,\Delta x,y)$ of resolution Δy. In theory, then, the result obtained after complete reconstruction should be the same as if the pulse RFS(0) shown in Figure 4 had been used to excite the scaling function profiles $\{\Phi_{0,n}\}$.

5 SIGNAL-TO-NOISE RATIOS

It has previously been noted[3,7] that wavelet encoded images suffer a disadvantage in terms of signal-to-noise ratio (ratio of signal energy to noise energy) when compared with standard phase-encoded images. The intuitive explanation for this is that all spins participate in each phase encode, whereas only a small portion of the spins are, on average, involved during each wavelet encode. When wavelet encoding, on $N/2$ of the encoding steps $2/N$ of the spins are excited, on $N/4$ of the steps $4/N$ are excited, on $N/8$ of the steps $8/N$ are excited, and so on. It can be shown[8,9] that the signal-to-noise ratio for wavelet encoding is only $3/N$ of that obtained when phase encoding. Of course, since wavelet encoding generally involves only a small portion of the spins, there is a possibility that many wavelet encodes can be interleaved in a *TR* period, and the loss in signal-to-noise ratio may be compensated by an increase in the information rate.

If wavelet packets[5] are used instead of standard wavelets then the signal-to-noise ratio in the wavelet case can be made to approach the signal-to-noise ratio of Fourier imaging.[7,9] Wavelet packets are built from linear combinations of the standard wavelets, and can be designed to involve much larger portions of the field-of-view than the standard wavelets. When wavelet packet functions are used for encoding, more spins on average will be involved on each encoding step, and the signal-to-noise ratio increases accordingly. Of course, since the wavelet packet functions cover larger portions of the field-of-view, spatial selectivity is traded for the increase in signal-to-noise ratio.

6 LIMITATIONS AND APPLICATIONS

Since wavelet encoding is implemented by exciting wavelet-shaped profiles, the resolution when encoding along more than one dimension is limited by the ability to excite multidimensional profiles of sufficient spatial selectivity. Even at the maximum gradient strengths available on commercial imaging systems, the required rf pulse lengths for acceptable resolution will be prohibitively long. A further limitation of wavelet encoding is that many fast imaging approaches may not be compatible. For example, there is no means (that we are aware of) to wavelet-encode in echo planar or RARE imaging except as a method of encoding the third (normally slice-select) dimension. Fast gradient echo with low-flip-angle pulses should be well suited for wavelet encoding; however, finding a simple method of slice selection may be problematic.

The most promising application that we see for wavelet encoding of MRI images is in the development of adaptive imaging methods.[8] An imaging method is adaptive if the image acquisition strategy is modified according to that which is learned about the image during data acquisition. Adaptive methods can be very effective in reducing redundancy in data acquisition and increasing response time in dynamic imaging applications. As an example of a dynamically adaptive imaging strategy based on wavelet encoding,[8] data are acquired beginning with the coarsest resolution. Acquired data values are then examined to determine if a change has occurred since the last time the same coefficients were encoded. Only if change occurs at the coarser resolution are the finer detail coefficients then acquired in a region. Since the number of wavelet encodes grows (by a factor of 2) at each level of detail, this strategy leads to a very significant reduction in the number of encodes necessary to update an image when spatially localized change occurs.

Wavelet-based adaptive methods exploit the ability, when wavelet encoding, to produce images with spatially variable resolution. This ability is useful in any application, adaptive or not, where it is desired to resolve highly only a small region of interest leaving the rest of the field of view less resolved. The ability to resolve images selectively in this way, which derives from the nature of the wavelet transform as a spatially selective multiscale decomposition, is a feature that is unavailable with standard Fourier encoding.

7 RELATED ARTICLES

Echo-Planar Imaging; Fourier Transform & Linear Prediction Methods; Image Formation Methods; Imaging: A Historical Overview; Selective Excitation in MRI; Shaped Pulses; Whole Body Magnetic Resonance: Fast Low-Angle Acquisition Methods.

8 REFERENCES

1. L. P. Panych, P. D. Jakab, and F. A. Jolesz, *J. Magn. Reson. Imaging*, 1993, **3**, 649.
2. D. M. Healy and J. B. Weaver, *IEEE Trans. Inf. Theory*, 1992, **38**, 840.
3. J. B. Weaver, Y. S. Xu, D. M. Healy, and J. R. Driscoll, *Magn. Reson. Med.*, 1992, **24**, 275.
4. A. H. Tewfik, H. Garnaoui, and X. Hu, *Proc. SPIE*, 1992, 112.
5. C. K. Chui, 'An Introduction to Wavelets', Academic Press, San Diego, 1992.
6. S. G. Mallat, *IEEE Trans. Pattern Anal. Machine Intell.*, 1989, **11**, 674.
7. D. M. Healy and J. B. Weaver, in 'Proceedings of the IEEE International Symposium on Time–Frequency and Time-Scale Analysis', Victoria, British Columbia, Canada, October 1992, p. 133.
8. L. P. Panych, 'Adaptive magnetic resonance imaging by wavelet transform encoding'. Ph.D. thesis, Massachusetts Institute of Technology, 1993.
9. J. B. Weaver and D. M. Healy, *J. Magn. Reson. A*, 1995, **113**, 10.

Biographical Sketches

Lawrence P. Panych *b* 1950. B.Eng., 1979, McGill University; M.S., 1983, University of British Columbia; Ph.D., 1993, Massachusetts Institute of Technology. Research fellow in radiology at Harvard Medical School and Brigham and Women's Hospital, 1993–present. Approx. 10 publications. Research specialties: dynamic MRI and novel encoding techniques.

Ferenc A. Jolesz *b* 1946. M.D., 1971, Semmelweis Medical School, Budapest; M.S., 1975, Kando College of Electrical Engineering, Budapest. Research fellowships in neurology at Massachusetts General Hospital and in physiology at Harvard Medical School, 1979–1982. Associate Professor of Radiology at Harvard Medical School and Director of the Division of MRI at Brigham and Women's Hospital, Boston 1988–present. Approx. 85 publications. Research specialties: interventional MRI, image processing, demyelination, multiple sclerosis and Alzheimer's disease.

Well Logging

Robert L. Kleinberg
Schlumberger-Doll Research, Ridgefield, CT, USA

1 INTRODUCTION

Well logging is the means by which physical properties of subsurface earth formations are measured in situ. The most important, and the most technically challenging, application of well logging is to the characterization of hydrocarbon reservoirs. Oil and gas are found up to 10 km underground in beds of sedimentary or other porous rock. Only a part of a typical sedimentary rock is solid mineral matter. The pore space, which accounts for up to 30% of the volume, can be filled by combinations of oil, water, or natural gas. Well logging is directed toward understanding these fluids and their relationship to the solid mineral matrix. A large variety of electromagnetic, acoustic, and nuclear borehole instruments are used for various purposes. Each technique has drawbacks and limitations, and no one logging device ('tool') is adequate to give a complete description of an earth formation.[1,2]

Measuring properties of earth formations in situ by NMR obviously requires apparatus very different than that commonly used in the laboratory. Instead of placing the sample inside the apparatus, the apparatus is placed inside the 'sample', which is in fact the earth. Thus 'inside-out' NMR equipment is required: large static magnetic fields and high-frequency oscillatory magnetic fields must be projected outside of the apparatus and into the surrounding rock formations. Moreover, the apparatus is normally in motion during the measurement.

The amplitude of the proton NMR signal is proportional to the fluid content of the rock, and relaxation times give information on the pore size distribution and, under favorable conditions, the amount and type of oil. These measurements are used by the oil industry to estimate the quantity of hydrocarbon in a reservoir and the rate at which it can be economically extracted. The application of NMR to well logging is the subject of an extensive bibliography, covering 1946–1990, compiled by Jackson and Mathews.[3] The reader also is referred to review articles[4,5] and proceedings of international conferences.[6]

Owing to limitations of signal-to-noise ratio, the only nucleus accessible to borehole NMR is hydrogen. Throughout this article, all statements refer to proton NMR. Nonuniformity of the static magnetic field precludes the acquisition of chemical shift spectra.

2 NMR RELAXATION PROCESSES IN ROCKS

2.1 Surface-Limited and Diffusion-Limited Regimes

NMR relaxation rates of fluids in porous media are enhanced by relaxation at the pore–grain interface.[7,8] Fluid molecules diffuse, eventually reaching a grain surface, where they have a finite probability of being relaxed. The rate-limiting step can either be the relaxation process at the surface or the transport of unrelaxed spins to the surface.

If the rate-limiting step is relaxation at the surface, the nuclear magnetization in a pore is uniform. Therefore the magnetization decay in the pore is monoexponential and does not depend on pore shape but only on the surface-to-volume ratio of the pore. This is referred to as the 'fast diffusion' or 'surface-limited' regime.[9]

In the opposite case, magnetic relaxation occurs at the grain surface, but the decay of macroscopic magnetization is controlled by the transport of molecules to the surface. This is possible when pores are relatively large and/or surface relaxation is strong. This is called the 'slow diffusion' or 'diffusion-limited' regime. In this regime, magnetization in the pore is not uniform. This gives rise to a magnetization decay, which, in each pore, has multiexponential character, and which depends on the shape of the pore.[10]

Support for the surface-limited hypothesis comes from measurements of the temperature dependence of relaxation times in rocks.[11] Remarkably, T_1 and T_2 are nearly independent of temperature over the range 25–175 °C, at least for the modest number of rock samples studied. If NMR relaxation were in the diffusion-limited regime, the relaxation time would depend on the diffusion coefficient of the pore fluid, which is very temperature-dependent. The lack of temperature dependence is a mark of the surface-limited regime.

2.2 Surface Relaxation Mechanisms

The basic principles of NMR relaxation of fluids at solid surfaces in porous media were established by Korringa, Seevers, and Torrey (KST).[12] Two surface processes were identified. One occurs at all sites on the surface and the other is associated with dilute paramagnetic metal ion impurities on the surface. The relaxation process associated with nonmagnetic sites is much too weak to account for the observed relaxation of water in rocks.[13] Paramagnetic ions, such as iron, manganese, nickel, and chromium, are particularly powerful relaxers, and tend to control the rate of relaxation whenever they are present. Rocks primarily composed of silica (sandstones) generally have an iron content of about 1%, and also frequently contain manganese in significant concentrations; these ions make fluid proton relaxation fairly efficient. Occasionally, sandstones contain grains of magnetic minerals such as magnetite (Fe_3O_4), which are sites of relaxation. In carbonate rocks, which are composed of calcite ($CaCO_3$) and related minerals, the rate of fluid relaxation tends to be lower than in sandstones.

Considering only relaxation by dilute paramagnetic ions, the KST equations for surface relaxation in a single pore reduce to

$$\frac{1}{T_1} = \frac{Sh}{V}\frac{n_{\mathrm{M}}}{T_{1\mathrm{M}}} = \rho_1\left(\frac{S}{V}\right)_{\mathrm{pore}} \tag{1}$$

where S and V are the surface area and volume of the pore, h is the thickness of the surface layer within which relaxation can take place, n_{M} is the proportion of surface sites occupied by paramagnetic metal ions, and $T_{1\mathrm{M}}$ is the relaxation time of protons in molecules coordinated with paramagnetic ions. All material constants are included in the surface relaxivity parameter ρ_1, which has dimensions of $\mathrm{m\,s^{-1}}$.

KST do not discuss T_2 in porous media. T_2 is shortened by >molecular diffusion in the inhomogeneous magnetic field arising from the magnetic susceptibility contrast between grains and pore fluid (see Section 2.3), which is unrelated to surface relaxation. However, for proton Larmor frequencies below 5 MHz and for Carr–Purcell echo spacings less than about one millisecond, the enhancement of T_2 decay coming from diffusion in inhomogeneous local fields is often negligible compared with the surface relaxation mechanism. Then[14]

$$\frac{1}{T_2} = \frac{Sh}{V}\frac{n_{\mathrm{M}}}{T_{2\mathrm{M}}} = \rho_2\left(\frac{S}{V}\right)_{\mathrm{pore}} \tag{2}$$

The surface relaxation mechanism is treated in detail by Kleinberg, Kenyon, and Mitra.[14] Briefly, the relaxation of fluid protons at paramagnetic ion sites on the surface of rock grains is controlled by the dipolar and scalar interactions between the ion and nuclear moments. Among other things, the theory explains why T_1 and T_2 are correlated at low frequency: $T_1 \approx 1.6\,T_2$.

2.3 Molecular Diffusion in Magnetic Field Gradients

A second contribution to transverse relaxation is the diffusion of spins in static magnetic field gradients. In a uniform B_0 gradient, the contribution of this mechanism is

$$\left(\frac{1}{T_2}\right)_{\mathrm{D}} = \tfrac{1}{12}D(\gamma G T_{\mathrm{E}})^2 \tag{3}$$

where D is the molecular diffusion coefficient, γ is the gyromagnetic ratio of the proton, G is the gradient strength in $\mathrm{T\,m^{-1}}$, and T_{E} is the Carr–Parcell echo spacing. This equation is applicable to bulk liquids, for which the diffusion coefficient is a constant. In porous media, the diffusion coefficient is time-dependent,[15] and equation (3) must be modified.

A further complication is the presence of local magnetic field gradients, which result from the magnetic susceptibility contrast between grain material and pore fluid. Rocks typically have paramagnetic ion contents of about 1%, and grain volumetric susceptibilities (SI dimensionless) are typically $\chi_{\mathrm{g}} = +10^{-4}$; grain susceptibilities of the magnetic minerals are much higher. Pore fluids are usually diamagnetic. The internal gradient is proportional to B_0. It also depends on the details of the pore geometry, and therefore varies from point to point within the pore space; some realistic models have been developed.[16,17]

The Carr–Purcell–Meiboom–Gill (CPMG) method mitigates the effect of diffusion in a magnetic field gradient. Keeping both the CPMG echo spacing and the applied magnetic field to a minimum reduces the contribution of diffusion to T_2 relaxation. At low values of B_0 typical of borehole logging tools, in the absence of an applied field gradient, and for closely spaced pulses in the CPMG measurement sequence, T_2 is dominated by surface relaxation. Diffusion effects become important when substantial field gradients are applied, and/or longer pulse spacings are used.

2.4 Bulk Fluid Processes

The last relaxation process is that of the bulk fluid itself. This occurs in the absence of grain surfaces and internal field gradients, and is unaffected by them. Bulk relaxation often can be neglected, but it is important when fluid is in very large pores and therefore is little affected by a surface. Bulk relaxation can also be important for very viscous fluids, such as heavy oils, and also can be the dominant process when the pore fluid has a high concentration of paramagnetic ions. For example, chromium lignosulfonate is a common additive in drilling fluids; chromium ion that enters the pore fluid from the borehole can dramatically reduce the observed relaxation time.

Fine particulates from the borehole can also enter the pore space. When in suspension they also reduce the fluid relaxation time, by making 'floating' solid surface area available for fluid molecules to encounter. However, it has been shown, using a novel NMR imaging technique, that this is not likely to be a problem for NMR logging tools.[18]

2.5 Pore Size Distributions and Distributions of Relaxation Times

Rocks tend to have very broad distributions of pore sizes, and therefore magnetization decays of fluids in rocks are nonexponential. In early work,[19] NMR data were analyzed in terms of two- or three-exponential decays. Later, magnetization decays of water in rock were fitted to stretched exponentials,[20] which assumes a particular form of pore size distribution. The most general way to analyze relaxation data is with a distribution (spectrum) of relaxation times: a plot of signal intensity versus relaxation time. In the following discussion, we assume that the transverse relaxation is controlled by the surface relaxation mechanism and that relaxation is surface-limited. The magnetization decay in an individual pore is then monoexponential, and does not depend on pore shape but only on the surface-to-volume ratio:

$$M(t) = M_0 \exp\left(-\frac{t}{T_2}\right) = M_0 \exp\left[-\rho_2\left(\frac{S}{V}\right)_{\mathrm{pore}} t\right] \tag{4}$$

The time evolution of the magnetization of a sample having a distribution of pore sizes can be expressed as a sum of exponential decays:

$$M(t) = \sum_i m_i \exp\left(-\frac{t}{T_{2i}}\right) = \sum_i m_i \exp\left[-\rho_2\left(\frac{S}{V}\right)_i t\right] \tag{5}$$

where m_i is proportional to the volume of fluid relaxing at the rate $1/T_{2i}$. The sum of the volumes is proportional to the porosity ϕ:

$$M_0 = \sum_i m_i \propto \phi \tag{6}$$

For References see p. 4968

Thus, in this model, there is a direct mapping from the spectrum of pore sizes, or more precisely the spectrum of surface-to-volume ratios, to the spectrum of relaxation times.

Implicit in this discussion is the assumption that pores exist as identifiable entities. On the basis of steady state measurements such as hydraulic permeability and electrical conductivity, it is known that the pore space in rocks is well connected. However, pulse NMR is not a steady state measurement. Molecules only sample a limited volume of pore space before they are relaxed.[21] The radius of the volume sampled is approximately the root mean square distance the molecule diffuses in the NMR relaxation time, $\sqrt{6DT_1}$, the diffusion coefficient, is itself a function of observation time in porous media.[15] If there is significant coupling of pores of different sizes, the relaxation time distributions will be narrowed.[22]

A confirmation of the connection between pore size and NMR relaxation time was provided by Straley et al.[23] Rock specimens were fully saturated with water, and the spectrum of longitudinal relaxation times was determined. Then each rock was centrifuged at a number of rotor speeds. At successively higher speeds, water was progressively expelled from the samples as the centrifugal pressure overcame the capillary pressure of successively smaller pore spaces. At each step, the NMR relaxation spectrum was remeasured. At low centrifuge speed, the components with the longest T_1 disappeared from the spectrum, while the short-T_1 components were unaffected, indicating that only the largest pores had drained. As the centrifuge speed was increased, smaller pores drained and shorter components disappeared. A typical result of a similar T_2 measurement is shown in Figure 1.

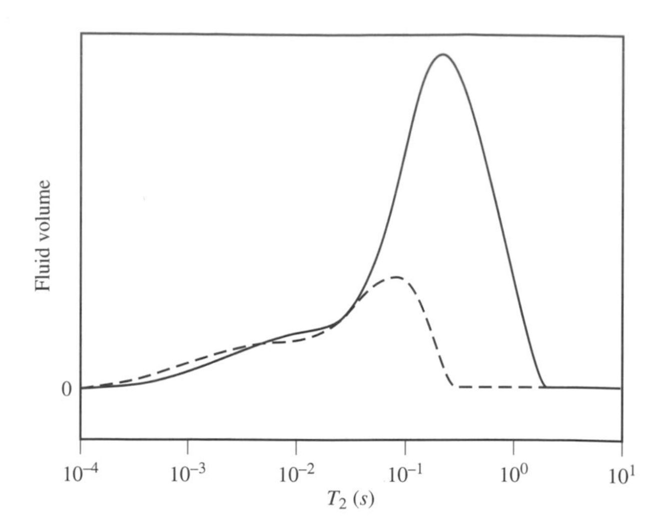

Figure 1 Distribution of relaxation times for a sedimentary rock. The spectrum is a plot of fluid volume as a function of relaxation time. Solid curve: rock fully saturated with water. Dashed curve: the same rock after centrifuging. Centrifugation drains the largest pores, and the corresponding part of the relaxation time distribution disappears. Fluid content before and after centrifuging is directly proportional to the area under the respective curve

The extraction of the values of m_i from the measurement data $M(t)$ poses significant mathematical problems. The inversion is not unique when $M(t)$ is noisy or when the mathematical computation has finite precision. In the past, either the number of terms in equation (5) was limited to two or three—a sparse spectrum indeed—or many terms were used, resulting in spiky, unphysical spectra. This problem can be solved by the use of regularization.[24] Regularization has the effect of smoothing the spectrum, which removes spikes and makes the inversions reproducible. The obvious disadvantages are that the distributions are artificially broadened and that real, narrow features can be filtered out of the results. To avoid problems of arbitrary smoothing, it is essential to use an algorithm that selects the regularization parameter in a data-dependent but objective manner. One such algorithm is that of Butler, Reeds, and Dawson.[25] This automatically selects a regularization parameter based on the signal-to-noise ratio of the measurement data to obtain a minimally broadened, but reproducible, inversion.

2.6 Determination of Surface Relaxivity

To relate quantitatively the NMR relaxation rate spectrum to the pore size spectrum, it is necessary to know the coefficient ρ in equations (1) and (2). Determination of the strength of the surface relaxation has proved to be elusive, because it is difficult to obtain independent information on pore size distributions of sedimentary rock.

The first problem is that the connection between a surface-to-volume distribution and a pore size distribution is poorly defined for rocks. The pore space of rocks has been modeled as spheres connected by throats, an interconnected network of tubes, and an interconnected network of sheets. None of these models has universal applicability. Thus the term 'pore size' should be understood to mean a quantity inversely proportional to the surface-to-volume ratio of a locality in the pore space.

Even more importantly, grain surfaces can have fractal character.[26] That is, the apparent surface area depends on the characteristic scale length η of the measurement. Relaxation time measurements also have characteristic scale lengths: the distance a molecule can diffuse in the shortest recovery time of an inversion-recovery measurement of T_1, or before the first spin echo of a CPMG determination of T_2.

Surface area determinations based on low-temperature adsorption of gas, e.g., the BET method,[27] probe the surface at molecular scale and are dominated by the finest features of grain surfaces. Optical microscopy of thin sections has a resolution of about $\eta = 5\,\mu$m: large pores appear to have smooth surfaces and small pores are invisible. It is not unusual to find that the surface-to-volume ratio determined by a combination of BET and buoyancy porosity methods is orders of magnitude larger than that determined by optical imaging measurements. An interesting new technique measures the surface-to-volume ratio by pulsed field gradient NMR measurements of diffusion.[28] The scale length is determined by the distance fluid molecules diffuse during the finite-length gradient pulses that tag the spins; in recent experiments,[29] $\eta = 2\,\mu$m.

The ambiguity in the determination of surface-to-volume ratio results in an ambiguity in the determination of ρ_1 and ρ_2. When the BET method is used to measure surface area and the relaxation time is determined by the initial slope method, for which $\eta < 1\,\mu$m,[21] ρ_1 is found to be in the range 0.1–0.8 μm s^{-1}. The pulsed field gradient diffusion measurements[29] yield ρ_1 in the range 3–40 μm s^{-1}. From optical imaging,[30] ρ_1 is 30–300 μm s^{-1}.

Yet another technique for determining ρ is based on capillary pressure measurements.[27,31] Differential mercury intrusion porosimetry measures a pore size distribution, on the assumption that the pore space is composed of a well-connected network of tubes of various diameters. For many sandstones, the pore size distribution deduced from mercury porosimetry overlays the NMR relaxation time distribution.[32] These comparisons indicate that $\rho_1 = 4\,\mu$m s^{-1} and $\rho_2 = 7\,\mu$m s^{-1}.

The relatively high values of ρ_1 found using the optical microscopy, PFG NMR, and capillary pressure methods reflect the fact that they measure not $(S/V)_{pore}$ but pore size, from which are calculated *effective* relaxivities. Sandstones can have significant amounts of clay or microporosity, which have large surface areas. BET measurements include this surface area while other methods miss it. Since large pore spaces are the dominant passages for fluid flow, the effective relaxivity ρ_{eff} is the appropriate quantity to use to convert relaxation times to pore sizes for many petrophysical applications.

3 PRACTICAL IMPORTANCE OF NMR MEASUREMENTS

3.1 T_1 versus T_2

Relaxation measurements can provide valuable information about rock formations. Unfortunately, measurements of T_1 and T_2 both have serious problems, though the problems are of very different natures. T_2 of fluids in rocks is dominated by molecular diffusion in random internal magnetic field gradients when the Larmor frequency is 10 MHz and above. This makes 'high-field' T_2 difficult to interpret. In contrast, T_1 is generally controlled by the surface relaxation mechanism, because it is unaffected by diffusion in gradients; thus interpretation of T_1 measurements is usually straightforward. However, determination of the longitudinal decay curve requires a time-consuming series of inversion–recovery measurements. For practical reasons, the time available for NMR measurements in an oil well is very limited (see Section 4.1). A CPMG determination of T_2 is much faster, and is therefore preferred.

Fortuitously, the borehole tools are not able to generate high static magnetic fields, and therefore do not generate large internal magnetic field gradients. Static field gradients are either known apparatus constants or are negligible (see Section 4). Under these conditions, transverse relaxation times are readily interpretable. Therefore, most recent work related to borehole NMR has focused on T_2 at frequencies of 2 MHz and below.

3.2 Porosity

Porosity, defined as the fraction of rock volume that can be occupied by fluid, is one of the most important parameters characterizing a hydrocarbon reservoir. In economically significant reservoirs, porosities generally range from 5% to 30%. There are many ways of measuring porosity in situ.[1,2] The speed of compressional sound waves, the scattering of neutrons, and the scattering of gamma rays are commonly measured for this purpose. However, these techniques require knowledge of the mineralogy of the solid matrix to make a determination of porosity.

NMR has the unique capability of measuring fluid-filled porosity without the need to know anything about the solid matrix. The amplitude of a proton NMR measurement is directly proportional to the amount of hydrogen in the material investigated. Protons are present in both oil and water (in approximately equal concentrations), in water associated with clay minerals, and in some matrix minerals such as gypsum (calcium sulfate hydrate). T_2 is sufficiently short in solid matrix materials, on the order of 10 μs, that the signal from protons in the matrix is lost in the deadtime of the borehole instruments. Protons in clay-bound water relax in a millisecond or less, so these generally are lost in the noise of the first one or two CPMG echoes of the borehole tools. On the other hand, relaxation times of protons in the fluids are usually greater than 10 ms, so these protons contribute significantly to the long echo trains collected by the downhole apparatus.

Pore fluids can be further subdivided into capillary-bound fluid and producible fluid. The partition depends on the pressure difference that drives fluids from the rock formation to the wellbore. This pressure is commonly around 0.7 MPa. It has been found that centrifuging sandstones at a differential pressure of 0.7 MPa depletes them of all water protons with $T_1 > 50$ ms.[23] Thus protons with $T_1 > 50$ ms (or $T_2 > 30$ ms) are associated with large pores containing 'unbound' or 'free' fluid. Although the total porosity measurement is independent of rock type, the 50 ms partition between bound and free fluid is merely typical, and depends on the surface relaxivity (see Section 2.2). Nonetheless, the estimation of producible fluid is a unique and valuable capability of NMR logging tools.

3.3 Permeability to Fluid Flow

Hydraulic permeability is one of the most difficult rock properties to estimate, because it depends sensitively on pore sizes and connectivities, which are themselves hard to measure. In simple models of porous media, the permeability is inversely proportional to the square of the surface-to-volume ratio and therefore directly proportional to the square of the NMR relaxation times, a hypothesis that has been substantiated by a large body of laboratory experiments on water-saturated sandstones.[20,33] The correlation between hydraulic permeability and NMR relaxation time appears to depend on limited variability of the surface relaxivity. Attempts to establish similar correlations for carbonates have been less successful, probably because their microgeometry is much more diverse than that of sandstones.

3.4 Measurement of Oil Content

Oil and water are generally found together in reservoir rocks. Many factors affect the NMR response of oil and water mixed in porous media:

1. the relative amounts of each phase;
2. the wettability of the grain surfaces;
3. the surface relaxivity;
4. the bulk relaxation times of each phase;

For References see p. 4968

5. the extent to which the original formation fluids have been replaced by borehole fluid filtrate.

When magnetic field gradients are present (see Section 4.5), the restricted diffusion coefficients of the phases are also a factor.

The water–oil interface does not assist relaxation, so if a fluid is not in contact with grain surface, it relaxes at its bulk liquid relaxation rate. The bulk relaxation time of oil depends on its viscosity, while the bulk relaxation time of water depends primarily on the concentration of paramagnetic ions. The proton density of an oil is generally within 10% of that of water, so its contribution to the relaxation time distribution is simply weighted by concentration.

Most reservoir rocks are thought to be hydrophilic. In such cases, droplets of oil sit in the center of pores and are unaffected by the surface. Therefore they relax at their bulk liquid relaxation time, which can be rather long if the oil has a low viscosity. In many cases, the water in contact with the grains will relax considerably faster, and there can be two distinct peaks in the relaxation time spectrum.[23] Then the amount of oil in the rock can be determined quantitatively.

When the oil is very viscous, its bulk relaxation time is shorter than the tool deadtime, and it will not be detected. Then the NMR porosity is just the water-filled porosity.[34] By subtracting the NMR porosity from a porosity measurement that is insensitive to the material properties of the oil, the quantity of oil can be obtained.

Borehole NMR measurements probe the formation to depths of 10 cm or less. In many cases, original formation fluids are flushed away from the rock near the well bore and replaced by borehole fluids ('invasion' by 'mud filtrate'). In these cases, NMR does not give information about the quantity of original oil in place. It can, however, estimate the amount of oil that will be retained in the formation after the usual recovery processes: water floods that, in fact, resemble invasion by mud filtrate, but on a larger scale. This information is useful in planning enhanced oil recovery projects.

4 INSTRUMENTATION FOR WELL LOGGING

4.1 The Borehole Environment and Field Operations

The borehole environment is unusually harsh. Boreholes drilled to extract oil are typically 20 cm in diameter and 1–10 km deep. The geothermal gradient of the Earth can give rise to temperatures of 175 °C or more, and pressures range to 140 MPa. Borehole logging tools must not only survive but must make quantitative measurements under these conditions. The requirements on electronic components exceed military specifications by a wide margin.

Borehole NMR tools must be rugged enough to survive transport in arctic, tropical, desert, and marine environments, and then a 1 m drop onto a steel surface, which typically produces a shock of 500*g*. They must survive the abrasion resulting from being dragged over kilometers of rough rock face in the well bore. They must comply with laws regulating transport by aircraft and helicopter, which is of particular significance for NMR equipment containing strong permanent magnets. The conditions and space constraints are in many respects more severe than those encountered in the exploration of outer space or the ocean bottom.

During well logging operations, the drilling rig is idle. Since rig charges can be $10 000 to $100 000 per day, oil companies wish to minimize logging time. This poses a significant constraint on practical logging operations. Typically, data might be required over a depth interval of 300 m or more, with a vertical resolution of 50 cm. To be economically viable, the NMR apparatus must move continuously past the formation at a rate of 300 m h^{-1}. This means that it is necessary to measure signal amplitude and a distribution of relaxation times in 6 s, while the sample and apparatus are in motion relative to each other, at low field, and with an unprecedentedly small coil filling factor.

Skilled NMR specialists are not available for logging operations. The borehole NMR tool must be maintained, set up, and operated by field engineers, who also are responsible for other electromagnetic, acoustic, and nuclear radiation instruments that are being run simultaneously. Thus there is a large premium on instruments that can operate autonomously under conditions of continuously changing environmental conditions, including radical excursions in temperature. The NMR and electromagnetic properties of the materials presented to the apparatus also vary radically, on a timescale of seconds.

Logging tools are generally transported to the wellsite on a truck, which also carries 10 km of 7-conductor armored cable ('wireline') used to lower a string of several tools into the borehole. In present practice, the cable carries nothing but 1 kW of power and 500 kilobits s^{-1} of digital telemetry. The power budget for an individual tool is normally around 100 W, which for an NMR tool must be apportioned among transmitter, receiver, auxiliary sensors, and a downhole computer.

Laboratory NMR apparatus consists of a magnet that provides a strong, steady magnetic field that is as uniform as possible, and a radiofrequency coil that produces an oscillating magnetic field perpendicular to the static field direction. A relatively small, compact sample (which may be a human being) is placed inside the coil. It requires a leap of imagination to see how NMR measurements can be made 'inside-out' on a sample external to the apparatus.

4.2 Early Work

It was recognized at an early date that nuclear magnetic resonance could contribute to the in situ investigation of earth formations. Laboratory NMR studies of fluids in rocks, of clays, and in other porous media started in the 1950s at a number of oil company research laboratories, most notably those of Chevron and Mobil. More than three dozen patent applications for borehole NMR devices were filed between 1950 and 1960, representing work sponsored by Chevron, Schlumberger, Mobil, Texaco, and Varian. Efforts accelerated in the 1960s and 1970s. A chronological bibliography[3] is an excellent guide to the early patents and publications.

4.3 Chevron Tool

The first practical NMR borehole logging device did not use magnets and an rf coil, but instead employed free precession in the Earth's magnetic field. It was invented at the Chevron research laboratory,[35] and was subsequently commercialized by Schlumberger as the Nuclear Magnetism Tool (NMT). The NMT contains a large coil of 1000 turns of wire energized by 1 kW direct current; the power consumption prevents other tools from being used at the same time as the NMT. The current produces a spatially inhomogeneous magnetic field in the formation. The current is left on for several seconds, long enough to polarize the proton spins of the formation fluids. The magnetization varies in magnitude and direction from point to point in the earth formation, but over a large volume the polarization is substantially greater than that associated with the Earth's field alone.

The coil current is then turned off rapidly, and the spins precess around the Earth's field direction.[36] A signal is coupled into the same coil used to polarize the spins. A bandpass filter selects the signal from hydrogen, which in any event is by far the largest signal of nuclear origin. The proton Larmor frequency in the Earth's magnetic field varies considerably with geographical location: 2.6 kHz in Canada and 1.0 kHz in Argentina. The free induction decay is often controlled by the T_2 of the rock. It can also be affected by small perturbations in the local Earth's field, such as might be caused by grains of magnetic minerals in the formation.

The deadtime of the NMT, controlled by the decay of coil current after the polarizing pulse, is about 25 ms. Most of the signal from clay-bound and capillary-bound fluid has decayed by then, so the amplitude of the remaining signal, extrapolated back to the turn-off time, is directly proportional to the producible fluid porosity ('free fluid index'). The formation response is spatially weighted and decreases rapidly with distance from the coil. Therefore the tool is calibrated in a large tank of water-saturated sand, representing a uniform formation of known porosity.

The borehole is always full of water or oil. In the absence of precautions, the received signal would be dominated by protons in the borehole, which are both closer and more numerous than those in the surrounding rock formation. To solve this problem, the borehole is 'doped' with magnetite (Fe_3O_4) grains. The magnetic grains rapidly relax protons in the wellbore. Since the grains cannot enter the formation, relaxation of protons in the pore fluid is unaffected.

The tool can make neither continuous wave nor pulsed NMR measurements. Neither T_2* not T_2 is measured with this device. T_1 can be measured by varying the polarization time, and measuring the signal amplitude at subsequent field turn-offs.

4.4 Los Alamos Tool

An 'inside-out NMR' device was designed and built at Los Alamos National Laboratory (LANL);[37] the basic idea is illustrated in Figure 2. The apparatus contains two axially magnetized cylindrical bar magnets. These are placed with like poles facing each other in a cylindrical housing ('sonde'). The radial magnetic field in the transverse symmetry plane is a maximum at a distance from the sonde axis that depends on the radius and spacing of the magnets. A toroidal region of relatively uniform static magnetic field strength is created. A prototype instrument generated a static field of 12 mT in a toroid having a major diameter of 28 cm.

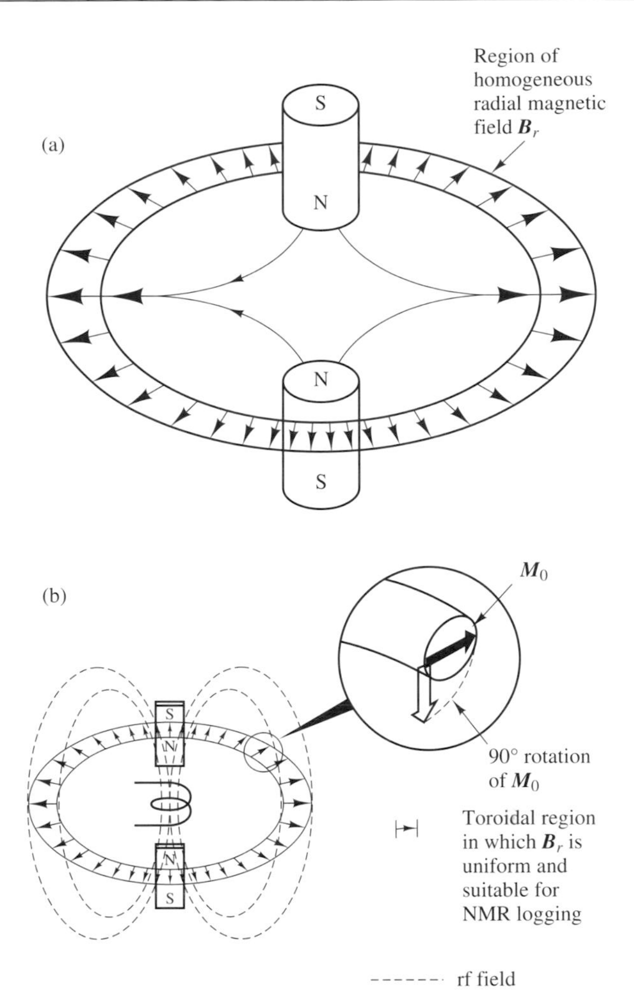

Figure 2 Schematic views of Los Alamos NMR well logging tool. (a) Static field generation. (b) rf field generation. The inset shows a cross section of the toroidal resonance region, where B_0 is perpendicular to B_1. (Reproduced by permission of Academic Press from J. A. Jackson et al., *J. Magn. Reson.*, 1980, **41**, 411)

A radiofrequency coil is placed between the magnets, and irradiates the resonance region. In the resonance region, the rf field is perpendicular to the static magnetic field. When driven by 1 kW peak power, the coil generates an oscillating magnetic field of 0.1 mT in the resonance region. The same coil detects the NMR response of the earth formation.

Unlike the field turn-off scheme of the NMT, the LANL tool is capable of performing a variety of pulsed NMR measurements. Both inversion–recovery and Carr–Purcell–Meiboom–Gill measurements are easy to implement. Pulse lengths are comparable to those of laboratory instruments, and the 200 μs deadtime achieved at Los Alamos is adequate for relaxation measurements on rock formations.

Although the Los Alamos design was never commercialized, it attracted considerable interest in the geophysical community, and was an inspiration for a number of other tool design efforts.

For References see p. 4968

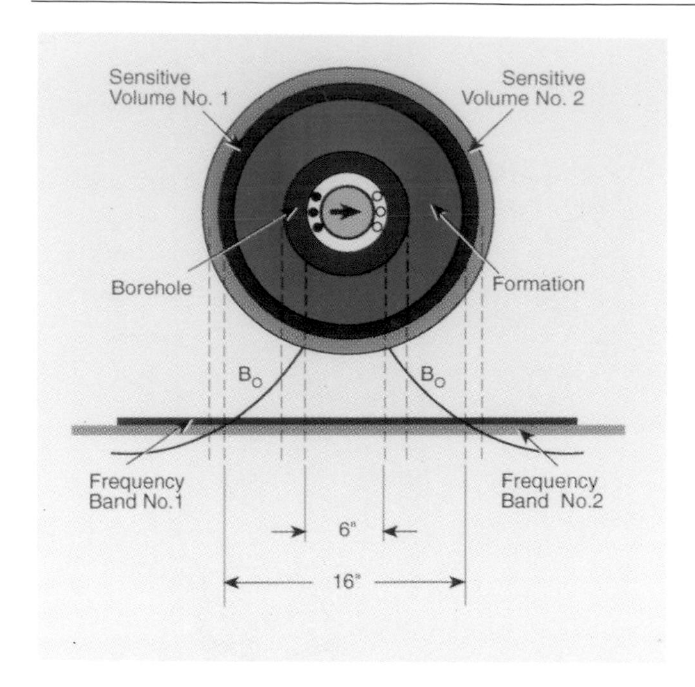

Figure 3 Cross-sectional view of the NUMAR tool. The permanent magnet is magnetized in the direction of the arrow. The tool is centered in the borehole, and the resonance region is a thin cylindrical shell. $\boldsymbol{B}_0$ is perpendicular to $\boldsymbol{B}_1$ everywhere in space. (Courtesy of NUMAR Corp.)

4.5 NUMAR Tool

Shtrickman and Taicher invented another NMR borehole logging tool,[38] and a small company (NUMAR) commercialized the technology. A cylindrical ferrite magnet roughly 1 m long is magnetized perpendicular to its long axis. A cross-sectional view is shown in Figure 3. The field outside the magnet is purely dipolar in the transverse plane; thus the magnitude of $\boldsymbol{B}_0$ is independent of azimuth and depends only on radius.

The radiofrequency field $\boldsymbol{B}_1$ is provided by a coil wound longitudinally around the magnet, so that $\boldsymbol{B}_1$ is also transverse to the tool's long axis. Whereas a metallic magnet would exclude the rf field from its interior, the ferrite magnet allows free penetration of the field, improving efficiency in both transmission and reception. The windings are spaced so as to produce a dipolar field, and so, like $\boldsymbol{B}_0$, the magnitude of $\boldsymbol{B}_1$ is independent of azimuth. Because the dipoles are perpendicular to each other, $\boldsymbol{B}_1$ is perpendicular to $\boldsymbol{B}_0$ at every point in space.

The NUMAR tool is unusual in that it does not produce a region of uniform static magnetic field; to be more precise, there is no place outside the magnet where all spatial derivatives of $\boldsymbol{B}_0$ vanish. The coil may be energized at any frequency: selecting a frequency selects a thin cylindrical shell on which the resonance condition is satisfied. The cylindrical resonant shell has thickness because spin tipping is effective where $B_1 > |\boldsymbol{B}_0(r) - \boldsymbol{B}_0(r_0)|$, where $\boldsymbol{B}_0(r)$ is the static magnetic field at radius r, and r_0 is the radius at which the resonance condition is exactly satisfied. Thus, the larger the B_1, the thicker the shell from which signal is obtained. The resonance frequency is selected to be between 0.5 and 1 MHz, with resonant shell radii $r_0 = 0.23$ and 0.18 m, respectively. $\mathrm{d}B_0(r_0)/\mathrm{d}r = 0.25\,\mathrm{T\,m^{-1}}$, and the shell is 2.5 mm thick. Volume selection allows CPMG sequences to be collected continuously without waiting for longitudinal recovery: the tool hops between two or more frequencies, resonating separate shells during successive sequences.

The applied gradient of $0.25\,\mathrm{T\,m^{-1}}$ is considerably larger than typical internal gradients due to magnetic susceptibility contrast between grains and pore fluids of rocks. Diffusion in the magnetic field gradient is an important contribution to transverse relaxation, and must be taken into account explicitly in understanding the measurements. Since the magnetic field gradient is known, an effective diffusion coefficient can be estimated by varying the CPMG pulse spacing. Extrapolating the pulse spacing to zero allows estimation of the part of T_2 that is due to surface relaxation.

4.6 Schlumberger Tool

Schlumberger has developed a novel tool design for NMR logging.[39] Unlike the other tools, the sonde is pressed against the borehole wall; the length of the wall-engaging section is about 40 cm. A cross sectional view of the Schlumberger logging sonde is shown in Figure 4. Two samarium cobalt magnets are magnetized in the same direction, transverse to the sonde axis. The static magnetic field is predominantly radial into the formation. The figure shows a contour plot of the total field strength in the plane transverse to the borehole axis. The field has a saddle point about 25 mm inside the formation; at this location, all three spatial derivatives of $\boldsymbol{B}_0$ are zero. The field strength in the resonant region is 55 mT, so the Larmor frequency is 2.3 MHz.

The antenna used to radiate the formation at the resonance frequency is placed in a semicircular cylindrical cavity on the face of the sonde. The antenna is essentially a half coaxial

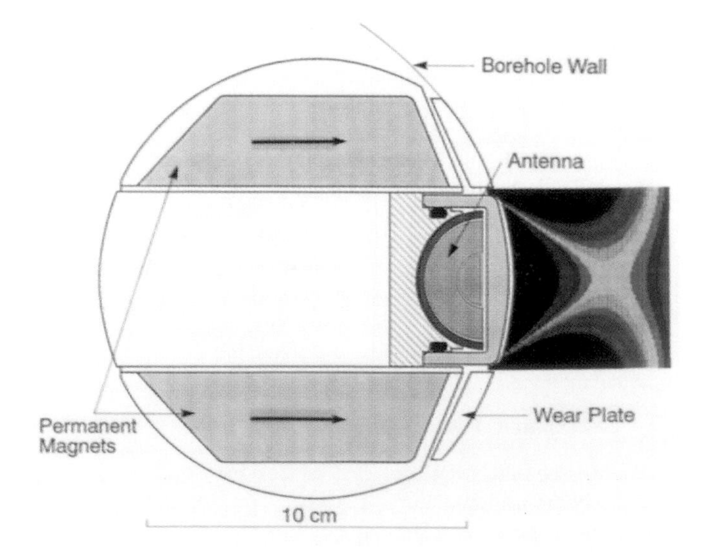

Figure 4 Cross-sectional view of the Schlumberger tool. The permanent magnets are magnetized in the direction of the arrows. The sonde is pressed against the borehole wall, and $\boldsymbol{B}_0$ is directed radially into the formation. A region of relatively homogeneous magnetic field is located inside the formation. Colors code for static field strengths: the resonance region is coded in yellow, stronger fields are coded in greens, weaker fields in reds. The antenna is in a semicircular cavity under a nonmetallic wear plate

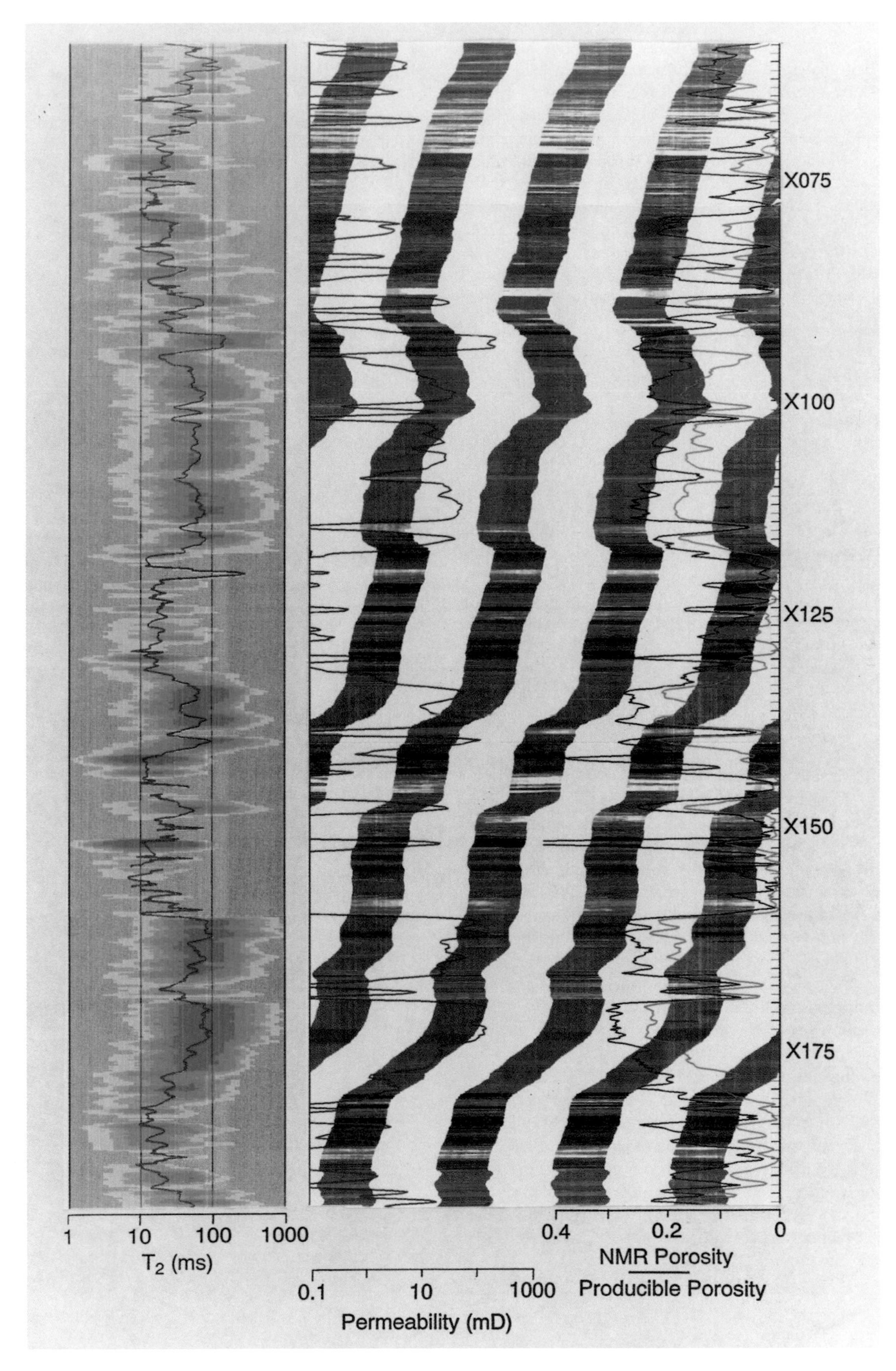

Figure 5 NMR well log. Left panel: A log of the transverse relaxation time distribution is shown in blue, darker color indicating higher amplitude. The red line is the logarithmic mean of T_2. Right panel: The variegated orange and brown background is an unwrapped electrical image of the borehole wall, showing fine statigraphic features. NMR porosity is shown as a black curve, producible fluid porosity is shown as a green curve, and permeability to fluid flow is shown as a blue curve. The curves are explained fully in the text. The depth scale is in feet; the most significant figure has been replaced by X (Reproduced by courtesy of Schlumberger)

cable 15 cm long that irradiates the formation with a field transverse to the static field. It is loaded with ferrite and has a high Q, which is desirable for the reception of weak signals from the nuclear spin system. The field produced by the antenna is not particularly homogeneous over the resonance region; inhomogeneity of $\boldsymbol{B}_1$ is acceptable, so long as error-correcting pulse sequences such as the Carr–Purcell–Meiboom–Gill sequence are used. The antenna resides underneath a wear plate, which is the only nonmetallic surface on an otherwise all-metal sonde. Magnetoacoustic ringing of metal structures, which would increase the deadtime of the instrument, is suppressed.

4.7 Log Example

A well log consists of the readings of one or more borehole logging tools plotted as a function of depth. Figure 5 is an example of an NMR well log. The vertical scale is depth, marked in feet. The left panel is a presentation of the spectrum of transverse relaxation times; the horizontal scale is T_2 in milliseconds. Higher amplitude is coded as darker color. This is an indicator of pore size distribution: for example, high amplitude at short relaxation times indicates a formation with relatively small pores. The red line is the logarithmic mean of T_2, $T_{2\text{LM}}$:

$$T_{2\text{LM}} = \exp\langle\log T_2\rangle = \exp\left(\frac{\sum_i m_i \log T_{2i}}{\sum_i m_i}\right) \qquad (7)$$

On the right panel is an electrical image of the borehole created by the Formation Microscanner (FMS). The FMS measures the local electrical conductivity using a number of small electrodes placed in contact with the borehole wall by retracting arms. Darker colors indicate high electrical conductivity, usually associated with the presence of saline water or clay. Lighter colors indicate low conductivity, and suggest low porosity or the presence of oil. The resulting electrical image of the borehole is unwrapped: the 0° azimuth is at the left edge of the panel and the 360° azimuth is at the right. The blank stripes represent areas of the borehole wall that are not in contact with the four arms of the tool. The stripes change azimuth slowly because the tool rotates as it is pulled up.

The NMR logs overlay the FMS image. By convention, porosity is plotted on the right-hand side of the panel and increases to the left, while hydraulic permeability is plotted on the left-hand side of the panel and increases to the right. The total NMR porosity is shown in black and the producible fraction of the porosity in green. Zones with small pores have low producible porosity, even when the total porosity is high. The permeability to fluid flow (measured in millidarcies, mD) is computed using the empirical equation

$$k = (2.25 \times 10^6 \text{ mD s}^{-2})\phi^4 T_{2\text{LM}}^2 \qquad (8)$$

where ϕ is the porosity. The permeability is plotted on a logarithmic scale, and changes by orders of magnitude from one zone to the next. In this powerful combination of measurements, the FMS gives a high-resolution stratigraphic image of the subsurface, while NMR provides quantitative producibility information.

For list of General Abbreviations see end-papers

5 RELATED ARTICLES

Diffusion in Porous Media; Microporous Materials & Xenon-129 NMR; Oil Reservoir Rocks Examined by MRI; Terrestrial Magnetic Field NMR.

6 REFERENCES

1. J. R. Hearst and P. H. Nelson, 'Well Logging for Physical Properties', McGraw-Hill, New York, 1985.
2. D. V. Ellis, 'Well Logging for Earth Scientists', Elsevier, New York, 1987.
3. J. A. Jackson and M. Mathews, *Log Anal.*, 1993, **34**(3), 35.
4. C. Chardaire-Riviere and J.C. Roussel, *Rev. Inst. Fr. Pet.*, 1992, **47**, 503.
5. W. E. Kenyon, *Nucl. Geophys.*, 1992, **6**, 153.
6. *Magn. Reson. Imaging*, 1991, **9**(5), entire issue; *Magn. Reson. Imaging*, 1994, **12**(2), entire issue.
7. F. Bloch, *Phys. Rev.*, 1951, **83**, 1062.
8. R. J. S. Brown and I. Fatt, *Petrol. Trans. AIME*, 1956, **207**, 262.
9. M. H. Cohen and K. S. Mendelson, *J. Appl. Phys.*, 1982, **53**, 1127.
10. K. R. Brownstein and C. E. Tarr, *Phys. Rev. A*, 1979, **19**, 2446.
11. L. L. Latour, R. L. Kleinberg, and A. Sezginer, *J. Colloid Interface Sci.*, 1992, **150**, 535.
12. J. Korringa, D. O. Seevers, and H. C. Torrey, *Phys. Rev.*, 1962, **127**, 1143.
13. F. D'Orazio, S. Bhattacharja, W. P. Halperin, K. Eguchi, and T. Mizusaki, *Phys. Rev. B*, 1990, **42**, 9810.
14. R. L. Kleinberg, W. E. Kenyon, and P. P. Mitra, *J. Magn. Reson., Ser. A*, 1994, **108**, 206.
15. L. L. Latour, P. P. Mitra, R. L. Kleinberg, and C. H. Sotak, *J. Magn. Reson., Ser. A*, 1993, **101**, 342.
16. P. Le Doussal and P. N. Sen, *Phys. Rev. B*, 1992, **46**, 3465.
17. R. J. S. Brown and P. Fantazzini, *Phys. Rev. B*, 1993, **47**, 14 823.
18. E. J. Fordham, A. Sezginer, and L. D. Hall, *J. Magn. Reson., Ser. A*, 1995, **113**, 139.
19. A. Timur, *J. Petrol. Tech.*, 1969, **21**, 775.
20. W. E. Kenyon, P. I. Day, C. Straley, and J. F. Willemsen, *Soc. Petrol. Eng. Form. Eval.*, 1988, **3**, 622; erratum: *Soc. Petrol. Eng. Form. Eval.*, 1989, **4**, 8.
21. W. P. Halperin, F. D'Orazio, S. Bhattacharja, and J. C. Tarczon, in 'Molecular Dynamics in Restricted Geometries', eds. J. Klafter and J. M. Drake, Wiley, New York, 1989, Chap. 11.
22. K. R. McCall, D. L. Johnson, and R. A., Guyer, *Phys. Rev. B*, 1991, **44**, 7344.
23. C. Straley, C. E. Morriss, W. E. Kenyon, and J. J. Howard, in 'Transactions of the SPWLA 32nd Annual Logging Symposium, 1991', Paper CC.
24. D. P. Gallegos and D. M. Smith, *J. Colloid Interface Sci.*, 1988, **122**, 143.
25. J. P. Butler, J. A. Reeds, and S. V. Dawson, *SIAM J. Numer. Anal.*, 1981, **18**, 381.
26. A. H. Thompson, *Annu. Rev. Earth Planet. Sci.*, 1991, **19**, 237.
27. S. Lowell and J. E. Shields, 'Powder Surface Area and Porosity', Chapman and Hall, New York, 1984.
28. P. P. Mitra, P. N. Sen, L. M. Schwartz, and P. Le Doussal, *Phys. Rev. Lett.*, 1992, **68**, 3555.
29. M. D. Hurlimann, K. G. Helmer, L. L. Latour, and C. H. Sotak, *J. Magn. Reson. Ser. A*, 1994, **111**, 169.
30. J. J. Howard, W. E. Kenyon, and C. Straley, *Soc. Petrol. Eng. Form. Eval.*, 1993, **8**, 194.
31. J. D. Loren and J. D. Robinson, *Soc. Petrol. Eng. J.*, 1970, **10**, 268.

32. C. E. Morriss, J. MacInnis, R. Freedman, J. Smaardyk, C. Straley, W. E. Kenyon, H. J. Vinegar, and P. N. Tutunjian, in 'Transactions of the SPWLA 34th Annual Logging Symposium, 1993', Paper GGG.
33. G. C. Borgia, G. Brighenti, P. Fantazzini, G. D. Fanti, and E. Mesini, *Soc. Petrol. Eng. Form. Eval.*, 1992, **7**, 206.
34. C. H. Neuman and R. J. S. Brown, *J. Petrol. Tech.*, 1982, **34**, 2853.
35. R. J. S. Brown and B. W. Gamson, *Petrol. Trans. AIME*, 1960, **219**, 199.
36. A. Abragam, 'Principles of Nuclear Magnetism', Clarendon Press, Oxford, 1961, p. 64–65.
37. R. K. Cooper and J. A. Jackson, *J. Magn. Reson.*, 1980, **41**, 400; L. J. Burnett and J. A. Jackson, *J. Magn. Reson.*, 1980, **41**, 406; J. A. Jackson, L. J. Burnett, and J. F. Harmon, *J. Magn. Reson.*, 1980, **41**, 411.
38. Z. Taicher, G. Coates, Y. Gitartz, and L. Berman, *Magn. Reson. Imaging*, 1994, **12**, 285.
39. R. L. Kleinberg, A. Sezginer, D. D. Griffin, and M. Fukuhara, *J. Magn. Reson.*, 1992, **97**, 466.

Biographical Sketch

Robert L. Kleinberg. *b* 1949. B.S., 1971, chemistry, University of California, Berkeley. Ph.D., 1978, physics, University of California, San Diego. Postdoctoral physicist, Exxon Research, 1978–80. Schlumberger-Doll Research, 1980–present. Approx. 50 publications, 10 US patents. Research specialties: borehole instrumentation (ultrasonics, electrical resistivity, nuclear magnetic resonance, gravimetry) and NMR properties of rocks and other porous media.

White Matter Disease MRI

Donald M. Hadley

Institute of Neurological Sciences, Glasgow, UK

1 INTRODUCTION

1.1 Characteristics

The white matter of the brain constitutes the core of the hemispheres, brainstem, and cerebellum. It is composed of axons, which transmit chemically mediated electrical signals, and glial supporting cells set in a mucopolysaccharide ground substance. The glial cells—oligodendrocytes, astrocytes, ependymal cells, and microglia—account for about half the brain's volume and 80–90% of its cells. The oligodendrocytes provide an insulating sheath of myelin by invagination, wrapping concentric layers of their cell membrane around the axons. Astrocytes are now known to influence and communicate through their long foot processes, which are in intimate contact with capillaries, neurones, synapses, and other astrocytes. The ependymal cells form the lining of the brain's internal cavities, while the microglia, normally relatively inconspicuous, are capable of enlarging and becoming active macrophages.

The white matter fibers are grouped into location-specific tracts, which can be divided into three main types:

(a) projection fibers that allow efferent and afferent communication between the cortex and target organ;
(b) long and short association fibers, which connect cortical regions in the same hemisphere; and
(c) commissural fibers, which connect similar cortical regions between hemispheres.

The formation and maturation of axons has been reviewed by Barkovich et al.[1] After development of the axons and their synapses, the final process of myelination occurs. This is crucial to the appearance on MRI.

1.2 Evolution and Imaging

The contrast obtained between gray and white matter is largely due to the myelination of the white matter tracts. Myelin is composed of a bilayer of lipids (phospholipids and glycolipids), cholesterol, and large proteins. In 1974, Parrish et al.[2] showed differences in the relaxation times of gray and white matter by spectroscopy before imaging was possible. These differences were later confirmed by MRI. In white matter, the

For References see p. 4981

myelin lipids themselves contain few mobile protons visible to routine MRI, but they are hydrophobic, and therefore, as myelination progresses, there is loss of brain water and a decrease in T_2 signal. Cholesterol tends to have a short T_1, and the increased protein also decreases the T_1 of water. This results in white matter having a reduced intensity on T_2-weighted images and increased intensity on T_1-weighted images compared with unmyelinated fibers or gray matter.

Myelination of the white matter is first noted in the cranial nerves during the fifth fetal month, and continues throughout life.[3] By birth, myelination is present in the medulla, dorsal midbrain, cerebellar peduncles, posterior limb of the internal capsule, and the ventrolateral thalamus. In general, myelination is completed from caudal to cephalad, from central to peripheral, and from dorsal to ventral. Key landmarks include the pre- and post-central gyri, which are myelinated at one month, with the motor tracts completed by three months. At this time, myelination is completed in the cerebellum, and progresses through the pons in the corticospinal tracts, cerebral peduncles, the posterior limb of the internal capsule, and up to the central portion of the centrum semiovale, to be completed by six months. The optic radiations and the anterior limb of the internal capsule are myelinated by three months. Myelination in the subcortical white matter is first noted in the occipital region at three months, and proceeds rostrally to the frontal lobes. This posterior-to-anterior maturation is noticeable in the corpus callosum, with the splenium first showing myelin at four months, progressing to the genu, where it is complete at six months.

This normal development is best visualized by MRI with age-related heavily T_1-weighted (e.g., inversion–recovery) sequences for the first six months, by which time the appearance is close to adult; after this, T_2-weighted sequences are most helpful, with all the major tracts assuming an adult appearance by 18 months. The cause of these differences is not fully understood, but is thought to be related to the initial hydrophilicity, with its associated increase in hydrogen bonding with water. Next, the T_2 shortening may be caused by the subsequent tightening of the myelin sheath, further redistributing the free water components. It has been shown that the very earliest changes of myelination are shown better by diffusion- or fluid-attenuated T_2-weighted sequences than by conventional T_1- or T_2-weighted sequences.[4,5] It must be noted, however, that some areas around the trigones of the lateral ventricles may not fully myelinate in normal children until they are 10 years old.

1.3 Classification of Abnormalities

There are a bewildering number of white matter diseases with multiple etiologies and pathological mechanisms. Although MRI is very sensitive to any white matter abnormality, it is rarely possible for the radiologist to make a specific diagnosis.[6] It is, however, useful to divide them into three main groups:

(a) a dysmyelinating group in which there is a biochemical defect in the production or maintenance of normal myelin; some of the individual enzyme deficiencies have been identified and will be discussed below;
(b) a demyelinating group in which myelin is formed normally but is later destroyed,
C(c) a vascular group in which normally myelinated white matter is destroyed by a critical reduction in blood flow to a particular region; this may also involve the adjacent gray matter or a large segmental part of the brain, depending on the extent and severity of the reduction in blood flow and the susceptibility of the cells involved.

In many of these conditions, the diagnosis is biochemical, but the radiologist has an important role in suggesting the diagnosis, and documenting progression, response to therapy, or complications.

2 DYSMYELINATION DISEASE

2.1 Leukodystrophies

These are diseases where dysmyelination occurs as a result of the production and maintenance of abnormal myelin. Becker[7] and Kendall[8] have produced excellent reviews of this subject.

2.1.1 *Alexander's Disease (Fibrinoid Leukodystrophy)*

This usually presents in the first few weeks of life with macrocephaly and failure to attain developmental milestones. There is progressive spastic quadriparesis and intellectual failure. Death ensues in infancy or early childhood, although cases have been reported in adolescents and adults. An enzyme defect has not yet been identified.

MRI shows increased T_1 and T_2 relaxation times, starting in the frontal lobes, and progressing to the parietal and capsular regions.[9,10] With the accumulation of Rosenthal fibers around blood vessels, there may be disruption of the blood–brain barrier, producing frontal periventricular enhancement. Frank cystic changes develop in the frontal lobes in the later stages, with atrophy of the corpus callosum.

2.1.2 *Canavan's Disease (Spongiform Leukodystrophy)*

This is a lethal autosomal recessive neurodegenerative disorder of Jewish infants caused by a deficiency of aspartoacylase. The disease progresses with marked hypotonia, macrocephaly, seizures, and failure to attain motor milestones in the first few months of life, although sometimes it starts as early as a few days, progressing to spasticity, intellectual failure, and optic atrophy. Death usually occurs in the second year of life. The radiological features may be seen before the full clinical picture has developed, but the diagnosis depends on the biochemical testing.

The demyelinated white matter shows increased T_1 and T_2 relaxation times, preferentially in the arcuate U fibers of the cerebral hemispheres. The occipital lobes are more involved than the frontal, parietal, and temporal lobes. Initially, it may spare the corpus callosum, deep white matter, and internal and external capsules, but, as it progresses, diffuse white matter involvement occurs, which leads to eventual cortical atrophy.[11]

2.1.3 *Krabbe's Disease (Globoid Cell Leukodystrophy)*

This is a rare, lethal, autosomal recessive leukodystrophy (locus now mapped to chromosome 14) due to a deficiency of the first of the two galactocerebroside β-galactosidases. This arrests the normal breakdown of cerebroside, disrupts the turnover of myelin, and results in the accumulation of galactosylsphingosine. This is toxic to oligodendrocytes, and causes a marked loss of myelin, although the minute amount of myelin remaining is normal. Changes become evident between one and six months old, although it is occasionally noted earlier, and leads to death within one to three years. The clinical diagnosis is based on an assay of β-galactosidase from leukocytes or skin fibroblasts.

Early in the disease process, MRI can be normal[12] and a spectrum of lesions then develops over several months. These are nonspecific symmetrical patchy changes in the periventricular white matter similar to many other demyelinating diseases such as multiple sclerosis, with increased T_1 and T_2 signals.[8,13] The thalami, central white matter, and cerebellar white matter may show decreased T_1 and normal or slightly decreased T_2 signals.[14] These changes are reflections of the increased attenuation sometimes seen on computerized tomography (CT), and are probably the result of paramagnetics such as crystalline calcification. In advanced disease, there is diffuse cerebral atrophy.[12]

2.1.4 *Pelizaeus–Merzbacher Disease*

This term has been used to cover the five subtypes of sudanophilic leukodystrophy,[15] but here it will be taken to mean the slowly progressive X-linked recessive leukodystrophy. The dysmyelination is now thought to be due to a point mutation in the PLP gene coding for the myelin–protein proteolipid protein. It presents in infancy, and runs a very chronic course leading to death in adolescence or early adulthood.

On MRI, there is a general lack of myelination without white matter destruction. The brain has the appearance of the newborn, with high signal intensity only appearing in the internal capsule, optic radiations, and proximal corona radiata on T_1-weighted images, and practically no low signal in the supratentorial region on T_2-weighted sequences.[16] A 'tigroid' pattern consisting of normal myelinated white matter within diffuse dysmyelination can be seen later on T_2-weighted sequences. When severe, there may be a complete absence of myelin. Cortical sulcal enlargement may be seen.

2.1.5 *Metachromatic Leukodystrophy*

The commonest of the sphingolipidoses is due to a deficiency in the activity of arylsulfatase A. This enzyme is responsible for normal metabolism of sulfatides, which are important constituents of the myelin sheath. The disease is subdivided into:

(a) neonatal, with a rapid downhill course leading to early death;
(b) infantile, presenting between one and four years with polyneuropathy, ataxia, progressive retardation, and spastic tetraparesis;
(c) juvenile, with dementia[17] and behavioral disorders progressing to spastic tetraparesis;
(d) the rare adult type, presenting at any age with dementia and spastic paraparesis.

The imaging findings are nonspecific, with symmetrical areas of increased T_1 and T_2 relaxation times in the centrum semiovale, representing progressive dysmyelination and gliosis within areas of normal myelination. The peripheral white matter, including the arcuate U fibers, is spared until late in the disease. As there is no inflammation, enhancement is not a feature. These appearances allow differentiation from the gross lack of myelination seen in Pelizaeus–Merzbacher disease. As the disease progresses, brain atrophy becomes more prominent than the white matter signal changes. Proton MRS may have a clinical role in the diagnosis.[18]

2.1.6 *Adrenoleukodystrophy (Childhood Type: X-Linked)*

This is seen exclusively in males, and was thought to be due to a deficiency of acyl-CoA synthetase. Long-chain fatty acids are incorporated in cholesterol esters, replacing the normal nonesterified cholesterol. It usually presents between the ages of five and ten years, with a disturbance of gait and intellectual impairment, with fairly rapid progression and the development of hypotonia, seizures, visual impairment, and bulbar symptoms. Neurological complaints are classically preceded by adrenal insufficiency and skin pigmentation, which may be precipitated by an intercurrent infection; however, they sometimes may never appear.

Symmetrical long T_1 and T_2 signals are usually first seen in the peritrigonal regions extending into the splenium of the corpus callosum. Although typical, these may rarely be seen in other white matter diseases such as Krabbe's. These signal changes gradually extend to involve the occipital lobes and more anterior regions such as the medial and lateral geniculate bodies, thalami, and the inferior brachia. The pyramidal tracts and the occipito-temporo-parieto-pontine fibers show progressive alteration, with sparing of the fronto-pontine fibers in the medial part of the crus cerebri. The lateral lemnisci and the cerebellar white matter may also become involved. Although this is the usual pattern, atypical symmetrical or asymmetrical involvement of other lobes sometimes occurs[19] (Figure 1).

Three zones of abnormal long T_1 and T_2 signals can be recognized:

(a) a central region of gliosis with necrosis and cavitation, next to
(b) an intermediate region of active inflammatory demyelination that shows enhancement due to blood–brain barrier breakdown, surrounded by
(c) a peripheral less marked zone of demyelination without inflammatory reaction.

As the disease progresses, atrophy becomes the more dominant feature on MRI.[8]

2.1.7 *Adrenoleukodystrophy (Adrenomyeloneuropathy—Adult Type)*

This often presents in the same family as the childhood type, but occurs in adult life. It is caused by a similar enzyme defect. The abnormal myelination is most marked in the corticospinal and spinocerebellar tracts, but can extend into the brainstem, involving the pyramidal tracts running into the posterior limb of the internal capsules, and the frontopontine and occipito-temporo-parieto-pontine fibers. The cerebellar white matter is usually affected, with sparing of the cerebral white matter. MR changes usually appear late in the course of the disease.

2.1.8 *Adrenoleukodystrophy (Neonatal)*

Several disorders have been grouped under this heading, but with active research at present underway, the classification of

For References see p. 4981

the enzyme defects may change. Severe progressive neurological impairment occurs, with psychomotor retardation, dysmorphic facial features, hypotonia, seizures, and defective liver function. In contradistinction to the childhood type, these abnormalities are present from birth. The enzyme defect may be confined to fatty acyl-CoA oxidase resulting in defective very long chain fatty acid oxidation.

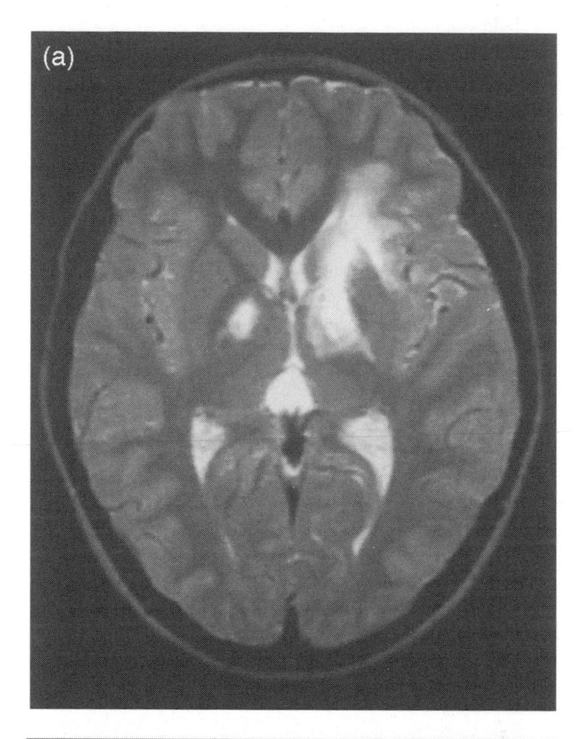

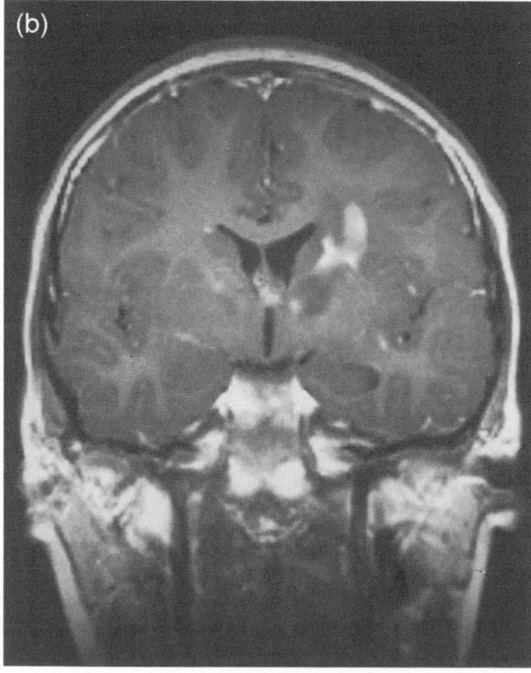

Figure 1 Adrenoleukodystrophy: (a) T_2-weighted and (*b*) gadolinium-enhanced T_1-weighted sections of a five-year-old boy showing bilateral focal areas of white matter abnormality with marginal enhancement (biopsy-proven)

There is diffuse degeneration of cerebral white matter, causing atrophy at a very early age. Progressive MRI changes have been described[20] in a single case followed for three years, with delayed myelination followed by symmetrical demyelination of the corona radiata, optic radiations, and pyramidal tracts.

2.1.9 Phenylketonuria

This is an autosomal recessive metabolic encephalopathy due to a defect in phenylalanine hydroxylase conversion of phenylalanine to tyrosine, resulting in hyperphenylalaninemia. Strict dietary restrictions must be maintained from early infancy to prevent profound mental retardation. Other cofactor defect variants may result in lesser or greater degrees of encephalopathy.[21] When there is a defect of dihydropteridine reductase, there are severe neurological and cognitive abnormalities in spite of adequate dietary restrictions. Severe white matter changes have been noted,[22] with cystic degeneration and loss of parenchyma.

Subtle abnormalities of white matter have been shown by MRI in older children and adults who have classical phenylketonuria despite having maintained a degree of dietary restriction.[23,24] This possibly provides some evidence for continuing the restrictive diet and phenylalanine-free protein supplements and into adulthood.[25]

Varying degrees of periventricular white matter abnormality have been shown, with focal and diffuse lengthening of T_1 and T_2 relaxation times most easily seen on the T_2-weighted images (Figure 2). In some studies, these changes were found to correlate loosely with the adequacy of the reduction and

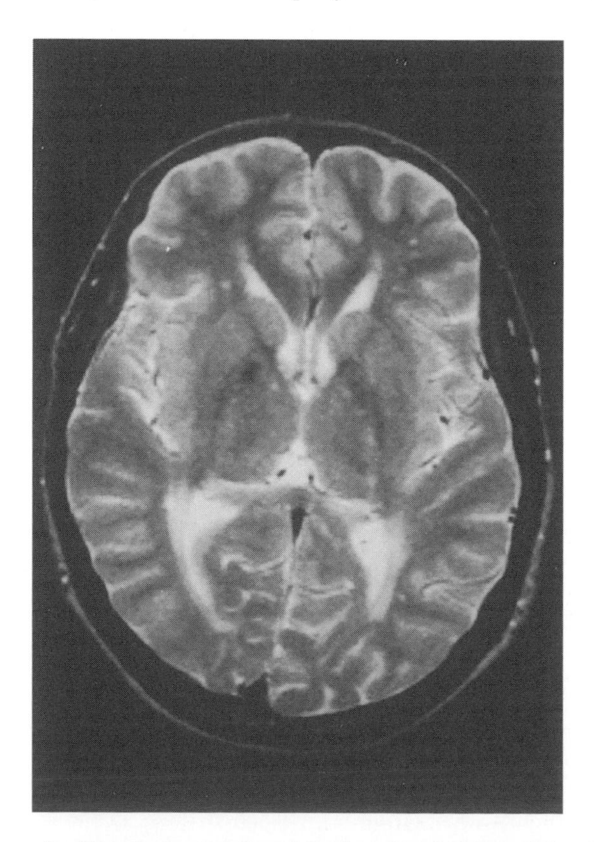

Figure 2 Phenylketonuria: T_2-weighted section showing subtle white matter hyperintensities in the optic radiations in spite of apparently adequate dietary control in a 13-year-old

maintenance of serum phenylalanine levels,[25,26] while other workers showed no clear relationship.[24]

2.2 Miscellaneous

Over 600 individual dysmyelination disorders that may affect MRI appearances have been identified in childhood alone. Continuing research is progressively isolating the individual enzyme or gene defects, which will in time allow more specific classification. Meanwhile, the following less well defined groups of disorders will be considered.

2.2.1 Neurodegenerative Disorders

These occur in a number of devastating developmental disorders of childhood, which are either congenital or acquired. Clinical findings are usually nonspecific, and laboratory tests have to be selected carefully. Imaging demonstrates the results of abnormal cellular function on parenchymal morphology. MRI is sensitive, but specificity is limited, and it must be integrated with the other clinical findings. Proton MRS may be able to detect abnormal metabolite levels and allow an earlier and more specific determination of neurodegeneration.[27]

2.2.2 Lysosomal Disorders

Several of these have been mentioned under specific enzyme defects above, such as metachromatic leukodystrophy and Krabbe's disease. All lack activity of a specific lysosome enzyme, which is inherited in an autosomal recessive manner. Abnormal materials build up in the lysosomes. The CNS is affected directly or secondary to the metabolic abnormality in adjacent structures.

Dysmyelination is shown as an increase in T_1 and T_2 relaxation times in the white matter, with a variable degree of involvement of the arcuate U fibers. In Fabry's disease, involvement of the small arteries may cause multifocal small infarcts visible on MRI. The gangliosidosis in addition may show focal decreases of T_2 relaxation time in the thalami, possibly reflecting the calcification seen on CT.[7,8]

2.2.3 Peroxisomal Disorders

These relate to deficiencies in the activity of respiratory enzymes and organelles of most cells. In most of these diseases, the CNS is involved. Adrenoleukodystrophy has already been discussed above, but there are multiple other rarer diseases that belong to this classification. In general, they produce dysmyelination. Several also show disturbances of neuronal migration. In some, there is an additional inflammatory response.

2.2.4 Mitochondrial Encephalopathies

These group of disorders are characterized by functionally or structurally abnormal mitochondria in the CNS or muscle. They are transmitted by non-Mendelian maternal inheritance, resulting in slowly progressive multisystem diseases with a wide range of clinical presentations, usually appearing in childhood but showing considerable variability depending on their severity.

Imaging is nonspecific.[28] There is diffuse but variable white matter atrophy and lengthened T_1 and T_2 signals in the basal ganglia. Focal infarcts may also be seen (Figure 3). Abnormal metabolites, including lactate, have been shown by MRS in the

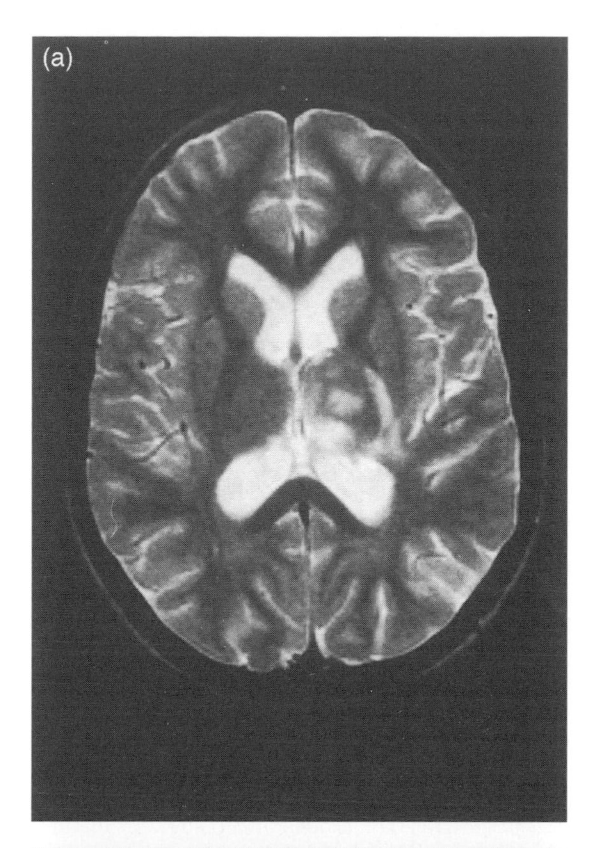

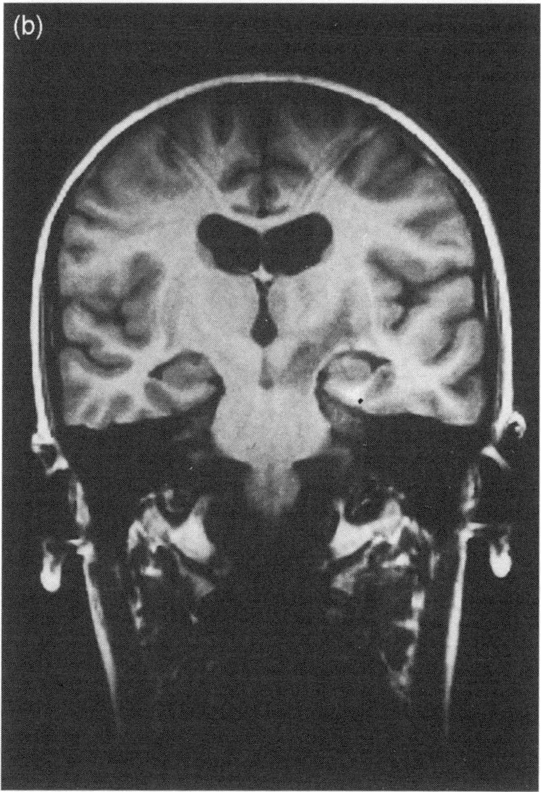

Figure 3 Mitochondrial encephalopathy: (a) T_2-weighted and (b) T_1-weighted sections in a nine-year-old boy showing increased T_1 and T_2 relaxation times representing focal infarction in the posterior limb of the left internal capsule and thalamus

For References see p. 4981

brain lesions. This is thought to be due to impaired aerobic metabolism of pyruvate.[29,30]

In Leigh's disease, there may be spongy degeneration with astroglial and microglial reaction, with vascular proliferation affecting the basal ganglia, brainstem, and spinal cord. Cerebellar and cerebral white matter undergoes demyelination, with preservation of the nerve cells and axons resulting in hyperintensity and hypointensity on T_2- and T_1-weighted images, respectively.

3 DEMYELINATION DISEASE

3.1 Idiopathic

3.1.1 Multiple Sclerosis

Multiple sclerosis (MS) is an idiopathic inflammatory and demyelinating disorder of the central nervous system (CNS). The definitive clinical and pathological features of the disease were established by Charcot[31] as long ago as 1868. Since then, the disorder's characteristics have been refined, with improvements in imaging giving the most recent insights into its pathophysiology. It is now one of the commonest reasons given for requesting MRI in the northern latitudes of the Western world, and this diagnosis has huge social and economic consequences. It is therefore considered in some depth in the following paragraphs.

Clinically, MS usually follows a fluctuating course, with symptoms varying from paroxysmal and brief to slowly progressive and chronic. The lesions affect single or multiple sites simultaneously, usually involving long white matter tracts, but clinical–pathological correlation is often poor. The disorder leads to visual loss, numbness, tingling in the limbs, spastic weakness, and ataxia.[32] Diagnosis is allowed when a combination of signs and symptoms localize lesions in separate and distinct areas of the CNS disseminated in time and space.[33] Supportive laboratory and imaging data can now be defined for research studies, dividing the disease into clinically definite and probable MS, with or without laboratory support.[34,35]

Although some CT studies using high-dose iodine-enhanced delayed imaging have reported sensitivities as high as 72% when patients are in an acute relapse,[36] generally MRI has proved to have considerably greater sensitivity, and, by using intravenous gadolinium-based contrast agents, can separate acute from subacute and chronic lesions.[37]

Initially, MS lesions were shown at low field on T_1-weighted (inversion–recovery) sequences,[38] but within four years several rigorously controlled studies demonstrated the effectiveness of spin echo sequences where the T_2-dependent contrast can be organized to maximize the sensitivity between normal and abnormal tissue while minimizing partial volume effects between cerebrospinal fluid (CSF) and adjacent lesions.[39,40] Although no single sequence will detect all lesions, multifocal supratentorial white matter abnormalities have been shown on moderately T_2-weighted images in 96.5% of a group of 200 consecutive patients with clinically definite MS.[41] One or more periventricular lesions were seen in 98%, and lesions discrete from the ventricle in 92.5%, with cerebellar lesions in just over half of the group. Normal scans were found in 1.5% of patients. The majority of these lesions were clinically silent; therefore MRI can produce the extra information that helps to fulfill the criteria of dissemination in space (Figure 4). It can also exclude other causes of the patients' signs and symptoms, such as Arnold Chiari malformations and spinocerebellar degeneration. Serial studies with careful repositioning may also fulfill the criterion of dissemination in time by showing new, often asymptomatic, lesions[42,43] This high sensitivity has now been confirmed by many workers.[36,42,44]

The T_1 of apparently normal white matter in patients with MS may be increased,[42,45] and an apparent increase in the iron content has been found in the thalamus and striatum at high field.[46] Post-mortem studies have confirmed that the long T_2 lesions found correspond to MS plaques.[42] This sensitivity makes MRI the most appropriate modality for examining a patient with suspected MS.

Unfortunately, these multifocal white matter lesions may be indistinguishable from other conditions that produce demyelination, gliosis, or periventricular effusions[47] (Table 1), and between 5 and 30% of apparently normal controls older than 50 years have been shown to have white matter lesions, probably due to asymptomatic cerebrovascular disease. The patient's age and pattern of lesions can help to improve the specificity of the MRI examination.[48] Fazekas et al.[49,50] have shown that if at least three areas of increased T_2 signal intensity are present with two of the following features—abutting the body of the lateral ventricles, infratentorial location, and size greater than 5 mm—then the sensitivity is decreased to 88%, but the specificity increases to 96% in patients with clinically definite multiple sclerosis. Although the relaxation times of acute plaques have been found to be longer,[51] the range was wide, and the age of an individual lesion could not be determined by T_1 or T_2 measurement alone. It is important to be able to define new and active lesions to differentiate between multiphasic (MS) and monophasic (ADEM) disease and to determine whether there is evidence of continuing disease progression (e.g., in clinical therapeutic trials).

The areas of perivenular inflammation and edema associated with the acute MS plaque[52] cause a transient disruption of the blood–brain barrier[53,54] and allow leakage of intravascular contrast agents. This is shown as enhancement on T_1-weighted images (Figure 5), and is safer and more effective than high-dose delayed contrast-enhanced CT.[53] Correlation with the lesions seen on T_2-weighted sections is good. Only a small number of cortical or subcortical plaques were seen solely on enhanced T_1-weighted images. Enhancement is now considered a consistent feature of recognizably new lesions or new parts of existing plaques,[54] although occasionally blood–brain barrier breakdown develops in older previously nonenhancing plaques associated with no increase in their size. The inflammatory demyelination has been shown pathologically to begin in a perivenular distribution and spread centrifugally, corresponding with the ring enhancement noted in several studies[37,54] (Figure 5). Gadolinium enhancement is particularly useful in the clear delineation of lesions in the spinal cord and optic nerves, especially if T_1-weighted fat saturation chemical shift sequences are used to reduce the high signal from surrounding periorbital fat.[55,56]

Differences in the enhancement pattern of primary and secondary progressive MS, the two major clinical patient groups, have been identified.[57] The secondary progressive group had more new lesions (18.2 lesions per patient per year), of which a larger proportion (87%) enhanced. In addition, there was enhancement at the edge of preexisting lesions. This compares with few new lesions (3.3 lesions per patient per year), of which only one enhanced in the primary progressive group at the time when there were no differences over six months in the rates of clinical deterioration between the two groups. This suggests a difference in the underlying dynamics of the inflammatory component of the disease. With improved imaging techniques, e.g., fast scanning methods, and particularly with

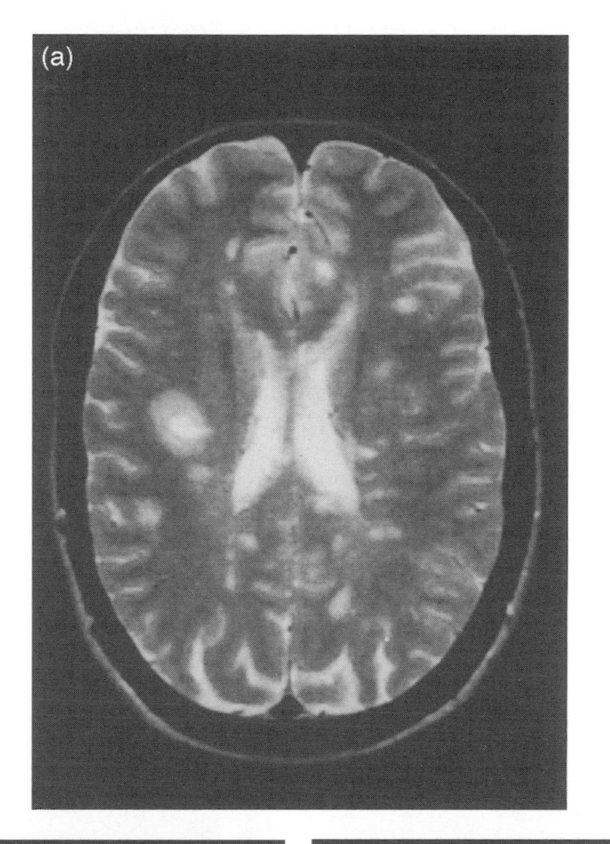

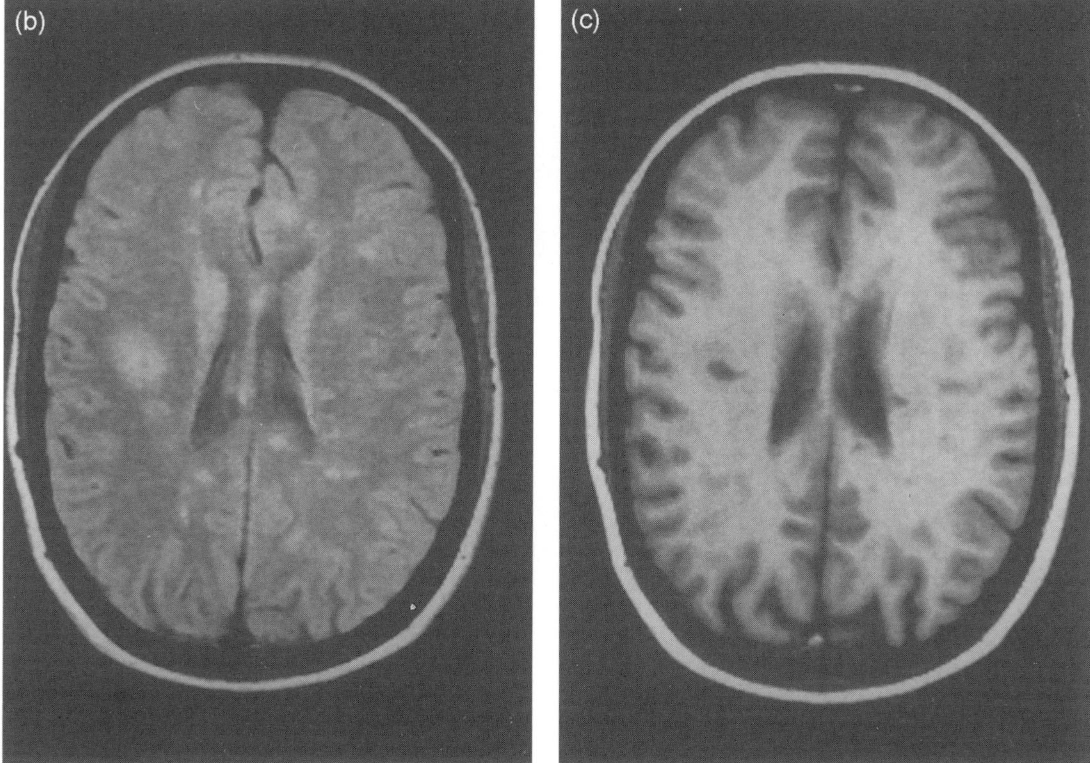

Figure 4 Multiple sclerosis: (a) T_2-weighted, (b) proton density, and (c) T_1-weighted sections showing the typical signal changes found in the multiple focal and coalescing acute and chronic plaques

For References see p. 4981

real-time echo planar imaging,[58] the complex morphology of the initial phase of gadolinium enhancement after intravenous injection may be further elucidated and related to the lesions on the unenhanced scan. Correlation with lipid imaging may allow study of the relationship between demyelination and inflammation. Advances in the use of MRI and MRS have been reviewed by Paty et al.[59]

Now that new and biologically active lesions can be identified routinely in clinic patients by gadolinium-enhanced T_1-weighted scans, the association of blood–brain barrier leakage in some but not all of multiple plaques shown on T_2-weighted images indicates the presence of dissemination in time[60] and refines the diagnosis of MS. The method can be used to subdivide clinical groups, and will be useful in monitoring and possibly shortening the time required for therapeutic trials (e.g., with steroids[61] and interferon[62]) in MS.

3.1.2 Schilder's Disease (Myelinoclastic Diffuse Sclerosis)

This is a rare but distinctive acute demyelinating condition, which can be defined by biochemical, pathological, and electrophysiological criteria, yet remains faithful to Schilder's original description of 1912.[63] There is a severe selective inflammatory demyelination with sparing of the subcortical U fibers, and extensive attempts at remyelination.

MR white matter changes have been described,[64] with bilateral involvement of the anterior hemispheres, extensive fluctuating increased relaxation times, mass effect, and varying partial ring enhancement indicating changes in blood–brain barrier breakdown. Other leukodystrophies with a similar appearance such as adrenoleukodystrophy and Pelizaeus–Merzbacher disease must be excluded by biochemical testing.

3.2 Postinflammatory: Viral, Allergic, Immune-Mediated Responses to Previous Infection

3.2.1 ADEM

Acute disseminated encephalomyelitis is a demyelinating disease that is thought to be an immune-mediated disorder secondary to a recent viral infection or more rarely to vaccination. It has an acute onset and a monophasic course, in contradistinction to MS. Most patients make a complete recovery, with no neurological sequelae. This makes it one of the most important differential diagnoses in the acute clinical situation.

Pathologically, there is a diffuse perivenous inflammatory process resulting in confluent areas of demyelination. These

Table 1 Differential Diagnosis of Multifocal White Matter Lesions

Multiple sclerosis
Aging, small vessel vascular disease, lacunar infarcts
Infarction
Acquired immune deficiency syndrome
Encephalitis (ADEM), (SSPE)
Progressive multifocal leukoencephalopathy
Metastases
Trauma
Radiation damage
Granulomatous disease (e.g., sarcoid)
Inherited white matter disease
Normal (in healthy elderly, especially hypertensive)
Hydrocephalus with CSF interstitial edema

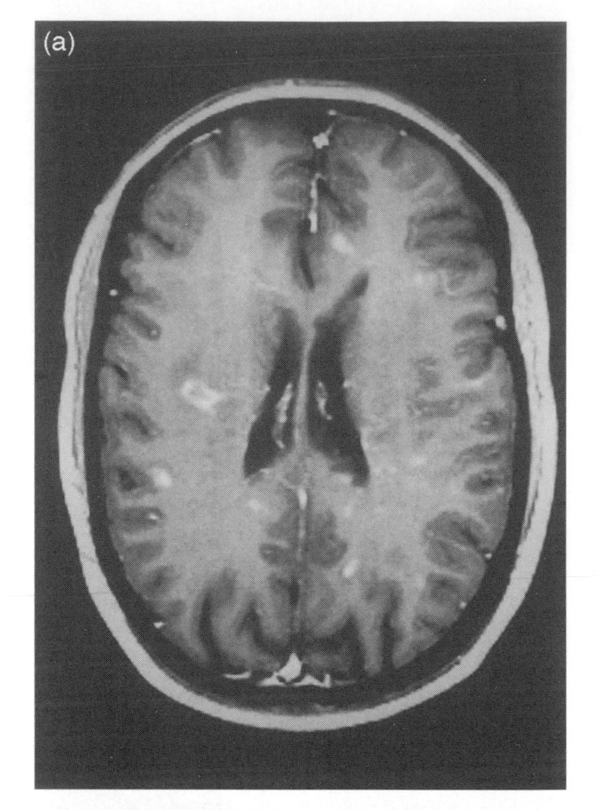

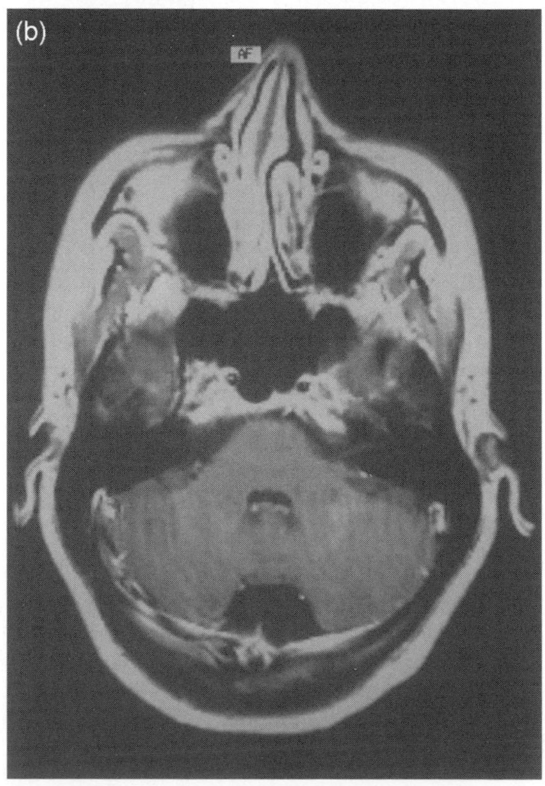

Figure 5 Multiple sclerosis: gadolinium-enhanced T_1-weighted sections showing (a) multiple enhancing acute lesions including a ring enhancing plaque and (b) a nonenhancing chronic cerebellar peduncular lesion

frequently occur at the corticomedullary junction, with gray matter much less often involved than white. Large areas of increased T_1 and T_2 relaxation time are found on MRI in both hemispheres, but the effects are usually asymmetrical.[65] Although in the acute stage, the demyelinating lesions may enhance, the blood–brain barrier quickly returns to normal. As with MS, MRI is much more sensitive than CT at demonstrating these lesions.

3.2.2 SSPE

Subacute sclerosing panencephalitis is a rare progressive demyelination resulting from reactivation of the measles virus due to a defect in immunity that allowed the virus to remain latent.[66] There is a variable rate of progression, with death between two months and several years after reactivation.

There is perivascular infiltration by inflammatory cells, cortical and subcortical gliosis, and white matter demyelination progressing from occipital to the frontal lobes and from the cerebellum to the brainstem and spinal cord. This is reflected in an increase in T_1 and T_2 relaxation times of the multifocal patchy white matter lesions.[67,68]

3.2.3 PML

Progressive multifocal leukoencephalopathy is a demyelinating disease probably caused by the papova viruses (e.g., JC and SV40-PML). These are universal childhood infections that are reactivated in the immunosuppressed patient. It is characterized by demyelination, with abnormalities of oligodendrocytes in the white matter. Initially, the lesions are widely disseminated, but later tend to become confluent, producing large lesions. It is now seen with increasing incidence in patients with AIDS[69] and those treated with immunosuppressive drugs. It most commonly involves the subcortical white matter of the posterior frontal and parietal lobes extending to the level of the trigones and occipital horns of the lateral ventricles (Figure 6). The lesions give an increased T_2 and slightly increased T_1 relaxation time, and because there is only occasional perivenular inflammation in the acute stage, they do not have mass effect and gadolinium enhancement is not usually a feature. These are useful differentiating features from lymphoma and toxoplasmosis.[70]

3.3 Posttherapy

3.3.1 Disseminated Necrotizing Leukoencephalopathy

When patients are treated with intrathecal antineoplastic agents such as methotrexate for disseminated lymphoma, leukemia, or carcinomatosis, a necrotizing leukoencephalopathy can occur despite the fact that the agent does not usually cross the blood–brain barrier.[71] Radiation therapy potentiates this neurotoxicity. There is endothelial injury, loss of oligodendroglial cells, coalescing foci of demyelination, and axonal swelling. The damaged endothelium responds by attempts at repair, resulting in hyalinization, fibrosis, and mineralization of the vessel walls. This causes a relative tissue ischemia, demyelination, and necrosis.

On MRI,[72] there is a diffuse increase in T_1 and T_2 relaxation times, with no mass effect and little or no gadolinium enhancement, reflecting only the edema and demyelination present.

The patterns of injury to the white matter from radiation therapy on its own are divided into three stages: (a) acute, (b) early, and (c) delayed. In the first days or weeks after therapy, vasogenic edema may be produced because of transient disruption of the blood–brain barrier, and some enhancement will

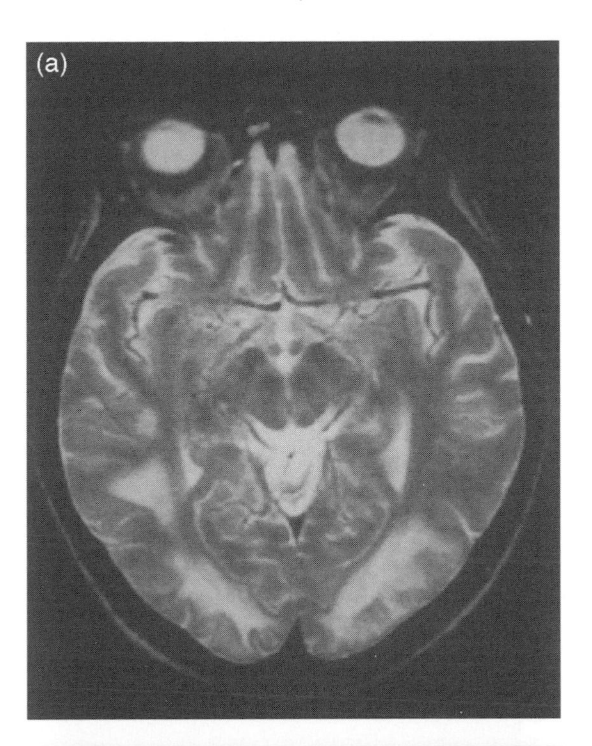

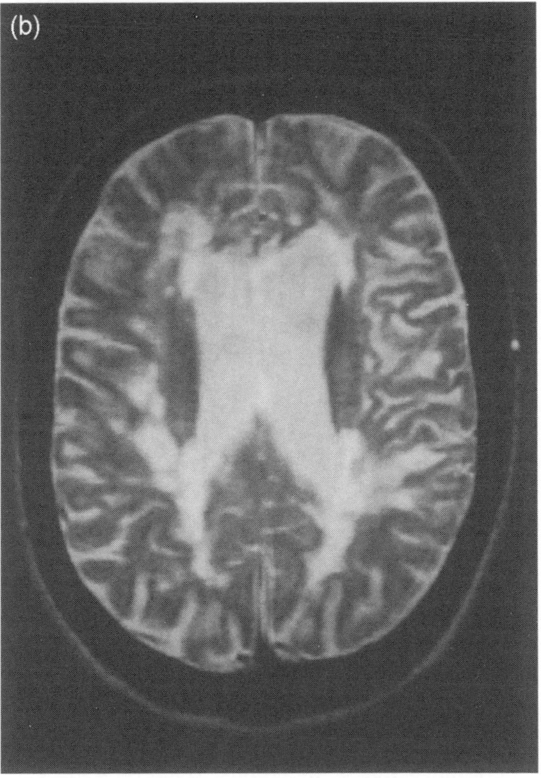

Figure 6 PML in AIDS: T_2-weighted images showing (a) extensive subcortical demyelination in the trigonal and occipital regions and, in a different patient, (b) gross loss of white matter substance with periventricular coalescing hyperintensities

For References see p. 4981

occur. In the weeks or months following radiation, demyelination may occur. MR may show focal or diffuse increases in T_1 and T_2 relaxation times in a patchy often periventricular distribution, which may be asymmetrical, affecting the white matter but generally sparing the corpus callosum, internal capsules, and basal ganglia.[73,74] The gray matter is only involved in severe cases. Delayed effects can occur months to years after therapeutic irradiation. These are less common, and develop later with hyperfractionated doses. There is endothelial hyperplasia, fibrinoid necrosis of perforating arterioles, and thrombosis. Cerebral necrosis supervenes, with blood–brain barrier disruption, edema, and mass effect. MRI at this stage shows mass effect, increased T_1 and T_2 relaxation times, and enhancement after intravenous gadolinium. Therefore this cannot be differentiated from recurrent tumor with MRI. There is some evidence that measures of the lesion's metabolism with ^{18}F-deoxyglucose positron emission tomography, ^{201}Tl or L-3-[^{123}I]iodo-α-methyltyrosine single photon emission computerized tomography will selectively differentiate the hypometabolic radiation necrosis from the hypermetabolic malignant tumor.[75,76]

A reversible acute cerebellar and cerebral syndrome has been reported[77] after systemic high-dose cytarabine therapy used for treatment of postremission and refractory leukemia treatment. Diffuse patchy areas of increased intensity on T_2-weighted images were shown in the deep white matter of the frontal, occipital, and parietal lobes. Punctate enhancement was observed in the occipital lobes. Over a month, the symptoms and white matter abnormalities resolved. At post-mortem later, there was no evidence of white matter disease.

3.4 Toxic and Degenerative

3.4.1 *Central Pontine Myelinolysis*

In this condition, there is loss of myelin and oligodendroglia in the central pons, which may extend to the lateral thalamus and mesencephalon, sparing the ventrolateral longitudinal fibers.[78] It is usually associated with rapidly corrected hyponatremia, often in alcoholics.[79,80] When there is only a tiny lesion or the patient is in coma due to the underlying disease process, it may be asymptomatic, but usually there is tetraparesis with a pseudobulbar palsy or a 'locked-in' syndrome. Mild cases with full recovery have now been reported.[81]

The demyelination is depicted as T_1 and T_2 prolongation with no mass effect, and although there is a single report of ring enhancement, there is usually no blood–brain barrier breakdown.[82]

3.4.2 *Marchiaflava–Bignami Disease*

This is characterized by a toxic demyelination of the corpus callosum in alcoholics.[83,84] A rapidly fatal form and a more chronic form have been recognized. In the acute form, extensive lesions have been reported in the centrum semiovale and corpus callosum, while at the chronic stage only corpus callosum lesions are seen, and occasionally there is a favorable outcome.[85] MRI shows these as small areas of increased T_1 and T_2 relaxation times, with no mass effect. Enhancement has not been reported.

3.4.3 *Carbon Monoxide Encephalopathy*

As carbon monoxide binds to the hemoglobin molecule and displaces oxygen, it induces hypoxia and vulnerable cells are destroyed. Although the gray matter structures are damaged first, the white matter is also involved, especially when there is episodic or chronic exposure.

MRI shows areas of increased T_1 and T_2 relaxation time in the thalamus, basal ganglia, hippocampus, and centrum semiovale. These areas may show enhancement with gadolinium in the acute stage. The lesions are usually bilateral and symmetrical, but can be patchy. Laminar necrosis has been reported as high signal cortical foci on T_2-weighted images.[86]

3.4.4 *Substance Abuse*

Inhalation of organic solvents and black market drugs such as heroin vapors (pyrolysate) and cocaine produce a wide variety of acute and chronic neurological signs and symptoms. The effects on the white matter will depend largely on the chemical constituent involved.

Xiong et al.[87] have shown that in toluene (one of the constituents of paint sprays) abuse, there is generalized cerebral, cerebellar, and corpus callosum atrophy, with a loss of gray–white matter contrast associated with diffuse multifocal hyperintensity of the cerebral white matter on T_2-weighted sequences. Additionally, hypointensity of the thalami are also seen.

Adulterated and synthetically produced drugs can produce severe leukoencephalopathy. Tan et al.[88] reported on four patients who inhaled contaminated heroin vapor and developed extensive, symmetrical lesions of the white matter of the cerebrum, cerebellum, and midbrain. Selective involvement of the corticospinal tract and lemniscus medialis was also found.

These have to be differentiated from the effects of cocaine abuse, where there is generally neurovascular damage with vasculitis, vasospasm, and thrombosis. Eventually, cerebral atrophy can be seen.[89]

3.4.5 *Hypoxic–Ischemic Encephalopathy*

This generally refers to brain damage in the fetus and infant. It may be focal or diffuse. When focal, it may be the cause of territorial infarcts such as can occur in cyanotic congenital heart disease when emboli can bypass the filtering effect of the lungs. This will be discussed in Section 3.4.6. In asphyxia, there is diffuse hypoxia hypercarbia, acidosis, and loss of the brain's normal vascular autoregulation, resulting in pressure-passive flow and reduced perfusion. Capillary permeability is also altered. Sudden reperfusion of these weakened capillaries can result in rupture and intracerebral hemorrhage. The periventricular white matter is particularly susceptible, lying at the distal end of the supply zone of the long narrow centripetal arteries that run from the cerebral surface.[90]

When 100 high-risk neonates of different gestational ages were followed prospectively with MRI, CT, and ultrasound examinations,[91,92] it was found that lesions associated with hypoxic–ischemic encephalopathy such as coagulative necrosis and germinal matrix hemorrhage were best shown on MRI. In the analysis, ultrasound showed 80%, while CT only showed 40% of those lesions depicted on MRI.

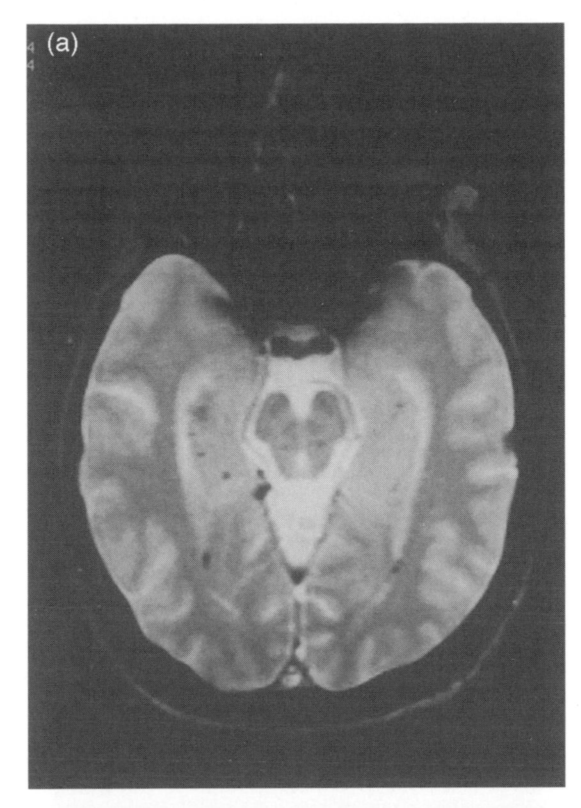

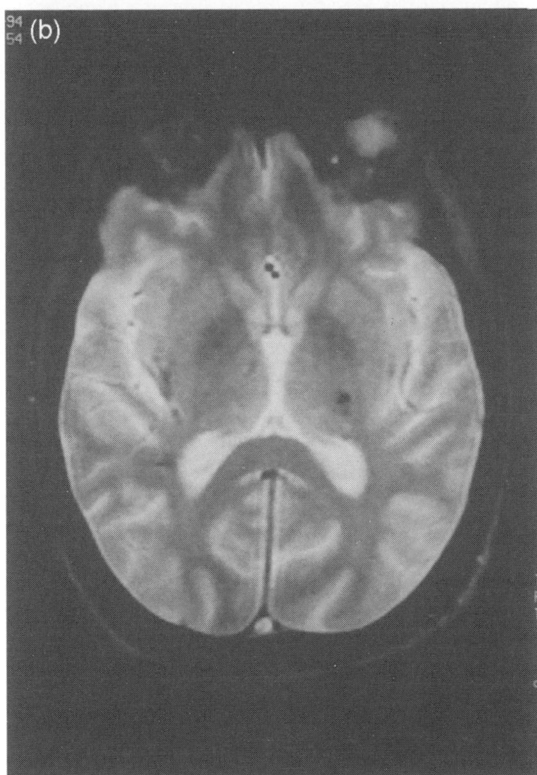

Figure 7 Head injury: T_2* gradient echo sections in an 25-year-old unconscious patient with a normal CT scan. Multiple focal hypointensities represent hemorrhage at white/gray matter interfaces—evidence of diffuse axonal injury

This has raised interest in medico-legal circles, since the timing of the insult may be more clearly defined. The appearances of the brain damage on MRI can now give important clues as to the time and nature of the asphyxia. Barkovich and Truwit[93] found that when the asphyxia occurred before the 26th week of gestation, there was dilatation of the ventricles without any signal changes, whereas in older fetuses there was increasing periventricular gliosis. Both periventricular and more peripheral white matter gliosis with associated general atrophy were found in cases who had been partially asphyxiated, or where asphyxia had occurred near term or in postmature fetuses. Total asphyxia involves the deep gray matter nuclei and the brainstem, and presents a different pattern.

These MRI patterns may have prognostic value, with initial studies[94] reporting good correlations between imaging findings at 8 months and neurodevelopmental outcome at 18 months.

3.4.6 *Trauma (Contusions–Shear Injuries)*

In children, the effects of asphyxia and mechanical trauma, be they accidental or nonaccidental, may initially produce the same imaging appearance, with generalized cerebral swelling resulting in blurring of the clear distinction between the gray matter and white matter boundaries, ventricular compression, and loss of CSF from the sulci and cisterns. This is due to a combination of edema and a failure of autoregulation producing an increase in cerebral blood volume. This can result in watershed ischemia and infarction, with eventual loss of white matter producing ventricular and sulcal dilatation.

Focal cerebral contusions involve the gyral crests, and can extend into the subcortical and deeper white matter regions, depending on their severity. There is edema and petechial perivascular hemorrhage, but tissue integrity is largely preserved in small lesions. With more severe contusions, the petechial hemorrhages coalesce into focal hematomas, which have some space-occupying effect. These are well shown by MRI at all stages,[95,96] although, in practice, CT is easier to perform and gives clinically adequate information in the acute stage.

Blunt trauma resulting in sudden acceleration or deceleration of the skull, especially when this is rotational, sets up shear–strain deformation at the moment of impact in response to the inertial differences between tissues of different density and viscosity.[97] This can cause immediate and irreversible structural damage to axons, and has been termed diffuse axonal injury.[98–100]

Diffuse axonal injury is a pathological diagnosis, and imaging may only show a few apparently focal lesions in the lobar white matter. It is, however, very important to recognize these as the 'hallmark' of associated widespread microscopic axonal disruption. Although CT is the most commonly conducted examination in the acute situation, MRI is much more sensitive[95,101] and essential when the clinical state is not explained by the CT imaging appearances (Figure 7). In the acute situation, foci of edema that may or may not contain macroscopic hematomas can be seen on T_2-weighted MRI in the corpus callosum, the parasagittal frontal white matter close to the gray–white matter interface, the basal ganglial regions and the dorso-lateral quadrant of the rostral brainstem.[96] At the subacute stage, hemorrhage will be better depicted on T_1-weighted

For References see p. 4981

images, but both acutely and in the chronic phase T_2* gradient echo sequences are most sensitive to deoxyhemoglobin and hemosiderin respectively. MRI can rarely appear entirely normal[102] in severe diffuse axonal injury, and it is only on follow-up that the widespread white matter damage is reflected in atrophy with ventricular and cortical sulcal enlargement.[100,103,104]

4 VASCULAR DISEASE

4.1 Infarction

Stroke remains one of the commonest causes of hospital admission in the developed world, has a high morbidity and mortality, and consumes more healthcare resources than any other single disease. Ninety percent of cases are due to ischemia (10% to cerebral hemorrhage) from thrombosis of a nutrient artery, with only a small number due to emboli from the heart or other vessels. Despite treatment advances, the mortality remains at around 50%. It is thought that only preventative public health measures and earlier thrombolytic therapy can improve this situation.[105] This requires the accurate early identification of patients with acute infarcts and those with transient ischemia at risk of completing their infarcts.

Routine unenhanced MRI can detect abnormalities within about 8 h of the onset of symptoms (although changes on MRI with a vessel occlusion stroke model were shown as early as 1–2 h without paramagnetic contrast),[106,107] whereas CT is normal for at least 14 h and if perfusion is not re-established remains 'bland' for several days. MRI initially shows subtle swelling and an increase in T_1 and T_2 signals due to failure of the 'sodium–potassium' pump and increasing intracellular water–cytotoxic edema. At this stage, function is lost but structure is maintained. It is only with continuing ischemia that blood–brain barrier breakdown occurs, structural integrity is lost, and vasogenic edema supervenes. Although there are anecdotal reports of MRI-defined cytotoxic edema being reversed on treatment of ischemia in humans,[108,109] and more rigorous demonstrations in cats using diffusion sequences,[107] it has not been established in routine clinical practice whether these MRI changes, unlike those on CT, are reversible.

Recent gadolinium-enhanced MRI studies[110–112] of the first 24–48 h after the ictus in the clinical population has shed light on this crucially important acute stage. Sato and colleagues[113] studied six patients within 8 h and a further two between 8 and 26 h. They showed areas of cerebral ischemia/infarction using gadolinium-enhanced T_1-weighted spin echo sequences. Abnormal curvilinear areas of enhancement thought to represent cortical arterial vessels with markedly slowed circulation were seen adjacent to affected brain. This tissue was shown to progress to frank infarction on follow-up CT and MRI. These features have been confirmed and extended by the Iowa group.[110,114] They demonstrated the vascular flow-related abnormalities with absence of normal flow voids and the presence of arterial enhancement detected within minutes of the onset of symptoms. Brain swelling on T_1-weighted images without signal changes on T_2-weighted images was detected within the first few hours. In contrast to the usual absence of parenchymal enhancement typically found in cortical infarctions in the first 24 h, a few lesions showed paradoxical early exaggerated enhancement. These were the transient or partial occlusions and isolated watershed infarcts. Longer term prospective observations through the first fortnight have defined the subacute appearances.[115] Three stages have been demonstrated:

(a) vascular enhancement—days 1–3, seen in 77% of cases;
(b) leptomeningeal enhancement—days 4–7, seen only in larger infarcts;
(c) brain parenchymal enhancement—days 7–14, seen in 100% of cases studied.

Enhancement is not noted after two to three months. These changing patterns of enhancement reflect the underlying pathophysiology, and may have prognostic significance; if this proves to be so then gadolinium enhancement will be crucial in the evaluation of early ischemia and its response to intervention.

Once the parenchymal long T_1 and T_2 signal changes are established, the differential diagnosis must be made in a similar way to conventional CT, following consideration of

(a) the site: vascular territory, watershed region, deep gray or white matter tracts;
(b) the shape: wedge, involving gray and white matter with subtle bowing of interfaces;
(c) the margins: sulcal effacement, blurring of the gray and white matter borders;
(d) the degree of edema;
(e) the sequence of resolution of mass effect over three to four weeks.[116]

Hypointensity on heavily T_2-weighted sequences and the use of gradient echoes or susceptibility mapping can often show petechial hemorrhage in the second week that is not seen on CT. While frank hemorrhage correlates with a worsening clinical state, fine interstitial bleeding mainly due to diapedesis relates neither to anticoagulation nor to a poorer clinical condition.[116]

The patency of the extracranial and major intracerebral arteries can be assessed on routine MRI sequences as a 'flow void' or with slower laminar flow even as echo rephasing. At present, projectional images produced by time-of-flight and phase contrast angiographic sequences are being evaluated, and may yet replace preoperative conventional cerebral angiography.[117,118]

Both diffusion imaging[119,120] and spectroscopy[121] are being used experimentally in clinical populations to try to gain an understanding of microscopic water shifts as the different types of edema develop and to give an insight into the progressive cycles of bioenergetic exchange as oxidative metabolism breaks down in the ischemic brain cells.

4.2 Ischemic White Matter Disease—Normal Aging

Focal and confluent white matter abnormalities seen on MRI do not necessarily represent actual necrosis and infarction, but can be due to a spectrum of chronic cerebrovascular insufficiency.[122] These merge with the changes found in as many as 30% of the normal aging population over 60 years of age who show no clinical cognitive deficit, but which are seen with increasing frequency in patients with hypertension, diabetes mellitus, and coronary artery disease (Table 1).[123]

Dilated perivascular spaces give a CSF signal, are usually smaller than lacunar infarcts, and occur in typical locations in the base, deep white matter, and cortex of the brain. Gliosis may become more confluent around these vessels as the vascular insufficiency progresses and produces an increased signal on proton density images in addition to the increased T_2 signal differentiating it from CSF. This has now been confirmed microscopically.

In a small post-mortem study,[124] histological examination showed that the larger lesions were characterized centrally by necrosis, axonal loss, and demyelination, and therefore represent true infarcts. Reactive astrocytes oriented along the degenerated axons were identified at distances of up to several centimeters from the central infarct. This isomorphic gliosis shows hyperintensity on T_2-weighted images, and increases the apparent size of the central lesion. Confirmation has recently been provided by Munoz et al.,[125] who investigated the pathological correlates of increased T_2 signal in the centrum ovale in an unselected series of 15 post-mortems. On the basis of size, greater than and less than 10 mm, two types of lesion were described, namely, extensive and punctate. The extensive areas of hyperintensity on T_2-weighted images were found to show myelin pallor that spared the subcortical U fibers. There was diffuse vacuolation and reduction in glial cell density. The punctate abnormalities were less well defined, and were found to be due to dilated Virchow–Robin spaces.

The white matter changes seen on MRI are therefore nonspecific, and although seen with increased frequency in ischemic brains, there is often little or no correlation with the clinical state in the elderly patient.

5 CONCLUSIONS

Over the last 10 years, MRI has become the main diagnostic tool in the investigation of white matter disease. In some conditions, its sensitivity is the key to selecting patients for further attention, while in others, it identifies more specific features that in turn lead to further laboratory investigations leading to a final diagnosis. MRI can be used to select patients for treatment and to monitor the effects of this treatment. The implementation of new sequences, faster scanning techniques, and better patient–machine ergonomics will ensure the preemptive position of MRI for the investigation of white matter diseases for the foreseeable future.

6 RELATED ARTICLES

Brain Neoplasms Studied by MRI; Cerebral Infection Monitored by MRI; Diffusion: Clinical Utility of MRI Studies; Echo-Planar Imaging; Focal Brain Lesions in Human Subjects Investigated Using MRS; Gadolinium Chelate Contrast Agents in MRI: Clinical Applications; Hemorrhage in the Brain and Neck Observed by MRI; Pediatric Brain MRI: Applications in Neonates & Infants.

7 REFERENCES

1. A. J. Barkovich, G. Lyon, and P. Evrard, *Am. J.N.R.*, 1992, **13**, 447.
2. R. G. Parrish, J. R. Kurland, W. W. Janese, and L. Bakay, *Science (Washington, DC)*, 1974, **483**, 349.
3. A. J. Barkovich and T. V. Maroldo, *Top. Magn. Reson. Imaging*, 1993, **5**, 96.
4. Y. Nomura, H. Sakuma, K. Takeda, T. Tagami, Y. Okuda, and T. Nakasasa, *Am. J.N.R.*, 1994, **15**, 231.
5. A. Oatridge, J. V. Hajnal, F. M. Cowan, C. J. Baudouin, I. R. Young, and G. R. Bydder, *Clin. Radiol.*, 1993, **47**, 82.
6. B. E. Kendall, *J. Inherited Metab. Dis.*, 1993, **16**, 771.
7. L. E. Becker, *Am. J.N.R.*, 1992, **13**, 609.
8. B. E. Kendall, *Am. J.N.R.*, 1992, **13**, 621.
9. G. B. Bobele, A. Garnica, G. B. Schaefer, J. C. Leonard, D. Wilson, W. A. Marks, R. W. Leech, and R. A. Brumback, *J. Child. Neurol.*, 1990, **5**, 253.
10. T. Ichiyama, T. Hayashi, and T. Ukita, *Brain Dev.*, 1993, **15**, 153.
11. J. Brismar, G. Brismar, G. Gascon, and P. Ozand, *Am. J.N.R.*, 1990, **11**, 805.
12. D. A. Finelli, R. W. Tarr, R. N. Sawyer, and S. J. Horwitz, *Am. J.N.R.*, 1994, **15**, 167.
13. T. J. Farley, L. M. Ketonen, J. B. Bodensteiner, and D. D. Wang, *Pediatr. Neurol.*, 1992, **8**, 455.
14. S. Choi and D. R. Enzmann, *Am. J.N.R.*, 1993, **14**, 1164.
15. J. C. Koetsveld-Baart, I. E. Glaudemans-van-Gelderen, J. Valk, and P. G. Barth, *Ned. Tijdschr. Geneeskd.*, 1993, **137**, 2494.
16. M. Ishii, J. Takanashi, K. Sugita, A. Suzuki, M. Goto, Y. Tanabe, K. Tamai, and H. Niimi, *No To Hattatsu*, 1993, **25**, 9.
17. E. G. Shapiro, L. A. Lockman, D. Knopman, and W. Krivit, *Neurology*, 1994, **44**, 662.
18. B. Kruse, F. Hanefeld, H. J. Christen, H. Bruhn, T. Michaels, W. Hanicke, J. Frahm, *J. Neurol.*, 1993, **241**, 68.
19. P. J. Close, S. J. Sinnott, and K. T. Nolan, *Pediatr. Radiol.*, 1993, **23**, 400.
20. M. S. van der Knaap and J. Valk, *Neuroradiology*, 1991, **33**, 30.
21. J. Brismar, A. Aqeel, G. Gascon, and P. Ozand, *Am. J.N.R.*, 1990, **11**, 135.
22. R. Sugita, I. Takahashi, K. Ishii, K. Matsumoto, T. Ishibashi, K. Sakamoto, and K. Narisawa, *J. Comput. Assist. Tomogr.*, 1990, **14**, 699.
23. D. W. W. Shaw, K. R. Maravilla, E. Weinberger, J. Garretson, C. M. Trahms, and C. R. Scott, *Am. J.N.R.*, 1991, **12**, 403.
24. K. D. Pearsen, A. D. Gean-Marton, H. L. Levy, and K. R. Davis, *Radiology*, 1990, **177**, 437.
25. A. J. Thompson, I. Smith, D. Brenton, B. D. Youl, G. Rylame, D. C. Davidson, B. Kordall, and A. J. Lees, *Lancet*, 1990, **336**, 602.
26. A. J. Thompson, S. Tillotson, I. Smith, B. Kendall, S. G. Moore, and D. P. Brenton, *Brain*, 1993, **116**, 811.
27. A. A. Tzika, W. S. Ball Jr., D. B. Vigneron, R. S. Dunn, and D. R. Kirks, *Am. J.N.R.*, 1993, **14**, 1267.
28. A. J. Barkovich, W. V. Good, T. K. Koch, and B. O. Berg, *Am. J.N.R.*, 1993, **14**, 1119.
29. P. M. Matthews, F. Andermann, K. Silver, G. Karpati, and D. L. Arnold, *Neurology*, 1993, **43**, 2484.
30. B. Barbiroli, P. Montagna, P. Martinelli, R. Lodi, S. Lotti, P. Cortelli, R. Fanicello, and P. Zaniol, *J. Cereb. Blood Flow Metab.*, 1993, **13**, 469.
31. J. M. Charcot, *Gaz. des Hôp. Civ. Mil., Paris*, 1868, **41**, 554.
32. W. I. McDonald and D. H. Silberberg, 'Multiple Sclerosis', Butterworths, London, 1986.
33. G. A. Schumacher, G. Beebe, R. F. Kibler, L. T. Kurland, J. F. Kurtzke, F. McDowell, B. Nagler, W. A. Sibley, W. W. Tourtellotte, and T. L. Willmon, *Ann. NY Acad. Sci.*, 1965, **122**, 552.

34. C. M. Poser, D. W. Paty, L. Scheinberg, W. I. MacDonald, F. A. Davis, G. C. Ebers, K. P. Johnson, W. A. Sibley, D. H. Silberberg, and W. H. Tourtellote, *Ann. Neurol.*, 1983, **13**, 227.
35. A. K. Asbury, R. M. Herndon, H. F. McFarland, W. I. McDonald, W. J. McIlroy, D. W. Paty, J. W. Prineas, L. C. Scheinberg, and J. S. Wolinsky, *Neuroradiology*, 1987, **29**, 119.
36. D. W. Paty and D. K. B. Li, in 'Clinical Neuroimaging', ed. W. H. Theodore, Alan R. Liss, New York, 1988, Vol. 4, Chap. 10.
37. R. I. Grossman, B. H. Braffman, J. R. Brorson, H. I. Goldberg, D. H. Silberberg, and F. Gonzalez-Scarano, *Radiology*, 1988, **169**, 117.
38. I. R. Young, A. S. Hall, C. A. Pallis, N. J. Legg, G. M. Bydder, and R. E. Steiner, *Lancet*, 1981, **ii**, 1063.
39. S. A. Lukes, L. E. Crooks, M. J. Aminoff, L. Kaufman, H. S. Panitch, C. Mills, and D. Norman, *Ann. Neurol.*, 1983, **13**, 592.
40. I. E. C. Ormerod, G. H. du Boulay, and W. I. McDonald, in 'Multiple Sclerosis', ed. W. I. McDonald and D. H. Silberberg, Butterworths, London, 1986.
41. D. H. Miller, *MRI Decis.*, 1988, **2**, 17.
42. I. E. C. Ormerod, D. H. Miller, W. I. McDonald, E. P. G. H. du Boulay, P. Rudge, B. E. Kendall, I. F. Moseley, G. Johnson, P. S. Tofts, and A. N. Halliday, *Brain*, 1987, **110**, 1579.
43. D. W. Paty, *Can. J. Neurol. Sci.*, 1988, **15**, 266.
44. D. W. Paty, J. J. F. Oger, L. F. Kastrukoff, S. A. Hashimoto, J. P. Haage, A. A. Eisen, K. A. Eisen, S. T. Purves, M. D. Low, and V. Brandejs, *Neurology*, 1988, **38**, 180.
45. D. Lacomis, M. D. Osbakken, and G. Gross, *Magn. Reson. Med.*, 1986, **3**, 194.
46. B. P. Drayer, P. Burger, B. Hurwitz, D. Dawson, and J. Cain, *Am. J.N.R.*, 1987, **8**, 413.
47. D. H. Miller, I. E. C. Ormerod, A. Gibson, E. P. G. H. du Bouley, P. Rudge, and W. I. McDonald, *Neuroradiology*, 1987, **29**, 226.
48. F. Z. Yetkin, V. M. Haughton, R. A. Papke, M. E. Fischer, and S. M. Rao, *Radiology*, 1991, **178**, 447.
49. F. Fazekas, H. Offenbacher, S. Fuchs, R. Schmidt, K. Niederkorn, S. Horners, and H. Lechner, *Neurology*, 1988, **38**, 1822.
50. H. Offenbacher, F. Fazekas, R. Schmidt, W. Freidl, E. Floch, F. Payer, and H. Lechner, *Neurology*, 1993, **43**, 905.
51. I. E. C. Ormerod, A. Bronstein, P. Rudge, G. Johnson, D. G. P. MacManus, A. M. Halliday, H. Barratt, E. P. du Boulay, B. E. Kendall, and I. F. Moseley, *J. Neurol. Neurosurg. Psychiatry*, 1986, **49**, 737.
52. J. Prineas, *Hum. Pathol.*, 1975, **6**, 531.
53. R. I. Grossman, F. Gonzalez-Scarano, S. W. Atlas, S. Galetta, and D. H. Silberberg, *Radiology*, 1986, **161**, 721.
54. H. Miller, P. Rudge, B. Johnson, B. E. Kendall, D. G. MacManus, I. F. Moseley, D. Barnes, and W. I. McDonald, *Brain*, 1988, **111**, 927.
55. E.-M. Larsson, S. Holas, and O. Nilsson, *Am. J.N.R.*, 1989, **10**, 1071.
56. S. F. Merandi, B. T. Kudryk, F. R. Murtagh, and J. A. Arrington, *Am. J.N.R.*, 1991, **12**, 923.
57. A. J. Thompson, A. J. Kermode, D. Wicks, D. G. MacManus, B. E. Kendall, D. P. Kingley, and W. I. McDonald, *Ann. Neurol.*, 1991, **29**, 53.
58. M. K. Stehling, P. Bullock, J. L. Firth, A. M. Blamire, R. J. Ordidge, B. Coxon, P. Gibbs, and P. Mansfield, *Proc. Soc. Magn. Reson. Med. 1989 Meet.*, p. 358.
59. D. W. Paty, *Curr. Opin. Neurol. Neurosurg.*, 1993, **6**, 202.
60. R. Heun, L. Kappos, S. Bittkau, D. Staedt, E. Rohrbach, and B. Schuknecht, *Lancet*, 1988, **ii**, 1202.
61. M. J. Kupersmith, D. Kaufman, D. W. Paty, G. Ebers, M. McFarland, K. Johnson, J. Reingold, and J. Whitaker, *Neurology*, 1994, **44**, 1.
62. D. W. Paty and D. K. Li, *Neurology*, 1993, **43**, 662.
63. P. Schilder, *Z. Gesamte Neurol. Psychiatr.*, 1912, **10**, 1.
64. M. F. Mehler and L. Rabinowich, *Am. J.N.R.*, 1989, **10**, 176.
65. S. W. Atlas, R. I. Grossman, H. I. Goldberg, D. B. Hackney, L. T. Bilanuck, and R. A. Zimmerman, *J. Comput. Assist. Tomogr.*, 1986, **10**, 798.
66. A. J. Barkovich, 'Pediatric Neuroimaging', Raven Press, New York, 1990, p. 312.
67. R. Murata, H. Hattori, O. Matsuoka, T. Nakajima, and H. Shintaku, *Brain. Dev.*, 1992, **14**, 391.
68. S. Yagi, Y. Miura, S. Mizuta, A. Wakunami, N. Kataoka, T. Morita, K. Morita, S. Ono, and M. Fukunaga, *Brain. Dev.*, 1993, **15**, 141.
69. A. S. Mark and S. W. Atlas, *Radiology*, 1989, **173**, 517.
70. L. Ketonen and M. Tuite, *Semin. Neurol*, 1992, **12**, 57.
71. F. Ebner, G. Ranner, I. Slavc, C. Urban, R. Kleinert, H. Roulner, R. Ernspieler, and E. Justich, *Am. J.N.R.*, 1989, **10**, 959.
72. R. Asato, Y. Akiyama, M. Ito, M. Kubota, R. Okumura, Y. Miki, J. Konishi, H. Mikaua, *Cancer*, 1992, **70**, 1997.
73. J. T. Curnes, D. W. Laster, M. R. Ball, T. D. Koubek, D. M. Moody, and R. L. Witcofski, *Am. J.N.R.*, 1986, **7**, 389.
74. W. J. Curran, C. Hecht-Leavitt, L. Schut, R. A. Zimmerman, and D. F. Nelson, *Int. J. Radiat. Oncol. Biol. Phys.*, 1987, **13**, 1093.
75. R. B. Schwartz, P. A. Carvalho, E. Alexander III, J. S. Loeffler, R. Folkerth, and B. L. Holman, *Am. J.N.R.*, 1991, **12**, 1187.
76. Karl-J. Langen, H. H. Coenen, N. Roosen, P. Kling, O. Muzik, H. Herzog, T. Kuwort, G. Stocklin, and L. E. Femendegen, *J. Nucl. Med.*, 1990, **31**, 281.
77. D. J. Vaughn, J. G. Jarvik, D. Hackney, S. Peters, and E. A. Stadtmauer, *Am. J.N.R.*, 1993, **14**, 1014.
78. Y. Korogi, M. Takahashi, J. Shinzato, Y. Sakamoto, K. Mitsuzaki, T. Hirai, and K. Yoshizumi, *Am. J.N.R.*, 1993, **14**, 651.
79. M. Mascalchi, M. Cincotta, and M. Piazzini, *Clin. Radiol.*, 1993, **47**, 137.
80. R. D. Laitt, M. Thornton, and P. Goddard, *Clin. Radiol.*, 1993, **48**, 432.
81. V. B. Ho, C. R. Fitz, C. C. Yoder, and C. A. Geyer, *Am. J.N.R.*, 1993, **14**, 163.
82. K. J. Koch and R. R. Smith, *Am. J.N.R.*, 1989, **10**, S58.
83. M. E. Charness, *Alcohol Clin. Exp. Res.*, 1993, **17**, 2.
84. P. Tomasini, D. Guillot, P. Sabbah, C. Brosset, P. Salamand, and J. F. Briant, *Ann. Radiol. (Paris)*, 1993, **36**, 319.
85. S. Canaple, A. Rosa, and J. P. Mizon, *Rev. Neurol. (Paris)*, 1992, **148**, 638.
86. A. L. Horowitz, R. Kaplan, and G. Sarpel, *Radiology*, 1987, **162**, 787.
87. L. Xiong, J. D. Matthes, J. Li, and R. Jinkins, *Am. J.N.R.*, 1993, **14**, 1195.
88. T. P. Tan, P. R. Algra, J. Valk, and E. C. Wolters, *Am. J.N.R.*, 1994, **15**, 175.
89. E. Brown, J. Prager, H. Y. Lee, and R. G. Ramsey, *Am. J. Roentgenol.*, 1992, **159**, 137.
90. M. D. Nelson, I. Gonzalez-Gomez, and F. H. Gilles, *Am. J.N.R.*, 1991, **12**, 215.
91. S. E. Keeney, E. W. Adcock, and C. B. McArdle, *Pediatrics*, 1991, **87**, 421.
92. S. E. Keeney, E. W. Adcock, and C. B. McArdle, *Pediatrics*, 1991, **87**, 431.
93. A. J. Barkovich and C. L. Truwit, *Am. J.N.R.*, 1990, **11**, 1087.

94. P. Byrne, R. Welch, M. A. Johnson, J. Darrah, and M. Piper, *J. Pediatr.*, 1990, **117**, 694.
95. E. Teasdale and D. M. Hadley, in 'Handbook of Clinical Neurology, 2nd Series: Head Injury', ed. R. Braakman, Elsevier, Amsterdam, 1990, Vol. 13, Chap. 7.
96. D. M. Hadley, *Curr. Imaging*, 1991, **3**, 64.
97. A. H. S. Holbourn, *Lancet*, 1943, **ii**, 438.
98. J. H. Adams, D. I. Graham, L. S. Murray, and G. Scott, *Ann. Neurol.*, 1982, **12**, 557.
99. T. A. Gennarelli, G. M. Spielman, T. W. Langfitt, P. L. Gildenberg, T. Harrington, J. A. Jane, L. F. Marshall, J. D. Miller, and L. H. Pitts, *J. Neurosurg.*, 1982, **56**, 26.
100. A. D. Gean, 'Imaging of Head Trauma', Raven Press, New York, 1994.
101. A. Jenkins, G. M. Teasdale, D. M. Hadley, P. Macpherson, and J. O. Rowan, *Lancet*, 1986, **ii**, 445.
102. D. M. Hadley, P. Macpherson, D. A. Lang, and G. M. Teasdale, *Neuroradiology*, 1991, **33**, 86.
103. K. D. Wiedmann, J. T. L. Wilson, D. Wyper, D. M. Hadley, G. M. Teasdale, and D. N. Brooks, *Neuropsychology*, 1990, **3**, 267.
104. J. T. L. Wilson, K. D. Wiedmann, D. M. Hadley, B. Condon, G. M. Teasdale, and J. D. N. Brooks, *J. Neurol. Neurosurg. Psychiatry*, 1988, **51**, 391.
105. C. D. Forbes, *Scot. Med. J.*, 1991, **36**, 163.
106. M. Brant-Zawadzki, B. Pereira, P. Weinstein, S. Moore, W. Kusharczyk, I. Berry, M. McNamara, and N. Derugin, *Am. J.N.R.*, 1986, **7**, 7.
107. M. E. Moseley, Y. Cohen, J. Mintorovitch, L. Chileuitt, H. Shimizer, W. Kueharczyk, M. F. Wendland, and P. R. Weinstein, *Magn. Reson. Med.*, 1990, **14**, 330.
108. A. M. Aisen, T. O. Gabrielsen, and W. J. McCune, *Am. J.N.R.*, 1985, **6**, 197.
109. W. G. Bradley, *Neurol. Res.*, 1984, **6**, 91.
110. W. T. C. Yuh, M. R. Crain, D. J. Loes, G. M. Greene, T. J. Ryals, and Y. Sato, *Am. J.N.R.*, 1991, **12**, 621.
111. S. Warach, W. Li, M. Ronthal, and R. R. Edelman, *Radiology*, 1992, **182**, 41.
112. R. N. Bryan, L. M. Levy, W. D. Whitlow, J. M. Killian, T. J. Preziosi, and J. A. Rosario, *Am. J.N.R.*, 1991, **12**, 611.
113. A. Sato, S. Takahashi, Y. Soma, K. Ishii, T. Watanabe, and K. Sakamoto, *Radiology*, 1991, **178**, 433.
114. M. R. Crain, W. T. C. Yuh, G. M. Greene, D. J. Loes, T. J. Ryals, Y. Sato, and M. N. Hart, *Am. J.N.R.*, 1991, **12**, 631.
115. A. D. Elster and D. M. Moody, *Radiology*, 1990, **177**, 627.
116. W. G. Bradley, in 'MRI Atlas of the Brain', eds. W. G. Bradley and G. Bydder, Martin Dunitz, London, 1990, Chap. 5.
117. T. J. Masaryk, G. A. Laub, M. T. Modic, J. S. Ross, and E. M. Haacke, *Magn. Reson. Med.*, 1990, **14**, 308.
118. A. W. Litt, *Am. J.N.R.*, 1991, **12**, 1141.
119. M. Doran and G. M. Bydder, *Neuroradiology*, 1990, **32**, 392.
120. R. M. Henkelman, *Am. J.N.R.*, 1990, **11**, 932.
121. M. Brant-Zawadzki, P. R. Weinstein, H. Bartkowski, and M. Moseley, *Am. J. Roentgenol.*, 1987, **148**, 579.
122. M. L. Bots, J. C. van-Swieten, M. M. Breteler, P. T. de-Jong, J. Van Gijn, A. Hofman, and D. E. Grobbee, *Lancet*, 1993, **341**, 1232.
123. T. Horikoshi, S. Yagi, and A. Fukamachi, *Neuroradiology*, 1993, **35**, 151.
124. V. G. Marshall, W. G. Bradley, C. E. Marshall, T. Bhoopat, and R. H. Rhodes, *Radiology*, 1988, **167**, 517.
125. D. G. Munoz, S. M. Hastak, B. Harper, D. Lee, and V. C. Hachinski, *Arch. Neurol.*, 1993, **50**, 492.

Biographical Sketch

Donald M. Hadley. *b* 1950. M.B.Ch.B., 1974, Ph.D., 1980, D.M.R.D., 1981, Aberdeen University, Scotland; F.R.C.R., 1983, London, UK. Introduced to NMR by Professor John Mallard and Dr Francis Smith while carrying out postdoctoral work in the Department of Bio-medical Physics, University of Aberdeen 1981. MRC research fellow, Glasgow University 1984, consultant and director of Neuroradiology 1992, Institute of Neurological Sciences, Glasgow, UK. Approx. 200 publications. Current research interests: MRI and MRS investigation of acute trauma, epilepsy, metabolic white matter diseases and stroke.

Whole Body Machines: NMR Phased Array Coil Systems

Peter B. Roemer
Advanced NMR, Wilmington, MA, USA

William A. Edelstein
GE Corporate Research and Development, Schenectady, NY, USA

&

Cecil E. Hayes
University of Washington, Seattle, WA, USA

1 INTRODUCTION

A small surface coil gives a good MRI (magnetic resonance imaging) or MRS (magnetic resonance spectroscopy) signal-to-

For References see p. 4989

noise ratio (S/N) close to the coil for a region whose dimensions are comparable to the size of the coil. A small coil has the drawback, however, that its field-of-view (FOV) is therefore limited. A volume coil, which surrounds a substantial portion of the body such as the torso or head, has a large field-of-view but poorer S/N. The *NMR phased array* collects data simultaneously from a distribution of surface coils (one- or two-dimensional) to achieve the S/N of a surface coil with the FOV of a volume coil.

The present discussion generally focuses on MRI. However, the NMR phased array is equally applicable to providing good S/N over a large FOV for spatially resolved MRS, which is accomplished using imaging techniques.

There are a number of technical difficulties that must be overcome in order to use closely spaced surface coils. There is an electrical interaction, which can be eliminated for adjacent coils by overlapping just the right amount to produce zero mutual inductance. Next-nearest neighbors and more distantly positioned coils can be decoupled by attaching a low-input-impedance preamplifier to one of the resonating capacitors for each surface coil.

The simplest realization of the NMR phased array acquires and processes separate images for each surface coil in the array, and produces a final image that is the optimized weighted sum of the images from all the coils. A further advance convolves a few Fourier coefficients from the Fourier transform of the coil sensitivity profiles with the NMR signal in the time domain, and yields a single, small data set that can be processed into images by straightforward Fourier transformation. While this latter procedure has been proven by computer modeling, a true implementation of the convolution scheme will require the development of sophisticated hardware using dedicated high-speed signal processing electronics.

2 OPTIMUM S/N FOR SURFACE COILS: APERTURE SYNTHESIS

The best S/N for a circular-loop surface coil is obtained when the loop diameter is approximately equal to the voxel depth.[1,2] (The actual relationship is optimum coil diameter = 0.89 × voxel depth.) This relationship is derived by considering the signal from the target voxel versus the electrical noise from the body as the loop size is varied.

Roemer et al.[3] calculated the S/N for arbitrary, abstract current distributions on the body surface, and concluded that the best possible S/N consistent with the laws of electromagnetism (the 'ultimate' S/N) would be achieved by a current distribution that approximates the current in a circular loop. A real circular loop has an S/N over 90% of the ultimate S/N.[3] Thus a simple circular loop is a useful building block in trying to maximize the S/N quality of MRI and MRS.

With an array of small coils, it is possible to synthesize a larger coil and achieve the optimum S/N for that larger coil. Roughly speaking, if one wishes to look at a voxel at a depth large compared to one of the coils in the array, the optimum SNR is obtained by adding the signals from a collection of circular loops whose total diameter is approximately equal to the depth of the voxel in question.

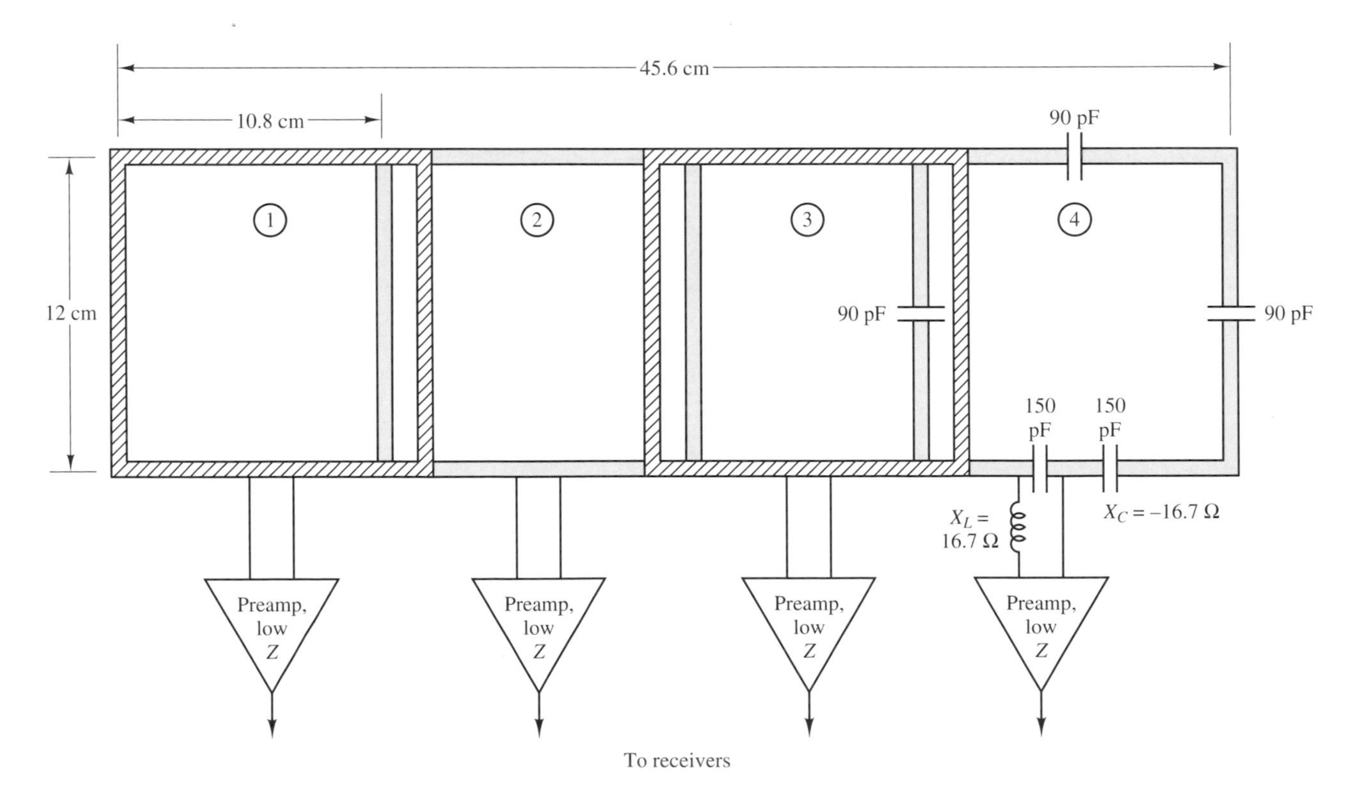

Figure 1 Four-element linear phased array coil for spine imaging. Nearest-neighbor interactions are eliminated by overlap. Next-nearest-neighbor and more distant coil interactions are reduced to negligible levels by connection to low-input-impedance preamplifiers, which, with a series inductor, act as a blocking network

3 NMR PHASED ARRAY TECHNIQUES: OVERLAPPING COILS AND LOW-INPUT-IMPEDANCE PREAMPLIFIERS

Figure 1 shows a schematic diagram of a four-element phased array for imaging of the spine that illustrates some technical aspects of the phased array assembly.[4] The basic difficulty of using an assembly of surface coils is their interactions. If two resonant circuits (i.e., surface coils) tuned to the same frequency are brought close together and have significant mutual inductance, the pair develops two new resonances: one above and one below the desired frequency. There are two steps that can correct this problem. First, as the coils actually overlap and the degree of overlap varies, the sign of the mutual inductance changes. There is an overlap for which the mutual inductance is exactly zero, and both coils can be tuned independently to the same frequency. Thus, in Figure 1, coil 1 and coil 2 are overlapped to have zero mutual inductance, as are the coil pairs 2 and 3, and 3 and 4.

The interaction between next-nearest neighbors, for example, coils 1 and 3, and 2 and 4, in Figure 1, can be alleviated by the electrical circuitry of the preamplifier attached to each coil in the array. If one of the several capacitors that tune each coil (see coil 4 in Figure 1) is shorted out, that coil is no longer resonant at its original frequency and no longer disturbs the operation of other nearby coils at the target frequency. Alternatively, if one of the capacitors were resonated with an inductor then current would be prevented from flowing in the coil, and again that coil would not interfere with its neighbors.

In the phased array of Roemer et al., preamplifiers with low input impedance (a few ohms or less) are connected, in series with an inductor, across one of the capacitors in each coil to receive the NMR signal. The inductance plus series low impedance of the preamplifier resonates the coupling capacitor and forms a blocking network that decreases currents in the surface coil at the NMR frequency, thereby destroying interactions with other coils in the array. The preamplifier–surface coil combination still works as an efficient, low-noise receiver at the NMR frequency, although a complete explanation of this point is beyond the scope of this article.

4 CONVENTIONAL NMR PHASED ARRAY SYSTEM ARCHITECTURE

Figure 2 shows a conventional system architecture for the NMR phased array that is present commercial practice. Each separate surface coil making up the array has its own receiver chain, consisting of preamplifier, mixers, analog-to-digital converter (A/D), and computer buffer. The signals from each coil are digitized and individually converted into images by a two- or three-dimensional FFT (fast Fourier transform). The individual coil images are then combined to form the final image (see Section 5). The drawback of this arrangement is the large amount of data and consequent data manipulation required if there are more than a few coils in the phased array. Present commercial implementation is limited to simultaneous imaging from four coils.

A more advanced concept, time domain processing of phased array data, which makes coil arrays with many individual coils practical, will be discussed below; realization of that scheme requires a very different system architecture.

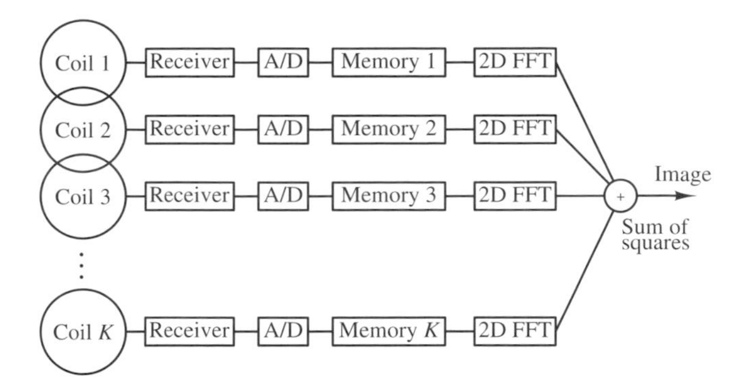

Figure 2 NMR phased array system architecture for sum-of-squares reconstruction. Each coil in the array has its own receiver chain. The signal from each coil is digitized and FFT-processed. The result is combined to give the final image by adding the sums of squares of the individual signals. All receivers operate from the same local oscillator, and the data on all channels are sampled simultaneously

5 PHASED ARRAY IMAGE RECONSTRUCTION

The present method of image reconstruction takes images from individual surface coils in the NMR phased array and combines them into a final image using weights that maximize the S/N of the result. In a radar phased array, signals in the antennas composing the array have the same amplitude but differ in phase. An image is produced by adjusting each signal phase to a common value and then adding the signals constructively. For the NMR phased array, each voxel induces a signal in each coil that varies in both phase and amplitude, depending on the relative location of coil and voxel. Individual coil signals produced by a particular voxel are not of equal quality, so a simple summing of the signals is not adequate.

Intuitively, it is evident from Figure 1 that a voxel above, for example, coil 1, will, in the final image, be mostly composed of the image signal for that voxel from coil 1. Similarly, a voxel above coil 2 will, in the end, be mostly composed of the image signal for that voxel from coil 2. A voxel midway between coil 1 and coil 2 will consist of about half of the image signal from coil 1 and half the image signal from coil 2, with a small amount of additional information from other coils. The image intensity of the ith pixel would, in general, be given by[4,5]

$$I(i) = \sum_{j} w(i,j)S(i,j) \tag{1}$$

where $S(i,j)$ includes the phase and magnitude of the signal in coil j due to pixel i, and $w(i,j)$ is the weighting factor needed to compensate for the phase and magnitude differences between coils.

If all coils in the array are exactly the same, and have the same gain in their receiver chains and the same amount of noise from the body, the signal magnitude $S(i,j)$ is a good measure of signal quality and can be used as the weighting factor $w(i,j)$. Hence, to a good approximation, the best S/N for

For References see p. 4989

the final image is obtained if the image intensity $I(i)$ at pixel i is given by[4,5]

$$I(i) = \left[\sum_j S(i,j)^2\right]^{1/2} \tag{2}$$

where the square root is taken in order to restore the appropriate image contrast. This method of combining the images is called 'sum of squares', and is shown at the combining '+' in Figure 2. For arrays with different coil sizes and receiver chain gains, the inverse of the noise variance for each coil must be included in $w(i,j)$ in order to give the correct statistical weighting. Equation (2) becomes

$$I(i) = K\left[\sum_j \frac{S(i,j)^2}{\langle N(j)^2\rangle}\right]^{1/2} \tag{3}$$

where $\langle N(j)^2\rangle$ is a time average of the square of the noise at the output of the receiver chain for coil j, and K adjusts the dynamic range of the displayed image. The technique using equation (3) is known as the 'sum-of-squares' method, and is presently used in commercial phased arrays because it requires no prior knowledge of coil location or sensitivity and hence is easy to use. A drawback is that image artifacts (e.g., wraparound signal) may give false weights.

The above equations produce an image that retains the strong intensity variations characteristic of surface coil imaging: the image intensity decreases for objects farther from the coils.

The weighting can also be determined using the 'reciprocity principle'.[6] The signal $S(i,j)$ in each coil is proportional to a coil sensitivity factor times the nuclear magnetization $M(i)$, which contains the clinically relevant information. According to reciprocity, the sensitivity of coil j to the magnetization contained in a given voxel i is proportional to the magnitude of the magnetic field $B_1(i,j)$ that would be produced at voxel i by unit current in coil j. Since the spins are only sensitive to rf fields perpendicular to the static field, the component of the calculated $\boldsymbol{B}_1(i,j)$ perpendicular to the static magnetic field must be used.

A combined image without the surface coil variations in intensity can be constructed from a weighted average of each coil's estimate of $M(i)$, which is proportional to $S(i,j)/B_1(i,j)$ The appropriate weighting factor for each $M(i)$ is the inverse noise variance scaled by the relative signal strength, which is proportional to $B(i,j)$. After cancellation of some of the factors of $B_1(i,j)$, the intensity corrected image is calculated as

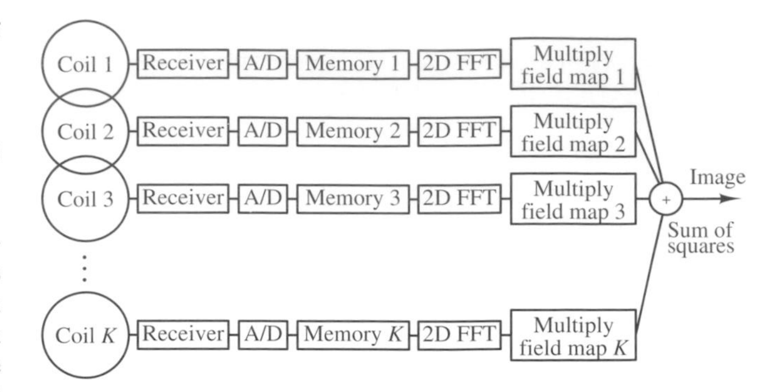

Figure 3 NMR phased array system architecture for field-map weighting reconstruction. Each coil in the array has its own receiver chain. The signal from each coil is digitized and FFT-processed. The result is combined to give the final image by dividing the image strength for a given voxel from a particular coil by the sensitivity of that coil to the voxel position. The sensitivity is proportional to the field strength at the voxel position caused by a unit current in the coil

Figure 4 Four sagittal images taken with a four-element linear phased array coil shown schematically in Figure 1. The four coils receive data simultaneously. The FOV is 48 cm, with a 512 × 512 pixel resolution, NEX = 2, 5 mm slice, TR = 400 ms, TE = 20 ms, imaging time = 6 min 50 s, and B = 1.5 T. Phase encoding is vertical. Only the central 256 points in the horizontal direction are shown. Artifacts at the top and bottom of the images are caused by wraparound

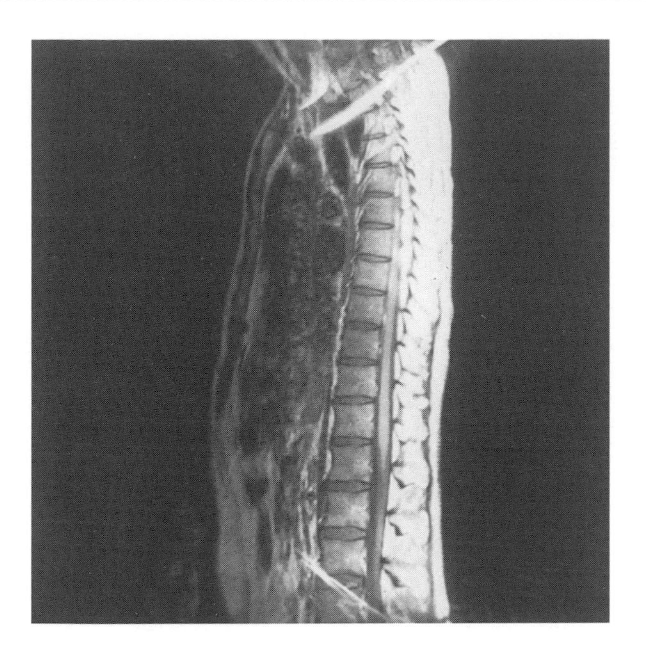

Figure 5 Image obtained by combining the images shown in Figure 4. Regions with weak signals have had their image intensity boosted so that the signal levels are approximately constant over the image. Areas with very poor S/N have been suppressed. The S/N gets worse away from the spine

$$I(i) = \frac{\sum_j S(i,j)B_1(i,j)/\langle N(j)^2\rangle}{\sum_j B_1(i,j)^2/\langle N(j)^2\rangle} \quad (4)$$

where the sum in the denominator normalizes the total weighting to unity. Figure 3 shows a system architecture for B_1 field-map weighting phased array image reconstruction.

The B_1 weighting in equation (4) requires detailed knowledge of the location of the rf coils and a stored array of calculated weights $B_1(i,j)$, and is harder to implement than the sum-of-squares method. Unlike the sum-of-squares method, B_1 weighting is not subject to wraparound artifacts.

The sum-of-squares method [equation (2) and (3)] does not achieve the highest possible S/N for a given reconstruction because the use of signal magnitudes does not produce the optimum combination of signals when the S/N is low or noise is correlated between coils. An additional refinement using the correct phase relationships makes the final expressions somewhat more complicated.[4,5] To take correlated noise fully into account, the optimum reconstruction would combine signals with a phase angle dependent on the amount of noise correlated in-phase and 90° out of phase with the NMR signal. In practice, the small gain in S/N, typically less[4] than 10%, is not worth the added complexity in keeping track of the coil positions.

Figures 4 and 5 show the initial images from a four-element spine phased array combined into a final image. In this case, the final image (Figure 5) has also been corrected for nonuniform sensitivity of the surface coils. Areas with low signal (and poor S/N) have been boosted so that all regions have approximately the same signal level. The noise levels in these regions are also increased, so areas that started with weak signals still have the same (poor) S/N.

6 NOISE CORRELATIONS AMONG SURFACE COILS IN THE NMR PHASED ARRAY

The NMR phased array receives both NMR signals and electrical noise from the imaging subject. The latter is the result of thermally generated noise currents in the subject that produce fluctuating rf magnetic fields picked up by the array. Thermally generated noise in the electrical components of the phased array system also contributes to noise in the final signal and resultant image or spectrum.

Noise generated by the subject can be received simultaneously by several coils in the array, which may produce noise correlations among the coils. The relative amount of correlated noise in two coils depends on the proximity of the coils—the closer the coils, the greater the correlation.

Although noise is one of the most difficult and fascinating theoretical aspects of the NMR phased array system, the practical effects of noise correlation are minor. The amount of correlated noise is sufficiently small that it cannot be used to reduce noise significantly in the reconstructed image or spectrum. Roemer et al.[4] show that, for a typical case such as adjacent overlapped coils in a spine array, the correlated noise power is calculated to be a maximum of 41%, assuming a dissipative, infinite half-space. The measured correlated noise power was 23%, because the body is finite.[4] After accounting for the uncorrelated noise in a preamplifier and receiver components and performing the optimum reconstruction, less than a 10% improvement in S/N could be realized by using knowledge of these correlations.

7 CLINICAL APPLICATIONS

A variety of phased array coil geometries have been used for clinical applications. The improved S/N obtained with the phased array has enabled established diagnostic procedures to be performed faster and/or with higher spatial resolution, and has made possible new procedures that were not practical with conventional coils.

Clinical phased arrays can be classified by configuration into surface coil arrays and volume arrays. The surface coil array is usually a linear chain of coils, and is best suited to observe a superficial structure whose length is greater than its depth. The coils of a volume array attempt to examine all the internal structures of the body part they circumscribe.

The spinal phased array is a prime example of a surface coil array used to improve an established MRI application. Spine imaging with conventional surface coils has been a time-consuming endeavor, because conventional coils have to be repositioned and the scan repeated for several locations along the spine.[7] The spinal phased array allows the acquisition of a 40–45 cm linear field-of-view in a single imaging period. Using 512 × 256 pixel resolution on a rectangular field-of-view can yield high-resolution sagittal images. One commercial spinal phased array design is made up of six coils spaced 12 cm apart along the spine. Three choices of four contiguous coils can be activated to image the cervical, thoracic, or lumbar spines. Phased array techniques reduce total examination time for spinal studies by a factor of two or more, and lead to greater patient acceptance and scanning efficiency.[7]

For References see p. 4989

Volume phased arrays were first applied to pelvic imaging[8] because it was assumed that improved resolution was only meaningful where respiratory motion was minimal. As an alternative to the more traditional volume imaging coils such as the birdcage resonator (see ***Birdcage & Other High Homogeneity Radiofrequency Coils for Whole Body Magnetic Resonance***), the volume phased array offers several advantages. First, the individual coils of the array can be tailored to fit the anatomy with fewer constraints than are needed for a standard volume coil. Therefore the array can provide a tighter coupling to the signal from the region of interest and can exclude noise from more distant tissue. Second, the array that circumscribes the volume of interest incorporates the benefits of quadrature reception without needing to satisfy the usual requirements of field orthogonality and mode isolation for a traditional quadrature coil.[5] Finally, superficial regions yield the high S/N of an individual surface coil, whereas the central region has the S/N resulting from averaging signals of the four coils forming the array. The S/N at the center is no better than that of a small quadrature volume coil that has the same overall dimensions as the array. Motion of subcutaneous fat can produce disproportionally larger motion artifacts from the surface array because of tight coupling to the array.

A four-coil pelvic array comprises two coils placed anteriorly and two posteriorly. The S/N for the pelvic array is approximately three times higher than the S/N for a full-sized body coil imaging the same region.[9] Such an S/N increase permits pelvic imaging with smaller fields-of-view (16 cm), thinner slices (3 mm), and/or higher resolution (512 × 512).

Phased arrays also facilitate the use of FSE (fast spin echo) techniques to obtain T_2-weighted images. Long effective echo times provide good contrast in pelvic structures, but at reduced signal strength. The enhanced S/N of the phased array permits good images to be acquired in a practical scan time using FSE.[10]

High-speed imaging sequences using low-flip-angle excitation and gradient-recalled echoes can produce images in about 1 s. Such images taken with a body coil have marginally useful S/N. With the volume phased array, low-flip-angle, gradient-recalled echo sequences can produce clinically useful, motion-free images of the liver[11] and pulmonary vasculature[12] during a single breath-hold.

High-resolution images of brain structures can be acquired by applying a volume phased array to the head.[13] The head array consists of two pairs of coils that cover about 16 cm of the circumference on each side of the head. When studying temporal lobe epilepsy, the coils are placed over both hippocampi, which lie about 5–6 cm from the lateral scalp surfaces in adults. At this depth, the S/N is about 1.67 times better than that of the standard quadrature head coil. Figure 6 is an example of a T_2-weighted fast spin echo image acquired for a 512 × 512 matrix on a 3 mm thick slice. The 0.3 mm × 0.3 mm in-plane resolution is degraded by pulsatile motion of the brain tissue unless flow compensation and cardiac gating are used with the pulse sequence.

MR neurography is an example of a new MR diagnostic technique whose development depends on the S/N advantages of phased arrays. MR neurography[14] employs pulse sequences tailored to highlight peripheral nerves and suppress signals from surrounding tissue such as fat, vessels, and muscle. Inflamed nerves appear hyperintensive compared with normal

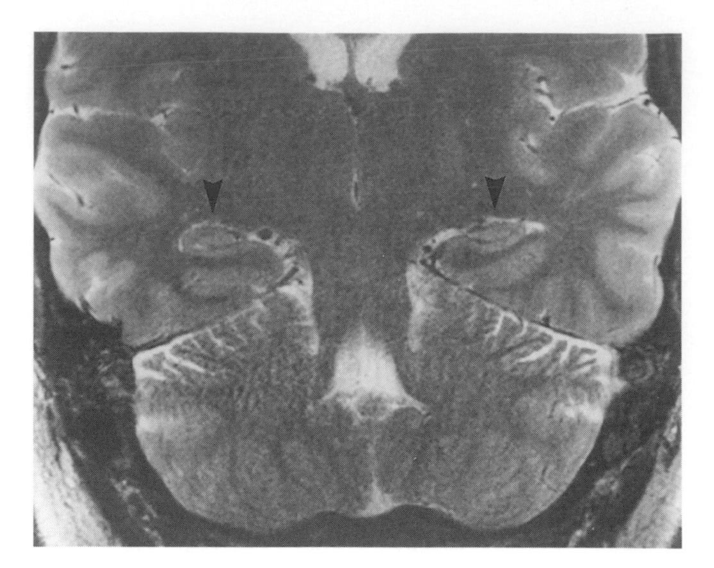

Figure 6 Portion of a high-resolution oblique coronal brain image, showing the internal architecture of the hippocampi (the almond-shaped structures indicated by the arrowheads)

nerves. Because the resulting signals are weak, a high S/N performance is required from the coil array in order to distinguish the nerve structures from the surrounding tissue.

A linear surface coil array has been built to track the brachial plexus, which runs from the neck to the axilla. The brachial plexus is a complex network of four cervical and one thoracic nerves that fuse, mix, and split into nerves supplying the arm, shoulder, and chest. The brachial plexus array has six coils contoured fit with one coil anterior to each axilla and each clavicle, and a lateral pair covering the volume of the lower neck. The three coils on one side plus the opposite neck coil are activated for a unilateral study. Selecting the middle four coils can give a bilateral study of the central region. The tight coupling of the array to the region of interest produces S/N not possible with conventional coils.

8 REAL-TIME PROCESSING

Real-time data processing techniques, similar to those used for radar and ultrasound, have been developed for NMR signals from the multiple receivers in the NMR phased array.[15] This approach has not been implemented on commercial imagers, since they have relatively few receive channels and since there is not presently a great demand for real-time imaging. As methods of building phased arrays with larger numbers of coils develop and the applications of fast imaging expand, the increased amount of data may make real-time processing attractive.

To see how a real-time processing method is implemented, we need only look at equation (4) and construct a mathematically equivalent expression in the time domain. According to equation (4), the reconstructed image from each coil is multiplied pixel-by-pixel with the associated field map from the rf coil. The results from each channel are then added together. The multiplication function in the image domain corresponds to convolution in the time domain. Hence the NMR signal from each receiver can be convolved with an appropriately

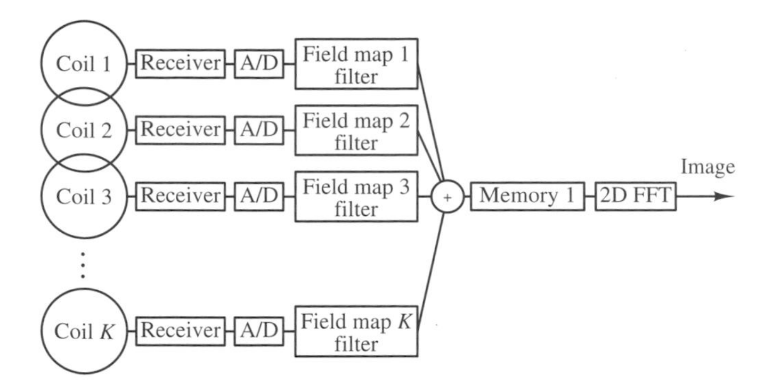

Figure 7 NMR phased array system architecture for time domain filtered reconstruction. Each coil in the array has its own receiver chain. The signal from each coil is digitized, multiplied by a filter kernel, and combined in a data array with signals from other array coils. The data array is FFTd to give the final image. So only one FFT needs to be done in this case, no matter how many individual coils in the array, in contrast with one FFT per coil in the conventional system (Figures 2 and 3)

designed filter, the results of which are added to form a composite time domain signal.

Figure 7 is a schematic diagram of a real-time processing system. The information from each coil is processed by its associated filter, and the results from all coils are combined into an array in memory as shown. The contents of the memory are then processed by a single FFT. This arrangement differs from that shown in Figure 2 where the data from each coil must be FFT-processed. The point is that the FFT is generally done in an array processor, which is relatively expensive. For a coil assembly consisting of a few surface coils, the FFTs of the coils can be done sequentially. For a coil assembly containing many coils, sequential FFTs would impose a significant time penalty. So the present scheme allows the processing of data from many coils in a short time.

The remaining difficulty is the filters, which, at least conceptually, are straightforward. They represent the Fourier transform of the field map generated by each coil. A substantial practical simplification results when one realizes that the coil field map is slowly varying in the image space and needs relatively few terms to describe it properly in the time domain. With this simplification, a practical implementation of the time domain processing uses a filter kernel derived by Fourier transforming the field map and truncating the result. It would be implemented by the application of programmed high-speed signal processing integrated circuits, which would multiply the incoming signal by the appropriate filter function and then combine into an array in memory. These methods are discussed in detail by Roemer.[15]

9 RELATED ARTICLES

Birdcage & Other High Homogeneity Radiofrequency Coils for Whole Body Magnetic Resonance; Surface & Other Local Coils for In Vivo Studies; Whole Body Magnetic Resonance Spectrometers: All-Digital Transmit/Receive Systems; Whole Body MRI: Local & Inserted Gradient Coils.

10 REFERENCES

1. W. A. Edelstein, T. H. Foster, and J. F. Schenck, *Proc. 4th Ann. Mtg. Soc. Magn. Reson. Med.*, London, 1985, p. 964.
2. C.-N. Chen and D. I. Hoult, 'Biomedical Magnetic Resonance Technology', Adam Hilger, Bristol, 1989.
3. P. B. Roemer and W. A. Edelstein, *Proc. 6th Ann. Mtg. Soc. Magn. Reson. Med.*, New York, 1987, p. 410.
4. P. B. Roemer, W. A. Edelstein, C. E. Hayes, S. P. Souza, and O. M. Mueller, *Magn. Reson. Med.*, 1990, **16**, 192.
5. C. E. Hayes and P. B. Roemer, *Magn. Reson. Med.*, 1990, **16**, 181.
6. D. I. Hoult and R. E. Richards, *J. Magn. Reson.*, 1976, **24**, 71.
7. D. M. Yousem and M. D. Schnall, *J. Comput. Assist. Tomogr.*, 1991, **15**, 598.
8. C. E. Hayes, N. Hattes, and P. B. Roemer, *Magn. Reson. Med.*, 1991, **18**, 309.
9. C. E. Hayes, M. J. Dietz, B. F. King, and R. L. Ehman, *J. Magn. Reson. Imaging*, 1992, **2**, 321.
10. R. C. Smith, C. Reinhold, T. R. McCauley, R. C. Lange, R. T. Constable, R. Kier, and S. McCarthy, *Radiology*, 1992, **184**, 671.
11. A. E. Holsinger-Bampton, S. J. Riederer, N. G. Campeau, R. L. Ehman, and C. D. Johnson, *Radiology*, 1991, **181**, 25.
12. T. K. Foo, J. R. MacFall, C. E. Hayes, H. D. Sostman, and B. E. Slayman, *Radiology*, 1992, **183**, 473.
13. C. E. Hayes, J. S. Tsuruda, and C. M. Mathis, *Radiology*, 1993, **189**, 918.
14. A. G. Filler, F. A. Howe, C. E. Hayes, M. Kliot, H. R. Winn, B. A. Bell, J. R. Griffiths, and J. S. Tsuruda, *Lancet*, 1993, **341**, 659.
15. P B. Roemer, US Patent 5 086 275 (1989).

Biographical Sketches

Peter B. Roemer. 1978, Ph.D., 1983, M.I.T., Cambridge, MA. Electrical engineer, GE Corporate Research and Development, 1983–1990; Manager, MRI, GE Corporate Research and Development, 1990–1994; Engineering Manager, Advanced NMR Systems, June 1994–present. Research specialities: MRI imaging systems, MRI gradient and RF coils.

William A. Edelstein. 1965, U. Illinois, Urbana, IL; A.M., 1967; Ph.D., 1974, Harvard University, Cambridge, MA. Research fellow, Glasgow University, 1974–77; Research Fellow, Aberdeen University, 1977–1980; physicist, GE Corporate Research and Development, Schenectady, NY, 1980–present. Research specialties: MR imaging, thermal remediation of environmental contamination.

Cecil E. Hayes. *b* 1941. B.Eng.Phys., 1964, Cornell University, Ithaca, NY; Ph.D., 1973, (supervisor Robert V. Pound), Harvard University, Cambridge, MA. Postdoctoral work at Rutgers University (with Herman Y. Carr) and at University of Utah (with David C. Ailion), 1973–1982. Senior physicist, GE Medical Systems 1982–91. Associate Professor of Radiology, University of Washington, 1991–present. Approximately 25 publications. Current research specialty: rf coils for MRI.

Whole Body Machines: Quality Control

Franca Podo
Istituto Superiore di Sanità, Rome, Italy

Wim Bovée
Delft University of Technology, Delft, The Netherlands

&

J. Stewart Orr
Hammersmith Hospital, London, UK

1 INTRODUCTION

A whole body system is one used for clinical investigations and studies of living human subjects. The studies may be of any part of the body. A whole body system includes the machine itself plus associated equipment and the mode of use. A whole body system may be used for magnetic resonance imaging (MRI), magnetic resonance spectroscopy (MRS), or both.

Quality control means

1. ensuring that an individual system performs in a way that consistently gives reproducible results that render good clinical value and provide a basis for improving techniques;
2. ensuring that any system gives results conveying the same clinical interpretation as other systems for the same mode of use, so that comparisons of diagnoses and clinical opinion can be made, trained staff can apply their skill generally, new techniques can be propagated, and the body of useful knowledge increased.

1.1 Necessity for Quality Control and its Evolution

Quality control is needed because whole body systems do not automatically perform to give the desirable results referred to above. Nor does any one system automatically convey the same interpretation as others.

Examples of poor results from checks on systems and comparisons between them are shown in Figure 1. Progress occurs through improvements in equipment and techniques, combined with local quality control, and the analysis of clinical results that are reproducible and made comparable by controlled calibration.

For list of General Abbreviations see end-papers

For NMR characteristics such as relaxation times, spin density, and chemical shift, pressures have grown for the development of methods for quality control. The pressures come from users (both clinical and scientific), from suppliers, and from government agencies. Users want quality control for improving performance and clinical usefulness, and so do suppliers, together with the public demonstration of high quality. Government agencies want quality control to ensure safety and value for money.

The evolution of quality control is an iterative process, because all the details of a method cannot be clear when work begins, and because practical advantages and disadvantages cannot all be foreseen. There have been several lines for developments in quality control of whole body systems.

Government agencies in individual countries were faced with responsibilities for approving the purchase of the new systems and judging performance. Progress was made in many countries, and in the United States the suppliers' organization (NEMA) produced standard methods for selected parameters.[1–4] Manufacturers needed methods for checking performance during development, before despatch, and on site. Such methods were the basis of the NEMA standards. The international standards body, IEC, is adopting some of the methods as tests of compliance with health and safety requirements.[5]

In the European Community, NMR Concerted Actions, mainly of users, ran from 1984 to 1992.[6–8] These Actions, which had useful interactions with individual manufacturers and with NEMA, went through the repeated iteration of steps for defining the parameters to be measured, creating test objects and substances, specifying protocols, and assessment in practical trials.[9–20] The methods were then tested in studies of selected organs.[21,22] The EC Concerted Action standard methods and test objects have been available and used in Europe and elsewhere.

1.2 Physical Basis and Practical Requirements for Quality Control

Whole body systems can use NMR features to define the position of tissues and also to establish their characteristics. The link between characteristics and positions is the NMR signal. These signals are dependent on basic parameters and functions, and the measurement of these is part of quality control. However, it has been found that tests using simulations of actual use are necessary.

The purpose of the test objects and methods is to define absolutely and separately the characteristic and the position. Quality control with test objects then assesses performance in stimulation and detection of the signal. However, the complexities of whole body systems makes clear separation difficult, and interpretation of tests is not straightforward.

A general practical requirement in quality control is that the results can be obtained in an acceptably short period of time, by using the normal clinical operating modes of the system. The methods may, however, require the use of additional computer software.

Quality control is of value only if the finding of any deficiency is followed up by effective action. The control tests, the recommended actions, and the outcomes should be systematically recorded.

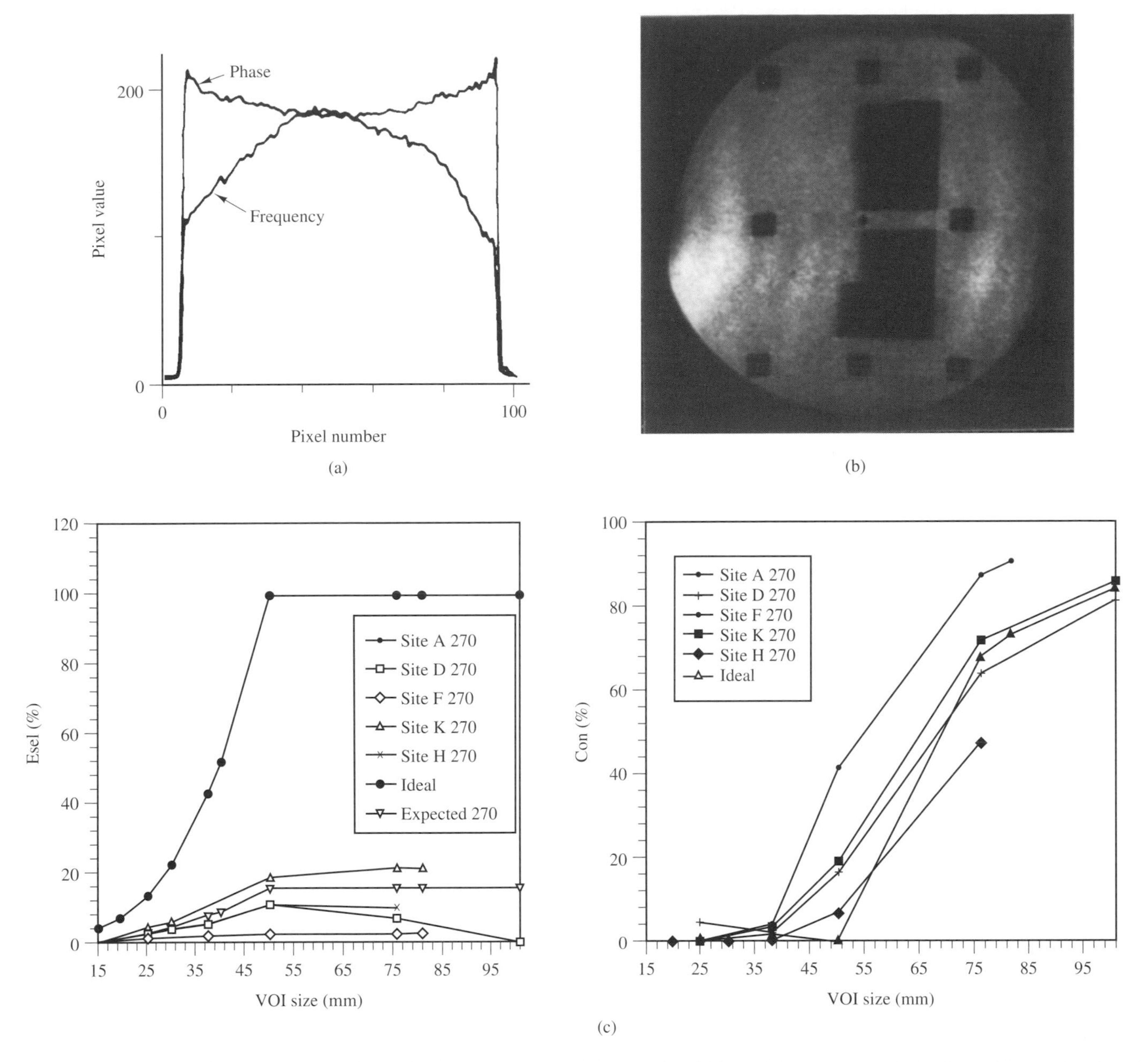

Figure 1 (a) Low-quality uniformity profiles and (b) gross geometric distortion in MRI. (c) Measurements of selection efficiency (Esel) and contamination (Con) in volume-selected proton MRS, using STEAM with $TE = 270$ ms. ((a) and (b) reprinted with permission of Pergamon Press, Oxford, from R. A. Lerski et al., *Magn. Reson. Imaging*, 1988, **6**, 203, 205. (c) reprinted with permission of Elsevier Science Ltd., New York, from S. F. Keevil et al., *Magn. Reson. Imaging*, 1995, **13**, 148)

2 MRI: TEST PARAMETERS AND PROTOCOLS

Most of the MRI test parameters need different features in the relevant test objects. The use of a standardized set of test objects[9,11] in combination with the test procedures offers some clear advantages to the users of MRI equipment, especially when these methods have been validated in multicenter trials performed on different MRI systems.[8,10,12] Some protocols (NEMA) provide only recommendations on the essential requirements for test objects. These requirements[1–4] are expressed primarily as the geometry and NMR characteristics of the signal-producing volume, together with definitions for the specification volume and area and the measurement region of interest (ROI).

Key parameters for assessing performance in quality control are signal uniformity, geometric distortion, signal-to noise ratio, slice thickness, and profile. Additional important parameters are spatial resolution, signal linearity, uniformity of image signal-to-noise ratio, slice position and warp, and image contrast-to-noise ratio. Of particular importance for quantitation and diagnosis of malfunction are T_1, T_2, and spin density precision and accuracy.[9,11] Deviations of these parameters from their expected values indicate malfunction or erroneous calibration of instrument components.

For References see p. 4995

Key parameters are described below, together with their value for assessing performance and the basics of measurement. It is worth noting that there are three aspects of variation of signal across the image of a uniform test object: uniformity, artifacts, and signal-to-noise ratio.

Uniformity of image signal refers to the ability of the equipment to produce an image with identical signal response over one plane from a test object with uniform NMR characteristics. This parameter, which characterizes the low-spatial-frequency nonuniformities (i.e., non-noise) typical of MRI, may be affected by main field and rf inhomogeneities, rf penetration and coil geometry, gradient field nonlinearities, inadequacies in gradient pulse calibration or eddy current corrections, irregular response of the rf receiver, and signal processing.[2,11]

Uniformity is given by pixel variations within the full ROI occupied during clinical use. The parameter can be quantified as the deviation in the image signal intensity relative to the midrange pixel value in the profiles derived from scans in both the phase-encoding and frequency-encoding directions.

Geometric distortion is the deviation between measured distances in an image and the actual corresponding dimensions in the object (allowance being made for magnification).

This parameter (produced by factors such as field inhomogeneity, gradient defects, or signal sampling imperfections[3,11]) is derived by measuring the distances between fixed points in the image of the test object and determining their deviations from the respective expected values. These deviations are measured in the three orthogonal planes centered at the isocenter.

Image *signal-to-noise ratio* (S/N) represents the mean of the pixel values over the ROI divided by their standard deviation. This noise is the random variations in pixel intensity. It affects the clarity of MR images and is also an index of hardware performance.[1,9–11] Changes in image S/N values may occur as a consequence of variations in system calibration, gain, coil tuning, rf shielding, etc.

To simulate image noise in a typical clinical situation, the rf receiver coil must be electrically loaded. Loading may be accomplished by the test objects themselves or with accessories, or by the use of inductive damping loops that can be adjusted for the loaded Q of individual machines.[1,11] The use of nonstandard test objects that provide some loading for a system may provide a convenient tool for routinely checking the S/N of equipment. Results acquired on different types of system may not, however, be comparable unless standardized test objects and loading procedures are applied.[11] The problems of loading, although very important, have not yet been completely solved.

Slice thickness is the width at half-maximum of the *slice profile*, which itself is the variation in contribution to the image signal as a function of position orthogonal to the plane. It expresses the variation in effectiveness of the selective excitation process through the slice. The slice thickness depends on the rf pulse shape and sequencing, rf field homogeneity, the selection gradient and other parameters. Although slice selection attempts to achieve reception of signal solely from a slab of rectangular profile, the selected slice may have a profile quite different from the ideal. The slice profile can be assessed either by direct measurement with a thin inclined slab of signal-producing material, or by numerical differentiation of the measured edge response function from an inclined surface of an inert wedge immersed in a signal-producing material.[4,9–11]

Spatial resolution is usually pixel-limited, but several features may lead to a loss of resolution. It is, for instance, degraded by filtering the raw data through a function used for reducing the image S/N. Eddy currents produced by rapid gradient switching will lead to uneven resolution loss throughout the imaging field. In 2D FT MRI, the frequency-encoding and phase-encoding resolutions will differ, since the sampling and filtering in these directions are independent. Spatial resolution can be measured by determining the minimum bar separation detectable in the image of a bar pattern of varying spacing.[9,11] The image profile available on most systems may be used to measure the modulation across a particular set of bars. Spatial resolution can also be assessed from an edge-containing test object, by calculating the modulation transfer function (MTF) associated with the obtained edge spread function.[11]

In addition to the above parameters, *signal linearity* represents the linearity of the relationship between image signal and the concentration of the excitation-sensitive material. This can be measured with the use of a test object including a series of tubes containing substances with different spin densities.

Uniformity of image signal-to-noise ratio measures the variation of image S/N over one imaging plane, obtained from the same test object utilized for S/N, while *slice position* and *slice flatness* or *warp* jointly describe defects in the selection and expression of planar slices. They can be measured with the same test object.[9,11]

T_1, T_2, *and spin density precision and accuracy* are the basis of the contrast in MRI images, without which they have no clinical value. Their long- and short-term precision is therefore worth study and control.

Many centers are also interested in the accuracy of T_1, T_2, and spin density, for both clinical and scientific reasons. The accuracy of measurements with the whole body system is the closeness with which parameter values correspond to the known values of standards that have been separately calibrated.[9,11]

2.1 MRI Test Objects

As mentioned above, NEMA gives recommendation on some of the features needed for suitable test objects. These recommendations are similar to the detailed final designs worked out and tested by the EC Concerted Action (Figure 2).

2.2 Levels of Quality Control in MRI

There are several levels of quality control appropriate for different circumstances. When receiving a new machine, complete acceptance tests should be done, preferably in the presence of the manufacturer or the supplier.

Periodic maintenance requires tests less detailed than acceptance tests, but more extensive than those for routine application by users. The latter need to be acceptably rapid and to provide control, particularly of signal quality and geometry.

One of the most critical performance indicators is S/N. It is necessary to measure this frequently, because it is easy to lose signal or gain noise. The quality of image is also critically dependent on optimal adjustment of rf pulses and sequence timings, and it is important that these should be checked on a

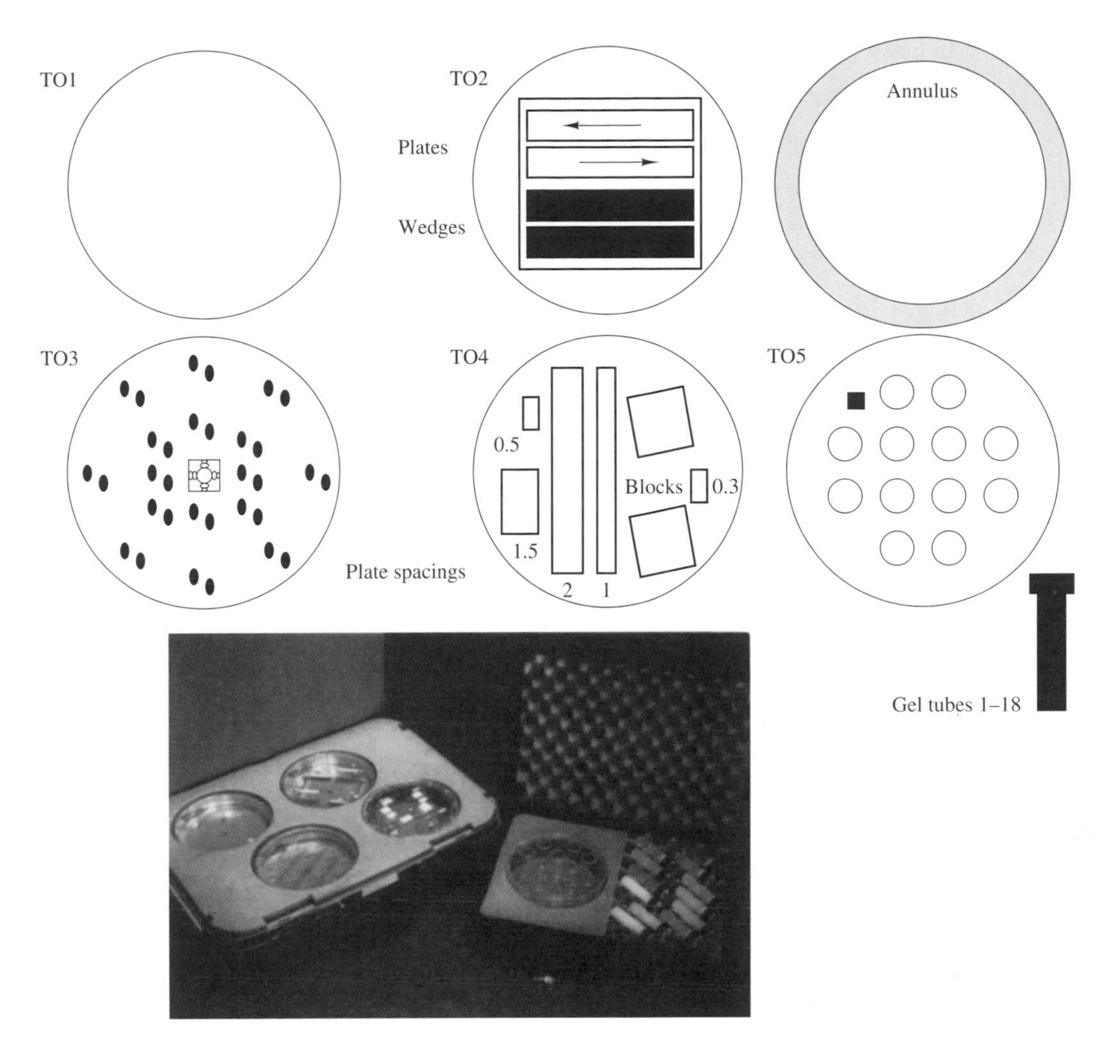

Figure 2 MRI test objects designed and constructed by the EC Concerted Action for quality control in whole body clinical systems:[9–12] TO1 [uniformity of image signal; image S/N; uniformity of image S/N; artifacts (ghosting)]; TO2 (geometric distortion; slice profile; slice thickness); TO3 (slice position; slice warp); TO4 (spatial resolution; modulation transfer function); TO5 (T_1, T_2, and spin density accuracy and precision; signal linearity; image contrast-to-noise ratio). The annulus represents one approach to providing appropriate loading for measuring S/N. (Schemes of test objects are reproduced with permission of Pergamon Press, Oxford, from R. A. Lerski et al., *Magn. Reson. Imaging*, 1993, **11**, 837. Photograph of test objects courtesy of R. A. Lerski)

regular basis. For all tests, it is necessary to have a series of techniques with standard protocols, so that each performance feature may be measured and compared with least effort and time.

3 MRS: TEST PARAMETERS, TEST OBJECTS, AND PROTOCOLS

While the same procedures in principle might be used for MRI and *magnetic resonance spectroscopic imaging* (MRSI), for single volume spectroscopy, where only one volume of tissue is under study, additional and different tests are needed.[13]

For single volume spectroscopy, the localization performance is the important factor for many of the parameters. *Localization* is the process of maximizing signal from the single volume that is the object of study and of restricting signals from outside. There are many useful techniques used for localization but none are ideal.

A simple approach that does not put much demand on the MRS equipment and software is to employ a test object having two compartments (an inner and outer one) for the localization testing,[13,15–18] and to use standard bottles and spheres for some other tests. The inner and outer compartments may each contain a substance with a single NMR line, but with different chemical shifts. The relative intensities of the two lines in the spectrum then vary as the size and position of the localized volume of study varies with respect to size and position of the two compartments.

It should be noted that some of the parameters below are also relevant to techniques that do not employ localization.

3.1 Test Parameters

Linearity is the linearity of the relations between signal strength and concentration, and signal strength and volume.

Spectral resolution is the full linewidths at one-half and at one-tenth of the maximum peak height for the spectral peaks

For References see p. 4995

from the inner and outer compartments. To assess the effect of the measuring sequence, this test can be performed with and without localization.

The *signal-to-noise ratio* (S/N) is measured per unit concentration in a fixed volume.

The *selection efficiency* is the ratio of the signal strengths from the inner compartment with and without localization. This parameter indicates signal loss due to the applied localization sequence. Ideally, its value equals one if the selected volume is isocentric with and equals or exceeds that of the inner container. In practice, this signal loss will depend on the ratio of the selected volume and the inner compartment volume, and should therefore be determined for various ratios.

The *extra volume suppression factor* is the ratio of the signal strengths from the outer compartment without and with localization. This parameter quantifies unwanted signal leakage from outside into the selected *volume of interest* (VOI). Ideally, its value is infinite if the selected volume is centered in and smaller than or equals the volume of the inner container. Owing to imperfect slice profiles, it will depend on the ratio of the selected and the inner compartment volumes. It should therefore also be measured for various ratios.

Contamination is the ratio of the unwanted signal from outside the VOI and the total signal in the spectrum. This is a less fundamental localization performance property than the suppression factor and selection efficiency. It depends on the relative volumes of and the concentrations in the two compartments, and it directly expresses the percentage of unwanted signal in the spectrum, using the actual pulse sequence and VOI.

The *VOI profile* characterizes the slice profile in one, two, or three dimensions. Ideally, the profile should be plotted and specified, for instance by the full width at 20, 50, and 80% of the maximum signal height. The profile can be determined using imaging methods.

The *accuracy of the VOI position* is the deviation between the position of the center of the VOI, as determined with a test object and procedure, and the true position. An imaging method can be used, after performing the imaging quality control tests.[9] The VOI center is estimated from the slice profile.

Selective suppression of signals in the spectrum is the ratio of the signal strength without and with suppression, obtained from two localized experiments, with and without suppression of the signal of the inner compartment. This parameter can be used to test the effectiveness of suppression methods (e.g., for water and fat), and with spectral editing.

For assessing the *stability* of localization, an experiment is repeated several times. The average value of the strengths of the signals from the two compartments and their standard deviations are determined. This test can be used to assess the short- and long-term stability of the equipment.

Precise descriptions of the implementation of the tests to determine the parameters indicated above, and more detailed information about the parameters themselves and other quantities involved, can be found in the literature.[8,13–18]

3.2 Test Objects and Substances

Substances are needed for constructing and filling test objects. For single volume spectroscopy, the two-compartment object is very useful (Figure 3). The inner compartment is usually a rectangular volume, since most methods localize this shape. For shimming reasons, a sphere as outer compartment is advisable for proton MRS, and a material with a favorable diamagnetic susceptibility should be chosen from which to construct the test objects. Perspex is therefore to be preferred to glass. Moreover, it is less fragile—but it is not inert, and therefore imposes restrictions to the choice of the substances for filling the compartments.

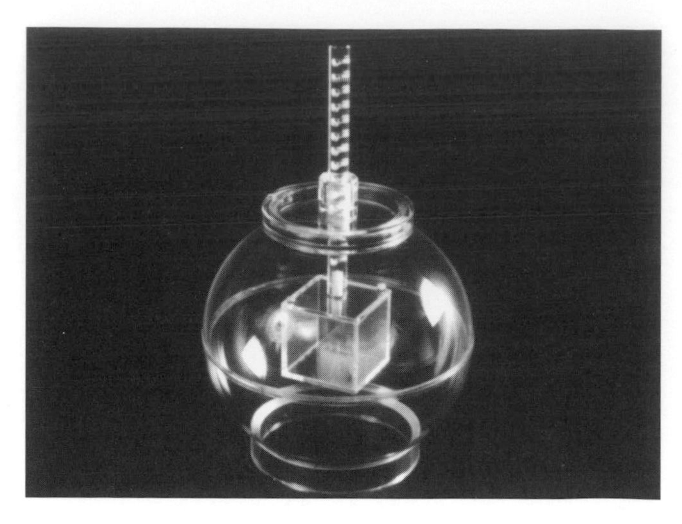

Figure 3 Test object designed and constructed by the EC Concerted Action for quality control of volume selection in whole body MRS systems.[13,15] (Courtesy of Dr. S. F. Keevil)

The substances used to fill the two compartments should, as far as possible, meet requirements that are sometimes conflicting.

1. The NMR spectrum of a substance should be a single line.
2. The chemical shift difference between the compounds in the two compartments should be sufficient to minimize line overlap in the spectrum, but also small enough to minimize localization artifacts when selecting slices or volumes by means of frequency-selective pulses and gradients.
3. The relaxation times T_1 of the two lines should be equal, as should be the T_2 values; both should be in the range of in vivo values. Doping with paramagnetic substances can sometimes give the desired relaxation values. If suitable values cannot be obtained, equal T_1 values are to be preferred for assessing the quality of localization obtained by using ISIS sequences in ^{31}P MRS. For ^{1}H MRS, equal T_2 values are preferable, because, in general, echo techniques are employed for localization.
4. If possible, the compounds should be easily available, not poisonous, not flammable, stable, and have relaxation times with a weak temperature dependence and chemical shift values with none.

Details are given by Keevil et al.[17]

The field of MRS is still in development, and new techniques evolve. Therefore quality control needs continued adaptation, and should also be applied to techniques like spectroscopic imaging that have not yet received much attention in this respect. Quality control should be used to check and calibrate apparatus and techniques regularly.

3.3 Data Analysis

In vivo MRS data do not have a high signal-to-noise ratio, and considerable line overlap occurs. As a consequence, data analysis methods have to be chosen very carefully. Several quantification methods exist, which may lead to widely varying results, owing to systematic and operator-dependent errors. Model function fitting is the best way to quantify the data and minimize these errors. The effect of the pulse sequence and the data analysis method can be checked using test objects and standard solutions. The results can be incorporated as prior information in the data analysis method, which significantly decreases the systematic errors and the standard deviation.[13,19]

4 RELATED ARTICLES

Data Processing; Quantitation in Whole Body MRS; Relaxation Measurements in Imaging Studies; Surface Coil NMR: Quantification with Inhomogeneous Radiofrequency Field Antennas.

5 REFERENCES

1. NEMA Standards Publication No. MS 1, 'Determination of Signal-to Noise Ratio (SNR) in Diagnostic Magnetic Resonance Images', National Electrical Manufacturers Association, Washington, DC, 1988.
2. NEMA Standards Publication No. MS 3, 'Determination of Image Uniformity in Diagnostic Magnetic Resonance Images', National Electrical Manufacturers Association, Washington, DC, 1989.
3. NEMA Standards Publication/No. MS2, 'Determination of Two-dimensional Geometric Distortion in Diagnostic Magnetic Resonance Imaging', National Electrical Manufacturers Association, Washington, DC, 1990.
4. NEMA Standards Publication/No. MS5, 'Determination of Slice Thickness in Diagnostic Magnetic Resonance Imaging', National Electrical Manufacturers Association, Washington, DC, 1992.
5. IEC/TC 62 B Medical Electrical Equipment. Part 2: Particular Requirements for the Safety of Magnetic Resonance Equipment for Medical Diagnosis. Committee Draft 62 B (Secretariat) 196, 1993.
6. F. Podo, J. S. Orr, K. H. Schmidt, and W. M. M. J. Bovée, *Magn. Reson. Imaging*, 1988, **6**, 175.
7. F. Podo, J. S. Orr, W. M. M. J. Bovée, J. D. de Certaines, and D. Leibfritz, *Magn. Reson. Imaging*, 1993, **11**, 809.
8. F. Podo and J. S. Orr, eds., 'Tissue Characterization by MRS and MRI', Istituto Superiore di Sanità, Rome, 1992.
9. EEC Concerted Research Project, *Magn. Reson. Imaging*, 1988, **6**, 195.
10. R. A. Lerski, D. W. McRobbie, K. Straughan, P. M. Walker, J. D. de Certaines, and A. M. Bernard, *Magn. Reson. Imaging*, 1988, **6**, 201.
11. R. A. Lerski and J. D. de Certaines, *Magn. Reson. Imaging*, 1993, **11**, 817.
12. R. A. Lerski, *Magn. Reson. Imaging*, 1993, **11**, 835.
13. W. M. M. J. Bovée, in 'Magnetic Resonance Spectroscopy in Biology and Medicine', eds. J. D. de Certaines, W. M. M. J. Bovée, and F. Podo, Pergamon Press, Oxford, 1992, Chap. 12.
14. F. Podo, W. M. M. J. Bovée, J. de Certaines, D. Leibfritz, and J. S. Orr, *Magn. Reson. Imaging*, 1995, **13**, 117.
15. EEC Concerted Action Project, *Magn. Reson. Imaging*, 1995, **13**, 123.
16. M. O. Leach, D. J. Collins, S. Keevil, I. Rowland, M. A. Smith, O. Henriksen, W. M. M. J. Bovée, and F. Podo, *Magn. Reson. Imaging*, 1995, **13**, 131.
17. S. F. Keevil, B. Barbiroli, D. J. Collins, E. R. Danielsen, J. Hennig, O. Henriksen, M. O. Leach, R Longo, M. Lowry, C Moore, E. Moser, C. Segebarth, W. M. M. J. Bovée, and F. Podo, *Magn. Reson. Imaging*, 1995, **13**, 139.
18. F. Howe, R. Canese, F. Podo, B. Vikhoff, J. Slootboom, J. R. Griffiths, O Henriksen, and W. M. M. J. Bovée, *Magn. Reson. Imaging*, 1995, **13**, 159.
19. R. de Beer, P. Bachert-Baumann, W. M. M. J. Bovée, E. Cady, J. Chambron, R. Dommisse, C. J. A. van Echteld, R. Mathur-De Vré, and S. R. Williams, *Magn. Reson. Imaging*, 1995, **13**, 169.
20. F. Podo, *Adv. Biomed. Eng.*, 1993, **7**, 304.
21. J. D. de Certaines, O. Henriksen, A. Spisni, M. Cortsen, and P. B. Ring, *Magn. Reson. Imaging*, 1993, **11**, 841.
22. O. Henriksen, J. D. de Certaines, A. Spisni, M. Cortsen, R. N. Müller, and P. B. Ring, *Magn. Reson. Imaging*, 1993, **11**, 851.

Acknowledgements

The authors are grateful to the Commission of European Communities, DG XII, for support to the Concerted Actions on MRI and MRS COMAC-BME II.1.3 and BIOMED 1 PL 920432. FP also acknowledges support by PF CNR ACRO.

Biographical Sketches

F. Podo. *b* 1944. Laurea Physics, 1967, University of Rome. Research associate, Rockfeller University, New York, 1971–1972 (Prof. G. Némethy) and University of Pennsylvania, 1972 (Professor B. Chance). Staff member, Istituto Superiore di Sanità, Rome, 1969–1980; Research Director 1980–present; Director of Physical Biochemistry, 1982–present. Introduced to NMR by Professor M. Ageno (1968). Over 100 publications. Research specialty: physical chemistry, biochemistry, cell biology and oncology.

W. M. M. J. Bovée. *b* 1938. M.Sc., chemistry, 1969, Ph.D., physics, 1975, Delft University. Billiton Research Laboratories, 1969. Delft University, 1970–present. Approx. 140 publications. Research specialty: in vivo NMR.

J. S. Orr. *b* 1930. B.Sc., physics, 1952, D.Sc., physics, 1972, University of Glasgow. FRSE, 1974. Introduced to MRI by Professor Robert Steiner, 1979; involved with early EMI imaging. Head of Radiotherapy Physics, Glasgow, 1962–1977. Professor of Medical Physics, Royal Postgraduate Medical School, Hammersmith Hospital, 1977–1986. Experience as Chairman of MRC Committee on MRI Technological Performance, and Member, Project Management Group of two Concerted Actions of the European Communities. Over 100 publications.

Whole Body Magnetic Resonance Artifacts

R. Mark Henkelman

Sunnybrook Health Science Centre and University of Toronto, Toronto, ON, Canada

1 INTRODUCTION

One would hope that an NMR image would show an exact one-to-one correspondence to the object being imaged. This, however, is not the case. There are many 'structures' that are observed in MR images for which there is no corresponding structure in the actual object. Because these unreal 'structures' are the product of human workmanship, they are called artifacts. Artifacts often arise from imperfections in the practical implementation of ideal imaging concepts. In this case, an understanding of the artifacts and their causes can often lead to improved imaging technique. Sometimes, artifacts are intrinsic to even an ideal imaging system, and are thus not correctable. In either case, an ability to recognize some of the wide variety of artifacts that occur in MR images protects against misinterpretation of their appearances as real structures. This is particularly important in clinical MR imaging, where artifacts have sometimes been interpreted as anatomical or pathological structures, leading to unwarranted therapeutic procedures.

Since artifacts are usually first encountered in the visual image, they are described in this article in terms of how they appear. In each case, the causes of the artifacts are identified with appropriate references to other articles of this Encyclopedia where the underlying science of the data acquisition and image formation are more fully described. Some prominent artifacts are subjects in their own right, and are not discussed in any detail in this article (see ***Respiratory Artifacts: Mechanism & Control***; ***Susceptibility Effects in Whole Body Experiments***; ***Pulsatility Artifacts due to Blood Flow & Tissue Motion & Their Control***; ***Tissue Water & Lipids: Chemical Shift Imaging & Other Methods***; ***Surface Coil NMR: Quantification with Inhomogeneous Radiofrequency Field Antennas***; ***Surface & Other Local Coils for In Vivo Studies***; ***Selective Excitation Methods: Artifacts***, and ***Functional Neuroimaging Artifacts***).

This overview of artifacts in whole body MRI is not exhaustive, since new artifacts are being identified whenever new pulse sequences and new applications are developed. However, this article introduces most of the classes of artifacts that are seen, and provides enough kinds of causes so that as new ones are identified, they can be readily understood. More detailed reviews of MRI artifacts have been published previously.[1–3]

2 IMAGE WRAP-AROUND

Figure 1 shows an image with wrap-around artifact. This is due to aliasing, in which precessional frequencies beyond the Nyquist sampling frequency appear as lower frequencies. (see ***Fourier Transform Spectroscopy***; ***Image Formation Methods***, and ***Spin Warp Data Acquisition***). Aliasing in the frequency-encode direction can be eliminated by over-sampling at a higher Nyquist frequency and digitally filtering the data as they are acquired (see ***Whole Body Magnetic Resonance Spectrometers: All-Digital Transmit/Receive Systems***). Over-sampling in the phase-encode direction requires extra data acquisition time, which often is not available. Selective saturation of spins that lie beyond the image boundary can be used to suppress aliased signal (see ***Selective Excitation Methods: Artifacts*** and ***Selective Excitation in MRI***). Aliasing can also occur in the slice selection direction when slices are phase encoded in three-dimensional acquisitions.

Images can also wrap over on themselves when attempts are made to image beyond the linear range of the gradient (see

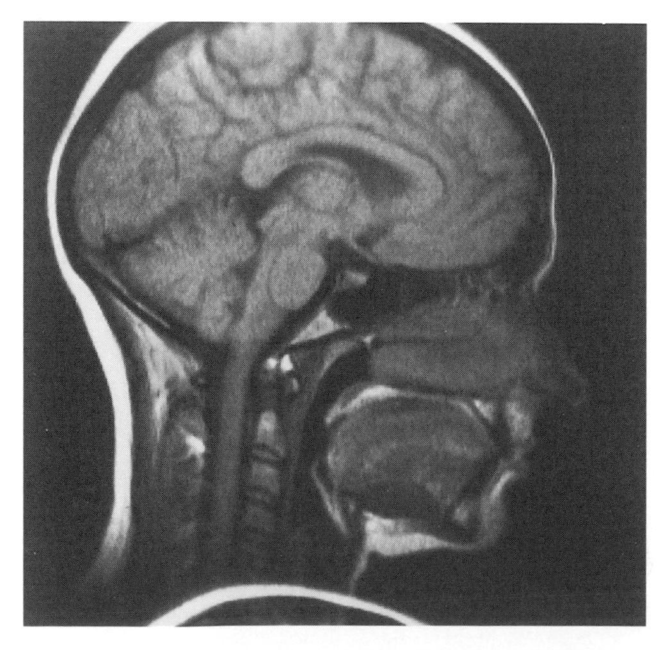

Figure 1 The top of the scalp extends beyond the upper boundary of the image frame. The image of this scalp is wrapped-around, and intrudes into the bottom of the image and overlaps the neck

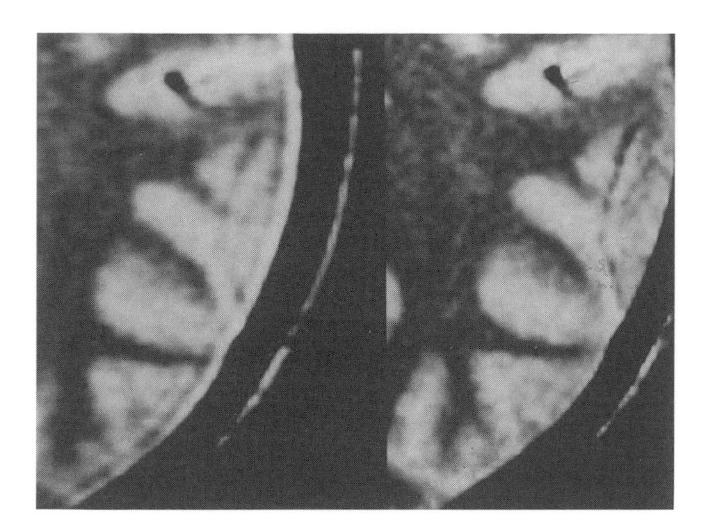

Figure 2 The left frame shows part of an axial brain image taken with 128 samples. The pattern of bright and dark rings propagating into the brain are edge ringing artifacts. On the right is the same image taken with 256 samples and then high-frequency-sampled to remove edge ringing. (Reproduced by permission of Dr. William Matthew Kelly, Department of Radiology, David Grant USAF Medical Center, CA)

Gradient Coil Systems). It is, therefore, important in whole body imaging to ensure that parts of the body that extend beyond the useful gradient are not excited by the rf system.

3 EDGE RINGING

Edge ringing appears as a series of ripples propagating away from any sharp discontinuity in signal intensity, as shown in Figure 2. Edge ringing is a truncation artifact due to inadequate sampling of the high frequencies of *k*-space (see ***Image Formation Methods***). It can be eliminated by sampling further in *k*-space with an associated increase in both image acquisition time and noise, or by low-pass filtering the acquired *k*-space data with an associated loss of image resolution. Alternatively, some postprocessing methods using predictive models of the missing data have been attempted with modest success[4,5] (see ***Fourier Transform & Linear Prediction Methods***).

Edge ringing can be particularly problematic when coarse voxel data are being acquired, as in chemical shift imaging. Here, spectral information from surrounding voxels may additively or subtractively contaminate the spectrum of the volume of interest (see ***Chemical Shift Imaging***).

Enhanced edge ringing can also arise from gradient problems. Eddy currents persisting into the data sampling window and nonlinear gradient amplifiers at high power can lead to enhanced apparent edge ringing. These are corrected by better amplifiers and self-shielded gradient designs (see ***Field Gradients & Their Application*** and ***Gradient Coil Systems***).

4 BLACK BOUNDARIES

Figure 3 shows an image with a well-defined, single-pixel-wide, black boundary between two adjacent tissues. This black line is an artifact. It results from the custom of showing MR images as magnitude images instead of as arrays of complex numbers (see ***Phase Contrast MRA***). Whenever two adjacent tissues have opposite phase, the magnitude image will present them with an artifactual black line between them. Such phase inversions can occur with inversion recovery images (see ***Inversion Recovery Pulse Sequence in MRI***), gradient echo images of fat and water, as seen in Figure 3, or motion shear between adjacent structures. Properly phased, real reconstructions serve to present the adjacent tissues in their true contrast relationship, and hence to eliminate the perception of a discrete black boundary.

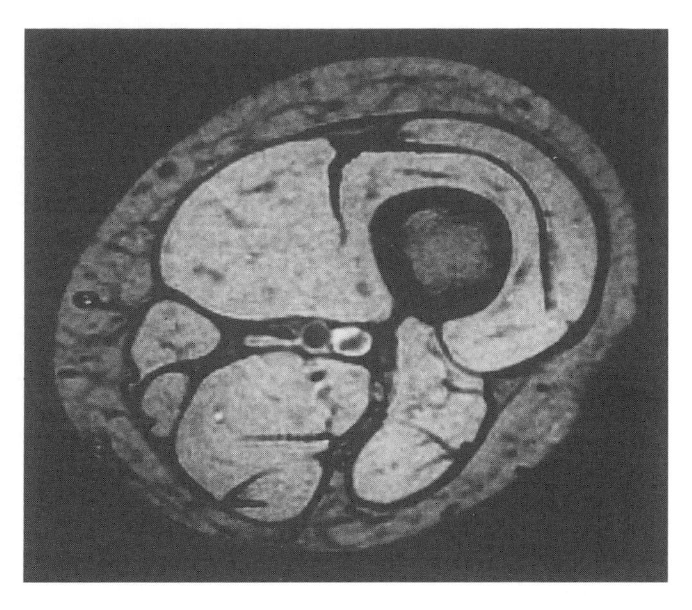

Figure 3 A gradient echo image through the thigh. The black lines between the muscle and subcutaneous fat and around the veins are artifactual black boundaries due to the fat and water signals being 180° out of phase. (Reproduced by permission of C. V. Mosby-Year Book, R. M. Henkelman, in 'Magnetic Resonance Imaging', eds. D. D. Stark and W. G. Bradley, Mosby-Year Book, St. Louis, MO, 1991, Chap. 10, p. 233)

These black boundary artifacts are different from the chemical shift artifact, which appears as a black boundary on one side and a bright boundary on the other side of fat/water interfaces (see ***Tissue Water & Lipids: Chemical Shift Imaging & Other Methods***).

5 STRIPING AND HERRINGBONE PATTERNS

Figure 4 shows an image with a pronounced striping artifact across it, which can be removed by correcting a single erroneous data point. When more than one data point is corrupted, the system of cross stripes gives an image with an imposed herringbone pattern. When many data points or whole data lines are corrupted, correction of the artifact using interpolation becomes less reliable.

Bad data points result from errors in the receiver analog-to-digital conversion (see ***Whole Body Magnetic Resonance Spectrometers: All-Digital Transmit/Receive Systems***). Such errors may result from ADC hardware failures, but more often are caused by electrostatic shocks into the rf receiver from the

For References see p. 5004

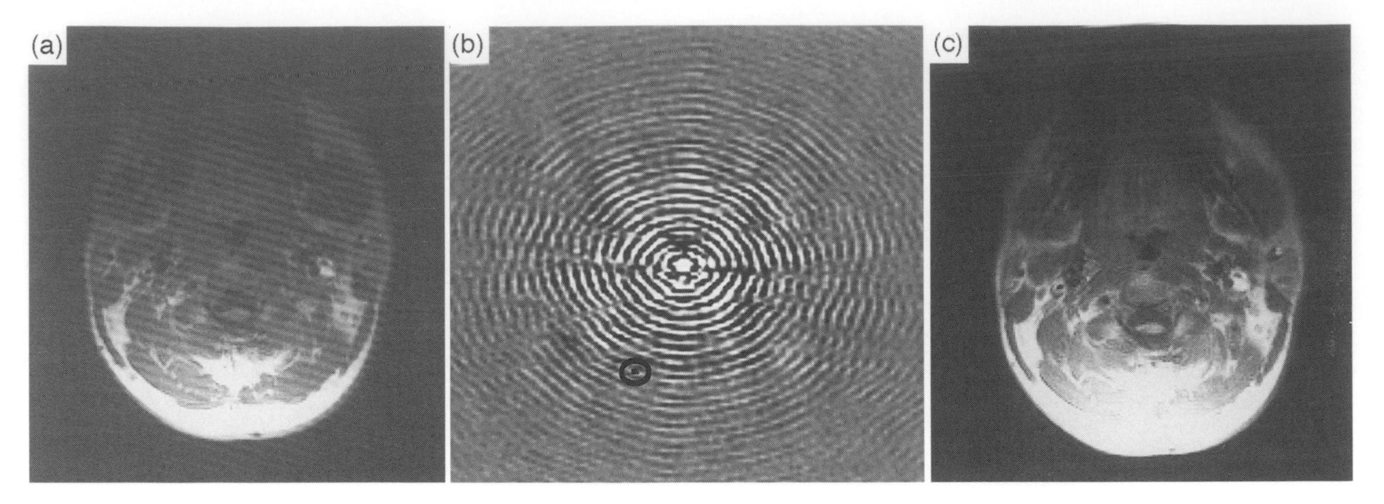

Figure 4 A bad data point. (a) An image with a pronounced diagonal striping across it. (b) Investigation of the *k*-space data from which this image was reconstructed shows a single anomalous data point (circled) at the *k*-space location corresponding to the frequency and orientation of the stripes. (c) Replacement of the bad data point by the average of its neighbors removes the artifact from the image. (Reproduced by permission of Pergamon Press, Elmsford, NY from R. M. Henkelman and M. J. Bronskill, *Rev. Magn. Reson. Med.*, 1987, **2**, 1)

patient or blankets (see ***Whole Body Machines: Quality Control***).

6 MOIRÉ FRINGES

Images can sometimes appear quite normal, except for an overlay of bright and dark alternating bands, often of irregular shape. These can be widely spaced, giving an appearance like stripes on a zebra, or can be closely spaced as in a moiré pattern of fringes. Such artifacts usually occur when two images overlap, such as occurs with aliasing, as shown in Figure 5. If the overlapping images have similar phase relationships, as in a spin echo image, they simply add. If, however, the images have a varying phase relationship, such as might occur with gradient echo imaging of anatomy in a large-B_0 inhomogeneity far from the center of the magnet, then the images will add constructively or destructively, depending on the relative phase. The fringing pattern is then an interference pattern between the two images with varying phase relationship.

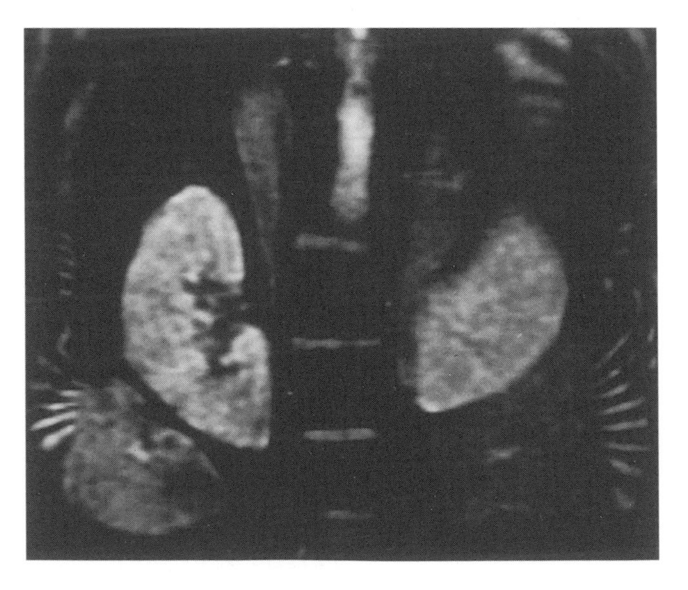

Figure 5 Moiré fringing patterns from interference effects between the abdominal image and aliased anatomy from beyond the boundary of the image. (Reproduced by permission of C. V. Mosby-Year Book, from R. M. Henkelman, in 'Magnetic Resonance Imaging', eds. D. D. Stark and W. G. Bradley, Mosby-Year Book, St. Louis, MO, 1991, Chap. 10, p. 233)

The situation is more subtle, but the concept is the same, when the two images are geometrically identical but are generated through different coherence pathways. This can occur, for example, in a late echo, multiple spin echo image, where the stimulated echoes arising from nonideal tip angles have been arranged to superimpose on the primary spin echo. If the accumulated phase is the same for the spin echo and stimulated echo over the whole image, the addition will be constructive; if there are regions where the phase is opposite, the image will interfere, leading to fringes (see ***Multi-Echo Acquisition Techniques Using Inverting Radiofrequency Pulses in MRI***; ***Selective Excitation Methods: Artifacts***, and ***Phase Cycling***).

Such interference fringes can be deliberately generated to provide a tagging grid for motion studies, particularly those of the heart (see ***Marker Grids for Observing Motion in MRI***).

7 CENTRAL POINT ARTIFACTS

The central point of the image is unique with respect to data acquired in *k*-space. For *k*-space acquisitions that begin acquiring data from the center of *k*-space (see ***Image Formation Methods***; ***Projection–Reconstruction in MRI***, and ***Spiral Scanning Imaging Techniques***), small errors in the time origin lead to bright or dark contrast pixels in the reconstructed image. Such central artifacts are almost the hallmark of radially acquired data. Removal of these artifacts requires careful time adjustment of the data acquisition or, equivalently, careful phasing of the reconstructed projection.

Even for rectangular data acquisitions (see ***Spin Warp Data Acquisition***; ***Partial Fourier Acquisition in MRI***, and ***Echo-Planar Imaging***), central artifacts occur if the rf transmitter allows the reference rf to leak through during data acquisition. Careful blanking of the rf transmitter and amplifier during all data acquisition effectively removes these central artifacts.

8 ZIPPERS

Zippers are lines of incorrect intensity that run across the image. They often appear with alternating bright and dark signal intensity, and thus can look like a 'zipper' crossing the image. The placement and orientation of such 'zippers' can be used to distinguish their various origins.

Zippers that run in the phase-encode direction (Figure 6) at any arbitrary location in the image are due to discrete rf noise. They can be identified as such by shifting the magnetic field and hence the central resonant frequency of the imager, causing the rf noise zipper to shift across the image. Elimination of rf noise requires better environmental rf shielding of the MR imager or better isolation of the imager electronics from the rf receiver chain. When better shielding is not an option, shifting the resonant frequency as described above can be used to search for a clean 'window' in which to operate the MR imager.

Another type of zipper can occur through the center of the image propagating in the frequency-encode direction. The origin of these zipper artifacts is different—they arise from free induction decay or stimulated echo signal that has not been phase-encoded and hence makes a consistent contribution to each acquired data line.[6] These spurious signals usually arise from nonideal selective excitation pulses (see ***Selective Excitation Methods: Artifacts***). As imaging methods become faster, there are greater numbers of rf pulses per T_1 interval, providing many more stimulated echo pathways that can give rise to zippers. There are three methods for reducing these artifacts:

1. design of better rf-generated slice profiles (see ***Selective Excitation in MRI***; ***Composite Pulses***);
2. phase cycling of alternate rf excitations to push the artifact to the boundary of the image (see ***Phase Cycling***);
3. the use of 'crusher' gradients to spoil any unwanted transverse coherence (see ***Field Gradients & Their Application***).

It should be recognized that, although the zippers discussed above are evident only in rectilinear acquisitions of k-space data, the physical principles from which they arise contribute unwanted signal to all data sampling schemes. For more elaborate data acquisition schemes, the resulting artifacts may be less recognizable and may even simply appear as 'noise'. This does not make it any less needful to carefully eliminate artifactual signal from the desired data.

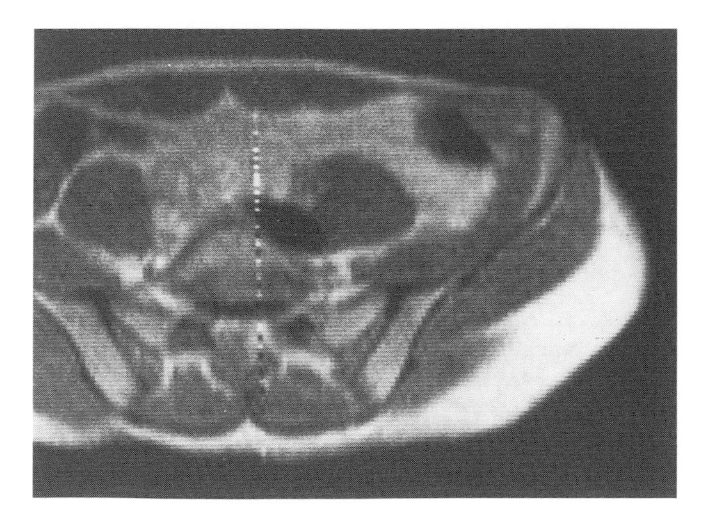

Figure 6 A 'zipper' of bright and dark signal intensity propagating in the phase-encode direction (vertical) is due to discrete rf noise from either the imager's computer systems or the environment that is being picked up by the NMR receiver coil. (Reproduced by permission of Pergamon Press, Elmsford, NY from R. M. Henkelman and M. J. Bronskill, *Rev. Magn. Reson. Med.*, 1987, **2**, 1)

9 IMAGE DISTORTIONS AND BLURRING

The spatial mapping in MR imaging is predicated on a fixed relationship between location and frequency. If either the magnetic field is not homogenous or the imaging gradients are not linear, this relationship is violated, and the image becomes distorted in some way.

A spin, expected to have a resonant frequency of γB_0, that has a resonant frequency of $\gamma(B_0 + \Delta B_0)$ because of a field inhomogeneity ΔB_0 will be shifted from its true location in the image along the frequency-encode direction toward higher frequency by $\Delta B_0/G_x$, where G_x is the strength of the frequency encoding gradient. It should be noted that phase encoding does not exhibit a similar displacement (see ***Spin Warp Data Acquisition***). Therefore, phase encoding in all three directions is an effective, although unrealistically slow, method to avoid magnetic field inhomogeneity distortions.

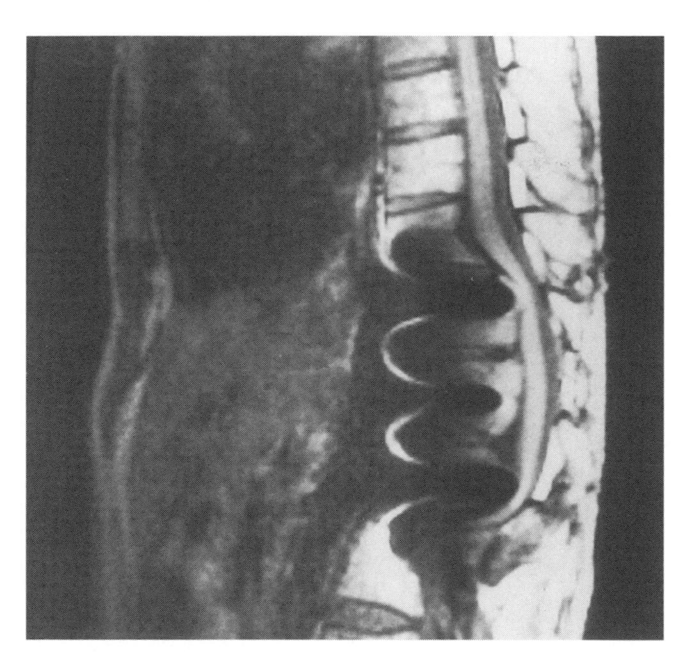

Figure 7 Sagittal image of the spine of a patient with a surgically implanted metal support rod. Displacement of the spinal cord posteriorly (in the frequency encode direction) results from distortion of the main magnetic field. (Reproduced by permission of C. V. Mosby-Year Book, from R. M. Henkelman, in 'Magnetic Resonance in Medicine', eds. D. D. Stark and W. G. Bradley, Mosby-Year Book, St. Louis, MO, 1991, Chap. 10, p. 233)

For References see p. 5004

Magnetic field inhomogeneities may also be caused by magnetic susceptibility variations within the patient, as seen in Figure 7. Distortions can be minimized with increased gradient strengths and decreased acquisition bandwidths. The associated loss in signal-to-noise ratio can be recovered by judicious averaging.

Even if the main field is uniform, gradient nonlinearities can produce image distortions in both the frequency *and* phase-encode directions, as well as in the slice selection direction. Gradient nonlinearities can be mapped, and images can be rerectified using postprocessing techniques.

Magnetic nonuniformities are more problematic in *k*-space acquisitions that are not unidirectional or are not rectilinear (see ***Image Formation Methods***; ***Echo-Planar Imaging***; ***Projection–Reconstruction in MRI***, and ***Spiral Scanning Imaging Techniques***). In these data acquisition strategies, points at field inhomogeneities will be mapped to several different points in the image resulting in image blur. Blur cannot be as readily corrected by postprocessing as can simple distortion, and hence elimination at the source should be attempted using stronger gradients, suppression of off-resonant spins, or some other technique.[7,8]

10 SIGNAL LOSS

Magnetic field variations not only result in image distortions, but also give rise to regions of signal loss (see ***Susceptibility Effects in Whole Body Experiments***).

In gradient echo imaging, if the phase dispersion of signal throughout an imaged voxel exceeds 2π radians, the net signal in the image will be significantly decreased. This signal loss is said to be due to intravoxel dephasing, and is illustrated in Figure 8.

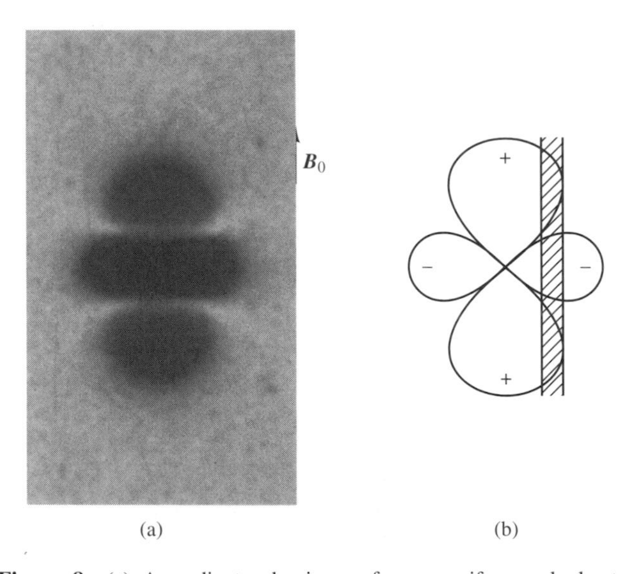

Figure 8 (a) A gradient echo image from a uniform gel phantom containing a 4 mm long wire of 60 μg mass. The wire produces a dipolar field in the gel as shown in (b). The image is taken in a plane (shaded) that is parallel to the applied magnetic field and displaced from the wire. The three regions of signal loss in the image are caused by intravoxel dephasing. (Reproduced by permission of RSNA Publications, Oak Brook, IL from J. K. Kim, W. Kucharczyk, and R. M. Henkelman, *Radiology*, 1993, **187**, 735)

Intravoxel dephasing can be reduced by using a spin echo acquisition that refocuses the spin phases at the time of echo acquisition. It can also be reduced by using short echo times (see ***Phase Contrast MRA***). It can also be improved by reducing the voxel size by moving to higher resolution images,[9] as shown in Figure 9.

Although signal loss is reduced with spin echo imaging, it cannot be eliminated when the spins are free to diffuse through the region of the inhomogeneity. Thus, diffusion-mediated signal loss in the presence of inhomogeneities cannot be refocused by a single spin echo. However, a train of π refocusing pulses will reduce the effects of diffusion-mediated dephasing, as in a Carr–Purcell sequence.

Finally, susceptibility-induced signal losses should not be thought of only as artifacts. These signal loss mechanisms are responsible for the contrast obtained in functional and diffusion imaging (see ***Brain: Sensory Activation Monitored by Induced Hemodynamic Changes with Echo Planar MRI***; ***Diffusion & Perfusion in MRI***, and ***Anisotropically Restricted Diffusion in MRI***).

11 IMAGE GHOSTS

Ghost images are probably the most prevalent of all MRI artifacts. Ghosts are faint replications of the image or parts of images that show the correct spatial structure, but occur at the wrong location within the field of view. Three different ghost mechanisms are discussed, which can be distinguished on the basis of the ghost's position.

11.1 Quadrature Imbalance Ghosts

Central symmetric ghosts, which are a rotation of the image by 180° about its center, result from imbalance in the two quadrature receiver channels. The quadrature signal that is recorded in MR images is intrinsically complex, consisting of real and imaginary parts that are recorded by two independent channels. If the real and the imaginary channels have slightly different gains or, if the phases of the two demodulating frequencies are not separated by exactly 90°, then quadrature imbalance ghosts may be seen. Current state-of-the-art all-digital receivers are able to eliminate these ghosts (see ***Whole Body Magnetic Resonance Spectrometers: All-Digital Transmit/Receive Systems***).

11.2 Stimulated Echo Ghosts

Stimulated echo ghosts appear as inversions of the image about the center of the image in the phase-encode direction. In a multiecho sequence, the primary image signal is obtained from echoes for which the spins have remained in the transverse plane for the whole *TE* time, but also from stimulated echoes that may have been stored longitudinally for part of the echo times. If the stimulated echo has received some odd number $(\ldots, -3, -1, +1, +3, \ldots)$ of addition phase encodes and refocusing pulses, the stimulated echo image will appear as an inverted ghost with respect to the primary image.

Such ghosts can be removed by crusher gradients (see Section 6), which spoil the stimulated echo signal. If it is desired to keep to stimulated echo signal, its polarity can be

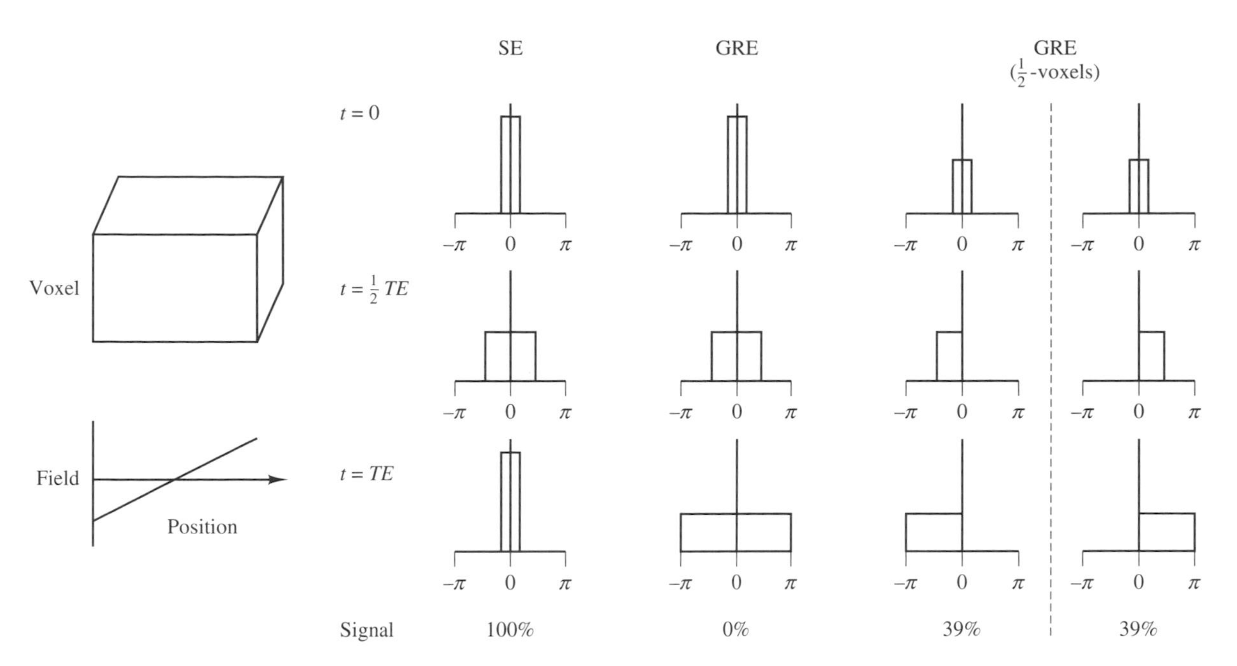

Figure 9 A single large voxel with a susceptibility-induced field gradient can be imaged in a spin echo (SE) sequence with 100% signal (left) because the dephasing at time $\frac{1}{2}TE$ is refocused at time TE. The same voxel imaged with a gradient-recalled echo (GRE) may show 0% signal due to intravoxel dephasing (middle). However, if the voxel size is halved with the same gradient echo sequence, each half-pixel shows 39% of the total signal. The intravoxel dephasing signal loss is diminished with higher resolution. (Reproduced by permission of C. V. Mosby-Year Book from R. M. Henkelman, in 'Magnetic Resonance in Medicine', eds. D. D. Stark and W. G. Bradley, Mosby-Year Book, St. Louis, MO, 1991, Chap. 10, p. 233)

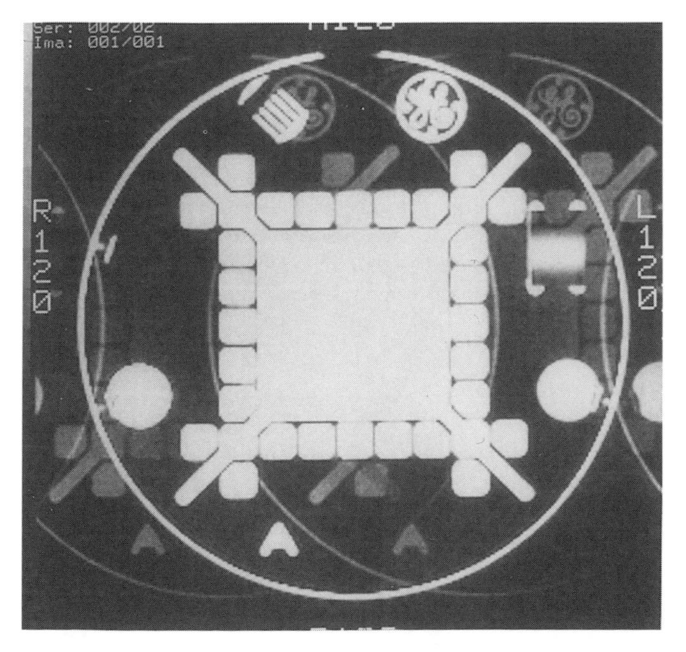

Figure 10 Ghost artifacts in an image of a quality control phantom. Ghost images are displaced in the phase-encoding direction (left and right in this case). The source of these ghosts was eventually identified as a cryopump. With each stroke, a slightly magnetic piston in the cryopump was periodically shifting the main magnet field strength. (Reproduced by permission of C. V. Mosby-Year Book, from R. M. Henkelman, in 'Magnetic Resonance Imaging', eds. D. D. Stark and W. G. Bradley, Mosby-Year Book, St. Louis, MO, 1991, Chap. 10, p. 233)

managed by ensuring that any applied phase encoding is rewound within the same interval between refocusing pulses[10] (see ***Multi-Echo Acquisition Techniques Using Inverting Radiofrequency Pulses in MRI***).

11.3 Periodic Modulation Ghosts

Displacement ghosts are replicas of parts of the image displaced in the phase-encode direction. They are the commonest type of ghost, and remain a problem in the imaging of living systems. Ghost artifacts from periodic physiological motions are discussed elsewhere (see ***Respiratory Artifacts: Mechanism & Control*** and ***Pulsatility Artifacts due to Blood Flow & Tissue Motion & Their Control***).

Besides physiological motions, displacement ghosts can arise from any other type of periodic variation concurrent with the imaging process, as shown in Figure 10.

Examples include building vibration, alternating line current pickup, and amplifier oscillation, as well as cryopumps. Elimination of such ghost artifacts requires an identification of the causative periodic variation and its subsequent correction. Such a detective process can be very long and expensive.

Even if a temporal variation imposed on the imaging process is not periodic, it will still propagate signal into the phase-encoding direction, as indicated in Figure 11—but it will not be recognized as ghosts.

In nonrectilinear acquisitions of k-space data, there is no preferred phase-encoding direction, and misassigned signal is scattered all over the image, where it can seldom be recognized

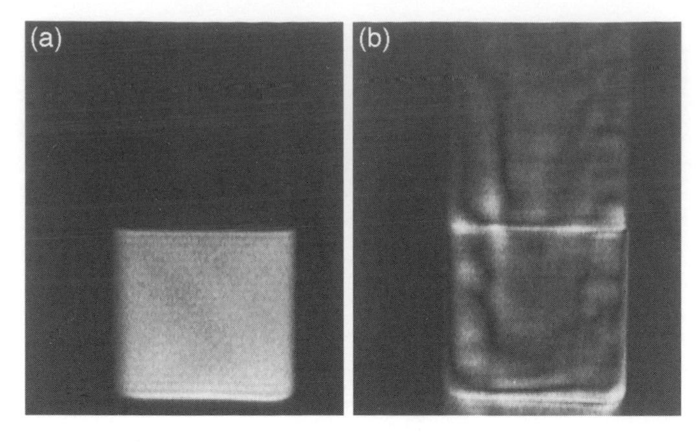

Figure 11 (a) An image of a beaker of water. (b) The same image acquisition, but after the water was stirred. Even though the motion of the water is not periodic, but rather is turbulent, signal intensity is displaced in the phase-encoding direction (up and down in this case). (Reproduced by permission of Pergamon Press, Elmsford, NY from R. M. Henkelman and M. J. Bronskill, *Rev. Magn. Reson. Med.*, 1987, **2**, 1)

as ghosts and may simply look like noise. Nonetheless, it is still artifactual signal that must be removed if the maximum image quality is to be achieved.

12 WASHED-OUT CONTRAST

There is a very characteristic artifact that presents as an image with good spatial definition and no obviously displaced image structures, but for which the contrast is washed out and there appears to be a weak aura around the imaged structure, as seen in Figure 12. The cause is simple. The artifact occurs when the data acquisitions at the center of *k*-space exceed the range of the analog-to-digital converter (ADC). This is most likely to occur in three-dimensional data acquisitions, where the integrated signal intensity over a 256^3 acquisition typically exceeds the high-order encoded data points by 2,[11] pushing the dynamic range of conventionally employed ADCs.

The artifact can be corrected by repeating the acquisition with increased receiver attenuation. Alternatively, the data can be retrospectively corrected if they are simply aliased as in Figure 12. If the central data are lost, the image can still be salvaged by imposing reconstruction conditions of compact support.[11]

13 INTENSITY VARIATIONS IN THE IMAGE

There are many reasons why the eventual signal intensity in an image does not correspond directly to the known spin density in the object. It is the wide variety of these artifacts that has led to the very slow development of quantitative methods in MRI (see ***Relaxation Measurements in Whole Body MRI: Clinical Utility***). Three causes of intensity nonuniformity are discussed.

13.1 Slice Profile Effects

Slice-selective rf pulses are never perfect, and always exhibit some transition region through which the effective tip angle is less than that which is prescribed. While this is not a large problem for small tip angle pulses, it becomes a major source of signal intensity artifact for 180° refocusing pulses. Thus, in a slice-selective multi-echo sequence and for a phantom with an infinite T_2, there is typically a 5–10% loss in signal for each subsequent echo due only to sequential degradation of the slice profile. This needs to be borne in mind when attempting to estimate T_2 values, since there is an apparent multiecho decay rate due to slice profile degradation.

Another cause of signal loss can come from eddy currents. If the readout gradient compensation pulse immediately precedes the refocusing 180° pulse in a spin echo sequence, and if the gradient shows an eddy current that persists throughout the 180° pulse, then the slice-selective plane of the 180° pulse will be rotated with respect to the 90° slice-selective excitation. Such a situation will provide ideal refocusing at the center of the slice, but degraded signal intensity as one moves away from the center in the frequency-encode direction.

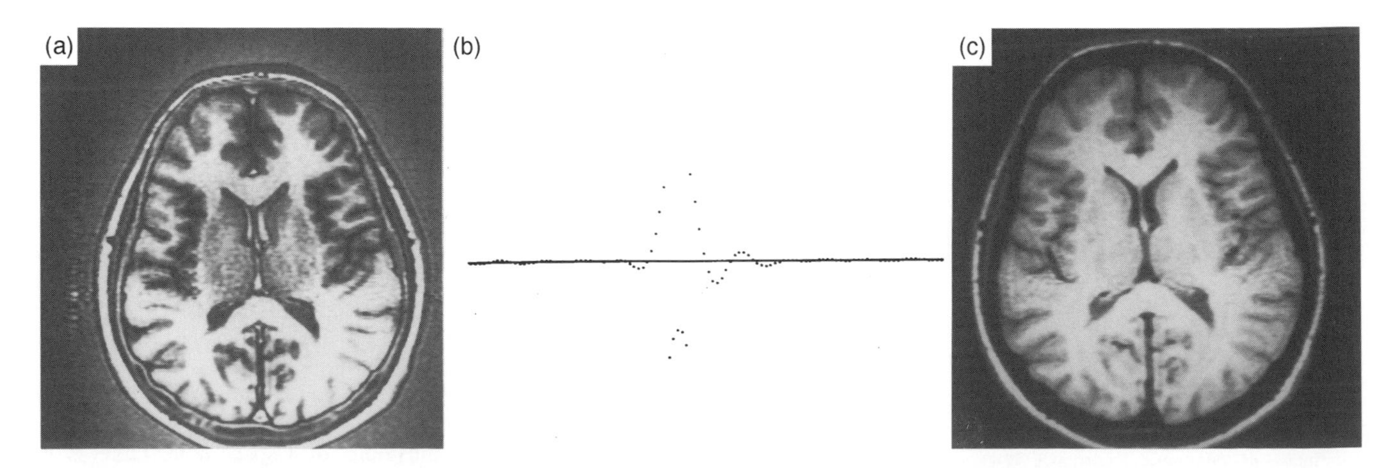

Figure 12 An axial brain image (a) with washed-out contrast and aura results from acquired *k*-space data that have exceeded the range of the analog-to-digital converter. (b) Correction of the aliased data and subsequent reconstruction results in an artifact-free image (c)

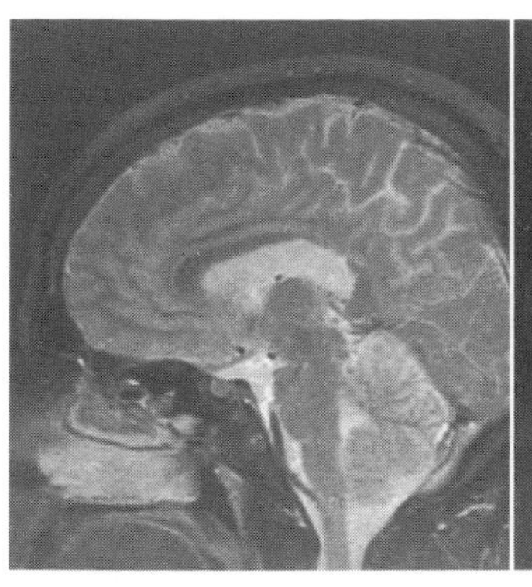
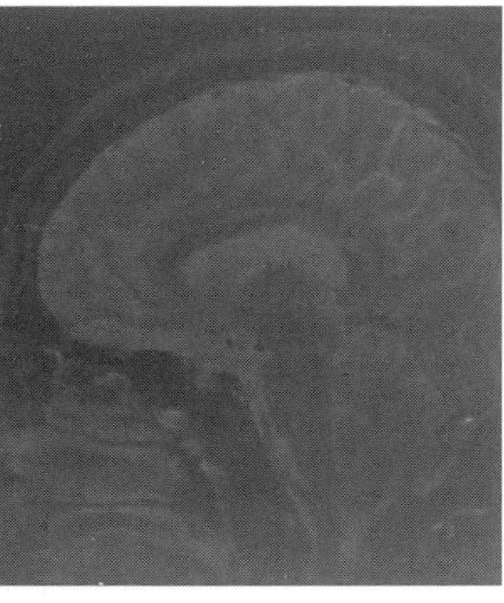

Figure 13 Sagittal head images acquired with a T_2-weighted spin echo sequence. The image on the left is a single slice whereas that on the right is the central image from a nine-slice contiguous multislice acquisition displayed with the same window and level. The multislice image shows the decreased signal intensity from slice interference and also the degraded contrast. (Reproduced by permission from W. Kucharczyk, A. P. Crawley, W. M. Kelly and R. M. Henkelman, *Am. J.N.R.* 1988, **9**, 443, American Society of Neuroradiology, Oak Brook, IL, USA)

13.2 Slice Interference Effects

The single slice nonuniformities become more complicated when multiple slices in close proximity are employed.

Because selective excitation is not perfect, selective pulses affect spins beyond the nominal boundaries of the slice. This has two consequences:

1. direct interference from neighboring slices causes signal reduction in each slice and variation in the average signal intensity of slices throughout a multislice block, depending on the order of slice acquisition and the gap between the slices;
2. the real-time contrast within a slice is degraded owing to interference from adjacent slices, particularly for T_2-weighted images.[12]

This effect is illustrated in Figure 13. These consequences can be diminished with better selective rf pulses (see ***Selective Excitation Methods: Artifacts***) or by leaving a slice gap between slices.

A slice interference intensity artifact that is not improved with better selectivity or slice gap is that arising from magnetization transfer effects. It has been realized that excitation and refocusing rf pulses from all neighboring slices serve to saturate the spins of the macromolecules within the slice of interest. While these spins are not directly observed in MRI because of their short T_2 relaxation times, their saturation diminishes the signal from the liquid pool that is imaged (see ***Phosphorus-31 Magnetization Transfer in Whole Body Studies***; ***Magnetization Transfer Contrast: Clinical Applications***, and ***Magnetization Transfer & Cross Relaxation in Tissue***).

13.3 Nonuniform Radiofrequency Fields

Image intensity variations also arise from nonuniformities in both the transmitted rf field and the receive sensitivities of the receiver coils.

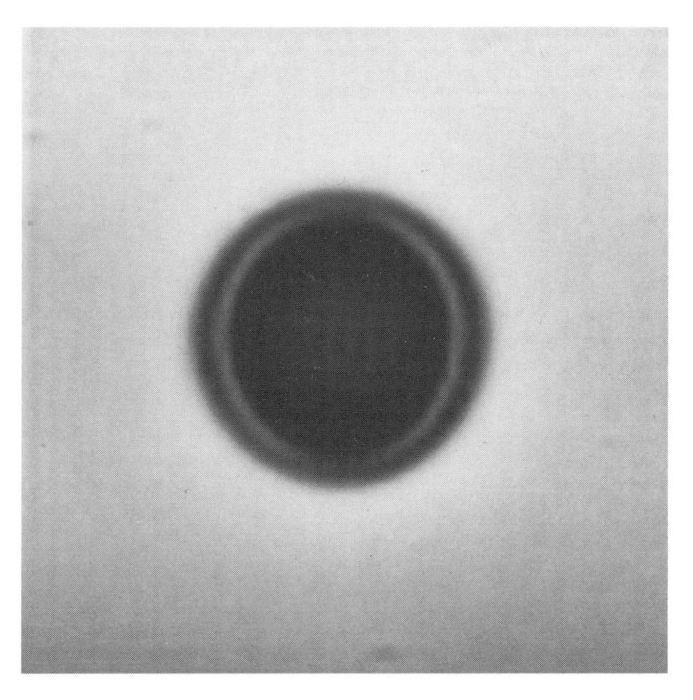

Figure 14 Spin echo image of a homogeneous doped water phantom containing a simple closed circular copper loop of 24 mm diameter. The loop serves as a secondary rf transmitter, and hence perturbs the uniformity of the transmitted rf field. The loop also plays the part of a secondary receiver, destroying the homogeneity of the rf receiver sensitivity.[13]

Nonuniformities in the transmitted rf field were originally believed to impose the fundamental limit to high-field imaging (see ***High Field Whole Body Systems***). However, although the images show slight rf shading, excellent MR images have been obtained at 4 T in whole body systems. Slight nonuniform rf intensities are seen as a left–right asymmetry on transverse abdominal images of large patients at 1.5 T.

Nonuniformities in the sensitivities of rf receiver coils are a common experience in MRI. The image nonuniformity can hardly be considered artifactual, since the smaller localized receiver coils are used deliberately to increase the signal-to-noise ratio (see ***Surface & Other Local Coils for In Vivo Studies***; ***Whole Body Machines: NMR Phased Array Coil Systems***, and ***Surface Coil NMR: Quantification with Inhomogeneous Radiofrequency Field Antennas***).

Finally, nonuniformities in signal intensity can arise from both changes in the transmitted rf field and the receive sensitivity when secondary rf conductors are included in the imaging field, as shown in Figure 14. Such secondary effects may become more of an issue with the further development of interventional MRI.

14 SUMMARY

MR imaging methods display a wide variety of image artifacts, some of which have been introduced in this article. All the artifactual signals arise from well-behaved and well-understood laws of physics, taking into account instrumental imperfection. Understanding the causes of MRI artifacts usually

For References see p. 5004

leads to ways of correcting, or at least avoiding them, so that images can be reliably interpreted with confidence.

15 RELATED ARTICLES

Anisotropically Restricted Diffusion in MRI; Birdcage & Other High Homogeneity Radiofrequency Coils for Whole Body Magnetic Resonance; Chemical Shift Imaging; Composite Pulses; Cryogenic Magnets for Whole Body Magnetic Resonance Systems; Diffusion & Perfusion in MRI; Echo-Planar Imaging; Field Gradients & Their Application; Fourier Transform & Linear Prediction Methods; Fourier Transform Spectroscopy; Functional Neuroimaging Artifacts; Gradient Coil Systems; High Field Whole Body Systems; Image Formation Methods; Magnetization Transfer Contrast: Clinical Applications; Magnetization Transfer & Cross Relaxation in Tissue; Marker Grids for Observing Motion in MRI; Multi-Echo Acquisition Techniques Using Inverting Radiofrequency Pulses in MRI; Partial Fourier Acquisition in MRI; Phase Contrast MRA; Phase Cycling; Phosphorus-31 Magnetization Transfer in Whole Body Studies; Projection–Reconstruction in MRI; Pulsatility Artifacts due to Blood Flow & Tissue Motion & Their Control; Radiofrequency Systems & Coils for MRI & MRS; Relaxation Measurements in Imaging Studies; Relaxation Measurements in Whole Body MRI: Clinical Utility; Respiratory Artifacts: Mechanism & Control; Selective Excitation Methods: Artifacts; Selective Excitation in MRI; Shimming of Superconducting Magnets; Spin Warp Data Acquisition; Spiral Scanning Imaging Techniques; Surface Coil NMR: Quantification with Inhomogeneous Radiofrequency Field Antennas; Susceptibility Effects in Whole Body Experiments; Tissue Water & Lipids: Chemical Shift Imaging & Other Methods; Whole Body Machines: NMR Phased Array Coil Systems; Whole Body Machines: Quality Control; Whole Body Magnetic Resonance: Fast Low-Angle Acquisition Methods; Whole Body Magnetic Resonance Spectrometers: All-Digital Transmit/Receive Systems.

16 REFERENCES

1. R. M. Henkelman and M. J. Bronskill, *Rev. Magn. Reson. Med.*, 1987, **2**, 1.
2. R. Harris and G. Wesbey, in *Magn. Reson. Annu.*, 1988, p. 71.
3. R. M. Henkelman, in 'Magnetic Resonance Imaging', eds. D. D. Stark and W. G. Bradley, Mosby-Year Book, St. Louis, MO, 1991, Chap. 10, p. 233.
4. R. T. Constable and R. M. Henkelman, *Magn. Reson. Med.*, 1991, **17**, 108.
5. Z. P. Liang, F. E. Boada, R. T. Constable, E. M. Haacke, P. C. Lauterbur, and M. R. Smith, *Rev. Magn. Reson. Med.*, 1992, **4**, 67.
6. A. P. Crawley and R. M. Henkelman, *Med., Phys.*, 1987, **14**, 842.
7. D. C. Noll, J. M. Peuly, C. H. Meyer, D. G. Nishimura, and A. Macovski, *Magn. Reson. Med.*, 1992, **25**, 319.
8. J. K. Kim, W. Kucharczyk, and R. M. Henkelman, *Radiology*, 1993, **187**, 735.
9. I. R. Young, I. J. Cox, D. J. Bryant, and G. M. Bydeler, *Magn. Reson. Imaging*, 1988, **6**, 585.
10. J. Hennig, *J. Magn. Reson.*, 1988, **78**, 397.
11. J. Jackson, A. Macovski, and D. Nishimura, *Magn. Reson. Med.*, 1989, **11**, 248.
12. W. Kucharczyk, A. P. Crawley, W. M. Kelly, and R. M. Henkelman, *Am. J.N.R.*, 1988, **9**, 443.
13. C. R. Camacho, D. B. Plewes, and R. M. Henkelman, submitted to *J. Magn. Reson. Imaging*, 1994.

Biographical Sketch

R. Mark Henkelman. *b* 1946. B.Sc., 1969; Ph.D., 1973, University of Toronto. Faculty at the University of British Columbia, Canada, 1973–1979, working on π-meson radiotherapy at TRIUMF. Professor of Medical Biophysics, University of Toronto 1979–present; Vice-President, Research for Sunnybrook Health Science Centre 1989–present. Approx. 175 publications. Research specialties: MRI, image acquisition techniques, mechanisms of tissue contrast.

Whole Body Magnetic Resonance: Fast Low-Angle Acquisition Methods

Axel Haase
Universität Würzburg, Würzburg, Germany

1 INTRODUCTION

The most frequent applications of NMR imaging are within the field of biology and medical diagnosis. The inherent problem in early clinical practice was that NMR imaging was supposed to be a time-consuming technique. However, as early as 1977, a fast NMR imaging method, echo planar imaging (EPI), was described[1] (see ***Echo-Planar Imaging***). During the first decade of clinical NMR imaging, EPI was not used owing to major limitations in technical implementation. In 1985, a further fast imaging method was found, which could be immediately applied in many clinical NMR imagers, the FLASH (fast low angle shot) technique.[2] The two methods rely on different approaches to the fast sampling of NMR image data. The purpose of this work is to present the physical principles and different versions of FLASH-based fast NMR imaging.

A clear definition of 'fast NMR imaging' cannot be found in the literature. It is even more confusing that measuring times of 'fast NMR imaging techniques' cover several orders of magnitude from a few milliseconds to a few minutes. A definition of fast imaging can only be given in the context of the object and its time constants under investigation. Therefore, 'fast imaging' is understood to be 'fast' compared with the time constant of the physiological process of the biological object of interest. A fast imaging technique should freeze a time-dependent process to give a 'snapshot' of the object or the time-dependent biological function. This is the case for heart studies, when the measuring times should be less than 50 ms, or for abdominal studies, where a measuring interval of several seconds during one breath-hold is needed. For other investigations, this definition is useless (e.g., for images of joints). Here, fast imaging has a practical or economic meaning to optimize the clinical investigation and increase patient comfort (see ***Whole Body MRI: Strategies Designed to Improve Patient Throughput***). Therefore, a 'fast NMR imaging' method has to cope with all the different needs of medical diagnosis, and cannot be restricted to a single measuring time interval. FLASH-based NMR imaging techniques address this goal.

The purpose of FLASH NMR imaging is to reduce all kinds of motional artifacts, for example, heart motion, breathing, and random motion by the patient. It will offer the possibility for the study of organ functions, and organ motion.[3,4] A further benefit is the ability to acquire multiparameter data sets within shorter time intervals, for example, three-dimensional imaging,[5] flow velocity information, and NMR parameter imaging.[6] A few years after its first conception in 1985, it has become generally accepted in routine clinical practice.

2 IMAGING TECHNIQUE

The physical principle of all FLASH-based techniques is the two- or three-dimensional Fourier transform, first proposed, as 'Fourier zeugmatography' in 1975[7] (see ***Image Formation Methods***). The practical question is, how to sample all necessary lines in $\boldsymbol{k}$-space as fast and efficiently as possible. First approaches for dealing with faster imaging were focused on a reduction of the number of $\boldsymbol{k}$-space lines in partial Fourier imaging (see ***Partial Fourier Acquisition in MRI***). However, this reduces the spatial resolution to a certain extent and produces (Gibbs ringing) truncation artifacts[8] (see ***Whole Body Magnetic Resonance Artifacts***). A factor of approximately two can be gained using 'half-Fourier' imaging, where only half of the data is sampled and the second half is filled later by software.[9] Partial Fourier imaging can be applied in addition to all fast techniques to reduce the measuring time further.

In principle, two methods are possible to reduce the time interval of data sampling: a reduction of the repetition time TR between the acquisition of different $\boldsymbol{k}$-space lines, or the measurement of the whole $\boldsymbol{k}$-space using a single NMR excitation followed by a series of NMR echoes. The second possibility includes the EPI methods. We focus on the first technique. However, a reduction of the repetition time has a considerable drawback. Every excitation of an NMR signal for a special $\boldsymbol{k}$-space line decreases the longitudinal magnetization to a lower value M_z. During the time interval TR, M_z recovers towards its equilibrium value M_0 by spin–lattice relaxation T_1, according to

$$M_z = M_0(1 - e^{-TR/T_1}) \tag{1}$$

It can be seen that this recovery is most inefficient, as long as TR is short compared with T_1. M_z tends to zero when the excitation flip angle is 90°. Furthermore, spin echo experiments having 180° rf refocusing pulses and separated by short TR intervals give almost zero values of M_z.

The idea of reducing the repetition time TR to short values is, however, most promising. According to the above discussion, the goal can be reached when spin echo experiments are avoided and lower flip angles are used. In 1D NMR spectroscopy low flip angles are used to speed up the experiment while retaining a high level of NMR signal. It is known that, for given TR and T_1 values, an optimum flip angle, the 'Ernst angle' α_E exists where the maximum signal strength per measuring time can be expected:[10]

$$\cos\alpha_E = e^{-TR/T_1} \tag{2}$$

Although an 'Ernst angle' cannot be defined under imaging conditions, where many different T_1 values can be found, the idea of using a reduced flip angle is promising.

The second need resulting from the above discussion is to avoid spin echo experiments. This is done in conventional 1D NMR spectroscopy by the acquisition of the FID signal. Under imaging conditions, the FID signal is acquired with the help of the inversion of one magnetic field gradient.[11] This signal is called a 'gradient echo'. The physical content of FLASH-based fast NMR imaging sequences is therefore fast acquisitions of gradient echoes excited by low-flip-angle NMR excitation (FLASH ≡ fast low angle shot).[2]

3 GRADIENT ECHOES

An FID signal rapidly decreases within a few milliseconds to zero in the presence of a magnetic field gradient. A gradient is needed to encode the spatial domain in the form of frequency information. Since the gradient switching process needs a certain time interval, the FID signal is already dephased at the time when the signal should be acquired in the presence of the gradient. There exists the possibility of shifting the NMR signal to a later time interval. This is achieved by an inversion of the polarity of the gradient. This gradient inversion produces a 'gradient echo'. The gradient echo appears at the time when the time integral of the negative $[G_-(t)]$ and positive $[G_+(t)]$ gradient pulses become equal:

$$\int_0^\tau G_+(t)\,dt = \int_\tau^\delta G_-(t)\,dt \tag{3}$$

Here the positive gradient pulse has a duration τ, and the negative gradient pulse until the appearance of the gradient echo a duration δ. Since this time integral can be changed by experimental parameters (gradient strength and timing), the time at which the gradient echo appears is under the control of the

For References see p. 5009

NMR expert. The time interval between the NMR excitation and the gradient echo is called the 'gradient echo time' *TG*.

The signal strength of the gradient echo is dependent on *TG* and the relaxation time responsible for the FID decay, T_2^*. T_2^* is much shorter than T_2. It is known that the relaxivity $1/T_2^*$ is composed of the sum of a number of contributions:

$$\frac{1}{T_2^*} = \frac{1}{T_2} + \frac{1}{T_{2i}} + \frac{1}{T_{2s}} + \frac{1}{T_{2c}} + \cdots \quad (4)$$

Here T_2 is the spin–spin relaxation time, T_{2i} is given by inhomogeneities of the main magnetic field, T_{2s} is produced by magnetic susceptibility gradients, and T_{2c} is due to the chemical shift dispersion of the NMR signal. Other field inhomogeneities, and motions within the magnet, can cause further relaxation paths.

It should be emphasized that a gradient echo signal is strongly affected by magnetic field inhomogeneities. Furthermore, a gradient echo acquired at *TG* often exhibits lower signal strength than a spin echo at the same spin echo time *TE*.

In two-dimensional Fourier gradient echo imaging, the phase-encoding gradient pulse is switched on during the negative polarity of the gradient needed for the gradient echo. The theoretical minimum gradient echo time *TG* is the sum of the time intervals of the slice-selective excitation pulse, inversion of the read gradient, and half of the time interval for acquisition of the gradient echo. At least two time intervals have to be reserved for the switching processes of gradients.

In summary, the minimum *TG* is totally hardware-dependent. Minimum values of approximately 1 ms have been achieved so far. The minimum repetition time of a gradient echo imaging experiment is approximately twice as long as *TG*. Therefore, measuring times of a 128 × 128 pixel image of the order of 250 ms are possible, and have been reached by many groups. However, it should be noted that the measuring time, or the minimum repetition time *TR*, is hardware-dependent, as can be seen in Figure 1. Increasing the gradient strength results in a decreased measuring time owing to shorter data acquisition, phase encoding, and slice selection intervals.

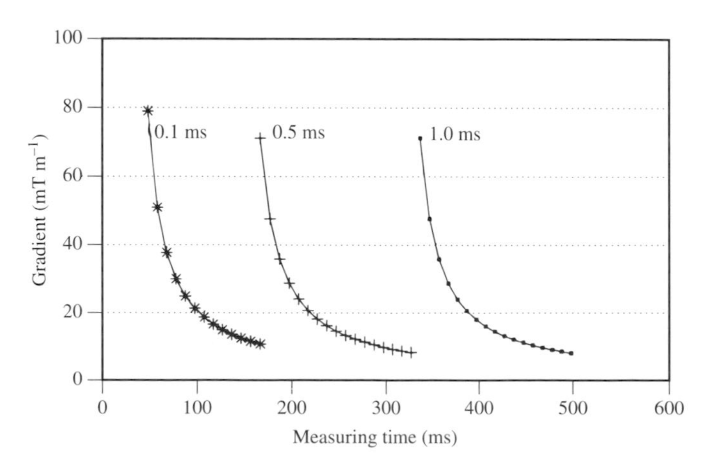

Figure 1 Dependence of the measuring time of FLASH MRI on the applied magnetic field gradient strength and switching time of the gradient pulses. Three switching time intervals are shown: 1 ms, 0.5 ms, and 0.1 ms. The curves were calculated using a 64 × 128 data matrix, slice thickness 5 mm and field of view 200 mm

Fast switching gradient power amplifiers further reduce *TG*, *TR*, and the measuring time.

Typical measurement parameters in whole body clinical scanners are a gradient switching time of 1 ms and a maximum gradient strength of 10 mT m^{-1}. Under these conditions, minimum repetition times of approximately 10 ms and gradient echo times of 5 ms are achieved with fast gradient echo imaging having a 20 cm field of view. In technologically advanced scanners repetition times of 3 ms and gradient echo times of 1.5 ms are possible. However, here, the gradient switching time must be in the range of 150 μs and maximum gradient strengths of at least 20 mT m^{-1} are needed.

4 SIGNAL INTENSITY

Fast repetitions of NMR excitations with short repetition times *TR* and low-flip-angle rf pulses need reconsideration of the signal behavior. It is known that a train of rf pulses with $TR < T_2$ generates a constant NMR signal level with an FID contribution following an rf pulse and an echo-like signal before each pulse.[12] The echo part of the signal can be destroyed by various kinds of 'spoiling' methods using gradient pulses[13] or phase scrambling of the rf pulses.[14] The echo part can be refocused, when all magnetic field gradient pulses are inverted in polarity after the acquisition of the gradient echo and before the next pulse.[15,16] The signal strength is totally dependent on the method, i.e., as between the use of 'spoiled' or 'refocused' fast gradient echo imaging.

If we assume that any transverse magnetization has been spoiled between the rf pulses, the signal will behave like

$$S \propto \frac{(1 - e^{-TR/T_1})\, e^{-TG/T_2^*} \sin\alpha}{1 - e^{-TR/T_1} \cos\alpha} \quad (5)$$

If we consider refocused experiments then the signal will behave like

$$S \propto \frac{(1 - e^{-TR/T_1})\, e^{-TG/T_2^*} \sin\alpha}{1 - e^{-TR/T_1} e^{TG/T_2} - (e^{-TR/T_1} - e^{-TG/T_2}) \cos\alpha} \quad (6)$$

It is interesting to note how this signal behavior changes under fast imaging conditions. Here *TR* and *TG* will be much smaller than T_1 and T_2.

For spoiled fast gradient echo imaging and large flip angles, the signal strength will behave like

$$S \propto TR/T_1 \quad (7)$$

For low-flip-angle pulses, the signal is no longer dependent on relaxation time differences, and will be dominated by spin density.

In refocused fast gradient echo imaging using large flip angles, the signal strength will behave like

$$S \propto T_2/T_1 \quad (8)$$

This parameter is relatively low in soft tissues compared with body fluids. Therefore, the most intense regions contain fluids—like CSF in the brain. Again, the signal will be dominated by the spin density contrast when using very low flip angles.

The application of fast gradient echo imaging allows great possibilities for changing the image contrast. Thus, 'refocused' gradient echo imaging can produce T_2-weighted images, according to equation (8), or 'spoiled' gradient echo images can give T_1-weighted images, according to equation (7), and both techniques result in 'spin density' images when low-flip-angle pulses are used.

The situation in the literature is rather complex, mainly because many acronyms have been used for the different contrast mechanisms in fast gradient echo imaging. Starting from the originally proposed fast gradient echo imaging using low-flip-angle pulses, FLASH, several other pulse sequences have been described using this concept:

1. 'spoiled' FLASH methods: (a) SPGR (spoiled GRASS); (b) spoiled FAST (Fourier acquired steady state); (c) FFE (fast field echo);
2. MPGR (multiplexed gradient refocusing) turbo methods: (a) Snapshot-FLASH, (b) Turbo-FLASH, (c) MPRAGE (magnetization prepared rapid acquired gradient echoes, (d) FSPGR (fast spoiled GRASS), (e) TFE (turbo FFE), (f) PFI (partial flip angle imaging);'refocused' FLASH methods: (i) FISP (fast imaging with steady precession); (ii) GRASS (gradient recalled acquisition in steady state); (iii) FAST (Fourier acquired steady state); (iv) GFE (gradient field echo); (v) CE-FAST (contrast enhanced FAST); (vi) FEER (fast field echo with even echo rephasing); (vii) GREC (gradient field echo with contrast).

It should be kept in mind that the principal idea of all these sequences is that fast NMR imaging can be achieved using gradient inversion, low-flip-angle excitations, and the acquisition of a gradient echo either with or without refocusing of transverse magnetization. The experiment has to be performed as fast as possible. The repetition time is only given by experimental or hardware constraints.

5 ARTIFACTS

Numerous artifacts can appear, but can be avoided in fast gradient echo imaging. The first, and not always visible, artifact can be a severe distortion of the slice profile. It has been known for many years that the slice profile is completely changed when using large flip angles and short repetition times.[17] In practical slice-selective imaging experiments, the flip angle varies across the slice, resulting in an often 'Gaussian-shaped' slice profile. As a consequence of this effect, large flip angles are found in the center and low flip angles at the edges of the slice profile. Higher flip angles and short repetition times *TR* will result in a heavy saturation in the center and almost no saturation at the edges of the slice profile. Therefore, high signal intensity will be found at the borders of the slice, giving strong partial volume artifacts.

This is the case when strongly T_1-weighted fast gradient echo images are needed, which can be measured according to the above discussion using short *TR* values and high flip angles in 'spoiled' imaging methods. It is therefore unrealistic to calculate T_1 maps from a series of strongly T_1-weighted images.

The most prominent feature of fast gradient echo imaging is the high-intensity signal of blood flowing perpendicular to the slice orientation. This is due to an inflow of fully relaxed magnetization during *TR*.[4] The enhancement of the signal intensity of flowing blood depends on imaging parameters (*TR*, flip angle, and slice thickness), T_1 of blood, and the flow velocity. Although this effect can be used to image blood vessels (MR angiography), it has a negative feature. Flowing spins accumulate a velocity (v)-dependent phase when flowing along a magnetic field gradient G, according to

$$\phi = \tfrac{1}{2}\gamma G v t^2 \qquad (9)$$

where t is the time interval between the NMR excitation and read-out of the signal. As long as the velocity remains constant, a constant phase is obtained. However, if the velocity changes during the imaging experiment (e.g., owing to pulsatile flow), different phases are accumulated, leading to severe 'flow artifacts' of bright intensity in the phase-encoding direction. This artifact can be reduced by rephasing gradients, giving no net phase shift due to flow velocity ('flow-compensated' gradient pulses). A detailed discussion of flow NMR imaging and flow compensation can be found in ***Flow in Whole Body Magnetic Resonance***.

All gradient echo imaging experiments suffer from the common feature of FID signals, namely, T_2* relaxation. The above discussion shows that owing to this effect, the signal intensity depends on magnetic field inhomogeneities from various sources, and chemical shift effects. If we have a magnetic field inhomogeneity given by the gradient G within a pixel of a diameter dr, a phase spread will appear during the gradient echo time *TG*:

$$\phi = \gamma G\,\mathrm{d}r\,TE \qquad (10)$$

High values of ϕ give low signal intensity in gradient echo imaging. It is clear that magnetic field homogeneity effects decrease when the spatial resolution increases and/or the gradient echo time *TG* decreases. Of course, this has technological limitations, because the gradient strength and gradient switching time has to be improved. This effect, often called the 'susceptibility' effect, appears near air–soft tissue interphases (e.g., the nasopharynx region, air-filled bowels, and lungs) (see ***Susceptibility Effects in Whole Body Experiments***).

A similar effect of signal change is observed when different chemical shift values are present in one image element. From inspection of an FID signal containing two close frequency components, we know that a beat is observed, having oscillations of the signal during T_2* decay. At periodic times, the signal vanishes—this is often called the opposed-phase signal. If the gradient echo time *TG* is exactly matched to this time, no signal can be measured from a pixel where two chemical shift components with equal intensity are present. Therefore, by a careful selection of the gradient echo time *TG*, 'in-phase' or 'opposed-phase' gradient echo images are measured.

Since transverse magnetization remains in fast gradient echo imaging without spoiling, a further 'center' or 'stripe' artifact appears. This is because residual transverse magnetization accumulates phase from step to step in 2D FT imaging. This gives a phase-encoding modulation of the signal, with low-frequency components giving a bright phase-encoding artifact in the middle of the image. The artifact can be reduced when low flip angles are used to minimize the transverse magnetization. Otherwise, good spoiling experiments should be selected, either with random gradient spoiling, or random phase of the rf excitation pulse.

For References see p. 5009

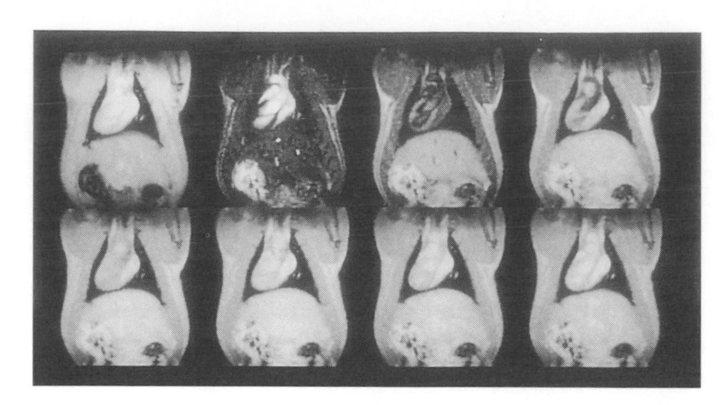

Figure 2 A series of 7 T inversion–recovery T_1-weighted snapshot-FLASH MR images of the thorax of a rat, showing the heart, liver, stomach, and skeletal muscle. The slice thickness was 2 mm, the data matrix 128 × 128, and the field-of-view 50 mm. The repetition time was 3 ms and the gradient echo time 1.5 ms. The left-hand image in the upper line is the first image taken immediately after magnetization inversion. The other images have been acquired successively

6 QUANTITATIVE IMAGING

Fast NMR imaging should have a well-defined image contrast, and should give quantitative NMR images of relaxation times, diffusion constants, and flow velocities. However, as pointed out above, it is difficult to observe well-defined image contrast, and even more difficult to image quantitative data using fast gradient echo sequences. For example, T_1 contrast suffers from slice profile artifacts, T_2 contrast is not obtainable using 'spoiled sequences', and a mixture of T_1 and T_2 information is given by 'refocused methods'. Furthermore, flow and diffusion information needs longer gradient echo times, with adverse effects arising due to 'susceptibility artifacts'. This problematic situation changed completely when 'magnetization prepared' fast gradient echo sequences appeared. These are called 'snapshot', 'turbo', or 'MP (magnetization prepared)' gradient echo sequences.[6]

These methods gain from the fact that, because of hardware improvements, short repetition times of 5 ms or less have become possible in whole body scanners. Now, the total measuring time is less than 1 s, and therefore comparable to an average T_1 relaxation time in high magnetic fields. Magnetization preparation means that the longitudinal magnetization can be changed in a definite way before the acquisition of the whole snapshot image. The image contrast will then be dependent on the 'magnetization preparation', because this lasts for a few T_1 intervals (typically, one to three times T_1).

Magnetization preparation can be performed using a single inversion pulse for IR T_1-weighted images, a $90°–\tau–180°–\tau–90°$ (DEFT) pulse for spin echo T_2-weighted images ($\tau = \frac{1}{2}TE$), or a DEFT pulse combined with diffusion or flow-encoding gradients for diffusion- or flow-weighted images. This technique does not suffer from all the artifacts described previously, and gives true quantitative image data when a series of images are acquired.[6]

Figure 2 shows a series of inversion–recovery T_1-weighted snapshot-FLASH cross-sectional images of the thorax of a rat measured at 7 T field strength. These images present a sufficient data base for quantitative evaluation of a relaxation time (T_1) image. This is further demonstrated in Figure 3, where calculated T_2 images are displayed. The left T_2 image was calculated from 16 T_2-weighted snapshot-FLASH images with a $90°–\tau–180°–\tau–90°$ magnetization preparation period of 16

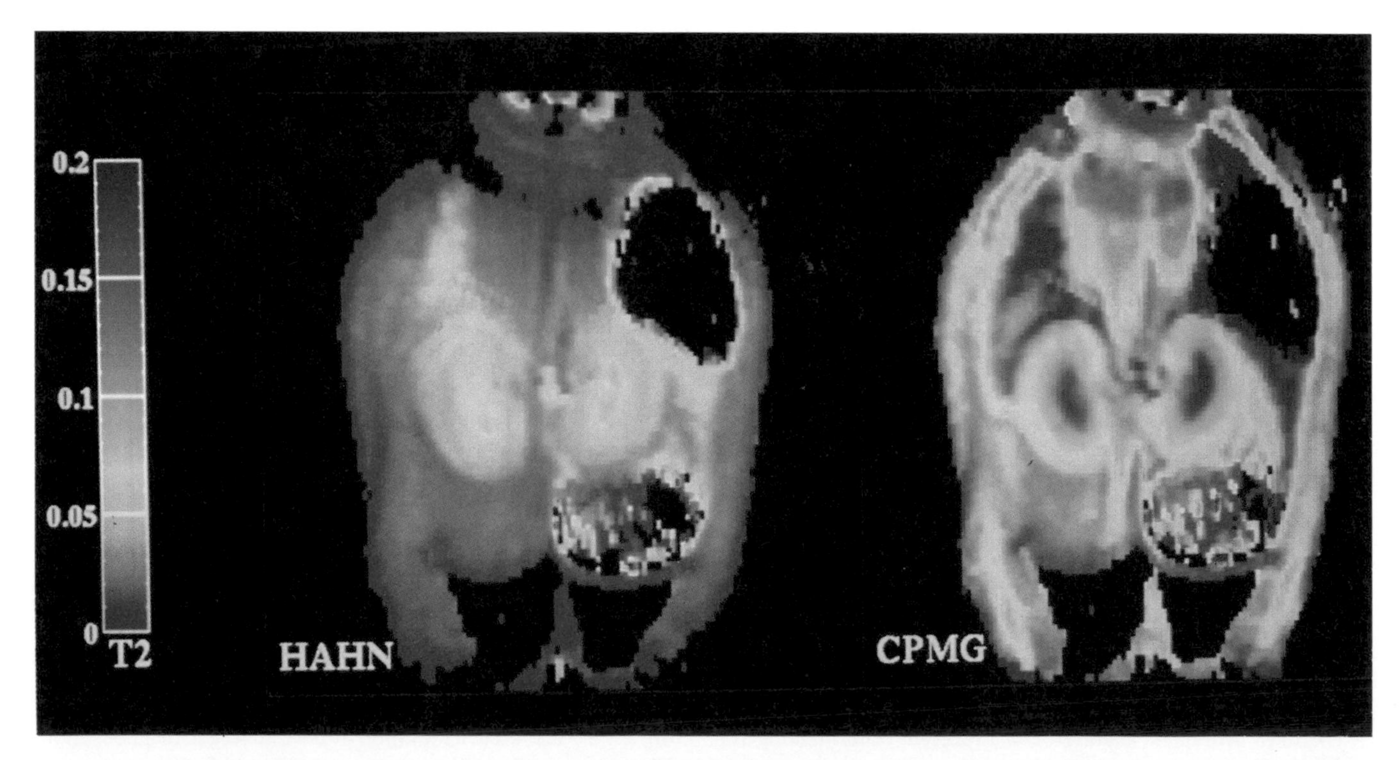

Figure 3 Calculated T_2 images, measured at 7 T, of the rat abdomen, showing the kidneys, liver, and stomach. The images were calculated from 16 T_2-weighted snapshot-FLASH MR tomograms. Images for the left-hand tomogram were obtained using Hahn spin echo magnetization preparation, and for the right-hand tomogram using Carr–Purcell–Meiboom Gill magnetization preparation. T_2 data are in seconds

different τ values (Hahn spin echo preparation). The right image was calculated from 16 T_2-weighted snapshot-FLASH images with $90°–\tau–(180°–\tau)_n–90°$ magnetization preparation ($n = 1,\ldots,16$) (CPMG preparation). Here, higher T_2 values have been obtained because of the decreased influence of molecular diffusion.

7 APPLICATIONS AND CONCLUSIONS

A multitude of different applications have been described in the past few years since the beginning of fast gradient echo experiments. One of the most prominent applications is in the field of rapid 3D NMR imaging. Here, the total measuring time can be limited to a few minutes, and it remains a practical method for medical diagnosis. Recent software and computer improvements have helped to increase the importance of 3D NMR imaging.

A further principal application is in the field of 'functional' NMR imaging. From the beginning, dynamic contrast media studies were very helpful in organ studies. The time resolution in fast gradient echo imaging is of the order of a few seconds. The time course of a bolus injection of an NMR contrast medium can be followed easily, and kinetic data calculated. This was first done for kidney and liver studies, and later for functional NMR imaging in the brain. However, it should be emphasized that dynamic fast gradient echo imaging using contrast media acquires a time-dependent dynamic signal change. As discussed above, the signal of fast gradient echo images has a complex dependence on various parameters—relaxation times, susceptibility, etc. Therefore, the signal intensity is not clearly related to the local concentration of a contrast medium, and is strongly dependent on the imaging experiment and parameters used. It should be emphasized that quantitative data have to be acquired for quantitative contrast media studies.

The importance and applications of fast gradient echo sequences will—as in the past—gain further from hardware improvements. An important parameter in this respect is the gradient echo time. All adverse artifacts can be reduced and the signal intensity improved when this parameter is minimized. This improvement should also help with the implementation of more 'hybrid' fast imaging sequences. In a few cases, hybrids between EPI and fast gradient echo imaging can be of advantage. The ultimate limit of all fast imaging sequences, however, is not given by technological constraints but by biological effects due to nerve stimulation in rapidly switched magnetic fields.

8 RELATED ARTICLES

Contrast Agents in Whole Body Magnetic Resonance: Operating Mechanisms; Echo-Planar Imaging; Flow in Whole Body Magnetic Resonance; Image Formation Methods; Partial Fourier Acquisition in MRI; Relaxation Measurements in Imaging Studies; Relaxation Measurements in Whole Body MRI: Clinical Utility; Spin Warp Data Acquisition; Susceptibility Effects in Whole Body Experiments; Whole Body Magnetic Resonance Artifacts; Whole Body MRI: Strategies Designed to Improve Patient Throughput.

9 REFERENCES

1. P. Mansfield, *J. Phys. C*, 1977, **10**, L55.
2. A. Haase, J. Frahm, D. Matthaei, W. Hänicke, and K.-D. Merboldt, *J. Magn. Reson.*, 1986, **67**, 258.
3. D. Matthaei, J. Frahm, A. Haase, and W. Hänicke, *Lancet*, 1985, **ii**, 893.
4. J. Frahm, A. Haase, D. Matthaei, K.-D. Merboldt, and W. Hänicke, *Magn. Reson. Med.*, 1986, **4**, 48.
5. J. Frahm, A. Haase, and D. Matthaei, *J. Comput. Assist. Tomogr.*, 1986, **10**, 363.
6. A. Haase, *Magn. Reson. Med.*, 1990, **13**, 77.
7. A. Kumar, D. Welti, and R. R. Ernst, *J. Magn. Reson.*, 1975, **18**, 69.
8. R. M. Henkelman and M. J. Bronskill, *Rev. Magn. Reson. Med.*, 1987, **2**, 126.
9. P. Margosian, F. Schmitt, and D. Purdy, *Health Care Instrum.*, 1986, **1**, 195.
10. R. R. Ernst and W. A. Anderson, *Rev. Sci. Instrum.*, 1966, **37**, 93.
11. W. A. Edelstein, J. M. S. Hutchison, G. Johnson, and T. Redpath, *Phys. Med. Biol.*, 1980, **25**, 756.
12. S. Patz, *Adv. Magn. Reson. Imaging* 1989, **1**, 73.
13. J. Frahm, K.-D. Merboldt and W. Hänicke, *J. Magn. Reson.*, 1987, **27**, 307.
14. Y. Zur, H. Israel, and P. Bendel, *Radiology*, 1987, **165**(P), 154.
15. A. Oppelt, R. Graumann, H. Barfuss, H. Fischer, W. Hartl, and W. Schajor, *Electromedica*, 1986, **1**, 15.
16. M. L. Gyngell, *Magn. Reson. Imaging*, 1988, **6**, 415.
17. I. R. Young, D. J. Bryant, and J. A. Payne, *Magn. Reson. Med.*, 1985, **2**, 355.

Biographical Sketch

Axel Haase. *b* 1952. Diploma in Physics, 1977, University of Giessen, Germany; Ph.D., 1980, working at Max-Planck-Institute, Göttingen and submitted to University of Giessen. Postdoc position, University of Oxford, with Prof. G. K. Radda, 1982. Scientist, Max-Planck-Institute in Göttingen, 1983–88. Chair of Biophysics at the University of Würzburg, Germany 1988–present. Published more than 60 articles. Present research interests are quantitative fast NMR imaging and NMR microscopy and NMR applications to cardiology and plant physiology.

Whole Body Magnetic Resonance Spectrometers: All-Digital Transmit/Receive Systems

G. Neil Holland

Otsuka Electronics, Fort Collins, CO, USA

1 INTRODUCTION

In a magnetic resonance (MR) imaging system, the spectrometer acts as the transmitter of radiofrequency (rf) waveforms and the receiver of the MR signal. Although MR was traditionally implemented by analog circuits,[1] advances in digital circuits have enabled rf transceivers in MR systems to be designed with digital devices for performing those functions previously implemented with analog circuitry.[2] As with the compact disk player and digital cellular telephone, when used in a spectrometer for MR imaging, digital technology can provide improvements in clarity and fidelity, since digital circuits are not subject to the imprecision and temporal drift inherent in analog equipment.

Implementation of spectrometer functions using digital electronics can provide benefits in the speed and precision of performing frequency shifts for multislice excitation and presaturation pulses, can give improved fidelity and linearity of the transmitted signal waveform resulting in improved slice profiles, can provide precise splitting of the received signal into real and imaginary parts, thus avoiding 'quadrature ghosts', and can give improved filtering for the prevention of aliasing or 'wraparound' artifacts. In addition, the digital spectrometer lends itself to being configured as a multichannel system, and thus is particularly important for implementation of phased array multiple receive coil technology.

2 MR TRANSMITTER AND RECEIVER

2.1 Transmitter

The transmitter forms the rf pulses required for imaging sequences. These rf pulses are fed to an rf power amplifier and ultimately the rf transmitter coil to excite the spin system and to generate a resonance signal. Slice-section techniques use spectrally tailored rf pulses for selective excitation, which is achieved by amplitude modulation of the transmitted rf pulse. The phase of the rf pulses is computer-controlled, and is used in artifact rejection schemes or advanced excitation schemes. To excite multiple slices, the transmitted rf pulse must vary in frequency for each excitation. Consequently, four forms of modulation of the rf carrier are used in MR imaging sequences: pulse, frequency, amplitude, and phase.

In a conventional transmitter, the rf frequencies are generated from a frequency synthesizer controlled by the system computer. Generally, two or more frequencies are mixed, whether in single or multiple stages, by devices known as double balanced mixers to form the final output frequency. Amplitude and phase modulation may also be accomplished with double balanced mixers. For amplitude modulation, the audiofrequency modulation envelope (or pulse shape) is fed as an analog waveform to one input of the mixer and fixed rf level (the 'carrier') to the other input. The modulation envelope is derived from a computerized waveform through a digital-to-analog converter (DAC).

On the transmitter side, several important performance parameters can affect image quality. The first of these, known as 'carrier feedthrough', is in the amplitude modulation section. All mixers have some signal leakage (carrier feedthrough) from input to output, which means that when an amplitude of zero is programmed, the modulator output is not truly zero. This small residual signal affects the quality of the rf pulses, particularly at lower modulation levels (such as for low-flip-angle sequences), and can affect slice profile. In addition, most imaging pulse sequences use a scheme in which the phase of successive rf pulses is alternated. Phase alternation works in conjunction with the data acquisition process to remove artifacts. This requires an exact balance between the 0° and 180° rf phases. Carrier feedthrough can contribute to phase imbalance, as can amplitude imbalance due to dc offset on the DAC that provides the audiofrequency modulation envelope. The exact manifestation of artifacts caused by carrier feedthrough and the complexity of design required to avoid them depend on the detailed implementation of the transmitter mixing scheme. However, the principal advantage of the digital implementation is that modulation artifacts of this form are eliminated entirely.

2.1.1 Digital Transmitter Implementation

A description of a typical digital transmitter is given below, and is illustrated in Figure 1. The major digital building block for this device is known as a numerically controlled oscillator (NCO), or as a direct digital synthesizer (DDS). This device is essentially a single chip frequency synthesizer, which can be interfaced to and controlled by the pulse programmer or host computer, and is set to produce an intermediate frequency (IF) (e.g., 3.5 MHz) via a DAC, which is mixed with a fixed frequency (e.g., 60 MHz) to generate the final output frequency (e.g., 63.5 MHz). The NCO architecture digitally divides the phase of a highly accurate clock by a digital tuning word (typically 32 bits) and uses a look-up table to convert the phase information to the desired frequency and sine wave output format. Structurally, the output of the NCO feeds a digital multiplier, which is used to perform amplitude modulation of the rf pulse. This is achieved by feeding one port of the multiplier with the digital modulation envelope while the frequency

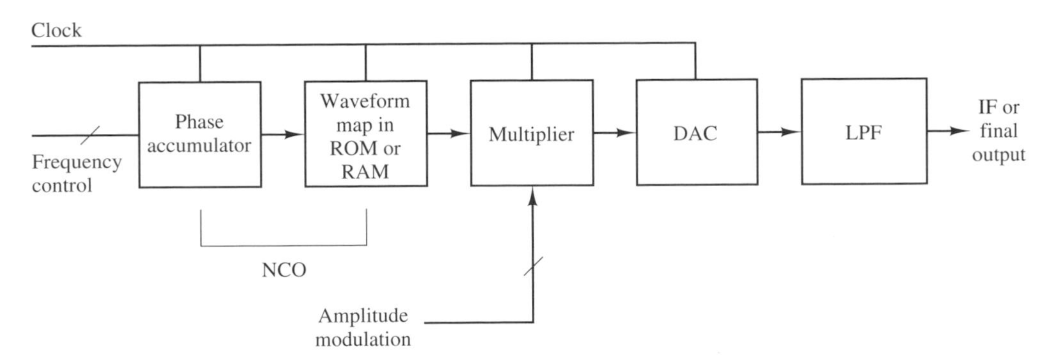

Figure 1 Digital transmitter representation showing the direct digital synthesis of the output frequency and digital modulation of the rf waveform. Depending on the final required Larmor frequency, an additional mixing step may or may not be required

data stream from the NCO feeds the other port of the multiplier. Note that with higher levels of digital integration now being achieved, the modulation may be internal to the NCO chip. In addition, the NCO has a phase port that allows coherent control of the phase of the output frequency. Consequently, all four forms of modulation referred to above can be accomplished with a single large-scale integrated (LSI) circuit. Not only does this significantly reduce parts count, and hence increase reliability, but the implementation also overcomes limitations of the analog design described earlier—basically, by avoiding the generation of the Larmor frequency during the time a zero profile is selected. In both the digital and analog implementations of the transmitter, the final output frequency is generated by mixing the modulated IF to the final output frequency. Although it is possible to get NCO devices that operate at frequencies as high as those used with MRI systems—thus, at least in principle, meaning that an MRI transmitter can be constructed from a single digital device—the output DAC cannot be greater than 8 bits owing to speed requirements, which places limitations on modulation accuracy and hence the quality of the resultant slice profile. The effect of DAC accuracy or number of bits on slice profile has been studied.[3] Evaluating the transverse magnetization profile following a small-tip-angle pulse shows that beyond 6 bits, the magnitude of unwanted signal is less than 1%. However, the DAC is also used to set the amplitude of the rf pulse for rf calibration, and must handle the range of tip angles from 180° to as low as approximately 10° for pulse sequences. This means about 5 bits of 'headroom' are required, thus meaning that an 11-bit converter would be recommended. The limiting factor in modulation accuracy is not, in fact, the word depth or number of bits used but rather the 'glitch' energy of the output DAC. Typically, this results in the loss of another bit of resolution, meaning that in practice a 12-bit DAC is required for rf pulse modulation.

2.2 Receiver

The receiver in an NMR spectrometer or MR imaging system takes the incoming NMR signal from the preamplifier and presents it to the computer system for data processing by FFT. The signal is preprocessed in the receiver to remove the rf carrier in a process known as demodulation or phase-sensitive detection, and split into phase quadrature. This is accomplished by splitting the incoming signal into two identical channels. The signals feed the rf ports of identical double balanced mixers. The mixer local oscillator (LO) ports are fed with reference frequencies at the same center frequency as the input signals, but the phase of the two reference signals differs by 90°. In a conventional receiver, programmable audiofrequency low-pass filters (usually Butterworth) operate on the quadrature signals to remove unwanted upper sidebands and to improve the signal-to-noise ratio (S/N) by matching the bandwidth of the filters to the known bandwidth of the NMR signal. After the filters, the quadrature NMR signals are digitized by a pair of analog-to-digital converters (ADCs). The conversion rate of the ADCs is determined by the bandwidth of the NMR signal. Typically, twofold oversampling is used. For example, for a 256^2 matrix image, 512 sampled points at 8 μs per point would result in a 62.5 kHz bandwidth. This operation is illustrated in Figure 2. This 'classical' implementation has a number of drawbacks. First, after the signal is split, the two channels must be exactly matched in both amplitude and phase. This is difficult to accomplish over a wide bandwidth. Second, the demodulation process places dc at the center of the signal bandwidth, with the result that any low-frequency noise causes a center frequency artifact (usually seen as a faint central line in an image). Last, quantization noise in the ADCs limits the overall S/N performance of the system.

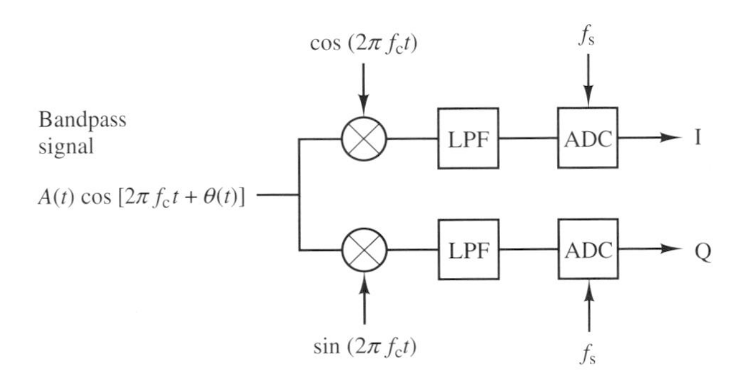

Figure 2 Classical receiver representation demonstrating the demodulation of incoming signal θ into in-phase and quadrature components by the use of parallel mixers to remove the rf carrier. After the mixers are low-pass filters (LPF), followed by analog-to-digital converters (ADCs)

For References see p. 5014

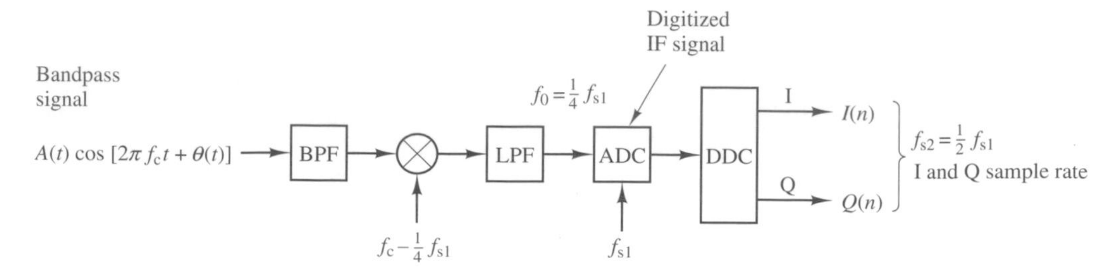

Figure 3 Digital receiver representation demonstrating the same operational steps as shown in Figure 2. The initial signal is filtered and mixed to an intermediate frequency which is one-fourth of the ADC sample frequency. A digital down-converter (DDC) performs final demodulation and low-pass filtering

2.2.1 Digital Receiver Implementation

Turning now to a digital receiver (Figure 3), the basic difference in design is that the NMR signal is mixed down to some intermediate frequency and digitized by a single ADC, so that final demodulation, separation into I and Q channels and low-pass filtering is performed in the digital domain.[4] The ADC is set to a fixed sampling frequency, operating at $N \times$ IF where IF is the intermediate frequency. Generally, $N = 4$, i.e., fourfold oversampling, is used. The choice of IF is determined by the maximum sampling rate of the ADC, but, in general, should be as high as possible, for two reasons. First, it dictates the minimum sampling interval, which then impacts the total sampling period, which then in turn can place a limitation on minimum echo time (*TE*). Second, to maximize S/N, the ratio between final subsampled baseband sampling rate and the IF should be as great as possible. This minimizes the ADC quantization noise per unit bandwidth.[5] There is another trade-off to be made when selecting the intermediate frequency, namely, the word depth of the ADC, i.e., the number of bits. As the number of bits decreases, conversion speed increases. For example, 8-bit converters can operate at speeds up to 100 MHz, whereas 16-bit converters only up to 1 MHz are readily commercially available. Quantization noise magnitude is determined by the voltage step size of the converter, and S/N is therefore proportional to $2^N/\sqrt{(\text{conversion rate})}$, where N is the number of bits in the converter. In other words, from a quantization noise perspective, a 14-bit converter operating at 4 MHz is equivalent to a 16-bit converter operating at 0.5 MHz.

To achieve the equivalent function of quadrature phase-sensitive detection for fourfold oversampled data, the input data stream is multiplied by the cosine and sine of the IF, which simplifies to multiplication by $+1, 0, -1, 0, \ldots$ for one channel and $0, +1, 0, -1, \ldots$ for the other. This process generates an exact 90° phase difference between channels, and centers the signals around dc. Low-frequency analog noise that in an analog receiver manifests self as a 'dc' artifact in the scheme just described is placed outside the imaging bandwidth so that the need for dc correction is eliminated. A second stage of digital filtering is implemented in each demodulated channel to provide the variable sampling required by the MR sequences. Finite-impulse-response (FIR) filters[6] perform this operation. These are multistage, linear phase filters, which are ideally suited to an implementation in silicon because of their repetitive pipelined operation. The selectable output sampling rate required by the NMR pulse sequence is achieved by choice of the decimation factor, which is the difference between input and output rates from the filters. FIR filter characteristics are more variable than their analog counterparts. Whereas the roll-off of an analog filter is determined by the number of poles (or filter stages), so that, for example, a six-pole filter attenuates at 36 dB per octave, the FIR filter design process allows tailoring of roll-off by trading final attenuation value (stopband attenuation) with rate of attenuation and ripple within the passband. FIR filters are defined in terms of the number of 'taps' they contain, where each tap represents a multiply–accumulate operation. With increasing decimation factor, a larger number of taps are needed to implement a filter with effective attenuation (> 80 dB at twice the corner frequency). Therefore, narrowband filters, such as would be used in spectroscopy sequences, require a greater number of taps, potentially up to 256, whereas for an imaging sequence with a decimation factor of say 32 only about 32 taps may be needed. While cut-off frequencies can be extremely sharp, with filters designed with many taps, actual implementation may require a large number of devices. In early digital receiver systems, each of the functions described above were implemented with individual digital devices. Today, the key building block for a current state of the art digital receiver is known as a digital down-converter (DDC). This is a single chip synthesizer, quadrature mixer, and lowpass filter, whose input is a sampled data stream taken directly from the ADC. Devices are available that can handle data up to 16-bits wide at a 52 MSPS (mega-samples-per-second) rate. The complex result of the demodulation process is lowpass-filtered with identical filters in the in-phase (I) and quadrature (Q) processing chains. Lowpass filtering is accomplished via a high decimation filter (HDF), followed by a fixed finite-impulse-response (FIR) filter. The devices provide automatically optimized matched bandwidth in that the lowpass filter inherently tracks the output (decimated) sampling rate. FIR filter coefficients are generated internally to the chip, so no preprogramming of filter parameters is required. Another advantage of the DDC is that automatic quadrature spectral reversal can be performed. In techniques such as EPI where echoes are taken alternately from read gradients of opposite sign, the frequency of each readout is therefore reversed and must be reordered. This is accomplished in the DDC by translating the lower sideband of the input to baseband rather than the upper sideband.

2.3 Experimental Validation of Digital Spectrometer S/N Improvements

An experimental validation of the relative performance between analog and digital receivers has been conducted (S. Mitchell, personal communication). The focus was to demonstrate that the reduced quantization noise of the digital receiver was reflected in actual image S/N improvement. An MR imaging system with identical 'front end' configuration (coils and preamplifiers), but different 'back end' equipment (spectrometer, data acquisition, and processing and display computers) was used. Images using identical scan parameters were taken first with the analog spectrometer and then with the digital spectrometer. Since the gains of the two spectrometers were different, the image pixel values and spectrometer input attenuator values were used to compute the gain difference between the two units. Then, a series of 'air scans' were taken at different receiver gain settings for the two spectrometers, with the standard deviation of the image noise being measured at each gain setting.

The results are shown in Figures 4(a) and (b). With increasing attenuation (i.e., as signal magnitude increases), the image noise as evidenced by standard deviation measurement exhibits progressively nonlinear behavior. In Figure 4(a), the response

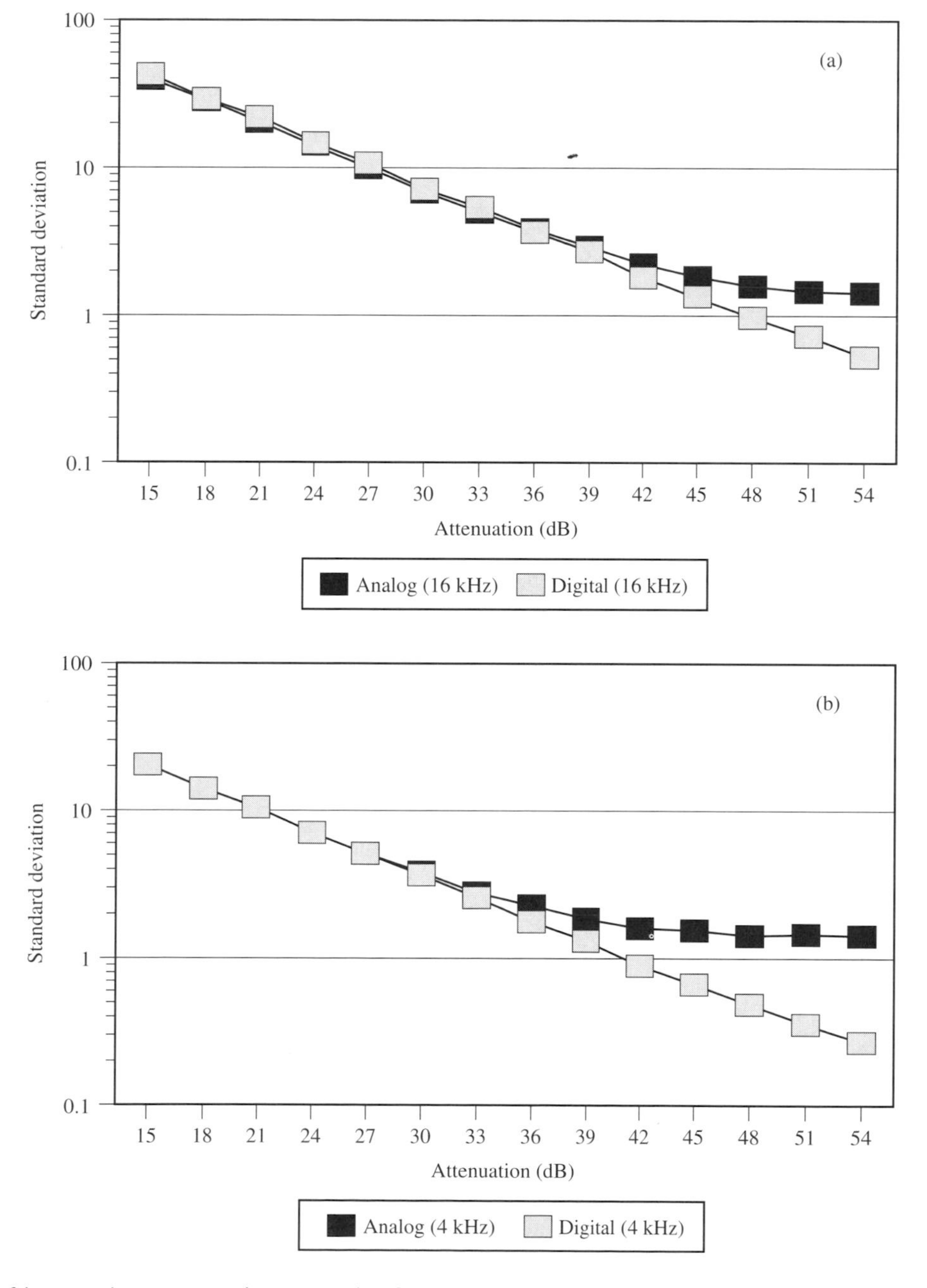

Figure 4 (a) A plot of image noise versus receiver attenuation for both analog and digital receivers operating at a 16 kHz bandwidth. Note the improved linearity at high attenuation factors (corresponding to high signal strength). (b) Data plotted for the same conditions as in (a), except with a 4 kHz bandwidth. The linearity and dynamic range of the digital system are unchanged, whereas the quantization noise per unit bandwidth of the analog system has increased fourfold

For References see p. 5014

for digital and analog receivers is plotted for effective sampling at 16 μs per point. This shows that the digital receiver exhibits about another 12 dB of dynamic range over its analog counterpart. The situation is demonstrated even more clearly for narrow bandwidth sequences. As shown in Figure 4(b), where data for a sampling rate of 64 μs per point are plotted, the analog receiver performs even more poorly relative to the digital design. The linearity of response for the digital system is unchanged by switching bandwidth whereas the analog system S/N drops by 6 dB. This because the quantization noise per unit bandwidth for the digital system is fixed, but the quantization noise generated by the ADCs in the analog implementation is now spread over a fourfold-narrower bandwidth, thus increasing the noise per unit bandwidth. The practical effect of this is to cause the signal magnitude at which reduced image S/N is visualized to be reduced.

2.4 Implementation of Multichannel Systems

Digital rf architecture is readily amenable to configuration in multichannel systems for both transmitter and receiver. In MRI, the most common use is in multichannel receive systems used with phased array rf coils. In such schemes, individual rf receive coil elements are formed into an array to cover the anatomical region of interest.[7] Each array element is connected to a separate receiver, and the resultant receive signals, either before or after Fourier transformation to the image domain, are combined to form a single image. The digital receiver implementation where the demodulator is connected to a digital signal processor (DSP) is particularly suited to the processing of phased array data. The DSP can be used for preprocessing of data to give weighted summation and correction for amplitude and/or phase variation between channels. The compact nature of a digital receiver design allows multiple receive channels to be configured on a single PC board, thus reducing the cost and complexity of multichannel systems and allowing the S/N advantages of phased array designs to be achieved without significant increase in system equipment. The use of a local NCO for receiver offset frequency allows each receiver in a multichannel array to have a unique demodulation frequency. This can be used to advantage for avoidance of aliasing and in the implementation of arrays where final composite images are not required, such as when used for imaging bilateral structures (e.g., TMJ). In addition, a multichannel system can be used advantageously with conventional quadrature coils by feeding each of the coil quadrature pair outputs to a separate receiver. This can improve performance by allowing for amplitude and phase balance as discussed above, but also can improve S/N by eliminating losses in analog hybrid combiner circuits, since these circuit components are not required.

3 CONCLUSIONS

Digital rf technology has now been widely adopted for use in MR imaging spectrometers, and is the technology of choice for new designs. Since initially demonstrated in the mid to late 1980s, advances in the speed and degree of integration of the digital devices required to implement spectrometer functions on both the transmitter and receiver side have increased dramatically. Other than the D/A converter (DAC) on the transmit side and the A/D converter (ADC) for the receiver, it is now possible to construct a spectrometer for a low- to mid-field system with only two devices—a single digital device for transmitter and a second for the receiver. With this degree of integration and forecasted improvements in speed of operation, the majority of whole body MR systems are expected to utilize direct-conversion digital spectrometers. In addition, because of the high degree of integration of digital spectrometers, it is anticipated that multichannel systems will become the operating standard in the near future. The inherent programmability and built-in self test features of the digital systems are also beneficial to reducing set-up and calibration time. The digital designs offer improvements in terms of accuracy and precision, and on the receive side can demonstrate improved dynamic range, which can be translated into improved S/N for high signal scans.

4 RELATED ARTICLES

Birdcage & Other High Homogeneity Radiofrequency Coils for Whole Body Magnetic Resonance; Coils for Insertion into the Human Body; Echo-Planar Imaging; Radiofrequency Systems & Coils for MRI & MRS; Sensitivity of Whole Body MRI Experiments; Surface & Other Local Coils for In Vivo Studies; Whole Body Machines: NMR Phased Array Coil Systems.

5 REFERENCES

1. D. I. Hoult, in 'Progress in Nuclear Magnetic Resonance Spectroscopy', eds. J. W. Emsley, J. Feeney, and L. H. Sutcliffe, Pergamon Press, Oxford, 1978, Vol. 12, p. 41.
2. G. N. Holland and J. R. McFall, *J. Magn. Reson. Imaging*, 1992, **2**, 241.
3. J. Slotboom, J. H. N. Creyghton, D. Korbee, A. F. Mehlkopf, and W. M. M. J. Bovee, *Magn. Res. Med.*, 1993, **30**, 732.
4. L. E. Pellon, *IEEE Trans. Sig. Process.*, 1992, **40**, 1670.
5. A. V. Oppenheim and R. W. Schafer, 'Discrete Time Signal Processing', Prentice-Hall, Englewood Cliffs, NJ, 1989, Chap. 3. Section 3.7.3
6. A. V. Oppenheim and R. W. Schafer, 'Digital Signal Processing', Prentice-Hall, Englewood Cliffs, NJ, 1975, Chap. 5.
7. P. B. Roemer, W. A. Edelstein, C. E. Hayes, S. P. Souza, and O. M. Mueller, *Magn. Res. Med.*, 1990, **16**, 192.

Biographical Sketch

G. Neil Holland. *b* 1952. M.Phil., 1979, University of Nottingham UK. Research fellow in Physics, Nottingham University, 1979–81. Picker International, Cleveland, Ohio, 1981–92; Otsuka Electronics 1992–present. Approximately 35 publications. Current interests include digital rf technology applied to MRI and MRI systems design.

Whole Body MRI: Local & Inserted Gradient Coils

Eric C. Wong
Medical College of Wisconsin, Milwaukee, WI, USA

1 INTRODUCTION

The gradient subsystem of the MRI scanner is responsible for all forms of spatial encoding, including image formation, encoding of flow information, encoding of diffusion information, and spatial modulation of magnetization for tissue tagging and selection of desired coherences (see also ***Gradient Coil Systems***). In many applications, the quality of the data obtained is limited by hardware-imposed limitations on the gradient amplitude or the gradient slew rate. Two examples are echo planar imaging (EPI),[1] which is typically limited in its spatial resolution by gradient amplitude and slew rate, and diffusion imaging,[2] which is limited primarily by gradient amplitude to very long echo times on conventional whole body systems.

Local gradient coils (LGCs) are gradient coils that are smaller than the whole body gradient coils built into clinical imaging systems, and are designed for specific purposes or areas of anatomy. Their smaller size gives them two advantages over whole body gradients.

The first advantage is that LGCs can produce stronger gradients per unit current, and higher gradient slew rates per unit voltage than whole body coils. For many clinical imaging systems, the use of local gradient coils makes more imaging techniques feasible (such as EPI), and improves the performance of many other techniques (i.e., higher resolution, shorter *TE*, etc.). This advantage is a technical and practical one. It brings higher performance gradients to clinical systems in a convenient and inexpensive way because local gradient coils are relatively simple add-on components, much like local rf coils, which are very widely used. However, there is active research at several institutions in the development of more powerful electronics to generate the current and voltage necessary to create strong, rapidly switched gradients using whole body gradient coils, and several commercial units are already available.

The second advantage is that LGCs can create higher gradient slew rates for a given limit on field slew rate. Because time-varying magnetic fields can induce electric currents and neuronal stimulation, there are limits on maximal field slew rates allowed in humans. In the United States, the FDA currently sets this limit at $20\,\mathrm{T\,s^{-1}}$. This field slew rate can be achieved not only using local gradient coils, but also by some of the higher performance whole body gradient systems available today. However, a local gradient coil will in general create a lower field slew rate for a given gradient slew rate, because the physical extent of the fields generated by a local gradient coil is smaller. For example, if a whole body Z gradient coil is used to image the head, the peak field will be in the area of the mid-chest, while the peak field of a local head gradient coil will be in the neck, and will be two to three times lower for a given gradient. This advantage is a more fundamental one in that local gradient coils can achieve stronger and faster switching gradients than whole body gradient systems, regardless of advances in whole body gradient system technology.

The primary disadvantage of using local gradient coils is that they add some complexity to the subject set-up procedure. The added time and effort are similar to that of using specialized rf coils, and, as local gradient coils become more common, commercialized, and integrated into gradient/rf coil units, the additional set-up time should be minimal. In some cases, patient comfort is decreased by the presence of local gradient coils owing to decreased visibility, restriction of patient movement, or acoustic noise. For a given imaging technique, LGCs are generally quieter in operation than whole body gradients. However, because LGCs allow the use of stronger and more rapidly switched gradients, acoustic noise at higher frequencies can be generated, and can attain high wound pressure levels (SPL). The SPL during EPI scanning in a local head gradient coil is approximately 100 dB, but is easily attenuated to tolerable levels using standard (30 dB) foam earplugs.

2 SAFETY

Two potential safety hazards accompany the use of LGCs: mechanical injury from forces and torques on the coil, and injury from electrical currents and voltages.

LCGs are usually designed so that there is no net force of torque on the coil when the coil is in place for imaging. However, if the coil is moved to an inhomogeneous area of the static field, or a portion of the coil fails without opening the circuit, then net forces and torques can result. If 100 A is applied to a coil in a 1.5 T field, torques on the order of 100 N m can be generated in a head gradient coil. If the coil is in an inhomogeneous portion of the field then forces on the order of 1000 N can result. Therefore, the system should be safeguarded so that current cannot be applied when the coil is not in place for imaging. The coil structure and support system should be strong enough to withstand the internal forces encountered during normal operation, and to prevent injury in case of an internal short.

The maximum voltage that drives most LGCs is on the order of 300 V. Although this is enough to cause injury, voltages of this magnitude are easily insulated from patients with only minimal measures such as plastic insulation. However,

For References see p. 5020

care must be taken, especially during installation and removal of the coil, that high voltage does not appear at any accessible terminals.

3 DESIGN CONSIDERATIONS

Many factors must be taken into account in the design of local gradient coils, including coil geometry, size, anatomy of interest, desired gradient strength, efficiency, inductance, desired switching rate, eddy currents, gradient uniformity, forces and torques, heat dissipation, and bioeffects. A large number of trade-offs exist between these factors, and only basic principles are discussed here.

The geometry of LGCs, like that of local rf coils, can be divided into three categories:

1. whole volume coils, which completely surround the anatomical structure of interest;
2. partial volume coils, which partially surround the structure;
3. surface gradient coils, which are placed against one side of the structure.

The primary deciding factor in the choice of coil geometry is the structure of the anatomical region of interest (ROI). If a whole volume coil can be placed around the ROI, as is usually the case for the head and the upper and lower extremities, this is generally preferable from the point of view of coil efficiency (gradient per unit current) and gradient field uniformity. The main drawbacks of whole volume coils are that they limit access to the patient, and they interfere more strongly with the rf fields. For large structures such as the chest and abdomen, a whole volume coil would be essentially a whole body coil. In order to obtain the advantages of smaller coils in these structures, partial volume or surface coils must be used. These will, in general, have smaller regions of usable gradient field uniformity, but can be useful for imaging of relatively small or superficial anatomical structures such as the heart.

The advantages gained by the use of a LGC have a strong dependence on the size of the coil. Two of the desirable properties of gradient coils are high efficiency and low inductance. For a coil of a given design, with a fixed number of turns,

$$\text{Efficiency} \propto \frac{1}{\text{size}^2} \tag{1}$$

$$\text{Inductance} \propto \text{size} \tag{2}$$

These factors indicate that the size of the coil should be as small as possible. The lower limit on the size of the coil is set by three other factors:

1. the coil must have a region of usable gradient field uniformity (the coil ROI) that encompasses the anatomical ROI;
2. the coil itself must be large enough to accommodate the anatomy of interest;
3. sufficient room must be left for the rf structure and return flux to allow dominant loading of the rf coil by the subject.

The largest of these three lower limits should, in general, determine the optimum size of the coil.

Once the desired gradient strength has been determined, the gradient coil should be designed to generate the required gradient and not more. If the available gradient strength is unnecessarily large then the inductance is higher and the maximum switching rate slower than they could otherwise have been. An exception to this principle can occur for the case of small coils in which the number of turns of wire required to produce the desired gradient is small, and may not be sufficient to produce the desired gradient field uniformity.

The inductance of a coil limits its maximum switching rate. A gradient coil can be thought of as a series resistance/inductance circuit. For a step voltage V_0 applied to the terminals of this circuit, the current will have the form

$$I = \frac{V_0}{R}(1 - \mathrm{e}^{-Rt/L}) \tag{3}$$

where L and R are the inductance and resistance of the circuit, respectively. Thus the current decays towards the final current with a time constant L/R. From this, it appears that either decreasing the inductance or increasing the resistance will decrease the switching time, but this is usually not the case. In practice, if an abrupt change in the current is desired, the voltage is not simply stepped to its new equilibrium value, but is given a 'pre-emphasis' pulse, which is a large overshoot of the voltage in order to increase the current slew rate. As the desired current is approached, the voltage is decreased to its equilibrium value. The current slew rate is limited by the maximum voltage output of the gradient amplifier (V_{max}), which is usually much higher than the voltage required to sustain the maximum current at steady state. The maximum current slew rate is calculated by differentiating the above equation with respect to time, replacing V_0 with V_{max}:

$$\left(\frac{\partial I}{\partial t}\right)_{max} = \frac{V_{max}}{L}\,\mathrm{e}^{-Rt/L} \tag{4}$$

For ramps that are short compared with L/R, as is usually the case, the current slew rate is simply V_{max}/L. Therefore, if fast switching is important to an application, the inductance of the gradient coil used should be minimized.

The importance of fast switching rates varies greatly, depending upon the imaging technique. For diffusion imaging using pulsed field gradients in a conventional spin echo sequence, the diffusion weighting gradient must have a large amplitude to minimize *TE*, but the switching rate of the gradient is, in general, not critical. For such an application, a gradient coil with many turns, a high efficiency, and a relatively high inductance would be practical. For fast imaging techniques such as echo planar, fast gradient switching speed is critical to the performance of the technique, and a coil with a minimum acceptable efficiency and low inductance should be used.

Eddy currents are currents induced in conducting structures due to time-varying magnetic fields. Eddy currents generate fields that oppose the field variations that induced the currents. Gradient eddy currents effectively limit the field slew rate by creating transient fields that usually oppose the desired gradient fields. Structures in an MRI scanner that may support eddy currents include the cryostat, the rf coil, and the rf shield. For whole body gradient coils, eddy currents generated in the cryostat are usually of the greatest concern, because the cryostat is physically close to the gradient coil and has very low resistance. In general, LGCs are much smaller than whole body gradient coils, and couple much less strongly to the cryostat. However, if rf coils or shields that have low-resistance closed current paths are used in close proximity to the gradient coil then these can support eddy currents and degrade the gradient field waveforms. Close attention to this problem must be paid in the design of rf coils and shields for use with local gradient coils.

Because the gradient fields directly map the spatial coordinates of physical space to the coordinates of the image, perfect gradient field uniformity is desired. However, if the gradient fields are not perfectly linear but are known then spatial transformations can be applied to the images to determine the location in space from which the signal in each pixel came. This can partially alleviate the problems associated with using nonlinear gradients, but several problems remain. In 2D imaging—the most common type—if the slice select gradient is nonuniform then information is obtained from tissue outside of the desired slice and the slice cannot be correctly reconstructed. If 3D data are collected, or if the slices are contiguous, then 3D spatial transformations can be used. Spatial transformations to correct for nonlinear gradients are straightforward, but increase image reconstruction time. Because nonlinearities in the gradients cause variations in the voxel sizes, the signal intensity, signal-to-noise ratio (S/N), and spatial resolution will be modulated by these variations, potentially adding to the difficulty of image interpretation. An interesting possibility is also raised by the availability of nonuniform voxel sizes. If gradient and rf coils with matched field nonuniformity are used then images with uniform S/N could be obtained using rf coils with high sensitivity but very nonuniform sensitivity profiles, such as surface rf coils.

If gradient pulses are used for functions other than spatial encoding then these functions will be affected by gradient nonlinearities as well. Nonlinear diffusion weighting or flow-encoding gradients will cause spatial variations in the diffusion or flow weighting. Gradient pulses that are used to dephase transverse magnetization will have an effectiveness that has spatial variations if the gradients are nonlinear.

The currents carried by a gradient coil interact with both the static field and self-generated fields. Because the static field is much stronger than the gradient fields for high- and mid-field scanners, interaction with the static field is of greater concern, and large forces on the current-carrying elements can result. In a uniform static field, a closed current loop cannot experience net forces, but can experience net torques. If these torques are not internally balanced then the coil must be mounted in a very strong and rigid mechanical structure that is tied to the structure of the main magnet in order to prevent large-amplitude vibration of the coil. Torques on an unbalanced head coil in a 1.5 T field are on the order of 100 N m. Internal balancing of the torque in a coil can be achieved by symmetry or by explicit nulling of the torque moment in an asymmetrical design.

4 DESIGN TECHNIQUES

Numerous mathematical techniques have been used to design gradient coils, many of which are described in a review article by Turner.[3] Most designs to date describe current patterns on the surface of one or two cylinders, since this is the most natural geometry for whole body patient access, and for LGCs a natural geometry for the head and extremities. Early designs consisted of simple counter-rotating pairs of loops for longitudinal gradients, and saddle coils for transverse gradients. Romeo and Hoult[4] used a spherical harmonic expansion of the fields to produce simple coils that approximate the lower order harmonics. These were then combined to approximate the desired fields. These designs give linear gradient fields in only a relatively small fraction of their enclosed volume, but are simple, easy to construct designs that can be useful if gradient uniformity is not critical.

Compton[5] introduced direct inversion of the field equations to derive a distribution of current density that would produce a desired field pattern. He defined an error function to describe the uniformity of the gradient fields over a region of interest, and used a least-squares approach to minimize this function. This approach can directly derive a current distribution that creates the desired fields for any coil geometry and any shape region of interest. The derived current density is approximated by a set of wires or etched current paths to arrive at a final coil design.

Turner[6] introduced an elegant approach to explicitly minimize inductance while creating the desired fields within the region of interest. Inductance is a quantity that is difficult to minimize using other techniques, because every element of the coil contributes not only self-inductance, but also mutual inductance with all other elements of the coil. For a given coil geometry, an appropriate transform applied to the current density linearizes the inductance so that each element of the design has zero mutual inductance with all other elements. For planar and cylindrical geometries, simple 2D Fourier transformation serves this purpose. In the transformed space, the inductance is a linear function of the transform components, and a Lagrange multiplier technique can be used to satisfy the desired field constraints while minimizing inductance. After deriving a current density pattern in transform space, a reverse transform is applied to produce a current density pattern in physical space, and again approximation by discrete current paths gives a final design. Several variations of the minimum inductance technique have been introduced to apply the technique to different geometries, and to use the same principles to minimize power dissipation instead of inductance.

Another class of design techniques uses numerical methods to optimize combinations of gradient uniformity, inductance, and coil efficiency. Using either a parameterized or a point-by-point description of the current path, mathematical techniques such as gradient descent, simulated annealing, Monte Carlo, and simplex can be used to minimize the chosen cost function. The author has used gradient descent techniques to design some of the coils that are shown in the examples below.[7] Numerical methods are, in general, more computationally intensive than analytical techniques, but they allow incorporation of arbitrary physical constraints in a simple manner, and they directly produce final designs, avoiding errors that may be introduced by the approximation of continuous current densities by discrete wires.

5 EXAMPLE COIL DESIGNS

Examples of whole volume, partial volume, and surface gradient coils are shown here.

For References see p. 5020

Figure 1 Three-axis local head gradient coil

The head coil shown in Figure 1 is an example of a whole volume local gradient coil. It is a three-axis design, torque-balanced by symmetry, and designed using gradient descent for gradient linearity. It was designed for EPI of the human brain, taking into account the specifications of the system in which it was to be used. The scanner it was designed for has a maximum sampling rate of 125 kHz and gradient amplifiers with a nominal current output of 85 A. The coil was designed to produce $20\,\mathrm{mT\,m^{-1}}$ gradients at 85 A, giving a minimum field of view of 16 cm at 125 kHz bandwidth, which was the smallest anticipated useful field-of-view in the human brain. Because the minimum coil efficiency for the application was specified, the number of turns of wire was kept to a minimum, and very low inductance was achieved. Inductances of 0.08–0.17 mH for the three axes allow switching times below $100\,\mu$s.

Figure 2 shows a biplanar cardiac gradient coil—an example of a partial volume coil. Although the coil structure encloses the chest for mechanical stability, the current-carrying elements occupy only two 30 cm × 40 cm rectangular regions, one above and one below the patient, making it in principle a partial volume coil. This coil produces only a single axis gradient in the vertical direction, and therefore requires the use of the whole body axial and horizontal gradient for imaging. It was also designed by gradient descent, and is used for EPI of the human heart. Because it only generates a vertical gradient field, only axial and saggittal EPI imaging, as well as oblique imaging intermediate to these, is possible.

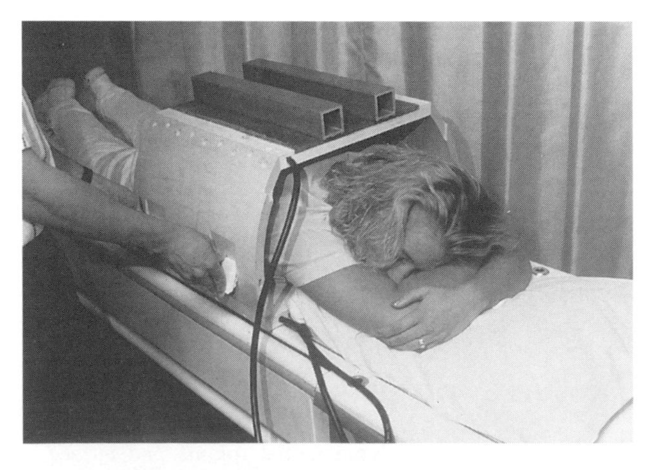

Figure 2 Single axis biplanar cardiac gradient coil

Figure 3 Three-axis surface cardiac gradient coil

Figure 3 shows a surface cardiac gradient coil. This coil is a three-axis design, upon which the patient lies prone, and generates nonlinear gradient fields that fall off by a factor of approximately two from the anterior surface to the posterior surface of the heart. The surface coil design in this case was chosen primarily for patient comfort, since the chest is not enclosed in the coil structure. This coil is used in conjunction with a surface rf coil, giving nearly uniform S/N across non-uniform voxel sizes, as described above.

6 APPLICATIONS

The most straightforward applications of LGCs are in high-resolution and short-*TE* conventional imaging techniques, which are usually severely limited by available gradient

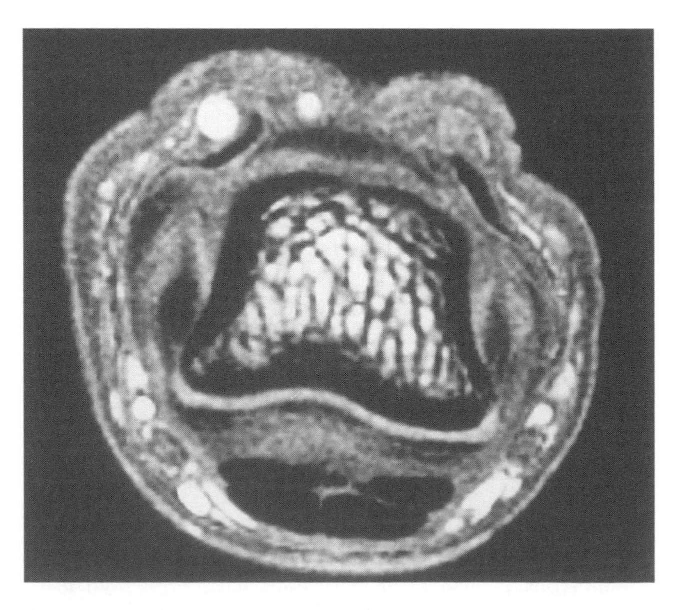

Figure 4 Axial image through the proximal interphalangeal joint of the middle finger. Acquisition parameters: gradient echo, FOV 3 cm × 1 mm, *TE* = 12 ms, matrix 256 × 256, *TR* = 500 ms, flip angle 90°, two excitations

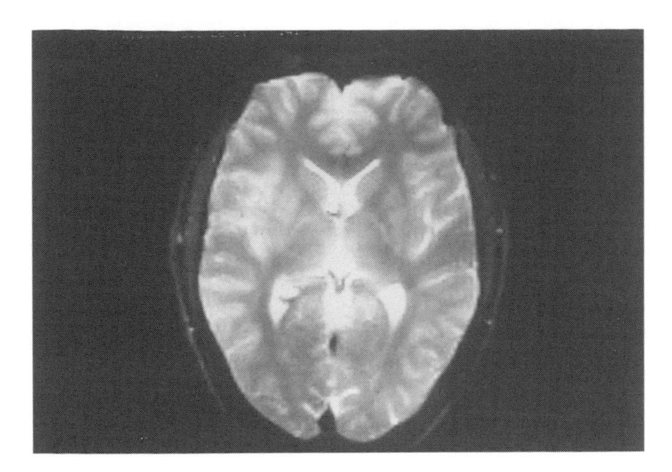

Figure 5 Single shot axial EPI image of the head. Acquisition parameters: single shot spin echo EPI, matrix 128 × 128 partial k-space, TE = 20 ms

strength in whole body systems. In principle, high resolution can be obtained using lower gradient strengths and longer pulse durations, but, in practice, the resolution will be limited by the T_2^* of the tissue. It is therefore often necessary to obtain high-resolution images with short TE and data acquisition times. An example is the image of the proximal interphalangeal joint of the human finger shown in Figure 4.[8] This image was acquired using a cylindrical LGC of 10 cm diameter and a small transmit–receive rf coil. The gradient strengths used were 60 mT m^{-1} in the slice select direction, and 27 mT m^{-1} in the in-plane direction.

An example of ultrafast imaging using a LGC is shown in Figure 5. This is a single shot EPI image of the human head acquired in 80 ms (see ***Echo-Planar Imaging***). For this image, the head gradient coil of Figure 1 was used, and no correction for gradient nonuniformity was employed. Only 15 mT m^{-1} gradients were used, but the coil was switched in just 112 μs using standard gradient amplifiers.

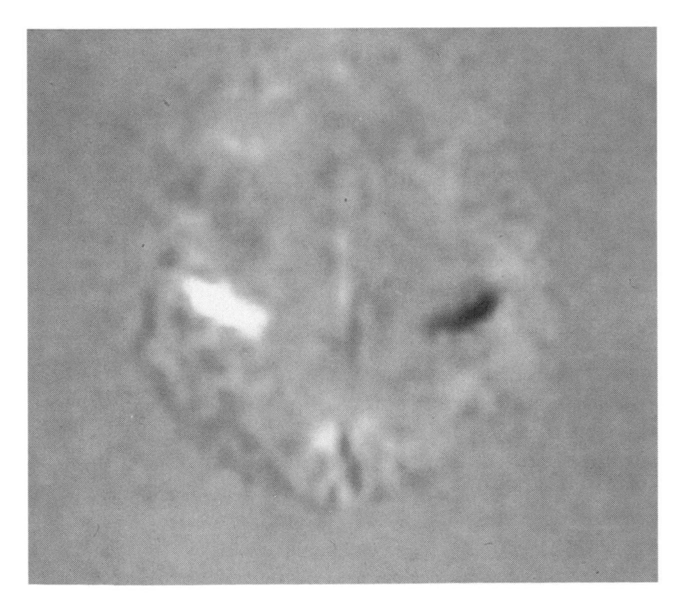

Figure 6 Calculated axial image of functional activation in the motor cortices during alternating right and left finger tapping. Acquisition parameters: single shot gradient-recalled EPI, 100 repetitions at 1 s intervals with five alternating periods of right and left finger tapping, matrix 64 × 64, TE = 40 ms. The displayed image is a subtraction of images collected during left finger tapping minus images collected during right finger tapping

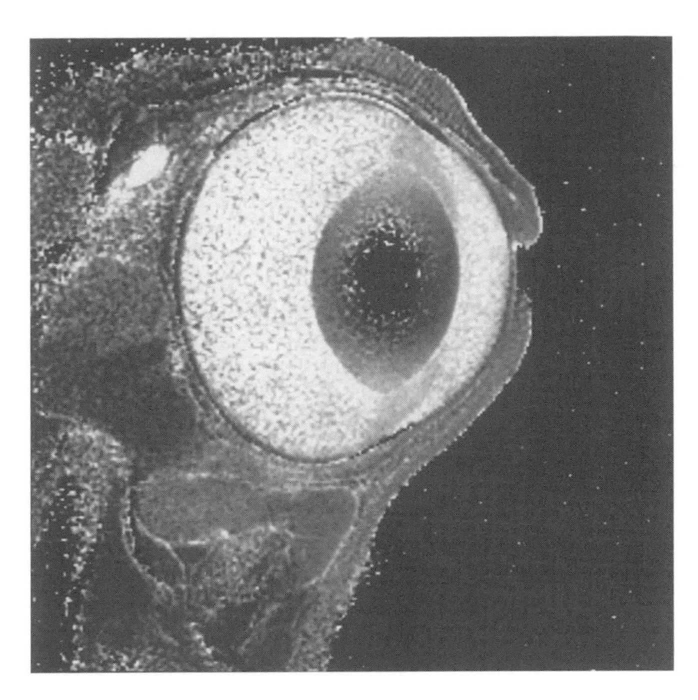

Figure 7 Calculated diffusion coefficient image of the normal rabbit lens. Acquisition parameters: spin echo with pulsed field gradients on either side of the refocusing pulse, b factor 400 s mm^{-2}, direction of diffusion sensitivity left/right on image, FOV 3 cm × 1 mm, TE = 30 ms, matrix 256 × 256, TR = 500 ms, two excitations

An extension of the EPI technique is functional neuroimaging using either BOLD contrast or flow-dependent contrast.[9–11] An example using BOLD contrast is shown in Figure 6. In this example, gradient-recalled EPI images through the motor cortex were collected repetitively while the subject alternately performed right and left finger tapping tasks. The figure shows a difference image between images acquired during right finger tapping and images acquired during left finger tapping, clearly demonstrating synchronous modulation of signal from the motor cortices with opposite phase.

Diffusion imaging is an example of the use of strong gradients for encoding of desired contrast, rather than for image formation (see ***Diffusion & Perfusion in MRI***). Figure 7 shows a calculated diffusion image of the eye of an anesthetized rabbit. This image was collected using the same 10 cm diameter LGC as that used to acquire Figure 4, inserted into the same clinical scanner. Diffusion weighting was achieved using pulsed gradients of 80 mT m^{-1} and 12 ms duration, resulting in a b factor of approximately 400 s mm^{-2}. With these gradients, the minimum TE for a conventional spin echo imaging sequence was 30 ms, similar to the average T_2 of the lens, which was the object of interest. If similar diffusion weighting is to be achieved using typical whole body gradients (10 mT m^{-1}), the minimum TE would be approximately 200 ms, which is many times longer than the T_2 of the tissue, and very little signal would be collected.

There are many other applications of LGCs for which examples are not shown. In angiography, stronger and faster

For References see p. 5020

gradients allow for shorter flow-compensating gradients and hence shorter *TE*. This decreases flow-related dephasing and increases the detected signal. In magnetic tissue tagging, such as myocardial tagging to observe cardiac wall motion, stronger gradients can improve tag profiles and decrease tagging time, thus generating sharper tagged grids in the final images. For imaging of very short T_2^* species, it is necessary to employ strong gradients in order to assure that the imaging gradients dominate the static field inhomogeneities. Short T_2^* occurs primarily in areas with strong susceptibility gradients, such as lung tissue. In localized spectroscopy, spectral resolution can be severely compromised by the presence of gradient eddy currents. For this reason, gradient switching rates are in general very conservative, resulting in relatively long minimum echo times. For shorter-T_2 species, LGCs can be used to improve switching rates with minimal eddy currents, and thereby decrease echo times.

7 SUMMARY

Many applications in MRI are either dependent upon or improved by the availability of stronger and more rapidly switched gradients than are typically found in whole body clinical systems. The use of local gradient coils is a simple and inexpensive means of obtaining high-performance gradients, and has the advantage of higher permissible gradient switching rate than whole body coils for a given limit on field switching rate.

8 RELATED ARTICLES

Diffusion & Perfusion in MRI; Echo-Planar Imaging; Eddy Currents & Their Control; Gradient Coil Systems.

9 REFERENCES

1. P. Mansfield, *J. Phys. C*, 1977, **10**, L55.
2. D. Le Bihan, E. Breton, D. Lallemand, P. Grenier, E. Cabonis, and M. Laval-Jeantet, *Radiology*, 1986, **161**, 401.
3. R. Turner, *Magn. Reson. Imaging*, 1993, **11**, 903.
4. F. Romeo and D. I. Hoult, *Magn. Reson. Med.*, 1984, **1**, 44.
5. R. A. Compton, US Patent 4 456 881 (1984).
6. R. Turner, *J. Phys. E: Sci. Instrum.*, 1988, **21**, 948.
7. E. C. Wong, A. Jesmanowicz, and J. S. Hyde, *Magn. Reson. Med.*, 1991, **21**, 39.
8. E. C. Wong, A. J. Jesmanowicz, and J. S. Hyde, *Radiology*, 1991, **181**, 393.
9. P. A. Bandettini, E. C. Wong, R. S. Hinks, R. S. Tikofsky, and J. S. Hyde, *Magn. Reson. Med.*, 1992, **25**, 390.
10. K. K. Kwong, J. W. Belliveau, D. A. Chesler, I. E. Goldberg, R. M. Weisshoff, B. P. Poncelet, D. N. Kennedy, B. E. Koppel, M. S. Cohen, and R. Turner, *Proc. Natl. Acad. Sci. U.S.A.*, 1992, **89**, 5675.
11. S. Ogawa, D. W. Tank, R. Menon, J. M. Ellerman, S. G. Kim, H. Merkle, and K. Ugurbil, *Proc. Natl. Acad. Sci. U.S.A.*, 1992, **89**, 5951.

Biographical Sketch

Eric C. Wong, *b* 1964. B.A., 1985, biophysics, University of California, San Diego. Ph.D., 1991, biophysics, M.D., 1994, Assistant Professor of Biophysics, 1992–present, Medical College of Wisconsin. Approx. 12 publications. Research specialties: design and applications of local gradient coils for MRI, pulse sequences for ultrafast MRI, contrast mechanisms in functional MRI.

Whole Body MRI: Strategies Designed to Improve Patient Throughput

Felix W. Wehrli

University of Pennsylvania Medical School, Philadelphia, PA, USA

1 INTRODUCTION

Since the inception of clinical MRI at the beginning of the 1980s, the rate at which a subject can be scanned—often denoted 'patient throughput'—has been reduced by nearly an order of magnitude. Some of the improved scan efficiency is unrelated to a shortening of the actual data acquisition time, and can be accounted for by improvements in some of the ancillary procedures such as patient set-up, rf coil placement, scan parameter entry, prescan (MRI jargon for rf carrier selection and adjustment of transmit and receive gain), and advances in digital system architecture (shortening in reconstruction times, concurrence of data acquisition and reconstruction, etc.). The focus of this article, however, will be on techniques for reducing scan time, since it is this that determines patient tolerance. Scan times in excess of 5–10 min typically lead to image degradation from subject motion.

Another strong motivation for shortening scan time is physiologic motion such as cardiac pulsation, respiration, and peristalsis. Clearly, if the data acquisition time could be lowered below the period of physiologic motion, the effect of motion unsharpness in the image data could effectively be suppressed. Many of the new applications, such as the assessment of organ function by monitoring the passage of a bolus of contrast agent[1] or the study of brain function by recording the response to stimuli,[2–4] demand temporal resolution on the order of 0.1–1 s (see also ***Contrast Agents in Whole Body Magnetic Resonance: An Overview***).

While many of the ideas for improved scan efficiency have been in existence for a decade or longer, they have become practical only recently, owing to engineering advances such as digital transceiver electronics (see also ***Whole Body Magnetic Resonance Spectrometers: All-Digital Transmit/Receive Systems***) and gradient technology (see ***Gradient Coil Systems***). A case in point is echo planar imaging (EPI), which was first conceived in 1977,[5] but has only recently become practical.[6] Finally, an increased data sampling rate typically exacts an unavoidable penalty in signal-to-noise ratio (S/N). Since the minimum tolerable S/N is usually dictated by diagnostic criteria such as lesion detectability, increases in temporal resolution will usually have to be traded against spatial resolution.

This article briefly reviews the concepts underlying the most important classes of rapid imaging techniques, each of which is discussed in more detail in other articles in this Encyclopedia, referenced later. For a simple introduction to rapid MRI imaging techniques, the reader is referred to Wehrli.[7]

2 *k*-SPACE FORMALISM

We shall in the following resort to the classical ***k***-space formalism[8] (see also ***Image Formation Methods***). In brief, data sampling occurs in reciprocal space, referred to as '***k***-space' or the 'spatial frequency domain'. For an object of spin density $\rho(\boldsymbol{r})$, ignoring relaxation, the spatial frequency signal $\rho'(\boldsymbol{k})$ is given by

$$\rho'(\boldsymbol{k}) = \int_{\boldsymbol{r}_1}^{\boldsymbol{r}_2} \rho(\boldsymbol{r}) \mathrm{e}^{-\mathrm{i}\boldsymbol{k}(t)\cdot\boldsymbol{r}}\, \mathrm{d}^3 r \tag{1}$$

Here $\boldsymbol{k}(t)$ is the spatial frequency vector, expressed in units of rad cm^{-1}:

$$\boldsymbol{k}(t) = \gamma \int_0^t \boldsymbol{G}(t')\, \mathrm{d}t' \tag{2}$$

where $\boldsymbol{G}(t')$ is the time-dependent spatial encoding gradient in vector format. Pictorially, the spatial frequency may be regarded as the phase rotation per unit length of the object experienced by the magnetization after being exposed to a gradient $\boldsymbol{G}(t')$ for some period t.

We further recognize from equation (1) that $\rho'(\boldsymbol{k})$ and $\rho(\boldsymbol{r})$ form a Fourier transform pair, and so

$$\rho(\boldsymbol{r}) = \int_{\boldsymbol{k}_1}^{\boldsymbol{k}_2} \rho'(\boldsymbol{k}) \mathrm{e}^{\mathrm{i}\boldsymbol{k}(t)\cdot\boldsymbol{r}}\, \mathrm{d}^3 k \tag{3}$$

The path of the spatial frequency vector during execution of the pulse sequence is determined by the time-dependent gradient waveforms.

Since ***k***-space is sampled discretely, sampling needs to satisfy the Nyquist criterion

$$\tfrac{1}{2} L_i k_{i,\max} = \tfrac{1}{2} N_i \pi \tag{4}$$

where $k_{i,\max}$ is the highest spatial frequency, L_i and N_i represent the field-of-view and number of data samples, respectively, and $i = x, y, z$. We can then write for the sampling interval Δk_i

$$\Delta k_i = \frac{k_{i\,\max}}{\tfrac{1}{2} N_i} = \frac{2\pi}{L_i} = \frac{2\pi}{N_i \Delta L_i} \tag{5}$$

with ΔL_i representing the pixel size. The factor of 2 in equation (5) arises because we collect N_i samples from $-k_{\max}$ to $+k_{\max}$ or $\tfrac{1}{2}N_i$ samples from 0 to $k_{\max}$.

Since the data sampling time increases with the number of data samples, one obvious approach toward shortening the sampling time is to reduce N_i. It is readily seen from equation (5) that this can be achieved in two different ways. If we wish to maintain the pixel size, this will result in a reduction in the field-of-view and thus an increase in Δk_i. Conversely, we may want to maintain the field-of-view, in which case Δk_i remains invariant but $k_{\max}$ is lowered, and thus pixel size is increased. In the latter case, we trade spatial resolution for a reduction in sampling time.

Further, since $\Delta k \approx \gamma \Delta t\, G$, the sampling interval Δt scales inversely with the amplitude of the spatial encoding gradients. Therefore, the duration of the spatial encoding process is governed significantly by the achievable gradient amplitudes. At a given resolution, the latter essentially determine what fraction of ***k***-space can be covered following a single excitation (since the signal decays with time constant T_2). Therefore, the properties of the gradients, notably their amplitudes, and for many imaging sequences also their slew rates, are paramount in determining ultimate imaging speed.

3 CLASSIFICATION OF IMAGING TECHNIQUES

Most of the currently used imaging techniques use rectilinear ***k***-space sampling, and as such may be regarded as descendants of the spin warp technique of Edelstein et al.,[9] which itself is an outgrowth of Fourier zeugmatography, pioneered by Kumar, Welti, and Ernst.[10] Since the latter is an embodiment of two-dimensional NMR spectroscopy,[11] conceptionally all rectilinear sampling techniques may be regarded as originating from 2D NMR. By contrast, projection–reconstruction imaging[12–14] or imaging with nonperiodic time-varying gradients[15] such as spiral scanning[16,17] are not rectilinear sampling techniques, since the ***k***-space path is non-Cartesian.

Common to all rectilinear ***k***-space sampling techniques are two distinct phases of spatial encoding, a first period during which a gradient G_y is applied whose time integral determines k_y and during which no sampling occurs, followed by a second period during which a gradient G_x is applied while the free precession signal is sampled, resulting in a line k_x. The former is typically called 'phase encoding' and the latter 'frequency encoding'. In this manner, a 2D ***k***-space grid is sampled by incrementally stepping the amplitude of the phase-encoding gradient. The concept can be extended to a third dimension by phase encoding along a second spatial coordinate to afford k_z, in which case ***k***-space is three-dimensional. In both back-projection and spiral scanning, two or all three spatial encoding gradients are simultaneously active while data are sampled. A generic 3D spin warp pulse sequence and ***k***-space map are shown in Figure 1.

4 VARIOUS EMBODIMENTS OF SPIN WARP IMAGING

4.1 Scan Time, Signal-to-Noise Ratio and Spatial Resolution

Except in those situations where ***k***-space can be scanned in a period on the order of T_2, such as in EPI[5] or multiple pulse techniques,[18] resulting in a series of consecutive echoes that can be individually encoded, only a fraction of ***k***-space is typically scanned upon creating transverse magnetization, and the process repeated several times, each cycle affording another ***k***-space segment. Longitudinal magnetization builds up in a recurrent fashion between cycles of period *TR* (also called 'pulse repetition time'). The data acquisition time T_{ac} thus scales with *TR* and the number of pulse sequence cycles, the latter being a function of the number of spatial encoding steps or ***k***-space samples. It is customary to denote the frequency-encoding axis by x and the phase-encoding axis by y (or y and z, respectively, if phase encoding in an additional dimension is effected), with N_i ($i = x, y, z$) representing the number of data samples. Assuming further that during a frequency-encoding period a single line $k_{x,i=1,...Nx}$ is sampled, T_{ac} to sample ***k***-space is given by

$$T_{ac} = N_y N_z n TR \tag{6}$$

where n represents the number of sequence repetitions (also called 'number of excitations' in imaging jargon) at given values of ***k***. At typical sampling rates, the sampling time for frequency encoding is on the order of 10 ms or less, and so does not appear in equation (6). It will, however, become critical in high-speed gradient echo imaging[19,20] and EPI and its derivatives.[21–23]

The scan time scales with the number of phase encodings (N_y in 2D, and N_y and N_z in 3D spin warp imaging), and thus a reduction in the number of phase-encoding samples entails a concomitant reduction in scan time. As previously pointed out, this can be achieved in two different ways: either by lowering k_{max} or decreasing sampling density $1/\Delta k$. The former lowers resolution, the latter leads to reduced field-of-view while preserving resolution [cf. equations (4) and (5)]. Thus, the term 'matrix size' is ambiguous, though in common parlance a change in matrix implicitly implies fixed field-of-view. The two different situations, each lowering data acquisition time of a 3D image acquisition by a factor of 4, are illustrated in Figure 2.

Let us now examine the S/N implications of the two different operations. The signal amplitude S scales with voxel size, i.e.,

$$S \propto \Delta x\, \Delta y\, \Delta z = \frac{L_x L_y L_z}{N_x N_y N_z} \tag{7}$$

where the product $\Delta x\, \Delta y\, \Delta z$ expresses the voxel volume, the remaining parameters have previously been defined. Further, the noise-reducing effect of sampling leads to

$$\mathrm{S/N} \propto \sqrt{N_x N_y N_z n} \tag{8}$$

and thus

$$\mathrm{S/N} \propto \frac{L_x L_y L_z \sqrt{n}}{\sqrt{N_x N_y N_z}} \tag{9}$$

From equation (9), we conclude that halving the number of data samples N_y and N_z by halving $k_{y,max}$ and $k_{z,max}$, increases S/N by a factor of 2 (Figure 2a). Skipping alternate data samples, on the other hand, penalizes S/N by a factor of 8

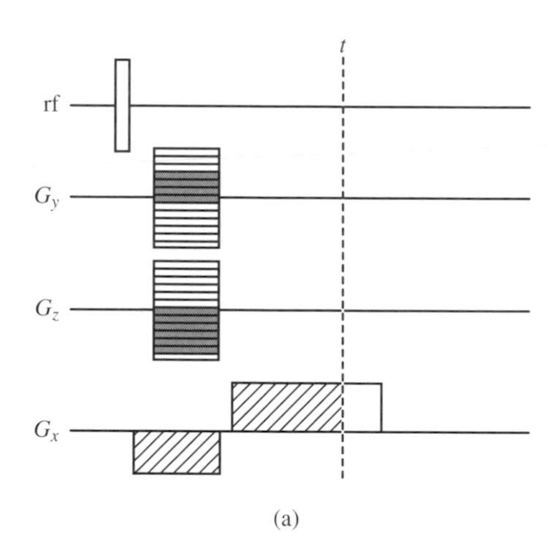

(a)

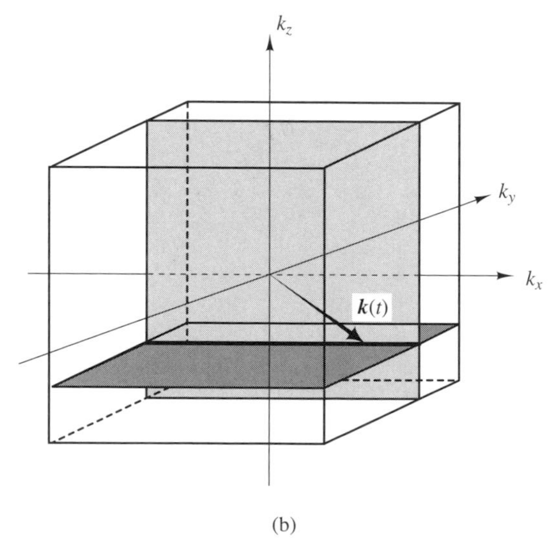

(b)

Figure 1 (a) Generic 3D spin warp imaging with phase encoding on the x and y axes and frequency encoding on z. Hatched areas indicate the spatial frequency vector at time t (b). Amplitudes of phase-encoding gradients are arbitrary. During readout, a line k_x is scanned corresponding to intersection between the (k_x,k_y) and (k_x,k_z) planes (shaded areas)

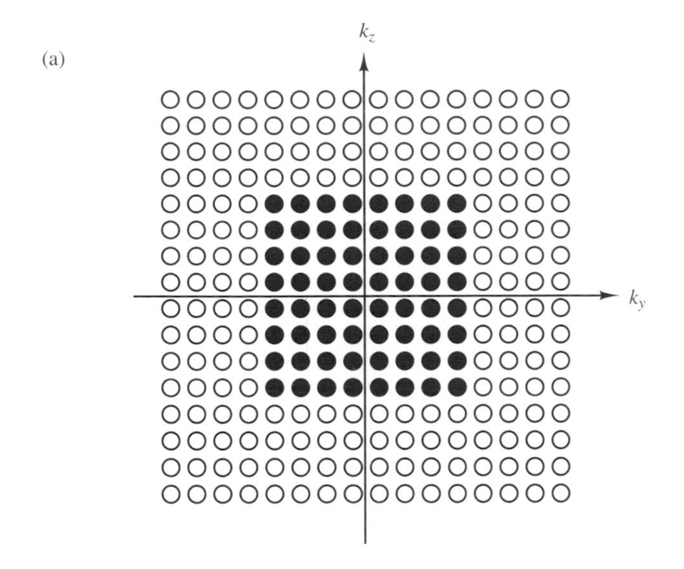

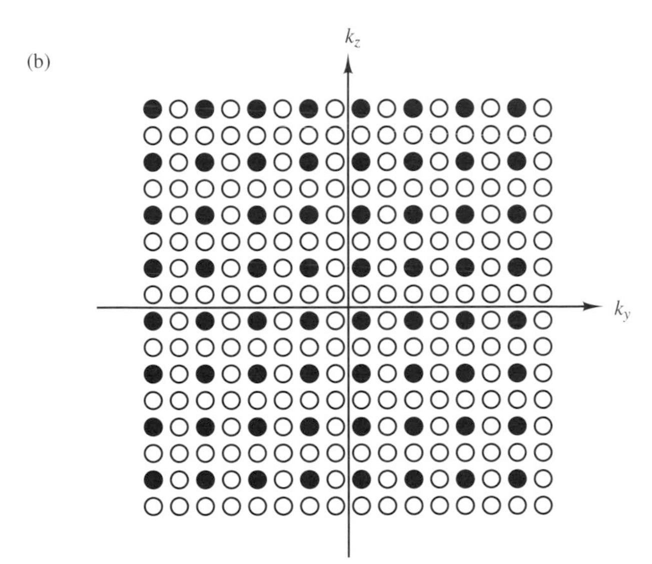

Figure 2 Halving the number of phase and slice encodings (N_y, N_z) in a 3D spin warp acquisition can be achieved by reducing either the ***k***-space area (a) or the sampling density (b), leading to lower resolution (a) or reduced field-of-view (b). Open circles represent skipped data samples

since this operation halves the field of view for both L_y and L_z [equation (5); Figure 2b].

4.2 Slice Multiplexing and Multiecho Acquisition

In most embodiments of 2D spin warp imaging, the period comprising slice selection, phase encoding, and frequency encoding accounts for only a fraction of the *TR* interval. Therefore, minimizing the dead time by shortening the pulse recycle time seems to be a straightforward means of optimizing scan time. However, *TR* is typically dictated by contrast requirements. Shortening of *TR* leads to progressive saturation, and thus, in spin echo imaging at least, is incompatible with proton density and T_2-weighted contrast A solution to this problem was first conceived by Crooks et al.,[24] who introduced what is commonly called '2D multislice imaging'. By making the rf pulses slice-selective, only the spins in the slice of interest are stimulated. Adjacent slices are then excited in sequence without perturbing the steady state. The gain in efficiency achieved in this manner depends on *TR*, the echo time *TE*, and the time required for applying the spatial encoding gradients, and typically ranges from about 10 to 50. The second important innovation was the incorporation of Carr–Purcell echoes into spin warp imaging. In this manner, images with different contrast[25] can be derived from a single acquisition. It is obvious that adding echoes reduces the number of slices that can be interrogated in a given *TR* period. Both 2D multislice and multiecho spin warp imaging, perhaps appearing trivial in hindsight, are, in historic perspective, among the most significant milestones in the quest for imaging efficiency.

4.3 Conjugate Synthesis

A salient property of ***k***-space with implications for imaging speed is its conjugate symmetry, i.e.,

$$\rho(\boldsymbol{k}) = \rho(-\boldsymbol{k}) \tag{10}$$

This implies that two data samples corresponding to spatial frequencies $\boldsymbol{k}$ and $-\boldsymbol{k}$ are identical. In principle, therefore, it suffices to acquire only half of ***k***-space. In reality, however, the conjugate symmetry is often not perfect. Causes of deviations are phase shifts from magnetic field inhomogeneity and motion, which require appropriate correction of the raw data before Fourier reconstruction[26] (see also ***Partial Fourier Acquisition in MRI***).

As already pointed out, sampling in the k_y and k_z directions is much slower than sampling in the k_x direction. Therefore, by exploiting the redundancy in phase-encoding lines, scan time is halved. Halving imaging time by conjugation was described by Margosian et al.[27] and also by Feinberg et al.[28] Because the phase correction requires sampling of a small number of data lines on the conjugate half, the time saving is usually somewhat less than 50%.

4.4 Variable Flip Angle Imaging

In their seminal work on pulse Fourier transform spectroscopy Ernst and Anderson[29] showed that for a train of equidistant rf pulses, the extent of saturation can be controlled by adjusting the rf pulse flip angle α and further that the signal peaks for the condition

$$\alpha_{\text{opt}} = \cos^{-1} \exp(-TR/T_1) \tag{11}$$

where α_{opt} is the optimum pulse flip angle (also called the 'Ernst angle'), which is valid provided that $TR >> T_2$. It is readily recognized that flip angles $\alpha < 90°$ are not compatible with the simple Hahn or Carr–Purcell imaging pulse sequence, since inversion of the residual longitudinal magnetization by the subsequent phase reversal pulse would lead to a steady state of very low magnetization. Haase et al.[30] modified the original spin warp technique in which a gradient echo is sampled and showed that in this manner *TR* could be reduced at least 10-fold relative to corresponding spin echo implementations, which may have prompted the authors to the acronym 'FLASH' (*f*ast *l*ow *a*ngle *sh*ot). Other solutions to the problem

For References see p. 5027

involve the idea to restore the longitudinal magnetization with a second 180° pulse.[31] An alternative solution is to select a flip angle so that $90° < \alpha < 180°$, followed by the usual phase-reversal 180° pulse.[32,33] In contrast to FLASH, the latter techniques provide true spin echo images in which the signal is attenuated as $\exp(-TE/T_2)$ rather than $\exp(-TE/T_2^*)$ as in FLASH.

The FLASH signal can readily be calculated from the Bloch equations to yield

$$S(TR, \alpha, TE) \propto \frac{1 - \exp(-TR/T_1)}{1 - \cos\alpha \exp(-TR/T_1)} \sin\alpha \times \exp(-TE/T_2^*) \quad (12)$$

For $\alpha = \alpha_{opt}$, equation (12) simplifies to

$$S_{\alpha_{opt}}(TE) \propto \tan(\tfrac{1}{2}\alpha) \exp(-TE/T_2^*) \quad (13)$$

In Figure 3(a), the relative S/N is plotted as a function of TR/T_1 for $\alpha = 90°$ and $\alpha = \alpha_E$, equation (12). In either case, a shortened data acquisition time exacts an S/N penalty, which, however, is less severe if the flip angle is optimized. At typical imaging field strengths (1 ± 0.5 T), mammalian tissue T_1 relaxation times span a range of about one order of magnitude (from about 0.3 s for fat to about 3–4 s for extracellular fluids). Hence, a compromise setting will have to be chosen for α. Figure 3(b) shows a series of images for which the flip angle was gradually increased, demonstrating the effect of increasing saturation and the progression from proton density to T_1-weighting. Low flip angles emphasize structures with long T_1, such as cerebrospinal fluid in the ventricles. It is noteworthy, however, that the flip angle for optimum contrast between two tissues of differing T_1 is greater than α_{opt}.[34]

The assumption implicit in equation (12), i.e., $TR >> T_2$ between successive rf pulses, may be invalid at high pulse repetition rates. As first described by Carr,[35] a steady-state situation arises for spins subjected to a train of rf pulses in inhomogeneous magnetic fields, with magnetization building up in a recurrent fashion from cycle to cycle. A similar situation applies in imaging, at least as long as the gradient moments are constant during successive cycles. In spin warp imaging, however, this latter condition is not satisfied, since the phase-encoding gradient is stepped in an incremental fashion. The ensuing cycle-dependent phase shifts cause intensity artifacts that have a positional dependence, since the phase depends on the spins' location on the phase-encoding axis.[36] If, on the other hand, the phase imparted by the phase-encoding gradient is unwound by a gradient of opposite moment prior to the next-following rf pulse,[36,37] the spatial dependence of the steady state vanishes. Common acronyms for this modification to the generic FLASH pulse sequence are 'GRASS' (*g*radient *r*ecalled *a*cquisition in the *s*teady *s*tate) and 'FISP' (*f*ast *i*maging with *s*teady state *p*recession).

From a tissue contrast point of view, it is often desirable to suppress transverse coherences. Application of spoiler gradients of varying amplitude has been found to be relatively ineffective for this purpose,[38] though schemes were later reported that provide some spoiling efficiency.[39] Nevertheless, a more effective means of eliminating the carry-over of transverse magnetization from previous cycles is to step the rf transmit and receive phase in constant increments from cycle to cycle. Although predicted by Crawley et al.[38] rf-spoiled gradient echo imaging was developed by Matt Bernstein and his colleagues in 1989, but, unfortunately, never published. The contrast characteristics of the two embodiments of FLASH are substantially different. In the high-flip-angle regime, GRASS affords images in which the signal scales as $T_2/(T_2 + T_1)$, hence favoring fluids, where $T_2 \approx T_1$ (as opposed to tissues, where $T_1 >> T_2$). By contrast, with spoiling, the signal obeys equation (12).

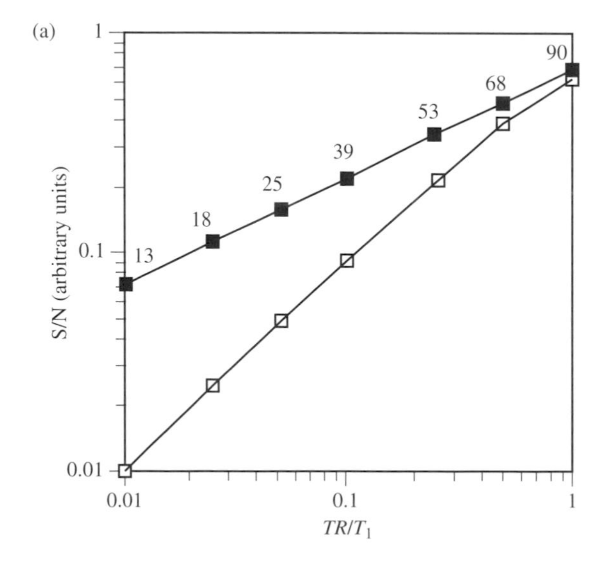

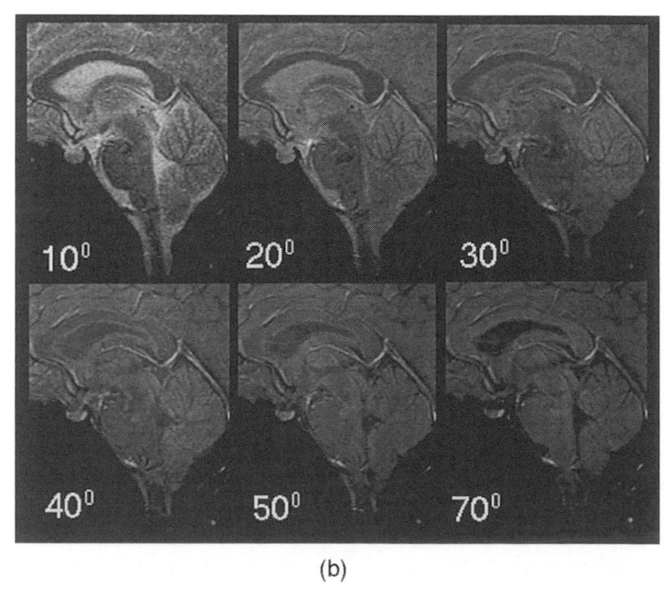

Figure 3 (a) Signal amplitude as a function of TR/T_1 for repetitive pulses after attaining steady state: $\alpha = \alpha_{opt}$ (upper curve, with optimum flip angle indicated), and $\alpha = 90°$ (lower curve), assuming negligible residual transverse magnetization. (b) Effect of increasing pulse flip angle in 2D gradient echo images (TR = 0.3 s) showing effect of gradual differential saturation

4.5 High-Speed Gradient Echo Techniques

Combination of increased sampling frequency bandwidth and fractional echo sampling with partial Fourier processing,[40] reduction in the number of frequency-encoding samples, and other strategies such as truncated rf pulses permit shortening of pulse repetition times to 5 ms or less.[19,20] The impact of reduced pulse repetition time on S/N is illustrated in Figure 3(a). These methods therefore need to be assessed in terms of the benefits and trade-offs they entail. Besides S/N, contrast is perhaps the single most important criterion for clinical utility. The typically unsatisfactory contrast inherent in high-speed steady-state techniques can be enhanced by means of magnetization preparation,[19] discussed in more detail in ***Whole Body Magnetic Resonance: Fast Low-Angle Acquisition Methods***. The underlying idea is to create a nonequilibrium situation by preceding the train of low angle rf pulses by an inversion pulse for T_1-dependent contrast[41–43] or driven-equilibrium pulses to achieve T_2-weighted contrast.[44] Since, in this case, the magnetization evolves during scanning of ***k***-space, effectively filtering the data in a manner determined by the spin dynamics, the usual sequential scanning from $-k_{\max}$ to $+k_{\max}$ is not optimal. Remembering that the signal components pertaining to $k \approx 0$ contribute most to contrast, it is necessary to acquire these data at the desired time point during magnetization evolution (e.g., when the signal from a particular tissue is nulled). These requirements can, for example, be reconciled with a ***k***-space ordering scheme that has been called 'centric phase encoding',[42] in which data collection starts at ***k***-space center and proceeds outward.

Finally, real-time implementations of fast gradient echo scanning are possible by confining images to a single spatial dimension. In this manner, projections can be displayed in real time at a frame rate of $100\,\mathrm{s}^{-1}$ as a scrolling bar, analogous to M-mode ultrasound, with possible applications in cardiology.[45] Another interesting idea based on fast gradient echo imaging is a method designed to track interventional devices. It consists of the consecutive generation of three orthogonal projections from a small pointlike object, followed by computation and display of its coordinates in a real-time mode.[46]

A comprehensive review of short-*TR* imaging techniques has been given by Haacke et al.[47]

4.6 Cardiac Gated Techniques

Except when the entire ***k***-space can be covered in a small fraction of the cardiac cycle, such as in EPI,[48] data acquisition in cardiac imaging needs to be synchronized to the cardiac cycle. Therefore, it appears that the minimum scan time in 2D spin warp imaging of the heart is given as $N_y T_c$, with N_y and T_c being the number of phase-encoding samples and cardiac period, respectively. The inherent speed of the gradient echo then can be exploited to subdividing the cardiac period by collecting data at multiple cardiac phases, resulting in a series of temporally resolved images that can be played back in a movie loop, a technique that has been called 'cardiac cine' imaging.[49,50] The minimum scan time can be considerably shortened (albeit at some cost in temporal resolution) if, instead of a single line k_y, multiple lines are scanned per cardiac phase, as illustrated in Figure 4. This modification, also termed 'segmented ***k***-space acquisition'[43] permits shortening of the

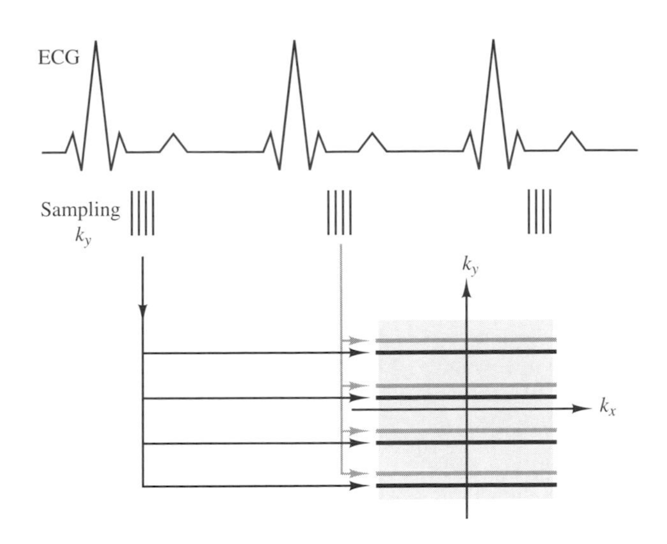

Figure 4 Principle of segmented ***k***-space acquisition in cardiac cine imaging. The RR interval is divided into time periods ('segments') of short duration, during which several lines k_y are encoded. The process is repeated during subsequent cycles until all lines are scanned

data acquisition time in cardiac cine imaging to 16–24 heartbeats,[51] which is within the range of a breath-hold period, thus considerably improving cardiac image quality.

5 MULTILINE SCANNING TECHNIQUES

All previously discussed embodiments of spin warp imaging have in common that within one pulse sequence cycle, a single data line is sampled. Additional echoes are merely a replication of the procedure affording images of increasing T_2-weighting. An inherently more efficient approach makes use of the idea to exploit a plurality of echoes in such a manner that each echo affords a separate k_y line.

5.1 Echo Planar Imaging

In 1977, Mansfield first proposed what he termed 'echo planar imaging' (EPI), in which the entire 2D encoding process could be completed during a single FID. The fundamental principle underlying the technique is to oscillate the read gradient G_x in the presence of a constant gradient G_y.[5] It is readily seen that in this manner, half of ***k***-space is traversed in a zig–zag trajectory. The difficulties in reconstructing images from only half the data and nonorthogonal sampling grids were remedied with the introduction of pre-encoding gradients and discontinuous phase encoding.[22,23] In this manner, entire images of 64 × 128 data samples could be obtained in as little as 50 ms at 2 T field strength.[23,48] For an overview of the subject, the reader is referred to a review by Cohen and Weisskoff[6] (see also ***Echo-Planar Imaging***).

An interesting corollary of single shot EPI is that the pulse repetition time loses its meaning, since the data are collected in single sequence cycle. However, concatenation of individual acquisitions for the purpose of generating movies, for example, underlies the same restrictions as the conventional FLASH method, demanding adjustment of the rf flip angle to control saturation. Cohen and Weisskoff[6] showed 16-frame cardiac

For References see p. 5027

movies obtained by acquiring complete images every 67 ms within a single heart beat. Unlike gated techniques, where each image data set, or fraction thereof, is collected during consecutive heart beats, the method is insensitive to variations in heart rate. In contrast to the high-speed FLASH-type methods, however, echo planar imaging puts far more stringent demands on instrumentation, in particular to amplitudes and slew rates of the gradients, which explains its currently limited diffusion. Some of these conditions are relaxed if only a fraction of ***k***-space is scanned from a single echo train. For a detailed discussion of the subject, see ***Echo-Planar Imaging***.

Finally, one wonders what S/N penalty this extraordinary scan speed exacts. At a given resolution, three parameters contribute to S/N:

1. sampling time (S/N $\propto \sqrt{t_s}$);
2. transverse relaxation losses (S/N $\propto \exp(-TE/T_2)$);
3. the fraction of magnetization available for signal creation (determined by saturation) (S/N $\propto f(TR, T_1, \alpha)$.

The following comparison may be instructive. Suppose a typical tissue such as neural white matter is scanned at 1.5 T field strength ($T_1 \approx$ 700 ms, $T_2 \approx$ 75 ms). Then consider a high-speed gradient echo acquisition with 128 phase encodings and 2 ms readout time, TR = 5 ms, $TE \approx 0$, operated at the optimum flip angle, which is 6.8°, and compare it with a single shot EPI with 50 ms total sampling time, performed such that $k_y = 0$ is sampled at TE = 30 ms. The results are summarized in Table 1. They imply a net advantage of about a factor of 5 in favor of EPI, defeating the common notion that shortened scan time inevitably exacts a S/N penalty.

Similar arguments in favor of EPI have been put forward by Cohen and Weisskoff.[6] Such a comparison, however, does not withstand more rigorous scrutiny. Suppose we were to repeat the EPI scan every 640 ms, i.e., the equivalent of the FLASH data acquisition time. This would mean that the longitudinal magnetization has only incompletely recovered, therefore requiring operation at the Ernst angle, which is 65°. Doing so reduces the transverse magnetization following the rf pulse to 0.64 [from equation (13)], thus lowering the relative S/N advantage of EPI to (a still substantial) factor of 3.

5.2 RARE

In 1986, Hennig introduced the spin echo counterpart of EPI,[52,53] initiating a development with significant impact on clinical efficiency. Surprisingly, this work did initially not elicit broad interest. The perception that the images emphasized structures with long T_2, thereby producing significantly T_2-weighted images, prompted Hennig to dub his technique 'RARE', short for *r*apid *a*cquisition with *r*elaxation *e*nhancement. While obviously highly efficient, the extreme T_2 weighting appeared to limit the method to specialized applications such as imaging of cerebrospinal fluid.

Table 1 Comparison of S/N in Single Shot EPI and High-Speed FLASH; For Assumptions, See Text

	EPI	FLASH	$(S/N)_{EPI}/(S/N)_{FLASH}$
Sampling time (ms)	50	256	0.44
Magnetization	1	0.06	16.7
exp ($-TE/T_2$)	0.67	1	0.67
Net gain			4.9

Besides scan time reduction by up to an order of magnitude, the current widespread clinical use of the method owes its popularity to its spin echo nature, i.e., being insensitive to magnetic field inhomogeneities. Equally important, however, is the almost unlimited latitude in image contrast RARE provides if implemented in a hybrid mode with only a fraction of ***k***-space covered per echo train. It is interesting that the possibilities for contrast manipulation of the RARE sequence were already recognized by Hennig in his 1986 paper.[52] By ***k***-space weighting the echoes appropriately, Melki et al.[54] showed that virtually any arbitrary contrast is possible. For this purpose, they devised an algorithm for phase-encoding ordering while minimizing adverse effects from discontinuous sampling of k_y-space. For example, assigning early echoes to low spatial frequencies de-emphasizes T_2 weighting. The convolution of the data with the relaxation function, of course, may result in point spread function blurring, since high-k signals are attenuated.[55]

Since each of the n echoes generates n lines k_y, the minimum data acquisition time is lowered n-fold. Nevertheless, when assessing the method's efficiency in terms of scan time per image, one has to consider that the prolonged echo train reduces the number of slice locations that can be interleaved during the dead time following data collection. For example, at an echo train length of 250 ms consisting of 16 echoes, a TR = 3 s acquisition provides for fewer than 12 slice locations, rather than about 30 as for a conventional dual echo scan with a second echo time of 100 ms. Finally, Hennig showed that the per-unit-time S/N can be up to a factor of 100 higher in single shot RARE, which again emphasizes what we have seen for EPI, i.e., that increased scan speed does not inevitably penalize S/N. Therefore, the greatly improved efficiency can be traded for improved resolution (by a factor of four in linear resolution) without adversely affecting S/N when compared with single line spin echo imaging. Further, it enables 3D imaging with all its benefits such as retrospective data rearrangement with image display in secondary planes, which was hitherto impractical in spin echo imaging because of excessive scan time. Examples of 3D RARE images are given in Figure 5. Finally, it has also been shown that RARE and EPI principles can be combined in such a manner that gradient echoes are interweaved into a spin echo train,[56,57] and thus the efficiency is further augmented. (See also ***Multi-Echo Acquisition Techniques Using Inverting Radiofrequency Pulses in MRI***.)

6 OTHER METHODS OF SCANNING *k*-SPACE

An alternative ***k***-space scanning technique that, like EPI, uses oscillating gradients is spiral scanning. As implied by the term, the ***k***-space trajectory in this case is a spiral originating at the ***k***-space center. Ahn et al.,[16] who published the first spiral scan images, used a back-projection algorithm for reconstruction. Most of the more recent work is based on constant-velocity interleaved ***k***-space spirals (as opposed to constant angular frequency), an embodiment that is credited to Macovski's laboratory[17] (see ***Spiral Scanning Imaging Techniques***). This scanning mode has the advantage that ***k***-space is covered more uniformly and that the amplitude of the oscillating gradients is constant. Compared with EPI, the gradient amplitude

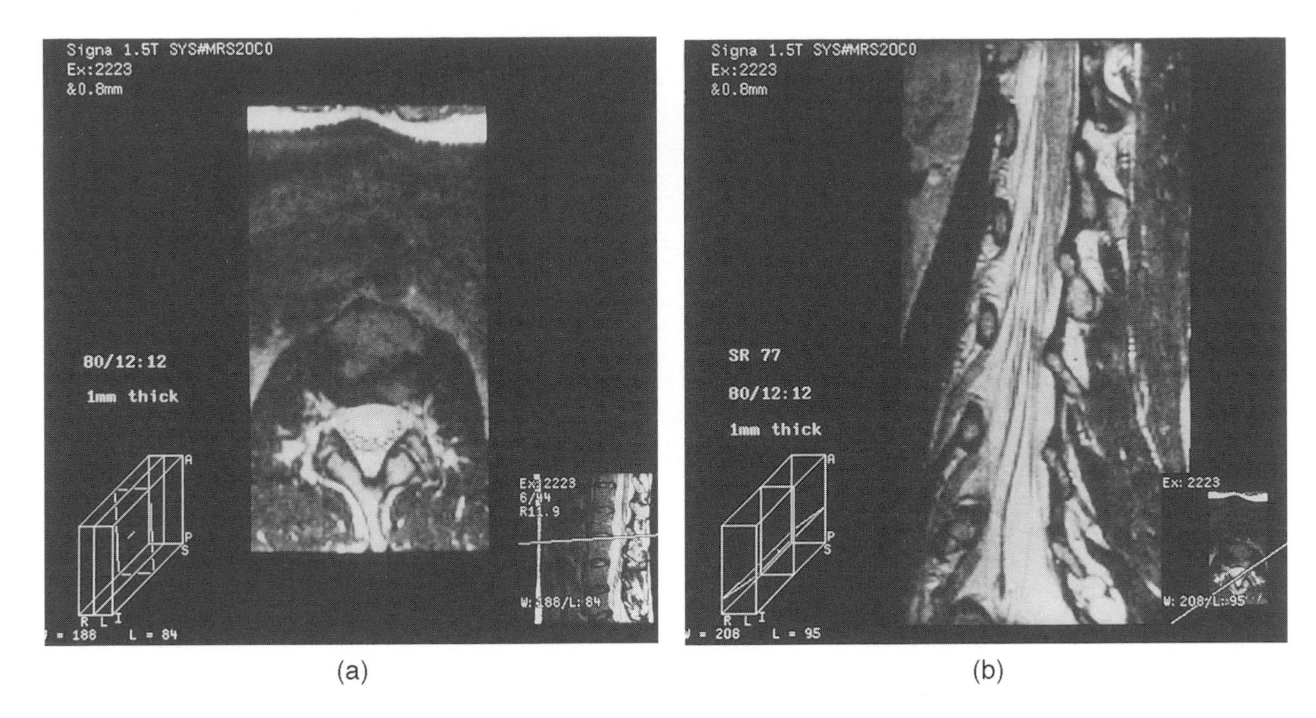

Figure 5 Three-dimensional RARE images of the lumbar spine: (a) axial section; (b) oblique reformation showing nerve roots

and slew rate requirements are less demanding, and the technique has been found to be tolerant to flow-related dephasing, an observation that was attributed to the low gradient moments near the ***k***-space center and the periodic simultaneous return to zero of all moments.[17]

Spiral scans require different reconstruction algorithms. Fourier transformation along the radial coordinate affords a series of projections, from which the image is obtained by filtered back-projection.[16] Alternatively, the data can be rearranged into a rectilinear grid by interpolation, thus allowing Fourier reconstruction[17] (see also ***Spiral Scanning Imaging Techniques***).

Another nonrectilinear sampling technique suited for high-speed imaging is projection imaging, where ***k***-space is sampled radially in two or three dimensions by simultaneous application of two or three constant-amplitude gradients that are proportional to the sine and cosine of the projection angle. In this manner, a radial path is traced, starting at the center of ***k***-space.[13,14] ***k***-space is covered by stepping the gradient amplitudes by incrementing the projection angle in successive sequence cycles. By selecting shallow rf pulses, the pulse sequence can be run at rates comparable to gradient echo spin warp imaging. Since sampling can start almost immediately following excitation, the technique is particularly suited for imaging short-T_2 tissues such as lung parenchyma[58] or imaging of nuclei with inherently short T_2 such as ^{11}B.[59] (See also ***Projection–Reconstruction in MRI***.)

In summary, during the past two decades since its inception, MRI has revealed a wide spectrum of technical approaches for covering data space. Scan speed is a continuum ranging from milliseconds to minutes, with trade-offs between temporal resolution, S/N and spatial resolution. The rate at which ***k***-space can be scanned is determined significantly by the imager's hardware, in particular the achievable amplitudes of the spatial encoding gradients and their switching rates, and the receiver's bandwidth. Besides intrinsic S/N, the ultimate scan speed may be limited by adverse bioeffects, notably peripheral nerve stimulation caused at increased dB/dt of the switched gradients.

7 RELATED ARTICLES

Echo-Planar Imaging; Gradient Coil Systems; Image Formation Methods; Multi-Echo Acquisition Techniques Using Inverting Radiofrequency Pulses in MRI; Partial Fourier Acquisition in MRI; Projection–Reconstruction in MRI; Spiral Scanning Imaging Techniques; Whole Body Magnetic Resonance: Fast Low-Angle Acquisition Methods; Whole Body Magnetic Resonance Spectrometers: All-Digital Transmit/Receive Systems.

8 REFERENCES

1. A. Villringer, B. R. Rosen, J. W. Belliveau, J. L. Ackerman, R. B. Lauffer, R. B. Buxton, Y. S. Chao, V. J. Wedeen, and T. J. Brady, *Magn. Reson. Med.*, 1988, **6**, 164.
2. J. W. Belliveau, D. N. Kennedy, R. C. McKinstry, B. R. Buchbinder, R. M. Weisskoff, M. S. Cohen, J. M. Vevea, T. J. Brady, and B. R. Rosen, *Science (Washington, DC)*, 1991, 716.
3. K. K. Kwong, J. W. Belliveau, D. A. Chesler, I. E. Goldberg, R. M. Weisskoff, B. Poncelet, D. N. Kennedy, B. E. Hoppel, M. S. Cohen, R. Turner, H. Cheng, T. Brady, and B. Rosen, *Proc. Natl. Acad. Sci. USA*, 1992, **89**, 5675.
4. S. Ogawa, D. W. Tank, R. Menon, J. M. Ellerman, S. G. Kim, H. Merkle, and K. Ugurbil, *Proc. Natl. Acad. Sci. USA*, 1992, **89**, 5951.
5. P. Mansfield, *J. Phys. C*, 1977, **10**, L55.
6. M. S. Cohen and R. M. Weisskoff, *Magn. Reson. Imaging*, 1991, **9**, 1.

7. F. W. Wehrli, 'Fast-Scan Magnetic Resonance: Principles and Applications', Raven Press, New York, 1991, p. 1.
8. S. Ljunggren, *J. Magn. Reson.*, 1983, **54**, 338.
9. W. A. Edelstein, J. M. S. Hutchison, G. Johnson, and T. Redpath, *Phys. Med. Biol.*, 1980, **25**, 751.
10. A. Kumar, D. Welti, and R. Ernst, *J. Magn. Reson.*, 1975, **18**, 69.
11. W. Aue, E. Bartholdi, and R. Ernst, *J. Chem. Phys.*, 1976, **64**, 2229.
12. P. C. Lauterbur, *Nature (London)*, 1973, **243**, 190.
13. C.-M. Lai and P. C. Lauterbur, *Phys. Med. Biol.*, 1981, **26**, 851.
14. I. R. Young, D. R. Bailes, M. Burl, A. G. Collins, D. T. Smith, M. J. McDonnell, J. S. Orr, M. L. Banks, G. M. Bydder, R. H. Greenspan, and R. E. Steiner, *J. Comput. Assist. Tomogr.*, 1982, **6**, 1.
15. A. Macovski, *Magn. Reson. Med.*, 1985, **2**, 29.
16. C. Ahn, J. Kim, and Z. Cho, *IEEE Trans. Med. Imaging*, 1986, **5**, 2.
17. C. H. Meyer, B. S. Hu, D. G. Nishimura, and A. Macovski, *Magn. Reson. Med.*, 1992, **28**, 202.
18. O. Heid, M. Deimling, and W. Huk, *Magn. Reson. Med.*, 1993, **29**, 280.
19. A. Haase, *Magn. Reson. Med.*, 1990, **13**, 77.
20. J. Frahm, K. D. Merboldt, H. Bruhn, M. L. Gyngell, W. Hänicke, and D. Chien, *Magn. Reson. Med.*, 1990, **13**, 150.
21. R. Rzedzian, B. Chapman, P. Mansfield, R. E. Coupland, M. Doyle, A. Crispin, D. Guilfoyle, and P. Small, *Lancet*, 1983, **2**, 1281.
22. B. Chapman, R. Turner, R. Ordidge, M. Doyle, M. Cawley, R. Coxon, P. Glover, and P. Mansfield, *Magn. Reson. Med.*, 1987, **5**, 246.
23. I. L. Pykett and R. R. Rzedzian, *Magn. Reson. Med.*, 1987, **5**, 563.
24. L. Crooks, M. Arakawa, J. Hoenninger, J. Watts, R. McRee, L. Kaufman, P. L. Davis, A. R. Margulis, and J. DeGroot, *Radiology*, 1982, **143**, 169.
25. L. E. Crooks, D. A. Ortendahl, L. Kaufman, J. Hoenninger, M. Arakawa, J. Watts, E. R. Cannon, M. Brant-Zawadzki, P. L. Davis, and A. R. Margulis, *Radiology*, 1983, **146**, 123.
26. J. R. MacFall, N. J. Pelc, and R. M. Vavrek, *Magn. Reson. Imaging*, 1988, **6**, 143.
27. P. Margosian, F. Schmitt, and D. Purdy, *Health Care Instrum.*, 1986, **1**, 195.
28. D. A. Feinberg, J. D. Hale, J. C. Watts, L. Kaufman, and A. Mark, *Radiology*, 1986, **161**, 527.
29. R. R. Ernst and W. A. Anderson, *Rev. Sci. Instrum.*, 1966, **37**, 93.
30. A. Haase, J. Frahm, and D. Matthaei, *J. Magn. Reson.*, 1986, **67**, 258.
31. J. A. Tkach and E. M. Haacke, *Magn. Reson. Imaging*, 1988, **6**, 373.
32. A. R. Bogdan and P. M. Joseph, *Magn. Reson. Imaging*, 1990, **8**, 13.
33. H. Jara, F. W. Wehrli, H. Chung, and J. C. Ford, *Magn. Reson. Med.*, 1993, **29**, 528.
34. N. J. Pelc, *Magn. Reson. Med.*, 1993, **29**, 695.
35. H. Carr, *Phys. Rev.*, 1958, **112**, 1693.
36. J. Frahm, K. D. Merboldt, and W. Hanicke, *J. Magn. Reson.*, 1987, **27**, 307.
37. K. Sekihara, *IEEE Trans. Med. Imaging*, 1987, **6**, 157.
38. A. P. Crawley, M. L. Wood, and R. M. Henkelman, *Magn. Reson. Med.*, 1988, **8**, 248.
39. H. Z. Wang and S. J. Riederer, *Magn. Reson. Med.*, 1990, **15**, 175.
40. T. K. Foo, F. G. Shellock, C. E. Hayes, J. F. Schenck, and B. E. Slayman, *Radiology*, 1992, **183**, 277.
41. J. P. Mugler III and J. R. Brookeman, *Magn. Reson. Med.*, 1990, **15**, 152.
42. A. E. Holsinger and S. J. Riederer, *Magn. Reson. Med.*, 1990, **16**, 481.
43. R. R. Edelman, B. Wallner, A. Singer, D. J. Atkinson, and S. Saini, *Radiology*, 1990, **177**, 515.
44. J. P. Mugler, III, T. A. Spraggins, and J. R. Brookeman, *J. Magn. Reson. Imaging*, 1991, **1**, 731.
45. J. D. Pearlman, C. J. Hardy, and H. E. Cline, *Radiology*, 1990, **175**, 369.
46. C. L. Dumoulin, S. P. Souza, and R. D. Darrow, *Magn. Reson. Med.*, 1993, **29**, 411.
47. E. M. Haacke, P. A. Wielopolski, and J. A. Tkach, *Rev. Magn. Reson. Med.*, 1991, **3**, 53.
48. R. R. Rzedzian and I. L. Pykett, *Am. J. Roentgenol.*, 1987, **149**, 245.
49. G. Nayler, D. N. Firmin, and D. B. Longmore, *J. Comput. Assist. Tomogr.*, 1986, **10**, 715.
50. G. H. Glover and N. J. Pelc, in 'Magnetic Resonance Annual', ed. H. Y. Kressel, Raven Press, New York, 1988, p. 299.
51. R. R. Edelman, W. J. Manning, D. Burstein, and S. Paulin, *Radiology*, 1991, **181**, 641.
52. J. Hennig, A. Nauerth, and H. Friedburg, *Magn. Reson. Med.*, 1986, **3**, 823.
53. J. Hennig, H. Friedburg, and D. Ott, *Magn. Reson. Med.*, 1987, **5**, 380.
54. P. S. Melki, R. V. Mulkern, L. P. Panych, and F. A. Jolesz, *Magn. Reson. Imaging*, 1991, **1**, 319.
55. P. S. Melki, F. A. Jolesz, and R. V. Mulkern, *Magn. Reson. Med.*, 1992, **26**, 328.
56. K. Oshio and D. A. Feinberg, *Magn. Reson. Med.*, 1991, **20**, 344.
57. D. A. Feinberg and K. Oshio, *Radiology*, 1991, **181**, 597.
58. C. J. Bergin, G. H. Glover, and J. M. Pauly, *Radiology*, 1991, **180**, 845.
59. G. H. Glover, J. M. Pauly, and K. M. Bradshaw, *J. Magn. Reson. Imaging*, 1992, **2**, 47.

Biographical Sketch

Felix W. Wehrli. *b* 1941. M.S., 1967, Ph.D., 1970, chemistry, Swiss Federal Institute of Technology, Switzerland. NMR application scientist, Varian AG, 1970–79; Executive Vice President, Bruker Instruments, Billerica, 1979–82; NMR Application Manager, General Electric Medical Systems, Milwaukee, 1982–88. Currently Professor of Radiologic Science and Biophysics, University of Pennsylvania Medical School. Approximately 95 publications. Current research specialty: NMR imaging of biomaterials, specifically trabecular bone.

Whole Body Studies: Impact of MRS

George K. Radda

University of Oxford and The John Radcliffe Hospital, Oxford, UK

1 INTRODUCTION

It is common to refer to MRS studies of humans as 'whole body' MRS, or, as indicated in the title of this article, 'whole body studies'. This is a misnomer, since much effort is expended in obtaining MR spectra from a defined organ or even a defined part of the given organ and not from the whole body. The term arose because generally, though not exclusively, this kind of measurement is taken in magnets that can accommodate the human body. In the study of limbs and the head, this is not always necessary, and much human MRS has been done using magnets that only accommodate part of the body. This article is therefore about the impact of MRS in human investigations, but excludes discussion of work in which MRS is used to study samples from biopsies or body fluids of humans. Much of the human data is inevitably backed up, extended, or elucidated by parallel animal investigations. Although such studies have played an essential role in the development and evaluation of human MRS, they are not included in this article for organizational rather than conceptual reasons.

1.1 The Varying Aims of Human MRS

Following from the success of MRI as a radiological tool in clinical diagnosis, it is not uncommon to view MRS as just another extension of the technique for the characterization and description of pathology. Thus one aim is to find clinical applications that can contribute to diagnosis in a routine examination. It will be shown below that this aim is far from having been achieved, and it will be argued that, given the nature and cost of the technique, the major justification for using MRS rests elsewhere. The second main aim of MRS is to investigate the dynamics of normal human biochemical processes. It is necessary to ask, however, why it is important to study fundamental biochemical processes in humans when the same cellular mechanisms are likely to be operative in all mammalian species. Significant special aspects of human biology, among others, include the wide range of polymorphism that results in the biochemical heterogeneity, the nature of the immune system, the advanced central nervous system, and the ease by which the effects of training, adaptation and environmental factors can be studied in humans. In addition, developmental processes at the structural and biochemical level require direct human investigations. The most important of the goals in MRS research is the study of the biochemical basis of human disease. Many human diseases have no useful animal models; the response to pharmacological interventions is often species-specific, and human morphology and anatomy has a unique influence on the expression and development of disease.

1.2 Approaches to Clinical MRS

A successful clinical study by MRS has several basic requirements:

(a) to obtain well-resolved spectra from defined regions of human organs and tissues in a time acceptable to patients;
(b) to obtain as much information as possible, and to assign and interpret the biochemical meaning of the spectra;
(c) to use the biochemical information derived to advance clinical understanding or management.

(a) Localization strategies fall into two major categories: single volume selection and spectroscopic images obtained in the form of one-dimensional slices or with lower frequency as two- or three-dimensional data acquisition. The choice of the method used is often limited by the instrumentation provided by the manufacturer (different manufacturers emphasize their own preferred options). It would be desirable if the localization could be tailored to the nature of the clinical condition to be studied. This is likely to become increasingly the case. Since the majority of current human spectrometers are based primarily on advanced MRI systems, image guided localization is the common practice.

(b) There are instances where a single spectroscopic measurement provides valuable clinical data. In most instances, however, the information content of the study is considerably increased by the measurement of reaction rates either directly, or more generally, in response to some intervention or stress test. For example, almost all the investigations of muscle metabolism and energetics in disease rely on following changes in the concentrations of high- and low-energy phosphates and in the pH values in response to exercise (dynamic or isometric), and during the recovery phase after exercise. In organs such as the liver, stress might be induced by the infusion or oral administration of substrates (glucose, fructose, etc.), while the intervention of most interest in tumors could well be the therapeutic method that is being used.

In the human brain, sensory stimulation may be used in some measurements (e.g., glucose uptake[1]) to provide regional activation. It is also possible that metabolic changes may be detected in responses to the administration of drugs that interfere with neurotransmitter release or function. Cardiac work can be increased by exercise (dynamic leg exercise or isometric hand grip) or by infusion of dobutamine.

The assignment of peaks in the spectra observed in vivo and their biochemical interpretation are largely based on extensive studies of cells, isolated organs, and animals. Spectral resolution and quantification often remain a problem, but progress is being made to replace measurements of metabolite

For References see p. 5037

ratios with derivation of absolute amounts or intracellular concentrations.

(c) Increasingly, the understanding of mechanisms of human disease should lead to the design and evaluation of new therapies. MRS, in the first instance, contributes to this process in a general sense. There are, at present, only a handful of specific cases where a spectroscopic evaluation altered the management of a particular patient. Indeed, the view has been advanced by the present author that the success of MRS will be measured by how readily the expensive and technically demanding MRS study can be replaced by a relatively simple but perhaps less informative and more empirical test for following the clinical status of a patient. This view is based on the belief that if we can define the nature and extent of the disease from MRS studies, it should be possible to design tests that can be correlated.

1.3 What Nuclei?

The nuclei that have been used so far in human investigations are ^{31}P, ^{1}H, ^{13}C, ^{19}F, ^{23}Na, and ^{7}Li. This choice is often governed by the technology, i.e., relative sensitivities, ease of detection, resolution, and quantification. Since we now know that high-resolution spectra can be obtained from all these nuclei in human organs (with the exception of ^{23}Na, where only total, largely extracellular, Na^+ is measured), it is essential that their choice should be governed by the biomedical problems we wish to solve.

1.3.1 ^{31}P MRS

This is widely used to investigate cellular bioenergetics in vivo.[2] Essentially, the following parameters that relate to the energetics are measured: the relative and, more seldom, absolute concentrations of ATP, PCr, Pi, and hexose phosphate, and the value of intracellular pH, which is often equated with the production of lactic acid. The concentration of ADP is often calculated assuming that creatine kinase is at equilibrium. Following changes in these values with stress (e.g., exercise in skeletal muscle and heart), substrate infusion, or administration (e.g., fructose, galactose, and glucose) yields direct or indirect measures of specific enzyme activities or of rates of transport of substrates or ions (e.g., Na^+/H^+ exchange rates). Additionally, magnetization transfer has been used to measure the activities of creatine kinase in human muscle,[3] brain,[4] and heart,[5] and in principle the activity of ATP synthase, so far only achieved in animal studies in vivo.[6]

In addition to the study of energetics, ^{31}P MRS can be used to follow pathways associated with the synthesis and degradation of phospholipids. Key molecules often detected include phosphoethanolamine and phosphocholine, which are intermediates in the Kennedy pathway of phospholipid biosynthesis, but may also be generated by specific phospholipase C catalyzed degradation of membrane phospholipids, perhaps involved in signaling pathways during cell proliferation. Other degradation products observed are glycerophosphorylcholine and glycerophosphorylethanolamine. In vivo, the resonances of these small metabolites may be obscured by the much larger and broader phosphodiester (PDE) signal arising from the phospholipids in the bilayers of membranes.[7]

1.3.2 ^{13}C MRS

This provides an extremely powerful approach, since, through the use of ^{13}C-enriched substrates, it can lead to measurements of individual reaction rates or fluxes through specific pathways.[8] The formation and breakdown of muscle and liver glycogen can be studied in this way.[9] The sensitivity of ^{13}C detection is considerably enhanced by the proton observe carbon edit (POCE) method. This allows detection of the ^{13}C label in the ^{1}H spectrum with heteronuclear editing, thus obtaining the information in the ^{13}C spectra with the sensitivity of ^{1}H MRS. This technique, as well as direct ^{13}C measurements, has been used to observe the flow from [1-^{13}C]glucose into the C-4 of glutamate via the tricarboxylic acid cycle in human brain.[10]

1.3.3 ^{1}H MRS

This is now widely used in clinical and physiological investigations, particularly of the human brain. Special techniques are required to suppress the water resonance, and in addition it is sometimes necessary to edit the spectra if specific components (e.g., lactate) are to be observed. The major metabolites that can be detected include *N*-acetylaspartate, glutamate and glutamine, creatine + phosphocreatine, and, under some conditions, lactate.[11] Additional signals are obtained from the so-called 'choline-containing compounds' (likely to represent a mixture of several substances), taurine and inositol phosphates.

Unlike the metabolites that are detected in the ^{31}P MR spectra, the components seen in ^{1}H NMR are not directly linked through major pathways, or in several instances have no established functional role. The lactate signal, however, represents a sensitive handle on anaerobic glycolysis. Knowledge of total creatine (from creatine and phosphocreatine) provides useful data when combined with measurements of phosphocreatine by ^{31}P MRS. *N*-acetylaspartate has been used empirically as a 'neuronal marker', though this remains to be established.

1.3.4 ^{19}F MRS

This is used for the measurement of the tissue concentration and metabolism of fluorinated drugs (e.g., the antidepressant fluoxetine in the brain)[12] or fluorinated agents like halothane.

1.3.5 ^{7}Li MRS

The unique use of lithium therapy in psychiatric patients is beginning to be studied by this technique, allowing the determination of brain concentrations and pharmacokinetics.[13]

In the discussion below, studies will not be classified according to the nuclei observed, but rather will be presented in relation to the investigation of physiological and pathological problems.

2 NORMAL HUMAN BIOCHEMISTRY

The distinguished Harvard physiologist Walter B. Cannon (1871–1945) coined the term 'homeostasis' to describe the coordinated physiological processes that maintain most of the steady state in an organism. The structural and functional features of the components of the organism are optimized to provide adjustments to the required performance of the system (input/output balance). Ultimately, cellular biochemical regulatory processes perform this function. In addition to short-term molecular control, the long-term response to chronically increased needs involves adjustments in the design properties of the system. Weibel[14] introduced the principle of symmorphosis and defined it 'as a state of structural design commensurate to functional needs resulting from regulated morphogenesis'.

MRS in vivo provides a unique way of studying control, development and adaptation, since it provides quantitative data about 'molecular integrative physiology' through the measurement of dynamics, structure and interactions within any part of the whole organism. While traditionally medicine is 'organ-oriented' and MRS is used to examine organs, the exciting aspect of the method is that it is particularly suited for the observation of coordinated bodily functions at the molecular level.

2.1 Control of Bioenergetics In Vivo

2.1.1 Skeletal Muscle

One of the fundamental questions in the understanding of bioenergetics in vivo is how the demand for ATP utilization is linked to increased supply of energy via oxidative metabolism and therefore to increased rate of oxygen delivery to the tissue. Oxygen delivery to the mitochondrion within the intact cells has four components:

1. the oxygen-carrying capacity of blood, i.e., the nature and amount of hemoglobin present;
2. the rate of flow of blood through major blood vessels;
3. capillary density and diffusion from the capillary to the mitochondrion;
4. the ability of oxyhemoglobin to release its oxygen, i.e., the dissociation of curve for oxyhemoglobin.

^{1}H NMR has been used to observe the ratio of oxy/deoxymyoglobin, since the chemical shifts of the proximal histidine resonance of the two forms are different.[15] Indirectly, blood oxygenation can be measured through the effect of paramagnetic deoxyhemoglobin on the T_2 relaxation time of blood,[16] although this effect has not yet been explored in studies of human muscle. It is believed to be partly responsible for the changes seen in the brain during sensory stimulation.[1]

It is possible to link NMR studies to measurement of tissue oxygenation using infrared detection of oxy/deoxymyoglobin and hemoglobin. During exercise, the decrease in the oxy forms can be related to oxygenation, and the rate of recovery is a measure of the oxidative capacity of the system.[17]

While the various pathways involved in the energetics of skeletal and cardiac muscle are well mapped out, the way in which they are controlled in the intact cell is less well understood. NMR studies in vivo have contributed a great deal to our understanding of some of the control functions, particularly in skeletal and cardiac muscle.

In skeletal muscle, mitochondrial oxidation in vivo was shown to depend, in a hyperbolic manner, on the concentration of ADP, with a K_m of about 30 μM. This value is very close to that reported by Chance and Williams for isolated mitochondria. Three types of experiments have been performed to arrive at this conclusion.

1. Chance and his colleagues have shown that in a graded exercise protocol in which the work is measured with an ergometer, during performance going from rest (state 4) to exercise (state 3), the transfer function, as defined in Chance et al.,[18] approximates a rectangular hyperbola (i.e., the plot of work against Pi/PCr is hyperbolic), giving a K_m for ADP of 28 μM. This follows from the fact that if there are no pH changes, as is the case in the relatively submaximal protocol used, the Pi/PCr ratio is proportional to the concentration of ADP.[18]
2. It was shown that the ADP concentration reached at the end of exercise and the initial rate of phosphocreatine resynthesis during recovery in human arm muscle also follow the same relationship, giving a K_m value of 27 mM and a V_{max} for the ATP synthesis rate of 43 mM min^{-1}.[19]
3. Measurements of the flux between ATP and Pi during steady state isometric muscle contraction using magnetization transfer in animal muscle give a direct indication of the relationship between ADP and mitochondrial oxidative phosphorylation.[6]

Recovery from exercise is a function of mitochondrial ATP synthesis.[20] It can therefore be thought of as an aerobic work jump, in which the rate constant of PCr recovery is a function of V_{max} for ATP synthesis, with the complication that the pH may also be recovering from a low value. As no work is done, the absolute rate of PCr resynthesis is an estimate of the oxidative ATP synthesis rate minus a small component of basal ATP turnover. The hyperbolic relationship between oxidation rate and [ADP] can be used to estimate apparent V_{max} and mitochondrial K_m in the same way as in oxidative exercise.

The apparent V_{max} is a function of the density and capacity of working mitochondria and the supply of substrate and oxygen and is independent of muscle mass. A preliminary calculation for human forearm muscle suggests that the V_{max} calculated from PCr recovery kinetics is 60–70% of that predicted from the intrinsic activity and muscle content of mitochondria. A reduction in apparent V_{max} can be caused by an intrinsic mitochondrial defect or a reduction in mitochondrial density (e.g., heart failure) in arterial oxygen carrying capacity or arterial pO_2 or in muscle blood flow.[21]

The quantitative analysis of the energetic processes during exercise is more complex, since both oxidative and anaerobic glycolytic processes contribute to ATP synthesis while the ATP used against the work done depends on muscle mass and other parameters. In a detailed analysis of exercise performed at three levels of mechanical power output under ischemic and aerobic conditions, it was suggested that oxidative ATP synthesis was negligible during the first half-minute of aerobic exercise, and increased with a half-time of around 0.75 min, although the mitochondrial controller [ADP] increased much more rapidly.[22] Initial PCr resynthesis after exercise appeared to be an adequate estimate of oxidative synthesis at the end of aerobic exercise. The pH dependence of proton efflux inferred from analysis of recovery could be used to estimate ATP synthesis by glycogenolysis/glycolysis to lactic acid during aerobic exercise. Measurement of lactic acid production by ^{1}H NMR, which has been demonstrated as a possibility,[23] might be used to confirm these conclusions. The quantitative interpretation of ^{31}P MRS data from human skeletal muscle, which has many applications to the study of physiology and pathophysiology,

For References see p. 5037

and of muscle bioenergetics and proton handling, has been analyzed in detail.[20]

^{31}P NMR is particularly suited for the examination of the relationship between work output and energetics in vivo. This has been achieved in animal models using, preferentially, sciatic nerve stimulation inducing tetani at different frequencies and with different time intervals. In humans, a variety of experimental protocols have been worked out. Some studies have used the steady state or graded work rate from a constant load or a Cybex ergometer, while others have used force produced in an isometric contraction. Chance and co-workers have examined the relationship between reaction velocities (e.g., ATP utilization) and tissue work rates to the concentration of regulatory molecules in skeletal muscle.[18] They examined the forearm muscle and measured values of Pi/PCr and the work by an ergometer in a graded metabolic load situation. They expressed the relationship between work and velocities as a 'transfer function', showing the hyperbolic relationship between velocity and Pi/PCr. The form of this transfer function varies between normal individuals and well-trained athletes, and can be used as an indication of exercise performance. Boska studied the ATP cost of force production in the human *gastrocnemius* muscle using ^{31}P NMR and an exercise protocol of isometric maximum voluntary contraction, and estimated the contributions to ATP production from three different processes.[24] The rate of change of PCr was used to estimate ATP production rates from creatine kinase rates during exercise and from oxidative phosphorylation during the first 10 s of recovery. The anaerobic glycolysis was estimated from the rate of change of pH, from an assumed buffering capacity and hydrogen ion production stoichiometry. The results showed that by the end of 30 s of exercise, the total ATP production and ATP cost of force production had stabilized and remained constant until the end of a 2 min period. It was also shown that the ATP cost of force production is lower in the first second than at any time, or possibly that ATP production is underestimated at later time points.

2.1.2 *^{13}C NMR and Glycogen*

A major new approach to the study of substrate utilization in bioenergetics is provided by the use of ^{13}C NMR and in particular in the use of ^{13}C-enriched substrates, which can lead to measurements of individual reaction rates or fluxes through specific pathways (for a recent review of ^{13}C NMR, see Cerdan and Seelig[25]).

An important new observation for muscle investigations was that, in spite of its high molecular weight, glycogen is fully visible by ^{13}C NMR in organs and tissues.[26] Thus the synthesis and breakdown of glycogen during muscle exercise can now be followed for the first time noninvasively in human muscles. There are many opportunities to study fundamental questions of the regulation of carbohydrate metabolism under a variety of conditions. One example of this was the pioneering study in which the rate of human muscle glycogen formation was measured in normal and diabetic subjects, using infused, isotopically labeled [1-^{13}C]glucose and ^{13}C NMR.[27] The important conclusions from these measurements were that

(i) synthesis of muscle glycogen accounted for most of the total body glucose uptake and all of the nonoxidative glucose metabolism;
(ii) in subjects with non insulin-dependent diabetes, the rate of glycogen formation was decreased by 60% compared with the rate in normal controls.

2.1.3 *Human Heart*

Many of the same parameters can, in principle, be measured in heart muscle, and have indeed been studied in isolated and in vivo animal hearts. So far, only the PCr/ATP ratio as an index of energetics has been quantitatively measured in the human heart, using special localization or spectroscopic imaging techniques (for a review see Conway and Radda[28]). In normal subjects, the PCr/ATP ratio remained constant when the heart was stressed, either by an isometric hand grip exercise[29] or by a dynamic leg exercise performed in the prone position.[28] Thus, in the normal situation, the heart can adequately regulate its phosphates over a range of workloads. The results show that free ADP is not the primary regulator of increased ATP synthesis in the normal heart in these situations.

2.1.4 *Brain Metabolism*

Glucose is the main, and generally the sole, energy source for the brain. Thus measurement of the regional cerebral rate of glucose consumption together with that of oxygen consumption and regional blood flow has been used to describe brain energetics. Using radiolabeled compounds, in humans [2-^{11}F]fluorodeoxyglucose accumulation and ^{17}O uptake are followed by positron emission tomography.

Recently, largely as a result of work by Shulman and colleagues (for a review, see Shulman et al.[1]) NMR spectroscopic methods have been developed for measuring regional metabolic parameters.

The kinetics of glucose transport in the human brain has been determined by measuring brain glucose concentration by ^{13}C NMR at steady state as a function of plasma glucose concentration. Using a model for facilitated transport values of $K_m = 4.9$ mM and $V_{max} = 1.1\ \mu mol\ g^{-1}\ min^{-1}$ were derived from the transporter. The maximal glucose influx was 3.6 times the normal glycolytic rate, showing that increases in energy demand can be accommodated by glucose transport. In addition, the concentration of glucose in the brain (about 1.2 mM) is well above the K_m for hexokinase, so that changes in plasma glucose will not change the hexokinase flux except at very low concentrations. The turnover rate of the brain glutamate pool can be measured by following the incorporation of ^{13}C isotope from [1-^{13}C]glucose into C-4 of glutamate on the first timing of the tricarboxylic acid (TCA) cycle. The isotopic flux has been measured directly by ^{13}C NMR and by the proton observe carbon edit method. By fitting the results to a metabolic model, the rates of the TCA cycle can be determined, which gives a measure for the O_2 consumption rates. For the human occipital lobe/visual cortex, this was found to be $1.5\ \mu mol\ O_2\ g^{-1}\ min^{-1}$, a value close to that reported from PET studies.[1]

Visual stimulation results in observable metabolic changes in the visual cortex that include increased lactate production (observed by ^{1}H NMR),[1] decreased brain glucose concentration,[30] and a change in the PCr/ATP ratio[31] as seen from ^{31}P NMR.

2.1.5 *Magnetization Transfer and Reaction Fluxes*

One of the many advantages of NMR in cellular studies is that it can provide dynamic information about intracellular events, and give a measure of reaction fluxes catalyzed by enzymes, whether in the steady state or at equilibrium. The role of creatine kinase in skeletal muscle, heart, and brain has been debated for some years, and has been studied in these organs (in animal preparations) by magnetization transfer techniques.[2] In humans, creatine kinase fluxes have been measured in skeletal muscle, where it was shown that the PCr → ATP flux decreases during exercise,[3] and in the brain, where the activity was shown to be twice as high in gray as in white matter.[4] These kinds of measurements as well as studies of ATP-synthase activities have not yet been explored further in human investigations, largely because of the time required for obtaining reliable data.

3 HUMAN DISEASE

In broad terms, human diseases that have been studied using MRS can be conveniently divided into four major categories (Sections 3.1–3.4 below).

3.1 Impaired Oxygen and Substrate Delivery

Oxygen delivery to tissues and organs depends on three factors: blood flow, hemoglobin concentration in the blood, and oxyhemoglobin dissociation characteristics. Alterations in any one of these parameters may impair oxidative phosphorylation, with a concomitant increase in anaerobic glycolysis leading to the production of lactic acid.

3.1.1 *Muscle*

Stenosis in limb arteries produces cramp and pain in patients, with intermittent claudication. ^{31}P MRS has been used by several groups to investigate the metabolic consequences and severity of peripheral vascular disease. While at rest changes in metabolites and intracellular pH (pH_i) were observed in severe claudicants (the pH being significantly alkali), the most characteristic and specific changes were detected during exercise and during the following recovery phase. During exercise, PCr utilization and intracellular acidosis were enhanced, indicating a greater dependence on glycolytic mechanisms for energy production. The reduced rate of PCr resynthesis is consistent with abnormal mitochondrial oxidation, and the slow pH recovery is indicative of substantially reduced blood flow and hence a decreased oxygen delivery (for a review, see Radda et al.[32]).

The hyperbolic relationship between cytosolic [ADP] and rate of PCr resynthesis after exercise has been used to estimate the apparent maximum rate of oxidative ATP synthesis (Q_{max}) in several human diseases in which mitochondrial oxidation may be impaired (see Table 1).[21]

Muscle responds to impaired oxidation by stimulating anaerobic ATP synthesis and/or by increasing [ADP], the stimulus to the mitochondrion. However, these responses interact: [ADP] depends on pH and [PCr], and lactic acid production tends to lower [ADP] (by lowering pH), while proton efflux has the opposite effect. Four patterns were identified:

(d) in mitochondrial myopathy, apparent Q_{max} is reduced and [ADP] is appropriately increased, because increased proton efflux reduces the pH change in exercise despite increased lactic acid production;
(e) in some conditions (e.g., cyanotic congenital heart disease), apparent Q_{max} is reduced, but there is no compensatory rise in [ADP], probably because anaerobic ATP synthesis during exercise is increased without increase in proton efflux;
(f) in other conditions (e.g., myotonic dystrophy) [ADP] is increased during exercise, but apparent Q_{max} is normal, suggesting either an increase in proton efflux and/or decrease in anaerobic ATP synthesis during exercise;

Table 1 Conditions with Probable or Possible Defects of Mitochondrial ATP Synthesis

Mechanism	Examples analyzed
Intrinsic mitochondrial defects	Mitochondrial myopathy Iron deficiency Cardiac failure Uremia
Impaired blood flow	Peripheral vascular disease
Reduced blood oxygen tension	Chronic respiratory failure Cyanotic congenital heart disease
Reduced blood oxygen content	Iron-deficient anemia Myelodysplastic anemia
Increased Hb oxygen affinity	Treatment with 12C79[a]
Other conditions with increased [ADP] in exercise	Hypertension Duchenne dystrophy carriers Myotonic dystrophy

[a] 12C79 is an experimental ‘left-shifter’ drug, 5-(2-formyl-3-hydroxyphenoxy)pentanoic acid.

For References see p. 5037

(g) there are also conditions (e.g., respiratory failure) where, despite impaired oxygen supply, both apparent Q_{max} and end-exercise [ADP] are normal.

Patients with heart failure might have been expected to have reduced muscle blood flow and, perhaps, reduced oxygen delivery due to arterial hypoxemia. Several groups, however, found no blood flow changes,[32] but demonstrated reduced mitochondrial content and activity and increased glycolytic capacity in the muscles of such patients, the pattern of response falling largely in group (b) above.

3.1.2 *Heart*

The elucidation of the metabolic abnormalities in ischemic heart disease has been a major objective of MR spectroscopists for some time. Recently Weiss et al.[29] have demonstrated a decrease in the PCr-to-ATP ratio during hand grip exercise in patients with coronary heart disease and ischemia, reflecting a transient imbalance between O_2 supply and demand in the myocardium with compromised blood flow. Exercise testing in five of these patients after revascularization showed no changes in high-energy phosphates, as was also observed in normal subjects and in nine patients with nonischemic heart disease. The possibility that an exercise-linked spectroscopic test may markedly improve the recognition and evaluation of myocardial ischemia is certainly of great interest.

3.1.3 *Brain*

Several important clinical problems arising out of impaired O_2 delivery have been investigated by both ^{31}P and ^{1}H MRS.

Perinatal asphyxia is the commonest cause of neurological handicap in full-term infants. Reynolds and colleagues[33] have studied infants with birth asphyxia from birth onwards. ^{31}P NMR spectra from neonatal brain differ from those from the adult brain in that a large monoester peak, identified as phosphoethanolamine, is present and the ratios PCr/ATP are lower. These workers found that birth asphyxia caused significant changes in the ratio PCr/Pi, the magnitude of which was helpful in prognosis. These changes occurred early, and the ratios returned to normal within a few days.

Several ^{1}H MRS studies have reported that the *N*-acetylaspartate (NAA) signal in neurologically abnormal infants was lower than in normal controls (for a review, see Howe et al.[34]).

Cerebrovascular disease and stroke are among the major causes of death in the adult population in the Western world. Numerous ^{1}H MRS studies in patients with stroke have suggested that NAA may be a marker for neuronal loss. Serial measurements of NAA following stroke demonstrated recovery of the NAA peak with neurological improvement.[34] Elevation of brain lactate during acute ischemia could be detected, but surprisingly lactate was also found within the infarcted area months after the event. The metabolic turnover of this lactate pool was demonstrated by the appearance of [^{13}C]lactate, following infusion of [1-^{13}C]glucose.[35] It has been suggested that brain macrophages, which begin to appear three days after infarction and gradually disappear over several months, could be a major source of elevated lactate signals that persists for months after stroke.[36] Levine et al. investigated early human focal ischemia with ^{31}P MRS to characterize the temporal evolution and relationship of brain pH and phosphate energy metabolism.[37] Serial ischemic brain pH levels indicated a progression from early acidosis to subacute alkalosis. When acidosis was present, there was a significant elevation in the relative signal intensity of inorganic phosphate and significant reductions in signal intensities of ATP compared with those of control subjects. Ischemic brain pH values correlated directly with the relative signal intensity of PCr and the PCr index, and correlated inversely with the signal intensity of Pi. There was a general lack of correlation between either ischemic brain pH or phosphate energy metabolism and the initial clinical stroke severity. The data suggested a link between high-energy phosphate metabolism and brain pH, especially during the period of ischemic brain acidosis, and the authors proposed that effective acute stroke therapy should be instituted during this period.

Subarachnoid hemorrhage may be complicated by cerebral ischemia, which, though reversible initially, can progress to an irreversible neurological deficit. ^{31}P MRS studies of 10 patients on 30 occasions at various times after hemorrhage showed in some of the patients ($n = 5$) focal areas of intracellular acidosis (pH < 6.8), which in most cases returned towards normal.[38] The recovery of pH_i to normal paralleled an improvement in clinical conditions in each case. These areas of acidosis probably affect ischemia, which is potentially reversible. The MRS measurement provides an opportunity for assessing methods of treatment.

3.2 Genetic and Metabolic Diseases

MRS is particularly valuable in the study of metabolic diseases and genetic conditions.

3.2.1 *Mitochondrial Disease*

Abnormal mitochondria are increasingly recognized as being responsible for muscle and brain disorders. In primary mitochondrial diseases, there are mutations in the nuclear or mitochondrial genes necessary for the synthesis of the components of the electron transport chain. The functional bioenergetic consequences of these abnormalities can be studied by ^{31}P MRS.[39] At rest, in both skeletal muscle and the brain, PCr/Pi and PCr/ATP are decreased, while the Pi/ATP ratio is often increased. This means that the phosphorylation potential, a measure of the available cellular energy, is decreased.

In skeletal muscle, oxidative ATP synthesis is impaired, and this often results in slow PCr resynthesis during recovery. As already pointed out (see Table 1), the reduction in mitochondrial Q_{max} is partly compensated for by an increase in free ADP during exercise and recovery, which stimulates the ATP synthesis rate. This arises from an apparent adaptation in the muscle to chronic mitochondrial insufficiency in which the rate of H^+ removal from the cell by the Na^+/H^+ exchange mechanism is enhanced. This provides a protection to the cell against low pH that would arise out of excess lactate production, and also results in an increase in ADP through the creatine kinase equilibrium.

3.2.2 *Diseases in Carbohydrate Metabolism*

Defects of glycogenolysis and glycolysis are inborn errors of metabolism mainly of muscle or liver. As in the mitochondrial myopathies, the biochemical defects result in impaired production of ATP and in reduced (or absent) amounts of pyruvate for mitochondrial oxidation. Not only are the MRS findings in this group of disorders quite distinct from the oxidative disorders, but glycogenolytic defects are distinguishable from glycolytic defects.[32]

Among several metabolic liver disorders, abnormal metabolism of fructose was studied in livers of patients with hereditary fructose intolerance.[40] In this autosomal recessive disease, aldolase B activity is greatly reduced in liver, kidney, and small intestine. Homozygotes were easily detectable by MRS. Heterozygotes for this disorder cannot be identified by conventional tolerance tests, but in the MRS investigations they showed impaired tolerance to fructose compared with control subjects. In the heterozygotes, the rise in plasma urate after fructose (attributed to the activation of AMP deaminase by the low Pi concentration) correlated significantly with the decrease in liver Pi. The drop in liver Pi after fructose was very marked in those heterozygotes who also suffered from gout, suggesting that some patients suffering from gouty attacks might benefit from a low-fructose diet. Indeed, subsequent studies on a group of patients with a family history of gout show that in some families this resulted from being carriers of hereditary fructose intolerance, and a fructose-free diet improved their condition.[41]

^{1}H MRS has been used to detect specific signals in metabolic brain diseases. For example, high lactate concentrations were found in the brains of patients with Leigh's disease, and *N*-acetylaspartate was consistently increased in Canavan's disease, an inborn error of *N*-acetylaspartate metabolism.[34]

3.2.3 Dystrophy

^{31}P MRS studies of skeletal muscle bioenergetics in patients with muscular dystrophies have given new clues as to the functional consequences of the genetic lesion. For example, in patients with Duchenne (DMD) and Becker (BMD) muscular dystrophy, there is an increase in intracellular Pi and decrease in PCr. Intracellular pH was raised substantially in DMD, moderately in BMD, and slightly but significantly in carriers of the disease.[2,42] Investigations on the mouse equivalent of DMD (the mdx mouse) have shown that the raised pH_i is linked to increased intracellular Na^+ and Ca^{2+}, suggesting that the absence of dystrophin in the muscle leads to a series of ionic abnormalities. Measurements of these biochemical parameters in vivo may provide a way of following the success of therapies such as might be achieved by gene replacement.

Dystrophin is also known to be absent in the brain (neuronal cells) of patients with DMD, and it is known that about 30% of such patients are mentally retarded. It is of interest, therefore, that abnormal Pi/PCr ratios are also observed in the brains of children with DMD and in the brain of the mdx mouse. Neuropsychological studies on such children have shown a relationship between full-scale IQ and the Pi/PCr in the brain.[43]

3.3 Abnormal Control and Disease

The control and coordination of metabolic processes require an intricate network of input functions. Some of these have been emphasized in Section 2, while others include the role of specific ions, general ionic environment, potentials across intra- and extracellular membranes, concentrations of substrates, and hormonal input.

We may consider diseases arising out of a 'control lesion' as being primary intracellular or extracellular, i.e., systemic in origin.

3.3.1 Intracellular Control

There are several examples where MRS has led to the identification of defects that affect intracellular control. These include a patient with abnormal intracellular substrate transport in skeletal muscle (malate aspartate shuttle defect) and the recognition of a sarcoplasmic reticulum Ca^{2+}-ATPase defect.[32]

As already mentioned, perturbations in the ionic balance occur in Duchenne and Becker muscular dystrophy. There are other conditions where the rates of ion-exchange processes (Na^+/H^+ antiport/Na^+/K^+-ATPase) are significantly increased. For example, in essential hypertension, skeletal and heart muscle as well as vascular smooth muscle are affected in this way, as are other peripheral cells (white cells and erythrocytes). It is not known in this case to what extent these changes are secondary or whether they are related to the fundamental mechanism of the disease.[2]

3.3.2 Extracellular Control

The relationships between the extracellular environment (blood biochemistry) and intracellular processes is central to our understanding of physiological regulation.

3.3.2.1 Acid/Base Regulation In Vivo. This is an important physiological problem, with profound clinical implications. Intracellular pH regulation in individual organs and its dependence on the physiological acid/base balance can be examined very well by ^{31}P MRS,[44] and many studies have been carried out on this problem in animal models and humans. Some illustrative examples are given below. Hood et al.[45] examined the effect of systemic pH on pH_i using ^{31}P MRS by measuring lactic acid generation in the forearm muscle of a group of volunteers who performed exhaustive arm exercise with arterial blood flow occluded. On ingestion of NH_4Cl (acidosis), blood pH decreased to 7.30 from 7.39 (control), whereas muscle pH_i did not change. During alkalosis induced by $NaHCO_3$ ingestion, blood pH increased, while pH_i was again well controlled. Systemic acidosis inhibited and alkalosis stimulated lactic acid output (during exercise), which suggests that systemic pH regulates cellular acid production, protecting muscle pH at the expense of energy availability. In contrast, in severe respiratory acidosis produced by CO_2 breathing, muscle pH is significantly reduced.[45] The adult mammalian brain is protected from oscillations in arterial blood pH by a different mechanism, namely the blood–brain barrier, but it is not protected from changes in arterial partial pressure of CO_2 that have been shown to produce changes in the pH of cerebrospinal fluid. ^{31}P MRS was used to measure intracellular pH in different parts of the human brain while the subject was breathing air, and during hypercapnea. The white matter responded with a greater decrease in pH_i than gray matter.[46] The response of liver to acidosis was examined in rats.[47] Varying severity of acidosis was induced by HCl infusion, NH_4Cl ingestion, or induction of experimental diabetic ketoacidosis. Whereas at the more severe degrees of systemic acidosis a marked decrease in hepatic pH_i was observed, there was little change in ketoacidosis. This explained why, in the latter condition, gluconeogenesis from lactate is not inhibited, despite

For References see p. 5037

evidence in vitro of inhibition of glyconeogenesis by systemic acidosis.

3.3.2.2 Cellular Inorganic Phosphate (Pi) Concentration and Total Intracellular Phosphate. These are dependent on blood Pi homeostasis, which in turn is linked to renal Pi clearance and whole body phosphate incorporation and regulation in bones. The relationship between extracellular and intracellular concentrations of Pi was studied in the muscles and erythrocytes of patients with vitamin-D-resistant rickets and patients with Paget's disease of bone before and after they had been made hyperphosphatemic by treatment with the drug ethylidene-1-hydroxy-1,1-biphosphonate. Even though the plasma Pi concentration in these patients spanned a fourfold range (0.5–2.0 mmol L^{-1}), the corresponding intramuscular Pi concentration increased by only 70%.[32]

Patients in renal failure also have elevated blood phosphate, and again this is accompanied by a proportional but smaller increase in muscle Pi. In spite of this, in this condition, the resting phosphorylation potential remains unaltered, implying that cellular levels of ADP and ATP are adjusted to compensate for increase in Pi.

3.3.2.3 Hormonal Regulation of Metabolism. This has long been recognized, and MRS has been used in many investigations of patients with abnormal hormonal levels on hormonal control. The study of diabetes using ^{13}C MRS has already been mentioned in Section 2.1.2. An extension of this work to insulin resistance is of some interest, since the latter may precede the diabetic condition.

Thyroid hormones affect metabolism in many tissues. Skeletal muscle metabolism has been examined in patients with both hypo- and hyperthyroid conditions.[2] The abnormalities found are totally reversible with treatment, and MRS is therefore particularly useful in the evaluation of the relationship between clinical symptoms (e.g., decreased exercise capacity in hypothyroidism) and the underlying metabolic abnormalities.

3.3.3 Tumor Biochemistry

Uncontrolled cellular growth and proliferation represent an extreme case of disease where normal molecular homeostasis is not observed. The study of tumor biochemistry by MRS represents a major area of clinical research, and advances in the field have been regularly reviewed since the 1980s and more recently examined in some detail.[34,48]

In the majority of studies, ^{31}P or ^{1}H MRS have been used but there are reports using ^{13}C, ^{19}F, and ^{23}Na as the nuclei of interest. Several specific developments have arisen out of the ^{31}P MRS measurements.

1. The fact that the majority of human tumors were shown to have alkaline intracellular pH values—and not acidic as was expected—focused attention on the role of ionic control in relation to cellular proliferation.
2. The observation that many tumors had prominent signals in the phosphomonoester (PME) region, which were mostly associated with increased phosphoethanolamine and phosphocholine, brought into the forefront research on phospholipid metabolism in tumors. While it is still not known whether such changes are the result of alterations in the biosynthetic pathways or are associated with phospholipid degeneration and cellular signaling, the possibility that they can be used as an index of cellular proliferation rates has generated much interest.
3. Some researchers hoped that the bioenergetic status of the tumour as reflected in the ^{31}P MR spectrum might be characteristic of tumor types. Others argued[32] that such measures of the metabolic state are more likely to provide an indication of the physiological state of a particular tumor in its specific location. The latter property could then be used to guide the choice of therapies that would utilize the state of oxygenation and activity of a given tumor at a defined stage of its development.
4. There is considerable evidence that response to therapy, particularly chemotherapy, might well be measurable by quantification of the phospholipid components.[48]

Recently, there has been a rapid expansion in the use of ^{1}H MRS in clinical investigations of tumors.[34] Given that in most ^{1}H MRS experiments, only four peaks are measured, namely those assigned to *N*-acetylaspartate, total creatine, choline-containing molecules, and lactate, much emphasis has been placed on empirical correlations with other measurements (histology, PET, etc.) or on pattern recognition methods.[34] It is perceived that multicenter clinical trials might clarify the applicability of such approaches to the grading and possible diagnosis of human tumors.

3.4 Disease and Foreign Invasion

Biochemical changes associated with the administration of toxic substances, viral infections, inflammation, and autoimmune disorders are readily observed by MRS. Alcoholic liver disease, the hepatic consequences of paracetamol poisoning, and viral hepatitis are just some of the conditions that have been studied by ^{31}P MRS.[49] Both the extent of damage and the liver regeneration process can be assessed from the MR measurements.

Acute liver failure can result in hepatic encephalopathy, possibly through the action of toxic substances like ammonia, mercaptans, and amino acids. Several ^{1}H MRS studies demonstrated an elevation of glutamine levels and the parallel decrease in brain myoinositol.[34,50] Liver transplantation completely reversed the ^{1}H MRS changes. Other forms of treatment (e.g., by neomycin or lactulose) were also examined by Kreis et al.[50]

4 CONCLUSIONS AND FUTURE PROSPECTS

This article has been written from the point of view of the author with emphasis on the biochemical aspects of human MRS. There can be no doubt that we have already learned a great deal from MRS about biochemical aspects of human diseases and control and fluxes in normal situations. The method is thus established as a truly new approach in clinical research. In some areas, such as muscle investigations and possibly the empirical uses of brain ^{1}H spectroscopy, it could be considered as part of clinical examination aimed at evaluation of the nature and severity of the disease. As the capability of obtaining spectra in a relatively routine manner becomes available on standard diagnostic imaging instruments, such examinations will become more common, though probably restricted to specialist centers. The author's bias remains, however, that, being a quantitative biochemical technique, the strength of MRS will remain in bringing new understanding about disease mechanisms. Once such understanding is derived from careful and focused studies on specific conditions, we should attempt to devise simpler, cheaper, and generally more readily available tests, probably with a less precise information content, that can be adopted in daily clinical use.

5 RELATED ARTICLES

Brain MRS of Human Subjects; Brain Neoplasms in Humans Studied by Phosphorus-31 MRS; Brain Infection & Degenerative Disease Studied by Proton MRS; Fluorine-19 MRS: General Overview & Anesthesia; Fluorine-19 MRS: Applications in Oncology; Focal Brain Lesions in Human Subjects Investigated Using MRS; Heart Studies Using MRS; Liver: in vivo MRS of Humans; Peripheral Muscle Metabolism Studied by MRS.

6 REFERENCES

1. R. G. Shulman, A. M. Blamire, D. L. Rothman, and G. McCarthy, *Proc. Natl. Acad. Sci. USA*, 1993, **90**, 3127.
2. G. K. Radda and D. J. Taylor, in 'Molecular Mechanisms in Bioenergetics', ed. L. Ernster, Elsevier, Amsterdam, 1992, Chap. 19, p. 463.
3. D. Rees, M. B. Smith, J. Harley, and G. K. Radda, *Magn. Reson. Med.*, 1988, **9**, 39.
4. T. A. D. Cadoux-Hudson, M. J. Blackledge, and G. K. Radda, *FASEB J.*, 1989, **3**, 2660.
5. P. A. Bottomley and C. Hardy, *J. Magn. Reson.*, 1992, **99**, 443.
6. K. M. Brindle, M. J. Blackledge, R. A. J. Challis, and G. K. Radda, *Biochemistry*, 1989, **28**, 4887.
7. P. M. Kilby, J. L. Allis and G. K. Radda, *FEBS Lett.*, 1990, **272**, 163.
8. R. G. Shulman, *Ann. NY Acad. Sci.*, 1987, **508**, 10.
9. T. Jue, D. L. Rothman, B. A. Tavitian, and R. G. Shulman, *Proc. Natl. Acad. Sci. USA*, 1989, **86**, 1439.
10. D. L. Rothman, E. J. Novotny, G. I. Shulman, A. M. Howseman, O. A. C. Petroff, G. Mason, T. Nixon, C. C. Hanstock, J. W. Pritchard, and R. G. Shulman, *Proc. Natl. Acad. Sci. USA*, 1992, **89**, 9603.
11. J. Frahm, T. Michaelis, K. D. Merboldt, W. Hanicke, M. L. Gyngell, D. Chien, and H. Bruhn, *NMR Biomed.*, 1989, **2**, 188.
12. P. F. Renshaw, A. R. Guimares, M. Fava, J. F. Rosenbaum, J. D. Pearlman, J. G. Flood, P. R. Puopolo, K. Clancy, and R. G. Gonzale, *Am. J. Psychiatry*, 1992, **149**, 1592.
13. R. A. Komorski, J. E. Newton, J. R. Sprigg, D. Cardwell, P. Mohanakrishnan, and C. N. Karson, *Psychiatry Res.*, 1993, **50**, 67.
14. E. R. Weibel, 'The Pathway of Oxygen', Harvard University Press, Cambridge, MA, 1984.
15. Z. Wang, E. A. Noyszewski, and J. S. Leigh Jr., *Magn. Reson. Med.*, 1990, **14**, 562.
16. K. R. Thulborn, J. C. Waterton, P. M. Matthews, and G. K. Radda, *Biochim. Biophys. Acta*, 1982, **714**, 265.
17. W. J. Bank, G. Tino and B. Chance, *Neurology*, 1992, **42**, 146.
18. B. Chance, B. J. Clark, S. Nioka, H. Subramanian, J. M. Maris, Z. Argov, and H. Bodle, *Circulation*, 1985, **72**, 103.
19. G. K. Radda, *Phil. Trans. R. Soc. Lond. Ser. A.*, 1990, **333**, 515.
20. G. J. Kemp and G. K. Radda, *Magn. Reson. Quart.*, 1994, **10**, 43.
21. G. J. Kemp, D. J. Taylor, C. H. Thompson, L. J. Hands, B. Rajagopalan, P. Styles, and G. K. Radda, *NMR Biomed.*, 1993, **6**, 302.
22. G. J. Kemp, C. H. Thompson, P. R. J. Barnes, and G. K. Radda, *Magn. Reson. Med.* 1994, **31**, 248.
23. H. P. Hetherington, J. R. Hamm, J. W. Pan, D. L. Rothman, and R. G. Shulman, *J. Magn. Reson.*, 1989, **82**, 86.
24. M. Boska, *NMR Biomed.*, 1991, **4**, 173.
25. S. Cerdan and J. Seelig, *Annu. Rev. Biophys. Chem.*, 1990, **19**, 43.
26. J. R. Alger, L. O. Sillerud, K. L. Behar, R. J. Gillies, R. G. Shulman, R. E. Gordon, D. Shaw, and P. E. Hanley, *Science*, 1981, **214**, 660.
27. G. I. Shulman, D. L. Rothman, T. Jue, P. Stein, R. A. De Fronzo, and R. G. Shulman *N. Engl. J. Med.*, 1990, **322**, 223.
28. M. A. Conway and G. K. Radda, *Trends Cardiovasc. Med.*, 1991, **1**, 300.
29. R. G. Weiss, P. A. Bottomley, C. J. Hardy, and G. Gerstenblith, *N. Engl. J. Med.*, 1990, **323**, 1593.
30. K.-D. Merboldt, H. Bruhn, W. Hanicke, T. Michaelis, and J. Frahm, *Magn. Reson. Med.*, 1992, **25**, 187.
31. D. Sappey-Marinier, G. Calabrese, G. Fein, J. W. Hugg, C. Biggins, and M. W. Weiner *J. Cereb. Blood Flow Metab.*, 1992, **12**, 584.
32. G. K. Radda, B. Rajagopalan, and D. J. Taylor, *Magn. Reson. Quart.*, 1989, **5**, 122.
33. P. L. Hope, A. M. Costello, E. B. Cady, D. T. Delpy, P. S. Tofts, A. Chu, P. A. Hamilton, E. O. R. Reynolds, and D. R. Wilkie, *Lancet*, 1984, **2**, 366.
34. F. A. Howe, R. J. Maxwell, D. E. Saunders, M. M. Brown, and J. R. Griffiths, *Magn. Reson. Quart.*, 1993, **9**, 31.
35. D. L. Rothman, A. M. Howseman, G. D. Graham, O. A. C. Petroff, G. Lantos, P. B. Fayad, L. M. Brass, G. I. Shulman, R. G. Shulman, and J. W. Pritchard, *Magn. Reson. Med.*, 1991, **21**, 302.
36. O. A. Petroff, G. D. Graham, A. M. Blamire, M. al-Rayess, D. L. Rothman, P. B. Fayad, L. M. Brass, R. G. Shulman, and J. W. Pritchard, *Neurology*, 1992, **42**, 1349.
37. S. R. Levine, J. A. Helpern, K. M. Welch, A. M. Vande-Linde, K. L. Sawaya, E. E. Brown, N. M. Ramadan, R. K. Deveshwar, and R. J. Ordidge, *Radiology*, 1992, **185**, 537.
38. N. S. R. Brooke, R. Ouwerkerk, C. B. T. Adams, G. K. Radda, J. G. G. Ledingham and B. Rajagopalan, *Proc. Natl. Acad. Sci. USA*, 1994, **91**, 1903.
39. D. J. Taylor and G. K. Radda, in 'Current Topics in Bioenergetics', ed. C. P. Lee, Academic Press, San Diego, 1994, Vol. 17, p 99.
40. R. D. Oberhaensli, B. Rajagopalan, D. J. Taylor, G. K. Radda, J. E. Collins, J. V. Leonard, H. Schwarz, and N. Herschkowitz, *Lancet*, 1987, **24**, 931.
41. J. E. Seegmiller, R. M. Dixon, G. J. Kemp. P. W. Angus, T. E. McAlindon, P. Dieppe, B. Rajagopalan, and G. K. Radda, *Proc. Natl. Acad. Sci. USA*, 1990, **87**, 8326.
42. G. J. Kemp, D. J. Taylor, J. F. Dunn, S. P. Frostick and G. K. Radda, *J. Neurol. Sci.*, 1993, **116**, 201.
43. I. Tracey, R. Scott, C. H. Thompson, J. F. Dunn, P. R. J. Barnes, P. Styles, G. J. Kemp, C. D. Rae, M. Pike, and G. K. Radda, in Proc. 2nd Ann. Mtg. Soc. Magn. Reson. Med., San Francisco, CA, 1994, p. 345.
44. G. K. Radda, *FASEB J.*, 1992, **6**, 3032.
45. V. L. Hood, C. Schubert, U. Keller, and S. Muller, *Am. J. Physiol.*, 1988, **255**, 479.
46. T. A. D. Cadoux-Hudson, B. Rajagopalan, J. G. G. Ledingham, and G. K. Radda, *Clin. Sci.*, 1990, **79**, 1.
47. J. S. Beech, S. R. Williams, R. D. Cohen, and R. A. Iles, *Biochem. J.*, 1989, **263**, 737.
48. W. Negendank, *NMR Biomed.*, 1992, **5**, 303.
49. G. K. Radda, R. M. Dixon, P. W. Angus, and B. Rajagopalan, in 'Regulation of Hepatic Function (Alfred Benzon Symposium 30)',

eds. N. Grunnet and B. Quistorff, Munksgaard, Copenhagen, 1990, p. 433.
50. R. Kreis, N. Farrow, and B. D. Ross, *Lancet*, 1990, **336**, 635.

Biographical Sketch

George K. Radda. *b* 1936. B.A. 1960, chemistry, M.A., D. Phil., 1962, University of Oxford, UK. Introduced to NMR by Rex Richards. Successively junior research fellow, lecturer (biochemistry), and Professor (Molecular Cardiology), University of Oxford, 1962–present. Approx. 650 publications. Research specialties: in vivo NMR, clinical spectroscopy, control of bioenergetics.

Whole Body Studies Involving Spin–Lattice Relaxation in the Rotating Frame

Raimo E. Sepponen
Picker Nordstar Inc., Helsinki, Finland

1 INTRODUCTION

The dependence of the longitudinal relaxation time T_1 of biological tissue on the strength of the polarizing magnetic field $\boldsymbol{B}_0$, i.e., on the NMR frequency ν_0 is often called T_1 dispersion.[1] The efficiency and the number of relaxation mechanisms related to water–macromolecule interaction increase with decreasing ν_0, and tissue T_1s get shorter. Longitudinal relaxation observed at low field strengths may therefore discriminate better between normal and pathological tissues than that observed at high field strengths. The spin lock (SL) technique allows studies of relaxation at very low field strengths with a high signal-to-noise ratio.

In an SL experiment, nuclear spins are locked with a radiofrequency (rf) field. The locked nuclear magnetization relaxes along the magnetic component of the locking rf field, $\boldsymbol{B}_{1\mathrm{L}}^{\nu_1}$, which rotates at an angular velocity ω_1. If the frequency ν_1 of the rf field is equal to ν_0, the relaxation is characterized by the relaxation time $T_{1\rho}$, which is approximately T_1 at $B_{1\mathrm{L}}$. If ν_1 differs from ν_0 then the relaxation time is called $T_{1\rho}^{\mathrm{off}}$, which approaches T_1 as the frequency offset increases; generally $T_2 < T_{1\rho} < T_{1\rho}^{\mathrm{off}} < T_1$.[2] At very low strengths of $\boldsymbol{B}_{1\mathrm{L}}^{\nu_1}$, $T_{1\rho}$, $T_{1\rho}^{\mathrm{off}} \approx T_2$. The strength of $\boldsymbol{B}_{1\mathrm{L}}^{\nu_1}$ may be varied by variation of the strength and/or frequency of the rf field. Hence, the SL technique allows studies of dispersion of relaxation in the rotating frame, i.e., $T_{1\rho}$ dispersion.

The relative efficiencies of different relaxation mechanisms under the locking conditions are still to be evaluated. It has been assumed that thermal motion of macromolecules and slow motions of water molecules such as chemical exchange are important sources of relaxation. However, it has been suggested that relaxation under the locking conditions is mainly a manifestation of magnetization transfer (MT) between mobile water protons and solid state broadened protein proton levels.[3–5]

The potential to provide unique relaxation information and tissue contrast has motivated the development of clinically feasible SL imaging techniques.[6,7]

2 SPIN LOCK IMAGING TECHNIQUES

In an SL experiment, the longitudinal nuclear magnetization M_z is tilted from the direction of the polarizing magnetic field $\boldsymbol{B}_0$ by an rf pulse or an adiabatic sweep. This is followed by a locking pulse, $\boldsymbol{B}_{1\mathrm{L}}^{\nu_1}$, which is applied during the locking time *TL*. This is illustrated in Figure 1, which shows the basic SL imaging sequence.[8] The phase of $\boldsymbol{B}_{1\mathrm{L}}^{\nu_1}$ is adjusted so that the magnetic component of the effective field is aligned with the tilted $\boldsymbol{M}$. The locked magnetization relaxes along $\boldsymbol{B}_{1\mathrm{L}}^{\nu_1}$ with a characteristic relaxation time $T_{1\rho}$ or $T_{1\rho}^{\mathrm{off}}$.

In imaging applications of the SL technique, $\boldsymbol{B}_{1\mathrm{L}}^{\nu_1}$ is followed by an excitation, gradient encoding, and signal collecting sequence (Figure 1).

The intensity S of the signal is given approximately by

$$S \propto M_0(1 - \mathrm{e}^{-TR/T_1})\,\mathrm{e}^{-TE/T_2}\,\mathrm{e}^{-TL/T_{1\rho}} \qquad (1)$$

Generally, tissues with a long T_2 also have a long $T_{1\rho}$, so the inevitable T_2 weighting of the final image does not decrease the contrast. This allows optimization of the signal-to-noise ratio by selecting *TE*. *TR* should be relatively long in order to minimize the antagonistic T_1 weighting.

If the strength of $\boldsymbol{B}_{1\mathrm{L}}^{\nu_1}$ is low, the contrast properties of $T_{1\rho}$-weighted images are similar to those of T_2-weighted images.[9] An increase in the strength of $\boldsymbol{B}_{1\mathrm{L}}^{\nu_1}$ makes the $T_{1\rho}$ of tissues longer and alters the contrast.[8]

$T_{1\rho}$ dispersion may be quantified by collecting two or more SL images with different $\boldsymbol{B}_{1\mathrm{L}}^{\nu_1}$ strengths. This unique information about relaxation processes may prove useful in tissue characterization.

For calculated $T_{1\rho}$ images, at least two images with different *TL* values need to be collected. The calculation is similar to the calculation of a T_2 image.

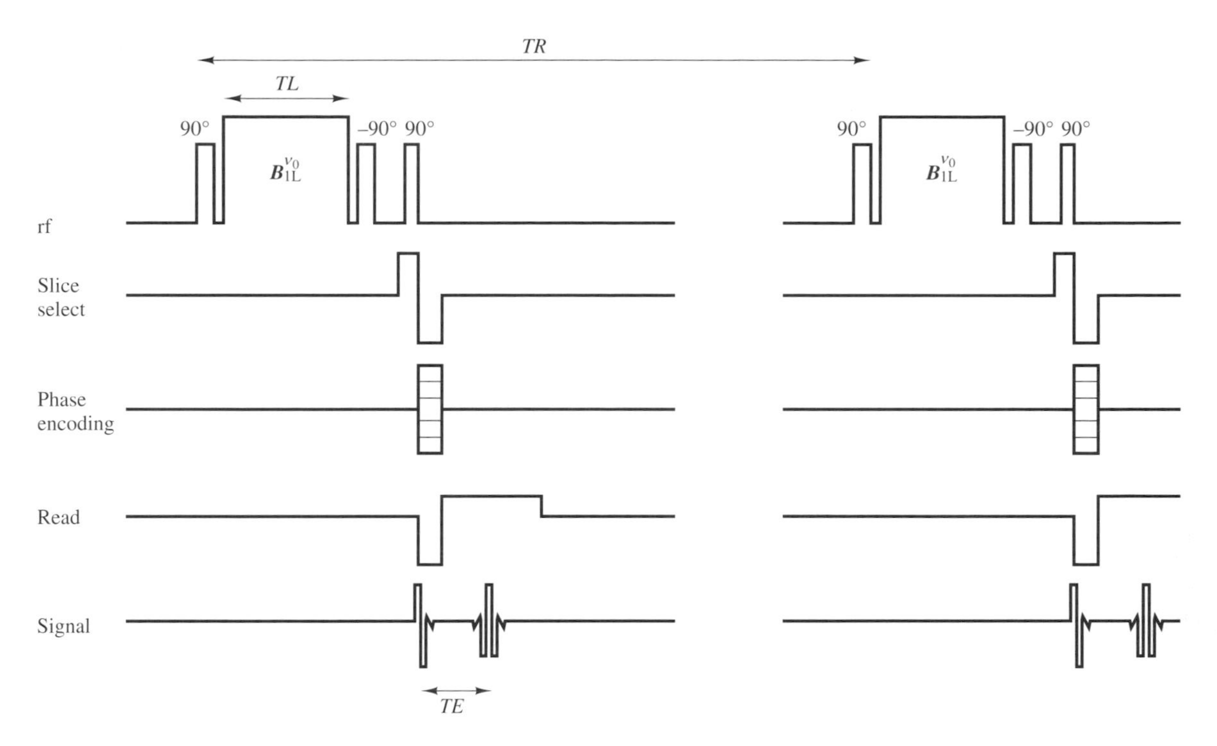

Figure 1 The basic pulse sequence for $T_{1\rho}$ imaging. *TR* is the repetition time, $B_{1L}^{\nu_0}$ is the locking radio frequency pulse applied at ν_0, *TL* is the length of the locking pulse, and *TE* is the time to echo: 'rf' indicates radiofrequency operations; 'Slice select' indicates slice-selection gradient operations, 'Phase encoding' indicates phase-encoding gradient operations; 'Read' indicates read gradient operations, and 'Signal' indicates NMR signals

A multiple slice SL imaging technique has been recently introduced.[10] The multiple slice SL imaging sequence is presented in Figure 2. Nonselective, short locking pulses are used for the generation of the $T_{1\rho}$ contrast in the whole imaging volume. After each locking pulse, the magnetization is returned to the direction of $\boldsymbol{B}_0$. In this way, the generated contrast is 'stored' in the longitudinal magnetization. The images are obtained with the multiple slice technique.

The intensity S of the signal is given by

$$S \propto M_0 \sin\theta \frac{E_2[1-(E_\mathrm{D}E_\mathrm{L})^N(1-E_\mathrm{D})E_\mathrm{L}]}{1-(E_\mathrm{D}E_\mathrm{L})^N(1-E_\mathrm{D}E_\mathrm{L})\cos\theta} \tag{2}$$

where

$$E_2 = \mathrm{e}^{-TE/T_2}, \quad E_\mathrm{D} = \mathrm{e}^{-TD/T_1}, \quad E_\mathrm{L} = \mathrm{e}^{-TL/T_{1\rho}} \tag{3}$$

The contrast also depends on the number N of slices and the time delay TD between successive locking pulses. A small θ

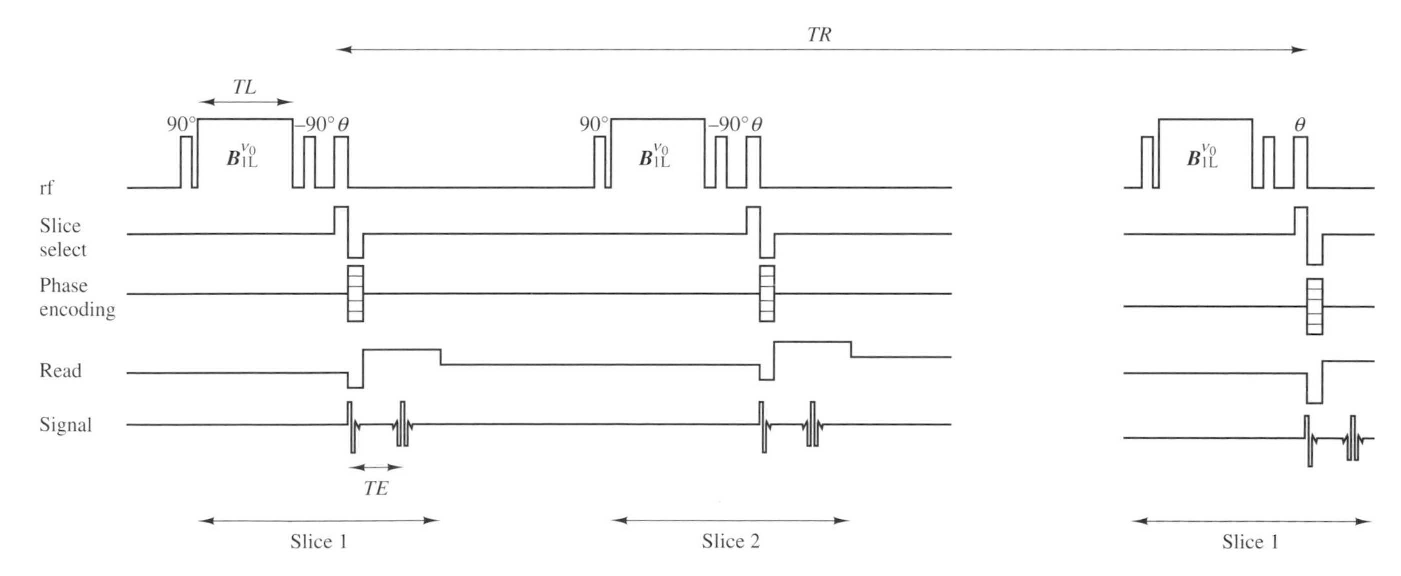

Figure 2 The multiple slice SL sequence. The notation is the same as in Figure 1; θ indicates a $\theta°$ excitation pulse

For References see p. 5042

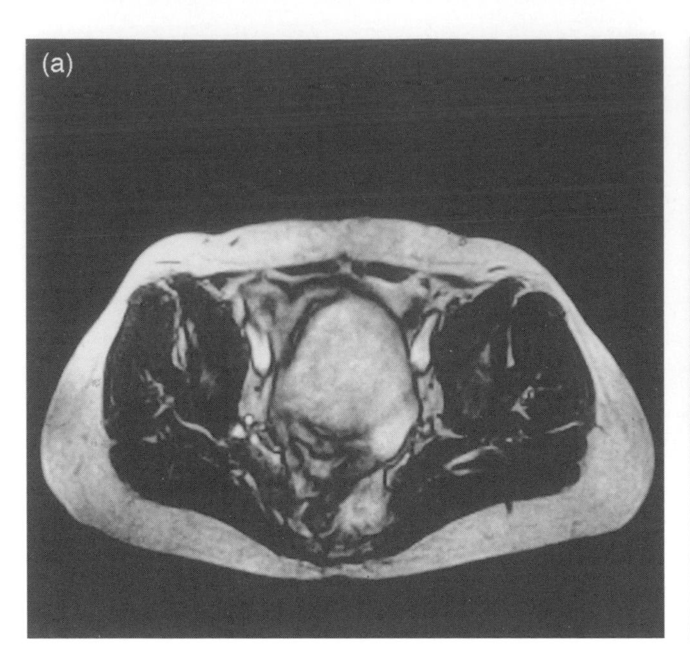

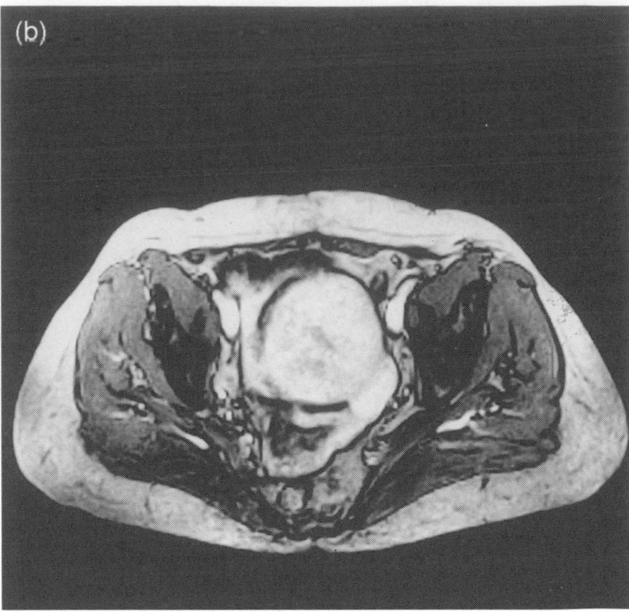

Figure 3 (a) A multiple slice SL image of a female pelvis. The imaging parameters are TR = 1500 ms, TE = 40 ms, B_{1L} = 60 μT, TL = 10 ms, θ = 60°, and B_0 = 0.1 T. The gradient echo technique has been used. A cyst is visible adjacent to the urinary bladder, with a high signal intensity. The wall between the urinary bladder and the cyst is also visible. (b) A gradient echo (GE) image of the same slice as in (a). The imaging parameters are TR = 1500 ms, TE = 40 ms, θ = 60°, and B_0 = 0.1 T. Note that the intensity of the adjacent intestine is as high as that of the cyst. The contrast between muscle and fat tissues is better in the SL image than in the GE image

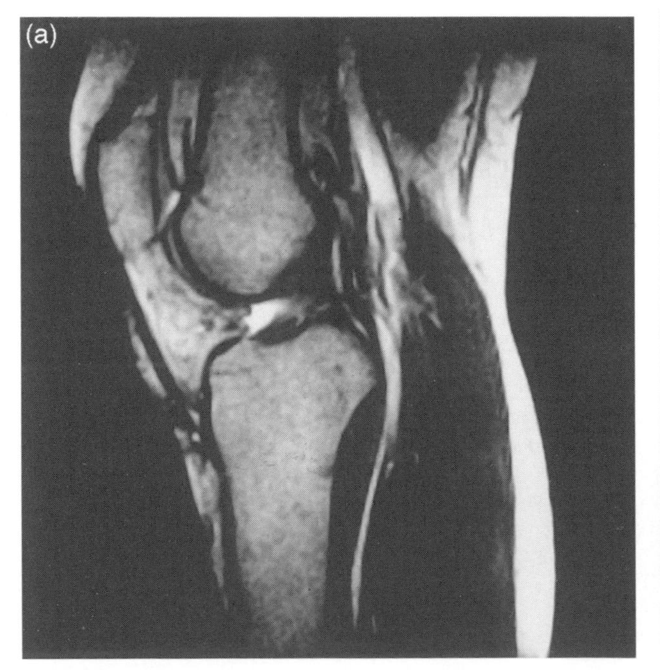

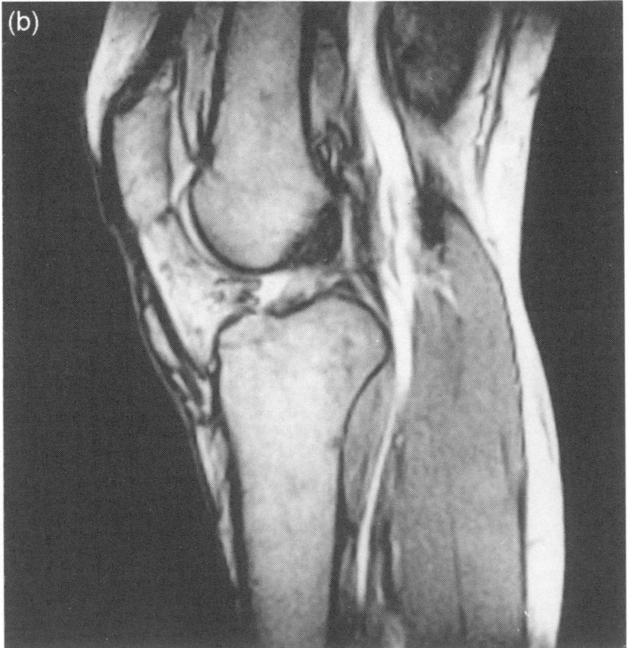

Figure 4 (a) A multiple slice SL image of an injured knee. The imaging parameters are TR = 1500 ms, TE = 40 ms, B_{1L} = 50 μT, TL = 8 ms, θ = 60°, and B_0 = 0.1 T. The gradient echo technique has been used. A small fluid collection adjacent to the patellar cartilage is clearly visible, with a high signal intensity. Muscle tissue shows a very low signal intensity. (b) A gradient echo (GE) image of the same slice as in (a). The imaging parameters are TR = 1500 ms, TE = 40 ms, θ = 60°, and B_0 = 0.1 T. Note that the contrast between the fluid collection and the patellar cartilage is low. Muscle tissue has a markedly higher signal intensity than in the SL image

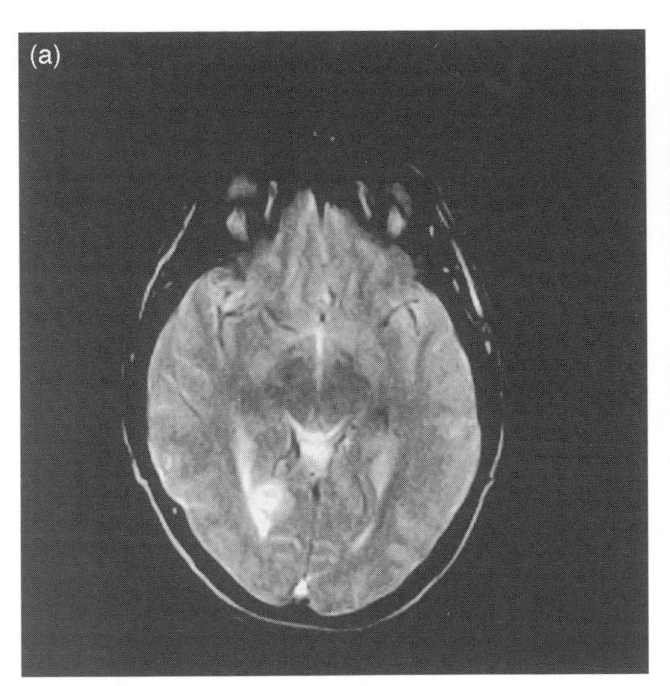

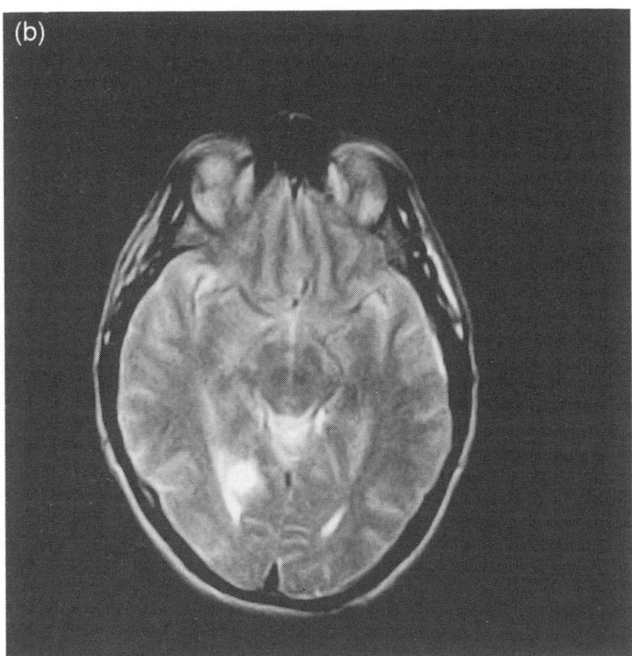

Figure 5 (a) A multiple slice SL image of a patient suffering from multiple sclerosis. A multiple sclerosis plaque is visible, with a high signal intensity. The center of the lesion has a slightly lower intensity. The imaging parameters are TR = 1500 ms, TE = 45 ms, B_{1L} = 60 μT, TL = 10 ms, θ = 60°, and B_0 = 0.1 T. (b) A T_2-weighted SE 2100/120 image of the same slice as in (a). The lesion has a high signal intensity. The imaging parameters are TR = 2100 ms, TE = 120 ms, θ = 90°, and B_0 = 0.1 T

generally has a synergistic effect on the contrast by reducing the antagonistic T_1 weighting.

With this sequence, high-contrast images may be obtained by using a short TE. This minimizes motion artifacts and allows a large number of slices to be accommodated within a given TR. This makes the sequence feasible for abdominal imaging, for example. TL may be short, because of the cumulative effect of the locking pulses. TD should be short compared with T_1 s of tissues.

3 RESULTS WITH T_1 AND $T_{1\rho}$ DISPERSION IMAGING

$T_{1\rho}$ imaging may prove to have a good capability for tissue differentiation. For example, $T_{1\rho}$ measured at B_{1L} = 0.23 mT discriminated tumor from normal fat and fibrous breast tissues. Conventional T_2 and T_1 measurements at B_0 = 0.15 T failed to demonstrate this difference.[11]

$T_{1\rho}$ dispersion imaging has been used to evaluate brain and muscle diseases. Fresh brain infarctions and plaques of multiple sclerosis have a long $T_{1\rho}$ and a slightly smaller $T_{1\rho}$ dispersion than normal brain tissue in the B_{1L} range 0.05–0.46 mT.[7]

A difference in $T_{1\rho}$ dispersion has been found between normal and diseased muscle tissue. The B_{1L} range used was 0.025–0.25 mT. The method seems to provide better identification of affected muscles than T_1- and T_2-weighted imaging.[12]

$T_{1\rho}$ dispersion imaging has been shown to be highly sensitive to the presence of paramagnetic substances. This has been demonstrated in a mouse tumor. $T_{1\rho}$ dispersion imaging seems to be superior to conventional T_1-weighted imaging in detecting the presence of a paramagnetic contrast agent.[13]

The SL technique may be used with the inversion–recovery method to improve the conspicuity of blood vessels in angiography. The length of the locking pulse and the inversion time may be adjusted in order simultaneously to suppress, for example, fat and muscle signals.[14]

The SL technique seems to provide a sensitive method to detect fluid collections, such as edema and cysts in different body areas. This is demonstrated in Figures 3 and 4. In Figure 3, a cyst adjacent to the urinary bladder is differentiated better in the SL 1500/30 (B_{1L} = 60 μT, TL = 10 ms, θ = 60°, N = 15, B_0 = 0.1 T) image (a) from intestines than in a gradient echo GE 1500/30 image (b). Note that the contrast between subcutaneous fat and muscle tissue is better in the SL image than in the conventional GE image.

In Figure 4, fluid collections in an injured knee are better discriminated in the SL 1500/40 (B_{1L} = 50 μT, TL = 8 ms, θ = 60°, N = 12, B_0 = 0.1 T) image (a) than in the GE 1500/40 image (b). Note the high contrast between fluid and cartilage tissue in the SL image.

As already mentioned, the SL technique may provide complementary information for tissue characterization to the conventional imaging methods. In Figure 5, the visualization of a multiple sclerosis plaque is demonstrated with the multiple slice SL technique (a) and with the conventional T_2-weighted spin echo (SE) technique (b). In the SL 1500/45 (B_{1L} = 60 μT, TL = 10 ms, θ = 60°, N = 12, B_0 = 0.1 T) image, the center of the lesion has a relatively low intensity. This feature is difficult to appreciate in the SE 2100/120 image.

For References see p. 5042

4 INSTRUMENTATION AND SAFETY

The instrumentation required for spin lock experiments is usually available in a standard MRI unit. However, the rf power amplifier for supplying the long, high-amplitude B^{ν}_{1L} pulse may be limited to ensure rf safety. The excitation and B^{ν}_{1L} pulses are emitted by the same coil system, which should be able to withstand the required rf power.

The rf absorption of biological tissues is described by the specific absorption rate (SAR). This is proportional to the product $(B^{\nu}_{1L})^2 B_0^2\, TL/TR$. The duty cycle TL/TR and amplitude of B^{ν}_{1L} at B_0 are limited by the maximum allowed SAR.[7]

The SL technique can be applied at medium and high fields by locking the magnetization with an off-resonance rf pulse B^{off}_{1L} and by using the multiple slice SL technique. The SAR in the multiple slice SL images presented earlier is less than $0.12\,\text{W kg}^{-1}$. This indicates that the technique may be safely used at any field strengths currently in clinical use.

5 CONCLUSIONS

The spin lock technique provides an effective method to generate MR images with a high tissue contrast. Magnetization transfer between protons of macromolecules and those of water molecules seems to be an important source of relaxation in the rotating frame. Slow motions of large molecules and paramagnetic substances also play an important role.

The technique may be used even at high field strengths via the utilization of the multiple slice technique or off-resonance locking pulses. The clinical potential of SL imaging is still to be evaluated. It may be assumed that it will find applications, especially in tissue characterization, abdominal imaging, and concurrent use with contrast agents and angiography.

6 RELATED ARTICLES

Biological Macromolecules: NMR Parameters; Brownian Motion & Correlation Times; Field Cycling Experiments; Magnetization Transfer Contrast: Clinical Applications; Magnetization Transfer & Cross Relaxation in Tissue; Magnetization Transfer between Water & Macromolecules in Proton MRI; Protein Dynamics from NMR Relaxation; Relaxometry of Tissue; Rotating Frame Spin–Lattice Relaxation Off-Resonance.

7 REFERENCES

1. S. H. Koenig, R. D. Brown III, D. Adams, D. Emerson, and C. G. Harrison, *Invest. Radiol.*, 1984, **19**, 76.
2. G. P. Jones, *Phys. Rev.*, 1966, **148**, 332.
3. G. H. Caines, T. Schleich, and J. M. Rydzewski, *J. Magn. Reson.*, 1991, **95**, 558.
4. J. H. Zhong, J. C. Gore, and I. M. Armitage, *Magn. Reson. Med.*, 1989, **11**, 295.
5. R. D. Brown III and S. H. Koenig, *Magn. Reson. Med.*, 1992, **28**, 145.
6. E. Rommel and R. Kimmich, *Magn. Reson. Med.*, 1989, **12**, 390.
7. R. E. Sepponen, in 'Magnetic Resonance Imaging', eds. D. D. Stark and W. G. Bradley, 2nd edn., Mosby-Year Book, St. Louis, 1992, Vol. 1, Chap. 8.
8. R. E. Sepponen, J. A. Pohjonen, J. T. Sipponen, and J. I. Tanttu, *J. Comput. Assist. Tomogr.*, 1985, **9**, 1007.
9. R. V. Mulkern, S. Patz, M. Brooks, P. C. Metcalf, and F. A. Jolesz, *Magn. Reson. Imaging*, 1989, **7**, 437.
10. R. E. Sepponen and J. I. Tanttu, *Radiology*, 1993, **189** (P), 320.
11. G. E. Santyr, R. M. Henkelman, and M. J. Bronskill, *Magn. Reson. Med.*, 1989, **12**, 25.
12. A. E. Lamminen, J. I. Tanttu, R. E. Sepponen, H. Pihko, and O. A. Korhola, *Br. J. Radiol.*, 1993, **66**, 783.
13. E. Rommel, R. Kimmich, H. Körperich, C. Kunze, and K. Gersonde, *Magn. Reson. Med.*, 1992, **24**, 149.
14. J. I. Tanttu and R. E. Sepponen, *Radiology*, 1993, **189** (P), 187.

Biographical Sketch

Raimo E. Sepponen. *b* 1950. M.S., 1974, D.Sc., 1986, Helsinki University of Technology. Instrumentarium Inc., Manager of Research & Development (MRI) 1978–1988, Manager of Clinical Applications 1988–1992. Picker Nordstar Inc., Consultant, 1993–1994. Helsinki University of Technology, Senior Lecturer, 1976–1994, Professor (applied electronics), 1994–present. 32 Publications. Research interests; tissue characterization with NMR and imaging techniques.

Wide Lines for Nonquadrupolar Nuclei

Cecil Dybowski
University of Delaware, Newark, DE, USA

&

Günther Neue
Universität Dortmund, Dortmund, Germany

1 INTRODUCTION

Since the inception of NMR, one hallmark of its use has been the ability to analyze materials specifically by selectively

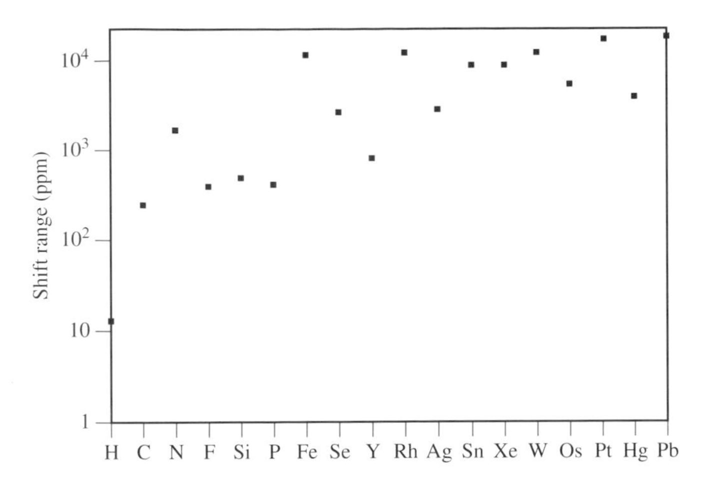

Figure 1 Shift ranges of some spin-$\frac{1}{2}$ nuclei ordered by increasing atomic number. The tendency of heavier nuclei to show larger shift ranges is clearly visible

detecting only resonances of certain nuclei in the material. Early on, the focus on protons by chemists resulted from the favorable NMR properties of this nucleus, but other spin-$\frac{1}{2}$ nuclei such as ^{13}C and ^{31}P have become routine targets because their spectroscopy addresses appropriate chemical problems with the same specificity as protons have for proton-containing organic materials.

An examination of the Periodic Table shows that approximately one-third of naturally occurring elements have at least one stable isotope with spin-$\frac{1}{2}$. Because the nuclear system of single spins $\frac{1}{2}$ has only two energy levels, these isotopes represent the simplest NMR spectroscopic systems. The proton occupies a unique position, being the spin-$\frac{1}{2}$ isotope of lowest atomic number. At the other extreme stands ^{207}Pb, the isotope of spin-$\frac{1}{2}$ having the highest atomic number. Thus, the NMR spectroscopy of this isotope logically represents the opposite extreme of NMR properties of spin-$\frac{1}{2}$ isotopes from the proton. Quite a few materials containing ^{207}Pb have been studied in solution, particularly the organolead compounds.[1] Despite its unique position, surprisingly little has been done to investigate the spectroscopy of ^{207}Pb in solid materials.

Figure 1 shows the shift ranges of several spin-$\frac{1}{2}$ nuclei. It is obvious that there is a loose correlation between the shift range of an isotope and its number of electrons. Elements with more electrons show a tendency to react more sensitively to their chemical environment than those with fewer electrons. It can be expected that similar arguments also hold for chemical shift anisotropies. The resulting broad spectra in noncubic environments lead to low signal-to-noise ratios and present problems for uniform excitation by pulses. The situation becomes even worse if the magnetogyric ratio is also low. Spin-$\frac{1}{2}$ nuclei of that type are sometimes nicknamed the 'Cinderella nuclei'.[2] Long T_1 values in solids may add to the problem.

Certainly, inherently difficult detection is a major reason for the comparatively fewer NMR studies of heavier isotopes. This article summarizes a few techniques for obtaining and analyzing spectra that consist of broad lines.

2 CW DETECTION

This old non-FT method scans a spectrum sequentially. The sensitivity is low, and therefore CW NMR has been replaced in most applications by the much more sensitive and more versatile pulse NMR. But this statement holds only for spectra consisting of narrow lines. Under such circumstances, a CW spectrometer scans detector noise of the baseline most of the time, while pulses excite all signals at once, as long as the pulse is strong enough. For liquids, the difference in the signal-to-noise ratio that can be obtained in a given time by both methods is striking (see e.g. Figure 4.3.4 on p. 157 of Ernst et al.[3]), and this characteristic boosted the development of FT NMR. However, this advantage becomes less important the broader the spectra are. For a faithful recording of very wide lines, CW spectrometers can be used. Swept CW detection always operates on resonance with a constant-amplitude rf signal. Among other things, this avoids offset effects and artifacts due to pulse imperfections. The detection of the ^{59}Co resonance in $KCoF_3$ gives an impression of what is possible with a CW wide line spectrometer.[4]

Essentially, there are two CW methods. The first is the slow passage experiment, which for a long time was the standard method for detecting NMR signals from liquids. Neglecting transient effects that result from noninfinitely slow scanning, the signal shape $g(\omega)$ depends on the amplitude of the rf field, B_1. For a Lorentzian peak with a linewidth described by T_2, one obtains[3]

$$g(\omega) \propto \gamma B_1 \frac{1/T_2}{(1+S)/T_2^2 + \omega^2} \qquad (1)$$

where $S = (\gamma B_1)^2 T_1 T_2$ is called the saturation parameter. This quantity depends on both T_1 and T_2, as well as on B_1. It can be seen that there is a conflict. To maximize the signal, B_1 has to be as big as possible, but to keep distortions low, B_1 has to be as small as possible. For solids with very long T_1 values, this may be a substantial problem. In such cases, a second CW method, the adiabatic fast passage through resonance, is useful.[5] This is based on the fact that, under certain circumstances, the nuclear magnetization always remains aligned with the effective field $\boldsymbol{B}_e = (B_1, 0, B_0 - \omega/\gamma)$ in the rotating frame. In solids, a signal can be observed if the B_1 field strength and the scanning rate $\mathrm{d}|B_0|/\mathrm{d}t$ fulfil the condition

$$\frac{1}{T_1} \ll \frac{1}{B_1}\frac{\mathrm{d}|B_0|}{\mathrm{d}t} \ll |\gamma B_1| \approx \frac{1}{T_2} \qquad (2)$$

But, since the last relationship means that the nutation frequency $\omega_1 = \gamma B_1$ must be comparable to the linewidth, one has a restriction similar to that for pulse methods.

For details on both CW methods, the reader is referred to Abragam.[5] It should be noted, however, that it is becoming more and more difficult to find a commercially produced CW NMR spectrometer to use in such experiments.

3 POINTWISE DETECTION BY SPIN ECHOES

Spectra representing extremely broad shift distributions cannot be excited evenly over the whole shift range if pulses are used. Instead, only a region, which may be small compared

For References see p. 5047

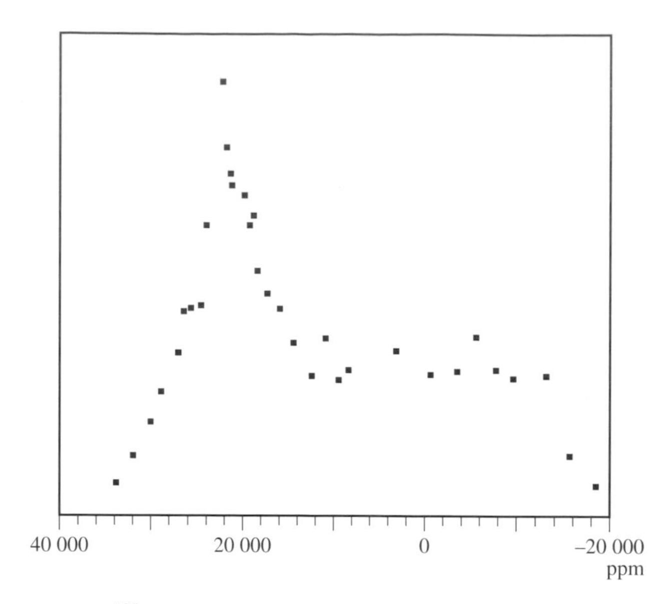

Figure 2 ^{195}Pt spin echo spectrum of small CO-covered Pt clusters on an alumina-supported catalyst. The data are taken from Makowka et al.[7]

with the total width of the spectrum, will be selected by the pulse. By using the basic idea of CW spectroscopy, sequential detection in different parts of the spectrum will outline the lineshape. In practice, one uses spin echoes to circumvent deadtime problems. The height of the echo is a direct measure of the intensity of the spectrum at the irradiation frequency. More precisely, it is a point on the true line smoothed with the excitation profile. Obviously, there are two possibilities to achieve the goal. Either the frequency or the field is changed from point to point. Field sweeps, if possible, are the better choice, because the tuning of the spectrometer can be kept constant.

With the limited availability of CW spectrometers, this method is becoming more and more the standard for observing extremely broad lines. It was, for example, extensively used for measuring the copper resonances in high-T_c superconductors.[6]

An instructive example to demonstrate the application and power of the spin echo technique is the study of the ^{195}Pt resonance of metallic Pt particles on an alumina-supported catalyst.[7,8]

Figure 2 shows a spin echo spectrum of such a catalyst with a Pt dispersion of 26% (= the percentage of surface atoms) covered by CO. The meaning of the structure of the spectrum is elucidated by a double resonance version of the spin echo experiment, where the ^{195}Pt resonance is observed while ^{13}C is irradiated. As only Pt atoms directly attached to CO can be observed, this method is surface-sensitive. By using this approach, it could be shown unambiguously that only the pronounced peak on the left side of the spectrum is due to surface atoms, while the right part belongs to resonances of ^{195}Pt atoms located within the small metal clusters.

4 EXCITATION BY COMPOSITE PULSES

To improve excitation over a wider range of frequencies, composite pulses have been used for many years.[9] The basic idea of this approach is to use a sequence of pulses that mimics the effect of a single pulse. The extra degrees of freedom in times (pulse durations, delays) and phases that are introduced by using several pulses are used to improve the properties of the equivalent pulse towards an ideal delta function pulse.

There is a wide variety of suggestions for composite pulses. Usually, they are optimized to achieve a particular property. Many pulse sequences are designed to be tolerant to missettings of the pulse length to allow a more uniform excitation over a large sample volume with B_1 inhomogeneities.[10] Another area of successful application of composite pulses is spin decoupling.[11]

More important in the context of exciting wide lines is an optimization with respect to a minimal phase dispersion as a function of frequency offset. A well-known sequence with that property is the 'spin knotting' sequence[12]

$$(\phi_1)_{+x}\text{–}\tau_1\text{–}(\phi_2)_{-x}\text{–}\tau_2\text{–}(\phi_3)_{+x} \tag{3}$$

The rotation angles ϕ_i have to fulfil the condition $\phi_1 - \phi_2 + \phi_3 = 90°$. The remaining four degrees of freedom can be used to optimize the performance of the sequence for each B_1 field strength by trial and error.[12] Figure 3 shows a simulated powder spectrum with shift anisotropy excited by a single $(\frac{1}{2}\pi)_{+x}$ pulse. The B_1 field strength was assumed to correspond to a nutation frequency of 10 kHz. Distortion of the lineshape is clearly visible. If the same spectrum is detected by applying the 'spin knotting' sequence with pulses of the same B_1 field strength, the result is an almost perfect powder pattern (Figure 4).

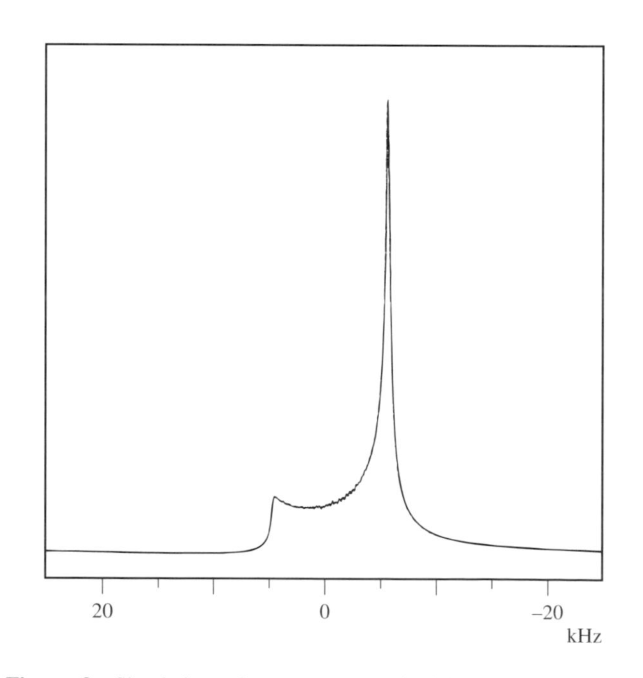

Figure 3 Simulation of a spectrum excited by a weak pulse and governed by chemical shift anisotropy. The rf pulse strength was assumed to correspond to a nutation frequency of 10 kHz

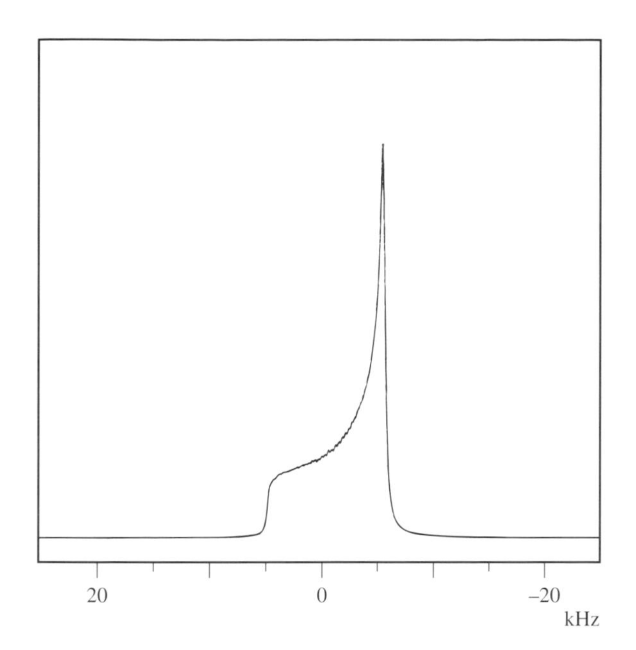

Figure 4 Simulation of the response of the same system as in Figure 3 to excitation by the spin knotting sequence with optimized parameters.[12] Again, the rf pulse strength was assumed to correspond to a nutation frequency of 10 kHz

A more direct and general approach to the construction of sequences that approximate ideal pulses[13] predicts that a composite pulse

$$(385^\circ)_{+x}(320^\circ)_{-x}(25^\circ)_{+x} \qquad (4)$$

is equivalent to a $(\frac{1}{2}\pi)_{+x}$ pulse with minimum phase dispersion. As this particular method uses a series expansion, higher order approximations can be found that consist of more and more pulses and that increase the spectral range over which the phases of the excitation are almost constant.

It may seem that cleverly designed composite pulses are the ultimate solution to the problem of exciting broad lines. But there are limitations. The power theorem of Fourier transforms,

$$\int_{-\infty}^{\infty} |p(t)|^2 \,\mathrm{d}t = \frac{1}{4\pi}\int_{-\infty}^{\infty} |P(\omega)|^2 \,\mathrm{d}\omega \qquad (5)$$

relates the power in a pulse train as a function of time [amplitudes $p(t)$] to the power available in the spectrum as a function of the frequency [amplitudes $P(\omega)$]. In general, both amplitudes are complex functions, since one has to include the phases. By assuming a constant amplitude $|A|$ for the pulses and a flat spectrum of the pulse sequence with a constant amplitude $|B|$ (disregarding its practical realization), one obtains

$$|A|^2 \Delta t = \frac{1}{4\pi}|B|^2 \Delta\omega \qquad (6)$$

where Δt is the total time during which the transmitter is on and $\Delta\omega$ is the bandwidth of the excitation. It is obvious that a larger frequency range $\Delta\omega$ can be obtained only by increasing the total lengths of the pulses, since the transmitter power $|A|^2$ is limited in practice. Note that the other quantity, $|B|$, is fixed by the condition to obtain a 90° pulse. Usually, in solids, there will be dipolar interactions that give rise to a T_2 decay of the order of a few tens of microseconds. On typical solids spectrometers $\frac{1}{2}\pi$ pulse lengths may be 5 μs. It should be observed that even simple composite pulse sequences such as those given above may last as long as typical decay times owing to dipolar interactions. Thus, in most solids, the bandwidth of ideal excitation cannot be increased much by more sophisticated and longer sequences than those given above. Similar bandwidth versus time arguments also hold for a newer development in the field of tailored excitation, namely, shaped pulses,[14] where extra degrees of freedom are obtained by modulating the amplitudes by analog signals rather than by pulses.

5 ANALYZING BROAD LINES USING TRANSFER FUNCTIONS

Despite the success of the methods described in the previous section, it might be inconvenient, difficult, or impractical to use them. Owing to the nonideal behavior of electronic circuits (finite rise times, power droop, etc.), one usually needs to adjust the pulses to optimize the operation of the sequence. Adjustments on the sample of interest may not be possible in cases of low intensities or long relaxation times. If one works with special probeheads that need insulation from the environment, like low- or high-temperature probes, then it may not be practical to substitute a calibration sample.

Very wide lines also present another problem. As discussed above, the possible bandwidth of excitation is subject to practical limitations like those imposed by relaxation. If the line extends over more than this range, application of the common composite pulse sequences destroys the wings of the line very efficiently. According to the power theorem, increasing the power in a given region of the spectrum means that, outside of that region, the excitation is reduced. In contrast to the almost rectangular excitation profile (transfer function) of a composite pulse, a simple single pulse spreads its energy over a much wider frequency range, though unevenly. This fact allows one to use another strategy for wide lines. As the transfer function is given by pulse timings, it can be calculated exactly and included in the data analysis. Under these circumstances, a simpler excitation has the advantage of giving a better signal-to-noise ratio far off resonance, allowing a better fit of data.[15] This is illustrated by Figure 5. It is clearly seen that if the lines are so broad that they cannot be covered by composite pulses, a simple pulse shows the singularities and shoulders of powder spectra better, and the extraction of tensor elements consequently becomes more precise.

Because of receiver dead time, one cannot use a single pulse, but has to rely on spin echo techniques. Furthermore, the detection of weak signals very often suffers from acoustic ringing. A simple pulse sequence that produces a spin echo and removes acoustic ringing, as well as some other imperfections is

$$\begin{array}{c} (\frac{1}{2}\pi)_{+x}\text{–}\tau\text{–}(\pi)_{+y}\text{–}\tau\text{–}(\mathrm{acq})_{+x} \\ (\pi)_{+x}\text{–}\tau_1\text{–}(\frac{1}{2}\pi)_{+x}\text{–}\tau\text{–}(\pi)_{+y}\text{–}\tau\text{–}(\mathrm{acq})_{-x} \\ (\frac{1}{2}\pi)_{-x}\text{–}\tau\text{–}(\pi)_{-y}\text{–}\tau\text{–}(\mathrm{acq})_{-x} \\ (\pi)_{+x}\text{–}\tau_1\text{–}(\frac{1}{2}\pi)_{-x}\text{–}\tau\text{–}(\pi)_{-y}\text{–}\tau\text{–}(\mathrm{acq})_{+x} \end{array} \qquad (7)$$

For References see p. 5047

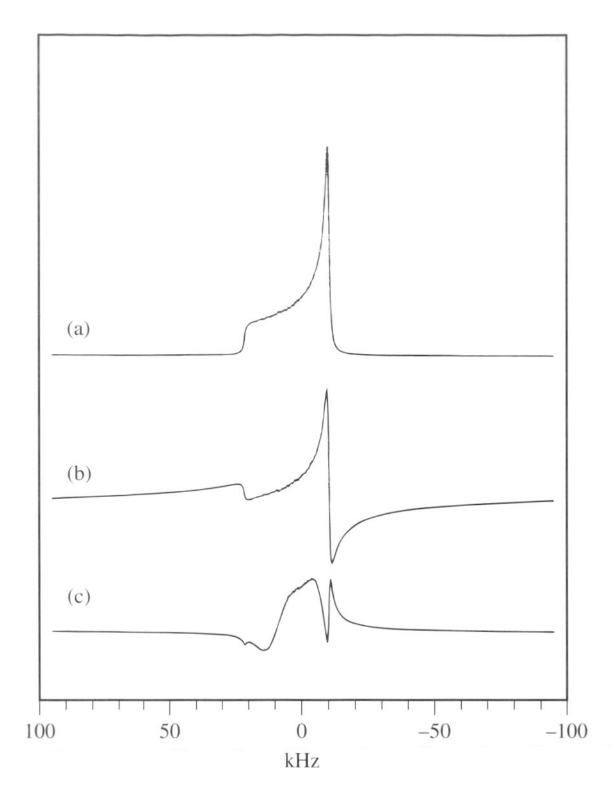

Figure 5 Simulation of the effects of different excitation methods on the appearance of spectra. (a) Ideal spectrum governed by chemical shift anisotropy. (b) Excitation by a single pulse. (c) Excitation by the spin knotting sequence. The rf pulse strength was always assumed to correspond to a nutation frequency of 10 kHz

where $(\theta)_\phi$ denotes a pulse with phase ϕ that produces a nutation by angle θ, and $(\text{acq})_\phi$ stands for acquisition with receiver phase ϕ.

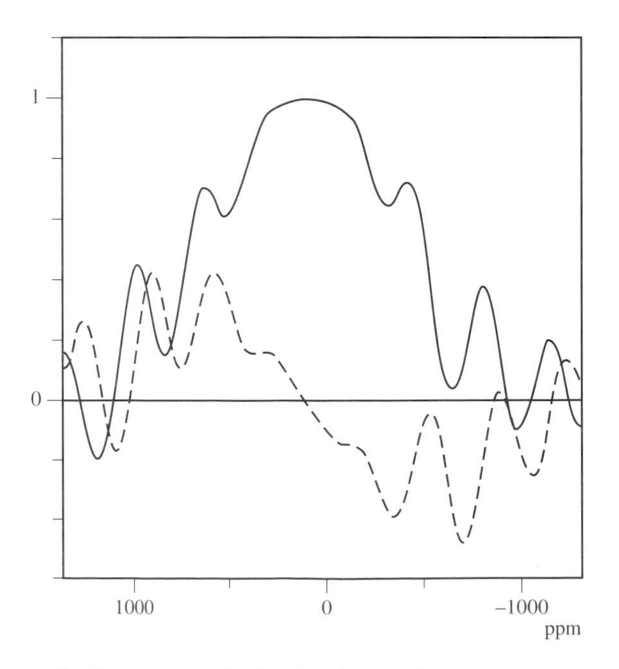

Figure 6 Complex transfer function for a pulse sequence using τ = 20 μs, $(\frac{1}{2}\pi)$ = 3.75 μs, and $\omega_0/2\pi$ = 62.60 MHz

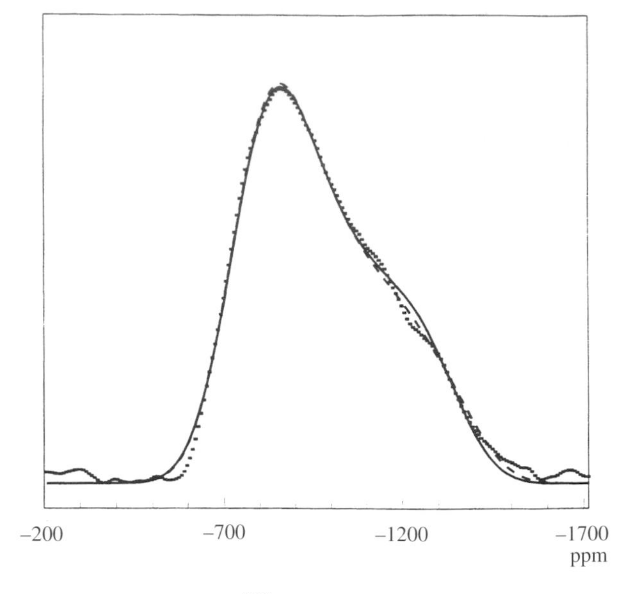

Figure 7 Experimental ^{207}Pb spectrum of $PbBr_2$ (squares) at 62.601 MHz. The rf pulse strength corresponds to a nutation frequency of 1000 ppm. The solid line represents a simple fit that assumes three shift tensor elements as well as a Gaussian broadening, while the dashed line is a best fit including the transfer function of the excitation. The ppm scale is referred to the chemical shift of $Pb(CH_3)_4$

The effect of this sequence can be calculated by standard density matrix theory.[5,8,16] For signals governed by chemical shift anisotropy (CSA), we define the Hamiltonian in the rotating frame as

$$\hat{\mathcal{H}} = \begin{cases} -\Delta\omega_0 \hat{I}_z & \text{no pulse} \\ -\Delta\omega_0 \hat{I}_z \mp \omega_1 \hat{I}_{x,y} & \text{during pulses} \end{cases} \tag{8}$$

where $-\omega_1 \hat{I}_x$ has to be used for a $(\theta)_x$ pulse, $+\omega_1 \hat{I}_x$ for a $(\theta)_{-x}$ pulse, etc. As this Hamiltonian is a piecewise-constant function of time, the evolution of the density matrix $\hat{\rho}$ is given simply by[5]

$$\hat{\rho}(t) = \mathrm{e}^{-\mathrm{i}\hat{\mathcal{H}}_n t_n} \cdots \mathrm{e}^{-\mathrm{i}\hat{\mathcal{H}}_1 t_1} \hat{\rho}(0)\, \mathrm{e}^{\mathrm{i}\hat{\mathcal{H}}_1 t_1} \cdots \mathrm{e}^{\mathrm{i}\hat{\mathcal{H}}_n t_n} \tag{9}$$

where subscripts n label subsequent time intervals.

The numerical calculation of the evolution of the density matrix is fairly easy, since all operators of a spin-$\frac{1}{2}$ can be expressed by 2 × 2 matrices. Figure 6 shows a calculated transfer function for the excitation by the above sequence. The effect on both receiver channels of a quadrature detector was calculated. One sees the complicated distortions of amplitudes and phases as a function of offset frequency.

Least-squares fits become significantly more precise if the transfer function is included. It should be noted that no new fitting parameters are introduced, but the experiment with its timings of pulses and delays determines the transfer function completely.

The possible improvement is demonstrated by Figure 7. Though, at first sight, the differences seem to be small, application of the transfer function leads to a better reproduction of the almost triangular lineshape. The set of best-fit parameters changes from σ_{11} = −710 ppm, σ_{22} = −849 ppm, and σ_{33} = −1350 ppm without correction to σ_{11} = −699 ppm, σ_{22} =

-845 ppm, and $\sigma_{33} = -1398$ ppm if the transfer function is included in the theoretical lineshape. The width of the line, $\sigma_{11} - \sigma_{33}$, changes by 10%, or 69 ppm, which is clearly outside the limit of statistical error. The mean-square deviation is also reduced by about 25%.

6 RELATED ARTICLES

Chemical Shift Tensor Measurement in Solids; Chemical Shift Tensors; Composite Pulses; Echoes in Solids; Fourier Transform & Linear Prediction Methods; Fourier Transform Spectroscopy; Magic Angle Turning & Hopping; Phase Cycling; Radiofrequency Pulses: Response of Nuclear Spins; Selective Pulses; Shaped Pulses; Shielding Tensor Calculations; Solid State Probe Design.

7 REFERENCES

1. B. Wrackmeyer and K. Horchler, in 'Annual Reports on NMR Spectroscopy', ed. G. A. Webb, Academic Press, London, 1990, Vol. 22, p. 249.
2. B. E. Mann, in 'Annual Reports on NMR Spectroscopy', ed. G. A. Webb, Academic Press, London, 1991, p. 141.
3. R. R. Ernst, G. Bodenhausen, and A. Wokaun, 'Principles of Nuclear Magnetic Resonance in One and Two Dimensions', Oxford University Press, Oxford, 1987.
4. R. G. Shulman, *Phys. Rev. Lett.*, 1959, **2**, 459.
5. A. Abragam, 'The Principles of Nuclear Magnetism', Oxford University Press, Oxford, 1961.
6. T. Shimizu, H. Yasuoka, T. Imai, T. Tsuda, T. Takabatake, Y. Nakazawa, and M. Ishikawa, *J. Phys. Soc. Jpn.*, 1988, **57**, 2494.
7. C. D. Makowka, C. P. Slichter, and J. H. Sinfelt *Phys. Rev. Lett.*, 1982, **49**, 379; *Phys. Rev. B*, 1985, **31**, 5663.
8. C. P. Slichter, 'Principles of Magnetic Resonance', 3rd edn., Springer-Verlag, Berlin, 1990.
9. M. H. Levitt, and R. Freeman, *J. Magn. Reson.*, 1979, **33**, 473.
10. M. H. Levitt, *J. Magn. Reson.*, 1982, **48**, 234.
11. M. H. Levitt, R. Freeman, and T. Frenkiel, in 'Advances in Magnetic Resonance', ed. J. S. Waugh, Academic Press, New York, 1983, Vol. 11, p. 48.
12. R. Freeman, S. P. Kampsell, and M. H. Levitt, *J. Magn. Reson.*, 1980, **38**, 453.
13. R. Tycko, H. M. Cho, E. Schneider, and A. Pines, *J. Magn. Reson.*, 1985, **61**, 90.
14. L. Emsley and G. Bodenhausen, *J. Magn. Reson.*, 1990, **87**, 1.
15. G. Neue, C. Dybowski, M. L. Smith, and D. H. Barich, *Solid State NMR*, 1994, **3**, 115.
16. M. Mehring, 'High Resolution NMR in Solids', 2nd edn, Springer-Verlag, Berlin, 1982.

Acknowledgements

Simulations were performed using the program ANTIOPE, which was kindly provided by J. S. Waugh.

Biographical Sketches

Cecil Dybowski. *b* 1946. B.S., 1969, Ph.D., 1973, Texas. Faculty at the University of Delaware, 1976–present. Over 100 publications, including the book "Transient Techniques in the NMR of Solids" (with B. C. Gerstein). Research interests include all facets of NMR spectroscopy, the analysis of heterogeneous catalysis and catalytic processes, materials structure and function.

Günther Neue. *b* 1955. Dipl.-Chem., 1978, Dr.rer.nat., 1983, Priv.-Doz., 1988, Dortmund. Hochschuldozent, Universität Dortmund, 1988–present. Over 25 research publications. Research interests include development of analytical techniques in the fields of surface science and environmental chemistry.

Wood & Wood Chars

Mark S. Solum

University of Utah, Salt Lake City, UT, USA

1 INTRODUCTION

Wood is a heterogeneous material composed roughly of 40–50% cellulose, 18–35% lignin, 15–35% hemicellulose, and a smaller amount, 1–10%, of miscellaneous material collectively called extractives.[1–4] The extractives are composed of tannins, starches, gums, fats, waxes, sugars, and resins. All of the carbohydrates, or what is left after removal of lignin and the extractives from wood, is called holocellulose. Whole wood has been studied by solid state NMR, and its components by both solid and liquid state NMR.[5] Chars or activated carbons made from pyrolysis of wood at temperatures of about 650 °C or less are also suitable for study by solid state NMR. This article will only deal with solid state NMR of wood and wood chars. Solid state NMR, because of its nondestructive nature, can be used to study wood without chemical modification of the wood components or with less modification than would be needed to solubilize them. Most solid state NMR studies have used the ^{13}C CP MAS technique for studies of whole woods or various components.[6] The usual problems of quantitation in CP MAS, which have been reviewed for coals, will also apply to woods and especially wood chars.[7] Quantitation problems have been discussed for wood with respect to the determination of carbohydrate and lignin content.[8,9]

Articles on NMR of wood or wood components in both liquid and solid states can be found in a variety of journals,

For References see p. 5051

some of which are hard to find, especially at nonagricultural schools. Some journals not referenced here in which NMR wood articles often appear are *Holzforschung*, *Tappi*, *Appita*, *Carbohydrate Research*, *The Journal of Wood Chemistry and Technology*, and *Mokuzai Gakkaishi*.

1.1 Experimental Considerations

Most solid state NMR data of wood or wood components have been obtained with a single-contact-time (τ_{cp}) CP MAS experiment[10] using contact times of 1 or 2 ms and pulse delays T_d of 0.5–10 s, with the longer times for pure cellulose. In the CP experiment, carbon magnetization is usually approximated as building up exponentially with a time constant T_{CH}, and dying out exponentially with a time constant $T^H_{1\rho}$. The experiment may be repeated, to increase the signal-to-noise ratio, with a time determined by T^H_1. For the data to represent the different carbon types fairly with one τ_{cp}, the following conditions should hold for each nucleus $T_{CH} << \tau_{cp} << T^H_{1\rho}$ and $T_d > 5T^H_1$. Several researchers[8,9,11,12] have measured these parameters for wood components, and in one case[9] a comparison has been made between the CP data and data from a single pulse experiment (SPE), which can be repeated in a time determined by the longer T^C_1.

The reported results for T_{CH} of the carbons in cellulose are 150 ms or shorter, with $T^H_{1\rho}$ values from 7 to 11 ms. For the nonprotonated aromatic carbons in lignin, T_{CH} values were found to be just under 0.5 ms, with $T^H_{1\rho}$ values from about 5 to 20 ms. The T^H_1 value in cellulose was found to be 369 ms,[8] and the values for lignin were less than 41 ms. The most slowly polarizing group in wood was found to be the carboxyl resonance from an acetate group in hemicellulose, where the T_{CH} was 0.8 ms.[9] These researchers also compared CP experiments with contact times of 1 and 2 ms and a 1.5 s T_d with single pulse experiment (SPE) data using a 25 s T_d, and in all three experiments virtually all of the carbon was detected. However, at a contact time of 1 ms, the carboxyl region was underestimated in comparison to SPE data. These results agree with our data on coals, where carbonyl and carboxyl groups are the main functional groups underrepresented in CP experiments as compared with SPE. A 2 ms contact time and a 2 s pulse delay appears to be the best choice for the most quantitative data, although this might not be the maximum signal intensity.

2 WHOLE WOOD AND ITS MAJOR COMPONENTS

Wood can be studied by solid state NMR by looking at the intact wood or by breaking it into components such as cellulose, hemicellulose, and lignin. The different types of lignin are in fact defined by the procedure used to isolate them from the whole wood. Studying whole wood has the advantage that there is no modification of components as is the case in separation procedures. The different components all show characteristic resonances in the NMR spectrum, with some overlap of signals. One early NMR study on lodgepole pine shows spectra of the whole wood, the effect of ball milling on wood, two types of hollocellulose, two hemicelluloses, cellulose and five types of lignin isolated from the wood.[6]

2.1 Cellulose

Cellulose, the major constituent of wood, is a linear homopolysaccharide composed of β-D-glucopyranose units. The empirical formula for cellulose is $(C_6H_{10}O_5)_n$, where n is the number of glucose monomer units in the polymer chain and is about 10 000 in wood cellulose and 15 000 in cotton cellulose.[3] Cellulose is found in the form of microfibrils, which are the smallest structural units seen by an electron microscope. These microfibrils consist of a central core of parallel cellulose chains forming a crystalline region that is surrounded by amorphous cellulose and hemicellulose and encrusted and bound in the wood cell wall by lignin.[1,13] The size of microfibrils varies, depending on the source, with the highly crystalline[14] *Valonia macrophysa* having about a 20 nm cross section and the crystalline region running at least 50 nm in length.[13] The exact details of the fibrillar and supramolecular crystal structure of cellulose is still open to debate.[1] All native celluloses have the same crystal structure called cellulose I, with four glucose units per unit cell and cellobiose (**1**) as the repeating unit in the chain direction with each glucose unit displaced 180° to give a twofold screw axis, (see Figure 1). There are at least three other crystal structures of cellulose (cellulose II, III, and IV). Cellulose II is more thermodynamically stable than cellulose I, because the crystal structure allows extra intermolecular hydrogen bonding between chains.

The ^{13}C CP MAS spectra of cellulose from three different sources is shown in Figure 2. The spectrum of amorphous cellulose has four broad lines. Lines in an amorphous material are usually broader than for crystalline material owing to chemical shift dispersion. The most deshielded line at 105.7 ppm is assigned to the anomeric C-1 carbon, and the next most deshielded line at 84.7 ppm can be assigned to the C-4 carbon of the glucosidic linkage. The large unresolved peak between 70 and 80 ppm is assigned to the three other ring carbons, C-2,3,5, and finally the nonring carbon, C-6, is assigned to the resonance at 63.1 ppm. The spectrum of powdered wood cellulose (Aldrich) is sharper than the amorphous cellulose, and two new sharp peaks at 89.6 and 65.9 ppm appear and can be assigned to carbons C-4 and C-6 of cellulose in the crystalline regions. The broad resonance between 70 and 80 ppm is now split also. The spectrum of cellulose from cotton is the most resolved of the three, with the anomeric carbon now showing a small splitting and with the sharp peaks at 90 and 66 ppm larger in comparison with broad shoulders at 84 and 64 ppm,

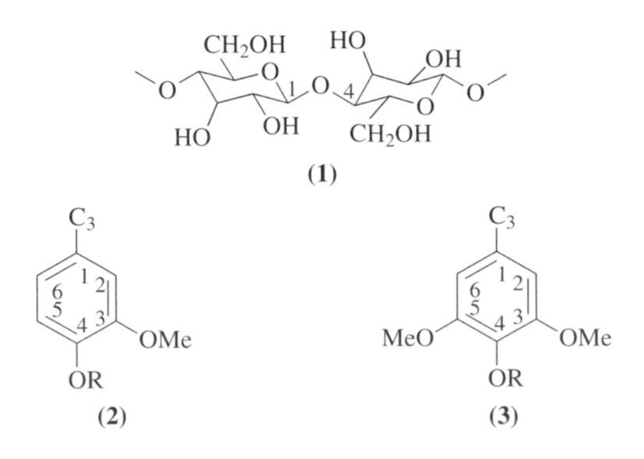

Figure 1 Basic structural units from wood components: cellobiose (**1**) unit in cellulose chain and guaiacyl (**2**) and syringyl (**3**) units from lignin

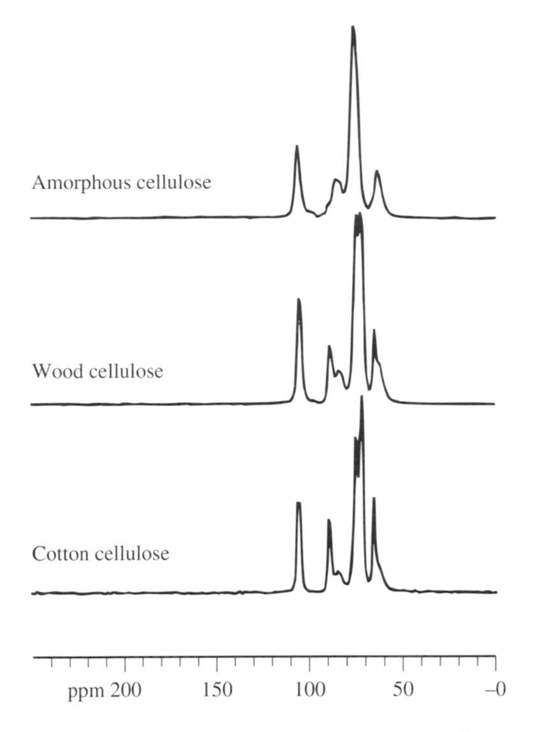

Figure 2 Carbon-13 CP MAS spectra of three types of cellulose. All spectra were run with a 2 ms contact time and 1 s pulse delay

showing this sample to be more crystalline than the wood cellulose. It should be noted that in cellulose the degree of crystallinity as measured by different techniques seems to vary with the technique[1,14] The NMR assignments for celluloses II and IV have also been given.[15]

2.2 Hemicellulose

Hemicelluloses are heterogeneous branched polysaccharides easily hydrolyzed by acid to their monomeric sugars consisting mainly of D-glucose, D-galactose, D-mannose, D-xylose, and L-arabinose. Also found are 4-*O*-methyl-D-glucuronic acid and galacturonic acid. The number of monomer units in a hemicellulose chain is about 200,[3] the chains are branched, and there are acetate, methoxy, and carboxylic acid functional groups not found in cellulose. The structures of hemicelluloses in hard woods are also different than in soft woods.[3] The ^{13}C CP MAS spectra of hemicelluloses A and B have also been assigned.[6] The general shape is somewhat similar to that of amorphous cellulose between 60 and 110 ppm, with the anomeric, C-1, carbon now at 103 ppm. Some additional resonances at 174 ppm are due to the acetate carboxyl groups, at 21.5 ppm from acetate methyls, and at 56 ppm from methoxy groups. Because clean separation of hemicellulose from lignin is difficult, small resonances from residual lignin can usually be seen in the aromatic region.

2.3 Lignin

Lignin is a highly aromatic irregular polymer of hydroxy- and methoxy-substituted phenylpropane units found in the middle lamella of plants, binding cells together, and in the secondary plant cell wall, encrusting cellulose microfibrils. Lignin in softwoods is composed mainly of guaiacyl units (**2**), while hardwood lignin is composed of syringyl (**3**) and guaiacyl units (**2**) (see Figure 1). Grass lignin is based on the *p*-hydroxyphenyl unit. However, most plants probably contain some of all three units. There are many methods used to isolate lignin from whole wood, and the name of the lignin is derived from the method (or the person who developed the method) used to isolate it. Because there are several types of ether and carbon–carbon linkages between phenylpropane units, there is great difficulty in isolating lignin without altering its chemical structure.

The ^{13}C CP MAS spectra of three different types of lignin (Aldrich) are shown in Figure 3. Kraft lignin is isolated from the spent liquor during Kraft pulping, organosolv lignin is isolated from a mixture of hardwood chips by aqueous ethanol in the presence of mineral acids, and hydrolytic lignin is isolated from bagasse with superheated steam in the presence of acetic acid catalyst. In the NMR spectra, one can see major similarities and minor differences between the three lignins. All three show a very strong resonance between 50 and 60 ppm from methoxy groups on aromatic rings. The hydrolytic lignin also has a strong aliphatic resonance at 33.5 ppm from CH_2 groups not significant in the other two lignins. All three lignins show a resonance from carbonyl and carboxyl groups at about 206 ppm and 175–181 ppm respectively. The phenolic region between 145 and 153 ppm is very important in lignins, because (**2**) and (**3**) have slightly different resonances.[5] The 153 ppm peak is assigned to the C-3 and C-5 carbons of (**3**), and the

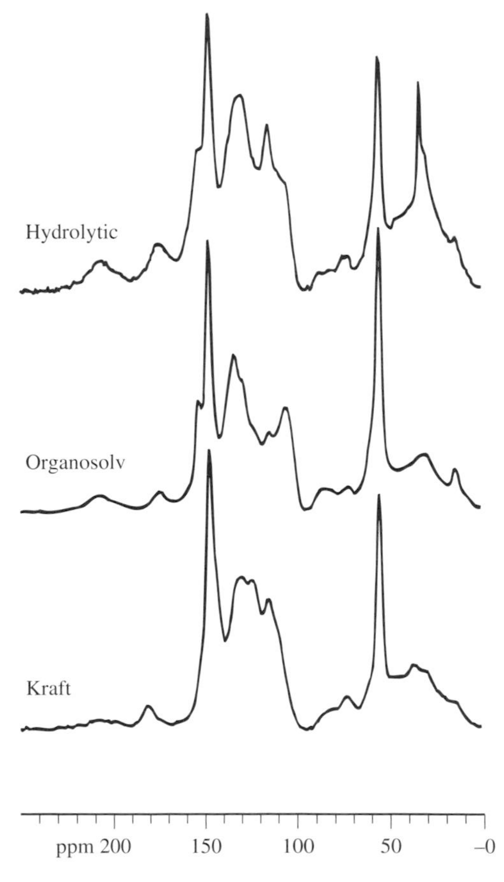

Figure 3 Carbon-13 CP MAS spectra of three types of lignin. All spectra were run with a 2 ms contact time and 1 s pulse delay

For References see p. 5051

145 ppm and the 148 ppm peaks are from the C-4 and C-3 carbons of (**2**), respectively. All three lignins in Figure 3 show a strong resonance at 148 ppm from (**2**), but only the organsolv and hydrolytic lignins show significient intensity from (**3**) at 153 ppm. Because of the large amount of oxygen substitution in the aromatic rings, *ortho* and *para* effects shift significant aromatic intensity from the protonated carbons to the more shielded region between 100 and 120 ppm. Spectra of Klason, dioxane, periodate, enzymatic, and Brauns native lignin have been discussed elsewhere,[6] along with different Kraft preparations,[11] and comparisons between solid and liquid NMR spectra have also been given.[5]

2.4 Whole Wood

Some ^{13}C CP MAS spectra of whole woods are shown in Figure 4 for three hardwoods (white oak, yellow poplar, and aspen) and a softwood (pine). All spectra are dominated by the resonances from cellulose (and hemicellulose) between 60 and 90 ppm and the anomeric carbon at 105 ppm. The one major notable difference in this region is associated with the yellow poplar sample. Judging by the size of the 89 ppm peak compared with the 84 ppm peak and the size of the shoulder on the 66 ppm peak, the cellulose in this sample may be more crystalline than in the other three samples. All four woods show acetate resonances at 174 and 21 ppm from carboxyl and methyl groups in the hemicellulose (and possibly lignin). There is no significant carbonyl intensity in the spectra of any of the woods, and all samples show a strong methoxy resonance at 56 ppm. Another difference among the four woods is in the phenolic region between 145 and 155 ppm. The hardwoods, white oak, and aspen have a larger resonance at 153 ppm from (**3**) than the 148–145 ppm resonance from (**2**). The opposite is true for the softwood pine sample, but also for the hardwood yellow poplar sample, showing the variability of lignin in hardwoods.

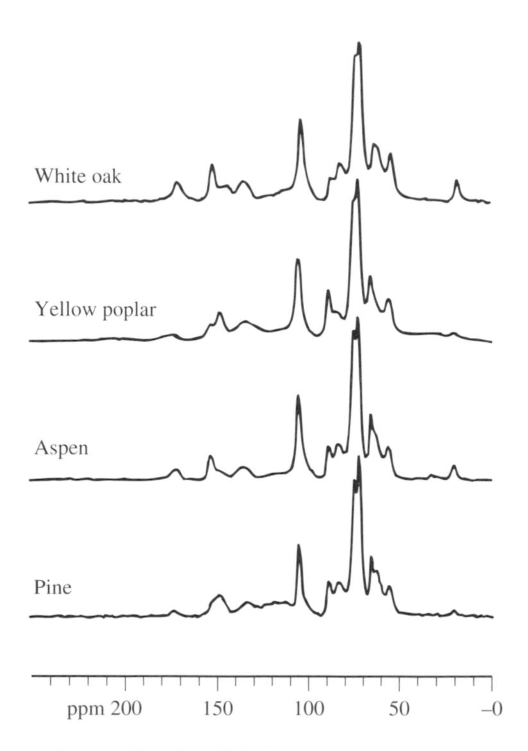

Figure 4 Carbon-13 CP MAS spectra of four whole woods, three hardwoods (white oak, yellow poplar, and aspen), and a softwood (pine). All spectra were run with a 2 ms contact time and 1 s pulse delay

If spectra are taken under quantitative conditions, solid state ^{13}C NMR can be used for the nondestructive determination of several structural parameters of whole (extractive free) woods. The parameters usually determined by solid state NMR are the total fraction of lignin in the wood,[7,11] OCH_3/R, the number of methoxy groups per aromatic ring, and $f_a^{a,H}$, the fraction of aromatic carbon that has a proton attached as determined from dipolar dephasing experiments.[8,16] These last two parameters give information on the basic lignin monomer unit and the ring substitution. One should note that in woods, the separation between the aromatic and aliphatic region is not as clean as in coals, owing to the carbohydrate C-1 carbons' deshielded resonance. While good results have been obtained by integrating the aromatic region from 109–110 ppm to the end of the aromatic range,[8,11] the most quantitative results require some estimate of the lignin contribution under the C-1 carbohydrate peak and also carbohydrate overlap of the lignin methoxy peak at 56 ppm.[8]

3 WOOD CHARS

Pyrolysis is the application of heat to a material in the absence of oxygen to degrade the material into solid, liquid, and gasous phases. The remaining solid material, called char (or charcoal if the material was wood), has been historically used as a fuel. Sometimes, with chemical activation (H_3PO_4 or KOH), the pore structure of these chars (or activated carbons) can be increased, with the chars becoming highly adsorptive. The properties of these chars will depend on both the temperature and the time the wood has been heated. The products of pyrolysis of whole wood are basically what one would expect from pyrolysis of components separately. There have been very few reported studies of pyrolysis of wood[17–20] or wood components[18,19,21] by solid state NMR spectroscopy.

Spectra from four activated carbons[20] made by pyrolysis of white oak (see Figure 4) are shown in Figure 5. Two of the samples were soaked in phosphoric acid for 1 h before pyrolysis. The 150 °C sample was slowly heated to temperature and held there for 3 h. The other samples were heated to 170 °C over 2 h, held there for 30 min, and then taken to their final temperature and held for 1 h. The 250 °C thermal-only sample has an NMR spectrum very similar to that of the starting material, except that the carboxyl and methyl groups from the acetate groups on hemicellulose have decreased in intensity and new resonances are starting to appear between 110 and 150 ppm from new aromatic (or alkenic) moieties and at 0–50 ppm from new aliphatic groups. In contrast, the acid-treated wood has lost all carbohydrate resonances by 150 °C, the spectrum is highly aromatic with a large new phenolic resonance at 147.5 ppm, a new carbonyl peak has appeared at 204 ppm, and there are significient new aliphatic bands but with the lignin methoxy peak still visible. At 250 °C, the methoxy groups have left, aliphatic material has been lost or aromatized, carbonyl and carboxyl resonances have decreased, and the phenolic shoulder at about 150 ppm has been reduced in intensity. At 650 °C, only an aromatic peak can be seen. Underneath this

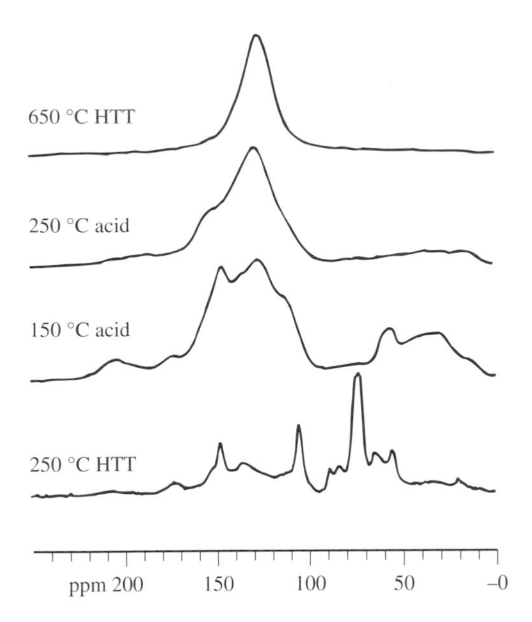

Figure 5 Carbon-13 CP MAS spectra of chars from white oak, showing final pyrolysis temperature. Two of the samples were treated with H_3PO_4 before pyrolysis. All spectra were run with a 2 ms contact time and 1 s pulse delay

peak, a very low intensity very wide band, probably from spins near free radicals, obscures any residual aliphatic material still present. The 650 °C spectrum is typical of that from the end product of pyrolysis of a carbonaceous material. On further heating, these types of material become too conductive to study by NMR.

4 RELATED ARTICLES

Amorphous Materials; Biological Macromolecules; Biological Macromolecules: NMR Parameters; Coal Structure from Solid State NMR; Cokes; Cross Polarization in Rotating Solids: Spin-1/2 Nuclei; Cross Polarization in Solids; Fossil Fuels; Magic Angle Spinning; Microporous Materials & Xenon-129 NMR; Solid Biopolymers.

5 REFERENCES

1. G. Tsoumis, 'Science and Technology of Wood: Structure, Properties, Utilization', Van Nostrand Reinhold, New York, 1991, Chap. 4.
2. R. C. Pettersen, in 'The Chemistry of Solid Wood', ed. R. M. Rowell, American Chemical Society, Washington, DC, 1984, Chap. 2.
3. E. Sjöström, 'Wood Chemistry: Fundamentals and Applications', 2nd edn., Academic Press, San Diego, 1993, Chaps. 3 and 4.
4. S. Saka, in 'Wood and Cellulosic Chemistry', eds. D. N.-S. Hon and N. Shiraishi, Marcel Dekker, New York, 1991, Chap. 2.
5. M. A. Wilson, 'NMR Techniques and Applications in Geochemistry and Soil Chemistry', Pergamon Press, Oxford, 1987, Chap. 6.
6. W. Kolodziejski, J. S. Frye, and G. E. Maciel, *Anal. Chem.*, 1982, **54**, 1419.
7. C. E. Snape, D. E. Axelson, R. E. Botto, J. J. Delpuech, P. Tekely, B. C. Gerstein, M. Pruski, G. E. Maciel, and M. A. Wilson, *Fuel*, 1989, **68**, 547.
8. A. L. Bates and P. G. Hatcher, *Org. Geochem.*, 1992, **18**, 407.
9. G. D. Love, C. E. Snape, and M. C. Jarvis, *Biopolymers*, 1992, **32**, 1187.
10. M. A. Wilson, 'NMR Techniques and Applications in Geochemistry and Soil Chemistry', Pergamon Press, Oxford, 1987, Chap. 4.
11. J. F. Haw, G. E. Maciel, and H. A. Schroeder, *Anal. Chem.*, 1984, **56**, 1323.
12. R. H. Newman, in 'Viscoelasticity of Biomaterials', eds. W. G. Glasser and H. Hatakeyama, ACS Symposium Series 489, American Chemical Society, Washington, DC, 1992, Chap. 20.
13. M. Fujita and H. Harada, in 'Wood and Cellulosic Chemistry', eds. D. N.-S. Hon and N. Shiraishi, Marcel Dekker, New York, 1991, Chap. 1.
14. W. L. Earl and D. L. VanderHart, *Macromolecules*, 1981, **14**, 570.
15. R. L. Dudley, C. A. Fyfe, P. J. Stephenson, Y. Deslandes, G. K. Hamer, and Marchessault, *J. Am. Chem. Soc.*, 1983, **105**, 2469.
16. P. G. Hatcher, *Energy & Fuels*, 1988, **2**, 48.
17. W. L. Earl, in 'Proceedings of 1981 International Conference on Residential Solid Fuels', eds. J. A. Copper and D. Malek, Oregon Graduate Center, Beaverton, OR, 1982, 772.
18. Y. Sekiguchi, J. S. Frye, and F. Shafizadeh, *J. Appl. Polym. Sci.*, 1983, **28**, 3513.
19. F. Shafizadeh, in 'The Chemistry of Solid Wood', ed. R. M. Rowell, American Chemical Society, Washington, DC, 1984, Chap. 13.
20. M. S. Solum, R. J. Pugmire, M. Jagtoyen, and F. Derbyshire, *Carbon*, 1995, **33**, 1247.
21. R. A. Wind, L. Li, G. E. Maciel, and J. B. Wooten, *J. Appl. Magn. Reson.*, 1993, **5**, 161.

Biographical Sketch

Mark S. Solum. *b* 1956. B.S., 1978, Ph.D., 1986, University of Utah. Research Associate Department of Chemistry and the Advanced Combustion Engineering Research Center (ACERC) at Brigham Young University & the University of Utah, 1986–present. Approx. 25 publications. Research interests include application of solid state NMR techniques for the study of coals, chars, woods, activated carbons and other fossil fuel materials with applications to modeling devolatilization processes and structure determinations.

Zero Field NMR

David B. Zax

Cornell University, Ithaca, NY, USA

1 INTRODUCTION

After 50 years of development, techniques for achieving high resolution in NMR are well developed for molecules in the liquid state, where subhertz linewidths are common. Achieving high resolution is a substantially different problem for materials in the solid state, where the 'natural' spectral linewidths (measured from the Fourier transform of the free induction decay following a single excitation pulse) are many kilohertz wide. Both theory and experiment have extensively demonstrated that manipulations of the spin degrees of freedom (using rf fields) and/or spatial degrees of freedom (typically via bulk sample rotation) make it possible to 'line-narrow'. In line-narrowing experiments, deterministic 'dynamics' imposed by the experimentalist efficiently substitute for Nature's chaotic efforts so as to eliminate those anisotropic interactions responsible for excess spectral breadth. What remains are the isotropic interactions only—a portion of the chemical shift tensor, some fraction of sufficiently large second-order interactions, and J couplings.

Where the molecular probes of greatest spectroscopic interest are the dipole–dipole or quadrupolar interactions (whose anisotropies are often responsible for the broad spectral lines observed in solids), techniques for achieving high resolution are substantially less well developed. Methods that focus on eliminating anisotropies are inappropriate, since in the limit of large applied magnetic fields, these interactions are entirely anisotropic. A fundamentally different approach is therefore required. Eliminating the anisotropy in the interaction is accomplished not by eliminating the interaction itself but by eliminating instead the laboratory-based, externally applied magnetic field. The resulting experiments are, therefore, zero field NMR experiments. In the absence of an applied field, NMR spectra of solids in their 'natural' spectroscopic state can be revealed.

In an era where research facilities rush to order spectrometers based on not yet operating superconducting magnets at newer and higher fields, it is far from obvious that some magnetic resonance experiments may be best executed in zero applied field. Yet zero field magnetic resonance has a long and important place in magnetic resonance, including experiments among the earliest[1] and of greatest impact.[2,3] Nonetheless, zero field NMR experiments occupy a relatively small niche in standard discussions of high resolution NMR in solids.[4] Thus one of the essential questions to be addressed below is the motivation for NMR experiments executed in low magnetic fields. A second question we shall address is how such experiments might be carried out without unacceptable sensitivity losses.

2 WHY DO HIGH FIELD NMR?

Why is NMR routinely performed in large applied magnetic fields? Two reasons come immediately to mind: the first is related to experimental sensitivity. Formulae for the signal amplitude in an NMR experiment predict an increase in that fundamental parameter by an amount that varies as the square of the transition frequency. This dependence can be traced directly to the Faraday law, which states that the electromotive force ϵ in a coil associated with a magnetic flux Φ is

$$|\varepsilon| \propto \frac{\mathrm{d}\Phi}{\mathrm{d}t} \propto \omega M_z \tag{1}$$

where ω is the transition frequency and M_z is the amplitude of the magnetization being detected. Experimentalists can control the transition frequency only where the Larmor interaction

$$\hat{\mathcal{H}}_{\mathrm{L}} = \gamma B_0 \hat{I}_z = \omega_{\mathrm{L}} \hat{I}_z \tag{2}$$

dominates the transition frequency so that $\omega \approx \omega_{\mathrm{L}}$. In the limit of high temperatures and Boltzmann statistics, the equilibrium magnetization M_z of equation (1) depends linearly on the same energy splitting ω. For most nuclei, and for all studies in liquid state NMR, the largest transition frequencies are found in the largest possible magnetic fields, and on combining equations (1) and (2) the signal amplitude is found to vary with $B_0{}^2$.

Large applied magnetic fields serve a second purpose. The chemical shift Hamiltonian can be written as a contraction

$$\hat{\mathcal{H}}_{\mathrm{cs}} = \gamma \hat{\boldsymbol{I}} \cdot \boldsymbol{\sigma} \cdot \hat{\boldsymbol{B}}_0 \tag{3}$$

and frequency shifts in equation (3) due to σ are proportional to B_0. Resolution of line shifts associated with two different chemical environments is therefore easiest in the largest available magnetic fields. Wherever chemical shift information is desired, the best measurements are made in the highest available fields.

In solids, the same considerations govern signal amplitudes, so that high field detection is always most sensitive. Chemical shifts scale in the same fashion in liquids and solids. But where the absorption spectrum is dominated by dipole–dipole coup-

For list of General Abbreviations see end-papers

lings between spins (for abundant spin-$\frac{1}{2}$ nuclei, especially ^{1}H) and/or quadrupolar couplings (for spin-1 or higher nuclei), resolution is not necessarily improved in higher fields—in fact, the opposite is true.

The direct coupling between the magnetic dipole moments of pairs of spins A and B, is represented by the dipole–dipole Hamiltonian

$$\hat{\mathcal{H}}_{\rm D} = \hat{\boldsymbol{I}}_{\rm A}\cdot\boldsymbol{D}\cdot\hat{\boldsymbol{I}}_{\rm B} \tag{4}$$

where $D_{\alpha\beta} = \gamma_i\gamma_j\hbar r_{\rm AB}{}^{-3}(\delta_{\alpha\beta} - 3e_\alpha e_\beta)$ and e_α is the αth component of the unit vector along $\boldsymbol{r}_{\rm AB}$. For collections of dipolar-coupled spins, the total dipolar Hamiltonian is the sum over all pairwise interactions. Typical dipole–dipole couplings are generally less than 50 kHz. (It is entirely possible that some of the dipole-like coupling between spins arises from anisotropic contributions to the J coupling tensor. As experiments provide little guidance as to how one might separate between the two, we shall ignore this possibility.)

The quadrupole Hamiltonian $\hat{\mathcal{H}}_{\rm Q}$ reflects the coupling between the nuclear electric quadrupole moment eQ (for $I \geqslant 1$ only; for $I = 0$ or $\frac{1}{2}$, the charge distribution in the nucleus is spherical and $eQ = 0$) and whatever local electric field gradients are found nearby, so that

$$\hat{\mathcal{H}}_{\rm Q} = \frac{eQ}{2I(2I-1)\hbar_i}\hat{\boldsymbol{I}}_i\cdot e\boldsymbol{q}\cdot\hat{\boldsymbol{I}}_i \tag{5}$$

where $\boldsymbol{q}$ is the field gradient tensor. Local field gradients reflect most strongly the distribution of electrons in valence orbitals, and somewhat less strongly the more distance ionic charges associated with a typical ionic lattice. Collectively, $\hat{\mathcal{H}}_{\rm D}$ and $\hat{\mathcal{H}}_{\rm Q}$ will be referred to as the local Hamiltonians.

Choosing the principal axis systems where these interactions are diagonal, the expanded version of the dipole–dipole Hamiltonian is

$$\hat{\mathcal{H}}_{\rm D} = \omega_{\rm D}(3\hat{I}_{z,i}\hat{I}_{z,j} - \hat{\boldsymbol{I}}_i\cdot\hat{\boldsymbol{I}}_j) \tag{6}$$

where $\omega_{\rm D} = \gamma_i\gamma_j\hbar/r_{ij}^3 = D_{xx}$. (In the absence of dynamics, the dipole–dipole Hamiltonian must satisfy $D_{zz} = -2D_{xx} = -2D_{yy}$.) The expanded version of the quadrupolar Hamiltonian is

$$\hat{\mathcal{H}}_{\rm Q} = \frac{\chi}{4I(2I-1)}[3\hat{I}_z^2 - I(I+1) + \tfrac{1}{2}\eta(\hat{I}_+^2 + \hat{I}_-^2)] \tag{7}$$

where q_{zz} is the largest principal value of $\boldsymbol{q}$ and $\chi = e^2q_{zz}Q/h$, with asymmetry parameter $\eta = (q_{yy} - q_{xx})/q_{zz}$. Typical values of χ range from 10 kHz to 100 MHz. Each of the interactions $\boldsymbol{D}$ and $\boldsymbol{q}$ is traceless, so that neither contributes to spectra of isotropically tumbling liquids.

Above, we have represented each of the local interactions $\hat{\mathcal{H}}_{\rm D}$ and $\hat{\mathcal{H}}_{\rm Q}$ in a locally defined reference frame. These interactions are often described (incorrectly) as field-independent because, unlike the chemical shift Hamiltonian of equation (3), the field appears nowhere in their description. This is a misnomer, which obscures the fact that these interactions may be best resolved in low applied fields. (In addition, second-order effects are largest in low fields. In DAS and DOR spectroscopy, for example, this leads to the surprising result that sensitivity to differing chemical environments may be best in low fields.[5])

The reference frames we have chosen are such that each of $\hat{\mathcal{H}}_{\rm D}$ and $\hat{\mathcal{H}}_{\rm Q}$ (or, collectively, $\hat{\mathcal{H}}_{\rm loc}$) is described in equations (6) and (7) in a consistent, molecule-based reference frame identical for both the spin vectors $\boldsymbol{I}$ and spatial interactions $\boldsymbol{D}$ or $\boldsymbol{q}$. A sufficiently large externally applied field $\boldsymbol{B}_0$ modifies the observable portions of $\hat{\mathcal{H}}_{\rm loc}$. Owing to the interaction of the field and spin moment of equation (2), the spin moment $\boldsymbol{I}$ is preferentially aligned parallel or antiparallel to the field. $\boldsymbol{D}$ and $\boldsymbol{q}$, however, are tied to their local frames of reference. Different orientations (with respect to the externally imposed field) of otherwise identical systems are differently affected. First-order perturbation theory predicts that the portions of $\hat{\mathcal{H}}_{\rm loc}$ observable in high field are

$$\hat{\mathcal{H}}'_{\rm D} = \tfrac{1}{2}\omega_{\rm D}(3\cos^2\theta - 1)(3\hat{I}_{z,i}\hat{I}_{z,j} - \hat{\boldsymbol{I}}_i\cdot\hat{\boldsymbol{I}}_j) \tag{8}$$

and

$$\hat{\mathcal{H}}'_{\rm Q} = \frac{\chi}{8I(2I-1)}(3\cos^2\theta - 1 + \eta\sin^2\theta\cos 2\phi)[3\hat{I}_z^2 - I(I+1)] \tag{9}$$

where the designation $\hat{\mathcal{H}}'$ in equations (8) and (9) indicates that these Hamiltonians are truncated by the external field, and $\hat{R}(\theta, \phi, 0)$ is the Euler transformation relating the molecular and laboratory frames. In $\hat{\mathcal{H}}'_{\rm loc}$, spectral energies depend explicitly on orientation, so that otherwise-equivalent local environments are spectroscopically distinguishable owing to the distribution of the angular parameters θ and ϕ. Where all orientations are equally probable, as in polycrystalline powders, this distribution manifests itself in 'powder pattern' lineshapes.[6] Where $\hat{\mathcal{H}}_{\rm loc}$ contains the information of greatest interest, $\hat{\mathcal{H}}'_{\rm loc}$ represents a more easily observed but lower resolution version of the original.

3 WHY DO ZERO FIELD NMR?

A deceptively simple answer to this simple question is that spectroscopy is performed in zero applied magnetic field because the information desired is less well observed in high magnetic fields. One of the first applications of zero field NMR techniques was to the measurement of the Earth's magnetic field, using the Larmor frequency of a suitably large sample of H_2O in zero applied field as the experimental observable.[7] Another early experiment suggested that the superior homogeneity of the Earth's field might yield longer-lived free induction decays than in available laboratory-generated fields, and this effect was demonstrated on a sample of fluorobenzene.[8] Nevertheless, applications focusing on liquid samples in low fields occupy a different scientific niche (see also ***Terrestrial Magnetic Field NMR***). Solid state applications of zero field spectroscopy most typically focus on measuring $\hat{\mathcal{H}}_{\rm loc}$, the low sensitivity spectrum, with high resolution so that all of the spectral information is available in its most readily interpreted form.

3.1 Relaxometry

In addition to revealing the interactions $\hat{\mathcal{H}}_{\rm loc}$ in their purest possible form, low field NMR experiments may reveal essential information inaccessible in high field experiments. A second application of low or zero field NMR is in the study of field-

For References see p. 5062

dependent spin–lattice relaxation times T_1 (see also ***Field Cycling Experiments***). Relaxation behavior depends sensitively on the spectral density functions $J(\omega)$, and the frequency dependence of dynamic processes may be best measured by varying the transition frequency of nuclear spin transitions over many orders of magnitude.[9] Low frequency fluctuations can be probed only at low Larmor frequencies.

In other systems, only the low field relaxation may be interesting. This is particularly true when a strong magnetic field changes the physical system in a fundamental, often undesireable, way. Early predictions that electron pairing at the metal-to-superconductor transition would lead to singularities in T_1 relaxation could only be verified by working in fields below the critical field strength where superconductivity is suppressed. Zero field relaxation data provided one of the earliest confirmations of the Bardeen–Cooper–Schrieffer theory of superconductivity.[2,3]

3.2 Overcoming Selection Rules

Where the applied magnetic field aligns nuclear dipole moments, high field nuclear magnetic resonance is characterized by spectral selection rules that determine which transitions are observable. Much of modern multidimensional NMR focuses on overcoming these limiting selection rules. In very low magnetic fields, similar opportunities exist—for example, where the distinction between homonuclear and heteronuclear spins (characterized by no-longer-differing Larmor frequencies) is lost.[10]

Expanded selection rules provide a mechanism for probing other problems only poorly suited to study in high magnetic field. The high field selection rule $\Delta m = \pm 1$ may be overcome in high field multidimensional NMR via indirect detection of multiple quantum transitions. In zero or low fields, these selection rules are loosened, and absorption lines at multiples of the Larmor frequency,[11] or combinations of quadrupolar transition frequencies between pairs of inequivalent, dipolar-coupled sites, are directly observable.[12]

Quantum mechanical tunneling in methyl groups can be associated with a tunneling Hamiltonian and a tunneling splitting superposed on the normal spin Hamiltonian.[13] A measurement of these additional splittings provides an experimental probe of the tunneling rates. Within a given rotational manifold of the methyl group, these tunneling splittings are unobservable, and in high field NMR transitions are allowed only where the rotational state is unchanged. Transitions between differing rotational manifolds are allowed in low field, where the spin-only transitions are flanked by satellite lines offset by the tunneling splitting.[14]

4 MAKING LOW FIELD DETECTION FEASIBLE

4.1 The NQR Regime

Pure NQR has a history nearly as old as NMR. In many covalent bonds to halogen atoms, the electric field gradients $\boldsymbol{q}$ are sufficiently large that pure quadrupolar transitions are found at 30 MHz or higher. For these systems, the sensitivity of NQR is quite comparable to that of NMR in electromagnets. With limited modifications, similar techniques are appropriate in either case, and spectra might be accumulated by either transient or continuous wave methods. Aspects of pure NQR are reviewed in some detail in various classic texts.[15–17]

4.2 The Field Cycling Regime

Other than for the halogens and a select few other nuclei, the sensitivity of pure NQR is insufficient for routine use as a spectroscopic probe. Pure quadrupolar transitions are typically found at frequencies below 5 MHz for such common nuclear spins as ^{17}O, ^{14}N, ^{27}Al, and ^{2}H, and all the alkali metals. And these splittings are quite large when compared with the dipole–dipole couplings between nuclei, which are always substantially less than 1 MHz and too small to be probed directly except in extraordinary circumstances.[1] In such cases, an idea derived from early attempts to understand spin temperature in NMR, where the $\boldsymbol{B}_0$ field amplitude was understood to be an experimental variable, is key.[18] Nuclear spin–lattice relaxation times T_1 are often sufficiently long that the applied magnetic field might be changed (including reduced to very low values) without substantial loss of spin order. As the coupling between spins and the lattice is quite weak, the initially prepared high field magnetization may be stable for sufficiently long times in low field that the spins may be manipulated and returned to high field without debilitating signal loss due to reequilibration. Such experiments were exploited in a number of different applications in the first decades of NMR.[1–3,19]

Under these conditions, the polarization and detection periods (where the signal intensity depends on $\boldsymbol{B}_0$) are separated from the period during which the field-free experiment is performed, as is illustrated in Figure 1. The experimental probe during the low field time period can be as simple as waiting for the spins to relax back toward equilibrium with the lattice in low field; this is the regime of relaxometry experiments. Spectroscopy can be probed by the application of fields, either CW or pulsed, that probe the energy levels of the low field spin systems. Reviews covering experimental aspects of field cycling NMR, and many of its applications, are available elsewhere.[20]

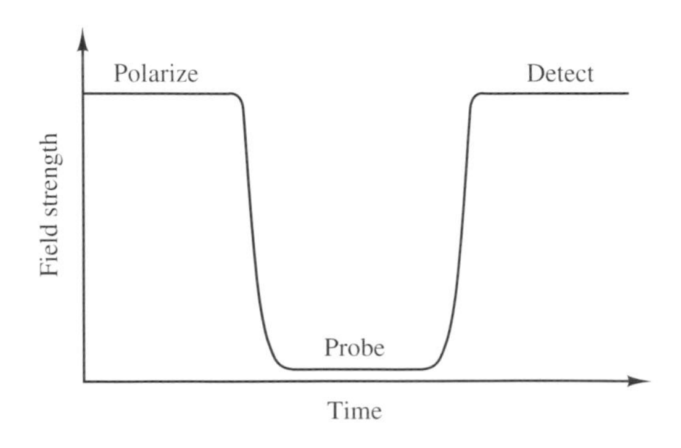

Figure 1 Generalized field cycling experiment. At the beginning of the field cycle, the sample sits in high field, so that an initially large spin polarization is prepared. In a time short compared with T_1, the field is reduced—either by draining current from an electromagnet or by removing the sample from the polarizing field. Features of the system in zero or low field are probed, and the sample returned to high field (often, but not necessarily, the same as the polarization field), where signal is detected

Two different experimental approaches can be taken to executing a field cycling NMR experiment. In the first, the field at the sample is reduced by varying the current applied to an electromagnet. An electronic field sweep between about 1 T and an arbitrarily small value can be performed in 1–10 ms. Alternatively, the field can be cycled by physically removing the sample into the fringe field of the polarizing magnet,[21] where the residual field can be nulled with relatively small electromagnets or suitable magnetic shielding.[22] This method appears more feasible, although somewhat slower (about 100 ms), when working in a superconducting magnet. When the field cycle cannot be safely completed in a short time period, varying the temperature may lengthen T_1.

4.3 Demagnetization Cycles

The motivation for probing $\hat{\mathcal{H}}_{\text{loc}}$ in zero applied field has been described above. Where no preferred direction can be identified because there is no externally imposed field, all orientations contribute to precisely the same sets of transition frequencies. Field cycling makes it possible to detect zero field transitions with the sensitivity of high field NMR. Precisely what is observed, however, depends on precisely how the field cycle is executed and how the spins are manipulated once they are in the low field regime.

The influence of demagnetization rate on the final state achieved can be illustrated by analogy to the results of more familiar but entirely analogous rotating frame demagnetization experiments executed in high field.[23] Imagine a 90° pulse transforming longitudinal Zeeman polarization into transverse magnetization. A phase-shifted rf field of strength much greater than typical internal fields will spin lock transverse magnetization so that its decay is associated with a time constant $T_{1\rho}$ and not the much shorter time constant associated with the disappearance of the free induction decay. Demagnetization in the laboratory frame is accomplished by removing the static magnetic field along which longitudinal magnetization is 'locked'.

When the locking field is removed, the response of the spin system depends critically on the rate at which that field is turned off, and in particular on the rate of transition between where the locking fields are larger than the local fields, to where the locking fields are smaller. Where the transition is either very slow (adiabatic) or very rapid (sudden), the state achieved at the end of the demagnetization can be cleanly evaluated. (Necessary conditions for transitions to qualify as 'adiabatic' or 'sudden' are available in most textbooks on quantum mechanics; a particularly lucid presentation is given by Messiah.[24])

Where the field is reduced adiabatically (rotating frame or longitudinal) Zeeman spin order is preserved as spin order at each point during the field cycle, and when the demagnetization is complete, the original spin magnetization is fully transformed into nonevolving dipolar- or quadrupolar-ordered states. Where the field is reduced suddenly, rotating frame Zeeman order is left unchanged—but the associated locking field is absent. The no-longer-locked magnetization evolves at the natural frequencies of the system. Remagnetization, adiabatically or suddenly, reverses these processes.

5 FREQUENCY VERSUS TIME DOMAIN

5.1 Frequency Domain Zero Field NMR

The adiabatically demagnetized state preserves all of the polarization initially prepared in high field, stored as spin order with respect to the eigenstates of $\mathcal{H}_{\text{loc}}$. Spin transitions in these nonevolving states can be probed directly by analogs to conventional absorption spectroscopy. Variable frequency irradiation will occasionally resonate with transitions of the spins in zero applied field. When the spin system is returned to high field, the saturation of the spins so effected in zero field is reflected by a decrease in the signal observed in high field, as illustrated in Figure 2.

Often, spin systems contain both rare, generally quadrupolar, nuclear spins S as well as more abundant high γ spin-$\frac{1}{2}$ nuclei. In such systems, substantially higher sensitivity is achieved if the 'interesting' S spins can be probed indirectly via polarization transfer to the easily observed abundant species.[23] As an example, assume the high γ nucleus is ^{1}H. In reasonably high magnetic fields, the ^{1}H Larmor frequency substantially exceeds the resonance frequency of all other spins. In low fields, where only the internal fields remain, the resonance frequencies of the ^{1}H spins are generally much lower than those associated with transitions in quadrupolar systems. At some intermediate field, the energy levels of the two spin systems (primarily determined by the Zeeman energies for the ^{1}H spins, and the quadrupolar energies for the S spins) must cross. When the level crossing field is traversed sufficiently slowly, dipole–dipole couplings transfer spin polarization from the more ordered ^{1}H spins to the less ordered S spins.

Once S spin order is destroyed via direct absorption in zero field, further ^{1}H spin order can be drained during the remagnetization process in the level crossing region. Zero field resonances of the S spins are observed as decreases in ^{1}H signal observed in high field. Where the S spins are rare, sensitivity can be substantially enhanced by repeated cycles through the level crossing region so that the contact between

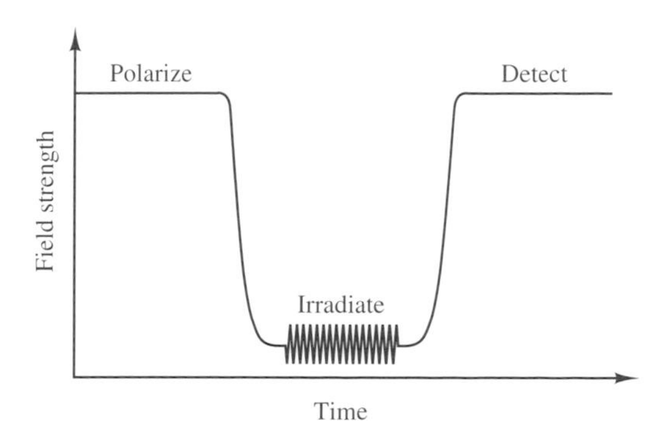

Figure 2 Frequency domain field cycling experiment. While the sample sits in zero applied field, it is irradiated with a low-frequency rf field. Where the frequency of the field matches a transition in the spin system, energy is absorbed until the transition is saturated. When the sample is returned to high field, this corresponds to a decrease in the detectable magnetization. A zero field spectrum is measured by repeating the experiment (polarization–demagnetization–irradiate–return to high field–probe the signal amplitude) with a continuously varied set of low field irradiation frequencies

For References see p. 5062

the S spins and the ^{1}H bath is established many times during a single high field polarization cycle.

This method has the substantial advantage of superior sensitivity to small numbers of S spins, and other modifications lead to yet greater signal-to-noise increases.[25,26] Its disadvantages are all those associated with absorption spectroscopy, namely, the trade-off between linewidths and sensitivity, decreasing sensitivity at low frequencies, and the inability to establish correlations of the sort common in high field multidimensional NMR.

5.2 Time Domain Zero Field NMR

An alternate approach that yields high resolution at very low frequencies is to perform the experiment entirely in the time domain. The advantages of time domain experiments are comparable to the advantages of pulsed high field NMR as compared with the older continuous wave techniques. Amongst the most important is that the resolution and sensitivity are uncoupled, so that linewidths are limited in the time domain version to only those naturally associated with the experiment. Furthermore, the time domain experiment provides the possibility of performing correlation experiments of the sort that are routinely executed in high field, high-resolution NMR. These correlations may be between low field and high field parameters, or between low field and low field parameters. Furthermore, with some additional technical and theoretical effort, spin Hamiltonians can be selectively averaged in low as well as high field.

As in frequency domain zero field NMR experiments, the sample is polarized in high field, demagnetized, and once the zero field frequencies have been probed, returned to high field. As in time domain high field NMR, a time domain zero field NMR experiment begins when a field pulse disturbs a previously nonevolving spin system so as to establish coherence between eigenstates.

In zero field, these pulses can take a variety of forms—for instance, the sudden field transient described above and shown in Figure 3(a),[27] a broadband dc pulse at the nominal Larmor frequency of zero as shown in Figure 3(b),[28,29] or a resonant rf pulse tuned to one or more of the expected transition frequencies as shown in Figure 3(c).[22,30] The evolution period t_1 can be terminated at some later time in like fashion, and, as in multidimensional high field NMR experiments, the evolved signal is sampled indirectly once the the sample has been returned to high field. A complete zero field spectrum is evaluated by repeating the experiment for continuously incremented values of t_1, followed by Fourier transformation so as to reveal the zero field frequencies.

Only in a limited number of cases can an exact calculation of the expected signal be provided. The primary theoretical stumbling block is the difficulty of obtaining a detailed understanding of $\hat{\rho}$ at the end of an adiabatic field transition. Where the zero field spectrum is continuous, as is often the case for dense networks of spin-$\frac{1}{2}$ nuclei, the spin-temperature hypothesis may apply,[31] so that the demagnetized state satisfies $\hat{\rho}(0) \propto \hat{\mathcal{H}}_{\mathrm{D}}$. Where the zero field spectrum is discrete so that only a small number of well-separated lines are found, no similarly simple theory of the demagnetized state exists. An analysis of isolated spin-1 nuclei has suggested that the demagnetized state is nearly orientation-independent,[32] though

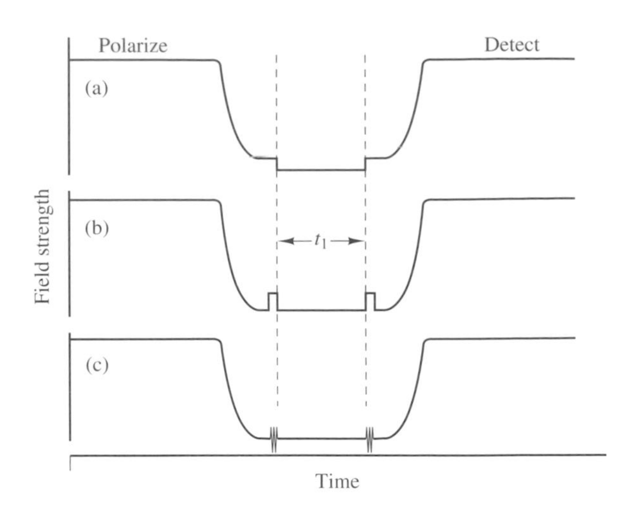

Figure 3 Time domain field cycling experiments. Polarization and detection occur in high magnetic fields. Low field evolution occurs during the time period t_1. A zero field interferogram is mapped out by repeating the experiment for regularly spaced increments in t_1. (a) Field cycling with sudden transients. The sample is partially demagnetized to an intermediate field strength where $\hat{\mathcal{H}}_{\mathrm{L}} >> \hat{\mathcal{H}}_{\mathrm{loc}}$. Typical values of the intermediate field are 0.01–0.05 T. Fields of this magnitude can be turned off and on in 1–5 μs. The evolution period t_1 is initiated when the intermediate field is rapidly turned off, and ends when it is suddenly turned back on. In high field, the amplitude of the evolved ^{1}H magnetization is probed following a 90° pulse. (b) Field cycling with zero frequency pulses. The sample is demagnetized to zero field. Evolution is initiated and terminated with a short, strong zero frequency field pulse. (c) Field cycling with low-frequency pulses. The sample is demagnetized to zero field. Evolution is initiated and terminated with a short, low-frequency field pulse tuned to one or more of the spin system transition frequencies

dipole–dipole couplings to other nuclei may substantially complicate the situation. A numerical simulation is certainly possible, but accurate simulations require complete knowledge of all the parameters of all quadrupolar and dipolar Hamiltonians in the system—presumably the goal of the experimental effort. Furthermore, the results of such simulations will generally be orientation-dependent.

Theory can provide substantial guidance where evolution is initiated and terminated by field transients that take the spin system from high field to zero field and back 'suddenly', so that both the prepared and detected states are well defined. While the conditions for applicability of the sudden approximation are impossible to satisfy for typical polarizing fields of several tesla, practical experiments need only switch rapidly fields sufficiently large that $\hat{\mathcal{H}}_{\mathrm{L}} >> \hat{\mathcal{H}}_{\mathrm{loc}}$. A practical scheme is illustrated in Figure 3(a). Under these conditions, a quantitative comparison between experiment and theory can be attempted.

In high magnetic field, the total system Hamiltonian is

$$\hat{\mathcal{H}}_{\mathrm{tot}} = \hat{\mathcal{H}}_{\mathrm{L}} + \hat{\mathcal{H}}'_{\mathrm{loc}} \tag{10}$$

where $\hat{\mathcal{H}}'_{\mathrm{loc}}$ represents whatever orientation-dependent local interactions exist, truncated as is appropriate for observation in high field. Any long-lived state prepared in high field must satisfy

$$[\hat{\rho}, \hat{\mathcal{H}}_{\rm tot}] = [\hat{\rho}, \hat{\mathcal{H}}_{\rm L}] = [\hat{\rho}, \hat{\mathcal{H}}'_{\rm loc}] = 0 \tag{11}$$

Equilibrium, which is the most easily prepared state satisfying equation (11), corresponds to $\hat{\rho}(0) \propto \hat{I}_z$.

In zero field, the total Hamiltonian is $\hat{\mathcal{H}}_{\rm loc}$. Assuming the field is turned off sufficiently rapidly, $\hat{\rho}$ is unchanged during the field transient. Other than in exceptional cases, $[\hat{\rho}(0),\hat{\mathcal{H}}_{\rm loc}] \neq 0$, and $\hat{\rho}$ evolves in time such that $\hat{\rho}(t) = \exp(-{\rm i}\hat{\mathcal{H}}_{\rm loc}t)\,\hat{\rho}(0)\exp({\rm i}\hat{\mathcal{H}}_{\rm loc}t)$. At some arbitrary time t_1 later, evolution under $\hat{\mathcal{H}}_{\rm loc}$ can be interrupted by reapplying the field. Within a short time, the evolved density matrix again satisfies $[\hat{\rho}(t_1), \hat{\mathcal{H}}_{\rm tot}] = 0$.

In high field, a single rf pulse converts only longitudinal magnetization back to observable transverse magnetization, and, under most circumstances, only that portion of $\hat{\rho}(t_1)$ proportional to $\hat{I}_z$ is observable. The detectable signal is therefore

$$\begin{aligned} S(t_1) &= {\rm Tr}[\hat{\rho}(t_1)\hat{I}_z] \\ &= {\rm Tr}[\exp({\rm i}\hat{\mathcal{H}}_{\rm loc}t_1)\hat{\rho}(0)\exp(-{\rm i}\hat{\mathcal{H}}_{\rm loc}t_1)\hat{I}_z] \end{aligned} \tag{12}$$

A model calculation illustrates many of the features of spectral simulations in time domain zero field NMR initiated with sudden transients,[27,32,33] based on an evaluation of equation (12). We assume longitudinal magnetization is prepared and, ultimately, detected. In a laboratory frame of reference, $\hat{\rho}(0) = \hat{I}_{z,{\rm lab}}$, where, for simplicity, the proportionality constant has been arbitrarily set to one. The same density operator represents the system after the sudden field transient. While $\hat{\rho}(0)$ is most easily defined in the same laboratory frame, $\hat{\mathcal{H}}_{\rm loc}$ is most easily represented in a consistently chosen local frame of reference. Computational efficiency demands that all calculations be carried out in the local frame where $\hat{\mathcal{H}}_{\rm loc}$ is orientation-independent, so that $\hat{\rho}(0)$ must itself be transformed from its natural representation in the laboratory frame into the local frame. This precisely inverts the common 'high field' transformation through Euler angles θ and ϕ that takes $\hat{\mathcal{H}}_{\rm loc}$ from its natural frame of reference into the laboratory frame. Rotating via $-\phi$ and $-\theta$,

$$\begin{aligned} \hat{\rho}_{\rm loc}(0,\theta,\phi) &= \hat{R}(-\phi,-\theta,0)\hat{I}_{z,{\rm lab}}\hat{R}^{-1}(-\phi,-\theta,0) \\ &= \exp({\rm i}\phi\hat{I}_{z,{\rm lab}})\exp({\rm i}\theta\hat{I}_{y,{\rm lab}})\hat{I}_{z,{\rm lab}} \\ &\quad \times \exp(-{\rm i}\theta\hat{I}_{y,{\rm lab}})\exp(-{\rm i}\phi\hat{I}_{z,{\rm lab}}) \\ &= -\hat{I}_{x,{\rm loc}}\sin\theta\cos\phi + \hat{I}_{y,{\rm loc}}\sin\theta\sin\phi \\ &\quad + \hat{I}_{z,{\rm loc}}\cos\theta \end{aligned} \tag{13}$$

Any laboratory frame operator that might be prepared is transformed, in analogy to equation (13), via arbitrary angles θ and ϕ into all local frame operators of similar tensor rank.

After a period of free evolution, using the shorthand notation $\hat{I}_{j,{\rm loc}}(t_1) \equiv \exp({\rm i}\hat{\mathcal{H}}_{\rm loc}t_1)\,\hat{I}_{j,{\rm loc}}\exp(-{\rm i}\hat{\mathcal{H}}_{\rm loc}t_1)$,

$$\begin{aligned} \hat{\rho}(t_1,\theta,\phi) = &-\hat{I}_{x,{\rm loc}}(t_1)\sin\theta\cos\phi \\ &+ \hat{I}_{y,{\rm loc}}(t_1)\sin\theta\sin\phi + \hat{I}_{z,{\rm loc}}(t_1)\cos\theta \end{aligned} \tag{14}$$

and the observable signal is

$$\begin{aligned} S(t_1,\theta,\phi) = {\rm Tr}\{&[-\hat{I}_{x,{\rm loc}}(t_1)\sin\theta\cos\phi \\ &+ \hat{I}_{y,{\rm loc}}(t_1)\sin\theta\sin\phi + \hat{I}_{z,{\rm loc}}(t_1)\cos\theta] \\ &\times(-\hat{I}_{x,{\rm loc}}\sin\theta\cos\phi \\ &+ \hat{I}_{y,{\rm loc}}\sin\theta\sin\phi + \hat{I}_{z,{\rm loc}}\cos\theta)\} \end{aligned} \tag{15}$$

A complete evaluation of equations (14) and (15) requires a complete specification of the Hamiltonian governing evolution, as well as θ and ϕ. However, some general statements that substantially simplify calculations are possible.

Application of zero field NMR is largely restricted to disordered systems, so the observable signal is

$$S(t_1) = \iint S(t_1,\theta,\phi)P(\theta,\phi)\,{\rm d}\theta\,{\rm d}\phi \tag{16}$$

In the common case where all orientations are equally probable (i.e., a polycrystalline powder), the distribution function of equation (16) is $P(\theta,\phi)\,{\rm d}\theta\,{\rm d}\phi = -(4\pi)^{-1}\,{\rm d}(\cos\theta)\,{\rm d}\phi$. Owing to the orthogonality relations between geometric factors (e.g., $\iint\cos\theta\sin\theta\cos\phi\,{\rm d}(\cos\theta)\,{\rm d}\phi = 0$), only autocorrelations of the angular momentum functions yield observable signals, and, to within a proportionality constant,

$$S(t_1) = \tfrac{1}{3}{\rm Tr}[\hat{I}_{x,{\rm loc}}(t_1)\hat{I}_{x,{\rm loc}} + \hat{I}_{y,{\rm loc}}(t_1)\hat{I}_{y,{\rm loc}} + \hat{I}_{z,{\rm loc}}(t_1)\hat{I}_{z,{\rm loc}}] \tag{17}$$

or, in matrix form,

$$S(t_1) = \frac{1}{3}\sum_{l,k}\sum_{p=x,y,z}|\langle l|\hat{I}_{p,{\rm loc}}|k\rangle|^2\cos\omega_{kl}t_1 \tag{18}$$

where $|k\rangle$ and $|l\rangle$ are eigenstates of $\hat{\mathcal{H}}_{\rm loc}$ and ω_{kl} is the frequency difference between the eigenstates. As in all autocorrelations, equations (17) and (18) predict that $S(t_1)$ is a maximum at zero time, and all lines appear in phase. Furthermore, no quadrature information is available.

For spin 1, the fictitious spin-$\frac{1}{2}$ formalism[34] provides a particularly compact and elegant framework. In zero field and where $\eta \neq 0$, the three nondegenerate eigenstates are conventionally labeled $|x\rangle$, $|y\rangle$, and $|z\rangle$. Each of the operators $\hat{I}_j$ couples one of the three pairs of levels so that

$$\hat{I}_j(t) = \hat{I}_j\cos\omega_{kl}t - (\hat{I}_k\hat{I}_l + \hat{I}_l\hat{I}_k)\sin\omega_{kl}t \tag{19}$$

Equation (19) states that the initially prepared dipolar magnetization $\hat{I}_j$ oscillates between itself and a quadrupolar operator $\hat{I}_k\hat{I}_l + \hat{I}_l\hat{I}_k$ at a rate proportional to the quadrupolar transition frequencies $\omega_{xy} = \eta\pi\chi$, $\omega_{yz} = -\frac{1}{2}(3 + \eta)\pi\chi$ and $\omega_{zx} = \frac{1}{2}(3 - \eta)\pi\chi$. Owing to the orthogonality of the dipolar and quadrupolar operators [i.e., ${\rm Tr}(\hat{I}_j\hat{I}_k\hat{I}_l) = 0$], the signal at the end of the time period t_1 is proportional to

$$\begin{aligned} S(t_1,\theta,\phi) = &\cos[\tfrac{1}{2}(3+\eta)\pi\chi t_1]\sin^2\theta\cos^2\phi \\ &+ \cos[\tfrac{1}{2}(3-\eta)\pi\chi t_1]\sin^2\theta\sin^2\phi \\ &+ \cos(\eta\pi\chi t_1)\cos^2\theta \end{aligned} \tag{20}$$

Equation (20) predicts that for any orientation of a spin-1 system in high field, one, two, or as many as three transitions might be observed, with varying intensities but well-defined frequencies simply related to the quadrupolar parameters χ and η. Differently oriented but otherwise equivalent systems have all spectral lines at identical zero field frequencies. Only the

For References see p. 5062

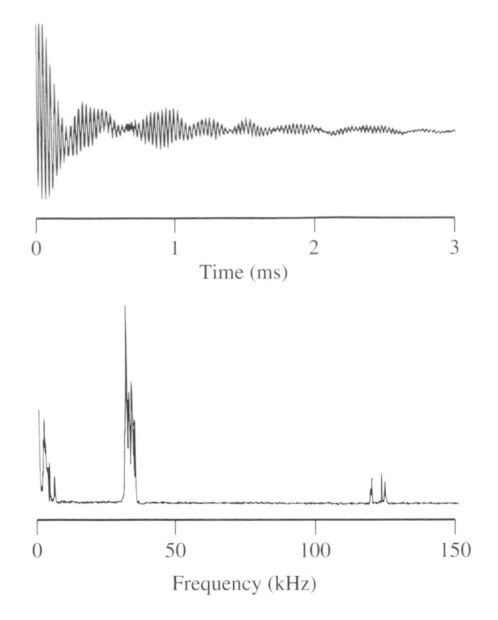

Figure 4 Zero field interferogram and associated zero field spectrum of perdeuterated 1,4-dimethoxybenzene, acquired with the field cycling sequence of Figure 3(a). The zero field spectrum reveals two inequivalent ring sites (with two lines each between 130 and 138 kHz, and one additional line each between 3 and 6 kHz) and two complex patterns (approx. 33–38 kHz and 1–3 kHz) associated with the dipole–dipole couplings within the $-C^2H_3$ groups. Intensity variations between the differing ^2H sites are due to vast differences in high field T_1 values. Reproduced by permission of D. B. Zax, A. Bielecki, K. W. Zilm, A. Pines, and D. P. Weitekamp, *J. Chem. Phys.*, 1985, **83**, 4877, ©American Institute of Physics, 1985

line intensities vary with orientation. In the common case where all orientations are equally probable, $\langle \sin^2\theta\cos^2\phi\rangle = \langle \sin^2\theta\sin^2\phi\rangle = \langle\cos^2\theta\rangle = \frac{1}{3}$, so that

$$S(t_1) = \tfrac{1}{3}\{\cos[\tfrac{1}{2}(3+\eta)\pi\chi t_1] + \cos[\tfrac{1}{2}(3-\eta)\pi\chi t_1] + \cos(\eta\pi\chi t_1)\} \tag{21}$$

An experimental verification of the predictions of equation (21) is provided in Figure 4, showing the zero field interferogram and spectrum of perdeuterated dimethoxybenzene. What is perhaps most striking in this spectrum is that at each of the ^2H sites (two inequivalent sites on the ring and three equivalent sites on the $-C^2H_3$ group) η is substantial ($\approx$ 0.05 or larger), and is greatest at the methyl group sites. Differences in intensities between different chemical sites reflect the substantial differences in high field T_1 values, which can vary by several orders of magnitude. Intensity variations amongst the three lines associated with a single ^2H environment are much smaller, since the T_1 anisotropies associated with a given environment are substantially less.

In homonuclear spin pairs, transitions between the singlet and triplet states for the nuclear spins are forbidden if $\hat{\rho}(0) = \hat{I}_{z,\mathrm{lab}}$. The three-level triplet system is formally identical to the three-level system of spin 1. Owing to the symmetry of the static dipole–dipole tensor, only a single nonzero evolution frequency ($\frac{3}{2}\omega_D$) is expected. Spectra of isolated water molecules, as are found in the crystalline forms of a number of inorganic salts, show the structure predicted from this two-spin model, and examples are shown in Figure 5. While the high field Pake doublet may be obscured by longer range dipole–dipole couplings and/or chemical shift anisotropies, the zero field spectrum is much better resolved, and clearly shows the dipole–dipole coupling frequency.

In many systems, dipolar resolution is limited because coupling networks are not well isolated from one another. More information can often be derived using isotopic dilution (e.g., replacing many of the ^1H nuclear spins with ^2H), so as to reduce longer range dipole–dipole couplings.[27,35] In the ^1H zero field spectrum of $Ba(ClO_3)_2\cdot H_2O$ (see Figure 5), the second moment of the line at $\frac{3}{2}\omega_D$ substantially exceeds calculated values[34] based on the known crystal structure. Figure 6 shows the zero field spectra as ^1H spins are successively replaced by increasing numbers of ^2H spins. At low ^1H pair densities, the dipolar spectrum shows sufficient resolution to reveal that the single line was actually a pair of lines due to dynamical averaging of the dipole–dipole coupling. The splitting between the two zero field frequencies is consistent with an effective asymmetry parameter $\eta \approx 0.05$.

Where local dynamics cause $\hat{\mathcal{H}}_{\mathrm{loc}}$ to be time-dependent, calculations of the zero field lineshapes are more involved. Dynamical averaging in zero field differs fundamentally from the same spectroscopic problem in high field. In pure dipole–dipole or quadrupolar spectroscopy, $\hat{\mathcal{H}}_{\mathrm{loc}}(t)$ is the main Hamiltonian rather than a perturbation. Furthermore, the averaging makes the orientation of $\boldsymbol{D}$ or $\boldsymbol{q}$ time-dependent, rather than its magnitude.[36–39]

5.2.1 *Selection Rules?*

What are the allowed nuclear spin transitions in time-domain zero field NMR? Equation (18) suggests that an equivalent question would be to ask which eigenstates are coupled by the initially prepared density operator. As in much of the theory of zero field NMR, somewhat different answers apply to half-integer quadrupolar nuclear spin states and integer spins (or, typically, collections of dipolar-coupled spin $\frac{1}{2}$ nuclei).

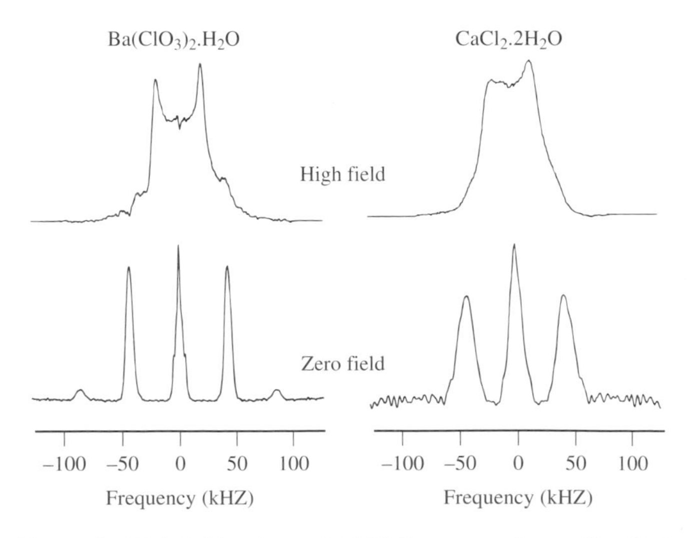

Figure 5 High field and zero field NMR spectra of crystalline H_2O molecules, in $CaCl_2\cdot 2H_2O$ and $Ba(ClO_3)_2\cdot H_2O$. In each, a three-line dipolar spectrum is observed. The high-frequency lines evolve at $\pm\frac{3}{2}\omega_D$; one-third of the total spin magnetization prepared in high field is nonevolving in a two-level system. Reproduced by permission of A. Bielecki, ©A. Bielecki, 1987

For half-integer quadrupolar nuclear spins, pairs of energy levels are always degenerate, and $\hat{I}_{z,\mathrm{loc}}$ is nearly a good quantum number. Under conditions where evolution is initiated by a sudden field transient, a portion of the initial magnetization is therefore nonevolving. Projections of $\hat{\rho}(0)$ corresponding to $\hat{I}_{x,\mathrm{loc}}$ and $\hat{I}_{y,\mathrm{loc}}$ couple transitions where $\Delta m = \pm 1$, and a series of $I - \frac{1}{2}$ zero field transitions will be observed. For nonzero η, additional transitions may be weakly allowed.[40]

Substantially different selection rules may be observed where field pulses generate and store coherence after adiabatic demagnetization. The demagnetized state satisfies $[\hat{\rho}, \hat{\mathcal{H}}_Q] = 0$. For half-integer quadrupolar spin systems, each of the operators that commute with $\hat{\mathcal{H}}_Q$ is proportional to $\hat{\mathcal{H}}_Q{}^n$, and $n_{\mathrm{max}} = I - \frac{1}{2}$. After a pulse initiates evolution, $\hat{\rho}$ contains operators that couple states with $\Delta m_{\mathrm{max}} = \pm(2I - 1)$, so new lines may appear at the sums of previously observed transition frequencies. These multiple quantum lines may prove helpful in making assignments in multisite systems.

Except in exceptional cases, no good quantum numbers are associated with $\hat{\mathcal{H}}_D$, or with $\hat{\mathcal{H}}_Q$ for integer-spin nuclei. (An exception is found where rapid motion in the molecular frame establishes a new quantization axis, as in rotating methyl groups.) Where dipole–dipole couplings dominate, essentially all eigenstates are coupled to all others; effectively, there are no selection rules. For N coupled spins, the number of transitions observable in zero field exceeds that found in a similarly sized spin system of specific orientation in high field. (Nevertheless, the absence of powder broadening generally ensures that the low field spectrum exhibits more structured features than the corresponding high field powder pattern.) Particularly when $N \leqslant 4$, experience shows that there is sufficient structure that a detailed analysis might be attempted.[27,41]

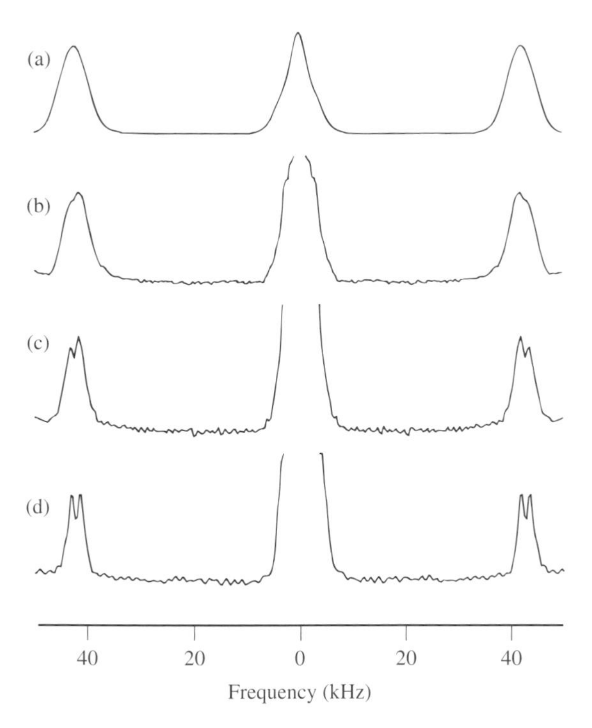

Figure 6 Zero field NMR spectra of progressively deuterated $Ba(ClO_3)_2 \cdot H_2O$. (a) ^1H zero field NMR spectrum in $Ba(ClO_3)_2 \cdot H_2O$, 100% ^1H. (b) ^1H zero field NMR spectrum in $Ba(ClO_3)_2 \cdot H_2O/^1H^2HO/^2H_2O$, 60% ^1H. (c) ^1H zero field NMR spectrum in $Ba(ClO_3)_2 \cdot H_2O/^1H^2HO/^2H_2O$, 31% ^1H. (d) ^1H zero field NMR spectrum in $Ba(ClO_3)_2 \cdot H_2O/^1H^2HO/^2H_2O$, 10% ^1H. Progressive deuteration reduces the numbers of substantial dipole–dipole couplings. Pairs of ^1H nuclei in an ^1H_2O water molecule evolve primarily at the frequency of the intramolecular dipole–dipole coupling of approximately ±42 kHz; ^1H nuclei in $^1H^2HO$ molecules evolve at very low frequencies associated with the weaker intermolecular couplings (and which overlap with the nonevolving magnetization from the spin pairs). As the magnitude and number of intramolecular couplings are decreased by replacement of ^1H by ^2H, sufficient additional resolution is achieved so as to observe a small motionally induced asymmetry parameter in the dipole–dipole tensor. Reproduced by permission of J. M. Millar, A. M. Thayer, D. B. Zax, and A. Pines, *J. Am. Chem. Soc.*, 1987, **91**, 2240, ©American Chemical Society, 1987

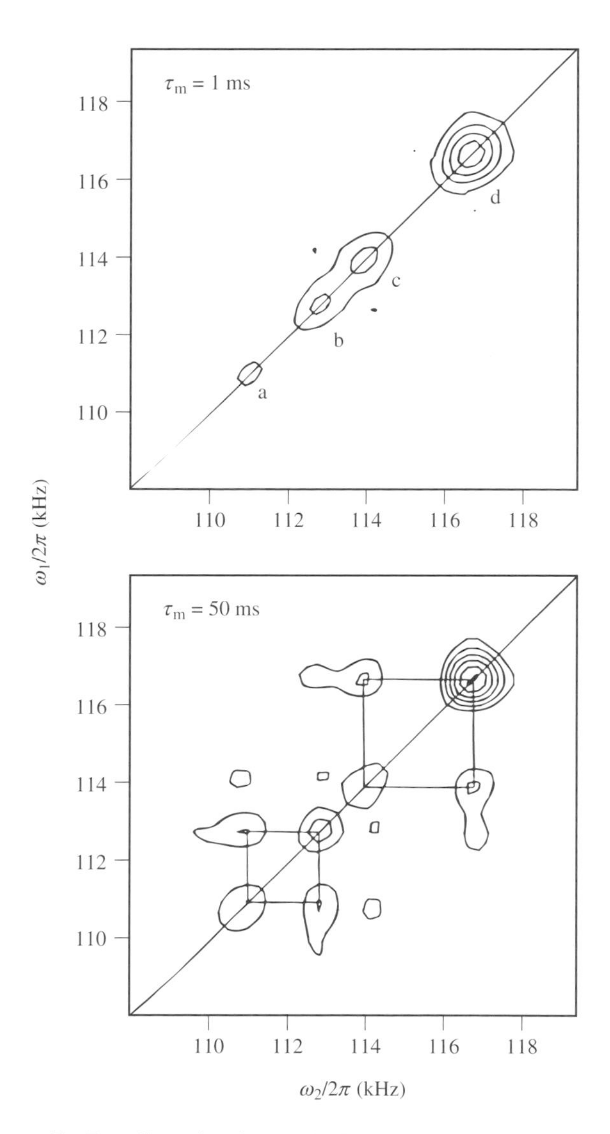

Figure 7 Two-dimensional zero field NMR experiment in diethyl terephthalate-d_4 (deuterated only at the methylene sites). The two time periods t_1 and t_2 are generated by short field pulses, in analogy to the one-dimensional experiment shown in Figure 3(b). The intervening mixing period τ_m allows for spin diffusion between deuterons. As can be seen, spin diffusion occurs on a timescale of a few milliseconds. Reproduced by permission of D. Suter, T. P. Jarvie, B. Sun, and A. Pines, *Phys. Rev. Lett.*, 1987, **59**, 106, ©American Physical Society, 1987

For References see p. 5062

For $N >> 4$, and where the dipolar fields are stationary, Gaussian and Markovian, the signal associated with $\hat{\mathcal{H}}_D$ is predicted to approach

$$S(t_1) = \tfrac{1}{3}[1 + 2(1 - \Delta^2 t_1^2)\exp(-\tfrac{1}{2}\Delta^2 t_1^2)] \tag{22}$$

where Δ^2 is a mean square measure of the magnitude of local dipolar fields.[36] Alternate initial conditions are unlikely to be of substantial use in dipolar-coupled systems, since successful spectral analyses for $N \geqslant 3$ require that both frequencies and intensities be matched. Initial and final conditions derived from adiabatic de/remagnetization cycles are too poorly characterized to yield predictable line intensities. In analyzing quadrupolar spin systems, a knowledge of transition frequencies only is generally sufficient, and adiabatic cycles share the sensitivity advantages of polarization transfer and/or of indirect detection via abundant, high γ nuclei,[28] as in zero field NQR with level crossings.

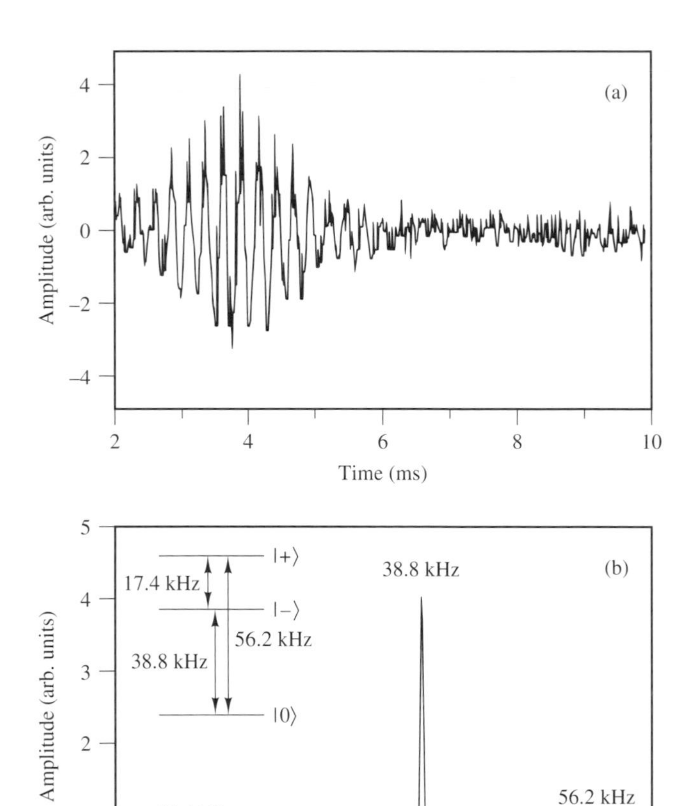

Figure 8 SQUID-detected zero field NQR in NH_4ClO_4 at 1.5 K. (a) Directly detected free induction decay and echo after two-pulse spin echo sequence (for purposes of display, demodulated with 35 kHz). (b) Fourier transform spectrum and energy levels in the three-level system corresponding to ^{14}N (I = 1). Reproduced by permission of N. Q. Fan and J. Clarke, *Rev. Sci. Instrum.*, 1991, **62**, 1453, ©American Institute of Physics, 1991

5.2.2 *Multipulse NMR Experiments in Zero Field*

A distinct advantage of experiments executed in the time domain is the possibility that the information content of the spectrum can be enhanced under the control of the experimentalist. In high field, rf pulses and/or sample rotation are the available tools. Qualitatively, it is a very different technical and theoretical problem to generate useful pulse sequences in zero field. Rapid sample rotation reintroduces powder broadening and is of limited value. In contrast to experience based on high field experiments, zero frequency pulses of different phases must be generated in physically orthogonal coils. As in traditional NQR experiments, pulse flip angles are only poorly defined, since the same pulse affects differently oriented molecular systems differently. Where the pulses are applied at nonzero frequencies, they must be carefully timed to maintain phase coherence between the spins and the rf field.[22,30]

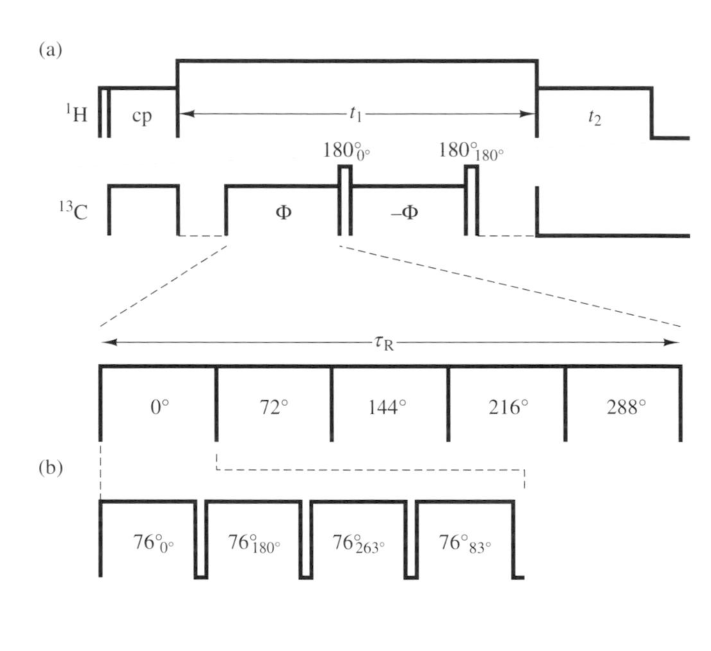

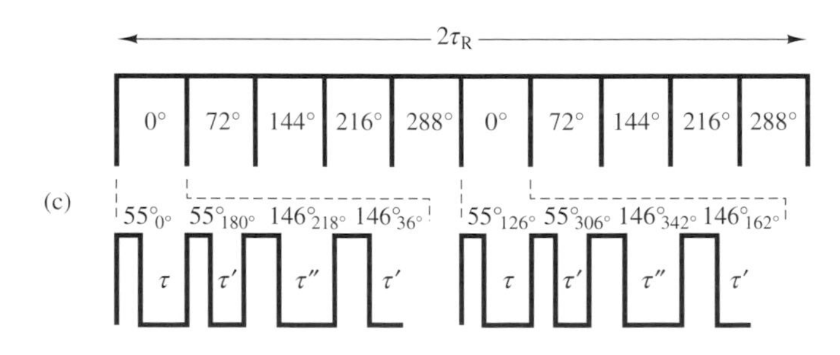

Figure 9 Two-dimensional pulse and rotation sequences for measuring zero-field-like spectra in high magnetic fields. Decoupling may be applied to abundant spins (i.e., ^1H) so as effectively to isolate small numbers of rare spin pairs. In sequence (b), the sample rotation axis is tilted by 64° with respect to the externally applied field; in sequence (c), the sample rotation axis is tilted by 66° with respect to the externally applied field. Angles such as 0°, 72°, etc. represent the relative phase of irradiation block, while θ_ϕ indicates a pulse with flip angle of θ and phase ϕ. Reproduced by permission of R. Tycko, G. Dabbagh, J. C. Duchamp, and K. W. Zilm, *J. Magn. Reson.*, 1990, **89**, 205, ©Academic Press, 1990

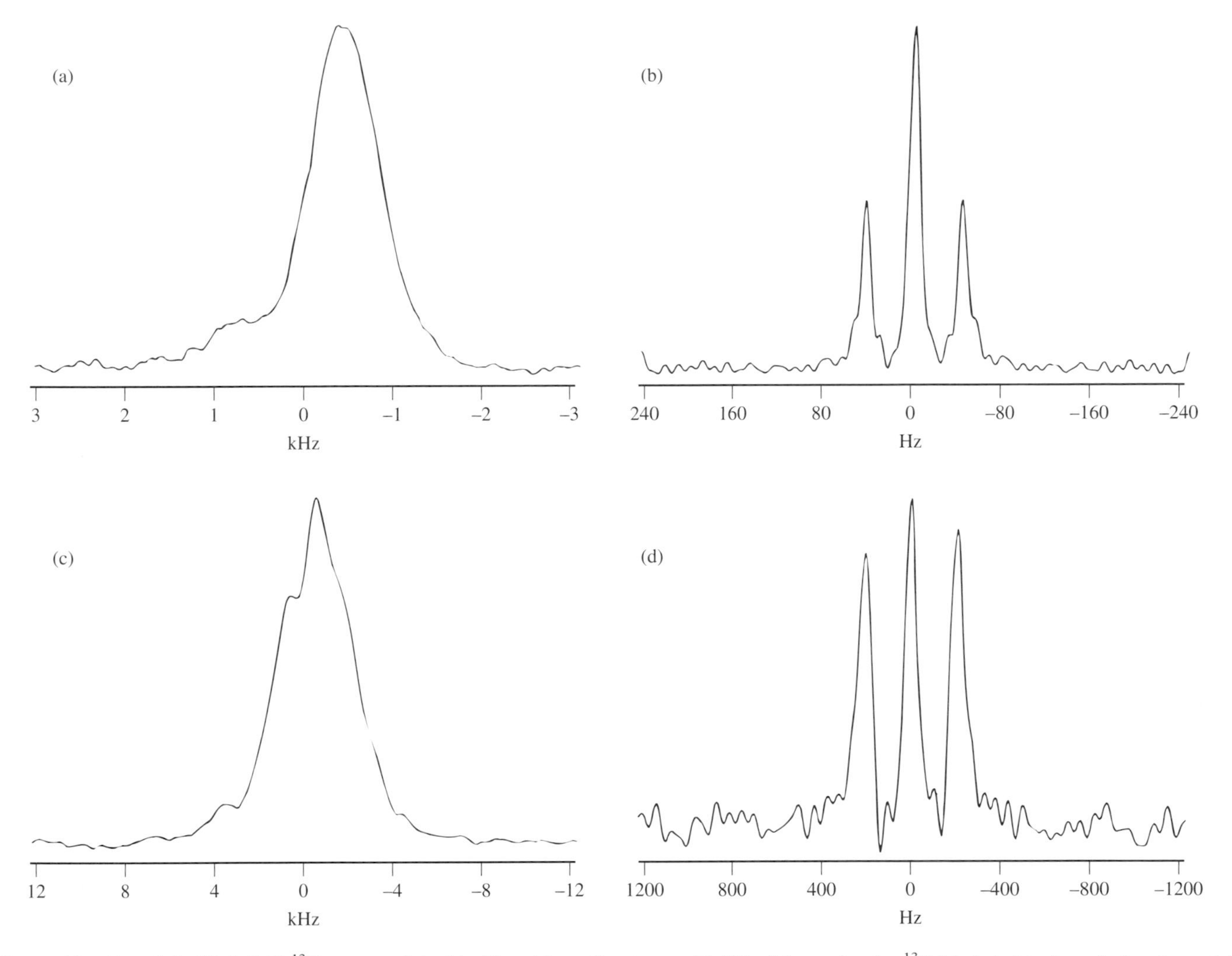

Figure 10 (a) and (b) High field ^{13}C spectra of the bisulfite adduct of acetone, with 5% of the molecules ^{13}C-labeled at both methyl carbons. (c) and (d) High field ^{13}C spectra of diammonium succinate, with 13% of the molecules labeled at both methylene carbons. (a) and (c) represent the normal high field powder pattern spectra, while (b) and (d) represent the spectra achieved using the zero field in high field sequences of Figures 9(c) and (a), respectively. Reproduced by permission of R. Tycko, G. Dabbagh, J. C. Duchamp, and K. W. Zilm, *J. Magn. Reson.*, 1990, **89**, 205, ©Academic Press, 1990

Despite these complicating factors, substantial effort has been devoted to experiments using pulsed dc fields in zero field,[28,29] including composite pulses selective for specific nuclear spins[42] and multiple pulse NMR experiments in zero field.[43,44] Multidimensional zero field NMR experiments have demonstrated how correlations between multiple lines associated with a single site might be identified,[45] and how small dipole–dipole couplings between quadrupolar nuclear spins might be used to establish through-space connectivities in crystals.[46] An example of a two-dimensional zero field correlation experiment is shown in Figure 7. Two-dimensional zero field experiments with field cycling are, of course, three-dimensional NMR experiments, since the signal $S(t_1, t_2)$ can only be detected in a third, high field dimension (not shown in Figure 7).

6 OTHER APPROACHES TO ZERO FIELD RESULTS

Two promising new approaches to achieving high-resolution measurements of $\hat{\mathcal{H}}_{loc}$ have been demonstrated, and each has appealing advantages over the field cycling methods described above. The first relies on a new technical accomplishment that makes direct low-frequency detection with reasonable sensi-

For References see p. 5062

tivity feasible. The second relies, instead, on a tremendously original application of traditional multiple pulse NMR and average Hamiltonian theory working together so as to probe zero-field-like coupling patterns directly in high field.

6.1 Zero Field NMR with SQUID Detection

Superconducting Quantum Interference Devices (SQUIDs) are based on Josephson junction technology, and have long been an area of interest to low temperature physicists. Applications of SQUIDs have focused on their sensitivity to small amounts of magnetic flux. In low-frequency magnetic resonance, the advantage of a SQUID-based detector is that its output is proportional to the magnetic flux enclosed, and therefore directly monitors the nuclear spin magnetization (rather than its time derivative.) This removes one of the frequency-dependent factors in equation (1), and it is possible to monitor directly such slow processes as T_1-driven relaxation of longitudinal magnetization to equilibrium.[47]

Within the bandwidth limits of the associated circuitry, a SQUID detector is broadband, so that sensitivity is independent of frequency. Furthermore, the noise figure of a SQUID detector is substantially below that of typical room-temperature amplifiers. As a result, direct detection of zero field time domain signals is possible at arbitrarily low frequencies (see Figure 8), and an entire zero field free induction decay can be accumulated in a single experiment.[48] The ability continuously to detect evolving magnetization in real time is a substantial advance over point-by-point field cycling schemes.

6.2 Zero Field NMR in High Magnetic Fields

A remarkable experiment addressing the problem of 'zero field NMR in high magnetic field', has demonstrated that measurements of dipole–dipole couplings might be made with neither the complications associated with powder pattern lineshapes nor the technical difficulties associated with cycling the field strength at the sample. Instead, multiple pulse NMR combined with rapid sample spinning may be sufficient. While the frequencies found in the high field spectrum are orientation-dependent owing to the anisotropy of the spatial parameters in $\hat{\mathcal{H}}_{\text{loc}}$, where both spatial and spin parameters are time-dependent, a portion of the time-dependent Hamiltonian, which depends on the product of two anisotropic terms, is isotropic. Selective averaging based on average Hamiltonian theory can be used to predict which sequences, acting on both spaces, average away all orders of anisotropy so as to leave behind only the desired orientation-independent information. Neither bulk sample rotation nor rf pulses alone can accomplish this. Instead, a clever combination of sample spinning correlated with multiple-pulse NMR and extensive phase cycling is required, as shown in Figure 9.[49] An example of a purely high field 'zero field spectrum' is shown in Figure 10. For two-spin systems, the same three-line spectrum is observed in high field NMR and in zero field NMR (see Figure 5), although in the high field version the splittings are scaled by a calculable scaling factor. While demonstrations have been limited to pairs of coupled spins, this is less of a limitation in high field than in low field, since spin decoupling can be applied where it is necessary to isolate experimentally specifically labeled spin pairs.

7 CONCLUSIONS

The realm of zero field NMR is the easy identification of structural features based on high resolution spectroscopy of anisotropic, traceless spin interactions. While amongst the oldest of NMR techniques, new technical, and theoretical advances have guaranteed that it is also amongst its youngest.

8 RELATED ARTICLES

Field Cycling Experiments; Internal Spin Interactions & Rotations in Solids; Line Narrowing Methods in Solids; Low Spin Temperature NMR; Molecular Motions: T_1 Frequency Dispersion in Biological Systems; Quadrupolar Nuclei in Solids; Quantum Tunneling Spectroscopy; Relaxometry of Tissue; SQUIDs; Terrestrial Magnetic Field NMR.

9 REFERENCES

1. F. Reif and E. M. Purcell, *Phys. Rev.*, 1953, **91**, 631.
2. L. C. Hebel and C. P. Slichter, *Phys. Rev.*, 1957, **107**, 901; 1959, **113**, 1504.
3. A. G. Redfield, *Phys. Rev. Lett.*, 1959, **3**, 85.
4. M. Mehring, 'Principles of High Resolution NMR in Solids', 2nd edn., Springer-Verlag, Berlin, 1983.
5. J. H. Baltisberger, S. L. Gann, E. W. Wooten, T. H. Chang, K. T. Mueller, and A. Pines, *J. Am. Chem. Soc.*, 1992, **114**, 7489.
6. N. Bloembergen and T. J. Rowland, *Acta Metall.*, 1953, **1**, 731.
7. M. Packard and R. Varian, *Phys. Rev.*, 1954, **93**, 941.
8. D. F. Elliott and R. T. Schumacher, *J. Chem. Phys.*, 1957, **26**, 1350.
9. R. Kimmich and H. W. Weber, *J. Chem. Phys.*, 1993, **98**, 5847.
10. D. B. Zax, A. Bielecki, K. W. Zilm, and A. Pines, *Chem. Phys. Lett.*, 1984, **106**, 550; A. M. Thayer, M. Luzar, and A. Pines, *J. Magn. Reson.*, 1987, **72**, 567.
11. A. Bielecki and A. Pines, *J. Magn. Reson.*, 1987, **74**, 381.
12. J. Koo and Y.-N. Hsieh, *Chem. Phys. Lett.*, 1971, **9**, 238.
13. F. Apaydin and S. Clough, *J. Phys. C*, 1968, **1**, 932.
14. K. J. Abed and S. Clough, *Chem. Phys. Lett.*, 1987, **142**, 209; K. J. Abed, S. Clough, A. J. Horsewill, and M. A. Mohammed, *Chem. Phys. Lett.*, 1988, **147**, 624.
15. M. H. Cohen and F. Reif, *Solid State Phys.*, 1957, **5**, 321.
16. T. P. Das and E. L. Hahn, *Solid State Phys. Suppl.*, 1958, **1**.
17. A. Abragam, 'Principles of Nuclear Magnetism', Oxford University Press, Oxford, 1961, Chap. 7.
18. R. V. Pound, *Phys. Rev.*, 1951, **81**, 156; N. F. Ramsey and R. V. Pound, *Phys. Rev.*, 1951, **81**, 278; E. M. Purcell and R. V. Pound, *Phys. Rev.*, 1951, **81**, 279.
19. A. G. Andersen, *Phys. Rev.*, 1959, **115**, 863.
20. F. Noack, in 'Progress in Nuclear Magnetic Resonance Spectroscopy', eds. J. W. Emsley, J. Feeney, and L. N. Sutcliffe, Pergamon Press, Oxford, 1986, Vol. 18, p. 171; E. Rommel, K. Mischker, G. Osswald, K. H. Schweikert, and F. Noack, *J. Magn. Reson.*, 1986, **70**, 219.
21. A. Bielecki, D. B. Zax, K. W. Zilm, and A. Pines, *Rev. Sci. Instrum.*, 1986, **57**, 393.
22. R. Kreis, A. Thomas, W. Studer, and R. R. Ernst, *J. Chem. Phys.*, 1988, **89**, 6623.
23. A. G. Redfield, *Phys. Rev.*, 1963, **130**, 589.
24. A. Messiah, 'Quantum Mechanics', North-Holland, Amsterdam, 1961, Vol. 2, Chap. 17.
25. R. Blinc, *Adv. Nucl. Quad. Reson.*, 1975, **2**, 71.

26. D. Edmonds, *Phys. Rep.*, 1977, **29**, 233; *Int. Rev. Phys. Chem.*, 1982, **2**, 103.
27. D. P. Weitekamp, A. Bielecki, D. B. Zax, K. W. Zilm, and A. Pines, *Phys. Rev. Lett.*, 1983, **50**, 1807; D. B. Zax, A. Bielecki, K. W. Zilm, A. Pines, and D. P. Weitekamp, *J. Chem. Phys.*, 1985, **83**, 4877.
28. J. M. Millar, A. M. Thayer, A. Bielecki, D. B. Zax, and A. Pines, *J. Chem. Phys.*, 1985, **83**, 934.
29. R. Kreis, D. Suter, and R. R. Ernst, *Chem. Phys. Lett.*, 1985, **118**, 120.
30. R. Kreis, D. Suter, and R. R. Ernst, *Chem. Phys. Lett.*, 1986, **123**, 154.
31. M. Goldman, 'Spin Temperature and Nuclear Magnetic Resonance in Solids', Clarendon Press, Oxford, 1970.
32. J. W. Hennel, A. Birczyński, S. F. Sagnowski, and M. Stachurowa, *Z. Phys. B*, 1984, **56**, 133.
33. Yu. A. Serebrennikov, *Chem. Phys.*, 1987, **112**, 253; *Chem. Phys. Lett.*, 1987, **137**, 183.
34. S. Vega, in 'Advances in Magnetic Resonance', ed. J. S. Waugh, Academic Press, New York, 1973, Vol. 6, p. 259; *J. Chem. Phys.*, 1975, **63**, 3769.
35. J. M. Millar, A. M. Thayer, D. B. Zax, and A. Pines, *J. Am. Chem. Soc.*, 1986, **108**, 5113; T. P. Jarvie, A. M. Thayer, J. M. Millar, and A. Pines, *J. Phys. Chem.*, 1987, **91**, 2240.
36. R. Kubo and T. Toyabe, in 'Magnetic Resonance and Relaxation', ed. R. Blinc, North-Holland, Amsterdam, 1967, p. 810; R. Kubo, T. Endo, S. Kamohara, M. Shimizu, M. Fujii, and H. Takano, *J. Phys. Soc. Jpn.*, 1987, **56**, 1172.
37. P. Jonsen, M. Luzar, A. Pines, and M. Mehring, *J. Chem. Phys.*, 1986, **85**, 4873.
38. J. W. Hennel, A. Birczyński, S. F. Sagnowski, and M. Stachurowa, *Z. Phys. B*, 1985, **60**, 49.
39. Yu. A. Serebrennikov, 'Advances in Magnetic and Optical Resonance', ed. W. S. Warren, Academic Press, New York, 1992, Vol. 17, p. 47.
40. M. H. Cohen, *Phys. Rev.*, 1954, **96**, 1278.
41. D. B. Zax, A. Bielecki, M. A. Kulzick, E. L. Muetterties, and A. Pines, *J. Phys. Chem.*, 1986, **90**, 1065.
42. A. M. Thayer and A. Pines, *J. Magn. Reson.*, 1986, **70**, 518.
43. C. J. Lee, D. Suter, and A. Pines, *J. Magn. Reson.*, 1987, **75**, 110.
44. A. Llor, Z. Olejniczak, J. Sachleben, and A. Pines, *Phys. Rev. Lett.*, 1991, **67**, 1989.
45. A. M. Thayer, J. M. Millar, and A. Pines, *Chem. Phys. Lett.*, 1986, **129**, 55.
46. D. Suter, T. P. Jarvie, B. Sun, and A. Pines, *Phys. Rev. Lett.*, 1987, **59**, 106.
47. C. Connor, J. Chang, and A. Pines, *J. Chem. Phys.*, 1990, **93**, 7639; *Rev. Sci. Instrum.*, 1990, **61**, 1059.
48. M. D. Hürlimann, C. H. Pennington, N. Q. Fan, J. Clarke, A. Pines, and E. L. Hahn, *Phys. Rev. Lett.*, 1992, **69**, 684.
49. R. Tycko, *Phys. Rev. Lett.*, 1988, **60**, 2734; *J. Chem. Phys.*, 1990, **92**, 5776; R. Tycko, G. Dabbagh, J. C. Duchamp, and K. W. Zilm, *J. Magn. Reson.*, 1990, **89**, 205.

Biographical Sketch

David B. Zax. *b* 1958. A.B., 1979, Harvard, USA; Ph.D., 1985, University of California at Berkeley, working with Professor A. Pines. Bantrell postdoctoral fellow with Professor S. Vega at the Weizmann Institute of Science, 1985–86. IBM Postdoctoral Fellow with Professor C. P. Slichter at the University of Illinois at Urbana–Champaign, 1986–87. Professor of Chemistry, Cornell, 1990–present. Approx. 40 publications in the field of magnetic resonance. Research interests include applications of solid state NMR to low-dimensional solids and confined solids, and of Floquet theory to problems in time-dependent quantum mechanics.

Zinc Fingers

Michael F. Summers

University of Maryland, Baltimore County, Baltimore, MD, USA

1 INTRODUCTION

The discovery in 1985 of 'zinc-binding fingers' in transcription factor IIIA from *Xenopus* oocytes[1,2] stimulated the development of a new field in bioinorganic chemistry that ultimately led to the identification of several novel classes of zinc-containing nucleic acid interactive protein domains (NAIDs). The term 'zinc finger' has been generally used to describe any miniglobular zinc-containing protein domain that functions in nucleic acid interactive processes. The 'classical **CCHH**' zinc finger[1,2] consists of a stretch of amino acids of the type: $\mathbf{C}\text{-}X_{2-5}\text{-}\mathbf{C}\text{-}X_{12}\text{-}\mathbf{H}\text{-}X_{2-4}\text{-}\mathbf{H}$ (**CCHH**). Other zinc-containing NAIDs are often referred to as 'nonclassical zinc fingers'. Examples of such nonclassical zinc finger domains that have been studied by NMR include

1. the **CCHC** type ($\mathbf{C}\text{-}X_2\text{-}\mathbf{C}\text{-}X_4\text{-}\mathbf{H}\text{-}X_4\text{-}\mathbf{C}$), observed originally in retroviral nucleocapsid proteins but more recently found in proteins from other organisms, including frogs and humans;[3–5]
2. the cysteine-rich DNA-binding domain of GAL-4 ($\mathbf{C}\text{-}X_2\text{-}\mathbf{C}\text{-}X_6\text{-}\mathbf{C}\text{-}X_6\text{-}\mathbf{C}\text{-}X_2\text{-}\mathbf{C}\text{-}X_6\text{-}\mathbf{C}$),[6] which coordinates two atoms of zinc in a binuclear cluster that contains two bridging cysteines;
3. the steroid hormone receptor (SHR) family of DNA binding domains (DBDs),[7] in which two zinc atoms are bound via cysteines in two structurally nonequivalent but interacting **CCCC** modules of the type, $\mathbf{C}\text{-}X_2\text{-}\mathbf{C}\text{-}X_{13}\text{-}\mathbf{C}\text{-}X_2\text{-}\mathbf{C}$ and $\mathbf{C}\text{-}X_5\text{-}\mathbf{C}\text{-}X_9\text{-}\mathbf{C}\text{-}X_2\text{-}\mathbf{C}$;
4. the DNA-binding domain of GATA-1, which binds zinc via a $\mathbf{C}\text{-}X_2\text{-}\mathbf{C}\text{-}X_{17}\text{-}\mathbf{C}\text{-}X_2\text{-}\mathbf{C}$ sequence;[8]

For References see p. 5070

5. the RING motif, which binds two atoms of zinc in independent sites via a **C**-X_2-**C**-X_{12}-**C**-X-**H**-X_2-**C**-X_2-**C**-X_{10}-**C**-X_2-**C** (or $\mathbf{C_3HC_4}$) sequence;[9]
6. the LIM motif, which also binds two equivalents of zinc in independent sites via a **C**-X_2-**C**-X_{16-23}-**H**-X_2-**C**-X_2-**C**-X_2-**C**-X_{16-21}-**C**-X_{2-3}-**C/H/D** ($\mathbf{C_2HC_5}$) sequence;[10,11]
7. the zinc ribbon motif observed in transcription elongation factor TFIIS that binds one atom of zinc in a **C**-X_2-**C**-X_{25}-**C**-X_2-**C** sequence;[12]
8. the zinc-containing domain of the tRNA synthetases;[13,14]
9. the zinc site of bacteriophage T4 gene-32 protein;[15,16]
10. the protein kinase C (PKC) zinc-binding domain;.[101]
11. the DNA methyl phosphotriester repair domain of Ada.[82]

Nuclear magnetic resonance spectroscopy has played a major role in advancing understanding of the structure and function of zinc-containing nucleic acid interactive domains. Indeed, the first structures of each class of zinc finger domain were determined originally by NMR methods, except for **GAL-4**, where the NMR[17,18] and X-ray structures[19] were reported simultaneously. Currently, there are at least 12 independent NMR structures reported for **CCHH** zinc finger domains,[20–25] six NMR structures of **CCHC** zinc finger domains and **CCHC**-containing proteins,[26–30,102] three NMR structures of the GAL-4-type DNA-binding domain,[17,18,103] and five NMR structures reported for steroid hormone receptor DNA-binding domains.[31–35] The structure of a **CCHC** zinc finger–nucleic acid complex has also been determined by NMR.[36] More recently, structures have been determined for the DNA-binding domain of GATA-1[8] and for the RING,[9] PKC,[101] Ada,[82] and LIM[37] zinc-binding domains by NMR methodologies. In addition, X-ray structures have been reported for DNA complexes with the **CCHH**-containing DNA-binding domains of the transcription factors Zif-286[38] and GLI[39] and with the DNA-binding domains of the GAL-4[19] and glucocorticoid receptor proteins.[40] In each of these domains, the essential zinc atoms do not participate directly in binding to nucleic acid, but rather function to stabilize a particular globular fold that allows amino acid side chains to make specific contacts with the nucleic acid.

The availability of structural data for a diverse class of zinc-containing NAIDs has enabled philosophical and experimental approaches to the question: why zinc fingers? Although current knowledge is not sufficient to address this question fully, one fact has become apparent; namely, all classes of Zn-NAIDs exist as highly stable, independently folded domains, despite the fact that the domains comprise a relatively small number of amino acid residues. Indeed, the **CCHC** retroviral-type motif comprises only 18 amino acid residues, and may represent the smallest known miniglobular protein domain. The 'zinc finger' motif thus appears to be Nature's choice for establishing stable, folded, miniglobular protein domains comprising a minimal number of amino acid residues. This prompts the following question: Why are zinc finger domains so stable?

By examination of the structures determined thus far for the different classes of Zn-NAIDs, common structural elements can be identified that may serve as the primary determinants of stability.[41–43] Specifically, all known Zn-NAIDS exhibit extensive hydrogen bonding between backbone amide protons and the cysteine sulfur atoms that are coordinated to zinc. NH···S hydrogen bonding was observed originally by Adman and co-workers[44] in the iron-binding domains of rubredoxins and ferredoxins, and, perhaps surprisingly, many of the structural elements observed originally in the iron domains of the iron–sulfur proteins are also observed in the zinc sites of the Zn-NAIDs.

2 NH···S HYDROGEN BONDING AS A DETERMINANT OF STABILITY IN MINIGLOBULAR ZINC-CONTAINING NUCLEIC ACID INTERACTIVE DOMAINS (NAIDs)

As indicated above, all of the known Zn-NAIDs that have been structurally characterized thus far contain NH···S hydrogen bonds between specific backbone amide protons and Zn-coordinated cysteine sulfurs. The local structural elements observed are often very similar to those observed in the iron-binding domain of rubredoxin (Rd). All rubredoxins contain two copies of a conserved iron-binding sequence, $X_{(i-2)}$-$X_{(i-1)}$-$\mathbf{C}_{(i)}$-$X_{(i+1)}$-$X_{(i+2)}$-$\mathbf{C}_{(i+3)}$-$\mathbf{G}_{(i+4)}$-$X_{(i+5)}$ (where X is a variable amino acid), and high-resolution X-ray and NMR structures of proteins from *Clostridium pasteurianum*, *Desulfovibrio gigas*, *D. vulgaris*, *D. desulfuricans*, and *Pyrococcus furiosus*, reveal that the four Cys residues are coordinated to the single iron atom. The backbone conformations of the N- and C-terminal X-X-**C**-X-X-**C**-**G**-X-X-X residues are essentially identical. Adman and co-workers[44] discovered that three of the backbone amide protons within each of the $\mathbf{C}_{(i)}$-$X_{(i+1)}$-$X_{(i+2)}$-$\mathbf{C}_{(i+3)}$-$\mathbf{G}_{(i+4)}$-$X_{(i+5)}$ loops are involved in amide-to-sulfur hydrogen bonding. The backbone amide protons of residues X(i + 2) and Cys(i + 3) are oriented in the direction of the Cys(i) side-chain sulfur atom, forming what are referred to as Types I and III tight turns, respectively (Figure 1). In addition, the backbone amide proton of residue X(i + 5) is oriented toward the side-chain sulfur atom of Cys(i + 3), forming a Type II NH···S tight turn. Rubredoxins also contain a conserved Gly residue at position (i + 4), and it was proposed that this residue is conserved in

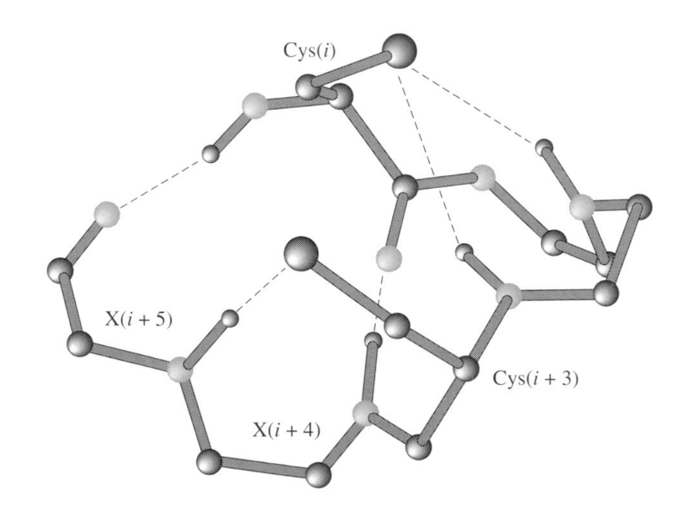

Figure 1 Structure of a 'rubredoxin knuckle' showing NH···S hydrogen bonds from the backbone amide protons of residues X(i + 2) and Cys(i + 3) to the S-γ sulfur of residue Cys(i), and from the backbone amide proton of residue X(i + 5) to the S-γ sulfur of residue Cys(i + 3). This folding pattern, which was observed originally in the iron-binding site of rubredoxin,[44,100] is commonly found in the zinc-binding sites of zinc fingers

order to stabilize the Type II NH···S tight turn.[44] The glycine NH is positioned to hydrogen bond with the carbonyl of Cys(i), forming a Type II NH···O bond. Throughout this article, this conserved folding pattern (Figure 1) will be referred to as an 'Rd knuckle'.

To gain insights into the nature of the NH–S hydrogen bonding, NMR studies have been performed on Zn-, ^{113}Cd-, and ^{199}Hg-substituted rubredoxin from the extreme hyperthermophile, *Pyrococcus furiosus* (pf), a bacterium that grows optimally at 100 °C.[41,45–47] The ^{1}H–^{113}Cd heteronuclear spin echo difference (HSED) spectrum obtained for ^{113}Cd–PfRd exhibited four backbone NH signals corresponding to the Ile(7), Cys(8), Ile(40), and Cys(41) backbone NH protons, of which Ile(40) and Cys(41) are part of the metal-binding **CXXCGX** knuckle (**C**$_{38}$-X-I$_{40}$-**C**$_{41}$-G$_{42}$-A$_{43}$). The observed scalar couplings between ^{113}Cd and the NH protons of Ile(40) and Cys(41) do not occur through the peptide backbone, since the NH protons are eight and five bonds removed from the metal, respectively. The observed amide HSED signals are thus attributed to two-bond $^{2}J(^{113}\text{Cd},^{1}\text{H})$ scalar coupling mediated by NH···S(Cys) hydrogen bonds. These findings provide direct evidence that Types I and III NH···S hydrogen bonds exist in PfRd, and further demonstrate that these hydrogen bonds contain significant covalent character.

^{1}H–^{199}Hg HSED NMR experiments were also performed on the ^{199}Hg adduct to evaluate the influence of this larger metal on the nature of the NH···S hydrogen bonds.[41,47] A portion of the ^{1}H–^{199}Hg HSED spectrum is shown in Figure 2. As observed for the ^{113}Cd adduct, HSED signals were observed for the Ile(7,40) and Cys(8,41) backbone NH protons,[45] confirming the presence of the Types I and III NH···S turns. Interestingly, the magnitudes of the scalar couplings, determined using a novel pulse sequence that affords pure-phase HSED spectra,[47] were approximately twofold greater than the coupling constants determined for ^{113}Cd–PfRd. This finding confirms that the HSED signals result from scalar coupling rather than from a dipolar cross-correlation mechanism. In addition, a weak ^{1}H–^{199}Hg HSED signal [$^{2}J(^{199}\text{Hg},^{1}\text{H}) = 0.4$ Hz] was detected for the Tyr(10) backbone amide proton. Although a similar signal for the analogous Ala(43) amide could not be distinguished owing to signal overlap, the observation of a Tyr(10) NH HSED signal confirmed the existence of the Type II NH···S turn. Thus, for the ^{199}Hg adduct, where the magnitudes of the scalar couplings are greatest, HSED spectroscopy provided confirmation for the existence of all three types of NH···S tight turns (Figure 1).

As a side note, ^{1}H–^{113}Cd and ^{1}H–^{199}Hg scalar coupling was also observed between the metal and the Ala(43) methyl protons (Figure 2).[41,45,47] This scalar coupling is probably due to direct orbital overlap between the methyl protons of Ala(43) and the Cys(42) sulfur atom, since 6- and 11-bond ^{1}H–^{113}Cd scalar coupling mediated by the Ala(43) NH···S hydrogen bond or by the covalent bonds, respectively, is implausible. This finding has important implications for understanding intraprotein electron transfer, since (a) it suggests that 'through-space electron jumps' may be energetically more favorable than previously expected,[48–52] and (b) CH_3··S(Cys) overlap provides a direct electron transfer pathway between the metal center and the hydrophobic core of the protein.

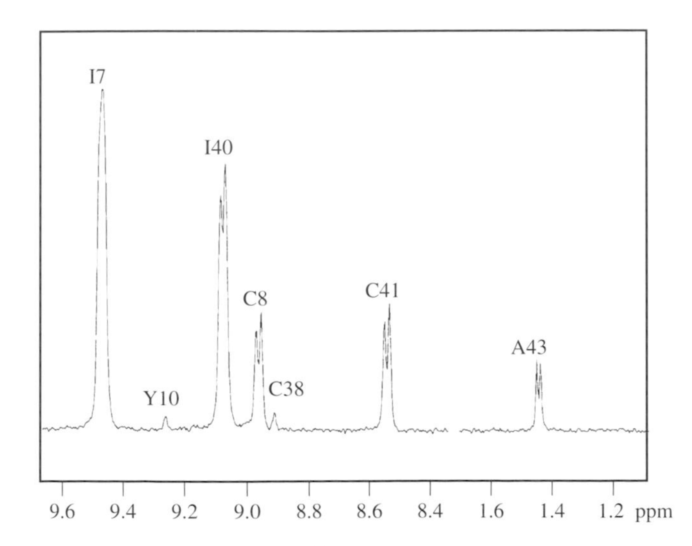

Figure 2 Portions of an ^{1}H–^{199}Hg heteronuclear spin echo difference spectrum obtained for ^{199}Hg-substituted rubredoxin from the hyperthermophile *P. furiosus*.[47] Difference signals are observed for backbone NH protons involved in Types I, II, and III NH···S hydrogen bonding, confirming the presence of these hydrogen bonds and demonstrating that these bonds contain significant covalent character

3 CLASSICAL CCHH ZINC FINGERS

To date, more than a dozen classical **CCHH** zinc finger domains have been structurally characterized by NMR and X-ray crystallographic methods.[20–25] Except for a few cases, the experimentally derived **CCHH** zinc finger structures are consistent with a structure predicted on the basis of sequence comparisons with other metalloproteins, including rubredoxin.[53] Klevit and co-workers[54] were the first to provide experimental NMR evidence in support of the predicted model, and shortly thereafter, atomic coordinates for a peptide with sequence of the 31st zinc finger of the *Xenopus* protein, Xfin, were published by Wright and co-workers.[20]

The **CCHH** zinc finger consists of a three-turn helix and a short antiparallel β-sheet. The strands of the β-sheet are linked by the **CXXCXX** metal-binding turn. The **CXXCXX** turns for five representative **CCHH** zinc finger structures are shown in Figure 3. For comparison, relevant residues of the N-terminal **CXXCXX** turn of *C. pasteurianum* Rd, and the associated NH···S and NH···O hydrogen bonds, are included in this figure. These residues within zinc fingers Xfin-31, ADR-1 and Zif286-2,3 are folded in a manner similar to the folding of analogous residues in rubredoxins. All of these structures exhibit the Types I and III NH···S hydrogen bonding found in rubredoxins. However, only the zinc fingers of ADR1b and Zif-268 exhibit the Type II NH···S turn after the second Cys. The **C**$_{(i+3)}$-X-X residues of the Xfin-31 structure were disordered, precluding classification of the potential NH···S hydrogen bonding for the X(i + 5) amide proton.

The **CXXCXX** loops of the Zif-268 and ADR-1 zinc fingers contain Type II NH···S turns after the second cysteine, whereas finger-3 of Zif-268 contains a Gly residue at position (i + 4). Finger-2 of Zif-268 and the ADR-1 zinc finger contain Met and Thr residues, respectively, at position (i + 4). Type II folding results in a positive ϕ value for the residue at position X(i + 4), and, as such, it was proposed early on that Type II

For References see p. 5070

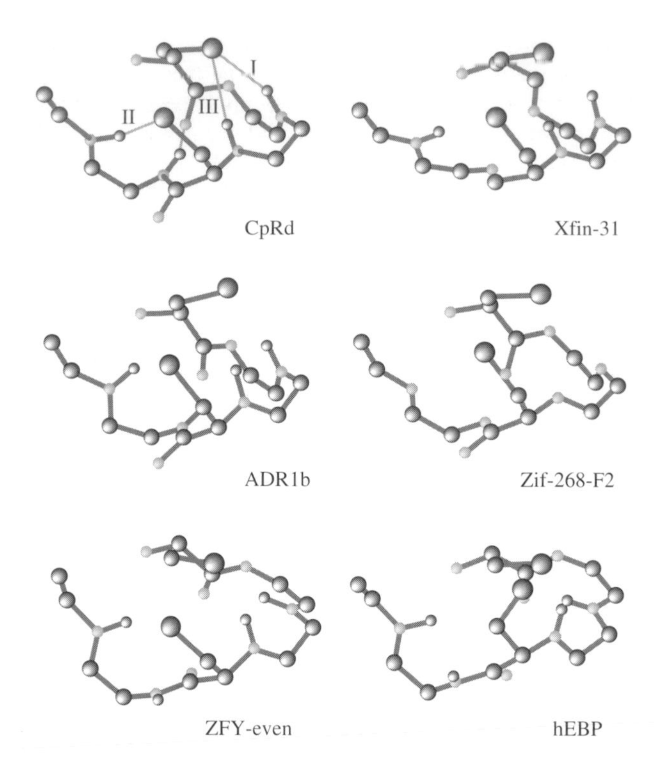

Figure 3 Comparison of the folding of the **CXXCXX** residues of representative classical-type **CCHH** zinc fingers with related residues of rubredoxin.[100] The folding of these residues in the Xfin-31,[20] ADR1b,[25] and Zif-268-2[38] zinc fingers is essentially identical to that observed in the X-ray structure of *C. pasteurianum* Rd. The ZFY-even finger domain[22–24] is similar, except that residues $\mathbf{C}_{(i+3)}$-$X_{(i+4)}$-$X_{(i+5)}$ form a Type I NH···S turn. The zinc fingers of the human enhancer DNA-binding domain[21] are unusual in that (i) the χ_1 value of the Cys(i + 3) residue precludes the formation of an NH···S hydrogen bond between the amide proton of X(i + 5) and the Cys(i + 3) sulfur, and (ii) the amide proton of X(i + 2) does not appear to form a hydrogen bond to the Cys(i) sulfur

NH···S folding should be stabilized by a glycine residue at position X(i + 4). This provided a rationale for the conservation of glycine residues at position X(i + 4) in the **CXXCXX** loops of all known rubredoxins.[55] As shown in Figure 1, Type II NH···S folding leads to hydrogen bonds between the X(i + 4) amide proton and the Cys(i) carbonyl and between the X(i + 5) amide proton and the Cys(i + 3) sulfur. Thus, the unfavorable steric interactions associated with the positive ϕ value may be offset by the additional hydrogen bond formed between the X(i + 4) amide and the Cys(i) carbonyl.[44] A positive ϕ value has also been observed for the Type II NH···S turn of residues **C**-A-G in the iron site of ferredoxin, where the Ala methyl group eclipses the Cys carbonyl.[44]

The amide proton of the X(i + 5) residue of the ZFY-even finger and an even finger mutant are directed toward the β-carbon of the Cys(i) residue and not toward the sulfur atoms of residue Cys(i + 3).[22–24] In addition, the Cys(i + 3)-to-X(i + 4) peptide bond is oriented in a manner that would preclude hydrogen bonding to the Cys(i) carbonyl (Figure 3). Thus, although the folding observed for residues **C**-X-X-**C** of the ZFY-even finger domains are consistent with the rubredoxin fold, the conformation of the $\mathbf{C}_{(i+3)}$-X-X residues differs from that observed in rubredoxins. The folding observed for the zinc finger domains of the human enhancer protein DNA-binding domain (hEBP)[21,56] are unusual in that the side-chain orientation of the Cys(i + 3) residues precludes formation of a Type II NH···S turn, and amide orientations are generally inconsistent with NH···S hydrogen bonding.

In summary, with the exception of the human enhancer protein DNA binding domain, all of the classical **CCHH** zinc finger structures exhibit Types I and III NH···S hydrogen bonds within the **C**-X-X-**C** knuckles. In addition, all **CCHH** zinc fingers except those of the hEBP and ZFY-even zinc finger domains exhibit an NH···S hydrogen bond from the X(i + 5) backbone amide to the Cys(i + 3) sulfur, forming either a Type I or II NH···S turn.

4 RETROVIRAL CCHC ZINC FINGERS

Retroviral genes encode a gag precursor polyprotein that functions in the recognition of viral RNA and in the assembly of virus particles.[57,58] Subsequent to assembly and budding, the gag polyproteins are cleaved by the viral protease to smaller proteins, including the nucleocapsid (NC) proteins. In 1981, Henderson and co-workers[5] at the Frederick Cancer Research and Development Center showed that retroviral gag and NC proteins contain at least one copy of an invariant amino acid sequence, **C**-X_2-**C**-X_4-**H**-X_4-**C**. This conserved sequence has recently been found in proteins ranging from plant viruses to frogs to humans, and thus may represent a common motif for binding single stranded nucleic acids.[59] For example, a nine tandem repeat of the conserved sequence is found in a protein encoded by the *Leishmania major* gene of the protozoan parasite *Leishmania*;[60] a seven-tandem repeat is found in human cellular nucleic acid binding protein (hCNBP) that binds to the sterol regulatory element;[61] and a single motif is present in a protein encoded by the *Xenopus-Posterior* (*Xpo*) gene of frog embryos.[62]

NMR structures have been determined for peptides with sequences of several **CCHC** zinc finger domains, as well as for the intact NC protein from HIV-1.[27–30,63] Like TFIIIA, overall folding of the NC protein consists of the independent **CCHC** zinc finger domains separated by a labile linker region. The global folding **CCHC** Zn-NAIDs consists of a short antiparallel β-sheet with an Rd knuckle, followed by a short β-like stretch that precedes a 3_{10} helix. Like **CCHH** zinc fingers and rubredoxin, the metal binding site of **CCHC** NAIDs fold with a number of NH···S hydrogen bonds, including a hydrogen bond between the side-chain NH of Asn(i + 2) and the Cys(i + 3) sulfur, and between the backbone NH protein of residue X(i + 15) and the Cys(i + 13) sulfur. In addition to these five NH···S hydrogen bonds, there are a total of six NH···O hydrogen bonds that further stabilize the **CCHC** zinc finger structure.[42]

5 STEROID HORMONE RECEPTOR DNA-BINDING DOMAINS

The steroid hormone receptor (SHR) superfamily is a class of transcription regulatory proteins whose activity is controlled by the binding of specific hormones. This class so far comprises more than a dozen proteins, including the glucocorticoid receptor, retinoic acid receptor β, progesterone receptor, estrogen receptor, and retinoid X receptor-α proteins. Typically, members of the SHR family bind as dimers to specific DNA hormone response elements (HREs),[64] forming a hormone–receptor complex that activates transcription.[65] To date, three-dimensional structures have been determined by NMR for five steroid hormone receptor DNA-binding domains,[31–35] and an X-ray structure of a DNA complex with the glucocorticoid receptor DNA binding domain has been reported.[40]

The steroid hormone receptor (SHR) DNA-binding domains contain two independent zinc sites, with each of the zinc atoms coordinated via four cysteine residues. Unlike classical **CCHH** zinc fingers, the two zinc atoms are incorporated into a single domain, with several contacts between residues that comprise the independent zinc sites. The N-terminal zinc coordination site of the SHR proteins contains two **CXXC** arrays, and the C-terminal zinc site contains one **CXXC** array. Luisi et al.[40] indicated that several backbone amide protons are involved in NH···S hydrogen bonding to the metal-coordinating sulfurs. Comparison of the three **CXXC** segments reveals a variation in χ_1 values of the cysteine side chains, with only the N-terminal knuckle exhibiting a Cys(i) side-chain orientation similar to that of rubredoxins (Figure 4). The backbone folding within each of the **CXXC** knuckles is generally consistent with the folding observed in rubredoxins, with backbone amide protons participating in Types I and III NH···S hydrogen bonding. The $\mathbf{C}_{(i+3)}$-X-X residues of the N-terminal **CXXC** knuckle form a Type II NH···S turn, with the X(i + 4) amide forming a hydrogen bond to the Cys(i) backbone carbonyl. The $\mathbf{C}_{(i+3)}$-X-X residues of the two subsequent **CXXC** segments do not form NH···S turns, owing to the fact that these residues reside within an α-helix (Figure 4). To summarize, the N-terminal **CXXCXX** array forms an Rd knuckle with Types I, II, and III NH···S turns, and the two subsequent arrays are similar in that they contain the Types I and III NH···S turns within the **CXXC** segment, but differ in that (i) the χ values of the Cys(i + 3) side chains are significantly different, and (ii) the $\mathbf{C}_{(i+3)}$-X-X residues do not form an NH···S turn.

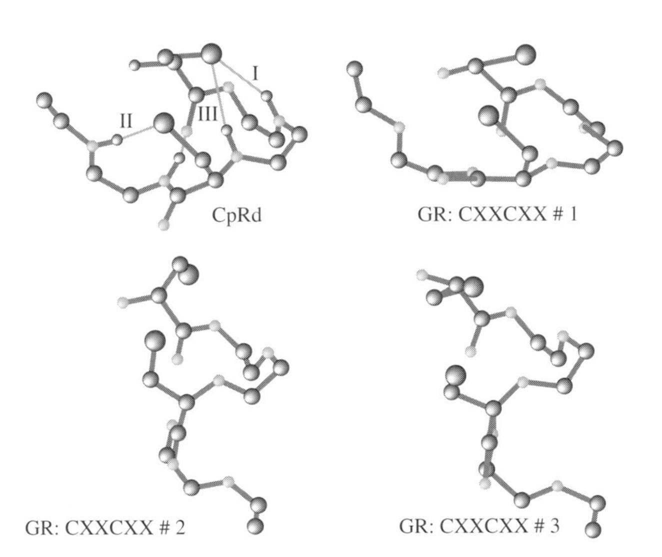

Figure 4 Comparison of the folding of the **CXXCXX** residues of the glucocorticoid receptor DNA-binding domain with related residues of *C. pasteurianum* Rd.[100] The GR protein contains three **CXXCXX** arrays. The folding of the N-terminal array (GR: CXXCXX # 1) is very similar to that observed in rubredoxins. The remaining arrays contain Types I and III NH···S turns within the **CXXC** knuckles common to rubredoxins; however, the $\mathbf{C}_{(i+3)}$-X-X residues do not form an NH···S turn, because of their involvement in α-helical conformations. Interestingly, the amide proton of residue X(i + 4) participates in a hydrogen bond to the Cys(i) carbonyl—an interaction that also occurs in rubredoxins

6 GAL-4 DNA-BINDING DOMAIN

GAL-4 is a yeast transcription activator that binds to an upstream activating sequence (UAS_G) for the genes (*GAL-1* and *GAL-10*) encoding galactose-metabolizing enzymes.[19] The 881-amino-acid residue GAL-4 protein binds as a dimer to the UAS_G-DNA site.[66] In the absence of galactose, GAL-4 remains on the DNA, but is inactivated by a complex formed with another protein, GAL-80.[67,68] In the presence of galactose, a metabolite derived from galactose binds to and dissociates GAL-80, thus activating GAL-4.[66] The DNA-binding domain of GAL-4 is a 62-amino-acid sequence located on the N-terminus of the protein.[17] This sequence contains a cysteine-rich array, **C**-X_2-**C**-X_6-**C**-X_6-**C**-X_2-**C**-X_6-**C**, which is conserved among 13 known fungal transcription factors.[17] Unlike other Zn-NAIDs, the two zinc atoms of GAL-4 form a binuclear cluster that contains two bridging cysteines.[17,69]

Structures of the GAL-4 NAIDs have been determined by NMR.[17,18] and X-ray crystallographic[19] methods. The GAL-4 DBD contains two helix–turn–strand motifs that are packed around the $Zn(II)_2Cys_6$ binuclear cluster.[17] The structure has a twofold pseudosymmetry axis that passes between the two helices and the center of the $Zn(II)_2Cys_6$ binuclear cluster. The GAL-4 zinc-binding domain contains two **CXXC** arrays that participate in metal coordination. Both the N- and C-terminal **CXXC** arrays exhibit backbone folding that is generally consistent with the folding observed in rubredoxins. Amide protons of residues X(i + 2) and Cys(i + 3) are oriented in a manner consistent with Types I and III NH···S hydrogen bonding to the Cys(i) sulfur (Figure 5). However, the χ_1 values of the Cys side chains differ from those of Rd by almost 180°. The $\mathbf{C}_{(i+3)}$-X-X residues comprise the N-terminal portions of α-helical stretches, and the associated backbone conformations preclude the formation of Type II NH···S turns. The folding of these residues is similar to that observed for analogous residues

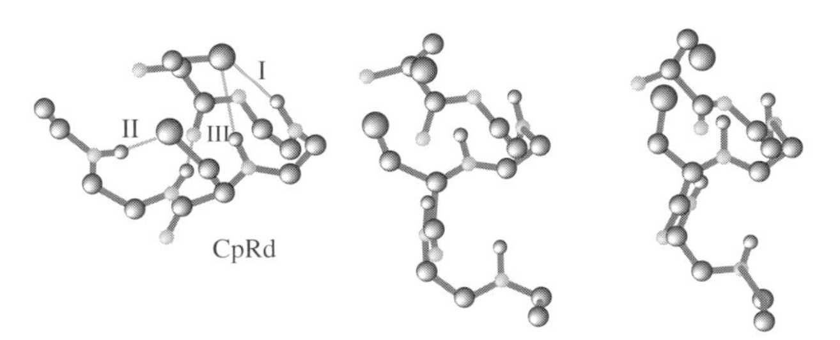

Figure 5 Comparison of the folding of the **CXXCXX** residues of the GAL-4 DNA-binding domain with related residues in *C. pasteurianum* Rd.[100] The folding of these residues is actually very similar to that observed for the two N-terminal **CXXCXX** stretches in the glucocorticoid receptor DNA-binding domain (see Figure 4)

For References see p. 5070

in the steroid hormone receptor proteins where an α-helical conformation exists (Figure 4).

7 GATA-1 DNA-BINDING DOMAIN

The DNA-binding domain of the transcription factor GATA-1 contains two conserved zinc binding sequences of the form **C**-X_2-**C**-X_{17}-**C**-X_2-**C**. Mutational studies indicated that the N-terminal motif may be unimportant for specific DNA binding,[70,71] although more recent studies indicate that sequence specificity is a function of both domains.[72]

The solution structure of the C-terminal GATA-DBD (residues 158–223) was determined recently by NMR methods.[8] The global folding of the peptide consists of a core (residues 2–51) that contains the zinc binding region, and an extended C-terminal tail (residues 52–59).[8] The folding of the majority of the core (residues 3–39) is very similar to the folding exhibited by the N-terminal **CCCC** zinc module of glucocorticoid receptor protein.[8] Interestingly, both the GATA-1 domain and the N-terminal **CCCC** module of GR participate directly in protein–nucleic acid interactions.

The N-terminal **CXXCXX** array of GATA-1 forms a classical Rd knuckle that contains Types I, II, and III NH···S turns. The Type II turn is present despite the fact that the array lacks a Gly residue at position X(i + 4). In addition, the backbone NH of residue Cys(i + 21) forms a hydrogen bond to the Cys(i) sulfur, and the sulfur of Cys(i + 21) accepts a hydrogen bond from the backbone amide proton of Cys(i + 24).

8 Met-tRNA SYNTHETASE

Aminoacyl-tRNA synthetases are responsible for the correct esterification of tRNAs with the corresponding amino acid during protein translation.[73] A number of aminoacyl-tRNA synthetases have been analyzed and shown to display a zinc finger domain of the type **C**-X_2-**C**-X_9-**C**-X_2-**C**.[74] Among these enzymes are the methionyl-tRNA (met-tRNA), alanyl-tRNA, and isoleucyl-tRNA synthetases, all of which have been shown to contain zinc.[75–77] On the other hand, tyrosyl- and phenylalanyl-tRNA synthetases, which do not contain zinc finger domains, apparently do not contain zinc.[78]

Dardel and co-workers[73] have recently determined the three-dimensional structure of a peptide corresponding to the zinc coordination site of met-tRNA synthetase by NMR methods. The structure differs significantly from that determined by X-ray crystallography for the intact protein. The zinc binding site is located within residues 138–163 of the protein, and contains a single zinc atom coordinated via four cysteine residues. Unlike other Zn-NAIDs, the zinc binding region of met-tRNA synthetase does not contain any distinguishable helices or extended β-sheet structure. Instead, the domain contains a series of four consecutive tight turns separated by short loops. Turns 2 and 4 form Rd knuckles. Superposition of the backbone and Cys side-chain atoms of the **XCXXCX** residues of the met-tRNA synthetase metallopeptide with analogous residues of *C. pasteurianum* Rd and Zn(HIV-1–F1) afforded RMSD values of 65–90 pm.[73] The backbone amide protons of residues X(i + 2) and Cys(i + 3) are oriented in a manner consistent with Types I and III NH···S hydrogen bonding. The N- and C-terminal Rd knuckles contain Gly and Lys residues, respectively, at position (i + 4). As expected, the backbone amide of residue X(i + 5) from the **CXXCGX** array forms a hydrogen bond to the Cys(i + 3) sulfur, with the backbone amide proton of Gly(i + 4) forming a hydrogen bond to the Cys(i) carbonyl. Thus, the $\mathbf{C}_{(i+3)}$-$\mathbf{G}_{(i+4)}$-$\mathbf{X}_{(i+5)}$ residues exist in a Type II NH···S turn common to rubredoxins.

The backbone amide proton of the X(i + 5) residue from the N-terminal **CXXCXX** array also appears to form a hydrogen bond with the Cys(i + 3) sulfur. In addition, the backbone amide proton of the lysine at position (i + 4) is oriented in the general direction of the Cys(i) backbone carbonyl. As such, the lysine at position (i + 4) possesses a positive ϕ value characteristic of Type II NH···S tight turns. Thus, as observed in the zinc finger of Zif-268-F2 and ADR1b, the presence of a non-glycine residue at position X(i + 4) does not preclude the formation of a Type II NH···S turn.

9 TRANSCRIPTION ELONGATION FACTOR (TFIIS)

Transcription elongation factor TFIIS is a protein that binds to the large subunit of RNA polymerase II in a paused elongation complex, and allows elongation through DNA pause and termination sites. It also facilitates transcription of sites blocked by sequence-specific DNA-binding proteins.[79] The 50-amino-acid zinc binding domain contains a **C**-X_2-**C**-X_{25}-**C**-X_2-**C** motif. The structure of a peptide with sequence of the TFIIS zinc binding site (residues 235–280) was determined recently by NMR methods.[80] Unlike other Zn-DBDs, the domain does not contain α-helical structure, but consists of a three-stranded antiparallel β-sheet. Two of the sheets are located between the **C**(15)-X_{25}-**C**(40) region, while the third exists after the second **C**(40)-X_2-**C**(43) array. TFIIS also contains three β-turns (residues 14–16, 30–32, and 42–44) two of which contain the zinc-binding cysteines.[80] The zinc atom is chelated via two Rd knuckles, both of which contain Types I, II, and III NH···S turns [residues $\mathbf{C}_{(12)}$-X_2-$\mathbf{C}_{(15)}$-X-X and $\mathbf{C}_{(40)}$-X_2-$\mathbf{C}_{(43)}$-**G**-Asp(45)]. The N-terminal Rd knuckle contains a Type II NH···S turn, even though the sequence lacks a glycine at residue 16.

10 DNA METHYL PHOSPHOTRIESTER REPAIR DOMAIN OF Ada

The Ada protein from *Escherichia coli* repairs methylated DNA by direct, irreversible transfer of a phosphotriester methyl group to a cysteine residue. The nucleophilic activity of the cysteine sulfur appears to be promoted by coordination to zinc.[81] Upon methylation, Ada binds specifically to DNA genes that confer resistance to methylating agents. The structure of an N-terminal 10 kDa fragment of Ada was determined by NMR methods.[82] This fragment, which retains DNA methyl phosphotriester repair activity, contains a single atom of zinc and exhibits a folding pattern that is unique by comparison with other known zinc finger domains. The structure consists of a three-stranded antiparallel β-sheet sandwiched between two α-helices. The atom of zinc is bound tetrahedrally by four cysteines, including the active-site Cys(69). Interestingly, no evidence was found for the presence of NH···S hydrogen bonds. Thus, the backbone NH protons of residues near the sulfur atoms exhibited disorder and/or NH···S distances incompatible with NH···S hydrogen bonding, and none of these protons exhibited slow chemical exchange with water protons. Such hydrogen bonding might be expected to decrease the negative charge density on the cysteine sulfurs and render the active-site cysteine a poorer nucleophile.[82] It will be very interesting to determine if the methylated form of the protein contains NH···S hydrogen bonds, since it is this form that exhibits sequence-specific DNA-binding properties.

11 RING FINGER

Over the past several years, more than 40 proteins have been identified from organisms including plants, fungi, viruses, and vertebrates that contain a '$\mathbf{C_3HC_4}$' or '**RING** finger' motif (**C**-X_2-**C**-X_{12}-**C**-X-**H**-X_2-**C**-X_2-**C**-X_{10}-**C**-X_2-**C**).[83,84] Proteins that contain a $\mathbf{C_3HC_4}$ motif are involved in numerous processes, including transcription, site-specific recombination, differentiation, and oncogenesis.[85]

The three-dimensional solution structure of a peptide of the sequence encompassing the $\mathbf{C_3HC_4}$ domain from a herpes virus protein was recently solved by Barlow and co-workers.[86] The two zinc atoms are bound in separate sites, with Cys(8) and (11) and Cys(29) and (32) coordinating to one zinc, and Cys(29), (43), (46) and His(26) coordinating to the second zinc atom. The peptide binds two zinc atoms and adopts a $\beta\beta\alpha\beta$ structure. Barlow and co-workers showed that the amphipathic α-helix is essential for *trans*- activation of gene expression. Interestingly, the N-terminal $\beta\beta\alpha$ portion of the protein is structurally very similar to that of classical **CCHH** zinc fingers.[86] Several backbone NH protons appear to be oriented to allow NH···S hydrogen bonding with the metal-coordinating cysteines.

12 LIM MOTIF

A number of proteins with diverse functions and subcellular distributions display from one to three copies of a cysteine- and histidine-containing sequence called the **LIM** motif (**C**-X_2-**C**-X_{16-23}-**H**-X_2-**C**-X_2-**C**-X_2-**C**-X_{16-21}-**C**-X_{2-3}-**C/H/D**). This motif was first identified in three developmentally regulated transcription factors, *C. elegans* Lin-11, rat Isl-1, and *C. elegans* Mec-3.[10,11] Each of these gene products exhibits two copies of the **LIM** motif plus a DNA-binding homeodomain and potential transcriptional activation domains.[10,11] In addition to a growing collection of LIM homeodomain proteins,[10,11,87–90] the **LIM** motif is also found in a number of proteins that lack additional DNA-binding sequences.[91–96] For example, a cytoskeletal cysteine-rich protein (CRP) that has been implicated in cell growth control and differentiation contains two LIM domains as predominant sequence elements.[95,97,98] CRP interacts with another LIM protein called zyxin, which is co-localized with CRP in association with the actin cytoskeleton, particularly at sites of cell–substratum adhesion.[93,95] It has been postulated that zyxin and CRP may mediate communication between the cell membrane and the nucleus.[93]

The $\mathbf{C_2HC_5}$ LIM domain comprises tightly packed **CCHC** and **CCCC** zinc-binding modules. The folding and metal binding mode differ significantly from that of the $\mathbf{C_3HC_4}$ **RING** finger motif, which also binds two atoms of zinc via seven cysteines and a histidine residue.[9,99] The C-terminal **CCCC** module of the LIM domain folds in a manner essentially identical to that observed for the DNA-interactive **CCCC** modules of the GATA-1 and GR DBDs. Although previous biochemical studies have suggested that LIM domains may participate in protein–protein rather than protein–DNA interactions,[93] the demonstration that the LIM domain contains a substructure that is closely related to those of well-characterized DNA-binding domains raises the possibility that the LIM domain may function in nucleic acid binding as well.

Heteronuclear spin echo difference (HSED) spectra were obtained for ^{113}Cd-substituted LIM to probe for NH···S hydrogen bonding. Scalar coupling was observed between the backbone NH protons of residues Lys(147), Cys(148) and Lys(150) and ^{113}Cd in the **CCCC** site, and between the backbone NH proton of Arg(120) and ^{113}Cd in the CCHC site, supporting the NOE evidence that these residues form Rd knuckles. These and additional NH···S hydrogen bonding patterns are similar to those observed in the GATA-1 DNA-binding domain.

13 SUMMARY

All of the zinc finger structures reviewed in this article contain at least one copy of a **CXXC** sequence at the metal binding site. As indicated above, many of these segments adopt similar backbone conformations when bound to zinc, with the NH protons of residues X(i + 2) and Cys(i + 3) oriented in a manner consistent with hydrogen bonding to the Cys(i) sulfur. In addition, many of the structures exhibit Type II NH···S folding for the subsequent $\mathbf{C}_{(i+3)}$-$X_{(i+4)}$-$X_{(i+5)}$ residues. In fact, for most Zn-NAID structures, the folding of at least one **CXXCXX** segment matches that of the analogous iron-chelating residues in rubredoxins. The formation of a Type II NH···S turn by residues $\mathbf{C}_{(i+3)}$-$X_{(i+4)}$-$X_{(i+5)}$ clearly does not depend on the presence of a Gly at position X(i + 4), even though the ϕ value for the X(i + 4) residue in a Type II NH···S turn is positive. The NH···O hydrogen bond between the X(i + 4) amide and the Cys(i) backbone carbonyl apparently compensates for the unfavorable steric interactions between the X(i + 4) side chain and the Cys(i + 3) carbonyl.[44] In several cases, the **CXXCXX** residues form a 'helix cap' at the N-terminal end of a helix. These structures still exhibit Types I and III NH···S turns within the **CXXC** segment, although the conformation of the subsequent residues precludes formation of a Type II NH···S hydrogen bond.

Heteronuclear ^{1}H–metal correlated NMR studies of ^{113}Cd- and ^{199}Hg-substituted forms of *P. furiosus* rubredoxin have provided the most detailed insights into the nature of the NH···S hydrogen bonds in the metal center. Protons that participate in Types I, II, and III NH···S hydrogen bonding exhibit NH···S hydrogen-bond-mediated scalar coupling, providing direct evidence that these protons are indeed hydrogen bonded to the cysteine sulfurs and demonstrating that the NH···S hydrogen bonds contain significant covalent character. NH···S hydrogen-bond-mediated scalar coupling has also been

For References see p. 5070

observed in a ^{113}Cd-substituted LIM domain for NH protons with sufficiently long T_2 values.[37] NH···S hydrogen-bond-mediated scalar coupling has been observed by Gardner, Coleman, and co-workers (personal communication) for ^{113}Cd-GAL-4.[103] Future studies with other classes of zinc domains are likely to lead to similar findings for what appear to be important determinants of stability in miniglobular Zn-NAIDs.

14 RELATED ARTICLES

Cadmium-113 NMR: A Surrogate Probe for Zinc & Calcium in Proteins; Coupling Through Space in Organic Chemistry; Inorganic Chemistry Applications; Iron–Sulfur Proteins.

15 REFERENCES

1. R. S. Brown, C. Sander, and P. Argos, *FEBS Lett.*, 1985, **186**, 271.
2. J. Miller, A. D. McLachlan, and A. Klug, *EMBO J.*, 1985, **4**, 1609.
3. T. B. Rajavashisth, A. K. Tayler, A. Andalibi, K. L. Svenson, and A. J. Lusis, *Science (Washington, DC)*, 1989, **245**, 640.
4. S. M. Sato and T. D. Sargent, *Development*, 1991, **112**, 747.
5. L. E. Henderson, T. D. Copeland, R. C. Sowder, G. W. Smythers, and S. Oroszlan, *J. Biol. Chem.*, 1981, **256**, 8400.
6. L. Keegan, G. Gill, and M. Ptashne, *Science (Washington, DC)*, 1986, **231**, 699.
7. P. H. Lohmar and D. O. Toft, *Biochem. Biophys. Res. Commun.*, 1975, **67**, 8.
8. J. G. Omichinski, G. M. Clore, O. Schaad, G. Felsenfeld, C. Trainor, E. Appella, S. J. Stahl, and A. M. Gronenborn, *Science (Washington, DC)*, 1993, **261**, 438.
9. P. N. Barlow, B. Luisi, A. Milner, M. Elliott, and R. Everett, *J. Mol. Biol.*, 1994, **237**, 201.
10. O. Karlsson, S. Thor, T. Norberg, H. Ohlsson, and T. Edlund, *Nature (London)*, 1990, **344**, 879.
11. G. Freyd, S. K. Kim, and H. R. Horvitz, *Nature (London)*, 1990, **344**, 876.
12. X. Qian, S. N. Gozani, H. Yoon, C. Jeon, K. Agarwal, and M. A. Weiss, *Biochemistry*, 1993, **32**, 9944.
13. L. H. Schulman, *Prog. Nucleic Acids Res. Mol. Biol.*, 1991, **41**, 23.
14. P. Schimmel, *Curr. Opin. Struct. Biol.*, 1992, **1**, 811.
15. D. P. Giedroc, B. A. Johnson, I. M. Armitage, and J. E. Coleman, *Biochemistry*, 1989, **28**, 2410.
16. D. P. Giedroc, H. Qiu, R. Kahn, G. C. King, and K. Chen, *Biochemistry*, 1992, **31**, 765.
17. P. J. Kraulis, A. R. C. Raine, P. L. Gadhavi, and E. D. Laue, *Nature (London)*, 1992, **356**, 448.
18. J. D. Baleja, R. Marmorstein, S. C. Harrison, and G. Wagner, *Nature (London)*, 1992, **356**, 450.
19. R. Marmorstein, M. Carey, M. Ptashne, and S. C. Harrison, *Nature (London)*, 1992, **356**, 408.
20. M. S. Lee, G. P. Gippert, K. A. Soman, D. A. Case, and P. E. Wright, *Science (Washington, DC)*, 1989, **245**, 635.
21. J. G. Omichinski, G. M. Clore, E. Appella, K. Sakaguchi, and A. M. Gronenborn, *Biochemistry*, 1990, **29**, 9324.
22. M. Kochoyan, T. F. Havel, D. T. Nguyen, C. E. Dahl, H. T. Keutmann, and M. A. Weiss, *Biochemistry*, 1991, **30**, 3371.
23. M. Kochoyan, H. T. Keutmann, and M. A. Weiss, *Biochemistry*, 1991, **30**, 7063.
24. M. Kochoyan, H. T. Keutmann, and M. A. Weiss, *Proc. Natl. Acad. Sci. USA*, 1991, **88**, 8455.
25. R. C. Hoffman, S. C. Horvath, and R. E. Klevit, *Protein Sci.*, 1993, **2**, 951.
26. M. F. Summers, T. L. South, B. Kim, and D. R. Hare, *Biochemistry*, 1990, **29**, 329.
27. T. L. South, P. R. Blake, D. R. Hare, and M. F. Summers, *Biochemistry*, 1991, **30**, 6342.
28. M. F. Summers, L. E. Henderson, M. R. Chance, J. W. Bess, Jr., T. L. South, P. R. Blake, I. Sagi, G. Perez-Alvarado, R. C. Sowder, III, D. R. Hare, and L. O. Arthur, *Protein Sci.*, 1992, **1**, 563.
29. J. G. Omichinski, G. M. Clore, K. Sakaguchi, E. Appella, and A. M. Gronenborn, *FEBS Lett.*, 1991, **292**, 25.
30. N. Morellet, N. Jullian, H. De Rocquigny, B. Maigret, J.-L. Darlix, and B. P. Roques, *EMBO J.*, 1992, **11**, 3059.
31. T. Hard, E. Kellenbach, R. Boelens, B. A. Maler, K. Dahlman, L. P. Freedman, J. Carlstedt-Duke, K. R. Yamamoto, J.-Å. Gustafsson, and R. Kaptein, *Science (Washington, DC)*, 1990, **249**, 157.
32. J. W. R. Schwabe, D. Neuhaus, and D. Rhodes, *Nature (London)*, 1990, **348**, 458.
33. R. M. A. Knegtel, M. Katahira, J. G. Schilthuis, A. M. J. J. Bonvin, R. Boelens, D. Eib, P. T. van der Saag, and R. Kaptein, *J. Biomol. NMR*, 1993, **3**, 1.
34. H. Baumann, K. Paulsen, H. Kovács, H. Berglund, A. P. H. Wright, J.-Å. Gustafsson, and T. Härd, *Biochemistry*, 1993, **32**, 13 463.
35. M. S. Lee, S. A. Kliewer, J. Provencal, P. E. Wright, and R. M. Evans, *Science (Washington, DC)*, 1993, **260**, 1117.
36. T. L. South and M. F. Summers, *Protein Sci.*, 1993, **2**, 3.
37. G. C. Perez-Alvarado, C. Miles, J. W. Michelsen, H. A. Louis, D. R. Winge, M. C. Beckerle, and M. F. Summers, *Nature Struct. Biol.*, 1994, **1**, in press.
38. N. P. Pavletich and C. O. Pabo, *Science (Washington, DC)*, 1991, **252**, 809.
39. N. P. Pavletich and C. O. Pabo, *Science (Washington, DC)*, 1993, **261**, 1701.
40. B. F. Luisi, W. X. Xu, Z. Otwinowshi, L. P. Freedman, K. R. Yamamoto, and P. B. Sigler, *Nature (London)*, 1991, **352**, 497.
41. P. B. Blake, B. Lee, M. F. Summers, J.-B. Park, Z. H. Zhou, and M. W. W. Adams, *New J. Chem.*, 1994, **18**, 387.
42. P. R. Blake and M. F. Summers, *Adv. Biophys. Chem.*, 1994, **4**, 1.
43. P. R. Blake and M. F. Summers, *Adv. Inorg. Biochem.*, 1994, **10**, 201.
44. E. Adman, E. D. Watenpaugh and L. H. Jensen, *Proc. Natl. Acad. Sci. USA*, 1975, **72**, 4854.
45. P. R. Blake, J.-B. Park, M. W. W. Adams, and M. F. Summers, *J. Am. Chem. Soc.*, 1992, **114**, 4931.
46. P. R. Blake, J.-B. Park, Z. H. Zhou, D. R. Hare, M. W. W. Adams, and M. F. Summers, *Protein Sci.*, 1992, **1**, 1508.
47. P. R. Blake, B. Lee, M. F. Summers, M. W. W. Adams, J.-B. Park, Z. H. Zhou, and A. Bax, *J. Biomol. NMR*, 1992, **2**, 527.
48. P. Dauber-Osguthorpe and D. J. Osguthorpe, *J. Am. Chem. Soc.*, 1990, **112**, 7921.
49. C. C. Moser, J. M. Keske, K. Warncke, R. S. Farid, and P. L. Dutton, *Nature (London)*, 1992, **355**, 796.
50. D. N. Beratan, J. N. Onuchic, J. N. Betts, B. E. Bowler, and H. B. Gray, *J. Am. Chem. Soc.*, 1990, **112**, 7915.
51. D. N. Beratan, J. N. Betts, and J. N. Onuchic, *Science (Washington, DC)*, 1991, **252**, 1285.

52. D. N. Beratan, J. N. Onuchic, J. R. Winkler, and H. B. Gray, *Science (Washington, DC)*, 1992, **258**, 1740.
53. L. M. Green and J. M. Berg, *Proc. Natl. Acad. Sci. USA*, 1990, **87**, 6403.
54. R. Klevit, J. R. Herriott, and S. J. Horvath, *Proteins: Struct., Funct., & Genet.*, 1990, **7**, 215.
55. E. Adman, K. D. Watenpaugh, and L. H. Jensen, *Proc. Natl. Acad. Sci. USA*, 1975, **72**, 4854.
56. J. G. Omichinski, G. M. Clore, M. Robien, K. Sakaguchi, E. Appella, and A. M. Gronenborn, *Biochemistry*, 1992, **31**, 3907.
57. D. P. Bolognesi, R. C. Montelaro, H. Frank, and W. Schäfer, *Science (Washington, DC)*, 1978, **199**, 183.
58. C. Dickson, R. Eisnman, R. Fan, E. Hunter, and N. Teich, in 'The Molecular Biology of Tumor Viruses: RNA Tumor Viruses', 2nd edn., eds. R. A. Weiss, N. Teich, H. Varmus, and J. Coffin, Cold Spring Harbor Laboratory, Cold Spring Harbor, NY, 1985.
59. M. F. Summers, *J. Cell. Biochem.*, 1991, **45**, 41.
60. J. R. Webb and W. R. McMaster, *J. Biol. Chem.*, 1993, **268**, 13 994.
61. T. B. Rajavashisth, A. K. Taylor, A. Andalibi, K. L. Svenson, L. Karen, and A. J. Lusis, *Science (Washington, DC)*, 1989, **245**, 640.
62. S. M. Sato and T. D. Sargent, *Development*, 1991, **112**, 747.
63. M. F. Summers, T. L. South, B. Kim, and D. R. Hare, *Biochemistry*, 1990, **29**, 329.
64. M. S. Lee, S. A. Kliewer, J. Provencal, P. E. Wright, and R. M. Evans, *Science (Washington, DC)*, 1993, **260**, 1117.
65. J. W. R. Schwabe, D. Neuhaus, and D. Rhodes, *Nature (London)*, 1990, **348**, 458.
66. K. H. Gardner, T. Pan, S. Narula, E. Rivera, and J. E. Coleman, *Biochemistry*, 1991, **30**, 11 292.
67. J. Ma and M. Ptashne, *Cell*, 1987, **48**, 847.
68. M. Johnston, *Nature (London)*, 1987, **328**, 353.
69. T. Pan and J. E. Coleman, *Biochemistry*, 1991, **30**, 4212.
70. H.-Y. Yang and T. Evans, *Mol. Cell. Biol.*, 1992, **12**, 4562.
71. D. I. K. Martin and S. H. Orkin, *Genes & Devel.*, 1990, **4**, 1886.
72. D. J. Whyatt, E. deBoer, and F. Grosveld, *EMBO J.*, 1993, **12**, 4993.
73. D. Fourmy, F. Dardel, and F. Blanquet, *J. Mol. Biol.*, 1993, **231**, 1078.
74. D. Fourmy, T. Meinnel, Y. Mechulan, and S. Blanquet, *J. Mol. Biol.*, 1993, **231**, 1068.
75. L. H. Posorske, M. Cohn, N. Yanagisawa, and D. S. Auld, *Biochim. Biophys. Acta*, 1979, **576**, 128.
76. J.-F. Mayaux and S. Blanquet, *Biochemistry*, 1981, **20**, 4647.
77. W. T. Miller, K. A. W. Hill, and P. Schimmel, *Biochemistry*, 1991, **30**, 6970.
78. J.-F. Mayaux and S. Blanquet, *Biochemistry*, 1981, **20**, 4647.
79. D. Reines and J. Mote Jr., *Proc. Natl. Acad. Sci. USA*, 1993, **90**, 1917.
80. X. Qian, S. N. Gozani, H. Yoon, C. Jeon, K. Agarwal, and M. A. Weiss, *Biochemistry*, 1993, **32**, 9944.
81. L. C. Myers, M. P. Terranova, A. E. Ferentz, G. Wagner, and G. L. Verdine, *Science (Washington, DC)*, 1993, **261**, 1164.
82. L. C. Myers, G. L. Verdine and G. Wagner, *Biochemistry*, 1993, **32**, 14 089.
83. P. S. Freemont, *Ann. NY Acad. Sci.*, 1993, **684**, 174.
84. R. D. Everett, P. Barlow, A. Milner, B. Luisi, A. Orr, G. Hope, and D. Lyon, *J. Mol. Biol.*, 1993, **234**, 1038.
85. R. D. Everett, P. Barlow, A. Milner, B. Luisi, A. Orr, G. Hope, and D. Lyon, *J. Mol. Biol.*, 1993, **234**, 1.
86. P. N. Barlow, B. Luisi, A. Milner, M. Elliott, and R. Everett, *J. Mol. Biol.*, 1994, **237**, 1201.
87. Y. Xu, M. Baldassare, P. Fisher, G. Rathbun, E. M. Oltz, G. D. Yancopoulos, T. M. Jessell, and F. W. Alt, *Proc. Natl. Acad. Sci. USA*, 1993, **90**, 227.
88. C. Bourgouin, S. E. Lundgren, and J. B. Thomas, *Neuron*, 1992, **9**, 549.
89. B. Cohen, M. E. McGuffin, C. Pfeifle, D. Segal, and S. M. Cohen, *Genes Devel.*, 1992, **6**, 715.
90. M. Taira, M. Jamrich, P. J. Good, and I. B. Dawid, *Genes Devel.*, 1992, **6**, 356.
91. T. Boehm, J. M. Greenberg, L. Buluwela, I. Lavenir, A. Forster, and T. H. Rabbitts, *EMBO J.*, 1990, **9**, 857.
92. E. H. Birkenmeier and J. I. Gordon, *Proc. Natl. Acad. Sci. USA*, 1986, **83**, 2516.
93. I. Sadler, A. W. Crawford, J. W. Michelsen, and M. C. Beckerle, *J. Cell Biol.*, 1992, **119**, 1573.
94. S. A. Liebhaber, J. G. Emery, M. Urbanek, X. Wang, and N. E. Cook, *Nucleic Acids Res.*, 1990, **18**, 3871.
95. A. W. Crawford, J. D. Pino, and M. C. Beckerle, *J. Cell. Biol.*, 1994, **124**, 117.
96. G. A. McGuire, R. D. Hockett, K. M. Pollock, M. F. Bartholdi, S. J. O'Brien, and S. J. Korsmeyer, *Mol. Cell. Biol.*, 1989, **9**, 2124.
97. X. Wang, G. Lee, S. A. Liebhaber, and N. E. Cooke, *J. Biol. Chem.*, 1992, **267**, 9176.
98. R. Weiskirchen and K. Bister, *Oncogene*, 1993, **8**, 2317.
99. R. D. Everett, P. Barlow, A. Milner, B. Luisi, A. Orr, G. Hope, and D. Lyon, *J. Mol. Biol.*, 1994, **234**, in press.
100. K. D. Watenpaugh, L. C. Sieker, and L. H. Jensen, *J. Mol. Biol.*, 1979, **131**, 509.
101. U. Hommel, M. Zurini, and M. Luyten, *Nature Struct. Biol.*, 1994, **1**, 383.
102. H. Déméné, N. Jullian, N. Morellet, H. de Rocquigny, F. Cornille, B. Maigret, and B. P. Roques, *J. Biomol. NMR*, 1994, **4**, 153.
103. K. H. Gardner, S. F. Anderson, and J. E. Coleman, *Nature Struct. Biol.*, 1995, **2**, 898.

Biographical Sketch

Michael F. Summers. *b* 1958. B.S., 1980, University of West Florida, Ph.D., 1984, Emory University. Postdoctoral training at NIH with Ad Bax, R. Andrew Byrd, William Egan. Introduced to NMR by Luigi G. Marzilli (Emory). Assistant professor, 1987–91, associate professor, 1991–present, University of Maryland, Baltimore County. Associate investigator, Howard Hughes Medical Institute, 1994–present. Approximately 70 publications in the areas of NMR and bioinorganic chemistry. Research interests include applications of NMR spectroscopy to studies of metalloproteins and macromolecular interactions.

Imaging and Medical Glossary

Pulse Sequence Notation

The sequence notation is based on the recommendations of the American College of Radiology (American College of Radiology Glossary of NMR Terms, ACR, Chicago, 1983). The notation has the form SE 1500/50, where the first two letters describe the sequence (PS = partial saturation, SE = spin echo, etc). The first number is the repetition time (TR) in milliseconds; the second is the time (TE) to echo formation; the third (if present) is the inversion time in an inversion recovery (IR) sequence. Additional, other, time intervals can be defined by third and other numbers, depending on the sequence.

The chemical shift between aqueous and lipid components in tissue is generally taken as 3.4–3.5 ppm.

Image Orientation

Transverse (transaxial) images are normally oriented in publications as if the slice was being viewed by an observer standing beyond the patient's feet. In a few non-North American instances, with little consistency even inside any one country, body images are oriented as if viewed from the feet, while the head and neck are viewed from above the top of the head.

Sagittal images are conventionally displayed as if the patient is lying on his/her back, with the observer looking down on the subject. The images are oriented as if viewed from the observer's left hand side (i.e. he/she is looking at the right hand wall of the slice).

Coronal images are oriented as if the patient were lying on his/her back with the observer looking down at the subject. The right side of the patient then appears at the left of the image.

Glossary terms

AA	Ascending Aorta
Acquisitions	See NSA
ADC	Apparent Diffusion Coefficient; experimentally observed value of D, in the presence of likely confusing factors such as tissue structural anisotropy; cf. D* [not to be confused with ADC, meaning analog-to-digital converter]
Adhesion	Fibrous band or structure linking two tissues not normally joined
ADP	Adenosine DiPhosphate
AE	Arterial Enhancement
Aerobic	Metabolism (usually), etc., operating using oxygen
Agenesis	Absence of an organ (from development failure onward)
AIDS	Acquired Immune Deficiency Syndrome
AMP	Adenosine MonoPhosphate
Anaerobic	Metabolism (usually), etc., operating without oxygen
Aneurysm	Region of a blood vessel in which a blockage has resulted in a swollen, weakened, and life threatening segment of the vessel
Angioplasty	Reconstruction or restructuring of a blood vessel by operation or interventional technique such as balloon dilator or laser
Anterior	Frontal; towards the front (notation used particularly in sagittal imaging)
Anteromedial	Towards the front on the center line of the body
Anteroposterior (movement)	Back to front movement towards and away from the spinal column (back/front motion); syn. AP
AP	See AnteroPosterior
Arrhythmia	Irregularity in a fixed rhythm; usually associated with heart beat irregularities (cardiac arrythmia)
Arteriole	Minute arterial branch; distribution level before, in circulatory terms, a capillary
Ataxia	Failure of muscular control and/or coordination; description of disease in other sites leading to muscular misfixation
ATP	Adenosine TriPhosphate
Atrophy	Shrinkage, loss of bulk, of an organ (e.g. atrophy of the brain)
Avascular	Without conventional blood supply; description applied to many lesions
Averages	See NSA
AVM	ArterioVenous Malformation; lesion innovating distortion and disruption of its normal vascular system
BBB	Blood Brain Barrier; largely impermeable vessel wall membrane in the brain, allowing passage of small molecules such as O_2 and CO_2; disruption of this membrane is frequently symptomatic of disease

BFAST	See Saturation
Biopsy	Procedure in which a small piece of tissue is extracted from a suspect region for histological examination
Birdcage coil (or birdcage)	Very common multiconductor cylindrical coil used in whole body MR
'Black blood'	Process of acquiring and reconstructing angiographic data which leaves vascular structure dark against a light background; cf. Bright or White blood
BOLD	Blood Oxygen Level Dependent contrast; mechanism used in describing changes detected in functional MRI (fMRI)
BPH	Benign Prostatic Hyperplasia; common older male nonmalignant disease; symptoms are blockage of urethra and difficulty in urinating
Breath-hold	Imaging method in which a patient is asked to hold his/her breath for the duration of data acquisition (usually a few seconds)
'Bright blood'	Process of acquiring and reconstructuring data such that the vasculature appears bright against a dark ground; syn. 'White blood'; cf. Black blood
BSA	Bovine Serum Albumen; extract used in MRI, particularly for relaxation behavior studies
Ca	Carcinoma; abbreviation used with localizing description of tumor
CABS	Coronary Artery Bypass Surgery
CAD	Coronary Artery Disease (see syn. CHD)
Calcification	Large or small ('microcalcification') solid deposits visible in MRI only through lack of signal; observed in a variety of lesions, and particularly associated with some tumors
Catabolite	Product of a destructive metabolic process
Caudal	Towards the foot; associated with direction of motion, flow; cf. Cephalad
CBF	Cerebral Blood Flow
CBV	Cerebral Blood Volume
CCT	Cranial Computed Tomography
CE	Contrast Enhanced; frequently used in the context 'CE images', meaning images acquired after a contrast agent has been given to the patient
CE-FAST	T_2 weighted rapid sequence, with analogues with FISP (see below) and direct connection with STEAM (see below)
Cephalad	Towards the head; associated with direction of motion, flow; cf. Caudal
CHD	Coronary Heart Disease (see syn. CAD)
Chemsat	See FATSAT
Cho	Choline; metabolite primarily observed in proton spectroscopy but with associated metabolites also visible elsewhere
CK	Creatine Kinase; enzyme controlling major step in Krebs' metabolic cycle
CNR	Contrast-to-Noise Ratio; in practice, the signal difference between two tissues (or other entitites) being considered divided by the RMS noise
CNS	Central Nervous System; brain, brain stem, and spinal cord
Coarctation	Restriction of structure (or contraction of component)
Collateral	Used in imaging in the sense of incidental, attendant (usually unwanted) effects of a process or procedure; thus 'collateral damage' is associated with unwanted damage to normal tissue resulting from the application of therapy
Contrast	Difference in magnitude between two signals; sometimes contrast is stated as a fractional difference, at other times as an absolute difference
COPE	See ROPE
Coronal	Slice along long axis of the body parallel to its back (or front)
Corpus callosum	White matter structure in the brain, towards the centre, which provides the link between the two hemispheres
Cortex (adj. cortical)	Outer layer of brain or kidney; in the case of the former, constitutes much of the gray matter of brain
CP angle	Cerebello Pontine angle; feature of the cerebellum (lower rear portion of the brain at the level of the brain stem) [CP not to be confused with cross polarization]
Cr	Creatine; metabolite seen, as such, in the proton spectrum; in combination, also detected in other forms of spectroscopy
Cranial	To head direction
Craniocaudal (movement)	Movement up and down the body, parallel to the spine
CRF	Chronic Renal Failure
CSF	Cerebro-Spinal Fluid; water-like fluid surrounding the brain, spinal cord, and filling ventricles in the brain
CSI (CS Imaging)	Chemical Shift Imaging; class of acquisitions in which spectral and positional data are encoded together and acquired at the same time

CT	(X-ray) Computerized Tomography (often abbreviated to CT X-ray, the original computer-based imaging method)
Cytology	Study of cells, including origin, function, structure, and pathology
Cytotoxicity	Ability to produce a specific toxic action in cells in various organs
D*	See ADC
DA	Descending Aorta [not to be confused with DA meaning data acquisition]
DAC	DiAcylglycerol; metabolite seen in spectroscopy [not to be confused with DAC meaning digital-to-analog converter]
Data sets	See NSA
Diastole	Period of cardiac cycle during which the heart expands (cf. Systole, shorter period of contraction)
Diffuse (disease)	Distributed disease affecting a significant fraction of an organ without specific localizing boundaries (cf. Focal)
Dilation	Swollen or expanded (of tissue)
Distal	Remote from, far away from, any reference point (cf. Proximal, adjacent to)
Dorsal	Pertaining to the back, or rear, of anything referred to
Double oblique	Slice, the plane of which is parallel to no major axis of the machine
DPDE	DiPhospho DiEster
DRESS	Slice-selective spectroscopic method used for phosphorus spatial localization
DTPA	DiethyleneTriaminePentaAcetic acid; chelate used to minimize toxicity of otherwise desirable agent (see Gd-DTPA)
DWI	Diffusion Weighted Imaging
Dysplasia	Abnormality of development; in pathology, alteration in size, shape, and organization of cells
EC	ExtraCellular
ECG	ElectroCardioGram (loosely used to describe both the equipment used, and its output)
Ectopic (beat)	Abnormal pattern or location of organ or structure
Edema	Additional extra and extracellular fluid associated with tissue which is damaged in some way
EDTA	EthyleneDiamineTetraAcetic acid; another common chelating agent (cf. DTPA)
Effusion	Undesirable escape of fluid into tissue or structure
End-diastole	Part of cardiac cycle at end of heart expansion
Endogenous	Developing, or growing, from within a tissue or structure
EPI	Echo Planar Imaging; single-shot fast acquisition introduced by Mansfield
Epidural	Structured upon, or outside, the dura mater (the outermost layer covering the brain and spinal canal); an epidural injection, for example, is one delivered, typically, into the spinal cord, outside the cord
EPRI	Electron Paramagnetic Resonance Imaging
Erythrocytes	Components of blood, known also as red blood cells, or corpuscles
ESP	Echo SPacing in multiple acquisitions such as RARE (see below) (stated as a time interval)
Etiology	Study of factors that cause disease, and their introduction to the host
ETL	Echo Train Length; number of echoes in RARE-type sequences (see RARE)
EVI	Echo Volumnar Imaging; three-dimensional version of echo planar imaging
Excitations	See NSA
Exogenous	Produced or otherwise originating outside the organism or region
Extracapsular	Situated or occurring outside a capsule (such as that, for example, surrounding a cyst)
Extradural	Situated or occurring outside the dura mater (the outer lining of the brain and spinal cord)
Extramural	Situated or occurring outside the wall of an organ
Extravasate	Escape from the vascular system
FAISE	See RARE
FATSAT	Sequence designed for the spectrally selective saturation of signals in MRI (normally used to saturate lipid components, as in breast imaging); there are a variety of methods employed for this purpose; syn. Chemsat
Fat saturation	Process of destroying the signal from lipid components in MRI to avoid chemical shift artifacts and improve lesion conspicuity in regions such as the breast
FE	Field Echo; see also GRE
Fibrillation	Uncoordinated quivering of the heart without normal blood-pumping beat
Fibrosis	Formation of fibrous tissue, or fibrous degeneration

FISP	Fast Imaging with Steady Precession; essentially a steady state derivative of FLASH (see below) in which magnetization is refocused after data acquisition, rather than being destroyed
Fistula	Abnormal passage or channel between two internal organs, or between an internal location and the body surface
FLAIR	Fluid Attenuated Inversion Recovery; inversion recovery sequence times so that signals from CSF (see above) are suppressed
FLASH	Fast Low Angle SHot; rapidly repeated (short TR) sequence with reduced precession angle and gradient recalled echo data acquisition
fMRI	Functional Magnetic Resonance Imaging; generic description of procedures in which a subject undergoes sensory stimulation, while responses are sought from brain imaging
FMRI	See fMRI
FNA	Fine Needle Aspiration
Focal (lesion)	A localized lesion with defined boundaries
FOV	Field Of View
FREEZE	See ROPE
Frequency encoding	Gradient applied to disperse spin frequencies during acquisition of data
FSE	Fast Spin Echo; see RARE
FSW	Fourier Series Windowing
fwhm	Full width half maximum; definition of linewidth, or, in context of point spread function, of resolution in imaging
$f(x, y, z)$	Spin density at point x, y, z
Gadopentate	See Gd-DTPA
Gag	An instrument for forcing open or holding open the mouth of the unconscious patient, particularly when under general anesthetic
Gating	Cardiac, respiratory; process by which machine operation is synchronized with the patient's behavior
Gd-DTPA	Gadolinium DiethyleneTriaminePentaAcetic acid; gadolinium chelate used as an in vivo contrast agent (also known as Gadopentate)
Gd-DTPA-BMA	Non-ionic variant of Gd-DTPA (last three letters represent 'BisMethylAmide')
GE	Gradient Echo; see GRE
Ghost	Artifact of motion (generally); normally an unwanted single or multiple repetition of image (or other) data in the phase encode direction(s) of a 2(3)D data set
Gibb's ringing	Artifact of the Fourier Transform process which is obvious where the resolution of the data is too coarse; analogous to effects when the data are truncated in some way; appears in images as ripples of reducing signal amplitude, adjacent to edges between sharp intensity changes
Gliosis	Excess of astroglia in damaged areas of the central nervous system
GLX	Glutamine plus glutamate; convenient code for components which are as yet unresolved in in vivo proton spectroscopy
GM	Gray Matter
GMC	Gradient Moment Compensation; see MAST
GMN	Gradient Moment Nulling; see MAST
GMR	Gradient Moment Rephasing; see MAST
Gradient echo	See GRE
GRASE	Combination of RARE and Gradient Recalled Acquisitions in a single multi-echo sequence (see RARE, EPI)
GRASS	Gradient Recalled Echo in Steady State; see GRE
GRE	Gradient Recalled Echo; commonly used acronym for 'spin warp' data acquisition; following a selective or nonselective pulse of significantly less than 180°, the magnetization is coded and an echo formed using gradients only; see also Spin warp
Great vessels	Large vessels entering the heart, including the aorta, pulmonary arteries and veins, and the venae cavae
Hc_t	Hematocrit (see below)
HE	Hepatic Encephalopathy; disease secondary (usually) to advanced liver disease
Hematocrit	Volume percentage of erythrocytes in whole blood
Hemorrhage	Deposit of blood products arising from a leak in the walls of, or break or damage to, a blood vessel
Heterozygote	Individual with different alleles with respect to a given characteristic
HIV	Human Immunodeficiency Virus

HMPAO-SPECT	HexaMethylPropyleneAmine-Oxide-SPECT (single photon emission computed tomography); lipid-soluble nuclear agent widely used in cardiac diagnosis
Homeostasis	Tendency to stability in the normal body states of an organism
Homozygote	Individual possessing a pair of identical alleles at a given location
Hyperintense	The component being described is brighter on the image than the region with which it is being compared (cf. Hypointense)
Hyperplasia	Abnormal multiplication of normal cells in an otherwise normal tissue structure
Hyperthermia	Class of methods in which tissue temperature is raised, particularly as a treatment for cancer; various thermal sources (rf, microwave, laser, ultrasound) are used, but the name is perhaps most closely associated with rf as the source of energy
Hypertrophy	Increase (swelling), particularly of an organ (cf. Hypotrophy)
Hypointense	The component being described is less bright on the image than that with which it is being compared (cf. Hyperintense)
Hypoplasia	Incomplete development or underdevelopment of an organ or tissue
Hypotrophy	Shrinkage, loss of size, of an organ (cf. Hypertrophy)
Hypoxia	Deficient in oxygen (in the sense of tissue being hypoxic)
IADSA	Intra Arterial Digital Subtraction Angiography; method of X-ray angiography
IC	IntraCellular
Ictus	Event, occurrence, usually in sense of a traumatic event clinically leading to consequential affects
Idiopathic	Self-originated; of unknown origin
Infarction	Result in tissue of loss of blood supply which has led to cell death; considered irreversible
Infusion	Therapeutic introduction of a fluid other than blood into a vein
Interleaving	Process (in spiral scanning, EPI, etc.) in which part of *k*-space is traversed in each of a series of acquisitions where the lines acquired at any one time are not contiguous
Intracranial	Inside the skull
Intramural	Within the wall of an organ
'Intrinsic' SNR	Signal-to-noise ratio calculated ignoring signal strength variations arising from the method of data acquisition (and, so, avoiding considerations such as the dependency of the relaxation time constants on field strength), i.e. signal-to-noise ratio of a fully recovered spin population with the signal measured immediately post excitation
IOP	Iron Oxide Particle; particularly in the context of superparamagnetic Fe_2O_3 particles used as contrast agent material (these have a typical diameter of 20–30 nm)
Ischemia	Process in which a region of tissue receives an inadequate blood supply for its needs; usually considered reversible
ISIS	Image Selected In vivo Spectroscopy
Isointense	There exists no contrast between two tissues which are of equal signal intensity
IV	IntraVenous; particularly used to qualify an injection, for example, of contrast agents
IVC	Inferior Vena Cava
k-space	Spatial frequency space, and so that space in which the time resolved data are recovered
k_x, k_y, k_z	Coordinates of *k*-space along major axes
La	Lactate
LGC	Local Gradient Coil; typically a small, temporarily inserted, coil structure
Lip	In positional sense, marginal part, or section
Locular	Pertaining to a small space or cavity
Lumen	Cavity within a tube or organ
LV	Left Ventricle
Lysis	Dissolution, loosening, destruction (e.g. of cells by a specific agency); also, gradual abatement of symptoms of a disease
M_0	Fully recovered magnetization, usually taken to be the observable magnetization in whole body experiments, thus allowing for the invisibility to normal MRI and MRS methods of some nuclei and reduced contribution from others
Macrophage	Any of the large phagocytic cells occurring in walls of blood vessels and loose connective tissue
Magnevist	Trade name for the widely used MRI contrast agent Gd-DTPA (see above)
MARF	Magic Angle in the Rotating Frame
Mass effect	Shift or distortion of the normal anatomical tissue pattern due to a lesion which is occupying space otherwise available for it
MAST	Motion Artifact Suppression Technique; method in which multiple gradient lobes are used to control motion artifacts

Mediolateral	At midline level towards the side (as could be a location in a coronal image)
Meninges	Three membranes enveloping the brain and spinal cord
MESA	Multiple Echo Spectroscopy Acquisition; method in which the first 90° pulse is chemical shift selective, which is followed by three slice-selective 180° pulses to define a volume from which the signal is obtained
Metabolite	Any chemical component present in the tissue which may or may not take part in biochemical reactions whether or not they are observable
MI	*Myo*-Inositol; metabolic component seen in proton spectroscopy; assume this usage in spectroscopy articles
MI	Myocardial Infarction; assume this usage in clinical articles
Microvasculature	Portion of circulatory system comprising the finer vessels (typically those with an internal diameter of less than 100 μm)
MIP	Maximum Intensity Projection; method in which the highest intensity signal along a line through an image data set is taken to be the value along that line; the method is frequently used in formatting angiography data
MnDPDP	Experimental, near clinical, contrast agent, with potential liver applications, using manganese as the paramagnetic moiety (actually Manganese DiPyridoxyl DiPhosphonate)
Morbidity	Death or other grossly unfavorable outcome
Morphology	Anatomy or anatomical structure
Motion refocusing	See MAST
MPR	MultiPlanar Reconstruction
MRA	Magnetic Resonance Angiography
MRI	Magnetic Resonance Imaging
MRM	Magnetic Resonance Microscopy
MRS	Magnetic Resonance Spectroscopy
MRSI	Magnetic Resonance Spectroscopic Imaging
MRV	Magnetic Resonance Venography
MS	Multiple Sclerosis
MT	Magnetization Transfer (see magnetization transfer contrast, MTC)
MTC	Magnetization Transfer Contrast
MTF	Modulation Transfer Function
Multi-coil	Coil system comprising more than one coil assembly, each with its own preamplifiers, matching network, etc.
Multi-echo	Method in which a series of spin-echo images are acquired following a single first excitation; economical of time when different image contrasts are required quickly; cf. RARE, as the most highly developed form of the process in routine imaging, or the CPMG (Carr–Purcell–Meiboom–Gill) sequence as a more extreme case
Multi-slice	Method in which interleaved parallel slices are acquired at the same time, so employing the recovery periods (TR) in sequences profitably to reduce the time needed to cover a region of the body
Myocardial	Of, or pertaining to, the myocardium (the thickest muscle layer of the heart)
Myopathy	Any disease of a muscle
NAA	*N*-AcetylAspartate; brain metabolite dominating normal adult white matter proton brain spectrum; function is unclear, but it is considered to be associated with the presence of neurons
NAD	Nicotinamide-Adenine Dinucleotide (NADP = nicotinamide-adenine dinucleotide phosphate)
NAD^+	Oxidized form of NAD ($NADP^+$ = oxidized form of NADP)
NADH	Reduced form of NAD (NAD(P)H also called NADPH = reduced form of NADP)
NCO	Numerically Controlled Oscillator (or DDS = direct digital synthesizer)
NDP	Nucleotide DiPhosphate (i.e. any diphosphate)
Necrosis	See necrotic
Necrotic	Dead (noun necrosis); description of a region where cells present have died and tissue has degenerated
Neoplasia	Formation of a neoplasm (tumor)
NEX	See NSA
N(H)	Symbol for proton density; see PD
Nidus	Point of origin, or focus, of a morbid process
NPH	Normal Pressure Hydrocephalus
NPV	Negative Predictive Value

NSA	Number of Signal Averages; number of repetitions of each encoding step through *k*-space which are integrated; syn. acquisitions, data sets, averages, NEX, excitations, repetitions, projections, views, and profiles
NTP	Nucleotide TriPhosphate (i.e. any triphosphate)
NV	Number of Views (see NSA)
Oblique	Slice, the plane of which is parallel to one major axis only of the body
Oblique-sagittal	Slice orientation but visualized as the notation of a sagittal slice about its vertical axis; oblique-coronal, by analogy, describes a coronal slice rotated about its transverse axis (i.e. that normal to the height of the patient)
Occlusion (occluded)	Blockage in a blood vessel (partially occluded = partially blocked)
Omniscate	Trade name for the contrast agent GdDTPA-BMA
OPE	See ROPE
OR	Operating Room (theater)
PA	Pulmonary Artery
Parenchyma	Normal tissue, excluding the vasculature and other fluid channels
Paresis	Slight or incomplete paralysis
Pathogen	Any agent which is inimical to the well being of tissue
Pb EDTA	Lead-EDTA (lead chelate; see EDTA)
PC	PhosphorylCholine, if in spectroscopy article
PC	Phase Contrast, if in imaging, particularly flow-related, article
PCa	Carcinoma of the prostate; this is a common type of abbreviation needing individual interpretation in context; see Ca
PCA extract	PerChloric Acid extract
PCr	PhosphoCreatine
PD	Proton Density; frequently used as in 'PD-weighted' in clinical articles
PDCE	Proton Detected Carbon Edited; technique for editing proton spectra dependent on the presence of carbon metabolites (used to detect latter)
PDE	PhosphoDiEsters; various phosphorus metabolites with a common double ester structure; normally seen in the phosphorus spectrum as a single broad peak, resolved in extracts, and, to some degree, by proton decoupling
PDW	Proton Density Weighted; frequently as 'pD or PD weighted'; see Proton density weighted
PEACH	Paramagnetic Enhancement Accentuated by CHemical shift; method in which a contrast agent is used with fat suppression using chemical shift selective methods to visualize (particularly) breast cancer
PEAR	See ROPE
PE	Parenchymal Enhancement
PEDRI	Proton–Electron Double Resonance Imaging
Peristalsis	Peristaltic movement; random, uncontrollable, motion of the gut
Periventricular	Round, and adjacent to, the ventricles
PEt	PhosphorylEthanolamine; phosphorus metabolite which is part of the phosphorus monoester complex seen in ^{31}P spectroscopy
PET	Positron Emission Tomography
Phase encoding	Process by which, using relatively short gradient pulses, a signal pattern is imposed through *k*-space which is present throughout the subsequent data acquisition; it is normally changed before the next data acquisition (except where multiple averages of the same data are required)
Phase image	Image in which the gray scale is a saw tooth plot of the signal for the phase angle θ in the range $-\pi < \theta \leqslant \pi$; if unconstrained it may vary through a number of consecutive cycles, with an appropriate pattern of increasing change from dark to light, then sudden reversion to dark; if the real and imaginary components of the signal of a voxel (i,j) are $x_{i,j}$ and $y_{i,j}$, then $\theta = \tan^{-1} y_{ij}/x_{ij}$
PhCh	PhosphatidylCholine
Phe	Phenylalanine
pH_i	Intracellular pH
Pi	Inorganic phosphate
Pixel	Two-dimensional cell which is the basic unit of an image
PM	Post Mortem
PME	PhosphoMonoEster; in phosphorus spectroscopy, large peak of in vivo spectrum contributed to by a number of metabolites which are frequently not resolved; proton decoupling can be used to assist in line resolution

pO_2	Tissue oxygen partial pressure
POPE	See ROPE
PPV	Positive Predictive Value
PR	Projection Reconstruction; method of acquisition directly analogous to the operation of the original Hounsfield translate–rotate CT X-ray scanner; has potential virtue of allowing very rapid start to data acquisition
PRE	Proton Relaxation Enhancement; typically, the result of a reduction of T_1 in situations where sequence repetition times are of the same order or less than T_1
PRESAT	See Saturation
Presaturation	Saturation of all or part of an imaging volume prior to mean sequence operation; often used to describe the situation in which a region upstream (in a blood flow sense) of a region to be studied is saturated so as to avoid artifacts due to flow effects in the region, but also used to eliminate risk of aliasing when very small fields of view, which are less than the body's natural size, are in use
Prescan	Operation of scanning sequence prior to start of data acquisition to stabilize magnetization signal
Present	In the sense 'the patient presented with such and such a symptom'; this means he/she was first seen, and investigated, when the symptoms noted were recorded
PRESS	Spectroscopic spatial localization method (particularly for protons) in which a selective 90° pulse is followed by two selective 180° pulses and data acquisition; each selective excitation defines one plane, and the target region is the volume common to all three
PRI	Projection Reconstruction Imaging
Profiles	See NSA
Prohance	Proprietary name for the contrast agent Gd-DOTA
Projections	See NSA
Prospective cardiac gating	Method of gating in which the position of the patient's abdomen (as it is particularly relevant to abdominal imaging) is recorded at the time of the acquisition of each line of *k*-space, and, by using multiple acquisitions, or convolution, attempts are made to modify the data so as to reorder it
Proton density weighted	Sequence resulting in data where the resultant contrast is mainly a function of proton density; typical examples are long TR (TR > 2000 ms), short TE (TE < 30 ms) brain imaging sequences; this can also be known as 'balanced weighting'
Proximal	Adjacent to, beside, same point of reference
PS	Partial Saturation; another name for a 90-TR-90 sequence with slice-selective 90° pulses and following spatial encoding and data acquisition
PSF	Point Spread Function
PSIF	See CE-FAST
PW	Proton Weighted; see Proton density weighted
QCT	Quantitative Computerized Tomography
QRS complex	Series of three consecutive peaks and troughs (including the large 'R' one) in an EKG wave form
r_1, r_2	Relaxivities of moieties (units s^{-1} $mmol^{-1}$)
r_t	Transmitter coil radius (specifically of a surface coil)
RARE	Rapid Acquisition with Relaxation Enhancement
RBC	Red Blood Cell
Refocused FLASH Methods	Derivation of FLASH (see above) in which magnetization is refocused, and the signal is pseudo steady state (acronyms, inter alia, FISP, GRASS, FAST, CE-FAST)
Relaxometry (field cycling relaxometry)	Method developed by Koenig in which the relaxation times of components are measured at a series of different field levels
Repetitions	See NSA
RESCOMP	See ROPE
REST	See Saturation
RFZ	Rotating Frame Zeugmatography
ROC (curve)	Receiver Operating Characteristics (curve); method used to measure the performance of those interpreting radiological images and other similar data
RODEO	Method pioneered by the Dallas group specifically for the suppression of lipid signals in breast imaging (cf. FATSAT)
ROI	Region Of Interest
ROPE	Respiratory Ordered Phase Encoding; method of correcting for patient (particularly) abdominal motion; though there are slight variations, the following synonyms use essen-

	tially the same concept: COPE, POPE, RESCOMP, RSPF, PEAR, phase reordering, FREEZE
ROS	Region Of Sensitivity
R–R interval (R-wave)	Interval between successive R wave peaks in an EKG waveform
RSPF	See ROPE
RT	Radiation Therapy
R-wave	Large peak in electrocardiogram measurement of heart operation
SaO_2	Oxygen saturation
Sagittal	Slice along the long axis of the body, with its plane lying from front to back
SAR	Specific Absorption Rate (in the context, usually, of rf energy deposition in tissue)
SAT	See Saturation
Saturation	Applied to saturation of magnetization of regions outside a region of interest, either to remove inflowing material (e.g. blood) to avoid artifacts or to avoid aliasing when a small field of view is used (many acronyms are used for this); note that the name is also used for many other processes (see also Presaturation)
SCHE	Sub-Clinical Hepatic Encephalopathy
SE	Spin Echo; echo formed as the result of an inverting rf pulse (selective or otherwise) and in the presence or absence of spatial encoding gradients
Section	Alternative word for slice
Sensitivity	In the clinical statistical sense, the sensitivity of a test is the ratio of the number of patients diagnosed with a disease to the total of those who have it (i.e. those correctly giving a positive result plus the false negative tests)
Shunt	Diversion or by-pass (usually channel)
SI	Signal Intensity [not to be confused with SI referring to the international system of units]
Slab	Block of tissue subdivided into slices (usually by phase encoding parallel to the slab selection direction), but falling well short of a volume acquisition; multi-slab (cf. multi-slice), series of parallel slabs acquired in an interleaved manner
SLIT DRESS	Version of DRESS in which multiple slices are interleaved
SMA	Superior Mesenteric Artery (also known as Arteria Mesenterica Superior)
SMV	Superior Mesenteric Vein (also known as Vena Mesenterica Superior)
SPAMM	SPAtial Modulation of Magnetization; technique for visualizing tissue motion
Specificity	In the clinical statistical sense, the fraction of normal subjects who give a correct negative result in a test to the total of normal subjects evaluated (i.e. the total of those correctly defined as normal plus the false positive results)
SPECT	Single Photon Emission Computed Tomography
Spin warp	Original Aberdeen-developed sequence using a gradient followed by its reversal to form an echo, through the development of which data are acquired; the method used a set of incremented pulses in a second direction to traverse *k*-space; see also GRE
SPIR	See FATSAT
Spoiled FFE	See spoiled FLASH
Spoiled FLASH	FLASH sequence with additional gradients or rf pulses to destroy residual coherent magnetization
Spoiled gradient echo	Variant of spoiled FLASH
SSFP	Steady State Free Precession; successful early method of image data acquisition; little used internationally nowadays
SSI	Solid State Imaging
ST	Saturation Transfer
ST	An abbreviation of the name of the Stejskal and Tanner pulsed gradient spin echo sequence (see PGSE) (has to be distinguished from previous entry by context)
STE	STimulated Echo (signal); see STEAM
STEAM (steam sequence)	STimulated Echo Acquisition Method of spatial localization
Stenosis	Narrowing or stricture of a duct or canal
STIR	Short T_1 Inversion Recovery; method primarily used to eliminate fat signals from images
Stroma	Supporting tissue or matrix of an organ (as distinguished from its functional component)
Subarachnoid	Located or occurring between the arachnoid and the pia motor (i.e. inner and mid-layer of meninges)
Subdural	Situated between the dura mater and the arachnoid (i.e. outer and central of the three layers of the meninges)
Supratentorial	Above the tentorium of the cerebellum
Susceptibility sensitive	In a pulse sequence, typically refers to gradient recalled echo sequences
SVC	Superior Vena Cava (large thoracic vein returning abdominal and lower blood to heart)

Synapse	(noun) Location of functional transfer between neurons; (verb) refers to the making of a connection
Systole	The part of the cardiac cycle during the contraction of the heart
$t_{1/2}$	The half-life of a process
T1W = T_1 weighted	Sequence resulting in data where the contrast is mainly a function of T_1; typical examples are short TR (300–500 ms), short TE (say 10–30 ms) brain imaging sequences
T2W = T_2 weighted	Sequence resulting in data where the contrast is mainly a function of T_2; typical examples are long TR (2000–3000 ms), long TE (TE > 60 ms) brain imaging sequences
T_2^* weighting	Conventionally, the contrast dependency found in gradient recalled acquisition sequences (or spin warp)
TCA-cycle	TriCarboxylic Acid (cycle); Krebs cycle
TE	Time to echo; time from the center of the rf pulse exciting transverse magnetization to the time at which the central point in image space is acquired (at some stage of the image acquisition process); if multiple lines are being acquired with individual excitations, TE is the time at which the central point of *k*-space is recovered in the data recovery without phase encoding; note that this is a more formal statement than that normally given, which simply defines TE as the time to the peak of the echo which is found at the center of the data acquisition (this version is not precise enough, however, in a number of situations)
TG	Gradient echo time; see TE
Thrombus (thrombosis)	Aggregate of blood components, primarily platelets and fibrin, frequently causing vascular obstruction
TI	Inversion time in an inversion recovery experiment; time between the center of the 180° inverting pulse and the pulse rotating magnetization into the *x*/*y* plane
Tip angle	Flip angle
TL	Locking time (in $T_{1\rho}$ experiments)
T_m	Interfacial interaction lifetime
TM	STEAM sequence time interval, between second and third rf pulses
TMA	TriMethylAmine (related to choline resonances)
TMJ	Temporo-Mandibular Joint
TMR	Topical Magnetic Resonance; early method of spatial location in spectroscopy using a carefully shaped inhomogeneous B_0 field
TOF	Time Of Flight; particularly used to describe weighting interval in some angiographic procedures where blood outside the region is preconditioned prior to the delay to allow it to reach its expected location
TOMROP	T One (T_1) by Multiple Read Out Pulses
TONE	Variable flip angle method for time-of-flight angiography
TR	Repetition time of a sequence (interval between the same stages of successive applications of a sequence to a single set of data)
Transaxial	See Transverse
Transmural	Extending through the wall of an organ
Transverse	Slice normal to the long axis of the body; syn. Transaxial
TRCF	Tilted Rotating Coordinate Frame
TSE	See RARE (means Turbo SE)
Turbo methods	Any very rapidly repeated sequence (where TR << T_1); there are a series of equivalent acronyms, e.g. snapshot FLASH = Turbo FLASH = MPRAGE = FSPGR = TFE, which do have variations in magnetization preparation but a common concept
Twave	Section of EKG waveform at end of main complex
US	UltraSound, UltraSonography
Valvular	Affecting a valve
VEC	Velocity Encoded Cone; flow imaging property
Vegetation	In a clinical sense, plant-like fungoid neoplasm, or luxuriant fungus-like growth of pathologic tissues
Ventral	Pertaining to the belly (abdomen) or towards the abdomen relative to a reference
Venule	Any small vessel collecting blood from the capillary bed, and uniting to form veins
Views	See NSA
V_{nec}	Aliasing velocity; velocity of flow where the encoding range in use means that the phase of the flowing material exceeds the $\pm\pi$ range, resulting in ambiguity
VOI	Volume Of Interest
Voxel	Unit of space into which the body is subdivided by the image formation process
VSE	Volume Selective Excitation

Wash-in effects	Enhanced signal from a relatively long T_1 component such as blood which has been allowed full recovery flowing into a volume of interest
Wash-out effects	Cf. Wash-in effects; effects arising from removal of wholly or partly saturated material moving out of a volume of interest
White blood	See Bright Blood
Window	Period of data recovery or, alternatively, visible gray scale range on a display
WM	White Matter
2-DFT imaging	Two-dimensional Fourier Transform imaging in which there is one phase encoding direction and a second (frequency encoding) direction
3-DFT imaging	Three-dimensional Fourier transform imaging; image acquisition method without slice selection having two phase encoding directions, with data acquisition along the third (frequency encoding) direction

Contributors

A Abragam — *Collège de France, Paris, France*
- Nuclear Ferromagnetism & Antiferromagnetism

J J H Ackerman — *Washington University, St Louis, MO, USA*
- Surface Coil NMR: Quantification with Inhomogeneous Radiofrequency Field Antennas

P F Agris — *North Carolina State University, Raleigh, NC, USA*
- RNA Structure & Function: Modified Nucleosides

D C Ailion — *Department of Physics, University of Utah, Salt Lake City, UT, USA*
- Incommensurate Systems
- Lung and Mediastinum: A Discussion of the Relevant NMR Physics
- Ultraslow Motions in Solids

M Akke — *Lund University, Sweden*
- Calcium-Binding Proteins

R Altenburger — *Universität Bremen, Bremen, Germany*
- Cell Suspensions

C Altona — *Leiden University, Leiden, The Netherlands*
- Vicinal Coupling Constants & Conformation of Biomolecules

W A Anderson — *Varian Associates, Palo Alto, CA, USA*
- Fourier Transform Spectroscopy

C F Anderson, — *University of Wisconsin Madison, USA*
- DNA–Cation Interactions: Quadrupolar Studies

I Ando — *Tokyo Institute of Technology, Japan*
- Indirect Coupling: Intermolecular & Solvent Effects
- Solid Biopolymers

E R Andrew — *University of Florida, Gainesville, FL, USA*
- Imaging: A Historical Overview
- Magic Angle Spinning

Y Anzai — *UCL School of Medicine, Los Angeles, CA, USA*
- Therapy Monitoring by MRI

T M Apple — *Rensselaer Polytechnic Institute, Troy, NY, USA*
- Reactions in Zeolites

M Arakawa — *UCSF-RIL, San Francisco, CA, USA*
- Low-Field Whole Body Systems

I M Armitage — *University of Minnesota, Minneapolis, MN, USA*
- Metallothioneins

D L Arnold
Montreal Neurological Institute, McGill University, Montreal, Quebec, Canada
- Focal Brain Lesions in Human Subjects Investigated Using MRS

J M Asakura
Stanford University, Stanford, CA, USA
- Projection–Reconstruction in MRI

A G Avent
University of Sussex, Brighton, UK
- Spin Echo Spectroscopy of Liquid Samples

L Axel
University of Pennsylvania, Philadelphia, PA, USA
- Marker Grids for Observing Motion in MRI

P Bachert
Forschungsschwerpunkt Radiologische Diagnostik und Therapie, Deutsches Krebsforschungszentrum (DKFZ), D-69120 Heidelberg, Germany
- Brain Neoplasms in Humans Studied by Phosphorus-31 MRS

G Bačíc
Dartmouth Medical School, Hanover, NH, USA.
- EPR and In Vivo EPR: Roles for Experimental & Clinical NMR Studies

A D Bain
McMaster University, Hamilton, Ontario, Canada
- COSY Spectra: Quantitative Analysis
- Radiofrequency Pulses: Response of Nuclear Spins

R S Balaban
National Heart Lung and Blood Institute, Bethesda, MD, USA
- Magnetization Transfer between Water & Macromolecules in Proton MRI

L Banci
University of Florence, Florence, Italy
- Cobalt(II)- & Nickel(II)-Substituted Proteins

P A Bandettini,
Medical College of Wisconsin, Milwaukee, WI, USA
- Brain: Sensory Activation Monitored by Induced Hemodynamic Changes with Echo Planar MRI

M Barfield
Department of Chemistry, University of Arizona, Tucson, AZ, USA
- Indirect Coupling: Theory & Applications in Organic Chemistry

M L Barnard
Royal Postgraduate Medical School, Hammersmith Hospital, London, UK
- Dietary Changes Studied by MRS

L S Batchelder
Cambridge Isotope Laboratories, Inc., Andover, MA, USA
- Deuterium NMR in Solids

C J Baudouin
Freeman Hospital, Newcastle upon Tyne, UK
- Magnetization Transfer Contrast: Clinical Applications

A Bax
National Institutes of Health, Bethesda, MD, USA
- ROESY

E D Becker
National Institutes of Health, Bethesda, MD, USA
- Rapid Scan Correlation Spectroscopy

J D Bell
Hammersmith Hospital, London, UK
- Body Fluids
- Dietary Changes Studied by MRS

G J Béné
Université de Genève, Switzerland
- Terrestrial Magnetic Field NMR

S Berger
Philipps University Marburg, Germany
- Carbon-13 Studies of Deuterium Labeled Compounds
- Two-Dimensional Carbon–Heteroelement Correlation

M Bernard *CNRS, Marseille, France*
- Cation Movements across Cell Walls of Intact Tissues using MRS

M Bernardo *Exxon Research and Engineering Company, Annandale, NJ, USA*
- Electron–Nuclear Multiple Resonance Spectroscopy

I Bertini *University of Florence, Florence, Italy*
- Iron–Sulfur Proteins

M Bilde *University of Copenhagen, Denmark*
- Shielding Calculations: LORG & SOLO Approaches

J R Binder *Medical College of Wisconsin, Milwaukee, WI, USA*
- Brain: Sensory Activation Monitored by Induced Hemodynamic Changes with Echo Planar MRI

R Bittman *Queens College of The City University of New York, Flushing, NY, USA*
- Phospholipid–Cholesterol Bilayers

K L Black *UCL School of Medicine, Los Angeles, CA, USA*
- Therapy Monitoring by MRI

R Blinc *J. Stefan Institute, University of Ljubljana, Ljubljana, Slovenia*
- Ferroelectrics & Proton Glasses
- Incommensurate Systems

J L Bloem *Leiden University Hospital, The Netherlands*
- Musculoskeletal Neoplasms: Investigations by MRI

B Blümich *Rheinisch-Westfälische Technische Hochschule Aachen, Aachen, Germany*
- Stochastic Excitation

J C Böck *Freie Universität Berlin, Germany*
- Tissue Perfusion in MRI by Contrast Bolus Injection

G Bodenhausen *Université de Lausanne, Switzerland/Florida State University and National High Magnetic Field Laboratory, Tallahassee, FL, USA*
- Selective Hartmann–Hahn Transfer in Liquids
- Selective NOESY

R Boelens *Utrecht University, Utrecht, The Netherlands*
- Protein Structures: Relaxation Matrix Refinement

C Boesch *University of Bern, Switzerland*
- Patient Life Support & Monitoring Facilities for Whole Body MRI

P H Bolton *Wesleyan University, Middletown, CT, USA*
- Relayed Coherence Transfer Experiments

A M J J Bonvin *Utrecht University, Utrecht, The Netherlands*
- Protein Structures: Relaxation Matrix Refinement

F Borsa *Iowa State University, Ames, IA, USA and Universita di Pavia, Pavia, Italy*
- Phase Transitions & Critical Phenomena in Solids

C S Bosch *Washington University, St Louis, MO, USA*
- Surface Coil NMR: Quantification with Inhomogeneous Radiofrequency Field Antennas

A A Bothner-By *Carnegie Mellon University, Pittsburgh, PA, USA*
- Magnetic Field Induced Alignment of Molecules

R E Botto — *Argonne National Laboratory, IL, USA*
- Fossil Fuels

W M M J Bovée — *Delft University of Technology, Delft, The Netherlands*
- Quantitation in Whole Body MRS
- Whole Body Machines: Quality Control

J L Boxerman — *Massachusetts General Hospital, Charlestown, MA, USA*
- Susceptibility Effects in Whole Body Experiments

W G Bradley, Jr. — *Long Beach Memorial Medical Center, CA, USA*
- Cerebrospinal Fluid Dynamics Observed by MRI

W J Brady — *Beth Israel Hospital, Boston, MA, USA*
- Coronary Artery Disease Evaluated by MRI

M N Brant-Zawadzki — *Hoag Memorial Hospital Presbyterian, Newport Beach, CA, USA*
- Brain Neoplasms Studied by MRI

P J Bray — *Brown University, Providence, RI, USA*
- Quadrupolar Nuclei in Glasses

M L Brey — *University of Florida, Gainesville, FL, USA*
- Fluorine-19 NMR

W S Brey — *University of Florida, Gainesville, FL, USA*
- Fluorine-19 NMR

K M Brindle — *University of Cambridge, Cambridge, UK*
- Enzyme-Catalyzed Exchange: Magnetization Transfer Measurements

M F Brown — *University of Arizona, Tucson, AZ, USA*
- Bilayer Membranes: Deuterium & Carbon-13 NMR

R A Brown — *University of Utah, Salt Lake City, UT, USA*
- Relaxation of Coupled Spins from Rotational Diffusion

R D Brown III — *Field Cycling Systems, Inc., River Edge, NY, USA*
- Relaxometry of Tissue

T R Brown — *Fox Chase Cancer Center, Philadelphia, PA, USA*
- Chemical Shift Imaging

E Brun — *University of Zürich, Zürich, Switzerland*
- Laser Devices in NMR: Routes to Chaos
- Ruby NMR Laser

H Brunner — *Max-Planck-Institut, Heidelberg, Germany*
- Electron–Nuclear Hyperfine Interactions

D J Bryant — *GEC Hirst Research Centre, Borehamwood, Herts., UK*
- Proton Decoupling in Whole Body Carbon-13 MRS
- Spatial Localization Techniques for Human MRS

R G Bryant — *University of Virginia, Charlottesville, VA, USA*
- Magnetization Transfer & Cross Relaxation in Tissue

J Buddrus — *Institut für Spektrochemie, Dortmund, Germany*
- INADEQUATE Experiment

M Burl *Hammersmith Hospital, London, UK*
- Eddy Currents & Their Control
- Refrigerated & Superconducting Receiver Coils for Whole Body Magnetic Resonance

D P Burum *Bruker Instruments, Inc., Billerica, MA, USA*
- Cross Polarization in Solids

C A Bush *University of Maryland, Baltimore County, Baltimore, MD, USA*
- Polysaccharides & Complex Oligosaccharides

L G Butler *Louisiana State University, Baton Rouge, LA, USA*
- Inorganic Solids

G M Bydder *University of London, London, UK*
- Inversion Recovery Pulse Sequence in MRI

E B Cady *University College London Hospitals, UK*
- Pediatric Brain MRS: Clinical Utility

P T Callaghan *Massey University, Palmerston North, New Zealand*
- Susceptibility & Diffusion Effects in NMR Microscopy

D Canet *Université H. Poincaré, Vandoeuvre-Nancy, France*
- Radiofrequency Gradient Pulses
- Relaxation Mechanisms: Magnetization Modes

G W Canters *Leiden University, Leiden, The Netherlands*
- Copper Proteins

J Carlson *UCSF-RIL, San Francisco, CA, USA*
- Low-Field Whole Body Systems

W R Carper *Wichita State University, KS, USA*
- Molten Salts

D A Case *The Scripps Research Institute, La Jolla, CA, USA*
- Amino Acids, Peptides & Proteins: Chemical Shifts

C Catalano *University of Rome 'La Sapienza', Italy*
- Abdominal MRA

G Celebre *Università della Calabria, Cosenza, Italy*
- Liquid Crystalline Samples: Spectral Analysis

S I Chan *California Institute of Technology, Pasadena, CA, USA*
- Bilayer Membranes: Deuterium & Carbon-13 NMR

X Chen *Brandeis University, Waltham, MA, USA*
- Sideband Analysis in Magic Angle Spinning NMR of Solids

H N Cheng *Hercules Incorporated Research Center, Wilmington, DE, USA*
- Polymer Reactions

T Cheng *Tokyo University of Agriculture and Technology, Koganei, Tokyo, Japan*
- Polymers: Regio-Irregular Structure

G R Cherryman *University of Leicester, Leicester, UK*
- In Vivo ESR Imaging of Animals

T T P Cheung *Phillips Petroleum Company, Bartlesville, OK, USA*
- Spin Diffusion in Solids

M H Cho — *Korea Advanced Institute of Science, Cheongyangni, Seoul, Korea*
- Microscopy: Resolution

Z H Cho — *University of California, Irvine, CA, USA & Korea Advanced Institute of Science, Seoul, Korea*
- Microscopy: Resolution

K Clarke — *University of Oxford, UK*
- Cation Movements across Cell Walls of Intact Tissues using MRS

G M Clore — *National Institutes of Health, Bethesda, MD, USA*
- Structures of Larger Proteins, Protein–Ligand, & Protein–DNA Complexes by Multi-Dimensional Heteronuclear NMR
- Three- & Four-Dimensional Heteronuclear Magnetic Resonance

S Clough — *Department of Physics, University of Nottingham, UK*
- Quantum Tunneling Spectroscopy

S Confort-Gouny — *Centre de Résonance Magnétique Biologique et Médicale, Faculté de Médecine, Marseille, France*
- Tissue & Cell Extracts MRS

A Connelly — *Institute of child Health and Great Ormond Street Hospital for Children, NHS Trust, London, UK.*
- Epilepsy MR Imaging & Spectroscopy

W W Conover — *Acorn NMR, Fremont, CA, USA*
- Shimming of Superconducting Magnets

P L Corio — *Department of Chemistry, University of Kentucky, Lexington, KY, USA*
- Analysis of High Resolution Solution State Spectra

B A Cornell — *CSIRO Food Research Laboratory, Sydney, NSW, Australia*
- Membranes: Carbon-13 NMR

D G Cory — *Massachusetts Institute of Technology, Cambridge, MA, USA*
- Ceramics Imaging

T Cosgrove — *University of Bristol, UK*
- Colloidal Systems

R M Cotts — *Cornell University, Ithaca, NY, USA*
- Diffusion in Solids

J Courtieu — *University of Paris-Sud, Orsay, France*
- Spinning Liquid Crystalline Samples

G A Coutts — *GEC Hirst Research Centre, Borehamwood, Herts., UK*
- Proton Decoupling in Whole Body Carbon-13 MRS
- Spatial Localization Techniques for Human MRS

G A Coutts — *Royal Postgraduate Medical School, Hammersmith Hospital, London, UK*
- Interventional MRI: Specialist Facilities & Techniques

D Cowburn — *The Rockefeller University, New York, NY, USA*
- Nucleic Acids: Chemical Shifts
- SH2 Domain Structure

I J Cox — *Royal Postgraduate Medical School, London, UK*
- Liver: in vivo MRS of Humans

P J Cozzone — *Centre de Résonance Magnétique Biologique et Médicale, Faculté de Médecine, Marseille, France*
- Tissue & Cell Extracts MRS

G P Crawford *Kent State University, Kent, OH, USA*
- Polymer Dispersed Liquid Crystals

L E Crooks *University of California, San Francisco, CA, USA*
- Image Formation Methods

T A Cross *Florida State University, Tallahassee, FL, USA*
- Gramicidin Channels: Orientational Constraints for Defining High-Resolution Structures

J V Crues III *Cedars-Sinai Medical Center, Los Angeles, CA, USA*
- Peripheral Joints Studied by MRI

P R Cullis *University of British Columbia, Vancouver, Canada*
- Lipid Polymorphism

F W Dahlquist *University of Oregon, Eugene, OR, USA*
- Phage Lysozyme: Dynamics & Folding Pathway

M A Danielson *University of Colorado, Boulder, CO, USA*
- Bacterial Chemotaxis Proteins: Fluorine-19 NMR

A Davidoff *University of Massachusetts, USA*
- Liver, Pancreas, Spleen, & Kidney MRI

F Davies *Oxford Magnet Technology Limited*
- Resistive & Permanent Magnets for Whole Body MRI

J H Davis *University of Guelph, Guelph, ON, Canada*
- Membranes: Deuterium NMR

R de Beer *Delft University of Technology, Delft, The Netherlands*
- Quantitation in Whole Body MRS

A de Crespigny *Stanford University, CA, USA*
- Anisotropically Restricted Diffusion in MRI

A C de Dios *University of Illinois at Urbana-Champaign, IL, USA*
- Chemical Shifts in Biochemical Systems

F De Luca *Università di Roma "La Sapienza", Rome, Italy*
- Linewidth Manipulation in MRI

B C De Simone *Università di Roma "La Sapienza", Rome, Italy*
- Linewidth Manipulation in MRI

H DeMeester *Picker International, Highland Heights, OH, USA*
- Partial Fourier Acquisition in MRI

J A den Hollander *University of Alabama, Birmingham, AL, USA*
- Quantitation in Whole Body MRS

N M deSouza *Royal Postgraduate Medical School, Hammersmith Hospital, London, UK*
- Interventional MRI: Specialist Facilities & Techniques

E E DeYoe, *Medical College of Wisconsin, Milwaukee, WI, USA*
- Brain: Sensory Activation Monitored by Induced Hemodynamic Changes with Echo Planar MRI

V Di Carlo *University of Rome 'La Sapienza', Italy*
- Abdominal MRA

M Di Girolamo — *University of Rome 'La Sapienza', Italy*
- Abdominal MRA

P Diehl — *University of Basel, Basel, Switzerland*
- Structure of Rigid Molecules Dissolved in Liquid Crystalline Solvents

R B Dietrich — *University of California, Irvine, Orange, CA, USA*
- Pediatric Body MRI

K Dixon — *Universität München, München, Germany*
- Phosphorus-31 NMR

W T Dixon — *Emory University, Atlanta, GA, USA*
- Tissue Water & Lipids: Chemical Shift Imaging & Other Methods

J W Doane — *Kent State University, Kent, OH, USA*
- Polymer Dispersed Liquid Crystals

D M Doddrell — *Centre for Magnetic Resonance, University of Queensland, 4072 Australia*
- COSY Two-Dimensional Experiments
- Polarization Transfer Experiments via Scalar Coupling in Liquids

R Y Dong — *Brandon University, Brandon, Manitoba, Canada*
- Liquid Crystalline Samples: Deuterium NMR

C Dorémieux-Morin — *Université P. et M. Curie (Paris 6), Paris, France*
- Brønsted Acidity of Solids

H C Dorn — *Virginia Polytechnic Institute and State University, Blacksburg, VA, USA*
- Flow NMR

F D Doty — *Doty Scientific Inc., Columbia, SC, USA*
- Probe Design & Construction
- Solid State Probe Design

S L Duce — *University of Cambridge, UK*
- Foods & Grains: Studied by MRS & MRI

H Duddeck — *University of Hannover, Germany*
- Sulfur, Selenium, & Tellurium NMR

C L Dumoulin — *General Electric Research and Development Center, Schenectady, NY, USA*
- Phase Contrast MRA

R Dunkel — *Salt Lake City, UT, USA*
- Computer Assisted Structure Elucidation

R Dupree — *University of Warwick, Coventry, UK*
- Ceramics

C Dybowski — *University of Delaware, Newark, DE, USA*
- Wide Lines for Nonquadrupolar Nuclei

H J Dyson — *The Scripps Research Institute, La Jolla, CA, USA*
- Proteins & Protein Fragments: Folding
- Thioredoxin & Glutaredoxin

H Eckert — *University of California, Santa Barbara, CA, USA*
- Amorphous Materials

R R Edelman *Beth Israel Hospital, Boston, MA, USA*
- Coronary Artery Disease Evaluated by MRI

W A Edelstein *GE Corporate Research and Development, Schenectady, NY, USA*
- Whole Body Machines: NMR Phased Array Coil Systems
- Radiofrequency Systems & Coils for MRI & MRS

U Edlund *University of Umeå, Sweden*
- Aluminum-27 NMR of Solutions

R H T Edwards *Magnetic Resonance Research Centre and Muscle Research Centre, The University of Liverpool, Liverpool, UK*
- Peripheral Muscle Metabolism Studied by MRS

P D Ellis *Battelle, Pacific Northwest Laboratories, Richland, WA, USA*
- Cadmium-113 NMR: A Surrogate Probe for Zinc & Calcium in Proteins

J W Emsley *University of Southampton, Southampton, UK*
- Liquid Crystalline Samples: Structure of Nonrigid Molecules
- Liquid Crystals: General Considerations

L Emsley *Ecole Normal Supérieure de Lyon, Lyon, France*
- Selective Pulses

G Engelhardt *University of Stuttgart, Stuttgart, Germany*
- Silicon-29 NMR of Solid Silicates

F Engelke *Ames Laboratory, Ames, IA, USA*
- Chemical Exchange on Solid Metal Surfaces
- Cross Polarization in Rotating Solids: Spin-1/2 Nuclei

M Engelsberg *Universidade Federal de Pernambuco, Recife, Pernambuco, Brazil*
- Dipolar Spectroscopy: Transient Nutations & Other Techniques

R R Ernst *Laboratorium für Physikalische Chemie, Eidgenössische Technische Hochschule, 8092 Zürich, Switzerland*
- Multidimensional Spectroscopy: Concepts

J N S Evans *Washington State University, Pullman, WA, USA*
- Time-Resolved Solid-State NMR of Enzyme–Substrate Interactions

J C Facelli *The University of Utah, Salt Lake City, UT, USA*
- Indirect Coupling: Semiempirical Calculations
- Shielding Calculations: Perturbation Methods
- Shielding Tensor Calculations

J J Falke *University of Colorado, Boulder, CO, USA*
- Bacterial Chemotaxis Proteins: Fluorine-19 NMR

R B Farahani *UCL School of Medicine, Los Angeles, CA, USA*
- Therapy Monitoring by MRI

R Farb *University of Toronto, Toronto, ON, Canada*
- Pituitary Gland & Parasellar Region Studied by MRI

T C Farrar *University of Wisconsin, Madison, WI, USA*
- Relaxation of Transverse Magnetization for Coupled Spins

N A Farrow *California Institute of Technology, Pasadena, CA, USA*
- pH Measurement In Vivo in Whole Body Systems

P J Feenan *Magnex Scientific Ltd, Abingdon, Oxon, UK*
- Cryogenic Magnets for Whole Body Magnetic Resonance Systems

J Feigon *University of California, Los Angeles, CA, USA*
- DNA Triplexes, Quadruplexes, & Aptamers

R Felix *Freie Universität Berlin, Germany*
- Tissue Perfusion in MRI by Contrast Bolus Injection

D B Fenske *University of British Columbia, Vancouver, Canada*
- Lipid Polymorphism

A Ferrarini *University of Padova, Padova Italy*
- Liquid Crystalline Samples: Relaxation Mechanisms

J A Ferretti *National Institutes of Health, Bethesda, MD, USA*
- Rapid Scan Correlation Spectroscopy

L D Field *Sydney University, NSW, Australia*
- Multiple Quantum Spectroscopy in Liquid Crystalline Solvents

D N Firmin *National Heart and Lung Institute, University of London, UK*
- Blood Flow: Quantitative Measurement by MRI
- Cardiac Gating Practice

J L Fleckenstein *University of Texas Southwestern Medical Center, Dallas, TX, USA*
- Skeletal Muscle Evaluated by MRI

U Fleischer *Ruhr-Universität Bochum, Bochum, Germany*
- Shielding Calculations: IGLO Method

H G Floss *University of Washington, Seattle, WA, USA*
- Biosynthesis & Metabolic Pathways: Carbon-13 and Nitrogen-15 NMR

S Forsén *Lund University, Sweden*
- Calcium-Binding Proteins

D A Forsyth *Northeastern University, Boston, MA, USA*
- Isotope Effects in Carbocation Chemistry

M A Foster *University of Aberdeen, Aberdeen AB92ZD, UK*
- Relaxation Measurements in Imaging Studies

J Frahm *Biomedizinische NMR Forschungs GmbH am Max-Planck-Institut für biophysikalische Chemie, Göttingen, Germany*
- Single Voxel Proton NMR: Human Subjects

J Fraissard *Université P. et M. Curie (Paris 6), Paris, France*
- Brønsted Acidity of Solids
- Microporous Materials & Xenon-129 NMR

R Freeman *Cambridge University, Cambridge, UK*
- Double Resonance

P Fritzsche *Loma Linda University School of Medicine, CA, USA*
- Cranial Nerves Investigated by MRI

B M Fung *University of Oklahoma, Norman, OK, USA*
- Liquid Crystalline Samples: Carbon-13 NMR

M M Fuson — *Denison University, Granville, OH, USA*
- Coupled Spin Relaxation in Polymers

D G Gadian — *Royal College of Surgeons Unit of Biophysics, Institute of Child Health, London, UK*
- Animal Methods in MRS

A N Garroway — *Naval Research Laboratory, Washington, DC, USA*
- Polymer MRI

I D Gay — *Department of Chemistry, Simon Fraser University, Burnaby, B.C., Canada*
- Adsorbed Species: Spectroscopy & Dynamics

J T Gerig — *University of California, Santa Barbara, CA, USA*
- Proteases

I P Gerothanassis — *University of Ioannina, Greece*
- Oxygen-17 NMR

B C Gerstein — *Ames Laboratory of Iowa State University, Ames, IA, USA*
- CRAMPS
- Echoes in Solids
- Multiple Quantum Coherence in Spin-1/2 Dipolar Coupled Solids
- Nutation Spectroscopy of Quadrupolar Nuclei
- Rudimentary NMR: The Classical Picture

H Gesmar — *University of Copenhagen, Denmark*
- Fourier Transform & Linear Prediction Methods

H Gibson — *Magnetic Resonance Research Centre and Muscle Research Centre, The University of Liverpool, Liverpool, UK*
- Peripheral Muscle Metabolism Studied by MRS

L M Gierasch — *University of Texas Southwestern Medical Center, Dallas, TX, USA*
- Peptide & Protein Secondary Structural Elements

R J Gillies — *University of Arizona Health Sciences Center, Tucson, AZ, USA*
- Bioreactors and Perfusion

S J Glaser — *Universität Frankfurt, Frankfurt, Germany*
- TOCSY in ROESY & ROESY in TOCSY

K K Gleason — *Massachusetts Institute of Technology, Cambridge, MA, USA*
- Diamond Thin Films
- Multiple Quantum NMR in Solids

J D Glickson — *The Johns Hopkins University School of Medicine, Baltimore, MD, USA*
- Cells and Cell Systems MRS

G H Glover — *Stanford University, Stanford, CA, USA*
- Projection–Reconstruction in MRI

M Goldman — *Centre d'Etudes de Saclay, Gif-sur-Yvette, France*
- Low Spin Temperature NMR
- Thermodynamics of Nuclear Magnetic Ordering

M Goodman — *University of California, San Diego, La Jolla, CA, USA*
- Synthetic Peptides

D G Gorenstein — *Department of Chemistry, Purdue University, West Lafayette, IN, USA*
- Nucleic Acids: Phosphorus-31 NMR

P J Grandinetti *The Ohio State University, Columbus, OH, USA*
- Dynamic Angle Spinning

P Granger *Université Louis Pasteur, CNRS, Bruker, Strasbourg, France*
- Quadrupolar Transition Metal & Lanthanide Nuclei

D M Grant *University of Utah, Salt Lake City, UT, USA*
- Chemical Shift Tensors
- Relaxation of Coupled Spins from Rotational Diffusion

C Griesinger *Universität Frankfurt, Frankfurt, Germany*
- TOCSY in ROESY & ROESY in TOCSY

M T R J Griesinger *University of Wisconsin Madison, USA*
- DNA–Cation Interactions: Quadrupolar Studies

R G Griffin *Massachusetts Institute of Technology, Cambridge, MA, USA*
- Rotating Solids

J R Griffiths *St. George's Hospital Medical School, London, UK*
- Animal Tumor Models
- Fluorine-19 MRS: Applications in Oncology

A M Gronenborn *National Institutes of Health, Bethesda, MD, USA*
- Structures of Larger Proteins, Protein–Ligand, & Protein–DNA Complexes by Multi-Dimensional Heteronuclear NMR
- Three- & Four-Dimensional Heteronuclear Magnetic Resonance

S Grzesiek *National Institutes of Health, Bethesda, MD, USA*
- ROESY

Z Gu *Columbia University, New York City, NY, USA*
- Carbon & Nitrogen Chemical Shifts of Solid State Enzymes

M Guéron *Ecole Polytechnique, Palaiseau, France*
- Water Signal Suppression in NMR of Biomolecules

H Günther *University of Siegen, D-57068 Siegen, Germany*
- Lithium NMR
- Vicinal ^{1}H,^{1}H Coupling Constants in Cyclic π-Systems

R K Gupta *Albert Einstein College of Medicine, Bronx, NY, USA*
- Rapid Scan Correlation Spectroscopy

E M Haacke *Washington University, St. Louis, MO, USA*
- Flow in Whole Body Magnetic Resonance

A Haase *Physikalisches Institut, Am Hubland, Würzburg, Germany*
- Relaxation Measurements in Imaging Studies
- Whole Body Magnetic Resonance: Fast Low-Angle Acquisition Methods

D M Hadley *Institute of Neurological Sciences, Glasgow, UK*
- White Matter Disease MRI

E L Hahn *University of California, Berkeley, CA, USA*
- Quantum Optics: Concepts of NMR

H C Hailes *University of Cambridge, UK*
- Biosynthesis & Metabolic Pathways: Deuterium NMR

J V Hajnal *Hammersmith Hospital, London, UK*
- Functional Neuroimaging Artifacts

L D Hall *University of Cambridge, UK*
- Foods & Grains: Studied by MRS & MRI

B Halle *Lund University, Lund, Sweden*
- Amphiphilic Liquid Crystalline Samples: Nuclear Spin Relaxation

M R Halse *University of Kent, Canterbury, Kent, UK*
- Imaging Techniques for Solids and Quasi-solids

T K Halstead *York University, York, UK*
- Fast Ion Conductors

D R Hamilton *UCL School of Medicine, Los Angeles, CA, USA*
- Therapy Monitoring by MRI

J A Hamilton *Boston University School of Medicine, Boston, MA, USA*
- Lipoproteins

W Hänicke *Biomedizinische NMR Forschungs GmbH am Max-Planck-Institut für biophysikalische Chemie, Göttingen, Germany*
- Single Voxel Proton NMR: Human Subjects

A E Hansen *University of Copenhagen, Denmark*
- Shielding Calculations: LORG & SOLO Approaches

W Happer *Princeton University, Princeton, NJ, USA*
- Polarization of Noble Gas Nuclei with Optically Pumped Alkali Metal Vapors

S E Harms *Baylor University Medical Center, Dallas, TX, USA*
- Temporomandibular Joint MRI

D N F Harris *Royal Postgraduate Medical School, London, UK*
- Postoperative Trauma Observed by MRI

R K Harris *University of Durham, Durham, UK*
- Magic Angle Spinning: Effects of Quadrupolar Nuclei on Spin-1/2 Spectra
- Nuclear Spin Properties & Notation
- Polymorphism & Related Phenomena

A N Hasso *Loma Linda University School of Medicine, CA, USA*
- Cranial Nerves Investigated by MRI

K H Hausser *Max-Planck-Institut, Heidelberg, Germany*
- Electron–Nuclear Hyperfine Interactions

T F Havel *Harvard Medical School, Boston, MA, USA*
- Distance Geometry

J F Haw *Texas A&M University, College Station, TX, USA*
- Thermometry

J M Hawnaur *Department of Diagnostic Radiology, University of Manchester, UK*
- Midfield Magnetic Resonance Systems

C E Hayes *University of Washington, Seattle, WA, USA*
- Birdcage & Other High Homogeneity Radiofrequency Coils for Whole Body Magnetic Resonance
- Whole Body Machines: NMR Phased Array Coil Systems

F Heatley *University of Manchester, Manchester, UK*
- Polymers: Relaxation & Dynamics of Synthetic Polymers in Solution

N Heaton *Universität Stuttgart, Stuttgart, Germany*
- Slow & Ultraslow Motions in Biology

J A Helpern *Nathan Kline Institute, Orangeburg, NY, USA*
- Single Voxel Whole Body Phosphorus MRS

R M Henkelman *Sunnybrook Health Science Centre and University of Toronto, Toronto, ON, Canada*
- Whole Body Magnetic Resonance Artifacts

J Hennig *Universität Freiburg, Germany*
- Multi-Echo Acquisition Techniques Using Inverting Radiofrequency Pulses in MRI

O Henriksen *Danish Research Center of Magnetic Resonance, Hvidovre University Hospital, Denmark*
- Diffusion: Clinical Utility of MRI Studies

R J Herfkens *Stanford University School of Medicine, CA, USA*
- Lung & Mediastinum MRI

J Herzfeld *Brandeis University, Waltham, MA, USA*
- Bacteriorhodopsin & Rhodopsin
- Sideband Analysis in Magic Angle Spinning NMR of Solids

J R Hesselink *UCSD Medical Center, San Diego, CA, USA*
- Cerebral Infection Monitored by MRI

S H Heywang-Köbrunner *University of Halle, Germany*
- Breast MRI

C W Hilbers *University of Nijmegen, Nijmegen, The Netherlands*
- Nucleic Acids: Spectra, Structures, & Dynamics

H D W Hill *Varian Associates, Palo Alto, CA, USA*
- Probes for High Resolution
- Spectrometers: A General Overview

J F Hinton *University of Arkansas, Fayetteville, AR, USA*
- Shielding Theory: GIAO Method
- Thallium NMR

C Ho *Carnegie Mellon University, Pittsburgh, PA, USA*
- D-Lactate Dehydrogenase

G L Hoatson *College of William and Mary, Williamsburg, VA, USA.*
- Deuteron Relaxation Rates in Liquid Crystalline Samples: Experimental Methods

J C Hoch *Rowland Institute for Science, Cambridge, MA, USA*
- Maximum Entropy Reconstruction

G N Holland *Otsuka Electronics, Fort Collins, CO, USA*
- Whole Body Magnetic Resonance Spectrometers: All-Digital Transmit/Receive Systems

M Holz *Universität Karlsruhe, Germany*
- Electrolytes

J Homer *Aston University, Birmingham, UK*
- Ultrasonic Irradiation & NMR

I Horváth *Exxon Research & Engineering Company, Annandale, NJ, USA*
- Gases at High Pressure

D I Hoult *Institute for Biodiagnostics, National Research Council Canada, Winnipeg, MB, Canada*
- Sensitivity of the NMR Experiment
- Sensitivity of Whole Body MRI Experiments

O W Howarth *University of Warwick, Coventry, UK*
- Recording One-Dimensional High Resolution Spectra

H Hricak *University of California, San Francisco, CA, USA*
- Male Pelvis Studies Using MRI

V J Hruby *University of Arizona, Tucson, AZ, USA*
- Peptide Hormones

P S Hsieh *Cedars-Sinai Medical Center, Los Angeles, CA, USA*
- Peripheral Joints Studied by MRI

J G Hu *Brandeis University, Waltham, MA, USA*
- Bacteriorhodopsin & Rhodopsin

J Z Hu *University Of Utah, Salt Lake City, UT, USA, & Wuhan Institute of Physics, Chinese Academy of Sciences, China*
- Magic Angle Turning & Hopping

J W Hugg *University of Alabama, Birmingham, AL, USA*
- Single Voxel Whole Body Phosphorus MRS

C D Hughes *Los Alamos National Laboratory, Los Alamos, NM, USA*
- Two-Dimensional Powder Correlation Methods

R E Hurd *GE Medical Systems, Fremont, CA, USA*
- Field Gradients & Their Application

G C Hurst *MetroHealth Medical Center and Case Western Reserve University, Cleveland, OH, USA*
- Coils for Insertion into the Human Body

J M S Hutchison *University of Aberdeen, Aberdeen, UK*
- Spin Warp Data Acquisition
- Spin Warp Method: Artifacts

J S Hyde *Medical College of Wisconsin, Milwaukee, WI, USA*
- Brain: Sensory Activation Monitored by Induced Hemodynamic Changes with Echo Planar MRI
- Surface & Other Local Coils for In Vivo Studies

M Ikura *University of Toronto, Canada*
- Calmodulin

J S Ingwall *Harvard Medical School, Boston, MA, USA*
- Cation Movements across Cell Walls of Intact Tissues using MRS

I Isherwood *Department of Diagnostic Radiology, University of Manchester, UK*
- Midfield Magnetic Resonance Systems

T Iwashita *Suntory Institute for Bioorganic Research, Osaka, Japan*
- Natural Products

G D Jackson *Institute of child Health and Great Ormond Street Hospital for Children, NHS Trust, London, UK.*
- Epilepsy MR Imaging & Spectroscopy

T L James *University of California, San Francisco, CA, USA*
- Nucleic Acid Structures in Solution: Sequence Dependence

C J Jameson *University of Illinois, Chicago, IL, USA*
- Chemical Shift Scales on an Absolute Basis
- Gas Phase Studies of Intermolecular Interactions & Relaxation
- Isotope Effects on Chemical Shifts & Coupling Constants

O Jardetzky *Stanford University, CA, USA*
- Biological Macromolecules
- Biological Macromolecules: NMR Parameters

H C Jarrell *National Research Council, Ottawa, Ontario, Canada*
- Glycolipids

J R Jinkins *University of Texas Health Science Center, San Antonio, TX, USA*
- Ischemic Stroke

C Job *University of Arizona, Tucson, AZ, USA*
- Instrumentation for the Home Builder

C S Johnson, Jr *University of North Carolina, Chapel Hill, NC, USA*
- Diffusion Measurements by Magnetic Field Gradient Methods
- Electrophoretic NMR

J Jokisaari *Department of Physics, University of Oulu, FIN-90570 Oulu, Finland*
- Anisotropy of Shielding & Coupling in Liquid Crystalline Solutions

F A Jolesz *Brigham and Women's Hospital and Harvard Medical School, Boston, MA, USA*
- Wavelet Encoding of MRI Images

B-H Jonsson *Umeå University, Sweden*
- Carbonic Anhydrase

P M Joseph *Hospital of the University of Pennsylvania, Philadelphia, PA, USA*
- Sodium-23 Magnetic Resonance of Human Subjects

P C Jurs *The Pennsylvania State University, University Park, PA, USA*
- Carbon-13 Spectral Simulation

L T Kakalis *Yale University School of Medicine, New Haven, CT, USA*
- Metallothioneins

J I Kaplan *Indiana University-Purdue University at Indianapolis, IN, USA*
- Chemical Exchange Effects on Spectra

K Kaptein *UCL School of Medicine, Los Angeles, CA, USA*
- Therapy Monitoring by MRI

S K Karampekios *University Hospital of Crete, Heraklion, Greece*
- Cerebral Infection Monitored by MRI

J Kärger *Leipzig University, Germany*
- Diffusion in Porous Media

L Kaufman *UCSF-RIL, San Francisco, CA, USA*
- Low-Field Whole Body Systems

L E Kay *University of Toronto, Toronto, ON, Canada*
- Three-Dimensional HMQC–NOESY, NOESY–HMQC, & NOESY–HSQC

A P Kelly *Hoag Memorial Hospital Presbyterian, Newport Beach, CA, USA*
- Brain Neoplasms Studied by MRI

K M W Keough *Memorial University of Newfoundland, St. John's, NF, Canada*
- Phospholipid–Cholesterol Bilayers

D J Kerwood *Informatrix Inc., Ann Arbor, MI, USA*
- Carbon-13 Relaxation Measurements: Organic Chemistry Applications
- Data Processing

H Kessler *Technical University, Munich, Germany*
- Peptides & Polypeptides

C L Khetrapal *Indian Institute of Science, Bangalore, India*
- Liquid Crystals: Mixed Magnetic Susceptibility Solvents

L D Kimmich *Cambridge University, UK*
- Multiple Quantum Coherence Imaging

R J Kirkpatrick *University of Illinois, Urbana, IL, USA*
- Geological Applications

P G Klein *Leeds University, Leeds, UK*
- Polymer Physics

R L Kleinberg *Schlumberger-Doll Research, Ridgefield, CT, USA*
- Well Logging

J Klinowski *University of Cambridge, UK*
- Molecular Sieves: Crystalline Systems

W D Knight *University of California, Berkeley, CA, USA*
- Knight Shift

S-i Kobayashi *University of Tokyo, Tokyo, Japan*
- Knight Shift

P Koehl *UPR 9003, CNRS and Louis Pasteur University, Strasbourg, France*
- Relaxation Matrix Refinement of Nucleic Acids

S H Koenig *Relaxometry, Inc., Mahopac, NY, USA*
- Dynamics of Water in Biological Systems: Inferences from Relaxometry
- Relaxometry of Tissue

J Kördel *Lund University, Sweden*
- Calcium-Binding Proteins

G Kothe *Universität Stuttgart, Stuttgart, Germany*
- Slow & Ultraslow Motions in Biology

G Kowalewski *Picker International, Highland Heights, OH, USA*
- Partial Fourier Acquisition in MRI

D Kramer *UCSF-RIL, San Francisco, CA, USA*
- Low-Field Whole Body Systems

M G Kubinec *Lawrence Berkeley National Laboratory, Berkeley, CA, USA*
- Tritium NMR
- Tritium NMR in Biology

G A Kucharczyk — *Duke University Medical Center, Durham, NC, USA*
- Plants, Seeds, Roots, & Soils as Applications of Magnetic Resonance Microscopy

A Kumar — *Indian Institute of Science, Bangalore, India*
- Two-Dimensional NMR of Molecules Oriented in Liquid Crystalline Phases

K Kume — *Tokyo Metropolitan University, Japan*
- Conducting Polymers

J Kümmerlen — *Universität Bayreuth, Bayreuth, Germany*
- Reorientation in Crystalline Solids: Propeller-Like R_3M Species

S Kuroki — *Tokyo Institute of Technology, Japan*
- Solid Biopolymers

H Kurosu — *Tokyo Institute of Technology, Ookayama, Meguro-ku, Tokyo, Japan*
- Solid Polymers: Shielding & Electronic States

D M Kurtz, Jr. — *University of Georgia, Athens, GA, USA*
- Nonheme Iron Proteins

W Kutzelnigg — *Ruhr-Universität Bochum, Bochum, Germany*
- Shielding Calculations: IGLO Method

G N La Mar — *University of California, Davis, CA, USA*
- Myoglobin

S Lacelle — *Université de Sherbrooke, Quebec, Canada*
- Multiple Quantum Coherences in Extended Dipolar Coupled Spin Networks

A Laghi — *University of Rome 'La Sapienza', Italy*
- Abdominal MRA

R M J N Lamerichs — *Philips Medical Systems, Best, The Netherlands*
- Proton Decoupling During In Vivo Whole Body Phosphorus MRS

O B Lapina — *Boreskov Institute of Catalysis, Siberian Branch of Russian Academy of Sciences, Novosibirsk, Russia*
- Vanadium Catalysts: Solid State NMR

P Laszlo — *Université de Liège B6, B-4000 Liège, Belgium, and École polytechnique, F-91128 Palaiseau, France*
- Quadrupolar Nuclei in Liquid Samples
- Sodium-23 NMR

G Laub — *Siemens AG., Erlangen, Germany*
- Time-of-Flight Method of MRA

P Lazzeretti — *Università degli Studi di Modena, Modena, Italy*
- Shielding in Small Molecules

D Le Bihan — *Warren G. Magnuson Clinical Center, Bethesda, MD, USA*
- Diffusion & Perfusion in MRI

J J Led — *University of Copenhagen, Denmark*
- Fourier Transform & Linear Prediction Methods

S C Lee — *Korea Advanced Institute of Science, Cheongyangni, Seoul, Korea*
- Microscopy: Resolution

J-F Lefèvre — *UPR 9003, CNRS and Louis Pasteur University, Strasbourg, France*
- Relaxation Matrix Refinement of Nucleic Acids

D Leibfritz *Universität Bremen, Bremen, Germany*
- Cell Suspensions

R E Lenkinski *University of Pennsylvania, Philadelphia, PA, USA*
- Brain Infection & Degenerative Disease Studied by Proton MRS

M H Levitt *Stockholm University, Sweden*
- Composite Pulses

G C Levy *Informatrix Inc., Ann Arbor, MI, USA*
- Carbon-13 Relaxation Measurements: Organic Chemistry Applications
- Data Processing

R N A H Lewis *University of Alberta, Edmonton, AB, Canada*
- Membrane Lipids of *Acholeplasma laidlawii*

F J Lexa *Hospital of the University of Pennsylvania, Philadelphia, PA, USA*
- Central Nervous System Degenerative Disease Observed by MRI

D Li *Washington University, St. Louis, MO, USA*
- Flow in Whole Body Magnetic Resonance

W Lin *Washington University, St. Louis, MO, USA*
- Flow in Whole Body Magnetic Resonance

G Lindblom *University of Umeå, Sweden*
- Liquid Crystalline Samples: Diffusion

S Lindskog *Umeå University, Sweden*
- Carbonic Anhydrase

S Linse *Lund University, Sweden*
- Calcium-Binding Proteins

M J Lipton *The University of Chicago, IL, USA*
- Outcome & Benefit Analysis in MRI

A S Lipton, *Battelle, Pacific Northwest Laboratories, Richland, WA, USA*
- Cadmium-113 NMR: A Surrogate Probe for Zinc & Calcium in Proteins

M Longeri *Università della Calabria, Cosenza, Italy*
- Liquid Crystalline Samples: Spectral Analysis

D B Longmore *Royal Brompton Hospital, London, UK*
- Cardiovascular NMR to Study Function

D López-Villegas *University of Pennsylvania, Philadelphia, PA, USA*
- Brain Infection & Degenerative Disease Studied by Proton MRS

C Luchinat *University of Bologna, Bologna, Italy*
- Iron–Sulfur Proteins

C Luchinat *University of Bologna, Bologna, Italy*
- Transferrins

N Lugeri *Università di Roma "La Sapienza", Rome, Italy*
- Linewidth Manipulation in MRI

D J Lurie *University of Aberdeen, Aberdeen, UK*
- Imaging using the Electronic Overhauser Effect

P R Luyten *Philips Medical Systems, Best, The Netherlands*
- Proton Decoupling During In Vivo Whole Body Phosphorus MRS
- Quantitation in Whole Body MRS

Z Luz *Weizmann Institute of Science, Rehovot, Israel*
- Dynamic NMR in Liquid Crystalline Solvents

J S MacFall *Duke University Medical Center, Durham, NC, USA*
- Plants, Seeds, Roots, & Soils as Applications of Magnetic Resonance Microscopy

G E Maciel *Colorado State University, Fort Collins, CO, USA*
- Silica Surfaces: Characterization

A Macovski *Stanford University, CA, USA*
- Spiral Scanning Imaging Techniques

M F Mafee *University of Illinois at Chicago, Chicago, IL, USA*
- Eye, Orbit, Ear, Nose, & Throat Studies Using MRI

M Malagoli *Università degli Studi di Modena, Modena, Italy*
- Shielding in Small Molecules

C Maldjian *Mount Sinai Medical Center, New York, NY, USA*
- Peripheral Vasculature MRA

C W Mallory *University of Pennsylvania, Philadelphia, PA, USA*
- Coupling Through Space in Organic Chemistry

F B Mallory *Bryn Mawr College, PA, USA*
- Coupling Through Space in Organic Chemistry

P P Man *Université Pierre et Marie Curie, Paris, France*
- Quadrupolar Interactions

C E Mann *The University of Chicago, IL, USA*
- Outcome & Benefit Analysis in MRI

C Manning *Universität Frankfurt, Germany*
- Coupling Constants Determined by ECOSY

P Mansfield *University of Nottingham, UK*
- Echo-Planar Imaging

F M Marassi *University of Pennsylvania, Philadelphia, PA, USA*
- Membrane Proteins

B Maraviglia *Università di Roma "La Sapienza", Rome, Italy*
- Linewidth Manipulation in MRI

C D Marcus *Hôpital Robert Debré, Reims, France*
- Tissue Behavior Measurements Using Phosphorus-31 NMR

P M Margosian *Picker International, Highland Heights, OH, USA*
- Partial Fourier Acquisition in MRI

A R Margulis *University of California, San Francisco, CA, USA*
- Clinical Medicine

H C Marsmann *Universität-GH Paderborn, Paderborn, Germany*
- Silicon-29 NMR

P A Martin *Magnetic Resonance Research Centre and Muscle Research Centre, The University of Liverpool, Liverpool, UK*
- Peripheral Muscle Metabolism Studied by MRS

J Mason *The Open University, Milton Keynes, UK*
- Nitrogen NMR

V M Mastikhin *Boreskov Institute of Catalysis, Siberian Branch of Russian Academy of Sciences, Novosibirsk, Russia*
- Vanadium Catalysts: Solid State NMR

Y Masuda *Aichi Gakuin University, Aichi, Japan*
- Metallic Superconductors

G B Matson *University of California San Francisco and VA Medical Center, San Francisco, CA, USA*
- Selective Excitation in MRI
- Single Voxel Whole Body Phosphorus MRS

P M Matthews *Montreal Neurological Institute, McGill University, Montreal, Quebec, Canada*
- Focal Brain Lesions in Human Subjects Investigated Using MRS

A A Maudsley *University of California San Francisco and VA Medical Center, San Francisco, CA, USA*
- Selective Excitation in MRI

C L Mayne *University of Utah, Salt Lake City, UT, USA*
- Liouville Equation of Motion
- Relaxation Processes in Coupled-Spin Systems

S M Mayr *Princeton University, Princeton, NJ, USA*
- Shaped Pulses

K McAteer *Battelle, Pacific Northwest Laboratories, Richland, WA, USA*
- Cadmium-113 NMR: A Surrogate Probe for Zinc & Calcium in Proteins

S McCarthy *Yale University School of Medicine, New Haven, CT, USA*
- Female Pelvis Studied by MRI

R E D McClung *University of Alberta, Edmonton, Alberta, Canada*
- Spin–Rotation Relaxation Theory

C L McCoy, *St. George's Hospital Medical School, London, UK*
- Animal Tumor Models

A McDermott *Columbia University, New York City, NY, USA*
- Carbon & Nitrogen Chemical Shifts of Solid State Enzymes
- Proton Chemical Shift Measurements in Biological Solids

I L McDougall *Oxford Instruments plc, Eynsham, Witney, Oxfordshire, UK*
- Resistive & Permanent Magnets for Whole Body MRI

C A McDowell *University of British Columbia, Vancouver, Canada*
- Magic Angle Spinning Carbon-13 Lineshapes: Effect of Nitrogen-14

R N McElhaney *University of Alberta, Edmonton, AB, Canada*
- Membrane Lipids of *Acholeplasma laidlawii*

R A McKay *Washington University, St. Louis, MO, USA*
- Probes for Special Purposes

P M J McSheehy *St. George's Hospital Medical School, London, UK*
- Fluorine-19 MRS: Applications in Oncology

M Mehring *Universität Stuttgart, Stuttgart, Germany*
- Internal Spin Interactions & Rotations in Solids

D K Menon *University of Cambridge, UK*
- Fluorine-19 MRS: General Overview & Anesthesia

A E Merbach *University of Lausanne, Lausanne, Switzerland*
- Kinetics at High Pressure

L A Metz *The University of Chicago, IL, USA*
- Outcome & Benefit Analysis in MRI

C H Meyer *Stanford University, CA, USA*
- Spiral Scanning Imaging Techniques

D F Mierke *Clark University, Worcester, MA, USA*
- Synthetic Peptides

A S Mildvan *Johns Hopkins School of Medicine, Baltimore, MD, USA*
- Enzymes Utilizing ATP: Kinases, ATPases and Polymerases

J M Millar *Exxon Research & Engineering Company, Annandale, NJ, USA*
- Gases at High Pressure

V W Miner *Acorn NMR, Fremont, CA, USA*
- Shimming of Superconducting Magnets

G J Misic *Medrad, Inc., MRI Products, Indianola, PA, USA*
- Coils for Insertion into the Human Body

U Mocek *Panlabs, Bothell, WA, USA*
- Biosynthesis & Metabolic Pathways: Carbon-13 and Nitrogen-15 NMR

M T Modic *Cleveland Clinic Foundation, OH, USA*
- Degenerative Disk Disease Studied by MRI

R H Mohiaddin *National Heart and Lung Institute, University of London, UK and Royal Brompton Hospital, London, UK*
- Blood Flow: Quantitative Measurement by MRI

C T W Moonen *National Institutes of Health, Bethesda, MD, USA*
- Water Suppression in Proton MRS of Humans & Animals

P B Moore *Yale University, New Haven, CT, USA*
- Ribosomal RNA

G A Morris *University of Manchester, Manchester, UK*
- INEPT
- Two-Dimensional *J*-Resolved Spectroscopy

P G Morris *Nottingham University, UK*
- Complex Radiofrequency Pulses

J D Morrisett *Baylor College of Medicine, Houston, TX, USA*
- Lipoproteins

M R Morrow *Memorial University of Newfoundland, St. John's, NF, Canada*
- Phospholipid–Cholesterol Bilayers

M E Moseley *Stanford University, CA, USA*
- Anisotropically Restricted Diffusion in MRI

C E Mountford *University of Sydney, NSW, Australia*
- Tissue NMR Ex Vivo

R Muhandiram *University of Toronto, Toronto, ON, Canada*
- Three-Dimensional HMQC–NOESY, NOESY–HMQC, & NOESY–HSQC

A Mühler *Research Laboratories Schering AG, Berlin, Germany*
- Gadolinium Chelates: Chemistry, Safety, & Behavior

R N Muller *University of Mons Hainaut, Mons, Belgium*
- Contrast Agents in Whole Body Magnetic Resonance: Operating Mechanisms

W Müller-Warmuth *Institut für Physikalische Chemie der Westfälischen Wilhelms-Universität, Münster, Germany*
- Intercalation Compounds

M Munowitz *Naperville, IL, USA*
- Double Quantum Coherence

M D Murphy *St. Frances College, Loretta, PA, USA*
- Dynamic Spin Ordering of Matrix Isolated Methyl Rotors

J Murphy-Boesch *Fox Chase Cancer Center, Philadelphia, PA, USA*
- Multifrequency Coils for Whole Body Studies

K Nakanishi *Columbia University, New York City, NY, USA*
- Natural Products

G Neue *Universität Dortmund, Dortmund, Germany*
- Wide Lines for Nonquadrupolar Nuclei

D Neuhaus *MRC Laboratory of Molecular Biology, Cambridge, UK*
- Nuclear Overhauser Effect

R H Newman *Industrial Research Limited, Lower Hutt, New Zealand*
- Agriculture & Soils

F Noack *Universität Stuttgart, Physikalisches Institut, Stuttgart, Germany*
- Field Cycling Experiments

R E Norberg *Washington University, St. Louis, MO, USA*
- Diffusion in Rare Gas Solids

P L Nordio *University of Padova, Padova Italy*
- Liquid Crystalline Samples: Relaxation Mechanisms

T J Norwood *Cambridge University, UK*
- Multiple Quantum Coherence Imaging

T J Norwood *Leicester University, UK*
- Multiple Quantum Spectroscopy of Liquid Samples

T M Obey *University of Bristol, UK*
- Colloidal Systems

H Oellinger *Universitäts-Klinikum Rudolf-Virchow, Freie Universität Berlin, Germany*
- Breast MRI

L-O Öhman *University of Umeå, Sweden*
- Aluminum-27 NMR of Solutions

E Oldfield *University of Illinois at Urbana-Champaign, IL, USA*
- Chemical Shifts in Biochemical Systems
- Oxygen-17 NMR: Applications in Biochemistry

U Olsson *University of Lund, Sweden*
- Micellar Solutions & Microemulsions

S J Opella *University of Pennsylvania, Philadelphia, PA, USA*
- Filamentous Bacteriophage Coat Protein
- Membrane Proteins
- Protein Dynamics from Solid State NMR

G Orädd *University of Umeå, Sweden*
- Liquid Crystalline Samples: Diffusion

R J Ordidge *University College, London, UK*
- Single Voxel Whole Body Phosphorus MRS

A M Orendt *University of Utah, Salt Lake City, UT, USA*
- Chemical Shift Tensor Measurement in Solids

J S Orr *Hammersmith Hospital, London, UK*
- Whole Body Machines: Quality Control

K G Orrell *University of Exeter, Devon, UK*
- Fluxional Motion
- Two-Dimensional Methods of Monitoring Exchange

G Otting *Karolinska Institute, Stockholm, Sweden*
- Protein Hydration

J D Otvos *North Carolina State University, Raleigh, NC, USA*
- Metallothioneins

M Overduin *The Rockefeller University, New York, NY, USA*
- SH2 Domain Structure

L P. Lemaire *St. George's Hospital Medical School, London, UK*
- Fluorine-19 MRS: Applications in Oncology

R J Pace *Australian National University, Canberra, Australia*
- Bilayer Membranes: Proton & Fluorine-19 NMR

D Packer *Universität Dortmund, Germany*
- Optically Enhanced Magnetic Resonance

K J Packer *Department of Chemistry, University of Nottingham, UK*
- Diffusion & Flow in Fluids

L P Panych *Brigham and Women's Hospital and Harvard Medical School, Boston, MA, USA*
- Wavelet Encoding of MRI Images

R Passariello *University of Rome 'La Sapienza', Italy*
- Abdominal MRA

P M Pattany *University of Miami, Miami, FL, USA*
- Pulsatility Artifacts due to Blood Flow & Tissue Motion & Their Control

R Pauly *Utrecht University, Utrecht, The Netherlands*
- Protein Structures: Relaxation Matrix Refinement

P Pavone *University of Rome 'La Sapienza', Italy*
- Abdominal MRA

R M Pearson *Tri-Valley Research, Pleasanton, CA, USA*
- Instrumentation for the Home Builder

J M Pennock *Royal Postgraduate Medical School, Hammersmith Hospital, London, UK*
- Pediatric Brain MRI: Applications in Neonates & Infants

M A Petrich *Northwestern University, Evanston, IL, USA*
- Amorphous Silicon Alloys

M Piccioli *University of Florence, Florence, Italy*
- Cobalt(II)- & Nickel(II)-Substituted Proteins

P Plateau *Ecole Polytechnique, Palaiseau, France*
- Water Signal Suppression in NMR of Biomolecules

F Podo *Istituto Superiore di Sanità, Rome, Italy*
- Whole Body Machines: Quality Control

B Ponceleti *Massachusetts General Hospital, MA, USA*
- Brain Parenchyma Motion Observed by MRI

R Poupko *Weizmann Institute of Science, Rehovot, Israel*
- Dynamic NMR in Liquid Crystalline Solvents

D H Powell *University of Lausanne, Lausanne, Switzerland*
- Kinetics at High Pressure

P S Pregosin *ETH Zentrum, Zürich, Switzerland*
- Inorganic Nuclei: Low Sensitivity Transition Metals

J H Prestegard *Yale University, New Haven, CT, USA*
- Motional Effects on Protein Structure: Acyl Carriers

J W Prichard *Yale University, New Haven, CT, USA*
- Brain MRS of Human Subjects

M Pruski *Ames Laboratory and Iowa State University, Ames, IA, USA*
- Cokes
- Supported Metal Catalysts

R J Pugmire *University of Utah, Salt Lake City, UT, USA*
- Coal Structure from Solid State NMR
- Magic Angle Turning & Hopping

P Pulay *University of Arkansas, Fayetteville, AR, USA*
- Shielding Theory: GIAO Method

J Quant *Universität Frankfurt, Frankfurt, Germany*
- TOCSY in ROESY & ROESY in TOCSY

G K Radda *University of Oxford and The John Radcliffe Hospital, Oxford, UK*
- Whole Body Studies: Impact of MRS

R Radeglia *Federal Institute for Materials Research and Testing (BAM), Berlin, Germany*
- Semiempirical Chemical Shift Calculations

B Radüchel *Research Laboratories Schering AG, Berlin, Germany*
- Gadolinium Chelates: Chemistry, Safety, & Behavior

R A Rauch — *University of Texas Health Science Center, San Antonio, TX, USA*
- Ischemic Stroke

M Ravikumar — *Infomatrix Inc, Ann Arbor, MI, USA*
- Data Processing

D L Rayner — *Magnex Scientific Ltd, Abingdon, Oxon, UK*
- Cryogenic Magnets for Whole Body Magnetic Resonance Systems

W T Raynes — *University of Sheffield, UK*
- Electric Field Effects on Shielding Constants

A G Redfield — *Brandeis University, Waltham, MA, USA*
- Relaxation Theory: Density Matrix Formulation

D Reed — *Edinburgh University, Edinburgh, Scotland, UK*
- Boron NMR

P Reimer — *Institute for Clinical Radiology, Munster, Germany*
- Contrast Agents in Whole Body Magnetic Resonance: An Overview

E O R Reynolds — *University College London Medical School, UK*
- Pediatric Brain MRS: Clinical Utility

J H Richards — *California Institute of Technology, Pasadena, CA, USA*
- pH Measurement In Vivo in Whole Body Systems

W Richter — *Princeton University, NJ, USA*
- Concentrated Solution Effects

C F Ridenour — *Otsuka Electronics, Inc., Fort Collins, CO, USA*
- Proton Chemical Shift Measurements in Biological Solids

S J Riederer — *Mayo Clinic, Rochester, MN, USA*
- Image Segmentation, Texture Analysis, Data Extraction, & Measurement

P A Rinck — *University of Mons-Hainaut Medical Faculty, NMR Laboratory, Mons, Belgium*
- Relaxation Measurements in Whole Body MRI: Clinical Utility

J M Risley — *University of North Carolina at Charlotte, NC, USA*
- Oxygen-18 in Biological NMR

J Rizo — *University of Texas Southwestern Medical Center, Dallas, TX, USA*
- Peptide & Protein Secondary Structural Elements

B E Roberts — *University of Sheffield, Sheffield, UK*
- Organometallic Compounds

J K M Roberts — *University of California, Riverside, CA, USA*
- Plant Physiology

S P Robinson, — *St. George's Hospital Medical School, London, UK*
- Animal Tumor Models

P B Roemer — *Advanced NMR, Wilmington, MA, USA*
- Whole Body Machines: NMR Phased Array Coil Systems

B R Rosen — *Massachusetts General Hospital, Charlestown, MA, USA*
- Susceptibility Effects in Whole Body Experiments

B D Ross — *Huntington Medical Research Institutes, Pasadena, CA, USA*
- pH Measurement In Vivo in Whole Body Systems

C Rossi — *University of Siena, Siena, Italy*
- Selective Relaxation Techniques in Biological NMR

G M Roth — *University of California, Irvine, Orange, CA, USA*
- Pediatric Body MRI

H D Roth — *Rutgers University, New Brunswick, NJ, USA*
- Chemically Induced Dynamic Nuclear Polarization

T J Rowland — *University of Illinois, Urbana, IL, USA*
- Metals: Pure & Alloyed

D Rugar, — *Almaden Research Center, San Jose, CA, USA*
- Force Detection & Imaging in Magnetic Resonance

G S Rule — *University of Virginia School of Medicine, Charlottesville, VA, USA*
- D-Lactate Dehydrogenase

V M Runge — *University of Kentucky, Lexington, KY, USA*
- Gadolinium Chelate Contrast Agents in MRI: Clinical Applications

P J Sadler — *Birkbeck College, University of London, UK*
- Body Fluids

H Saitô — *Himeji Institute of Technology, Kamigori, Hyogo, Japan*
- Polysaccharide Solid State NMR

J Schaefer — *Department of Chemistry, Washington University, St. Louis, MO, USA*
- REDOR & TEDOR

T Schaefer — *University of Manitoba, Winnipeg, MB, Canada*
- Stereochemistry & Long Range Coupling Constants

T Schleich — *University of California, Santa Cruz, CA, USA*
- Rotating Frame Spin–Lattice Relaxation Off-Resonance

J Schleucher — *Universität Frankfurt, Frankfurt, Germany*
- TOCSY in ROESY & ROESY in TOCSY

P Schmidt — *Universität Frankfurt, Germany*
- Coupling Constants Determined by ECOSY

T H Schmitt — *Stanford University, CA, USA*
- Biological Macromolecules: NMR Parameters

W Schmitt — *Technical University, Munich, Germany*
- Peptides & Polypeptides

M D Schnall — *University of Pennsylvania Medical Center, Philadelphia, PA, USA*
- Peripheral Vasculature MRA

G J Schrobilgen — *McMaster University, Hamilton, Ontario, Canada*
- Noble Gas Elements

H Schwalbe — *Universität Frankfurt, Germany*
- Coupling Constants Determined by ECOSY

M P Schweizer
Departments of Medicinal Chemistry and Radiology, University of Utah, Salt Lake City, Utah, USA
- Nucleic Acids: Base Stacking & Base Pairing Interactions

A Schwenk
Universität Tübingen, Tübingen, Germany
- Steady-State Techniques for Low Sensitivity & Slowly Relaxing Nuclei

A I Scott
Texas A&M University, College Station, USA
- Enzymatic Transformations: Isotope Probes
- Vitamin B_{12}

A Sebald
Universität Bayreuth, Bayreuth, Germany
- Reorientation in Crystalline Solids: Propeller-Like R_3M Species

W Semmler
Institut für Diagnostikforschung (IDF) an der Freien Universität, D-14050 Berlin, Germany
- Brain Neoplasms in Humans Studied by Phosphorus-31 MRS

F Separovic
CSIRO Food Research Laboratory, Sydney, NSW, Australia
- Membranes: Carbon-13 NMR

R E Sepponen
Picker Nordstar Inc., Helsinki, Finland
- Whole Body Studies Involving Spin–Lattice Relaxation in the Rotating Frame

N K Sethi
Amoco Research Center, Naperville, IL, USA
- Variable Angle Sample Spinning

A J Shaka
University of California, Irvine, CA, USA
- Decoupling Methods

F G Shellock
American Health Services Corporation, Newport Beach, CA, USA; Future Diagnostics, Inc., Los Angeles, CA, USA; UCLA School of Medicine, Los Angeles, CA>Bioeffects & Safety of Radiofrequency Electromagnetic Fields
- Bioeffects & Safety of Radiofrequency Electromagnetic Fields

M H Sherwood
IBM Research Division, Almaden Research Center, San Jose, CA, USA
- Chemical Shift Tensors in Single Crystals

J N Shoolery
Varian Associates, Palo Alto CA, USA
- Quantitative Measurements

T H Siddall, III
University of New Orleans, New Orleans, LA, USA
- Magnetic Equivalence

J A Sidles
University of Washington, Seattle, WA, USA
- Force Detection & Imaging in Magnetic Resonance

T M Simonson
The University of Iowa College of Medicine, Iowa City, IA, USA
- Ischemic Stroke

C M Slupsky
University of Alberta, Edmonton, Alberta, Canada
- Muscle Proteins

S L Smith
Department of Chemistry, University of Kentucky, Lexington, KY, USA
- Analysis of High Resolution Solution State Spectra

C M Smith
University of Leicester, Leicester, UK
- In Vivo ESR Imaging of Animals

I C P Smith
National Research Council of Canada, Winnipeg, Canada
- Tissue NMR Ex Vivo

R C Smith *Yale University School of Medicine, New Haven, CT, USA*
- Female Pelvis Studied by MRI

S A Smith *University of Utah, Salt Lake City, UT, USA*
- Relaxation Processes in Coupled-Spin Systems

S O Smith *Yale University, New Haven, CT, USA*
- Rotational Resonance in Biology

O Söderman *University of Lund, Sweden*
- Micellar Solutions & Microemulsions

M Sola *University of Bologna, Bologna, Italy*
- Transferrins

M S Solum *University of Utah, Salt Lake City, UT, USA*
- Wood & Wood Chars

H W Spiess *Max-Planck-Institut für Polymerforschung, Mainz, Germany*
- Polymer Dynamics & Order from Multidimensional Solid State NMR

C S Springer, Jr. *Brookhaven National Laboratory, Upton, NY, and State University of New York, Stony Brook, NY, USA*
- Biological Systems: Spin-3/2 Nuclei

D Stark *University of Massachusetts, USA*
- Liver, Pancreas, Spleen, & Kidney MRI

J Staunton *University of Cambridge, UK*
- Biosynthesis & Metabolic Pathways: Deuterium NMR

V M Stein *University of Wisconsin Madison, USA*
- DNA–Cation Interactions: Quadrupolar Studies

D S Stephenson *University of Munich, Munich, Germany*
- Analysis of Spectra: Automatic Methods

A S Stern *Rowland Institute for Science, Cambridge, MA, USA*
- Maximum Entropy Reconstruction

A D Stevens *University of Leicester, Leicester, UK*
- In Vivo ESR Imaging of Animals

L C Stewart *National Research Council, Ottawa, Ontario, Canada*
- Cation Movements across Cell Walls of Intact Tissues using MRS

J H Strange *University of Kent, Canterbury, Kent, UK*
- Imaging Techniques for Solids and Quasi-solids

T C Stringfellow *University of Wisconsin, Madison, WI, USA*
- Relaxation of Transverse Magnetization for Coupled Spins

P Styles *MRC Biochemical and Clinical Magnetic Resonance Unit, Oxford, UK*
- Localization by Rotating Frame Techniques

N S Sullivan *University of Florida, Gainesville, FL, USA*
- *Ortho–Para* Hydrogen at Low Temperature

M F Summers *University of Maryland, Baltimore County, Baltimore, MD, USA*
- Zinc Fingers

H M Swartz *Dartmouth Medical School, Hanover, NH, USA.*
• EPR and In Vivo EPR: Roles for Experimental & Clinical NMR Studies

B D Sykes *University of Alberta, Edmonton, Alberta, Canada*
• Muscle Proteins

S D Taylor-Robinson *Hammersmith Hospital, London, UK*
• Tissue Behavior Measurements Using Phosphorus-31 NMR

T Terao *Kyoto University, Japan*
• Conducting Polymers

H Thomann *Exxon Research and Engineering Company, Annandale, NJ, USA*
• Electron–Nuclear Multiple Resonance Spectroscopy

D A Torchia *National Institutes of Health, Bethesda, MD, USA*
• Protein Dynamics from NMR Relaxation

A S Tracey *Simon Fraser University, Burnaby, BC, Canada*
• Lyotropic Liquid Crystalline Samples

D D Traficante *The University of Rhode Island, Kingston, RI, USA*
• Relaxation: An Introduction

N S True *University of California, Davis, CA, USA*
• Gas Phase Studies of Chemical Exchange Processes

S Tunlayadechanont *University of Pennsylvania, Philadelphia, PA, USA*
• Brain Infection & Degenerative Disease Studied by Proton MRS

D L Turner *University of Southampton, UK*
• Phase Cycling

R Turner *Institute of Neurology, London, UK*
• Gradient Coil Systems

R Turner *Institute of Neurology, London, UK*
• Functional MRI: Theory & Practice

P A Turski *University of Wisconsin, Madison, WI, USA*
• Phase Contrast MRA

R Tycko *National Institutes of Health, Bethesda, MD, USA*
• Overtone Spectroscopy of Quadrupolar Nuclei

M van de Kamp *Nijmegen University, Nijmegen, The Netherlands*
• Copper Proteins

H van Halbeek *The University of Georgia, Athens, GA, USA*
• Carbohydrates & Glycoconjugates

D van Ormondt *Delft University of Technology, Delft, The Netherlands*
• Quantitation in Whole Body MRS

C van Wüllen *Ruhr-Universität Bochum, Bochum, Germany*
• Shielding Calculations: IGLO Method

P C M van Zijl *Johns Hopkins University Medical School, Baltimore, MD, USA*
• Water Suppression in Proton MRS of Humans & Animals

D L VanderHart — *National Institute of Standards and Technology, Gaithersburg, MD, USA*
- Magnetic Susceptibility & High Resolution NMR of Liquids & Solids

W S Veeman — *Gerhard-Mercator-Universität Duisburg, Germany*
- Polymer Blends: Miscibility Studies

A J Vega — *DuPont, Wilmington, DE, USA*
- Quadrupolar Nuclei in Solids

S Vega — *Weizmann Institute of Science, Rehovot, Israel*
- Floquet Theory

J Vion-Dury — *Centre de Résonance Magnétique Biologique et Médicale, Faculté de Médecine, Marseille, France*
- Tissue & Cell Extracts MRS

J Virlet — *CEA Saclay, Service de Chimie Moléculaire, Gif-sur-Yvette, France*
- Line Narrowing Methods in Solids

R L Vold — *College of William and Mary, Williamsburg, VA, USA.*
- Deuteron Relaxation Rates in Liquid Crystalline Samples: Experimental Methods
- Nucleic Acid Flexibility & Dynamics: Deuterium NMR

R Wand — *Universität Ulm, Germany*
- Molecular Motions: T_1 Frequency Dispersion in Biological Systems

W Wang — *University of Utah, Salt Lake City, UT, USA*
- Magic Angle Turning & Hopping

I M Ward — *Leeds University, Leeds, UK*
- Polymer Physics

R J Warner — *Magnex Scientific Ltd, Abingdon, Oxon, UK*
- Cryogenic Magnets for Whole Body Magnetic Resonance Systems

W S Warren — *Princeton University, NJ, USA*
- Concentrated Solution Effects
- Shaped Pulses

R E Wasylishen — *Dalhousie University, Halifax, Nova Scotia, Canada*
- Dipolar & Indirect Coupling Tensors in Solids

J S Waugh — *Massachusetts Institute of Technology, Cambridge, MA, USA*
- Average Hamiltonian Theory

G A Webb — *University of Surrey, Guildford, Surrey, UK*
- Shielding: Overview of Theoretical Methods

V J Wedeen — *Harvard Medical School, USA*
- Brain Parenchyma Motion Observed by MRI

F W Wehrli — *University of Pennsylvania Medical School, Philadelphia, PA, USA*
- Trabecular Bone Imaging
- Whole Body MRI: Strategies Designed to Improve Patient Throughput

M W Weiner — *University of California, San Francisco, CA, USA*
- Kidney, Prostate, Testicle, & Uterus of Subjects Studied by MRS

H-J Weinmann — *Research Laboratories Schering AG, Berlin, Germany*
- Gadolinium Chelates: Chemistry, Safety, & Behavior

R M Weisskoff *Massachusetts General Hospital, Charlestown, MA, USA*
- Susceptibility Effects in Whole Body Experiments

T J Weissleder *Massachusetts General Hospital and Harvard Medical School, Charlestown, MA, USA*
- Contrast Agents in Whole Body Magnetic Resonance: An Overview

D E Wemmer *University of California, Berkeley, USA*
- DNA: A-, B-, & Z-

L G Werbelow *New Mexico Tech, Socono, NM, USA*
- Relaxation Theory for Quadrupolar Nuclei
- Relaxation Processes: Cross Correlation & Interference Terms

L G Werbelow *Universite d'Aix-Marseille I, Marseille, France*
- Dynamic Frequency Shift

U Werner-Zwanziger *Indiana University, Bloomington, IN, USA*
- SQUIDs

D White *University of Pennsylvania, Philadelphia, PA, USA*
- Dynamic Spin Ordering of Matrix Isolated Methyl Rotors

S S Wijmenga *University of Nijmegen, Nijmegen, The Netherlands*
- Nucleic Acids: Spectra, Structures, & Dynamics

P G Williams *Lawrence Berkeley National Laboratory, Berkeley, CA, USA*
- Tritium NMR
- Tritium NMR in Biology

M P Williamson *University of Sheffield, Sheffield, UK*
- NOESY

W D Wilson *Georgia State University, Atlanta, GA, USA*
- Drug–Nucleic Acid Interactions

S Wimperis *University of Oxford, UK*
- Relaxation of Quadrupolar Nuclei Measured via Multiple Quantum Filtration

R A Wind *Battelle/Pacific Northwest Laboratories, Richland, WA, USA*
- Dynamic Nuclear Polarization & High-Resolution NMR of Solids

D E Woessner *The University of Texas Southwestern Medical Center at Dallas and Department of Radiology, The Mary Nell and Ralph B. Rogers Magnetic Resonance Center*
- Brownian Motion & Correlation Times
- Relaxation Effects of Chemical Exchange

R Wolff *Federal Institute for Materials Research and Testing (BAM), Berlin, Germany*
- Semiempirical Chemical Shift Calculations

E C Wong *Medical College of Wisconsin, Milwaukee, WI, USA*
- Whole Body MRI: Local & Inserted Gradient Coils

M L Wood *University of Toronto, Toronto, Ontario, Canada*
- Respiratory Artifacts: Mechanism & Control

B S Worthington *University of Nottingham, Nottingham, UK*
- Image Quality & Perception

B Wrackmeyer *Universität Bayreuth, Bayreuth, Germany*
- Germanium, Tin, & Lead NMR
- Inorganic Chemistry Applications

L A Wright — *University of Wisconsin Madison, USA*
- DNA–Cation Interactions: Quadrupolar Studies

P E Wright — *The Scripps Research Institute, La Jolla, CA, USA*
- Proteins & Protein Fragments: Folding

W-g Wu — *National Tsing Hua University, Hsinchu, Taiwan*
- Sonicated Membrane Vesicles

Y Wu — *University of North Carolina, Chapel Hill, NC, USA*
- Double Rotation

J Wüthrich — *Stockholm University, Sweden*
- Paramagnetic Relaxation in Solution

K Wüthrich — *Eidgenössische Technische Hochschule Zürich, Switzerland*
- Biological Macromolecules: Structure Determination in Solution

T Yamanobe — *Gunma University, Kiryu, Gunma, Japan*
- Solid Polymers: Shielding & Electronic States

C S Yannoni — *Almaden Research Center, San Jose, CA, USA*
- Force Detection & Imaging in Magnetic Resonance

P L Yeagle — *University at Buffalo, Buffalo, NY, USA*
- Membranes: Phosphorus-31 NMR

I R Young — *Hammersmith Hospital, London, UK*
- Eddy Currents & Their Control
- Refrigerated & Superconducting Receiver Coils for Whole Body Magnetic Resonance
- Selective Excitation Methods: Artifacts
- Temperature Measurement for In Vivo MRI

W T C Yuh — *The University of Iowa College of Medicine, Iowa City, IA, USA*
- Ischemic Stroke

R Zanasi — *Università degli Studi di Modena, Modena, Italy*
- Shielding in Small Molecules

D B Zax — *Cornell University, Ithaca, NY, USA*
- Zero Field NMR

K W Zilm — *Yale University, New Haven, CT, USA*
- Quantum Exchange
- Spectral Editing Techniques: Hydrocarbon Solids

O Züger — *Almaden Research Center, San Jose, CA, USA*
- Force Detection & Imaging in Magnetic Resonance

Subject Index